PHOTOSYNTHESIS: PHOTOREACTIONS TO PLANT PRODUCTIVITY

Photosynthesis:
Photoreactions to
Plant Productivity

Edited by

YASH PAL ABROL
Indian Agricultural Research Institute,
New Delhi, India

PRASANNA MOHANTY
Jawaharlal Nehru University,
New Delhi, India

and

GOVINDJEE
University of Illinois,
Urbana, Illinois, U.S.A.

SPRINGER-SCIENCE+BUSINESS MEDIA, B.V.

Library of Congress Cataloging-in-Publication Data

```
Photosynthesis : photoreactions to plant productivity / edited by Yash
  Pal Abrol, Prasanna Mohanty, Govindjee.
       p.   cm.
    Includes index.
    ISBN 978-0-7923-1943-6      ISBN 978-94-011-2708-0 (eBook)
    DOI 10.1007/978-94-011-2708-0
    1. Photosynthesis.    I. Abrol, Yash Pal.   II. Mohanty, Prasanna.
  III. Govindjee, 1933-
  QK882.P5577  1992
  581.1'3342--dc20                                        92-34817
```

ISBN 978-0-7923-1943-6

Preface

All biomass is derived from photosynthesis. This provides us with food fuel, as well as fibre. This process involves conversion of solar energy, via photochemical reactions, into chemical energy. In plants and cyanobacteria, carbon dioxide and water are converted into carbohydrates and oxygen. It is the best studied research area of plant biology. We expect that this area will assume much greater importance in the future in view of the depleting resources of the Earth's fuel supply. Furthermore, we believe that the next large increase in plant productivity will come from applications of the newer findings about photosynthetic process, especially through manipulation by genetic engineering.

The current book covers an integrated range of subjects within the general field of photosynthesis. It is authored by international scientists from several countries (Australia, Canada, France, India, Israel, Japan, Netherlands, Russia, Spain, UK and USA).

It begins with a discussion of the genetic potential and the expression of the chloroplast genome that is responsible for several key proteins involved in the electron transport processes leading to O_2 evolution, proton release and the production of NADPH and ATP, needed for CO_2 fixation. The section on photosystems discusses how photosystem I functions to produce NADPH and how photosystem II oxidizes water and releases protons through an "oxygen clock" and how intermediates between the two photosystems are produced involving a "two electron gate". These basic chapters are followed by an in-depth discussion on the various processes involved in coordination and regulation of photosynthesis at several levels: Photosystem II (by excess light, herbicides, bicarbonate, other inhibitors), thylakoid membranes and chloroplasts (by light and other factors, especially senescence), and the leaves (by substrates and products).

An equally in-depth discussion is then given on carbon assimilation and partitioning with special emphasis on: the roles of the enzyme ribulose 1, 5-*bis*phosphate carboxylase; C_3, C_4 and C_3–C_4 plants, fruiting structures, source-sink relationships and interaction between carbon and nitrogen metabolism. Two important aspects of photosynthesis related to the overall productivity, the effects of "stress" (particularly water, heavy metals, light and CO_2 enrichment) and of genetic variability, are then discussed at length. A rather significant discussion in the last section deals with photosynthesis improvement and the importance of photosynthesis in low light towards crop productivity, a sometimes neglected aspect. A detailed discussion of some of these aspects in various crop plants, especially rice, wheat and tobacco makes it a very useful book.

Finally, this book brings to light the multidisciplinary character of photosynthesis which spans from (bio)physics to agronomy. This book will help prepare students with the necessary conceptual outlook for integrating information from the bioenergetic and enzymatic angles, obtained at the molecular level, to the physiology of chloroplasts, leaves and eventually the crops. Thus, in our opinion, this book would serve the larger interests of both students and researchers in the areas of agriculture, biotechnology, biochemistry, biophysics, plant physiology, and molecular biology, who are engaged in studying not only the basic aspects of photosynthesis, a major process determining biomass production, but its relationship to plant productivity. The possibility of using information in this book to improve biomass and grain yield is something that no one should ignore.

September, 1992
New Delhi and
Urbana, Illinois

Y.P. ABROL
P. MOHANTY
GOVINDJEE

Contents

CARBON ASSIMILATION AND PARTITIONING

STRESS; CO$_2$ ENRICHMENT

GENETIC VARIATION; PRODUCTIVITY

The Contributors

Yash Pal Abrol
   Division of Plant Physiology
   Indian Agricultural Research Institute, New Delhi 110 012, India

Kailash C. Bansal
   Division of Plant Physiology,
   Indian Agricultural Research Institute, New Delhi 110 012, India

Satinder N. Bhardwaj
   Division of Plant Physiology
   Indian Agricultural Research Institute, New Delhi 110 012, India

William J. Coleman
   Service de Biophysique (OBCM-SBE)
   Centre d'Etudes de Sacky, 911 91 Gif-sur-Yvette, Cedex, France

Christa Critchley
   Department of Botany, The University of Queensland
   QLD 4072, Australia

Jeroen den Hertog
   Department of Plant Biology, University of Groningen
   PO Box 14, 9750 AA Haren (Gn), The Netherlands

Marvin Edelman
   Department of Plant Genetics, The Weizmann Institute of Science
   Rehovot, Israel

Tedd Elich
   Plant Molecular Biology Laboratory
   ARS, USDA, Building 006, BARC-West, Beltsville, MD 20705-2350, USA

Christine H. Foyer
   Laboratoire du Metabolisme, INRA, Route de Saint-Cyr
   78026 Versailles, Cedex, France

Maria Lucia Ghirardi
   Photoconversion Research Branch, Basic Sciences Division
   National Renewable Energy Lab, 1617 Cole Blvd.
   Golden, CO 80401-3393, USA

Govindjee
Department of Plant Biology
University of Illinois at Urbana-Champaign
265 Morill Hall, 505 South Goodwin Avenue
Urbana, IL 61801-3707, USA

Bruce M. Greenberg
Department of Biology, The University of Waterloo
Waterloo, Ontario N2L3GI, Canada

Anil Grover
Department of Plant Molecular Biology
University of Delhi, South Campus
New Delhi 110 021, India

David O. Hall
Division of Biosphere Sciences, King's College
Campden Hill Road, London W8 7AH, UK

Satoshi Hoshina
Department of Biology, Faculty of Science
Kanazawa University
Kakuma, Kanazawa 920-11, Japan

Ryuichi Ishii
Laboratory of Crop Sciences, Faculty of Agriculture
The University of Tokyo, Yayoi, Bunkyo-ku, Tokyo 113, Japan

Shigeru Itoh
Division of Bioenergetics
National Institute for Basic Biology
38 Nishigonaka, Myodaiji Chyo, Okazaki 444, Japan

Boris N. Ivanov
Institute of Soil Sciences and Photosynthesis
Russian Academy of Sciences, Pushchino, Moscow Region 142292, Russia

Katrien Jansen
Department of Plant Biology, University of Groningen
PO Box 14, 9750 AA Haren (Gn), The Netherlands

Sanjay Kapoor
Department of Botany, University of Delhi
Delhi 110 007, India

Navassard V. Karapetyan
A.N. Bakh Institute of Biochemistry, Russian Academy of Sciences
Leninsky prospekt 33, Moscow, 117071, Russia

Nyay Kelkar
Department of Botany, University of Delhi
Delhi 110 007, India

P. Ananda Kumar
Nuclear Research Laboratory
Indian Agricultural Research Institute, New Delhi 110 012, India

David W. Lawlor
Institute of Arable Crops Research
Rothamsted Experimental Station, Harpenden, Herts, AL5 2JQ, UK

Stephen P. Long
Department of Biology, University of Essex
Colchester, CO4 3SQ, UK

Satish C. Maheshwari
Department of Plant Molecular Biology
University of Delhi, South Campus, New Delhi 110 021, India

Autar K. Mattoo
Plant Molecular Biology Laboratory, 10300 Baltimore Ave.
ARS, USDA, Building 006, BARC-West, Beltsville, MD 20705-2350, USA

Hipolito Medrano
Lab de Fisiologia Vegetal, Dept. de Biologia Ambiental
Institut d'Estudis Avançats — Universitat de les Illes Balears
07071 Palma de Mallorca, Spain

Surma Mitra
Division of Plant Physiology
Indian Agricultural Research Institute, New Delhi 110 012, India

Prasanna Mohanty
School of Life & Environmental Sciences
Jawaharlal Nehru University, New Delhi 110 067, India

P. Murali
Biochemistry and Molecular Biology Laboratory
Department of Botany, University of Delhi, Delhi 110 007, India

Raghuveer Polisetty
Division of Plant Physiology
Indian Agricultural Research Institute, New Delhi 110 012, India

Agepati S. Raghavendra
School of Life Sciences
University of Hyderabad, Hyderabad 500 134, India

V.S. Rama Das
School of Life Sciences
University of Hyderabad, Hyderabad 500 134, India

Jack J.S. van Rensen
Laboratory of Plant Physiological Research
Agricultural University,
Arboretumlaan 4, 6703 BD, Wageningen, The Netherlands

A. Wendy Russell
Department of Botany, The University of Queensland
QLD 4072, Australia

Ramesh C. Sachar
Biochemistry and Molecular Biology Laboratory
Department of Botany, University of Delhi, Delhi 110 007, India

Daman Saluja
Biochemistry and Molecular Biology Laboratory
Department of Botany, University of Delhi, Delhi 110 007, India

Udayan K. Sengupta
Division of Plant Physiology
Indian Agricultural Research Institute, New Delhi 110 012, India

Aruna Sharma
Division of Plant Physiology
Indian Agricultural Research Institute, New Delhi 110 012, India

Prabhat K. Sharma
Department of Botany, Goa University
Taleigaun Plateau, Goa 403 202, India

Inder S. Sheoran
Department of Chemistry and Biochemistry
Haryana Agricultural University, Hisar 125 004, India

Richard C. Sicher
Climate Stress Laboratory,
Beltsville Agricultural Research Center
USDA, Beltsville, MD 20705, USA

Randhir Singh
Department of Chemistry and Biochemistry
Haryana Agricultural University, Hisar 125 004, India

Sudhir K. Sopory
School of Life Sciences
Jawaharlal Nehru University, New Delhi 110 067, India

Girish C. Srivastava
Division of Plant Physiology
Indian Agricultural Research Institute, New Delhi 110 012, India

Ineke Stulen
Department of Plant Biology, University of Groningen,
PO Box 14, 9750 AA Haren (Gn), The Netherlands

Akilesh K. Tyagi
Department of Plant Molecular Biology
University of Delhi, South Campus, New Delhi 110 021, India

Dinesh C. Uprety
Division of Plant Physiology
Indian Agricultural Research Institute, New Delhi 110 012, India

J. Vadell
Lab de Fisiologia Vegetal, Dept. de Biologia Ambiental
Institut d'Estudis Avançats — Universitat de les Illes Balears
07071 Palma de Mallorca, Spain

CHLOROPLAST GENOME

1

The Chloroplast Genome: Genetic Potential and Its Expression

A.K. Tyagi[1], N. Kelkar[2], S. Kapoor[2] and S.C. Maheshwari[1, 2]

[1]Department of Plant Molecular Biology,
University of Delhi, South Campus,
New Delhi 110021, India
[2]Department of Botany,
University of Delhi,
Delhi 110007, India

CONTENTS

4

ABBREVIATIONS

bp	:	base pairs;
cp DNA	:	Chloroplast DNA;
cyt	:	cytochrome;
DALA	:	δ-amino-levulinic acid;
IRs	:	Inverted repeats;
kD	:	kilo Dalton;
LSC	:	large single copy region;
nt	:	nucleotide(s);
ORFs	:	open reading frames;
PS I	:	photosystem I;
PS II	:	photosystem II;
RuBPcase	:	ribulose 1, 5-bisphosphate carboxylase;
SSC	:	small single copy region;
TAC	:	transcriptionally active chromosome.

ABSTRACT

The chloroplasts carry out the vital process of photosynthesis. Though chloroplasts contain their own DNA, they are only semi-autonomous. The majority of their components are coded in the nucleus. The size of cpDNA in higher plants varies from 120 to 180 kbp, which consists of a large and a small single copy region separated by a pair of inverted repeats. More than a hundred genes encoded by the cpDNA include genes for photosynthetic machinery as well as those involved with genetic system. Several features of the chloroplast gene organization and regulation of expression are prokaryotic. Thus, genes are organized into clusters (operons) and co-transcribed. The genes or gene clusters are flanked by prokaryotic −10 and −35 like promoter sequences at 5′ end and stem-and-loop structures at 3′ end. Although chloroplast RNA polymerase has been found to contain a variable number of polypeptides, *Escherichia coli* RNA polymerase can also recognize the chloroplast promoter sequences. The primary transcript is processed further which includes modification at ends as well as processing in intergenic regions.

The expression of chloroplast genes is regulated by light, but the exact level of control is debatable. Generally, there is an increase in the steady-state levels of transcripts after illumination of dark-grown plants, but the rate of transcription *per se* seems to play only a limited role and post-transcriptional events have been found to be more important. These are related to transcripts as well as translation system. Alternative promoters can also be used in light and specific light quality may effect such a selection. Attempts are being made to dissect the components and mechanism of regulatory controls which should be greatly facilitated by recent work on chloroplast transformation.

I. Introduction

Chloroplasts represent the site of photosynthesis, which involves conversion of solar energy to chemical energy, utilizing carbon dioxide and water. As is well known, the process of photosynthesis can be split into two phases: the light reaction(s) and the light independent or dark reaction(s). The mature chloroplast represents an intricately built apparatus, composed of membranous grana or the thylakoids and the stroma. The components necessary for the light reaction(s) are organized in the form of four supra-molecular complexes: photosystem I (PS I), photosystem II (PS II), cytochrome b_6/f complex (cyt b_6/f) and ATP synthase in the thylakoids; in contrast the components of the dark reactions are present in the stroma. Additionally, in the stroma there exist the components of the genetic machinery of chloroplasts. Investigations on the molecular biology of chloroplasts are, therefore, of great importance not only from the applied angle of agriculture and plant biotechnology but also for fundamental research on gene expression and organelle biogenesis. There is much evidence pointing towards a high degree of coordination in the working of the genetic machinery of all the three components—chloroplasts, mitochondria, and nucleus—of a cell. The elucidation of this coordination provides a great challenge to researchers.

Studies on various aspects of chloroplast molecular biology started in the 1960s when the presence of chloroplast DNA (cpDNA) was first demonstrated. However, it is only after the availability of reliable methods to purify intact cpDNA (Kolodner and Tewari, 1975) and the advent of the powerful recombinant DNA technology that the new era of intensive study of chloroplast molecular biology has really been ushered in. The mid-1970s saw the development of the first restriction maps of cpDNAs (Bedbrook and Bogorad, 1976; Herrmann *et al.*, 1976). This was followed by localization of genes for RNAs and proteins on cloned DNA fragments. Recently, by the application of the recombinant DNA techniques, sequencing of the entire chloroplast genome of *Marchantia polymorpha* (Ohyama *et al.*, 1986), *Nicotiana tabacum* (Shinozaki *et al.*, 1986), and *Oryza sativa* (Hiratsuka *et al.*, 1989) has been achieved. The new techniques are unravelling details of the organization and structure of chloroplast-encoded genes which in turn are invaluable for a deep insight into the regulatory mechanisms of gene expression. In this regard, much attention has been paid to the regulation of gene expression during light-dependent differentiation of chloroplasts from their precursors, proplastids and etioplasts (a developmental stage of plastids present in dark-grown seedlings).

Despite the fact that chloroplasts are at best only semi-autonomous (the vast majority of their components being nuclear-encoded), the chloroplast genome plays a crucial role in the biogenesis of chloroplasts. Several reviews dealing with various aspects of chloroplast molecular biology have appeared in the literature. Some of the recent reviews dealing mainly with organization of the chloroplast genome are those by Ozeki *et al.* (1987), Sugiura (1989) and Palmer (1990). Expression of chloroplast genes has been dealt with by Hanley-Bowdoin and Chua (1987), Mullet (1988), Gruissem (1989), and Herrmann *et al.* (1992). Nevertheless, the subject is in need of constant updating as nearly a hundred papers are being published annually. In the beginning of this review, we shall deal with physical as well as transcriptional organization and structure of chloroplast-encoded genes for components of the chloroplast genetic machinery and photosynthesis. This is followed by a discussion of various aspects related to regulation of gene expression during chloroplast biogenesis. Because of space limitations, we shall be selective in citing references, restricting ourselves to the first one or two for a discovery, unless notable new information is available.

II. Genetic Potential of Chloroplast Genome

The cpDNA is circular with an average contour length of about 50 μm that corresponds to a DNA molecule of about 150 kbp. The number of molecules has been found to be more than a hundred per chloroplast depending on the developmental statge of plastids. One dominant structural feature of most cpDNAs (barring a few cases like broad bean and pea) is the presence of two large inverted repeats that range from 10 to 80 kbp in size and that divide the chloroplast chromosome into unequal parts, a large single copy region of 70–85 kbp and a small single copy region of 10–20 kbp. It is largely due to the variation in the sizes of inverted repeats that cpDNA in different species varies from about 120 to 180 kbp. In all, about 140 open reading frames (ORFs) have been found to be present on both the strands of cpDNA, as exemplified by tobacco (Fig. 1), which can code either for a polypeptide of more than 30 amino acids or for ribosomal and transfer RNAs. Since about 20 ORFs are present in inverted repeats, the true genetic potential of cpDNA is limited to about 120 products, out of which more than 80% have already been identified. It should, however, be mentioned that certain products have been identified on the basis of similarity of the deduced primary structure of the product with that of a known bacterial product while a real product from chloroplast still remains to be found. In this section, we first discuss the different strategies adopted to identify various genes followed by information on their organization and structure.

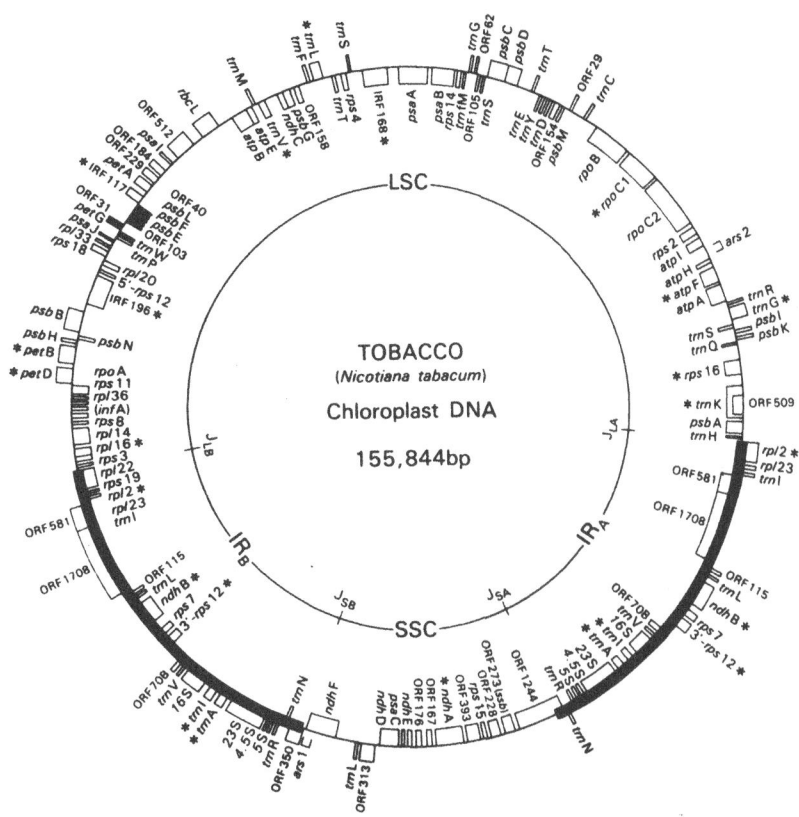

Figure 1. **The gene map of circular chloroplast DNA of tobacco.** The transcriptional polarities of the genes shown inside the circle are clockwise and those outside are anticlockwise. The nomenclature of genes follows the proposals of Hallick and Bottomley (1983) and Hallick (1989). Split genes are marked with asterisks. Also shown are major ORFs (open reading frames) and IRFs (interrupted reading frames) whose products remain unidentified. Other abbreviations used are: LSC, large single copy region (86, 684 bp); SSC, small single copy region (18, 482 bp); IR_A and IR_B, inverted repeats A and B (25, 339 bp); J_LA, junction between LSC and IR_A; J_LB, junction between LSC and IR_B; J_SA, junction between SSC and IR_A; J_SB, junction between SSC and IR_B; ars, autonomously replicating sequence in yeast; 23S, 16S, 5S, 4.5S, rRNA genes; trn, tRNA genes followed by single letter code for specific amino acids; rpo, genes for RNA polymerase subunits; rpl and rps, genes for 50S and 30S subunits of ribosomes, respectively; psb, pet, psa, atp, genes for thylakoid membrane supra-molecular complexes, photosystem II, cytochrome b_6/f complex, photosystem I, ATP synthase, respectively; ndh, genes for subunits of NADH dehydrogenases; rbcL, gene for the large subunit of ribulose-1, 5-*bis*phosphate carbosylase/oxygenase; infA, initiation factor-1. Reproduced with kind permission of Professor M. Sugiura.

a) Strategies for identification of genes

The first genes to be located on the chloroplast genome were those coding for rRNAs and tRNAs. This was achieved by hybridization of ^{32}P-labelled chloroplast rRNA (Thomas and Tewari, 1974; Bedbrook *et al.*, 1977) or ^{125}I-labelled chloroplast tRNAs (Steinmetz *et al.*, 1978) with cpDNA fragments resolved on agarose gels. Individual tRNA genes were identified by hybridization of individual tRNA species (separated on 2-D gels and confirmed on the basis of aminoacylation) to cpDNA fragments.

For identification of protein-coding genes, various strategies have been followed to identify and locate genes on the cpDNA fragments. The transcript level of many genes is increased by exposure to light. For identification of these genes, RNAs have been isolated from etiolated tissues exposed to light and then hybridized with isolated restriction fragments. Further characterization has been accomplished by selection of mRNA complementary to one of the strands of the fragment, translation of mRNA after dissociation from the hybrid, and then immunoselection of the proteins using antibodies for the known proteins (hybrid select translation). Similar selection can be done in another way—complementing the earlier approach—which involves addition of denatured restriction fragments to isolated chloroplast RNA that inhibit the synthesis of polypeptides coded by sequences in the fragment (hybrid arrest translation). Once a gene has been located on a DNA fragment, the exact position of an ORF can be fixed after determining the nucleotide sequence.

In vitro coupled transcription-translation of restriction fragments cloned in *E. coli* and then selection of particular clones, by using antibodies against known proteins, is another approach towards this goal.

A new approach developed recently for establishing the identity of genes is the use of antibodies against synthetic polypeptides. Once the nucleotide sequence of an ORF is determined, a synthetic polypeptide can be made from the deduced amino acid sequence, and antibodies raised against this polypeptide (either complete or partial) can then be used for identification of the desired protein in the chloroplast.

When a particular gene has already been characterized by various means including sequencing and determining the ORFs from one plant, it can be used as a heterologous probe for identifying the gene in chloroplast DNA fragments of another plant. The rapid advance in the last 10 years in our knowledge of organization of the chloroplast genome is indeed due to availability and exchange of such probes between various laboratories.

b) Genes for RNA polymerase

We may first discuss the genes coding for chloroplast RNA polymerase, the enzyme essential for maintaining the semi-autonomous nature of chloroplasts. The first indication that chloroplasts have an RNA polymerase came in the early 1960s from studies demonstrating synthesis of RNA by purified chloroplasts (Bandurski and Maheshwari, 1962). Since then, research on plastid RNA polymerase has been the subject of several investigations. The chloroplast RNA polymerase was first purified by Polya and Jagendorf (1971) from wheat and by Bottomley *et al.* (1971) from maize chloroplasts. Subsequently, the enzyme preparations have been shown to have a complex subunit composition of as many as 7–14 polypeptides of 25 to 180 kD (Kidd and Bogorad, 1979; Tewari and Goel, 1983; Lerbs *et al.*, 1983). However, from these investigations it could not be resolved whether RNA polymerase was coded by the nucleus or the chloroplast itself.

The sequencing of chloroplast DNA from the liverwort *Marchantia* and the higher plants, tobacco and rice, as also of relevant smaller fragments from spinach, maize, pea and wheat revealed the presence of ORFs (rpoA, rpoB, and rpoC) potentially coding for proteins homologous to α, β, and β′ subunits of *E. coli* RNA polymerase.

It may be recalled that *E. coli* RNA polymerase has the subunit composition α, β, β´ and σ. This indicates that chloroplast genome may encode at least all the principal RNA polymerase subunits if not the σ subunit (Ohme *et al.*, 1986; Sijben-Müller *et al.*, 1986; Shinozaki *et al.*, 1986; Ohyama *et al.*, 1986; Hudson *et al.*, 1988; Ruf and Kössel, 1988; Hird *et al.*, 1989; Igloi *et al.*, 1990; Shimada *et al.*, 1990). Although σ-factor-like activity has also been reported from chloroplasts (Jolly and Bogorad, 1980; Bülow and Link, 1988; Lerbs *et al.*, 1988), the factor is probably encoded in the nucleus.

The chloroplast gene corresponding to *E. coli* rpoC is divided into two cistrons, C1 and C2 (Hudson *et al.*, 1988; Hiratsuka *et al.*, 1989). Of these, rpoC1 is interrupted by an intron in the liverwort *Marchantia* (Ohyama *et al.*, 1986) as also in dicots like spinach (Hudson *et al.*, 1988) and tobacco (Shinozaki *et al.*, 1986). In monocots, the intron from rpoC1 is deleted though rpoC2 is interrupted by an intron as shown in maize (Igloi *et al.*, 1990) and rice (Shimada *et al.*, 1990).

The genes rpoB, C1 and C2 are clustered in an operon in all plants studied and can code for polypeptides of 1,070, 687, and 139 amino acids, respectively in tobacco. The region containing this operon in maize (Igloi *et al.*, 1990), rice (Shimada *et al.*, 1990), and wheat (Hird *et al.*, 1989) is inverted with respect to that of spinach or tobacco. A similar organization has also been found in *Euglena gracilis* where the rpoB gene is, however, interrupted by eight introns (Yepiz-Plascencia *et al.*, 1990). Recently, it has been found that several mRNA species hybridize with specific gene probes of rpo operon suggesting that the rpo gene cluster is an active transcription unit in *Marchantia* (in Ozeki *et al.*, 1987 cited as unpublished results of Nakashigashi and Umesono) and in *Euglena* (Yepiz-Plascencia *et al.*, 1990).

Using antibodies against either a synthetic polypeptide for the C-terminal sequence or the fusion proteins of α subunit of RNA polymerase with lacZ product, the presence of the α subunit of RNA polymerase has been shown in chloroplasts of maize (Ruf and Kössel, 1988) and pea (Purton and Gray, 1989). The protein has been shown to be synthesized in a transcriptionally active chloroplast extract. More recently, Hu and Bogorad (1990) have purified a transcriptionally active RNA polymerase containing four prominent polypeptides of 180, 120, 85, and 38 kD. The N-terminal amino-acid sequences of 180, 120, and 38 kD were found to be similar to the deduced sequences for rpoC2, rpoB, and rpoA respectively, thereby establishing that these subunits may indeed be coded and synthesized in the chloroplast.

c) Ribosomal genes

Bearing a similarity to the prokaryotic transcription-translation system, the chloroplasts, too, have 70S ribosomes, comprising 50S and 30S ribosomal subunits. In addition to several proteins, the 50S subunits have 23S, 5S, and 4.5S rRNAs, while a 16S rRNA is associated with the 30S subunit.

(i) Genes for ribosomal RNAs

The rRNA genes were among the first to be identified and located on chloroplast chromosome of maize by Tewari, Bogorad, and their co-workers (Thomas and Tewari, 1974; Bedbrook *et al.*, 1977). In *Zea mays*, chloroplast rDNA exists in two identical units as a part of a 22 kb sequence which is repeated in reverse orientation (Bedbrook *et al.* 1977). Sequencing of these DNA segments in maize by Schwarz and Kössel (1980) and in tobacco by Sugiura's group (Takaiwa and Sugiura, 1980b; Tohdoh *et al.*, 1981) indicated that the rRNA genes are arranged to form an operon in the order 16S, 23S, 4.5S and 5S and can code RNA of 1,489, 2,810, 103, and 121 nt, respectively, in tobacco. Certain tRNA genes were also found to inhabit the same region, in the spacer sequence between 16S and 23S rRNA genes of spinach, as demonstrated by Herrmann and co-workers (Bohnert *et al.*, 1979). However, *Chlamydomonas reinhardtii* has a slightly different composition of rRNA genes, i.e., it consists of 16S, 7S, 3S, 23S, and 5S rRNA genes in the given order (Rochaix and Malnoe, 1978). It is of interest that the first case of a split gene reported in the chloroplast genome was that of the 23S rRNA gene from *Chlamydomonas reinhardtii* (Allet and Rochaix, 1979). In another alga, *Chlorella ellipsoideae*, the rDNA cluster is split into two back-to-back operons, one operon comprising 16S rRNA-tRNA[Ile] (GAU) genes and the other operon comprising tRNA[Ala] (UGC)-23SrRNA-5S rRNA genes (Yamada and Shimaji, 1986).

The number of rRNA genes varies with the number of repeats in the chloroplast genome. Thus, in higher plants, there are two sets of rRNA genes per chloroplast genome, while in *Euglena,* where there are three to five direct repeats, there is a correspondingly higher number. In broad bean, pea and chickpea, where inverted repeats are absent, there is only one set of rRNA genes (Kollar and Delius, 1980; Chu *et al.*, 1981; Chu and Tewari, 1982). In *Euglena* strain Z, an additional 16S rRNA gene has been identified proximal to the first repeat by Jenni and Stutz (1979).

(ii) Organization of rRNA genes in operons

As discussed above, the rRNA genes are clustered in operons (rrn operons) along with tRNA genes located in the spacer regions. Transcripts of these operons are synthesized as long precursors which are then processed into smaller mature rRNAs and tRNAs. Thus, the transcription starts from 16S rRNA gene and continues till the end of 5S rRNA gene as shown in tobacco by Sugiura and co-workers (Kusuda *et al.*, 1980; Takaiwa and Sugiura, 1980a). Promoter sequences similar to those in prokaryotes at –10 and –35 nt were also detected by the same group at –326 to –319 and –337 to –331 nt upstream of 16S rRNA gene of tobacco (Tohdoh *et* al., 1981). This promoter has been cloned and characterized in pea (Sun *et al.*, 1989) employing a homologous *in vitro* transcription system and has been shown to have properties similar to those of prokaryotic promoters. However, like other chloroplast promoters but somewhat differently from prokaryotic examples, it promotes transcription in supercoiled form of template DNA.

(iii) Genes for ribosomal proteins

Chloroplasts contain around 60 ribosomal proteins, out of which, however, only about 20 are encoded in the chloroplasts. These proteins share a significant homology with amino-acid sequence of *E. coli* ribosomal proteins (in fact, the genes have been located with the use of probes from *E. coli*). Therefore, the same numbering system as per *E. coli* ribosomal proteins has been adopted for naming the choroplast ribosomal protein genes. A number prefixed with rpl refers to a gene coding for a large subunit protein wheras one with the prefix rps codes for a small subunit protein. Earlier in spinach, it was estimated that about 10 out of 24 proteins in the 30S subunit and 8 out of 33 proteins in the 50S subunit are chloroplast-encoded (Dorne *et al.*, 1984). The chloroplast genome of *Marchantia polymorpha* codes for 10 proteins of the 50S and 11 proteins of the 30S subunit (Ohyama *et al.*, 1986). The cpDNAs from *Nicotiana tabacum* and *Oryza sativa,* however, code for 12 proteins of the 30S subunit and 9 proteins of the large subunit (Shinozaki *et al.*, 1986; Hiratsuka *et al.*, 1989; Yokoi *et al.*, 1990). Clearly, from an evolutionary point of view, the genes are in a state of flux and a majority of the genes have already moved into the nucleus, if we believe in the endosymbiotic theory. A particularly interesting example of the dynamic flux is the rpl21 gene which is present in *Marchantia* cpDNA (Ohyama *et al.*, 1986), but is found in the nucleus in higher plants (Smooker *et al.*, 1990).

Introns are present in some of the ribosomal protein genes. In fact, the rpl2 gene from *Nicotiana debneyi* happens to be the first protein-coding gene from a higher plant chloroplast that is reported to contain an intron (Zurawski *et al.*, 1984). This intron was discovered as a result of a direct comparison between the nucleotide sequences of the rpl2 genes from *Spinacea oleracea* and *Nicotiana debneyi: N. debneyi* was found to contain a 666 bp insertion which was absent from the *Spinacea* gene. The genes rpl16, rps12 and rps16 are split (see Sugiura, 1989).

A few genes are of special interest from an organizational viewpoint. An extraordinary feature of the rps12 gene is that it is not only split by two introns but the first exon is also further separated from the remaining two exons which are located far away in the inverted repeat regions (and, therefore, duplicated) on the opposite strands as in the case of *Marchantia* (Fukuzawa *et al.*, 1986), *N. tabacum* (Torazawa *et al.*, 1986), and soybean (Von Allmen and Stutz, 1987). Such an organization requires that these two parts be transcribed separately and, yet, a mature mRNA be formed after *trans*-splicing which has been substantiated both by nucleic acid hybridization (Zaita *et al.*, 1987) and by electron microscopy of DNA-RNA hybrids (Koller *et al.*, 1987; Hildebrand *et al.*, 1988).

Additionally, the gene rpl23 (coding for protein, CL23, belonging to the 50S subunit of ribosomes) is present as a pseudogene with an internal deletion in spinach, rice, and wheat. The deletion in spinach splits it into two overlapping reading frames both of which are represented as specific transcripts, but strangely no protein is detected in an *in vitro* transcription-translation system. In rice, however, an intact copy of the gene is also present in addition to a pseudogene (Hiratsuka *et al.*, 1989) and it would be of interest to know if a normal plant gene for this protein would be present in the nucleus in case of the species having only a defective gene in the chloroplast.

To conclude, although chloroplast ribosomes and the organization of their genes are retained as of the prokaryotic type, the contribution of the chloroplast genome is so incomplete that without the aid of nuclear genes, chloroplast ribosomes cannot be constructed.

(iv) Organization of ribosomal protein genes in operons

Approximately half of the chloroplast genes coding for ribosomal proteins are clustered to form an operon in the order rpl23-rpl2-rps19-rpl22-rps3-rpl16-rpl14-rps8-infA-rpl36-rps11-rpoA, resembling the clustering of spc, S10, and alpha-operons, in *E. coli* (Tanaka *et al.*, 1986; Ohyama *et al.*, 1986; Sijben-Müller *et al.*, 1986; Hiratsuka *et al.*, 1989; Zhou *et al.*, 1989; see also Ozeki *et al.*, 1987). Similarly, rps12 and rps7 genes are also part of an operon resembling the str operon in *E. coli* (see Ozeki *et al.*, 1987). The other genes coding for remaining chloroplast-encoded proteins are scattered throughout the chloroplast genome.

Much less is known about the expression of the ribosomal protein genes. From the large cluster mentioned above consisting of almost a dozen genes, two to four genes were found to be co-transcribed in spinach (Thomas *et al.*, 1988; Zhou *et al.*, 1989) and *Euglena* (Christopher and Hallick, 1989), suggesting that the entire operon might be co-transcribed. Christopher and Hallick (1990) have recently confirmed that these genes are indeed co-transcribed as a pre-mRNA of 8.3 kb which is further processed in a stepwise manner to give rise to various processed products.

d) Genes for transfer RNAs

To date, there is no report of any RNA species having been found to be transported from cytosol to the chloroplasts. Chloroplasts have their own transcription-translation system. We can thus infer that all the tRNAs required for the translational machinery must be synthesized in the chloroplast itself. Haff and Bogorad (1976) first showed the presence of tRNA genes on chloroplast genome by RNA-DNA hybridization.

To identify and localize the genes for individual tRNAs on the circular DNA molecule of spinach, Weil, Herrmann, and their co-workers (Steinmetz *et al.*, 1978; Driesel *et al.*, 1979) separated tRNAs by 2-D gel electrophoresis into about 35 species. The individual ^{125}I-labelled tRNAs were hybridized to DNA fragments obtained by digestion of cpDNA with several restriction endonucleases.

The complete sequencing of the chloroplast genome has allowed definite identification of genes responsible for production of tRNA. It has now been found that the chloroplast genome does, in fact, contain genes for all the necessary tRNA species. About 30 species of tRNA genes from *Nicotiana tabacum* (Shinozaki *et al.*, 1986), 31 from *Marchantia polymorpha* (Ohyama *et al.*, 1986), and 30 from rice (Hiratsuka *et al.*, 1989) have been reported. The various species of tRNA have been identified by their anticodon triplet sequences in the familiar cloverleaf structure and the amino acids to be accepted are predicted from their anticodons according to the universal codon table. According to the "wobble" hypothesis, a minimum of 32 tRNA species are required; but "2 out of 3" mechanism or "extended wobbling" will allow all codons to be read by 30 tRNAs or even a smaller number. Duplication of tRNA genes, because of their location in the inverted repeats, however, makes the total number of tRNA genes 36 or 37. In contrast to *E. coli*, where the 3′ CCA sequence is coded by the tRNA genes themselves, in the chloroplast the 3′ CCA sequence is added post-transcriptionally, as in many cytoplasmic and mitochondrial tRNAs.

An extra tRNA gene present in *Marchantia*, but not in tobacco or rice, is trnR (CCG) which codes for arginyl tRNA. In addition, certain pseudogenes also exist. A possible pseudogene for tRNA/pro (CGG) showing a somewhat incomplete cloverleaf structure has been found in the liverwort (Ohyama *et al.*, 1986); in rice as many as five pseudogenes have been found for different tRNAs (Shimada and Sugiura, 1989).

Recently, it was found that a specific glutamyl tRNA, with an unusual anticodon modification (called "RNA^{DALA}" to distinguish it from tRNA^{Glu}), is involved in chlorophyll synthesis in the chloroplasts where it participates as a co-factor in the synthesis of δ-amino-levulinic acid (DALA) from glutamate (Schön et al., 1986). What is unusual is the finding that the substrate for synthesis of DALA is not free glutamate, but glutamate bound to tRNA. This is a case of a tRNA adapted for a process other than protein synthesis. The unmodified nucleotide sequence deduced from the trnE (UUC) coding for glutamyl tRNA shares a very high degree of homology with the reported RNA^{DALA} in higher plants (see also Ozeki et al., 1987). Since, however, there is only one gene for tRNA$^{\bar{G}lu}$ in the chloroplast genome, this gene must have a dual function for both chlorophyll and protein biosynthesis (i.e. for coding tRNA$^{\bar{G}lu}$), possibly through modification of the primary product.

The presence of introns in chloroplast tRNA genes was first demonstrated in maize trnA and trnI genes (coding for tRNA^{Ala} and tRNA^{Ile}, respectively) which are located in the spacer separating the 16S and 23S rRNA genes (Koch et al., 1981). In contrast, these genes are not interrupted in Euglena (Graf et al., 1980). In tobacco, however, six tRNA genes harbour rather long single introns of 503 to 2,526 bp (Shinozaki et al., 1986). Interestingly, trnG (UCC) coding for glycyl tRNA contains a 691 bp intron in the D-stem region (Deno and Sugiura, 1984) which seems to be a unique feature of the chloroplast-encoded tRNA genes. The trnL (UAA) gene, coding for leucyl tRNA, of maize contains an intron between the first and second nucleotide of the anticodon (Steinmetz et al., 1982). The gene coding for lysyl tRNA, trnK, has the longest intron (2.5 kb) as studied in tobacco (Sugita et al., 1985), mustard (Neuhaus and Link, 1987), and rice (Hiratsuka et al., 1989), and contains on ORF potentially capable of encoding a protein of 60 kD. The segment near the carboxyl-terminal of the deduced polypeptide from the mustard gene (at amino-acid positions 369–471) is structurally related to a portion of maturase-like polypeptides of mitochondrial introns (Neuhaus and Link, 1987). It may be noted here that a similar ORF coding for a polypeptide related to the mitochondrial maturase has been found in the intron of 23S rRNA gene of Chlamydomonas reinhardtii (Rochaix et al., 1985).

All tRNA species whose genes contain long introns are frequently used tRNAs. Therefore, the presence of long introns does not seem to interfere with the effective expression of these tRNA genes in Nicotiana (Wakasugi et al., 1986). Similarly, the presence of duplicated genes has no apparent relationship with codon usage frequency.

In higher plants, the tRNA genes are scattered all over the chloroplast genome, while in Euglena they are clustered at nine different loci on the genome (in groups of 2–6 genes). One of these loci in the rRNA operon is present three times per genome (Hallick et al., 1984).

e) Genes for photosynthesis-related proteins

(i) Gene coding for large subunit of RuBPcase/Oxygenase (rbcL)

Ribulose-1, 5-bisphosphate carboxylase/oxygenase (RuBPcase/oxygenase) is composed of eight large subunits and eight small subunits. Analysis of tryptic digests has shown that the amino-acid sequence may vary in different plant species. Because of the Mendelian pattern of inheritance of primary structure of polypeptides as also their synthesis by poly A^+ mRNAs, small subunits were thought to be coded by the nuclear genome (Kawashima and Wildman, 1972; Gray and Kekwick, 1974; Roy et al., 1976). Chan and Wildman (1972) showed, however, that the large subunit in *Nicotiana* species is inherited maternally and is thus likely to be coded in the chloroplasts.

In vitro linked transcription-translation of cpDNA was used to confirm that the large subunit is encoded by chloroplast DNA in *Zea mays* by Bogorad's group (Coen et al., 1977). It may be mentioned parenthetically that this was the first gene to be cloned and sequenced (coding capacity 475 amino acids) for a chloroplast protein concerned with photosynthesis by the same group (McIntosh et al., 1980). It is, therefore, not surprising that the gene has become the single most widely studied chloroplast gene allowing sequence comparisons to be made to determine even phylogenetic relationship among various plant species (see Zurawski and Clegg, 1987).

In land plants and *Chlamydomonas*, the rbcL gene of 1.8 kb contains no introns while in *Euglena* this gene is quite large in size (6.5 kb) and is interspersed with as many as nine introns of about 0.5 kb each (Koller et al., 1984).

Northern analysis in several plants like corn, spinach, mustard, and pea has revealed a transcript of about 1.6 kb (Link and Bogorad, 1980; Link, 1981; Zurawski et al., 1981). Later studies on mapping of the 5′ ends have revealed that at least two transcripts for rbcL exist *in vivo* in the same plant. In maize, their 5′ ends map at 300 and 65 bp upstream to translation start codon; but in spinach and pea the position of 5′ ends was found to be at 178 and 65 bp away from AUG codon (Crossland et al., 1984; Erion, 1985; Mullet et al., 1985). The larger transcript (1.8 kb) starting at 300 or 178 bp upstream is the primary transcript as could be established by labelling of 5′ end with guanylyltransferase as well as by *in vitro* transcription (Mullet et al., 1985; Hanley-Bowdoin et al., 1985), while the 1.6 kb transcript starting at 65 bp upstream is a processed product. The difference in the start-point of the larger transcript in different species was found to be due to an extra sequence of 130 nucleotides present in the gene in maize. The precise role of differential transcript processing is not known, but its possible significance in regulation of gene expression is discussed later.

While the general situation is summarized above, it is worthy of note that in mung bean and certain other legume species the transcripts are much longer, 2.6, 2.4, and 1.6 kb long (Palmer et al., 1982). Techniques such as S1 nuclease mapping and *in vitro* capping have not yet been applied in the investigations just cited. However, it is possible that recombinational events may have affected not merely the upstream regions of the gene but also the downstream regions, so that a much longer gene has arisen.

Sequences similar to bacterial −10 and −35 promoter regions have been found to be located upstream to the transcription start site of rbcL (see Hanley-Bowdoin and Chua, 1987). Possible stem-and-loop forming structures have been found to be located downstream to the coding region at 85–88 bp in spinach (Zurawski et al., 1981) and 111–153 bp in rice (Nishizawa and Hirai, 1987).

(ii) Genes coding for polypeptides of PS II

In higher plants, PS II is composed of the water-splitting or the oxygen-evolving apparatus, the P680 core complex, and the light harvesting complex. While all the proteins of oxygen-evolving system (33 kD, 23 kD, 16 kD, and 10 kD) and the light harvesting complex (chlorophyll a/b-binding proteins) are encoded in the nucleus, interestingly *all* of the 12 polypeptides of the core complex are coded for by the chloroplast genome in the LSC region.

The gene psbA codes for 32 kD Q_B-binding (D_1) protein. This gene has been studied extensively because of the herbicide-binding nature of the protein product and has been sequenced in a dozen different plants. It was first localized on the chloroplast genome by Bedbrook *et al.* (1978) and was also the first one to be sequenced among the various components of PS II, in *Nicotiana debneyi* (Zurawski *et al.*, 1982a), where it codes for a polypeptide of 353 amino acids. In *Chlamydomonas reinhardtii*, the psbA gene contains four introns and is entirely located in the inverted repeats; thus, two copies of psbA gene per chloroplast genome are present in this case (Erickson *et al.*, 1984). Similarly, the psbA gene in *Euglena* contains four introns (Karabin *et al.*, 1984; Keller and Stutz, 1984). However, an interesting point is that the introns seem to have been lost in the higher plants.

The genes for CP47 (psbB) and CP43 (psbC) were first localized on spinach plastome (Westhoff *et al.*, 1983a). The nucleotide sequence was also determined by the same group (of psbB with 508 codons by Morris and Herrmann, 1984 and of psbC with 473 codons by Alt *et al.*, 1984). A striking feature of gene organization is the tandem arrangement of psbC with psbD (coding for the D2 polypeptide of 353 amino acids) in a manner such that the 5´ end of psbC overlaps the 3´ end of psbD by 50 nucleotides (Alt *et al.*, 1984; Holschuh *et al.*, 1984; Rasmussen *et al.*, 1984). The significance of such an arrangement is not clear; however, this may reflect the consequence of a rather compact arrangement of genes in the chloroplast genome.

The genes psbE and psbF code for 9 kD and 4 kD polypeptides, respectively, which are subunits of cyt b_{559}. These genes have been characterized in spinach and sequenced by Herrmann *et al.* (1984). Bogorad and co-workers reported a gene, psbG, coding for a protein of 24 kD in maize and designated it as the G-protein (Steinmetz *et al.*, 1986), although subsequent studies failed to confirm that the product belongs to PS II (Gray *et al.*, 1990). Another gene, psbH coding for a phosphoprotein of 10 kD was discovered in spinach (Westhoff *et al.*, 1986) and wheat (Hird *et al.*, 1986). More recently, the genes for several smaller molecular weight proteins—of around 2–4 kD—have been revealed after the analysis of the remaining ORFs in the completely sequenced chloroplast genomes. Thus, psbI (36 codons) and psbJ (98 codons) from maize (Kato *et al.*, 1987), *Euglena* (Cushman *et al.*, 1988a), and cyanobacteria (Cantrell and Bryant, 1987), psbK (98 codons) from tobacco (Murata *et al.*, 1988), psbL (38 codons) from wheat (Webber *et al.*, 1989a), and psbM (34 codons) and psbN (43 codons) from a cyanobacterium (Ikeuchi *et al.*, 1989) have been reported. Among these, however, the association of the gene product of psbJ with PS II has not yet been established biochemically.

Certain genes coding for the polypeptides of PS II are clustered to form operons—sometimes even with the polypeptides of other supra-molecular complexes. While organization of most of these operons is similar in various species, variations also exist. Thus, the psbK-psbI genes in rice and barley are located adjacent to psbD (Hiratsuka *et al.*, 1989; Sexton *et al.*, 1990) forming a common operon whereas in tobacco and mustard these two genes are located away as part of a separate operon (see Sugiura, 1989; Neuhaus and Link, 1990). Some of the genes and operons which have been studied more intensively by various investigators are discussed.

psbA (*Gene coding for 32 kD herbicide-binding protein*): The gene psbA, also sometimes referred to as "photogene 32" (Rodermel and Bogorad, 1985), produces a single monocistronic transcript of 1.2 kb in most plants such as mustard (Link, 1982), pea, mung bean (Palmer *et al.*, 1982), and wheat (Hanley-Bowdoin and Chua, 1988). Rochaix and co-workers have shown, however, that as an exceptional case the psbA gene of *Chlamydomonas* (*C. moewusii* as also *C. reinhardtii*)—which is located in the inverted repeats—produces multiple transcripts which arise due to the presence of introns (Erickson *et al.*, 1984; Turmel *et al.*, 1989).

The transcription start site of psbA is supposed to be positioned at 85 to 93 bp upstream of the coding region in spinach (Zurawski *et al.*, 1982a), tobacco (Sugita and Sugiura, 1984), mustard (Link and Langridge, 1984), and pea (Boyer and Mullet, 1986). In pea, however, a processed transcript starting at 68 bp upstream has also been observed. The non-transcribed regions immediately upstream to 5′ ends of psbA are strongly conserved among these species and contain sequences homologous to the prokaryotic sequences at −10 and −35 bp (see Hanley-Bowdoin and Chua, 1987). On the other end of the gene sequences capable of forming stem-and-loop structure consisting of 22 bp stem and 6 bp loop ending in a run of 6Ts were found to be located 1100-1149 bp downstream of the ATG start codon in *Brassica napus* (Reith and Straus, 1987). These characteristics, typical of a prokaryotic transcriptional termination signal, have been found to be strongly conserved in plants.

psbD-psbC-operon: The genes coding for 32 kD-like protein (psbD) and 44 kD protein (psbC) of PS II were first shown to be co-transcribed in spinach (Alt *et al.*, 1984; Holschuh *et al.*, 1984) and barley (Berends *et al.*, 1987). The probes for individual genes hybridized with as many as 10 RNA species on Northern blots. This may be due to a multiple mode of processing of the primary transcripts as the operon encompasses a few other genes in addition to psbC and psbD. The detailed transcriptional patterns and the start sites of various transcripts have been investigated by S1 nuclease mapping in barley (Berends *et al.*, 1987; Gamble *et al.*, 1988) which shows that not only do different transcription start sites exist at the 5′ end of the genes but also multiple termination sites are present at the 3′ end . Illumination can cause newer transcripts to appear in barley. The two new transcripts share a common transcription start site (~ 600 bp upstream to psbD) but have separate termination sites at 175 and 1,175 bp downstream to psbC. The use of alternative start site under different conditions will be discussed later.

The latest sutdies on this operon are those of Sugiura and co-workers on tobacco (Yao *et al.*, 1989). Interestingly, analysis by capping as well as S1 nuclease mapping has clearly confirmed the presence of a separate transcription start site for psbC (earlier Mullet and co-workers could only give indirect evidence for this in barley) which lies within the coding region of psbD! This start site at 194 bp upstream from psbC as well as the other located upstream from psbD is preceded by typical prokaryotic promoter sequences. A similar additional transcription start site upstream of psbC gene has also been detected in pea (Woodbury *et al.*, 1989).

Finally, a point of interest is that in tobacco the start codon of psbC may not be ATG, which is the normal start codon for all the genes, but GTG. The comparison of Shine-Dalgarno sequences by Sugiura and co-workers (Yao *et al.*, 1989) in tobacco for psbC gene indicates that GTG is the start codon instead of ATG codon, which is located further upstream. But the fact that ATG codon is absent altogether in *Chlamydomonas* (Rochaix *et al.*, 1989) and a GTG codon is conserved in all species where psbC has been sequenced indicates that GTG is the start codon for this gene, perhaps in all plants.

psbE-psbF-psbL-ORF40 (genes coding for two subunits of cyt b_{559} *and L protein)*: Herrmann and co-workers in spinach (Westhoff *et al.*, 1985b), tobacco, and *Oenothera* (Carrillo *et al.*, 1986) and Gray, Dyer, and co-workers in wheat (Webber *et al.*, 1989a) recently found a major transcript of 1.1 kb hybridizing with the psbE gene cluster. The 5′ end of this multicistronic transcript is located upstream of psbE, the first gene in the cluster, at 90 and 105 bp upstream in spinach and *Oenothera* (Carrillo *et al.*, 1986), and at 136 bp upstream in wheat (Webber *et al.*, 1989a). Even more recently, Haley and Bogorad (1990) found a similar operon in maize and additionally suggested that two alternative promoters are used for transcription of this operon, one located immediately upstream to psbE and the other probably located further upstream (~ 1,000 nt) near the 5′ end of ORF31 (the exact site of the second promoter has not yet been determined). A stretch of 19 bp capable of forming a stem-and-loop structure has been detected about 1,000 bp downstream of transcription start site in wheat (Webber *et al.*, 1989a) and might be involved in termination of the transcript.

psbB-psbH-petB-petD (genes coding for 51 kD and 10 kD polypeptides of PS II and apocytochrome b_6 *and subunit IV of cyt* b_6/f *complex)*: So far, we have been considering operons comprising genes coding for polypeptides belonging to only one complex, for example PS II (the possibility that unrelated genes comprise the operon, however, cannot be excluded since certain ORFs are yet to be characterized). However, the psbB operon consists of genes coding for polypeptides of *two different* complexes, PS II and cyt b_6/f complex.

The expression of the genes from this operon has been studied in detail for *Marchantia polymorpha* (Fukuzawa *et al.*, 1987), maize (Rock *et al.*, 1987; Barkan, 1988), spinach (Westhoff and Herrmann, 1988), and tobacco (Tanaka *et al.*, 1987). The genes are transcribed as a large transcript of about 5.6 kb starting at about 175 bp before psbB and terminating beyond petD about 83 nt downstream, just behind a sequence capable of forming a stem-and-loop structure. Later, it is processed into 18–20 major species of mRNA varying in size from 400 to 5,600 nt which are products of stepwise processing to remove introns and produce gene-specific transcripts, as established by using smaller probes for Northern hybridization and by S1 nuclease protection anaylsis or reverse transcription. Such processing ultimately results in monocistronic mRNA for subunits of PS II and an intron-less bicistronic mRNA for polypeptides of the cyt b_6/f complex. Barkan (1988) has further shown that the removal of introns from petB and petD transcripts is necessary for translation to occur which can be carried out on both mono- and polycistronic mRNAs. Thus, complete post-transcriptional processing is not a prerequisite for translation to occur.

At the 5´end of the operon, sequences similar to prokaryotic –10 and –35 promoter sequences have been found upstream of the transcription initiation site (Morris and Herrmann, 1984). As for the 3´ end, in addition to the stem-and-loop structure beyond petD, various processing or termination points within the operon have also been recognized. Thus, a 39 bp sequence, 295 bp distal to psbB, which is capable of forming another stem-and-loop structure (Morris and Herrmann, 1984) can possibly be a processing site because this does not act as a termination signal during transcription (Tanaka et al., 1987; Barkan, 1988; Westhoff and Herrmann, 1988).

(iii) Genes coding for cytochrome b_6/f complex

The cyt b_6/f complex is composed of cyt f (33–34 kD), cyt b_6 (23 kD), Rieske FeS protein (20 kD), and subunit IV (17 kD). A new low molecular mass protein of 4 kD has recently been shown to be associated with this complex and has been named "subunit V" (Haley and Bogorad, 1989). Of these, the genes for Rieske FeS protein (petC) along with those of three other proteins of photosynthetic eletcron transport chains, namely, plastocyanin (petE), ferredoxin (petF), and FNR (petH), are present in the nuclear genome.

The genes of cyt f (petA), cty b_6 (petB), and subunit IV (petD) were first located in the spinach chloroplast genome by Herrmann and co-workers (Alt et al., 1983a). petA was first sequenced simultaneously in spinach by Alt and Herrmann (1984) and pea by Willey et al. (1984). This gene codes for a pre-apoprotein of 320 amino acids in several higher plants, except in Oenothera where it is shorter by two codons (Tyagi and Herrmann, 1986). It is close to rbcL. in most of the cases except in legumes (such as mung bean and pea) and Oenothera because of the large inversion that has taken place in these cpDNAs (see Palmer, 1990; Tyagi and Herrmann, 1986). Northern analysis has revealed that petA is transcribed as a part of a polycistronic operon as is evident from a complex pattern of transcripts hybridizing with gene-specific probes (Alt and Herrmann, 1984; Tyagi and Herrmann, 1986; Woodbury et al., 1989; Willey and Gray, 1990). Furthermore, various inverted repeats, capable of forming stem-and-loop structures, in the 5´ upstream and 3´ downstream regions can serve as transcript processing sites generating monocistronic transcripts for petA.

As to genes petB and petD, they are arranged close together (Alt et al., 1983a) and were first sequenced in spinach (Heinemeyer et al., 1984). These genes form a part of the psbB operon and code for polypeptides of 211 and 139 amino acids. Both genes are shown to contain a single intron (of 753 and 742 bp, respectively) which interrupts the open reading frame after the second and third codons in Marchantia (Fukuzawa et al., 1987), rice (Hiratsuka et al., 1989), spinach (Westhoff and Herrmann, 1988), and tobacco (Tanaka et al., 1987). A recent discovery of special interest is that the gene petD in the alga KS3/2 (probably a species of Ankistrodesmus or Monoraphidium) contains ORF in 3,533 bp long intron which can potentially code for an enzyme like reverse transcriptase (Kück, 1989). This is another example of a gene within another gene, and transcribing in the same direction. The transcriptional organization of these genes has already been discussed.

Finally, petG codes for a 37 amino-acid polypeptide—subunit V—reported by Haley and Bogorad (1989), but originally named "petE" by them. The polypeptide itself was discovered as a consequence of its recognition by an antiserum raised against a synthetic decapeptide representing the deduced C-terminus sequence of ORF37 of maize coded by the chloroplast genome. The same gene has also been discovered in rice by Sugiura and co-workers (Hiratsuka *et al.*,1989) and now correctly named "petG", since the name "petE" had meanwhile been assigned to the gene for plastocyanin (Hallick, 1989).

(iv) Genes coding for polypeptides of PS I

The core complex of PS I is constituted of at least 13 polypeptides, out of which genes for only five polypeptides have been localized on the chloroplast genome. These are psaA, psaB, psaC, psaI, and psaJ (Scheller and Møller, 1990). The genes psaA and psaB coding for the P700 chlorophyll apoproteins A and B were first localized in the large single copy region of spinach chloroplast genome (Westhoff *et al.*,1983b) and first sequenced in maize (Fish *et al.*, 1985). They encode proteins of 750 and 734 amino acids which have a fairly high degree of homology (~ 45%). The two genes were found to constitute an operon along with another ribosomal gene rps14 in maize (Fish *et al.*, 1985). Subsequently, a similar organization has been found in several other plants like spinach, barley, and tobacco, although some variations also exist, for example, in *Chlamydomonas*, where all genes are scattered (see Sugiura, 1989).

The gene psaC, i.e., the gene for apoprotein harbouring Fe-S centres A and B, was detected first in tobacco and located in the small single copy region (Hayashida *et al.*, 1987). This was done by comparison of N-terminal amino-acid sequences of the spinach 9 kD apoprotein with the appropriate sequences in tobacco chloroplast DNA. Later, a transcript corresponding to this gene was also detected.

The remaining genes, psaI and psaJ, are small genes and code for polypeptides of 36 and 45 amino acids. The gene psaI was first localized on barley chloroplast genome and sequenced by Scheller *et al.* (1989). This was achieved by first comparing the partial amino-acid sequence of the 4.0 kD polypeptide of PS I with deduced sequences from *Nicotiana* and *Marchantia* chloroplast genomes. Additionally, the sequence of a 4.1 kD polypeptide of PS I from a cyanobacterium (Koike *et al.*, 1989) has been found to be homologous to the deduced amino-acid sequence of an ORF of cpDNAs from higher plants which has been designated as psaJ.

Studies by several groups, for example, of Bogorad, Mullet, and Sugiura in maize, barley, and tobacco have shown a single transcript of about 4.9, 5.3, and 5.2 kb, respectively, which can potentially translate to give two proteins of 66–68 kD (Fish *et al.*, 1985; Berends *et al.*, 1987; Meng *et al.*, 1988). The 5′ ends of the transcripts in maize and tobacco have been localized at 183 and 194 nt upstream of psaA, respectively (Fish *et al.*, 1985; Meng *et al.*, 1988). Further upstream to the transcription initiation site are −10 and −35 canonical sequences of typical prokaryotic promoter. S1 nuclease mapping in barley has, interestingly, revealed two transcripts having the same 5′ ends but differing at their 3′ ends by 26 nt (Berends *et al.*, 1987). The significance of this observation is not yet clear.

In *Euglena*, there are several introns, three in psaA and six in psaB, and a complex array of transcripts hybridizes to gene-specific probes (Cushman *et al.*, 1988b). In

Chlamydomonas reinhardtii, two introns have been found, though only in psaA. However, the gene is split with coding regions widely scattered on the genome with thousands of bp distance between them. The transcripts from each exon, however, can recombine by *trans*-splicing to form a functional transcript (Kück *et al.*,1987; Choquet *et al.*, 1988). The significance of the exons of a gene being separated so far apart is not known. What is of interest is that, despite there being separate transcripts, there seems to be no obvious problem in synthesizing the protein product and having a functional photosynthetic machinery.

(v) Genes coding for ATP synthase complex

The proton-translocating ATPase of chloroplasts is an essential component of light-driven ATP synthesizing system and, as in mitochondria and *E. coli*, it also consists of two parts: one located on the outer surface of thylakoid membranes (CF_1) and the other located within the thylakoid membrane (CF_0). In *E. coli* all the eight genes coding for polypeptides of ATP synthase are arranged in a single (unc) operon (Saraste *et al.*, 1981). In chloroplasts, CF_1 is composed of five polypeptides (α, β, γ, δ, and ϵ) as in *E. coli*, while CF_0 consists of four polypeptides (I, II, III, and IV) instead of three in *E. coli* (Cozens *et al.*, 1986; Hennig and Herrmann, 1986). Of these, the genes for six polypeptides CF_1 α, β, ϵ, CF_0 I, III, and IV are located on the chloroplast genome (Westhoff *et al.*, 1981; Alt *et al.*, 1983b; Bird *et al.*, 1985; Westhoff *et al.*, 1985a; Cozens *et al.*, 1986; Hennig and Herrmann, 1986). The chloroplast-encoded genes of ATP synthase complex are part of two operons: one includes atpI, atpH, atpF, and atpA, while the other consists of atpB and atpE. The operons are further discussed below.

The smaller operon comprising atpB (498 codons) and atpE (137 codons) was sequenced for the first time, simultaneously, in maize (Krebbers *et al.*, 1982) and spinach (Zurawski *et al.*, 1982b). In most of the higher plants, atpB overlaps atpE, so that the last base of TGA stop codon of atpB forms a part of ATG start codon of atpE. However, in pea, the two genes are separated by a 25 bp spacer (Zurawski *et al.*, 1986).

The larger operon is separated from the smaller one by several thousand base pairs. The gene atpA was first localized in spinach (Westhoff *et al.*, 1981) and sequenced in tobacco (Deno *et al*, 1983). The entire gene cluster, atpI-atpH-atpF-atpA, has been characterized in spinach and pea (Hennig and Herrmann, 1986; Hudson *et al.*, 1987) and the genes code for polypeptides of about 247, 81, 183, and 504 amino acids, respectively. In higher plants, atpF gene contains a single intron of 823 bp (Bird *et al.*, 1985).

In higher plants, atpB and atpE are located near the rbcL and, with respect to the latter gene, are co-transcribed in an opposite direction, so that their promoter regions are located in very close proximity (Krebbers *et al.*, 1982; Zurawski *et al.*, 1982b). The transcription of the atpB-atpE genes was first studied in spinach, maize, and pea (Mullet *et al.*, 1985). Although in maize and pea, only one transcript— starting at 302/298 nt and 355/351 nt upstream, respectively—has been observed, a rather puzzling situation has been encountered in spinach, where at least four transcripts starting at 454/453, 273/272, 179, and 99 nt upstream have been found.

Very recently, Orozco *et al.* (1990) have also confirmed the existence of similar multiple transcripts in tobacco, the largest transcript being initiated only 20 bp away from the transcription start site of rbcL and transcribed in the opposite direction. The significance of these findings is not yet known. But *in vitro* transcription studies in spinach by Chen *et al.* (1990) confirm that three of them represent products of transcription from multiple initiation sites (multiple promoters) and not stages in processing of a single primary transcript. The same group has also shown that, in addition to the three transcripts, there is one additional transcript which is leaderless and which is also synthesized *in vivo* (Bennett *et al.*, 1990). It is to be noted that this is the first example of a plastid transcript that is totally devoid of any 5′ untranslated region.

The atpI-atpH-atpF-atpA operon also produces several transcripts (up to 20) of varying lengths (5.5 to 0.5 kb) which reflect polycistronic mRNAs at different stages of processsing (Bird *et al.*, 1985; Westhoff *et al.*, 1985a; Cozens *et al.*, 1986; Hudson *et al.*, 1987). The 5′ and 3′ ends of the transcripts from this operon are located in various intergenic regions, but the 5′ ends of some are located before rps2 (present before atpI) and the 3′ end of some others, after atpA. One wonders whether there is a very large transcript from rps2 to atpA gene encoding all genes. However, none has been detected so far, and one is left to speculate if transcription starts at more than one site.

f) Other putative genes (e.g., genes for NADH dehydrogenase, initiation and elongation fatcors)

The complete nucleotide sequences of cpDNA from tobacco, rice, and *Marchantia* have revealed numerous ORFs whose functions are still not assigned. Since the chloroplast genome is rather compactly arranged (with genes coding for components of both photosynthetic apparatus and protein synthesizing machinery), it seems likely that all the ORFs might be coding for some protein related to chloroplast organization and function.

Physiological studies in *Chlamydomonas* led Bennoun (1982) and Godde (1982) to suggest that a respiratory chain exists also in the chloroplasts. Interestingly, six ORFs—ndhA, B, C, D, E, F—in *Nicotiana tabacum* cpDNA can code for products whose predicted amino-acid sequences resemble those of the respective components—ND1, 2, 3, 4, 4L and 5— of NADH dehydrogenase in the respiratory chain of human mitochondria (Shinozaki *et al.*, 1986). All these ORFs have been shown to be highly expressed at mRNA level (Matsubayashi *et al.*, 1987). Similar sequences have also been found in other land plants (Meng *et al.*, 1986; Ohyama *et al.*, 1986; Schantz and Bogorad, 1988; Hiratsuka *et al.*, 1989). Additionally, another ORF, homologous to human mitochondrial URF6, has been discovered in *Marchantia* (Ohyama *et al.*, 1986) and is named ndhG. Also, a gene—ndhH—has been localized both in *Marchantia* (ORF392) and in tobacco (ORF393) coding for a polypeptide having homology with the nuclear-encoded 49 kD subunit of NADH-ubiquinone oxido-reductase (Fearnley *et al.*, 1989). The ndhA and ndhB genes from *Nicotiana* contain single introns of 1,148 bp and 67 bp, respectively (Matsubayashi *et al.*, 1987). The coded proteins, however, have not been detected in the chloroplasts so far.

Other than genes for NADH dehydrogenase, genes of initiation and elongation factors (with homology to bacterial proteins) are also coded for by the chloroplast genome. The spinach plastid chromosome has been shown to contain an ORF coding for a polypeptide of 77 amino acids which appears to be hydrophilic as deduced by hydropathy plot and homologous to *E. coli* initiation factor IF-1 (Sijben-Müller *et al.*, 1986).

A sequence similar to that coding for the *E. coli* elongation factor (EF-Tu) has been found in three algae, namely, *Chlamydomonas* (Watson and Surzycki, 1982), *Euglena* (Passavant *et al.*, 1983; Montandon and Stutz, 1984), and *Astasia longa* (Siemeister *et al.*, 1990), but is absent from cpDNAs sequenced so far from the land plants.

The ORF231, located immediately upstream to petA in pea, has been shown to encode a putative haem-binding polypeptide (Willey and Gray, 1990). Other ORFs from *Marchantia* (ORF434), tobacco (ORF228), and rice (ORF 230) can also code for a homologous polypeptide. This ORF is co-transcribed with petA and processed further, which results in multiple transcripts as discussed earlier.

Analysis of cpDNA from mutants of *Chlamydomonas reinhardtii* has revealed two other novel functions residing in a 4 kb region on plastome. These are *trans*-splicing of first and second exons of psaA mRNA and reduction of protochlorophyllide (Roitgrund and Mets, 1990).

g) Gene coding for a zinc finger-containing protein

A novel gene that may code for a protein containing Zn finger, has been found as an important structural feature of several *trans*-acting factors involved in regulation of gene activity in eukaryotes.

Sasaki *et al.* (1989) sequenced an ORF in far-upstream region of petA gene in pea which may code for such a putative protein. This gene is conserved also in tobacco and pea as in *Marchantia*. A point of further interest is that the gene can undergo transcription even in the absence of *de novo* protein synthesis while it is necessary for expression of certain other genes like rbcL. Apparently, the gene product is required ahead of chloroplast biogenesis, in a temporal sense, thus suggesting a regulatory role for the putative protein. The presence of this protein has not yet been shown in chloroplast of higher plants.

h) Transcription map

As we have discussed earlier, the chloroplast genome is compactly filled with genes coding for chloroplast products, and most of the genes are functional. Link (1984b) published the first studies on the hybridization of labelled chloroplast RNA with 50 mustard cpDNA fragments indicating that a rather large percentage of the DNA is indeed transcribed. Although the transcripts for certain genes like psbA and rbcL could be recognized, the main limitation of this study had been lack of information on precise location of various plastid genes on the chloroplast genome.

However, a detailed transcription map is now available for pea chloroplast genome in which the location of various genes whether arranged individually or in clusters is reflected. Palmer, Thompson, and co-workers (Woodbury *et al.*, 1988) used 53 cloned cpDNA fragments, representing 90% of the chloroplast genome, to analyse

Northern blots of pea RNA. They found that nearly 71% of pea chloroplast genome is represented as detectable transcripts. These transcripts were then assigned to particular regions of DNA by comparison of transcript patterns hybridizing to individual DNA probes. As was observed in the earlier studies of Link (1984b), these transcripts were also found to be more abundant in light-grown than in dark-grown seedling indicating that the overall expression of cpDNA is greatly enhanced by light.

The same group (Woodbury et al., 1989) more recently prepared a transcription initiation map of pea chloroplast genome by using a set of 73 clones containing about 90% of pea chloroplast genome. The chloroplast RNAs were capped in vitro with α-^{32}P-GTP (using guanylyltransferase) and hybridized to the cloned fragments. Ultimately, the DNA fragments capable of protecting the transcripts from RNase treatment were identified; these contained transcription initiation site(s), The results showed that there are about 33 such transcription sites located throughout the plastome representing the start-points of all polycistronic units. These studies have significantly enhanced our understanding of the transcriptional organization of the chloroplast genome.

III. Regulation of Chloroplast Gene Expression

Germinating seeds as also meristematic cells in an adult plant contain an immature form of plastids, the proplastid. Under conditions of darkness, the proplastids may develop only into etioplasts. On illumination, they develop into chloroplasts. Development of proplastids or etioplasts to any kind of mature plastid, such as a chloroplast, involves a dramatic change not only in terms of change in colour because of formation of pigments, but also in terms of enhanced DNA replication, changes in RNA and protein synthesis, membrane organization, and a number of other physiological processes.

What exactly controls such changes during development? There could be many alternative mechanisms, and more than one could work together. Major changes are brought about by light, which sets in a cascade of reactions and the etioplasts develop into functional chloroplasts. But again, at which step does light act, to bring about such changes? Is it at the level of RNA synthesis, or a post-transcriptional step such as RNA processing, or protein synthesis? Or are all these steps involved? Although the major control mechanisms operate in the nucleus, because of the restricted scope of this review, only the reports about chloroplast genes and the various factors that regulate their expression are discussed here.

a) Status of chloroplast RNA polymerase

(i) RNA polymerases of two types

The DNA-dependent RNA polymerases were first isolated in soluble form from chloroplasts of maize (Bottomley et al., 1971) and wheat (Polya and Jagendorf, 1971). Subsequently, more work appeared not only in maize (e.g., by Kidd and Bogorad, 1979; Orozco et al., 1985), but also in other plants such as spinach (Briat and Mache, 1980; Gruissem et al., 1983a; Lerbs et al., 1983, 1985), pea (Tewari and Goel, 1983), Euglena (Gruissem et al., 1983b), and mustard (Link, 1984a). Some of these enzyme preparations were sufficiently purified yielding a subunit composition of 7 to 14 polypeptides ranging from MW 25,000 to 180,000 (Kidd and Bogorad, 1979; Tewari and Goel, 1983; Lerbs *et al.*, 1983, 1985).

Another chloroplast RNA polymerase activity has been isolated in the form of DNA-protein complex, i.e., as transcriptionally active chromosome (TAC), from chloroplasts of *Euglena* (Hallick *et al.*, 1976), *Chlamydomonas* (Dron *et al.*, 1979), and spinach (Briat *et al.*, 1979). The enzyme is extremely difficult to dissociate from TAC; but the highly purified TAC enzyme from *Euglena* contains only two major polypeptide subunits of MW 118,000 and 85,000 as shown by Hallick and co-workers (Narita *et al.*, 1985).

The two types of RNA polymerase activities preferentially transcribe different chloroplast genes. The TAC enzyme, as studied in *Euglena* (Rushlow *et al.*, 1980) and spinach (Briat *et al.*, 1982), specifically transcribes the ribosomal RNA operon. This preference can be experimentally related to the presence of rrn operon in the inverted or tandem repeats. This is because although in pea chloroplasts—where inverted repeats have been lost—the TAC enzyme seems to be capable of transcribing the entire chloroplast genome (Tewari and Goel, 1983), the transcription initially starts at the rrn operon. As to the soluble enzyme isolated from the chloroplast extracts of spinach, mustard, maize, and *Euglena*, it transcribes tRNA and various mRNAs from supercoiled plastid DNA templates (see Gruissem, 1989).

The similarity of promoter sequences of chloroplast genes with those of *E. coli* consensus promoter sequences as well as the ability of *E. coli* RNA polymerase to transcribe chloroplast genes suggests the enzyme from both sources to be similar. As discussed earlier, the presence of similar polypeptide subunits coded by chloroplasts and their immunological cross-reactivity also support this view. However, the chloroplast enzyme differs from the *E. coli* one in some respects: (1) it does not efficiently recognize *E. coli* promoter regions (Gruissem and Zurawski, 1985); (2) it is insensitive to rifampicin, a potent inhibitor of the bacterial enzyme (Surzycki and Schellenberger, 1976); and (3) the sigma factor of *E. coli* RNA polymerase does not enhance the activity of chloroplast enzyme *in vitro* (Jolly and Bogorad, 1980).

On the other hand, the chloroplast enzyme differs from nuclear RNA polymerases II and III in that it is not inhibited by α–amanitin. Thus, the choroplast RNA polymerase is closer to the prokaryotic enzyme, but by no means similar or identical to it.

(ii) Sigma-like factors

As the first step towards gene expression, RNA polymerase holoenzyme has to recognize promoters such that transcription can be initiated precisely at specific sites and in a definite orientation. This is presumably accomplished through the sequence-specific binding of σ factors. Although an ORF showing homology with σ factor of *E. coli* has not been found so far, the possibility cannot be ruled out that a corresponding chloroplast gene, such as rpoD coding for σ factor, has not yet been detected due to very low homology with known prokaryotic protein or has moved during evolution to the nucleus. Nevertheless, a protein fraction implicated with sigma-like activity, on the basis of its ability to confer rifampicin insensitivity to core RNA polymerase, was obtained from *Chlamydomonas reinhardtii* (Surzycki and Shellenberger, 1976). Also, a maize chloroplast S factor was shown to preferentially stimulate transcription of cloned chloroplast gene by chloroplast RNA polymerase *in vitro* (Jolly and Bogorad, 1980).

More recently, σ-like activity has been detected also in chloroplast extracts of mustard (Bülow and Link, 1988) and spinach (Lerbs *et al.*, 1988) which may be responsible for enhanced transcription of chloroplast genes. If this subunit is nuclear-encoded, the nucleus may be capable of controlling the expression of chloroplast genes by σ factor supply, possibly differentially.

b) Gene dosage

When dark-grown plants are illuminated, there is an increase in the plastid number and simultaneously the DNA copy number per plastid. Thus, in developing chloroplasts the number of cpDNA molecules may increase from about 20 per plastid to 80–100 per plastid or even more (see Herrmann and Possingham, 1980). This increase obviously means increase in the copy number of every gene coded by the plastome.

Sasaki *et al.*, (1984, 1987) studied the expression of the gene for a large subunit of RuBPcase/oxygenase (rbcL) as controlled by light in pea and suggested that the gene dosage, in part, is responsible for the light-induced changes in mRNA level for rbcL.

The involvement of plastid DNA levels on the overall transcriptional activity of plastids was studied by Deng and Gruissem (1987). It was found that in spinach, with some decrease in plastid DNA level during chloroplast development, there is a three-fold increase in transcription in the initial phase in terms of transcripts made per plastid. In mature chloroplasts, however, there is five-fold decrease in transcription activity though DNA levels were still maintained high. From these results, they concluded that the transcriptional activity is independent of plastid DNA levels during developmental transitions. In other words, transcriptional activity does vary depending on the developmental stage.

Hosler *et al.* (1989) recently studied the relationship of gene dosage with the transcript levels and protein accumulation in the chloroplasts of *Chlamydomonas reinhardtii*. They used an analog of thymidine, 5-fluorodeoxyuridine which can reduce the cpDNA copy number preferentially as compared to the nuclear genome which remains constant during greening. They found that when the plastome copy number was lowered to one-fourth, the mRNA levels for rbcL, atpA, and rp12 did decrease, but there was no concomitant decrease in the protein levels. In fact, there was no change in the amount of translatable mRNA for these genes (as judged by immunological assay of *in vitro* translated products). They concluded that although a gene dosage effect may operate, there is a further step after transcription which may be specially subject to regulatory events (such that high levels of protein products are still maintained despite decrease in plastid DNA copy number).

c) Template topology

Another mechanism of regulation at (gene) DNA level could be availability of DNA template for RNA polymerase with respect to template topology, a subject which has been receiving much attention recently in relation to gene transcription (see McClure, 1985; Wang, 1985). Indeed, it may be that DNA topoisomerases represent an important class of enzymes which help in transcription of any gene by RNA polymerase, by relaxing supercoiled DNA or coiling DNA that is too relaxed. Presence of topoisomerases has been demonstrated not only in nuclei of higher organisms, but also in chloroplasts and mitochondria.

Bogorad and co-workers (Stirdivant *et al.*, 1985), using calf thymus topoisomerase I to generate different topoisomers of a plasmid containing two divergently transcribed chloroplast genes, i.e., rbcL and atpB, were the first to show that transcription was also regulated by the topology of the DNA *in vitro*. While fully relaxed template could not support transcription, increasing superhelicity increased their transcription. With more superhelical forms, however, rbcL could not be transcribed; this was in contrast to atpB, which could still be transcribed. The above results were also confirmed by Lam and Chua (1987), who could show dependence of transcription of these genes on the state of the template using a topoisomerase I inhibitor, novobiocin. The presence of DNA gyrase activity in addition to topoisomerase I, in chloroplast extracts further confirms that template topology may also be a regulatory factor.

An interesting observation has recently been made by Thompson and Mosig (1990) with respect to the torsional stress in the cpDNA of *Chlamydomonas*. They have used a UV-cross-linking assay based on the intercalation in DNA of HMT (4′ hydroxymethyl-4, 5′ 8′-trimethylpsoralen) and showed that, after exposure of cells to light, there is reduced intercalation indicating thereby that the torsional stress of cpDNA is reduced. Experiments with use of inhibitors such as atrazine and cycloheximide suggest that the effect of light is independent of photosynthesis and requires cytoplasmic protein synthesis. In summary, this study shows that light can affect not only transcription but also replication. Further studies, however, are required to prove this point.

d) Chloroplast promoters

As in bacteria, chloroplast genes contain at their 5′ upstream non-coding regions sequences similar to prokaryotic promoter sequences, the consensus being TATAAT at –10 nt and TTGACA at –35 nt upstream from the transcription initiation site; these can even be recognized by *E. coli* RNA polymerase. Of course, some genes are clustered to form operons, and therefore only the proximal genes of the operon in the upstream region contain these sequences.

The best approach to a study of regulatory or promoter sequences of any gene is via reverse genetics, i.e., of constructing a transgenic plant, with a modified promoter to examine its activity. This is because in higher organisms important sequences often exist far upstream and cannot be detected merely by comparison of consensus sequences. However, no efficient system has so far been developed for chloroplast transformation. Nevertheless, *in vitro* transcription systems have been developed to transcribe plastid DNAs (see Hanley-Bowdoin and Chua, 1987), which have greatly helped studies on the regulation of chloroplast genes.

Using such *in vitro* transcription system and employing *E. coli* RNA polymerase, Gruissem and Zurawski (1985) analysed the transcription mediated by promoters of chloroplast genes. They found that the –10 and –35 regions of chloroplast promoters show homology with the prokaryotic promoters not merely at the level of base sequence, but even in respect of the number of base pairs that separated them. Usually, –35 and –10 like sequences need to be separated by 17 or 18 bp. The type of nucleotides in this stretch does not matter, but an alteration in the number of nucleotides results in reduction in transcription. The strength of promoters also depends on the sequences comprising the –10 and –35 boxes. This was concluded from studies on psbA, rbcL and atpB promoters, when tagged with trnM2 gene. From such analysis, they arranged the –35 regions of the promoters according to their intrinsic strength as: TTGACA > TTGCTT > TTGCGC.

A system of special interest for studies on promoters and their interaction is provided by atpB and rbcL genes. These two genes, in higher plants, are located on opposite strands and transcribed in opposite directions, with their 5´ ends separated by 89–110 bp. Separation by such a limited stretch may not favour independent transcription of the two genes. In spite of this, the genes do express differentially. In light-grown plants, rbcL is generally strongly expressed in leaves while in the dark, both rbcL and atpB are expressed at similar levels (Deng and Gruissem, 1987, 1988). Various possibilities exist for the differential expression of these genes: variable promoter strength, or binding of specific protein factors possibly induced by a trigger such as light, or altered conformation of the template.

However, Hanley-Bowdoin and Chua (1989) while working on this region of pea found that the two promoters are somewhat interdependent. Insertional mutations showed that when the promoter of atpB gene was disrupted by a 8 bp insertion at –35 region, the transcription of rbcL remained unaltered; yet a 2 bp insertion in rbcL promoter (between –10 and –35 region) reduced the expression of atpB. When rbcL promoter was totally removed, atpB transcription increased. The results were quite different when they were separated by a long (500 bp) insert of psbA coding region which resulted in their independent transcription. This would mean that the expression of the genes is not merely dependent on promoter strength but also on their location; the motifs could possibly interact *in vitro* in *cis* and in a spacing-dependent manner.

From the studies on nuclear-encoded genes, it is known that transcription of genes is regulated in an important manner by *trans*-acting factors. Attempts have, therefore, been made to discover such factors involved in the regulation of chloroplast genes. Chua and co-workers (Lam *et al.*, 1988) studied interaction of chloroplast proteins with common upstream sequences of atpB and rbcL genes in maize, pea, spinach, and tobacco. They found that a protein CDF1 of about 115 kD does bind to sequences between –16 and –101 nt upstream of the rbcL transcription start site as shown by DNase footprinting. This activity is exclusive of binding of RNA polymerase since during the gel filtration step (which precedes other analyses), the enzyme is excluded in the other fraction. The protein possibly binds to an octanucleotide in this region as established by exonuclease III protection assay and sequence comparisons of promoter regions from various plants.

Recently, Link and co-workers studied the DNA-protein interactions in the promoter region of psbA (Eisermann *et al.*, 1990). By use of gel retardation assay, they found that the proteins from etioplasts and chloroplasts of mustard bind differentially to the promoter regions of psbA (the binding being stronger in chloroplasts), thus indicating that *trans*-acting factors may play an important role in regulation of gene expression in plastids.

e) Role of 3′ ends

The chloroplast genes or gene clusters are also flanked at the 3′ end of the coding region by inverted repeat sequences which are capable of forming stem-and-loop structures in the resultant transcript. Initially, it is only from homology with the prokaryotic transcription termination sites that these sequences were considered to be termination sites for the chloroplast transcripts (Zurawski *et al.*, 1981; Alt *et al.*, 1984; Heinemeyer *et al.*, 1984). Later, however, S1 nuclease mapping experiments conclusively proved them to be 3′ ends of the transcripts (Westhoff and Herrmann, 1988; Barkan, 1988).

Whether these sequences are really transcription termination sites or 3′ processing sites has been examined by Stern and Gruissem (1987) and Chen and Orozco (1988) in chloroplast extracts capable of *in vitro* transcription. Employing 3′ ends of psbA, rbcL, atpB-E, petD, and trnH, they showed that the inverted repeat sequences are not necessarily transcription termination sites because transcription is often carried forward from these sites. Instead, these sites may act as post-transcriptional processing sites in the RNA maturation. Stern and Gruissem (1987) concluded that these sequences also stabilize and protect RNA from nucleases. They found that of the various plasmid constructs made and introduced in *E. coli* for transcription, transcripts of those containing inverted repeat sequences were more stable than those lacking them. A similar suggestion has been made by Westhoff and Herrmann (1988) in case of psbB gene to the effect that the 3′ stem-and-loop forming sequences most probably prevent the nucleolytic cleavage of RNAs.

The most recent study of 3′ ends is by Adams and Stern (1990). They induced mutations in stem-loop structures at the 3′ end of psbA of spinach and studied the stability of the transcripts both *in vitro* as well as in intact chloroplasts after electroporation. They found that the thermodynamic stability of 3′ stem-and-loop structures alone is not responsible for the stability but their sequences and structure might also play an important role in determining the mRNA stability. The mutant RNAs are susceptible to RNase activity *in vitro*, but are stable in chloroplasts after electroporation suggesting that certain protein factors in the chloroplasts might also be involved in the stability.

Some questions remain unanswered, however, as to how the 3′ end processing occurs at all and how it confers stability to the transcript particularly in light-grown seedlings. As far as the mechanisms of termination of transcription and processing of 3′ ends are concerned, it may be recalled that in prokaryotic systems such as *E. coli*, proteins like the rho-factor are involved in termination of transcription and the stem-and-loop structure at the 3′ end of the RNA is considered to somehow help in this process. Attempts have been thus made to unravel the involvement of a protein in this reaction. Stern and Gruissem (1987) showed that in spinach, one or more proteins are required for the accurate processing to occur *in vitro*, since proteinase K or heat treatment inhibits the processing reaction. Nickelsen and Link (1989) studied the interaction of 3′ region of mustard trnK gene with proteins from chloroplast extracts. They have found three proteins of 70, 62, and 58 kD which could specifically bind and bring about termination of transcription to 3′ regions, 26 bp further downstream to stem-loop structure. However, the exact role of these proteins in processing the 3′ end is yet to be experimentally elucidated.

f) Effect of light

It is certain that light plays a very significant role in the development of chloroplasts. Exactly at which step it regulates expression is, however, a matter of debate. The earlier reports on effect of light on mRNA levels indicated that light is involved in regulation at the transcriptional level. Many authors found tremendous increase in mRNA levels for genes like rbcL in maize (Rodermel and Bogorad, 1985), mustard (Link, 1982, 1984b), pea (Palmer *et al.*, 1982; Thompson *et al.*, 1983; Sasaki *et al.*, 1984), spinach (Zurawski *et al.*, 1981) and psbA in maize (Bedbrook *et al.*, 1978; Rodermel and Bogorad, 1985), mustard (Link, 1982), pea, and mung bean (Palmer *et al.*, 1982; Thompson *et al.*, 1983) on exposure to light. But mere increase in mRNA levels does not necessarily mean that the transcription of the gene is specifically regulated by light. This is because the increase in the mRNA levels may be primarily due to increased template copy number, or RNA stability rather than transcription per se.

(i) *Light regulation: Transcription may have only a limited role and post-transcriptional mechanisms may be more important*

The realization that transcription may have only a limited role has come only in the last five years after detailed studies have been undertaken on selected systems with respect to particular genes. The genes on which much of the work has been focussed are: rbcL, psbA, psbB, psbD-psbC of PSII, psaA and psaB of PSI, petA, petB-petD of cyt b_6/f complex and atpA and atpB–atpE, of ATP synthase complex. In some cases, the investigations have been focussed on operons so that the information on all the constituent genes is available, e.g., psbD-psbC, psbB-psbH-petB-petD, and psaA-psaB. But in other cases expression of individual genes such as psbA, rbcL, and atpA has been studied.

Some of these studies show that light can indeed regulate the gene expression and recently as much as 20- to 30-fold increase in the level of transcription has been found. Generally, however, the increases in steady-state levels of mRNA are smaller—only two- to three-fold—which is far too low to account for the significant increase in translation products. Many workers think, therefore, that the major influence of light must be on translation or some post-transcriptional event such as processing of primary transcripts or stability of transcripts.

Having given the broad conclusions of recent studies, we can go into the cases of certain genes in more detail so that we can have a better idea of the regulatory processes involved. Because of limited space, the main discussion will be confined to only two genes, while others will be covered more generally.

(ii) The rbcL gene

Several investigations have been made on light-dependent expression of the rbcL gene as mentioned earlier. But, as an example of detailed studies, we may first take up those made by Mullet and co-workers (Klein and Mullet, 1986, 1987, 1990; Klein *et al.*, 1988; Mullet and Klein, 1987), who analysed the expression of rbcL at both mRNA and protein levels per plastid in young seedlings of barley, transferred from darkness to light. They found that high levels of rbcL transcripts are already present in dark. Though there is no significant increase in transcripts within one hour, they do increase during the first 16 hours of illumination, and then decline as a function of plant age. As to protein synthesis, rbcL transcripts from etioplasts do undergo some translation both *in vivo* and *in vitro;* but there is a significant increase after illumination (protein synthesis, however, also declines in consonance with the decrease in transcript levels after prolonged illumination). To investigate why translation proceeds more efficiently in chloroplasts than in etioplasts, thése workers fractionated the polysomes from the lysed plastids of etiolated and illuminated seedlings and found that the rbcL transcripts from etioplasts were distributed throughout the polysome gradients with the largest fraction associated with small polysomes. On the other hand, the rbcL transcripts from illuminated seedlings had larger proportion bound to large polysomes, consistent with increase in protein synthesis. Thus, these studies show that although light does have some effect on transcription, the major control may be at the translational level.

In other studies, employing transcription "run-on" system in spinach, it was found that when dark-grown plants are illuminated, there is indeed an increase in the transcriptional activity (by around three-fold) of all chloroplast genes studied, which declines to one-fifth during leaf development (Deng and Gruissem, 1987, 1988). The increase reflects a general enhancement in transcriptional activity of the chloroplast genome. The specific modulation of steady-state levels of a particular gene such as rbcL would, therefore, be a result of some post-transcriptional step like enhanced processing of primary transcript or controlling the stability of transcript.

The work of Klessig and co-workers on amaranth is also worth a special mention. In initial studies, they found that when the dark-grown seedlings are illuminated, there is almost a 20-fold increase in mRNA levels for rbcL during greening (Berry *et al.*, 1985, 1986). However, in the first four hours there is a two- to four-fold increase in protein levels, while increase in mRNA levels starts only after five hours. This means that protein synthesis starts even before any increase in mRNA level is triggered. Moreover, when the light-grown plants were shifted to darkness, there was a sudden drop in protein levels (approximately 20-fold by two hours and more than 50-fold by six hours). The decline in protein levels, however, was not accompanied by decrease in transcript levels, demonstrating that some translational block is indeed operative. The authors propose that for large subunits the rapid responses to the changes in the environment would be accomplished largely at the translational level, whereas long-term changes may be regulated at the level of transcription.

To ascertain how exactly such regulation is achieved, Berry *et al.* (1988, 1990) isolated the polysome fractions and found that in light-grown plants the rbcL transcripts are attached to the polysomes (they remain so even if plants are shifted for limited period to darkness), while in etioplasts the transcripts are not associated with polysomes. Translational run-off analysis of these polysomes *in vitro* indicated that large subunit polypeptide could be synthesized with the polysomes from light-grown or dark-shifted cotyledons, but not with polysomes from etiolated seedlings. These results indicate that translation can be down- or up-regulated by some mechanism such as enhancement of binding of mRNA to ribosomes which is, however, yet to be unravelled.

An interesting observation has been made in case of rbcL gene of spinach. As discussed earlier, in this case there is processing of 5´ ends of the transcripts (Mullet *et al.*, 1985), but the processing occurs in such a way that transcripts with two distinct 5´ ends accumulate differentially in light- and dark-grown plants. Thus, the transcripts with 5´ ends at 65 nt upstream to AUG codon accumulate in etiolated seedlings. On the other hand, in illuminated seedlings the transcripts starting at 178 bp upstream predominate. What exactly mediates such differential processing in light and dark is not yet clear and we also do not know of what benefit this is to plants.

(iii) The psaA-psaB operon

In the case of psaA-psaB operon, too, there is much evidence for a significant translational control on production of ultimate gene products. As in rbcL, Mullet and co-workers (Klein and Mullet, 1986, 1987; Mullet and Klein, 1987) found that here too fairly high amounts of transcripts are present in the dark-grown seedlings. However, at this stage there are very low or undetectable levels of corresponding polypeptides, though—after illumination for one hour by white light—the synthesis of chlorophyll apoproteins is rapidly induced with a time course similar to that of chlorophyll biosynthesis from protochlorophyllide (unlike rbcL, here mRNA level does not increase even transiently and in fact declines).

The above results suggest that there is some block working at translational level which does not allow the translation of psaA-psaB transcripts in dark. Later, Klein *et al.* (1988) observed that, both in etioplasts and chloroplasts the psaA-psaB transcripts are associated with membrane-bound polysomes but in etioplasts they are not translated even when bound to the polysomes. The exact mechanism of translational block is still unclear. However, when the membrane-bound polysomes were treated with a detergent, polyoxyethylene 10-tridecyl ether, polypeptide synthesis *in vitro* was enhanced and these results indicate that the block may work in dark at the level of polypeptide chain elongation and is functional only when polysomes are membrane-bound.

Light-dependent accumulation of P700 chlorophyll *a* apoprotein in barley, as controlled at the level of translation, has been demonstrated also by Apel's group (Kreuz *et al.*, 1986). It was found that mRNAs for psaA-psaB genes are already present at high levels in the dark and they are bound to polysomes even in etioplasts. However, as demonstrated by immunogold labelling methods in ultrathin sections, the chlorophyll-binding protein is below the limit of detection in dark-grown barley seedlings. Upon illumination with white light, a rapid accumulation of protein occurs. The same group has also provided evidence that the light effect may be mediated through a new photoreceptor, protochlorophyllide, rather than phytochrome (Laing *et al.*, 1988). This matter will be further discussed in Section IIIg, on photoreceptors.

(iv) Other genes

The investigations on other genes highlight the following points:

In general, transcriptional activity of the plastid genome increases in etioplasts as well as chloroplasts as a function of the age and development (regardless of exposure to light, it decreases at later stages). But the activity increases relatively more in illuminated seedlings. Thus, transcription appears to be largely constitutive. However, the strength of the promoter may control transcriptional level of individual genes. Response to stimuli such as light and steady-state levels of transcripts seems to be controlled more at a post-transcriptional level, such as processing or difference in stability of RNAs.

Alternative promoters may be used in light and dark. Also, different types of transcripts may be made from the same promoter.

Importance of strength of promoters and transcriptional processing: That light has a stimulative effect on all genes and that the stimulation is general rather than gene-specific has been shown by Deng and Gruissem (1987). They analysed the transcriptional activities of 10 plastid genes (psbA, rbcL, rpoA, atpB, rp12, psaA, psbB, petB, petD, and rRNA) by run-on transcription assay in illuminated spinach seedlings and found that the general transcriptional activity of the whole chloroplast genome follows the same pattern (of initial increase followed by decrease) in both dark and light. On illumination with light, there was an increase in transcriptional activity, but there was no major change in *relative* transcriptional activities. The increase in steady-state levels of various mRNAs is, therefore, most likely due to an effect of light at a post-transcriptional level such as increased stability of transcripts (Stern and Gruissem, 1989). Similar observations were also made in mustard and barley (Dietrich *et al.*, 1987; Mullet and Klein, 1987; Klein and Mullet, 1986, 1987).

The relative abundance of steady-state transcripts of various genes may also be controlled by the strength of promoters of individual genes. Thus, it was shown that the promoter of psbA gene is relatively stronger and the gene gives rise to a relatively larger number of transcripts (Gruissem and Zurawski, 1985; Deng and Gruissem, 1987). Link and co-workers (Fiebig *et al.*, 1990) studied the expression of six genes clustered around psbA gene in mustard. To account for the variety of responses, the genes have been divided into three groups: (1) those showing light-enhanced expression (psbA as well as other genes such as psbB and psaA-psaB); (2) those showing light-independent or constitutive mode of expression (genes such as psbK, psbI, and genes for tRNA); and (3) those showing transient light-enhancement and then constitutive mode of expression e.g., trnK gene, which contains an intron that probably codes for a maturase-like polypeptide.

Significance of multiple promoter sites in regulation or production of multiple transcripts from the same promoter: Haley and Bogorad (1990) have shown that for the two adjacent but divergently transcribed operons of maize, psbE-psbF-psbL-ORF40 and ORF31-petG-ORF42, there are two alternative promoters. In the latter operon, it has been shown that the promoters are used differentially in light and dark. It is not clear how light triggers transcription from different start sites. A somewhat similar situation also occurs in psbD-psbC operon. As discovered in barley (Berends *et al.*, 1987; Gamble *et al.*, 1988; Sexton *et al.*, 1990) and tobacco (Yao *et al.*, 1989), two psbD-psbC polycistronic transcripts have different 5′ ends in light- and dark-grown seedlings. This results in the synthesis of two entirely new transcripts in light. In barley, the transcripts have recently been shown to arise from use of different promoters and not from processing of longer transcripts. Even though the transcriptional activity is reduced as a function of age, the light-induced transcripts are still maintained at high levels, because of the stability of these transcripts as compared to the others (from etiolated leaves), which are degraded faster. Even if transcription is carried out *in vitro*, employing extracts of chloroplasts or etioplasts, the capacity of the system to initiate transcription at specific sites is retained and it has been suggested that there may be some factor which allows selection of the appropriate initiation site. Till now, it is not clear, however, what the factor is and how it regulates the transcription of the operon.

In conclusion, it may be said that, in the majority of cases studied, expression of chloroplast genes by light is regulated predominantly at post-transcriptional or translational levels (Gruissem, 1989). One must, however, keep the option of transcriptional level regulation open, at present, since in some cases, light does seem to act at this level (Klein and Mullet, 1990; Schrubar *et al.*, 1991).

g) Light intensity, quality and photoreceptor(s)

(i) Light intensity

Some attention can be given to adaptational strategies in response to change in light intensity. Jordan *et al.* (1989) found that in lettuce, grown under both low and high light irradiances, the expression of rbcL, psaA, psbA, and rDNA remains the same but the mRNAs for atpB-E and petB-D genes increase under high light irradiances. The change in mRNA levels seems to be related to the alteration in the corresponding protein levels.

(ii) Effect of red light—Phytochrome or protochlorophyllide control?

As should be abundantly clear by now, light plays the most important role in chloroplast development. Chloroplasts of higher plants can adapt to different environmental light conditions that allow the plants to succeed in a rather wide variety of habitats. It is thus of great interest also to know which photoreceptor regulates the expression of these genes and at what level.

The early experiments of Link (1982), Thompson *et al.* (1983), and Zhu *et al.* (1985) clearly indicated that transcripts for some genes like *rbcL* and *psbA* increase after illumination of etiolated seedlings with red light, which on subsequent treatment by far-red light declined, thus indicating the involvement of phytochrome in regulation of expression of chloroplast genes. But in some cases, like mung bean, a phytochrome effect of rbcL transcripts was not so apparent (Thompson *et al.*, 1983). Recently, however, Apel and co-workers published results of a study which necessitate taking a second look at phytochrome involvement in the regulation of expression of psaA (Laing *et al.*, 1988). They showed that the accumulation of P700 apoprotein is light-dependent but phytochrome *independent* since far-red light does not reverse the effect of red light. According to them the actual photoreceptor may be protochlorophyllide, and not phytochrome. As is well known, on exposure to red or white light, protochlorophyllide is reduced to chlorophyllide and subsequently esterified to yield chlorophyll. Thus, the protochlorophyllide holochrome could be involved in the control of chlorophyll biosynthesis (Harpster and Apel, 1985). The authors found that in a mutant of barley, which lacked protochlorophyllide, P700 apoprotein did not accumulate. Although more definitive evidence is required in favour of a new role of the pigment, the suggestion has since been supported by other workers, for example, Klein *et al.* (1988) and Deng *et al.* (1989), in respect of accumulation of chlorophyll *a* apoprotein.

(iii) Evidence for a blue light control

Evidence for mediation of chloroplast development by blue light was first provided by Richter and co-workers (see Richter, 1984). The dark-grown suspension cultures of *Nicotiana tabacum*, when illuminated by blue light continuously, caused the development of chloroplasts. Simultaneously, there was an increase in steady-state levels of transcripts for rbcL and psbA as well as the proteins. Interestingly, this effect was negated by red light when given at early or later stages of chloroplast development (Richter and Wessel, 1985).

Recently, Gamble and Mullet (1989) have also shown that the appearance of two new psbD-psbC transcripts is controlled by blue light. The response to phytochrome transformations is not much, but continuous blue light considerably increases the transcript levels even in the presence of an inhibitor of phytochrome synthesis, 4-amino-5-fluoropentanoic acid (AFPA). Interestingly, the blue light response is reversible by far-red light. Thus, although the experiments indicate a major role of a blue light photoreceptor (cryptochrome) in regulation of this operon, phytochrome may also be involved to some extent and act in concert with the former.

(iv) Effect of spectral quality of light on biogenesis of photosystems

Studies have also been made of the effect of spectral quality of light regimes on differential assembly of photosystems, by Gruissem, Melis, and their co-workers. These workers had found that in pea, polypeptide profiles of thylakoid membrane proteins varied under different physiological conditions. Thus, the chloroplasts from plants grown in PS II-sensitizing orange (far-red-enriched) light are relatively enriched in PS I complexes, whereas plants grown in light preferentially absorbed by PS I (far-red depleted) are enriched in PS II complex (Melis, 1984; Glick *et al.*, 1985). An alteration in the photosystem stoichiometry in the chloroplasts occurs as a compensatory strategy to correct unbalanced absorption of light by the two photosystems and this has been a subject of detailed physiological study by Chow *et al.* (1990). Such adjustment allows the plants to maintain high quantum efficiency under a variety of light conditions. Although the biochemical mechanisms have not been inspected as yet, the authors speculate that both biosynthetic and degradative mechanisms may operate.

Analysing the process further, the workers found that ratio of psbB/psaA transcripts was 2.6 times in red light as compared to yellow light (Glick *et al.*, 1986), which is broadly in agreement with the analysis of PS II/PS I stoichiometry. More recently,

Gruissem and co-workers extended their work to 10 plastid genes, the transcriptional activities of which were determined by run-on transcription assays (Deng *et al.*, 1989). However, it was found that the relative transcriptional activities remain generally constant under different light conditions. Even for the psaA gene, where the transcript level decreases to only one-tenth, the run-on transcription assay showed only a 30–40% decrease. In other genes, there was practically no difference in transcript level. Therefore, it would seem that regulation occurs largely by post-transcriptional mechanisms.

h) Circadian and diurnal rhythms

The expression of plastid genes is also under the control of diurnal and circadian rhythms as shown in tomato by Piechulla and Gruissem (1987) and Piechulla (1988). In developing tomato fruits as well as leaves, there are characteristic alterations during a natural day/night cycle in the mRNA levels of rbcL and psbA, implying that the regulation of plastid gene expression involves certain diurnal or circadian rhythms. When the plants were shifted to continuous light or continuous darkness, the mRNA levels in leaves for rbcL and psbA fluctuated with a diurnal periodicity of up to five days. Since the resultant fluctuating patterns of the activity of these genes were similar, both the genes are apparently regulated by a common mode.

i) Effect of age and developmental stage

The earlier reports of Thompson *et al.* (1983) and Link (1984b) and the more recent reports, e.g., of Mullet and co-workers (Mullet and Klein, 1987; Klein and Mullet 1986, 1987; Klein *et al.*, 1988; Gamble *et al.*, 1988) and Link (Dietrich *et al.*, 1987), indicate that for many chloroplast genes the transcript levels are already very high in etiolated seedlings which further increase after illumination. But this increase is only transient (for 36–72 hours) and again the levels decline; only psbA mRNA is maintained at higher levels in light. This finding has led some of the authors to conclude that the prevailing RNA level is also a function of age.

IV. Conclusions and Future Prospects

The work on molecular biology of chloroplast started in an intensive way 15 to 20 years ago after chloroplasts were shown to have their own genetic machinery. The cpDNA was isolated in a pure form by Kolodner and Tewari in 1975 and the first restriction maps were reported a year later by Bedbrook and Bogorad (1976) and Herrmann *et al.* (1976).

Since the 1970s, restriction maps of a rather large number of species have been prepared and many genes localized on them (these data have also been used to establish relationships between different taxa). The complete nucleotide sequence of three plants has also been determined and a large number of genes have been identified on the chloroplast genome. The chloroplast genome has been shown to code for 4 rRNAs, 30 tRNAs, and about 60 proteins. After nucleotide sequence analysis and expression studies, the promoters, the transcription initiation, and termination sites for various genes have been determined. It has been found that there is a high degree of homology between the promoters of chloroplast genes with those of prokaryotes, and that

E. coli RNA polymerase can also transcribe chloroplast genes. Simultaneously, studies on transcription and translation of chloroplast genes have been greatly facilitated by the establishment of *in vitro* systems from chloroplast extracts as well as for purified chloroplast RNA polymerase. Thus, the stage is all set for studies of various aspects of chloroplast development and biogenesis.

Despite intensive research in this field, much still remains to be done. Even after complete sequencing of chloroplast genomes and study of the products of many genes, there are several ORFs yet to be characterized. Some ORFs, which code for proteins with homology to mitochondrial maturase, reverse transcriptase, zinc finger protein, and NADPH-dehydrogenase, have been identified. The proteins, however, have not been detected from the chloroplasts and the same is true for several ribosomal proteins. Characterization of the products of such ORFs is a task for the future.

The regulatory aspects of chloroplast molecular biology have yet to be explored intensively. It is clear that the chloroplast proteins are coded by both nuclear and chloroplast genome, the contribution of the former being of a rather dominant character. Thus, the development and assembly of functional chloroplasts is dependent on both genomes. How these two compartments interact or what governs the simultaneous expression of genes residing on the two genomes is, however, not known. The question is pertinent in respect of several major constituents in plants, for example, RuBPcase/oxygenase, where of the two subunits the smaller one is encoded in the nucleus whereas the larger catalytic subunit is coded for by the chloroplast genome. Signals such as light which trigger chloroplast development must, therefore, affect related genes both in the nucleus and in the chloroplasts. However, the control mechanisms may not always be similar. Thus, when transcripts for the rbcS and others are synthesized *de novo*, in the nucleus, in response to light, this is generally not the case in regulation of rbcL. Apparently, even the etioplasts contain a certain level of transcripts already and the control by light is exercised more at the post-transcriptional and translational levels. It seems very likely that the nuclear factors move in chloroplast and regulate the expression of chloroplast genes and vice versa but this is a subject that needs to be studied intensively in the future (see also Taylor, 1989). The nuclear factors may be proteins that may form subunits of the chloroplast DNA polymerase, or resemble σ-factor or be translation-initiation factors.

The knowledge gained from the study of the molecular biology of chloroplasts can later be applied to increase the efficiency of photosynthesis by genetic engiheering. Well-established and efficient transformation systems, unfortunately, are not yet developed for chloroplasts, even though success in transformation of chloroplasts of tobacco (Svab *et al.*, 1990) has been reported recently. Another approach towards this goal, of relocation of chloroplast genes in nucleus, has been attempted by some laboratories (Cheung *et al.*, 1988). For achieving this, the nuclear transformation system has been used and the special segments of nuclear genes coding for transit peptides are tagged to the gene that normally resides on the chloroplast genome. In this way, products of the gene can be targeted to chloroplasts. Whereas chloroplast genetic engineering is of great interest from the viewpoint of agriculture, one should not lose sight of the fact that an efficient chloroplast transformation system may pave the way for further study of basic molecular biology, such as analysis of promoters and regulation of gene activity.

Acknowledgements

We thank Dr. N. Maheshwari for critically reading the manuscript.

V. References

Adams CC and Stern DB (1990) Control of mRNA stability in chloroplasts by 3′ inverted repeats: Effect of stem and loop mutations on degradation of psbA mRNA *in vitro*. *Nucleic Acids Res* **18**: 6003–6010

Allet B and Rachaix J-D (1979) Structure analysis at the ends of the intervening DNA sequences in the chloroplast 23S ribosomal genes of *C. reinhardii. Cell* **18**: 55–60

Alt J and Herrmann RG (1984) Nucleotide sequence of the gene for pre-apocytochrome *f* in the spinach plastid chromosome. *Curr Genet* **8**: 551–557

Alt J, Morris J, Westhoff P and Herrmann RG (1984) Nucleotide sequence of the clustered genes for the 44 kd chlorophyll *a* apoprotein and the "32 kd"-like protein of the photosystem II reaction center in the spinach plastid chromosome. *Curr Genet* **8**: 597–606

Alt J, Westhoff P, Sears BB, Nelson N, Hurt E, Hauska G and Herrmann RG (1983a) Genes and transcripts for the polypeptides of the cytochrome b_6/f complex from spinach thylakoid membranes. *EMBO J* **2**: 979–986

Alt J, Winter P, Sebald W, Moser JG, Schedel R, Westhoff P, and Herrmann RG (1983b) Localization, nucleotide sequence of the gene for the ATP synthase proteolipid subunit on the spinach plastid chromosome. *Curr Genet* **7**: 129–138

Bandurski RS and Maheshwari SC (1962) Nucleotide incorporation into nucleic acids by tobacco leaf homogenates. *Plant Physiol* **37**: 556–560

Barkan A (1988) Proteins encoded by a complex chloroplast transcription unit are each translated from both monocistronic and polycistronic mRNAs. *EMBO J* **7**: 2637–2644

Bedbrook JR and Bogorad L (1976) Endonuclease recognition sites mapped on *Zea mays* chloroplast DNA. *Proc Natl Acad Sci (USA)* **73**: 4309–4313

Bedbrook JR, Kolodner R and Bogorad L (1977) *Zea mays* chloroplast ribosomal RNA genes are part of a 22,000 base pair inverted repeat. *Cell* **11**: 739–749

Bedbrook JR, Link G, Coen DM, Bogorad L and Rich A (1978) Maize plastid gene expressed during photoregulated development. *Proc Natl Acad Sci (USA)* **75**: 3060–3064

Bennett DC, Rogers SA, Chen L-J and Orozco EM Jr (1990) A primary transcript in spinach chloroplasts that completely lacks a 5′ untranslated leader region. *Plant Mol Biol* **15**: 111–119

Bennoun P (1982) Evidence for a respiratory chain in the chloroplast. *Proc Natl Acad Sci (USA)* **79**: 4352–4356

Berends T, Gamble PE and Mullet JE (1987) Characterization of the barley chloroplast transcription units containing psaA-psaB and psbD-psbC. *Nucleic Acids Res* **15**: 5217–5240

Berry JO, Breiding DE and Klessig DF (1990) Light-mediated control of translational initiation of ribulose-1, 5-bisphosphate carboxylase in amaranth cotyledons. *Plant Cell* **2**: 795–803

Berry JO, Carr JP and Klessig DF (1988) mRNAs encoding ribulose-1, 5-bisphosphate carboxylase remain bound to polysomes but are not translated in amaranth seedlings transferred to darkness. *Proc Natl Acad Sci (USA)* **85**: 4190–4194

Berry JO, Nikolau BJ, Carr JP and Klessig DF (1985) Transcriptional and post-transcriptional regulation of ribulose 1, 5-bisphosphate carboxylase gene expression in light- and dark-grown amaranth cotyledons. *Mol Cell Biol* **5**: 2238–2246

Berry JO, Nikolau BJ, Carr JP and Klessig DF (1986) Translational regulation of light-induced ribulose 1, 5-bisphosphate carboxylase gene expression in amaranth. *Mol Cell Biol* **6**: 2347–2353

Bird CR, Koller B, Auffret AD, Huttly AK, Howe CJ, Dyer TA and Gray JC (1985) The wheat chloroplast gene for CFo subunit I of ATP synthase contains a large intron. *EMBO J* **4**: 1381–1388

Bohnert HJ, Driesel AJ, Crouse EJ, Gordon K, Herrmann RG, Steinmetz A, Mubumbila M, Keller M, Burkard G and Weil JH (1979) Presence of a transfer RNA gene in the spacer sequence between the 16S and 23S rRNA genes of spinach chloroplast DNA. *FEBS Letts* **103**: 52–56

Bottomley W, Smith HJ and Bogorad L (1971) RNA polymerase of maize: Partial purification and properties of the chloroplast enzyme. *Proc Natl Acad Sci (USA)* **68**: 2412–2416

Boyer SK and Mullet JE (1986) Characterization of *P. sativum* chloroplast psbA transcripts produced *in vivo, in vitro,* and in *E. coli. Plant Mol Biol* **6**: 229–243

Briat J-F, Dron M, Loiseaux S and Mache R (1982) Structure and transcription of the spinach chloroplast rDNA leader region. *Nucleic Acids Res* **10**: 6865–6878

Briat J-F, Laulhere JP and Mache R (1979) Transcription activity of a DNA-protein complex isolated from spinach plastids. *Eur J Biochem* **98**: 285–292

Briat J-F and Mache R (1980) Properties and characterization of a spinach chloroplast RNA polymerase isolated from a transcriptionally active DNA-protein complex. *Eur J Biochem* **111**: 503–509

Bülow S and Link G (1988) Sigma-like activity from mustard (*Sinapis alba* L.) chloroplasts conferring DNA-binding and transcription specificity to *E. coli* core RNA polymerase. *Plant Mol Biol* **10**: 349–357

Cantrell A and Bryant DA (1987) Molecular cloning and nucleotide sequence of the psaA and psaB genes of the cyanobacterium *Synechococcus* sp. PCC 7002. *Plant Mol Biol* **9**: 453–468

Carrillo N, Seyer P, Tyagi A, and Herrmann RG (1986) Cytochrome b-559 genes from *Oenothera hookeri* and *Nicotiana tabacum* show a remarkably high degree of conservation as compared to spinach. *Curr Genet* **10**: 619–624

Chan P-H and Wildman SG (1972) Chloroplast DNA codes for the primary structure of the large subunit of fraction I protein. *Biochim Biophys Acta* **277**: 677–680

Chen L-J and Orozco EM Jr (1988) Recognition of prokaryotic transcription terminators by spinach chloroplast RNA polymerase. *Nucleic Acids Res* **16**: 8411–8420

Chen L-J, Rogers SA, Bennett DC, Hu M-C and Orozco EM Jr (1990) An *in vitro* transcription termination system to analyze chloroplast promoters: Identification of multiple promoters for the spinach atpB gene. *Curr Genet* **17**: 55–64

Cheung AY, Bogorad L, Montagu M Von and Schell J (1988) Relocating a gene for herbicide tolerance: A chloroplast gene is converted into a nuclear gene. *Proc Natl Acad Sci (USA)* **85**: 391–395

Choquet Y, Goldshmidt-Clermont M, Girard-Bascou J, Kück U, Bennoun P and Rochaix J-D (1988) Mutant phenotypes support a trans-splicing mechanism for the expression of the tripartite psaA gene in the *C. reinhardtii* chloroplast. *Cell* **52**: 903–913

Chow WS, Melis A, and Anderson JM (1990) Adjustments of photosystem stoichiometry in chloroplasts improve the quantum efficiency of photosynthesis. *Proc Natl Acad Sci (USA)* **87**: 7502–7506

Christopher DA and Hallick RB (1989) *Euglena gracilis* chloroplast ribosomal protein operon: A new chloroplast gene for ribosomal protein L5 and description of a novel organelle intron category designated group III. *Nucleic Acids Res* **17**: 7591–7608

Christopher DA and Hallick RB (1990) Complex RNA maturation pathway for a chloroplast ribosomal protein operon with an internal tRNA cistron. *Plant Cell* **2**: 659–671

Chu NM, Oishi KK and Tewari KK (1981) Physical mapping of the pea chloroplast DNA and localization of the ribosomal RNA genes. *Plasmid* **6**: 279–292

Chu NM and Tewari KK (1982) Arrangement of the ribosomal RNA genes in chloroplast DNA of leguminosae. *Mol Gen Genet* **186**: 23–32

Coen DM, Bedbrook JR, Bogorad L and Rich A (1977) Maize chloroplast DNA fragment encoding the large subunit of ribulose bisphosphate carboxylase. *Proc Natl Acad Sci (USA)* **74**: 5487–5491

Cozens AL, Walker JE, Phillips AL, Huttly AK and Gray JC (1986) A sixth subunit of ATP synthase, an F_0 component, is encoded in the pea chloroplast genome. *EMBO J* **5**: 217–222

Crossland LD, Rodermel SR and Bogorad L (1984) Single gene for the large subunit of ribulose bisphosphate carboxylase in maize yields two differentially regulated mRNAs. *Proc Natl Acad Sci (USA)* **81**: 4060–4064

Cushman JC, Christopher DA, Little MC, Hallick RB and Price CA (1988a) Organization of the psbE, psbF, orf38 and orf42 gene loci on the *Euglena gracilis* chloroplast genome. *Curr Genet* **13**: 173–180

Cushman JC, Hallick RB, and Price CA (1988b). The two genes for the P_{700} chlorophyll *a* apoproteins on the *Euglena gracilis* chloroplast genome contain multiple introns. *Curr Genet* **13**: 159–171

Deng X-W and Gruissem W (1987) Control of plastid gene expression during development: The limited role of transcriptional regulation. *Cell* **49**: 379–387

Deng X-W and Gruissem W (1988) Constitutive transcription and regulation of gene expression in non-photosynthetic plastids of higher plants. *EMBO J* **7**: 3301–3308

Deng X-W, Tonkyn JC, Peter GF, Thornber JP and Gruissem W (1989) Post-transcriptional control of plastid mRNA accumulation during adaptation of chloroplasts to different light quality environments. *Plant Cell* 1: 645–654

Deno H, Shinozaki K and Sugiura M (1983) Nucleiotide sequence of tobacco chloroplast gene for the subunit of proton translocating ATPase. *Nucleic Acids Res* 11: 2185–2191

Deno H and Sugiura M (1984) Chloroplast tRNA^{Gly} gene contains a long intron in the D stem: Nucleotide sequences of tobacco chloroplast genes for tRNA^{Gly} (UCC) and tRNA^{Arg} (UCU). *Proc Natl Acad Sci (USA)* 81: 405–408

Dietrich G, Detschey S, Neuhaus H and Link G (1987) Temporal, light control of plastid transcript levels for proteins involved in photosynthesis during mustard (*Sinapis alba* L.) seedling development. *Planta* 172: 393–399.

Dorne A-M, Lescure A-M and Mache R (1984) Site of synthesis of spinach chloroplast ribosomal proteins and formation of incomplete ribosomal particles in isolated chloroplasts. *Plant Mol Biol* 3: 83–90.

Driesel AJ, Crouse EJ, Gordon K, Bohnert HJ, Herrmann RG, Steinmetz A, Mubumbila M, Keller M, Burkard G and Weil JH (1979) Fractionation and identification of spinach chloroplast transfer RNAs and mapping of their genes on the restriction map of chloroplast DNA. *Gene* 6: 285–306

Dron M, Robreau G, and Gal YL (1979) Isolation of a chromoid from the chloroplast of *Chlamydomonas reinhardtii*. *Exp Cell Res* 119: 301–305

Eisermann A, Tiller K and Link G (1990) *In vitro* transcription and DNA binding characteristics of chloroplast and etioplast extracts from mustard (*Sinapis alba* L.) indicate differential usage of the psbA promoter. *EMBO J* 9: 3981–3987

Erickson JM, Rahire M and Rochaix JD (1984) *Chlamydomonas reinhardii* gene for the 32,000 mol wt. protein of photosystem II contains four large introns and is located entirely within the chloroplast inverted repeat. *EMBO J* 3: 2753–2762

Erion JL (1985) Characterization of the mRNA transcripts of the maize, ribulose-1, 5-bisphosphate carboxylase, large subunit gene. *Plant Mol Biol* 4: 169–179

Fearnley IM, Runswick MJ and Walker JE (1989) A homologue of the nuclear coded 49 kd subunit of bovine mitochondrial NADH-ubiquinone reductase is coded in chloroplast DNA. *EMBO J* 8: 665–672

Fiebig C, Neuhaus H, Teichert J, Röcher W, Degenhardt J and Link G (1990) Temporal and spatial pattern of plastid gene expression during crucifer seedling development and embryogenesis. *Planta* 181: 191–198

Fish LE, Kück U and Bogorad L (1985) Two partially homologous adjacent light-inducible maize chloroplast genes encoding polypeptides of the P700 chlorophyll *a*-protein complex of photosystem I. *J Biol Chem* 260: 1413–1421

Fukuzawa H, Kohchi T, Shirai H, Ohyama K, Umesono K, Inokuchi H and Ozeki H (1986) Coding sequences for chloroplast ribosomal protein S12 from the liverwort, *Marchantia polymorpha*, are separated far apart on the different DNA strands. *FEBS Letts* 198: 11–15

Fukuzawa H, Yoshida T, Kohchi T, Okumura T, Sawano Y, and Ohyama K (1987) Splicing of group II introns in mRNAs coding for cytochrome b$_6$ and subunit IV in the liverwort *Marchantia polymorpha* chloroplast genome. Exon specifying a region coding for two genes with the spacer region. *FEBS Letts* 220: 61–66

Gamble PE and Mullet JE (1989) Blue light regulates the accumulation of two psbD-psbC transcripts in barley chloroplasts. *EMBO J* 8: 2785–2794

Gamble PE, Sexton TB and Mullet JE (1988) Light-dependent changes in psbD and psbC transcripts of barley chloroplasts: Accumulation of two transcripts maintains psbD and psbC translation capability in mature chloroplasts. *EMBO J* 7: 1289–1297

Glick RE, McCauley SW, Gruissem W and Melis A (1986) Light quality regulates expression of chloroplast genes and assembly of photosynthetic membrane complexes. *Proc Natl Acad Sci (USA)* 83: 4287–4291

Glick RE, McCauley SW and Melis A (1985) Effect of light quality on chloroplast-membrane organization and function in pea. *Planta* 164: 487–494

Godde D (1982) Evidence for a membrane bound NADH-plastoquinone-oxidoreductase in *Chlamydomonas reinhardii* CW-15. *Arch Microbiol* 131: 197–202

Graf L, Kössel H and Stutz E (1980) Sequencing of 16S-23S spacer in a ribosomal RNA operon of *Euglena gracilis* chloroplast DNA reveals two tRNA genes. *Nature* 286: 908–910

Gray JC, and Kekwick RGO (1974) The synthesis of small subunit of ribulose 1,5-bisphosphate carboxylase in the french bean *Phaseolus vulgaris*. *Eur J Biochem* **44**: 491–500

Gray JC, Webber AN, Hird SM, Willey DL and Dyer TA (1990) Genes for photosystem II polypeptides. In: (M Baltscheffsky, ed.) *Current Research in Photosynthesis*, Vol III, Kluwer Academic Publishers, Netherlands, pp 461–468

Gruissem W (1989) Chloroplast gene expression: How plants turn their plastids on? *Cell* **56**: 161–170

Gruissem W, Greenberg BM, Zurawski G, Prescott DM and Hallick RB (1983a) Biosynthesis of chloroplast transfer RNA in a spinach chloroplast transcription system. *Cell* **35**: 815–828

Gruissem W, Narita JO, Greenberg BM, Prescott DM and Hallick RB (1983b) Selective *in vitro* transcription of chloroplast genes. *J Cell Biochem* **22**: 31–46

Gruissem W and Zurawski G (1985) Analysis of promoter regions for the spinach chloroplast rbcL, atpB, and psbA genes. *EMBO J* **4**: 3375–3383

Haff LA and Bogorad L (1976) Hybridization of maize chloroplast DNA with transfer ribonucleic acids. *Biochemistry* **15**: 4105–4109

Haley J and Bogorad L (1989) A 4-kDa maize chloroplast polypeptide associated with the cytochrome b_6-f complex: Subunit 5, encoded by the chloroplast petE gene. *Proc Natl Acad Sci (USA)* **86**: 1534–1538

Haley J and Bogorad L (1990) Alternative promoters are used for genes within maize chloroplast polycistronic transcription units. *Plant Cell* **2**: 323–333

Hallick RB (1989) Proposal for the naming of chloroplast genes. II. Update to the nomenclature of genes for thylakoid membrane polypeptides. *Plant Mol Biol Reporter* **7**: 266–275

Hallick RB and Bottomley W (1983) Proposal for the naming of chloroplast genes. *Plant Mol Biol Reporter* **1**: 38–43

Hallick RB, Hollingsworth MJ and Nickoloff JA (1984) Transfer RNA genes of *Euglena gracilis* chloroplast DNA. *Plant Mol Biol* **3**: 169–175

Hallick RB, Lipper C, Richards OC, and Rutter WJ (1976) Isolation of a transcriptionally active chromosome from chloroplasts of *Euglena gracilis*. *Biochemistry* **15**: 3039–3045

Hanley-Bowdoin L and Chua N-H (1987) Chloroplast promoters. *Trends Biochem Sci* **12**: 67–70

Hanley-Bowdoin L and Chua N-H (1988) Transcription of the wheat chloroplast gene that encodes the 32 kd polypeptide. *Plant Mol Biol* **10**: 303–310

Hanley-Bowdoin L and Chua N-H (1989) Transcriptional interaction between the promoters of the maize chloroplast genes which encode the β subunit of ATP synthase and the large subunit of ribulose 1, 5-bisphosphate carboxylase. *Mol Gen Genet* **215**: 217–224

Hanley-Bowdoin L, Orozco EM Jr. and Chua N-H (1985) *In vitro* synthesis and processing of a maize chloroplast-transcript encoded by the ribulose 1, 5-bisphosphate carboxylase large subunit gene. *Mol Cell Biol* **5**: 2733–2745

Harpster M and Apel K (1985) The light-dependent regulation of gene expression during plastid development in higher plants. *Physiol Plant* **64**: 147–152

Hayashida N, Matsubayashi T, Shinozaki K, Sugiura M, Inoue K and Hiyama T (1987) The gene for 9 kd polypeptide, a possible apoprotein for the iron sulfur centres A and B of the photosystem I complex, in tobacco chloroplast DNA. *Curr Genet* **12**: 247–250

Heinemeyer W, Alt J and Herrmann RG (1984) Nucleotide sequence of the clustered genes for apocytochrome b_6 and subunit 4 of the cytochrome b/f complex in the spinach plastid chromosome. *Curr Genet* **8**: 543–549

Hennig J and Herrmann RG (1986) Chloroplast ATP synthase of spinach contains nine nonidentical subunit species, six of which are encoded by plastid chromosomes in two operons in a phylogenetically conserved arrangement. *Mol Gen Genet* **203**: 117–128

Herrmann RG, Alt J, Schiller B, Widger WR and Cramer WA (1984) Nucleotide sequence of the gene for apocytochrome b-559 on the spinach plastid chromosome. Implications for the structure of the membrane protein. *FEBS Letts* **176**: 239–244

Herrmann RG, Bohnert H-J, Driesel A and Hobom G (1976) The location of rRNA genes on the restriction endonuclease map of the *Spinacia oleracea* chloroplast DNA. In: (TH Bucher, W Neupert, W Sebald, S Werner, eds.) *Genetics and Biogenesis of Chloroplast and Mitochondria*, Elsevier/North Holland Biomedical Press, Amsterdam, pp 351–359

Herrmann RG and Possingham JV (1980) Plastid DNA—The plastome. In: (J Reinert, ed.) *Chloroplasts,* Springer, Berlin, pp. 45–96

Herrmann RG, Westhoff P, and Link G (1992) Biogenesis of plastids in higher plants. In: (RG Herrmann, ed.) *Cell Organelles—Plastids, Mitochondria, Glyoxysomes and Peroxisomes,* Springer, Wien, New York, in press.

Hildebrand M, Hallick RB, Passavant CW and Bourque DP (1988) Trans-splicing in chloroplasts: The rps12 loci of *Nicotiana tabacum. Proc Natl Acad Sci (USA)* **85:** 372–376

Hiratsuka J, Shimada H, Whittier R, Ishibashi T, Sakamoto M, Mori M, Kondo C, Honji Y, Sun C-R, Meng B-Y, Li Y-Q, Kanno A, Nishizawa Y, Hirai A, Shinozaki K and Sugiura M (1989) The complete sequence of the rice (*Oryza sativa*) chloroplast genome: Intermolecular recombination between distinct tRNA genes accounts for a major plastid DNA inversion during the evolution of the cereals. *Mol Gen Genet* **217:** 185–194

Hird SM, Dyer TA and Gray JC (1986) The gene for the 10 kDa phosphoprotein of photosystem II is located in chloroplast DNA. *FEBS Letts* **209:** 181–186

Hird SM, Dyer TA and Gray JC (1989) Nucleotide sequence of the rpoA gene in wheat chloroplast DNA. *Nucleic Acids Res* **17:** 6394

Holschuh K, Bottomley W and Whitfeld PR (1984) Structure of the spinach chloroplast genes for the D_2 and 44 kd reaction center proteins of photosystem II and for tRNA^{Ser} (UGA). *Nucleic Acids Res* **12:** 8819–8834

Hosler JP, Wurtz EA, Harris EH, Gillham NW and Boynton JE (1989) Relationship between gene dosage and gene expression in the chloroplast of *Chlamydomonas reinhardtii. Plant physiol* **91:** 648–655

Hu J and Bogorad L (1990) Maize chloroplast RNA polymerase: The 180-, 120- and 38-kilodalton polypeptides are encoded in chloroplast genes. *Proc Natl Acad Sci (USA)* **87:** 1531–1535

Hudson GS, Holton TA, Whitfeld PR and Bottomley W (1988) Spinach chloroplast rpoBC genes encode three subunits of the chloroplast RNA polymerase. *J Mol Biol* **200:** 639–654

Hudson GS, Muson JG, Holton TA, Koller B, Cox GB, Whitfeld PR and Bottomley W (1987) A gene cluster in the spinach and pea chloroplast genomes encoding one CF_1 and three CF_0 subunits of the H^+-ATP synthase complex and the ribosomal protein S2. *J Mol Biol* **196:** 283–298

Igloi GL, Meinke A, Döry I and Kössel H (1990) Nucleotide sequence of the maize chloroplast rpo B/C_1/C_2 operon: Comparison between the derived protein primary structures from various organisms with respect to functional domains. *Mol Gen Genet* **221:** 379–394

Ikeuchi M, Koike H and Inoue Y (1989) N-terminal sequencing of low molecular mass components in cyanobacterial photosystem II core complex—Two components correspond to unidentified open reading frames of plant chloroplast DNA. *FEBS Letts* **253:** 178–182

Jenni B and Stutz E (1979) Physical mapping of the ribosomal DNA region of *Euglena gracilis* chloroplast DNA. *Eur J Biochem* **88:** 127–134

Jolly SO and Bogorad L (1980) Preferential transcription of cloned maize chloroplast DNA sequences by maize chloroplast RNA polymerase. *Proc Natl Acad Sci (USA)* **77:** 822–826

Jordan BR, Hopley JG, and Thompson WF (1989) Chloroplast gene expression in lettuce grown under different irradiances. *Planta* **178:** 69–75

Karabin GD, Farley M and Hallick RB (1984) Chloroplast gene for Mr 32000 polypeptide of photosystem II in *Euglena gracilis* is interrupted by four introns with conserved boundary sequences. *Nucleic Acids Res* **9:** 5801—5812

Kato K, Sayre RT and Bogorad L (1987) Expression of the PSII-I gene is maize chloroplasts. *Proc Annu Meet Jpn Soc Plant Physiol, Urwa.* pp 208

Kawashima N and Wildman SG (1972) Studies on fraction I protein. IV. Mode of inheritance of primary structure in relation to whether chloroplast or nuclear DNA contains the code for a chloroplast protein. *Biochim Biophys Acta* **262:** 42–49

Keller M and Stutz E (1984) Structure of the *Euglena gracilis* chloroplast gene (psbA) coding for the 32-kDa protein of photosystem II. *FEBS Letts* **175:** 173–177

Kidd GH and Bogorad L (1979) Peptide maps comparing subunits of maize chloroplast and type II nuclear DNA-independent RNA polymerases. *Proc Natl Acad Sci (USA)* **76:** 4890–4892

Klein RR, Mason HS, and Mullet JE (1988) Light-regulated translation of chloroplast proteins. I. Transcripts of psaA- psaB, psbA, and rbcL are associated with polysomes in dark-grown and illuminated barley seedlings. *J Cell Biol* **106:** 289–301

Klein RR and Mullet JE (1986) Regulation of chloroplast-encoded chlorophyll-binding protein translation during higher plant chloroplast biogenesis. *J Biol Chem* **261**: 11138–11145

Klein RR and Mullet JE (1987) Control of gene expression during higher plant chloroplast biogenesis. *J Biol Chem* **262**: 4341–4348

Klein RR and Mullet JE (1990) Light induced transcription of chloroplast genes. psbA transcription is differentially enhanced in illuminated barley. *J Biol Chem* **265**: 1895–1899

Koch W, Edwards K and Kossel H (1981) Sequencing of the 16S-23S spacer in a ribosomal RNA operon of *Zea mays* chloroplast DNA reveals two split tRNA genes. *Cell* **25**: 203–213

Koike H, Ikeuchi M, Hiyama T and Inoue Y (1989) Identification of photosystem I components from the cyanobacterium *Synechococcus vulcanus* by N-terminal sequencing. *FEBS Letts* **253**: 257–263

Koller B and Delius H (1980) *Vicia faba* chloroplast DNA has only one set of ribosomal RNA genes as shown by partial denaturation mapping and R-loop analysis. *Mol Gen Genet* **178**: 261-269

Koller B, Fromm H, Galun E and Edelman M (1987) Evidence for *in vivo trans* splicing of pre-mRNAs in tobacco chloroplasts. *Cell* **48**: 111–119

Koller B, Gingrich JC, Stiegler GL, Farley MA, Delius H and Hallick RB (1984) Nine introns with conserved boundary sequences in the *Euglena gracilis* chloroplast ribulose-1, 5-bisphosphate carboxylase gene. *Cell* **36**: 545–553

Kolodner R and Tewari KK (1975) The molecular size and conformation of the chloroplast DNA from higher plants. *Biochim Biophys Acta* **402**: 372–390

Krebbers ET, Larrinua IM, McIntosh L and Bogorad L (1982) The maize chloroplast genes for the β and ε subunits of photosynthetic coupling factor CF_1 are fused. *Nucleic Acids Res* **10**: 4985–5002

Kreuz K, Dehesh K and Apel K (1986) The light-dependent accumulation of the P700 chlorophyll *a* protein of the photosystem I reaction center in barley—Evidence for translational control. *Eur J Biochem* **159**: 459–467

Kück U (1989) The intron of a plastid gene from a green alga contains an open reading frame for a reverse transcriptase-like enzyme. *Mol Gen Genet* **218**: 257–265

Kück U, Choquet Y, Schneider M, Dron M and Bennoun P (1987) Structural and transcription analysis of two homologous genes for the P700 chlorophyll *a*-apoproteins in *Chlamydomonas reinhardtii*: Evidence for *in vivo* trans-splicing. *EMBO J* **6**: 2185–2195

Kusuda J, Shinozaki K, Takaiwa F and Sugiura M (1980) Characterization of the cloned ribosomal DNA of tobacco chloroplasts. *Mol Gen Genet* **178**: 1–7

Laing W, Kreuz K and Apel K (1988) Light-dependent, but phytochrome-independent, translational control of the accumulation of the P700 chlorophyll-a protein of photosystem I in barley (*Hordeum vulgare* L.). *Planta* **176**: 269–276

Lam E and Chua N-H (1987) Chloroplast DNA gyrase and *in vitro* regulation of transcription by template topology and novobiocin. *Plant Mol Biol* **8**: 415–424

Lam E, Hanley-Bowdoin L and Chua N-H (1988) Characterization of a chloroplast sequence-specific DNA binding factor. *J Biol Chem* **263**: 8288–8293

Lerbs S, Braütigam E and Mache R (1988) DNA-dependent RNA polymerase of spinach chloroplasts: Characterization of α-like and σ-like polypeptides. *Mol Gen Genet* **211**: 459–464

Lerbs S, Braütigam E and Parthier B (1985) Polypeptides of DNA-dependent RNA polymerase of spinach chloroplasts: Characterization by antibody-linked polymerase assay and determination of sites of synthesis. *EMBO J* **4**: 1661–1666

Lerbs S, Briat JF and Mache R (1983) Chloroplast RNA polymerase from spinach: Purification and DNA-binding proteins. *Plant Mol Biol* **2**: 67–74

Link G (1981) Enhanced expression of a distinct plastid DNA region in mustard seedlings by continuous far-red light. *Planta* **152**: 379–380

Link G (1982) Phytochrome control of plastid mRNA in mustard (*Sinapis alba* L.). *Planta* **154**: 81–86.

Link G (1984a) DNA sequence requirements for the accurate transcription of a protein coding plastid gene in a plastid *in vitro* system from mustard (*Sinapis alba* L.) *EMBO J* **3**: 1697–1704

Link G (1984b) Hybridization study of developmental plastid gene expression in mustard (*Sinapis alba* L.) with cloned probes for most plastid DNA regions. *Plant Mol Biol* **3**: 243–248

Link G and Bogorad L (1980) Sizes, locations, and directions of transcription of two genes on a cloned maize chloroplast DNA sequence. *Proc Natl Acad Sci (USA)* **77**: 1832–1836

Link G and Langridge U (1984) Structure of the chloroplast gene for the precursor of the M_r 32,000 photosystem II protein from mustard (*Sinapis alba* L.). *Nucleic Acids Res* 12: 945–958

Matsubayashi T, Wakasugi T, Shinozaki K, Yamaguchi-Shinozaki K, Zaita N, Hidaka T, Meng BY, Ohto C, Tanaka M, Kato A, Maruyama T and Sugiura M (1987) Six chloroplast genes (ndhA-F) homologous to human mitochondrial genes encoding components of the respiratory chain NADH dehydrogenase are actively expressed: Determination of the splice sites in ndhA and ndhB pre-mRNAs. *Mol Gen Genet* 210: 385–393

McClure WR (1985) Mechanism and control of transcription initiation in prokaryotes. *Annu Rev Biochem* 54: 171–204

McIntosh L, Poulsen C, and Bogorad L (1980) Chloroplast gene sequence for the large subunit of ribulose bisphosphate carboxylase of maize. *Nature* 288: 556–560

Melis A (1984) Light regulation of photosynthetic membrane structure, organization, and function. *J Cell Biochem* 24: 271–285

Meng BY, Matsubayashi T, Wakasugi T, Shinozaki K, Sugiura M, Hirai A, Mikami T, Kishima Y and Kinoshita T, (1986) Ubiquity of the genes for components of a NADH dehydrogenase in higher plant chloroplast genomes. *Plant Sci* 47: 181–184

Meng BY, Tanaka M, Wakasugi T, Ohme M, Shinozaki K and Sugiura M (1988) Co-transcription of the genes encoding two P700 chlorophyll *a* apoproteins with the gene for ribosomal protein CS14: Determination of the transcriptional initiation site by *in vitro* capping. *Curr Genet* 14: 395–400

Montandon P-E and Stutz E (1984) The genes for the ribosomal proteins S12 and S7 are clustered with the gene for the EF-Tu protein on the chloroplast genome of *Euglena gracilis*. *Nucleic Acids Res* 12: 2851–2859

Morris J and Herrmann RG (1984) Nucleotide sequence of the gene for the P_{680} chlorophyll *a* apoprotein of the photosystem II reaction center from spinach. *Nucleic Acids Res* 12: 2837–2850

Mullet JE (1988) Chloroplast development and gene expression. *Annu Rev Plant Physiol Plant Mol Biol* 39: 475–502

Mullet JE and Klein RR (1987) Transcription and RNA stability are important determinants of higher plant chloroplast RNA levels. *EMBO J* 6: 1571–1579

Mullet JE, Orozco EM Jr and Chua N-H (1985) Multiple transcripts for higher plant rbcL and atpB genes and localization of the transcription initiation site of the rbcL gene. *Plant Mol Biol* 4: 39–54

Murata N, Miyao M, Hayashida N, Hidaka T and Sugiura M (1988) Identification of a new gene in the chloroplast genome encoding a low-molecular-mass polypeptide of photosystem II complex. *FEBS Letts* 235: 283–288

Narita JO, Rushlow KE and Hallick RB (1985) Characterization of a *Euglena gracilis* chloroplast RNA polymerase specific for ribosomal RNA genes. *J Biol Chem* 260: 1194–1199

Neuhaus H and Link G (1987) The chloroplast tRNA^{Lys} (UUU) gene from mustard (*Sinapis alba*) contains a class II intron potentially coding for a maturase-related polypeptide. *Curr Genet* 11: 251–257

Neuhaus H and Link G (1990) The chloroplast psbK operon from mustard (*Sinapis alba* L.): Multiple transcripts during seedling development and evidence for divergent overlapping transcription. *Curr Genet* 18: 377–383

Nickelsen J and Link G (1989) Interaction of a 3′ RNA region of the mustard trnK gene with chloroplast proteins. *Nucleic Acids Res* 17: 9637–9648

Nishizawa Y and Hirai A (1987) Nucleotide sequence of the large subunit of rice ribulose 1, 5-bisphosphate carboxylase. *Jpn J Genet* 62: 389–395

Ohme M, Tanaka M , Chunwongse J, Shinozaki K and Sugiura M (1986) A tobacco chloroplast DNA sequence possibly coding for a polypeptide similar to *E. coli* RNA polymerase β-subunit. *FEBS Letts* 200: 87–90

Ohyama K, Fukuzawa H, Kohchi T, Shirai H, Sano T, Sano S, Umesono K, Shiki Y, Takeuchi M, Chang Z, Aota SI, Inokuchi H and Ozeki H (1986) Chloroplast gene organization deduced from complete sequence of liverwort *Marchantia polymorpha* chloroplast DNA. *Nature* 322: 572–574

Orozco EM Jr, Chen L-J and Eilevs RJ (1990) The divergently transcribed rbcL and atpB genes of tobacco plastid DNA are separated by nineteen base pairs. *Curr Genet* 17: 65–71

Orozco EM Jr, Mullet JE, and Chua N-H (1985) An *in vitro* system for accurate transcription initiation of chloroplast protein genes. *Nucleic Acids Res* 13: 1283–1302

Ozeki H, Ohyama K, Inokuchi H, Fukuzawa H, Kohchi T, Sano T, Nakashigashi K and Umesono K (1987) Genetic system of choroplasts. *Cold Spring Harbor Symposia on Quantitative Biol* **52**: 791–804

Palmer JD (1990) Contrasting modes and tempos of genome evolution in land plant organelles. *Trends Genet* **6**: 115–120

Palmer JD, Edwards H, Jorgensen RA and Thompson WF (1982) Novel evolutionary variation in transcription and location of two chloroplast genes. *Nucleic Acids Res* **10**: 6819–6832

Passavant CW, Stiegler GL and Hallick RB (1983) Location of the single gene for elongation factor Tu on the *Euglena gracilis* chloroplast chromosome. *J Biol Chem* **258**: 693–695

Piechulla B (1988) Plastid and nuclear mRNA fluctuations in tomato leaves—diurnal and circadian rhythms during extended dark and light periods. *Plant Mol Biol* **11**: 345–353

Piechulla B and Gruissem W (1987) Diurnal mRNA fluctuations of nuclear and plastid genes in developing tomato fruits. *EMBO J* **6**: 3593–3599

Polya GM and Jagendorf AT (1971) Wheat leaf RNA polymerases. I. Partial purification and characterization of nuclear, chloroplast and soluble DNA-dependent enzymes. *Arch Biochem Biophys* **146**: 635–648

Purton S and Gray JC (1989) The plastid rpoA gene encoding a protein homologous to the bacterial RNA polymerase alpha subunit is expressed in pea chloroplasts. *Mol Gen Genet* **217**: 77–84

Rasmussen OF, Bookjans SG, Stummann BM and Henningsen KW (1984) Localization and nucleotide sequence of the gene for the membrane polypeptide D_2 from pea chloroplast DNA. *Plant Mol Biol* **3**: 191–199

Reith M and Straus NA (1987) Nucleotide sequence of the chloroplast gene responsible for triazine resistance in canola. *Theor Appl Genet* **73**: 357–363

Richter G (1984) Blue light control of the level of two plastid mRNAs in cultured plant cells. *Plant Mol Biol* **3**: 271–276

Richter G and Wessel K (1985) Red light inhibits blue light-induced chloroplast development in cultured plant cells at the mRNA level. *Plant Mol Biol* **5**: 175–182

Rochaix J-D, Kuchka M, Mayfield S, Schirmer-Rahire M, Girard-Bascou J and Bennoun P (1989) Nuclear and chloroplast mutations affect the synthesis or stability of the chloroplast psbC gene product in *Chlamydomonas reinhardii*. *EMBO J* **8**: 1013–1021

Rochaix J-D and Malnoe P (1978) Anatomy of the chloroplast ribosomal DNA of *Chlamydomonas reinhardii*. *Cell* **15**: 661–670

Rochaix J-D, Rahire M and Michel F (1985) The chloroplast ribosomal intron of *Chlamydomonas reinhardtii* codes for a polypeptide related to mitochondrial maturases. *Nucleic Acids Res* **13**: 975–984

Rock CD, Barkan A and Taylor WC (1987) The maize plastid psbB-psbF-petB-petD gene cluster: Spliced and unspliced petB and petD RNAs encode alternative products. *Curr Genet* **12**: 69–77

Rodermel SR and Bogorad L (1985) Maize plastid photogenes: Mapping and photoregulation of transcript levels during light-induced development. *J Cell Biol* **100**: 463–476

Roitgrund C and Mets LJ (1990) Localization of two novel chloroplast genome functions: Trans-splicing of RNA and protochlorophyllide reduction. *Curr Genet* **17**: 147–153

Roy H, Patterson R and Jagendorf AT (1976) Identification of the small subunit of ribulose 1, 5-bisphosphate carboxylase as a product of wheat leaf cytoplasmic ribosomes. *Arch Biochem Biophys* **172**: 64–73

Ruf M and Kössel H (1988) Structure and expression of the gene coding for the α-subunit of DNA-dependent RNA polymerase from the chloroplast genome of *Zea mays*. *Nucleic Acids Res* **16**: 5741–5754

Rushlow KE, Orozco EM Jr, Lipper CJ and Hallick RB (1980) Selective *in vitro* transcription of *Euglena* chloroplast ribosomal RNA genes by a transcriptionally active chromosome. *J Biol Chem* **255**: 3786–3792

Saraste M, Gay NJ, Eberle A, Runswick MJ and Walker JE (1981) The atp operon: Nucleotide sequence of the genes for β and ε subunits of *Escherichia coli* ATP synthase. *Nucleic Acids Res* **9**: 5287–5296

Sasaki Y, Nagano Y, Morioka S, Ishikawa H and Matsuno R (1989) A chloroplast gene encoding a protein with one zinc finger. *Nucleic Acids Res* **17**: 6217–6227

Sasaki Y, Nakamura Y and Matsuno R (1987) Regulation of gene expression of ribulose bisphosphate carboxylase in greening pea leaves. *Plant Mol Biol* **8**: 375–382

Sasaki Y, Tomoda Y and Kamikubo T (1984) Light regulates the gene expression of ribulose bisphosphate carboxylase at the levels of transcription and gene dosage in greening pea leaves. *FEBS Letts* **173**: 31–35

Schantz R and Bogorad L (1988) Maize chloroplast genes ndhD, ndhE, and psaC. Sequences, transcripts and transcript pools. *Plant Mol Biol* **11**: 239–247

Scheller HV and Møller BL (1990) Photosystem I polypeptides. *Physiol Plant* **78**: 484–494

Scheller HV, Okkels JS, Høj PB, Svendsen I, Roepstroff P and Møller BL (1989) The primary structure of a 4.0-kDa photosystem I polypeptide encoded by the chloroplast psaI gene. *J Biol Chem* **264**: 18402–18406

Schön A, Krupp G, Gough S, Berry-Lowe S, Kannangara CG and Söll D (1986) The RNA required in the first step of chlorophyll biosynthesis is a chloroplast glutamate tRNA. *Nature* **322**: 281–284

Schrubar HV, Wanner G and Westhoff P (1991) Transcriptional control of plastid gene expression in greening *Sorghum* seedlings. *Planta* **183**: 101–111

Schwarz Z and Kössel H (1980) The primary structure of 16S rDNA from *Zea mays* chloroplast is homologous to *E. coli* 16S rRNA. *Nature* **283**: 739–742

Sexton TB, Jones JT and Mullet JE (1990) Sequence and transcriptional analysis of the barley ctDNA region upstream of psbD-psbC encoding trnk (UUU), rps16, trnQ (UUG), psbK, psbI, and trnS (GCU). *Curr Genet* **17**: 445–454

Shimada H, Fukuta M, Ishikawa M and Sugiura M (1990) Rice chloroplast RNA polymerase genes: The absence of an intron in rpoCl and the presence of an extra sequence in rpoC2. *Mol Gen Genet* **221**: 395–402

Shimada H and Sugiura M (1989) Pseudogenes and short repeated sequences in the rice chloroplast genome. *Curr Genet* **16**: 293–301

Shinozaki K, Ohme M, Tanaka M, Wakasugi T, Hayashida N, Matsubayashi T, Zaita N, Chunwongse J, Obokata J, Yamaguchi-Shinozaki K, Ohto C, Torazawa K, Meng BY, Sugita M, Deno H, Kamogashira T, Yamada K, Kusuda J, Takaiwa F, Kato A, Tohdoh N, Shimada H and Sugiura M (1986) The complete nucleotide sequence of the tobacco chloroplast genome: Its gene organization and expression. *EMBO J* **5**: 2043–2049

Siemeister G, Buchholz C and Hachtel W (1990) Genes for the plastid elongation factor Tu and ribosomal protein S7 and six tRNA genes on the 73 kb DNA from *Astasia longa* that resembles the chloroplast DNA of *Euglena*. *Mol Gen Genet* **220**: 425–432

Sijben-Müller G, Hallick RB, Alt J, Westhoff P and Herrmann RG (1986) Spinach plastid genes coding for initiation factor IF-1, ribosomal protein S11 and RNA polymerase α-subunit. *Nucleic Acids Res* **14**: 1029–1044

Smooker PM, Kruft V and Subramanian AR (1990) A ribosomal protein is encoded in the chloroplast DNA in a lower plant but in the nucleus in angiosperms. *J Biol Chem* **265**: 16699–16703

Steinmetz AA, Castroviejo M, Sayre RT and Bogorad L (1986) Protein PSII-G. An additional component of photosystem II identified through its plastid gene in maize. *J Biol Chem* **261**: 2485–2488

Steinmetz A, Gubbins EJ and Bogorad L (1982) The anticodon of the maize chloroplast gene for tRNA[Leu] UAA is split by a large intron. *Nucleic Acids Res* **10**: 3027–3037

Steinmetz A, Mubumbila M, Keller M, Burkard G, Weil JH, Driesel AJ, Crouse EJ, Gordon K, Bohnert HJ and Herrmann RG (1978) Mapping of tRNA genes on the circular DNA molecule of *Spinacia oleracea* chloroplasts. In: (G Akoyunoglou and JH Argyuoudi-Akoyunoglou, eds.) *Chloroplast Development*, Biomedical Press, Elsevier, Amsterdam, pp 573–580

Stern DB, and Gruissem W (1987) Control of plastid gene expression: 3′ inverted repeats act as'mRNA processing and stabilizing elements, but do not terminate transcription. *Cell* **51**: 1145–1157

Stern DB and Gruissem W (1989) Chloroplast mRNA 3′ end maturation is biochemically distinct from prokaryotic mRNA processing. *Plant Mol Biol* **13**: 615–625

Stirdivant SM, Crossland LD and Bogorad L (1985) DNA supercoiling affects *in vitro* transcription of two maize chloroplast genes differentially. *Proc Natl Acad Sci (USA)* **82**: 4886–4890

Sugita M, Shinozaki K and Sugiura M (1985) Tobacco chloroplast tRNA[Lys] (UUU) gene contains a 2.5-Kilobase-pair intron: An open reading frame and a conserved boundary sequence in the intron. *Proc Natl Acad Sci (USA)* .**82**: 3557–3561

Sugita M and Sugiura M (1984) Nucleotide sequence of and transcription of the gene for the 32,000 dalton thylakoid membrane protein from *Nicotiana tabacum*. *Mol Gen Genet* **195**: 308–313

Sugiura M (1989) The chloroplast chromosomes in land plants. *Annu Rev Cell Biol* **5**: 51–70

Sun F, Wu B-W and Tewari KK (1989) *In vitro* analysis of the pea chloroplast 16S rRNA gene promoter. *Mol Cell Biol* **9**: 5650–5659

Surzycki SJ and Schellenberger DL (1976) Purification and characterization of a putative sigma factor from *Chlamydomonas reinhardii*. *Proc Natl Acad Sci (USA)* **73**: 3961–3965

Svab Z, Hajdukiewicz P and Maliga P (1990) Stable transformation of plastids in higher plants. *Proc Natl Acad Sci (USA)* **87**: 8526–8530

Takaiwa F and Sugiura M (1980a) Cloning and characterization of 4.5S and 5S RNA genes in tobacco chloroplasts. *Gene* **10**: 95–103

Takaiwa F and Sugiura M (1980b) Nucleotide sequences of the 4.5S and 5S ribosomal RNA genes from tobacco chloroplasts. *Mol Gen Genet* **180**: 1–4

Tanaka M, Obokata J, Chunwongse J, Shinozaki K and Sugiura M (1987) Rapid splicing and stepwise processing of a transcript from the psbB operon in tobacco chloroplasts: Determination of the intron sites in petB and petD. *Mol Gen Genet* **209**: 427–431

Tanaka M, Wakasugi T, Sugita M, Shinozaki K and Sugiura M (1986) Genes for the eight ribosomal proteins are clustered on the chloroplast genome of tobacco (*Nicotiana tobacum*): Similarity to the S10 and *spc* operons of *Escherichia coli*. *Proc Natl Acad Sci (USA)* **83**: 6030–6034

Taylor WC (1989) Regulatory interactions between nuclear and plastid genomes. *Annu Rev Plant Physiol Plant Mol Biol* **40**: 211–233

Tewari KK and Goel A (1983) Solubilization and partial purification of RNA polymerase from pea chloroplasts. *Biochemistry* **22**: 2142–2148

Thomas F, Massenet O, Dorne A-M, Briat J-F and Mache R (1988) Expression of the rp123, rp12 and rps19 genes in spinach chloroplasts. *Nucleic Acids Res* **16**: 2461–2472

Thomas JR and Tewari KK (1974) Conservation of 70S ribosomal RNA genes in the chloroplast DNAs of higher plants. *Proc Natl Acad Sci (USA)* **71**: 3147–3151

Thompson WF, Everett M, Polans NO, Jorgensen RA and Palmer JD (1983) Phytochrome control of RNA levels in developing pea and mung-bean leaves. *Planta* **158**: 487–500

Thompson RJ and Mosig G (1990) Light affects the structure of *Chlamydomonas* chloroplast chromosomes. *Nucleic Acids Res* **18**: 2625–2631

Tohdoh N, Shinozaki K and Sugiura M (1981) Sequence of a putative promoter region for the rRNA genes of tobacco chloroplast DNA. *Nucleic Acids Res* **9**: 5399–5406

Torazawa K, Hayashida N, Obokata J, Shinozaki K and Sugiura M (1986) The 5′ part of the gene for ribosomal protein S12 is located 30 kbp downstream from its 3′ part in tobacco chloroplast genome. *Nucleic Acids Res* **14**: 31–43

Turmel M, Boulanger J and Lemieux C (1989) Two group I introns with long internal open reading frames in the chloroplast psbA gene of *Chlamydomonas moewusii*. *Nucleic Acids Res* **17**: 3875–3887

Tyagi AK and Herrmann RG (1986) Location and nucleotide sequence of the pre-apocytochrome f gene on the *Oenathera hookeri* plastid chromosome (*Euoenothera* plastome I). *Curr Genet* **10**: 481–486

Von Allmen J-M and Stutz E (1987) Complete sequence of "divided" rps12 (r-protein S12) and rps7 (r-protein S7) gene in soybean chloroplast DNA. *Nucleic Acids Res* **15**: 2387

Wakasugi T, Ohme M, Shinozaki K and Sugiura M (1986) Structure of tobacco chloroplast-genes for tRNA Ile (CAU), tRNA Leu (CAA), tRNA Cys (GCA), tRNA Ser (UGA) and tRNA Thr (GGU): A compilation of tRNA genes from tobacco chloroplasts. *Plant Mol Biol* **7**: 385–392

Wang, JC (1985) DNA topoisomerases. *Annu Rev Biochem* **54**: 665–697

Watson JC and Surzycki SJ (1982) Extensive sequence homology in the DNA coding for elongation factor TU from *Escherichia coli* and *Chlamydomonas reinhardtii* chloroplast. *Proc Natl Acad Sci (USA)* **79**: 2264–2267

Webber AN, Hird SM, Packman LC, Dyer TA and Gray JC (1989a) A photosystem II polypeptide is encoded by an open reading frame co-transcribed with genes for cytochrome b-559 in wheat chloroplast DNA. *Plant Mol Biol* **12**: 141–151

Webber AN, Packman L, Chapman DJ, Barber J and Gray JC (1989b) A fifth chloroplast-encoded polypeptide is present in the photosystem II reaction centre complex. *FEBS Letts* **242**: 259–262

Westhoff P, Alt J and Herrmann RG (1983a) Localization of the genes for the two chlorophyll *a*-conjugated polypeptides (mol. wt. 51 and 44 kd) of the photosystem II reaction center on the spinach plastid chromosome. *EMBO J* **2**: 2229–2237

Westhoff P, Alt J, Nelson N, Bottomley W, Bünemann H and Herrmann RG (1983b) Genes and transcripts for P_{700} chlorophyll *a* apoprotein and subunit 2 of the photosystem I reaction center complex from spinach thylakoid membranes. *Plant Mol Biol* **2**: 95–107

Westhoff P, Alt J, Nelson N and Herrmann RG (1985a) Genes and transcripts for the ATP synthase CFo subunits I and II from spinach thylakoid membranes. *Mol Gen Genet* **199**: 290–299

Westhoff P, Alt J, Widger WR, Cramer WA and Herrmann RG (1985b) Localization of the gene for apocytochrome b-559 on the plastid chromosome of spinach. *Plant Mol Biol* **4**: 103–110

Westhoff P, Farchaus JW and Herrmann RG (1986) The gene for the M_r 10,000 phosphoprotein associated with photosystem II is part of the psbB operon of the spinach plastid chromosome. *Curr Genet* **11**: 165–169

Westhoff P and Herrmann RG (1988) Complex RNA maturation in chloroplasts. The psbB operon from spinach. *Eur J Biochem* **171**: 551–564

Westhoff P, Nelson N, Bünemann H and Herrmann RG (1981) Localization of genes for coupling factor subunits on the spinach plastid chromosome. *Curr Genet* **4**: 109–120

Willey DL and Gray JC (1990) An open reading frame encoding a putative haem-binding polypeptide is contrascribed with the pea chloroplast gene for apocytochrome *f*. *Plant Mol Biol* **15**: 347–356

Willey DL, Howe CJ, Auffret AD, Bowman CM, Dyer TA and Gray JC (1984) Location and nucleotide sequence of the gene for cytochrome f in wheat chloroplast DNA. *Mol Gen Genet* **194**: 416–422

Woodbury NW, Dobres M and Thompson WF (1989) The identification and localization of 33 pea chloroplast transcription initiation sites. *Curr Genet* **16**: 433–445

Woodbury NW, Roberts LL, Palmer JD and Thompson WF (1988) A transcription map of the pea chloroplast genome. *Curr Genet* **14**: 75–89

Yamada T and Shimaji M (1986) Peculiar feature of the organization of rRNA genes of the *Chlorella* chloroplast DNA. *Nucleic Acids Res* **14**: 3827–3839

Yao WB, Meng BY, Tanaka M and Sugiura M (1989) An additional promoter within the protein-coding region of the psbD-psbC gene cluster in tobacco chloroplast DNA. *Nucleic Acids Res* **17**: 9583–9591

Yepiz-Plascencia GM, Radebaugh CA and Hallick RB (1990) The *Euglena gracilis* chloroplast rpoB gene. Novel gene organization and transcription of the RNA polymerase subunit operon. *Nucleic Acids Res* **18**: 1869–1878

Yokoi F, Vassileva A, Hayashida N, Torazawa K, Wakasugi T and Sugiura M (1990) Chloroplast ribosomal protein L32 is encoded in the chloroplast genome. *FEBS Letts* **276**: 88–90

Zaita N, Torazawa K, Shinozaki K and Sugiura M (1987) *Trans* splicing *in vivo*: joining of transcripts from the "divided" gene for ribosomal protein S12 in the chloroplasts of tobacco. *FEBS Letts* **210**: 153–156

Zhou D-X, Quigley F, Massenet O and Mache R (1989) Cotranscription of the S10- and spc-like operons in spinach chloroplast and identification of three of their gene products. *Mol Gen Genet* **216**: 439–445

Zhu YS, Kung SD and Bogorad L (1985) Phytochrome control of levels of mRNA complementary to plastid and nuclear genes of maize. *Plant Physiol* **79**: 371–376

Zurawski G, Bohnert HJ, Whitfeld PR and Bottomley W (1982a) Nucleotide sequence of the gene for the M_r 32,000 thylakoid membrane protein from *Spinacia oleraceae* and *Nicotiana debneyi* predicts a totally conserved primary translation product of M_r 38,950. Proc Natl Acad Sci (USA) **79**: 7699–7703

Zurawski G, Bottomley W and Whitfeld PR (1982b) Structure of the genes for the beta and epsilon subunits of spinach chloroplast ATPase indicate a dicistronic mRNA and an overlapping translation stop/start signal. *Proc Natl Acad Sci (USA)* **79**: 6260–6264

Zurawski G, Bottomley W and Whitfeld PR (1984) Structure of the large single copy region and the inverted repeats in *Spinacea oleracea* and *Nicotiana debneyi* chloroplast DNA: Sequence of the genes for tRNA^{His} and the the ribosomal proteins S19 and L2. *Nucleic Acids Res* **12**: 6547–6558

Zurawski G, Bottomley W and Whitfeld PR (1986) Sequence of the genes for the β and ε subunits of ATP synthase from pea chloroplasts. *Nucleic Acids Res* **14**: 3974

Zurawski G and Clegg MT (1987) Evolution of higher plant chloroplast DNA-encoded genes: Implications of structure, function and phylogenetic studies. *Annu Rev Plant Physiol* **38**: 391–418

Zurawski G, Perrot B, Bottomley W and Whitfeld PR (1981) The structure of the gene for the large subunit of ribulose 1, 5-bisphosphate carboxylase from spinach chloroplast DNA. *Nucleic Acids Res* **9**: 3251–3270

PHOTOSYSTEMS

2

Photosystem I Reaction Centre in Oxygenic Photosynthetic Organisms: Current Views and the Future

Satoshi Hoshina[1] and Shigeru Itoh[2]

[1]Department of Biology, Faculty of Science
Kanazawa University, Kakuma, Kanazawa 920-11, Japan
[2]Division of Bioenergetics National Institute for Basic Biology
38 Nishigonaka, Myodaiji Chyo, Okazaki 444, Japan

CONTENTS

52

ABBREVIATIONS

A_0	:	primary electron acceptor, chlorophyll *a*
A_1 ($Q\emptyset$)	:	secondary electron acceptor, phylloquinone
(B)Chl	:	(Bacterio) Chlorophyll
CD	:	circular dichroism
EDC	:	N-ethyl-3-[3-(dimethylamino) propyl] carbodiimide
EG	:	ethylene glycol
ENDOR	:	electron nuclear double resonance
EPR	:	electron paramagnetic resonance
ESR	:	electron spin resonance
EXAFS	:	extended X-ray absorption fine structure
F_A, F_B, F_X	:	iron-sulphur centre A, B, X
Fe-S	:	iron-sulphur centre
LDS	:	lauryl dodecyl sulphate
LHC	:	light harvesting complex
PAGE	:	polyacrylamide gel electrophoresis
PS I	:	photosystem I
P700	:	primary electron donor of PS I
RC	:	reaction center
SDS	:	sodium dodecyl sulphate
PSI-A, PSI-B, PSI-C etc.	:	proteins encoded by *psaA, psaB, psaC* genes, etc.

ABSTRACT

Recent advances in structure and function of photosystem I reaction centre (PS I RC) protein-pigment complex in the inner membranes of green plants and of cyanobacteria are discussed. The complex cooperates with photosystem II reaction centre (PS II RC) complex in the same membranes to convert light energy into electrochemical energy in photosynthetic reaction. The structure of the protein moiety of the PS I RC complex is discussed on the basis of recently identified subunit components and their amino-acid sequences. The sequences show very low homology to those of subunits of RC complexes of PS II or purple bacteria, and suggest that this complex has a structure and an evolutionary origin somewhat different from the latter complexes. The PS I RC produces extremely low redox potential and reduces $NADP^+$. Almost all the prosthetic groups, such as chlorophylls, quinone, and iron-sulphur centres, in this RC have recently been identified. Recent experimental findings on characteristics of the prosthetic groups and their reactions are reviewed. It has become clear that the protein moiety of this RC complex creates special environments for the function of prosthetic groups. Future studies, expected from the modification of structure of this RC now in progress in various laboratories, are discussed.

I. Introduction

Photosynthetic reaction in green plants is carried out by two types of light reactions, photosystems I and II (for discussion on PS II see Chapter by Govindjee and Coleman this volume). Both reactions require the organized multi-subunit protein-pigment complexes embedded in chloroplast inner thylakoid membranes. PS II RC complex oxidizes water and reduces plastoquinone, which transfers reducing power to cytochrome b_6/f complex in the same membrane. The cytochrome b_6/f complex pumps protons, generates membrane potential and reduces plastocyanin, which is the water-soluble protein localized in the inner thylakoidal space, and donates electrons to PS I. PS I RC complex reduces $NADP^+$, which gives reducing power for the carbon fixation reaction carried out by the water-soluble enzymes in the stroma space of chloroplasts. Membrane potential and proton gradient across the thylakoid membrane, produced by the activities of these three complexes, contribute to the synthesis of adenosine triphosphate (ATP) by ATP-synthetase on the same membrane (see Chapter by Ivanov, this volume).

Photosynthetic reactions in prokaryotes, which have no chloroplasts, are carried out on the inner cytoplasmic membranes. Their RC complexes are grouped into four types; (1) PS I and (2) PS II RCs in cyanobacteria or prochlorophytes, (3) RCs in purple and green non-sulphur photosynthetic bacteria (*Chloroflexus*), and (4) RCs in green sulphur (*Chlorobium*) and Gram positive bacteria (*Heliobacterium*) (Fig. 1). Photosynthetic systems in cyanobacteria and in prochlorophytes have essentially the same system as that of chloroplasts having PS I and PS II RCs, and cytochrome b_6/f complexes. The other photosynthetic prokaryotes have only one photosystem and have cytochrome b/c_1 complexes which are almost analogous to cytochrome b_6/f complex. Purple bacteria have RC complexes, whose tertiary structure is now known from X-ray crystallography (Deisenhofer *et al.*, 1985; Allen *et al.*, 1987); they have some homology with the PS II RC complex. On the other hand, green sulphur (*Chlorobium*) and Gram positive bacteria (*Heliobacterium*) contain RC complexes which are likely to be analogous to the PS I RC complex. The RCs of PS I, green sulphur bacteria, and Gram positive bacteria have quinone and iron-sulphur centres as the electron acceptors in the RC complexes, and make clear contrast

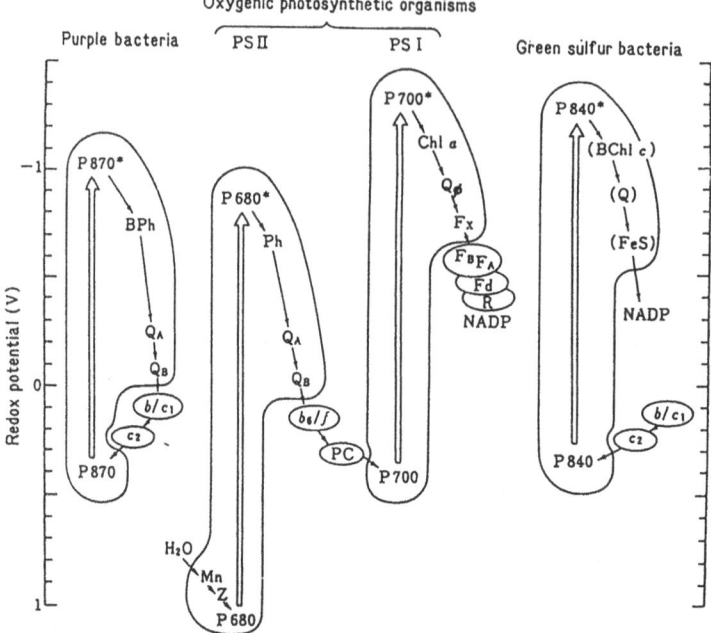

Figure 1. Energy diagram of photosynthetic electron transfer components in RC complexes of purple photosynthetic bacteria, oxygenic organisms and green sulphur bacteria. P870, P680, P700 and P840; RC electron donor chlorophylls. (B)Ph, electron acceptor (bacterio) pheophytin. (B)Chl, (bacterio) chlorophyll. Q, electron acceptor quinone. b/c_1 or b_6/f, cytochrome b/c_1 or b_6/f complex. PC, plastocyanin. Z, secondary electron donor tyrosine. Fd, ferredoxin. R, ferredoxin-NADP reductase. c_2, cytochrome c_{a2} complex.

to the RCs of PS II and purple bacteria, which have two electron acceptor quinones in a series and have no iron-sulphur centres. The PS I type RCs are called "iron-sulphur type RCs" and the PS II type, "quinone-type RCs". Amino-acid sequences of PS II and purple bacterial RC polypeptides (about 30 kD molecular mass) show homology with each other, but show very little homology to those of the PS I-RC polypeptides which have about 80 kD molecular mass. Green non-sulphur bacteria (*Chloroflexus*), although quite different from all systems, are similar in many respects to purple photosynthetic bacteria (Blankenship and Fuller, 1986). Comparison of L- and M-subunit of RCs of purple and green non-sulphur bacteria shows highly significant amino-acid sequence homologies (Ovchinnikov *et al.*, 1988a, b). RC polypeptides of green sulphur bacteria have not been well characterized.

The mystery in the evolution of the photosynthetic system is, therefore, the existence of two types of RCs. Purple and green sulphur bacteria, which are from an evolutionary viewpoint quite distant from each other, have only one RC, while cyanobacteria, prochlorophytes, and chloroplast have both RCs working in a series on the same membrane. Have the two types of RCs evolved separately from the one ancestral RC which is now lost? It is not easy to assume that a highly efficient and specific system like photosynthetic RC originated twice in the history of evolution of life. Did purple and green bacteria get RCs together by gene transfer? Have purple or green bacteria lost one of the RC complexes during evolution? Why are there these different types of RCs? These questions have not been fully answered so far (Olson and Pierson, 1987; Mathis, 1990). We need more information about different photosynthetic systems.

The structures of the purple bacterial RCs are now available by the X-ray crystallographic data on *Rhodopseudomonas viridis* (Diesenhofer *et al.*, 1985) and *Rhodobacter sphaeroides* (Allen *et al.*, 1987). These results have led to enormous progress in our understanding of the structure-function relationship in these RCs and in PS II RC. On the other hand, our knowledge about iron-sulphur type RCs (PS I, green sulphur bacteria, or Gram positive bacteria) is still limited. We describe here current views on PS I RC which, in some aspect, show common features with the other type of RCs.

The reader is referred to recent reviews on photochemistry and electron carriers of PS I (Rutherford and Heathcote, 1985; Ke and Shuvalov, 1987; Mathis and Rutherford, 1987; Andréasson and Vänngård, 1988; Lagoutte and Mathis, 1989; Evans and Bredenkamp, 1990), on PS I core complex (Malkin, 1987; Golbeck, 1988; Green, 1988; Margulies, 1989), and on polypeptides of PS I complex (Nelson, 1987; Møller *et al.*, 1990; Scheller and Møller, 1990). For peripheral light-harvesting antenna of PS I (LHC I), which will not be discussed here, see Thornber (1986), Zuber (1987), and Green (1988).

II. Current View of the Structure of PS I

a. General structure

PS I complex contains chlorophyll *a*, quinone, and three iron-sulphur centres, F_X, F_A, and F_B, shown in Fig. 2. The components other than F_A and F_B reside on the RC complex, and F_A/F_B on a small polypeptide, which is bound to the RC complex and shows homology to bacterial ferredoxins. Chlorophylls are grouped into three functional types: antenna, electron donor (P700), and electron acceptor (A_0 = Chl *a*-690). The function of chlorophyll *a* as an electron acceptor is analogous to the function of pheophytin in PS II or purple bacterial type RCs although the redox potential of A_0 is much lower. The existence of iron-sulphur centres in this RC has long been known and has been a reason to assume a very different structure for this RC from the PS II or purple bacterial type RCs. The function of quinone has only recently been confirmed in this RC. The quinone (phylloquinone = 2-methyl-3-phytyl-1, 4-naphthoquinone) has been known as vitamin K_1, which functions as the blood coagulation factor when fed to animals, but until recently its function in plants was not known. The discovery of quinone as the secondary electron acceptor in this RC has given the common electron transfer scheme in all the photosynthetic RCs as:

chlorophyll dimer → chlorophyll (or pheophytin) monomer → quinone

PS I RC produces reducing power which is much stronger than that produced by PS II RC. This seems to be related to the use of a series of electron acceptors, which have low redox potentials not very much separated from each other and which mediate reduction of ferredoxin weakly bound to the surface. This contrasts

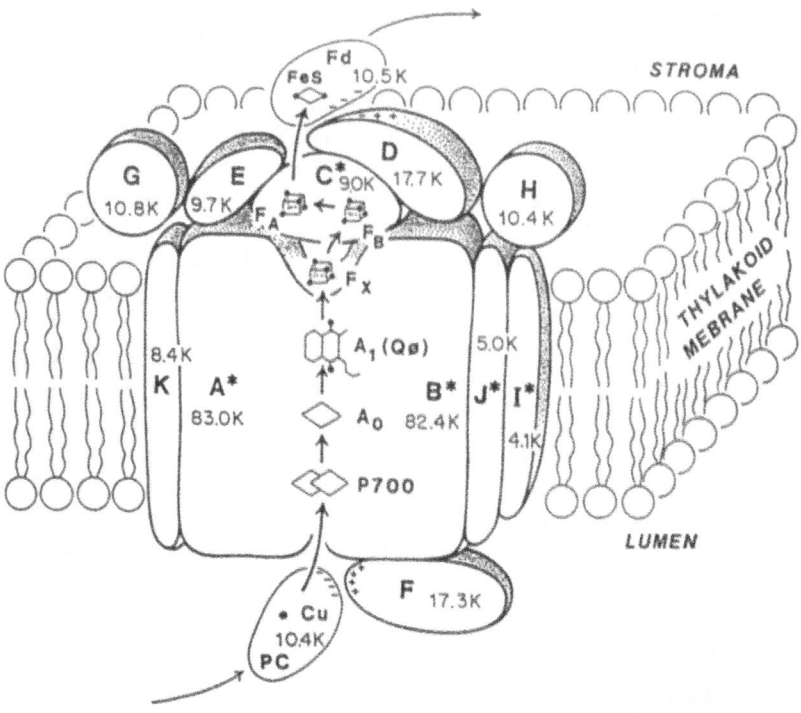

Figure 2. Model for the structure and organization of the PS I core complex. Polypeptides with asterisk are encoded by the chloroplast genes and the others by the nuclear genes. ◊, chlorophyll *a*; ⟶, direction of electron transfer. Details are given in the text (also for a discussions on chloroplast genes see chapter by Tyagi *et al.*, this volume)

with the situation in PS II or purple bacterial RCs in which the acceptor pheophytin and the Q_A quinone are separated with a large energy gap, which is interpreted to facilitate the fast (~ 200 ps) electron transfer rate (Gunner *et al.*, 1986). Phylloquinone plays an important role in PS I, since its semiquinone–/quinone redox potential is estimated to be much lower than that of other quinones in other RCs. This is presumably due to the unique nature of the protein environment inside this RC, which is very different from those in PS II (see chapter by Govindjee and Coleman, this volume) or purple bacterial RCs.

b. Purification and fractionation of PS I complex

PS I is a pigment protein complex localized mainly in non-appressed thylakoid membrane regions (Gounaris *et al.*, 1986; Staehelin, 1986; Golbeck, 1988). A variety of functional complexes with different pigment and polypeptide contents can be isolated by treatment of thylakoid membranes with detergents, chaotropic agents, or organic solvents, followed by density gradient ultracentrifugation, column chromatography, or gel electrophoresis. In this chapter, various PS I preparations are grouped following the nomenclature of Golbeck (1988).

(i) PS I complex

Almost native PS I complex is isolated from thylakoid membranes after treatment with Triton X-100, followed by sucrose density gradient ultracentrifugation. It shows Chl a/b ratio of 7-9 and contains approximately 200 Chl per P700 (Mullet *et al.*, 1980; Lam *et al.*, 1984; Hoshina and Itoh, 1987); in this complex, about 100 molecules of Chl bind to the central PS I core complex (see below) and the others to the peripheral light-harvesting Chl a/b complexes (LHC I) with a Chl a/b ratio of 3.5 (Thornber, 1986; Bassi and Simpson, 1987; Malkin, 1987; Green, 1988). The complex is also termed "PS I-200" to represent its chlorophyll content per P700. A similar PS I complex (CPIa) can be isolated by mild electrophoresis on polyacrylamide gels after treatment of thylakoids with sodium dodecyl sulphate (SDS) (Thornber, 1986; Malkin, 1987; Green, 1988).

(ii) PS I core complex

PS I core complex is the preparation in which LHC I is removed from the native PS I complex (Mullet *et al.*, 1980; Lam *et al.*, 1984) and is composed of PSI-A, PSI-B, PSI-C, and eight polypeptides (PSI-D, E, F, G, H, I, J, and K). This complex contains all the electron acceptors A_0, $A_{1(Q_\phi)}$, F_X, F_B, and F_A as well as P700. Polypeptides in the PS I core complex except PSI-A and PSI-B do not seem to bind any Chl. The PS I core complex also binds about 100 molecules of core antenna Chl a on the polypeptides PSI-A and PSI-B (Malkin, 1987; Golbeck, 1988; Green, 1988). PS I core complex catalyzes the photoreduction of $NADP^+$ if ferredoxin is added, and oxidizes added plastocyanin (Bengis and Nelson, 1975; Wynn and Malkin, 1988a). The number of chlorophylls associated with the complex varies from 40 to 100 per P700 depending on the method of preparation (Malkin, 1987; Golbeck, 1988; Green, 1988).

(iii) PS I RC complex

PS I RC complex is composed of PSI-A and PSI-B polypeptides which bind P700, A_0, A_1, F_X, and 40 to 100 molecules of Chl a, but lacks PSI-C polypeptide, which carries F_A/F_B. PS I RC complexes have been isolated from cyanobacteria and higher plant chloroplasts by treatment with urea (Golbeck *et al.*, 1988a; Parrett *et al.*, 1989). Hoshina *et al.*, (1989, 1990) prepared a PS I RC complex by Triton X-100 treatment after 60°C heat treatment of spinach PS I particles in the presence of 50% ethylene glycol (60°C/EG treatment). This preparation retained F_X but had lost almost all of F_A and F_B (Fig. 3). Both types of PS I RC preparations contained polypeptide(s) smaller than 10 kD in addition to PSI-A and PSI-B (Golbeck *et al.*, 1988a; Hoshina *et al.*, 1989, 1990). These complexes still contain more than 50 molecules of chlorophylls. This is in contrast to the situation in the PS II or purple bacterial RC complexes which bind only 5–6 Chl (BChl) per RC. The PS I RC preparations can be reconstituted with the PSI-C retaining F_A/F_B iron-sulphur centres (Golbeck *et al.*, 1988b; Hoshina and Oh-oka, unpublished data). Hiyama *et al.* (1990) reported the F_X-active PS I RC complex without small subunits. F_X cluster in the RC complex is easily destroyed when the complex is treated with SDS or LDS (Bengis and Nelson, 1977; Golbeck and Cornelius, 1986) or by treatment with urea

58

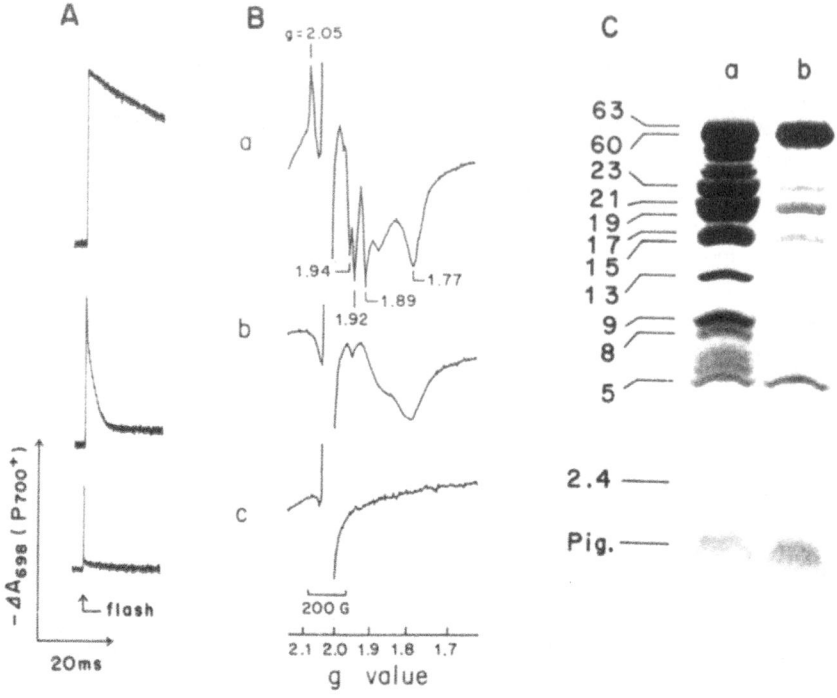

Figure 3. Characterization of PS I complex prepared by heat/EG treatment. (A) Flash-induced absorption changes at 698 nm measured at room temperature. (B) EPR spectra of the iron-sulphur centres measured at 10K. a. Native PS I complex. b. PS I RC complex prepared by 60°C/EG treatment. c. PS I RC protein by 70°C/EG treatment. (C) Pclypeptide compositions analysed by SDS-PAGE (reprinted, with permission, from Hoshina *et al.*, 1989).

and potassium ferricyanide (Parrett *et al.*, 1990). Incubation in the presence of $FeCl_3$, Na_2S, and beta-mercaptoethanol reconstitutes the F_X cluster (Parrett *et al.*, 1990).

Flash-induced absorption change of P700, electron paramagnetic resonance (EPR) signals of F_A, F_B, and F_X clusters, and polypetide compositions in PS I complex and PS I RC complex prepared by heat/EG treatment are shown in Fig. 3 (Hoshina *et al.*, 1989, 1990).

(iv) Other preparations

PS I complexes are also termed "PS I-200", "PS I-100", or "PS I-40", in which numbers represent their Chl content per P700. Efforts have been directed at isolating a PS I complex in which Chl content is as low as possible. Ikegami and Katoh (1975, 1989) reported the preparation of "P700-enriched PS I particles", which contain only 10 Chl per P700 after extraction of Chl with diethyl ether. This preparation retains all the electron transfer components except phylloquinone (A_1) and is suitable for physicochemical studies as reported elsewhere (Vacek *et al.*, 1977; Ikegami and Itoh 1986, 1988; Ikegami *et al.*, 1987; Breton and Ikegami, 1989).

c. Polypeptides in PS I

PS I of eukaryote (chloroplasts) is composed of RC core complex and LHC I. Bengis and Nelson (1975) reported that the core complex, prepared from higher plants, contained eight different polypeptides, designated as subunits Ia to VII. Similar core complexes were prepared from a wide variety of plants (Nelson, 1987, 1988; Münch *et al.*, 1988) and were shown to be composed of two large polypeptides (60–70 kD on the basis of SDS-polyacrylamide gel electrophoresis, or SDS-PAGE) and at least nine smaller polypeptides with apparent molecular masses between 1.5 and 22 kD when determined by the SDS-PAGE (Table 1). There is a considerable variation in the number and size of these small polypeptides in different preparations from different species. The subunit stoichiometry of the PS I complex has been investigated by densitometric scanning of Coomassie brilliant blue-stained bands of SDS-PAGE (Bengis and Nelson, 1977; Høj *et al.*, 1987), by the radioactivity associated with each band of SDS-PAGE from uniformly ^{14}C-labelled plants (Lundell *et al.*, 1985; Bruce and Malkin, 1988), and by the amino-acid composition of each polypeptide (Scheller *et al.*, 1989a). A stoichiometry of main subunits (five or six polypeptides) was shown to be 1:1 for the complex from Swiss chard (Bengis and Nelson, 1977), green alga, *Dunaliella salina* (Bruce and Malkin, 1988), and barley (Scheller *et al.*, 1989a). However, there is still some uncertainty.

The amino-acid sequences of all these subunits are now available from the corresponding gene sequences. The genes coding for these polypeptides were cloned and sequenced in higher plants, algae, and cyanobacteria. Amino-acid sequence of each subunit from various organisms shows high homology with each other and suggests conservation of their structure (Nelson, 1988). In this chapter, products (polypeptides) of genes *psaA, psaB, psaC*, etc., are denoted PSI-A, PSI-B, PSI-C, etc., according to Schantz and Bogorad (1988). The polypeptides associated with PS I complex may be grouped into four types, judging from their gene locations and functions:

1) Three subunits (PSI-A, PSI-B and PSI-C), encoded by the chloroplast genes, constitute the major part of PS I core complex (Table 1). The PSI-A and B polypeptides bind the primary electron donor P700, acceptors A_0, A_1, and F_X, and the core antenna pigments. The PSI-C polypeptide, bound to the stroma-side surface of the RC complex, carries iron-sulphur centres F_A and F_B.

2) Five subunits (PSI-D to H), encoded by nuclear DNA, bind to the surface of the core complex (Table 1). They do not seem to bind prosthetic groups. PSI-D and PSI-F seem to facilitate binding of ferredoxin and plastocyanin, respectively. The functions of PSI-E, PSI-G, and PSI-H remain to be determined.

3) Three small subunits (PSI-I, J and K) with apparent masses ranging from 1.5 kD to 7 kD (by SDS-PAGE) were identified only during 1989–1990. They are tightly bound to the core complex in various organisms (Hoshina *et al.*, 1989, 1990; Koike *et al.*, 1989; Scheller *et al.*, 1989a, b; Ikeuchi *et al.*, 1990; Wynn and Malkin, 1990). These small polypeptides are hydrophobic and seem to have membrane-spanning helices.

4) Three or four subunits with apparent molecular masses between 20 and 25 kD associate with LHC I (Haworth *et al.*, 1983; Thornber, 1986; Zuber, 1987; Green,

Table 1. Polypeptides of PS I core complex

Subunit	Protein	MWa (kD)	Gene location[b]	Gene	Amino-acid residues[c]	MWc[d] (kD)	PI[e]	Helices[f]	Function and prosthetic group content	Ref
IA	PSI-A	55–70	C	psaA	750	83.0	37	11	$P700$, A_0, A_1, F_X,	1–8
IB	PSI-B	55–70	C	psaB	734	82.4	38	11	Core antenna pigments	1–8
II	PSI-D	19–25	N	psaD	162	17.8	43	0	Fd-docking protein	9–14
III	PSI-F	16–20	N	psaF	154	17.3	42	0	PC-docking protein	15–17
IV	PSI-E	13–18	N	psaE	91	9.7	41	0	?	9, 13, 17, 18
V	PSI-G	11–16	N	psaG	98	10.8	41	0	?	15, 13, 19
VI	PSI-H	9–10	N	psaH	95	10.4	41	0	?	20, 13, 19, 21
VII	PSI-C	8–9	C	psaC	81	9.0	43	0	F_A, F_B-carrying protein	20, 4, 22–26
VIII	PSI-K	3–7	N	psaK	87	8.4	33	1–2	?	19, 27–30
IX	PSI-J	4.1	C	psaJ	44	5.0	30	1	?	3, 4, 28, 30, 31
X	PSI-I	1.5–3.9	C	psaI	36	4.0	28	1	?	32, 3, 4, 28, 31

[a] Apparent molecular mass calculated by SDS-PAGE. [b] C: chloroplast gene; N: nuclear gene. [c] Amino-acid residues of the first reference cited in the last column. [d] Molecular mass calculated from coding sequence of the first reference in the last column. [e] PI (polarity index) = D+N+E+Q+K+S+R+T+H/total amino-acids (mol %) according to Capaldi and Vanderkooi (1972). [f] Transmembrane helices predicted from sequences.

Sequences references PSI-A and PSI-B: 1. spinach (Kirsch et al., 1986); 2. maize (Fish et al., 1985); 3. tobacco (Shinozaki et al., 1986); 4. liverwort (Ohyama et al., 1986); 5. pea (Lehmbeck et al., 1986); 6. Euglena gracilis (Cushman et al., 1988); 7. Chlamydomonas reinhardtii (Küch et al., 1987); 8. Synechococcus (Cantrell and Bryant, 1987).

PSI-D: 9. spinach (Münch et al., 1988); 10. spinach (Lagoutte, 1988); 11. tomato (Hoffman et al., 1988); 12. barley (Scheller et al., 1988); 13. pea (Dunn et al., 1988); 14. Synechosystis (Reilly et al., 1988).

PSI-F: 15. spinach (Steppuhn et al., 1988); 16. barley (Scheller et al., 1990); 17. Chlamydomonas reinhardtii (Franzén et al., 1989a).

PSI-E: 9. spinach (Münch et al., 1988); 13. pea (Dunn et al., 1988); 18. barley (Okkels et al., 1988); 17. Chlamydomonas reinhardtii (Franzén et al., 1989a).

PSI-G: 15. spinach (Steppuhn et al., 1988); 13. pea (Dunn et al., 1988); 19. Chlamydomonas reinhardtii (Franzén et al., 1989b).

PSI-H: 20. spinach (Steppuhn et al., 1989); 21. barley (Okkels et al., 1989); 13. pea (Dunn et al., 1988); 19. Chlamydomonas reinhardtii (Franzén et al., 1989b).

PSI-C: 20. spinach (Steppuhn et al., 1989); 4. liverwort (Ohyama et al., 1986); 22. spinach (Oh-oka et al., 1988a); 23. pea and wheat (Dunn and Gray, 1988); 24. barley (Høj et al., 1987); 25. maize (Schantz and Bogorad, 1988); 26. tobacco (Hayashida et al. 1987).

PSI-K: 19. Chlamydomonas reinhardtii (Franzén et al., 1989b); 27. spinach (Hoshina et al., 1989); 28. spinach and pea (Ikeuchi et al., 1990); 29. spinach (Wynn and Malkin, 1990); 30. Synechococcus vulcanus (Koike et al., 1989).

PSI-J: 3. tobacco (Shinozaki et al., 1986); 4. liverwort (Ohyama et al., 1986); 28. spinach and pea (Ikeuchi et al., 1990); 31. rice (Hiratsuka et al., 1989); 30. Synechococcus vulcanus (Koike et al., 1989).

PSI-I: 32. barley (Scheller et al., 1989b); 3. tobacco (Shinozaki et al., 1986); 4. liverwort (Ohyama et al., 1986); 28. spinach and pea (Ikeuchi et al., 1990); 31. rice (Hiratsuka et al., 1989).

1988). These LHC I apoproteins are encoded by nuclear DNA and their sequences are available (Pichersky *et al.*, 1987).

Function and characterization of these polypeptides belonging to groups (1) and (3) are further discussed here. The other polypeptides will be briefly discussed in a later section. For recent review of PS I polypeptides and genes, see Scheller and Møller (1990).

(i) PSI-A and PSI-B

The two large RC polypeptides, PSI-A and PSI-B, are encoded by the chloroplast genes *psaA* and *psaB*, which are only 25 bp apart and are cotranscribed (Ohyama *et al.*, 1986; Shinozaki *et al.*, 1986). Molecular weights of these subunits are about 83 kD calculated from cording sequences (60–70 kD by SDS-PAGE) (Fish *et al.*, 1985; Kirsch *et al.*, 1986). PSI-A and PSI-B show a high, about 80% homology to each other, suggesting that they have evolved through gene duplication (Nelson, 1988). These two polypeptides form a heterodimer and constitute the RC complex containing the prosthetic groups. These polydpeptides are assumed to have eleven membrane-spanning alpha-helices which contain about 50 histidines, most of which presumably bind Mg atom of Chl (see Fig. 4). They retain the specific sequence

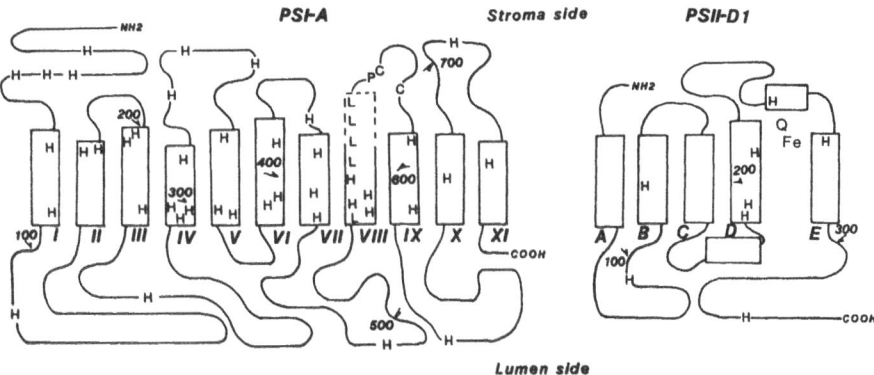

Figure 4. Schematic diagram of the primary structure of PSI-A and PSII-D1 polypeptides. Estimated membrane-spanning helix regions are shown as boxes numbered with roman numerals or large capitals. Arabic numbers indicate number of amino-acid residues. H, L, P, and C represent histidine, leucine, proline, and cysteine residues. The structure of PSI-A was modified from Kirsch *et al.* (1986) and Webber and Malkin (1990) and that of PS II-D1 was from Trebst (1987).

containing two cysteins (PCDGPGRGGTC), which is highly conserved in the extra-helix regions between VIII and IX helices of the PSI-A (residues at 572–582 in spinach) and the PSI-B (residues 558–568 in spinach) in several species (Fish *et al.*, 1985; Kirsch *et al.*, 1986). This sequence is a candidate for the binding of the unique iron-sulphur centre F_X, which is assumed to bridge PSI-A and PSI-B (Golbeck *et al.*, 1987) (Fig. 2). Webber and Malkin (1990) recently proposed that a number of leucine residues spaced every seventh amino-acid residue mediate the dimerization of these polypeptides forming a "leucine zipper" in helix VIII region preceding the above F_X candidate region.

The amino-acid sequence of PSI-A or PSI-B shows very low homology with RC polypeptides of PS II-D1 and D2 or of purple bacterial L and M (Fig. 4). However, existence of some sequence homology is suggested recently for some histidine residues (Robert and Moënne-Loccoz, 1990).

(ii) PSI-C

By the analysis of chloroplast DNA sequence, it was assumed that chloroplast has a polypeptide that contains two sets of amino-acid sequences of CXXCXXCXXXCP, corresponding to characteristic sequence of [4Fe-4S] type iron-sulphur clusters in bacterial ferredoxins, now known as the regions at 10–21 and 47–58 of PSI-C (Ohyama et al., 1986; Hayashida et al., 1987). PSI-C carries the two [4Fe-4S] clusters F_A and F_B and shows homology to bacterial ferredoxins (Fig. 5). The one [4Fe-4S] cluster is estimated to be coordinated by cysteins 10, 13, 16, and 57, and the other [4Fe-4S] cluster by cysteins 20, 47, 50, and 53 (Oh-oka et al., 1988a; Dunn and Gray, 1988) (Fig. 5). Apoprotein of F_A and F_B has been isolated from spinach (Oh-oka et al., 1987), tobacco (Hayashida et al., 1987), and barley (Høj et al., 1987). Spinach PSI-C, retaining the iron-sulphur clusters, was isolated and characterized (Oh-oka et al., 1988a, b, 1991; Wynn and Malkin, 1988b). The structure of the protein is assumed to change by the isolation of protein from PS I-RC complex (Oh-oka et al., 1991).

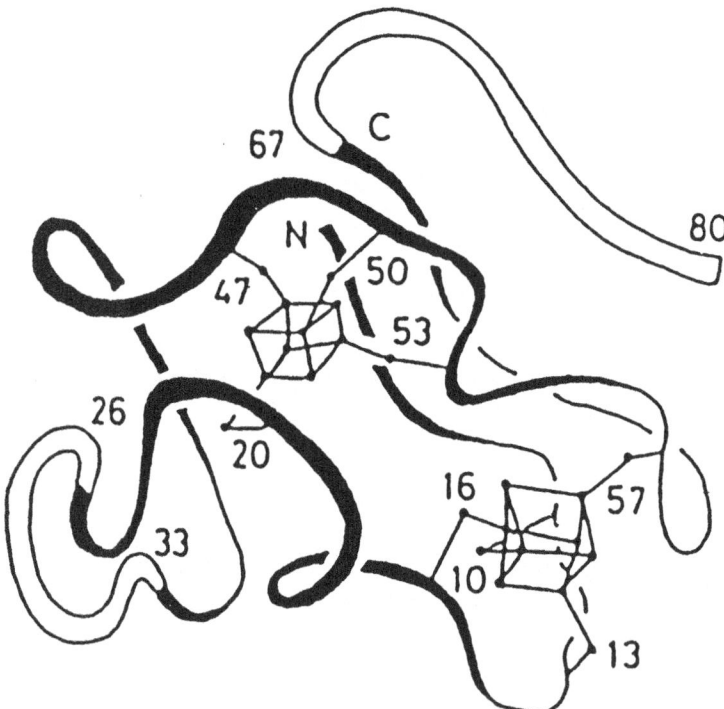

Figure 5. Simulated three-dimensional structure of PSI-C 9 kD polypeptide. The black ribbon (■) shows the main chain folding of *Peptococcus aerogenes* ferredoxin. The white (□) shows the insertion and the extended C-terminal region of this protein (reprinted, with permission, from Oh-oka *et al.*, 1988a).

(iii) PSI-K

When PS I particles were tested with Triton X-100 after 60°C/EG treatment, PSI-RC complex, mainly composed of PSI-A, PSI-B, and a polypeptide with an apparent molecular mass of 5 kD, was isolated (Hoshina *et al.*, 1989, 1990). This complex lacks PSI-C but contains F_X. The N-terminal sequence of the 5 kD polypeptide was determined (Hoshina *et al.*, 1989) and was shown to be 72% homologous to a 3 kD polypeptide (P37 subunit) from the green alga *Chlamydomonas reinhardtii*, encoded by the nuclear gene denoted *psaK* (Franzén *et al.*, 1989b) (Fig. 6). A mature protein encoded by *psaK* is made of 87 amino-acids with 8.4 kD molecular mass. A similar polypeptide was isolated from PS I complexes of spinach, pea (Ikeuchi *et al.*, 1990; Wynn and Malkin, 1990), and cyanobacterium *Synechococcus vulcanus* (Koike *et al.*, 1989). PSI-K from higher plants and green alga shows a highly conserved N-terminal sequence (Fig. 6). PSI-K is hydrophobic, can be extracted by organic solvent (Girard *et al.*, 1980), and presumably has one (or two) membrane-spanning alpha-helix at N- or C-terminal (Fig. 6).

(iv) PSI-J

The PS I core complex isolated from a thermophilic cyanobacterium, *S. vulcanus*, also contains a 4.1 kD polypeptide whose N-terminal sequence matches an open reading frame (ORF) 44 of tobacco (Shinozaki *et al.*, 1986) and liverwort ORF42 (Ohyama *et al.*, 1986) of chloroplast DNA. This gene is now designated *psaJ* (Koike *et al.*, 1989) (Fig. 6). PSI-J in tobacco is 98% and 83% homologous to the protein in rice and liverwort, respectively (Fig. 6), and seems to have a membrane-spanning domain (Ikeuchi *et al.*, 1990). PSI-J was found in the native PS I complex of spinach but was not retained in the PS I core complex (Ikeuchi *et al.*, 1990).

(v) PSI-I

Scheller *et al.* (1989a) reported PSI-I, the smallest subunit associated with the PS I complex, from barley with an apparent molecular weight of 1.5 kD. Such a polypeptide was also described as the 2.4 kD (Hoshina *et al.*, 1989) or 3.9 kD polypeptide in spinach (Ikeuchi *et al.*, 1990) and as the 4.4 kD polypeptide in pea PS I complex (Ikeuchi *et al.*, 1990). The gene sequence encoded by a chloroplast genome, designated *psaI*, encodes a polypeptide of 36 amino-acid residues with a deduced moleuclar mass of 4 kD which seems to have a membrane-spanning alpha-helix (Scheller *et al.*, 1989b) (Fig. 6). The PSI-I in barley is 89%, 64%, and 89% homologous to that in tobacco (Shinozaki *et al.*, 1986), liverwort (Ohyama *et al.*, 1986), and rice (Hiratsuka *et al.*, 1989), respectively (Fig. 6). Since PSI-I is partly homologous to transmembrane helix E of the D2 polypeptide of PS II RC (see Fig. 4), it is speculated that it participates in the binding of P700, A_0 and A_1 in a manner similar to that of D1 and D2 polypeptides in PS II RC (Scheller *et al.*, 1989b; Møller *et al.*, 1990; Scheller and Møller, 1990). The 2.4 kD polypeptide (presumably PSI-I), however, seems to be lost in the 60°C/EG-RC complex, although it still retains active P700, A_0, A_1, and F_X (Hoshina *et al.*, 1989). The function of PSI-I, therefore, is not known.

(A) PSI-K

```
                        1 ┌─────────────────────────┐                40
C. reinhardtii:         DGFIGSSTNLIMVASTTATLAAARFGLAPTVKKNTTAGLK
     (19)ᵃ                          :                    :
Spinach:                GD··········T·??LM?F·G····?·
     (27)                           :                    :
Spinach:                GD··········T···LM·F·G····
     (28)                           :
Pea:                    -D······V·······LM·F
     (28)                   :      :::      ::   : :::
S. vulcanus:            TLP·TTWTP·VG·VVIL·NLFAI·LG·YAI
     (30)
```

```
                        41            ┌────────────────────────┐        87
C. reinhardtii:         LVDSKNSAGVISNDPAGFTIVDVLAMGAAGHGLGVGIVLGLKGIGAL/
     (19)
```

(B) PSI-J

```
                        1  ┌───────────────────────────────┐           44
Tobacco:                MRDLKTYLSVAPVLSTLWFGALAGLLIEINRFFPDALTFPFFSF/
     (3)                     :              :               :
Rice:                   ···I·········V········R········L·····S······/
     (31)                 : :             :                          :
Liverwort:              ·Q·V·····T····A·····F···············VL···/
     (4)                     :
Spinach:                ···F··········?·
     (28)
Pea:                    ··········?····
     (28)                   : :
S. vulcanus:            ·KHFL····T····
     (30)
```

(C) PSI-I

```
                        1  ┌────────────────────────┐                  36
Barley:                 MTDLNLPSIFVPLVGLVFPAIAMTSLFLYVQKKKIV/
     (32)                    :                    :           :
Tobacco:                ··N·················A····H···N···/
     (3)                     :                    :           :
Rice:                   ·M·F················A·········N···/
     (31)                   : :           : :    : :::    :
Liverwort:              ··ASY···········I····T·A···I·IEQDE·L/
     (4)                     :
Spinach:                ·----·F················
     (28)                   :    :          :
Pea:                    ·--I····L········L·
     (28)
```

Figure 6. Alignment of amino-acid sequences for PSI-K, PSI-J, and PSI-I from different species (modified from Ikeuchi *et al.*, 1990). (a) Reference number, see Table 1. Single dot (•) indicates a residue identical to the first polypeptide; double dot (:), conservative substitutions; hyphen (–), gap; ?, unidentified; ⌐⌐⌐, putative transmembrane region.

d. Organization of PS I subunits

Topology of subunit polypeptides in the PS I complex has been studied using chemical labelling, cross-linking agents, and proteolytic enzymes (Andersson *et al.*,

1982; Ortiz *et al.*, 1985; Ryrie *et al.*, 1985; Enami *et al.*, 1986, 1987). At the stroma side of PS I, a 20 kD (or 22 kD) polypeptide (presumably PSI-D) was shown to insure the docking of ferredoxin for electron transfer from F_A and F_B by using a cross-linking agent EDC (N-ethyl-3-[3-(dimethylamino) propyl] carbodiimide) (Zanetti and Merati, 1987; Zilber and Malkin, 1988). Oh-oka *et al.* (1989) showed that PSI-C, PSI-D (19 kD), and PSI-E (14 kD), identified by their N-terminal amino-acid sequences, were situated close to each other on the stroma side of thylakoid membranes. Andersen *et al.* (1990) also showed that PSI-D cross-linked with both PSI-E and ferredoxin using EDC. These results suggest that PSI-C, D and E are located on the surface of the stroma side of PS I RC. On the other hand, ferredoxin did not directly cross-link with PSI-C.

A subunit PSI-F was shown to interact with the plastocyanin on the lumen side (intrathylakoidal space). A 19 kD polypeptide (as determined by SDS-PAGE; it is presumably PSI-F) of PS I was chemically cross-linked with the plastocyanin by EDC and facilitated the electron transfer between plastocyanin and the P700 (Wynn and Malkin, 1988a). The carboxyl groups of plastocyanin seem to be necessary for the cross-linking. Hippler *et al.* (1989) showed that PSI-F cross-linked with plastocyanin. These results are consistent with the earlier data that treatment of PS I core complex with 1% Triton X-100 removed a subunit III (presumably PSI-F) and led to a loss in the plastocyanin-dependent $NADP^+$ photoreduction (Bengis and Nelson, 1977). However, Scheller *et al.* (1990) suggested that PSI-F is not essential because no differences in the $NADP^+$ photoreduction activities were detected in the PS I preparations with different amounts of PSI-F.

In summary, a model for the organization of the PS I core complex is presented in Fig. 2. PS I complexes that have been crystallized from cyanobacteria form a trimer (Witt *et al.*, 1987; Ford *et al.*, 1987; Rögner *et al.*, 1990). It is hoped that high-resolution X-ray crystallography, now under progress, will provide a three-dimensional structure of this RC and will facilitate the understanding of the structure-function relationship.

III. Functional Components in PS I

a. Primary electron donor, P700

The primary electron donor of PS I which converts light energy into redox energy is a special chlorophyll called "P700". Upon excitation, P700 is oxidized and an electron is delivered to the primary electron acceptor A_0; this is followed by electron transfer from A_0^- to other acceptors (Table 2). The redox potential of P700/P700$^+$ couple is about +480 mV (at pH 7), which is more negative than that of Chl a/Chl a^+ in organic solvents; this difference suggests that P700 is located in a special environment. Because the reduced-minus-oxidized difference absorption spectrum of the primary electron donor has a peak at 700 nm, it was named P700 by its discoverer (Kok, 1959). The actual peak wavelength of the reduced form of P700 seems to be shifted by 3–4 nm towards shorter wavelengths (Schaffernicht and Junge, 1981; Ikegami and Itoh, 1988).

P700 is generally considered to be a special dimer of Chl a on the basis of its dimer-like circular dichroism (CD) and absorption spectrum (Ikegami and Itoh 1986, 1988; Schaffernicht and Junge, 1981). The photochemical hole burning experiments

Table 2. Characterization of donor and acceptors in PS I RC complex.

Electron carrier		Chemical identity	Polypeptide location	Redox potential (mV)	EPR characterization
Primary electron donor	P700	Chl *a* dimer	PSI-A and PSI-B (helix VIII?)	+480	oxidized form; g=2.0026, linewidth=8 gauss
Primary electron acceptor	A_0	Chl *a* monomer (Chl *a*-690)	PSI-A and PSI-B (helix VI?)	>−1,040	
Secondary electron acceptor	$A_{1\,(Q\emptyset)}$	phylloquinone	PSI-A and PSI-B	−830	
Tertiary electron acceptor	F_X (A_2)	[4Fe-4S] cluster	PSI-A and PSI-B (E9?)	−730	reduced form; g=2.08, 1.88, 1.78
	F_B	[4Fe-4S] cluster	PSI-C	−590	reduced form; g=2.05, 1.92, 1.89
	F_A	[4Fe-4S] cluster	PSI-C	−530	reduced form g=2.05, 1.94, 1.86

(Gillie *et al.*, 1987) and the EPR spectrum of its cationic form, whose bandwidth is narrower than that of monomeric chlorophyll *a* (Astashkin *et al.*, 1988), confirm its dimer-like character. Furthermore, low-temperature resonance Raman spectra suggest that P700 consists of two closely separated Chl *a* molecules which unequally interact through their keto-carbonyls (Moënne-Loccoz *et al.*, 1990). On the other hand, P700 shows features of monomeric chlorophyll, especially when oxidized, as shown by absorption spectroscopy (Schaffernicht and Junge, 1981; Ikegami and Itoh, 1988) or by electron nuclear double resonance (ENDOR) (O'Malley and Babcock, 1984). Its triplet state ESR signal also suggests monomeric nature (Rutherford and Mullet, 1981; Warden and Golbeck, 1987). It is highly likely that P700 is a dimer of Chl *a* and the interaction between these chlorophylls is weakened when they are oxidized.

C-epimer of Chl *a*(Chl *a*′) was proposed to constitute P700 since different PS I preparations containing different amounts of antenna chlorophylls (PS I-200 to PS I-10) were found to always contain two Chl *a*′ per P700 (Maeda *et al.*, 1992). Chl *a*′ introduced into Chl-depleted PS I-A/B proteins showed some optical similarity with those of P700, although the photochemical activity of P700 was not recovered (Hiyama *et al.*, 1987). Chlorinated-Chl *a* (RCI), once thought to constitute P700, is now shown to be an artefact produced after the extraction of chlorophyll (Senge *et al.*, 1988; Kobayashi *et al.*, 1988). Thus, the chemical nature of P700 remains to be established, although it is clear that it is a Chl *a* type pigment.

b. Primary electron acceptor, A_0

A_0 is presumably made of Chl *a* with an absorption maximum at 690–694 nm and a shoulder at 670 nm, as judged from picosecond and nanosecond laser spectroscopy (Nuijs *et al.*, 1986; Shuvalov *et al.*, 1986; Wasielewski *et al.*, 1987; Mathis *et al.*, 1988; also see Fenton *et al.*, 1979). It was suggested that A_0 was a dimer of Chl a on the basis of long wavelength peak position at around 694 nm (Nuijs et al., 1986; Shuvalov et al., 1986). However, no CD band corresponding to A_0 can be detected and, thus, A_0 seem to be a monomer (Ikegami and Itoh, 1988)

Another candidate for A_0 was a Chl a molecule peaking at 670 nm which photobleaches under the highly reducing condition at 210K (Ikegami and Ke, 1984; Mansfield and Evans, 1985). Under the same conditions, a radical named 'A_0^-' can be detected by EPR with a g value of 2.0024 and a linewidth of 8 gauss (Bonnerjea and Evans, 1982; Gast et al., 1983; Itoh et al., 1987; Smith et al., 1987; Warden and Golbeck, 1987), However, the 'A_0^-' was not rapidly photoaccumulated (Shuvalov et al., 1986; Wasielewski et al., 1987), so that the 'A_0^-' (Chl a-670) seems to be an antennae Chl a. The redox potential of A_0/A_0^- has been tentatively assumed to be-1.04 V (Nitsch et al., 1988).

c. Secondary electron acceptor, $A_1(Q\emptyset)$

The existence of A_1, which mediates electron transfer between A_0 and the iron-sulphur centre F_X, was first indicated by EPR studies (Bonnerjea and Evans, 1982; Gast et al., 1983) and then by absorption changes under highly reducing conditions at low temperature (Sétif et al., 1984).

The chemical identity of A_1 as phylloquinone has been shown by the extraction and reconstitution study of phylloquinone in PS I complex (Itoh et al., 1987; Itoh and Iwaki, 1988, 1989a b; Ikegami et al., 1987; Iwaki and Itoh, 1989; Biggins and Mathis, 1988) as well as by the optical measurement of A_1 (Mansfield and Evans, 1986; Brettel et al., 1986; Brettel, 1988; Mathis and Sétif, 1988). The usual PS I complex contains approximately two molecules of phylloquinone (2-methyl-3-phytyl-1,4- naphthoquinone = vitamin K-1) per P700, and one of them, tightly bound to the PS I RC preparations (Takahashi et al., 1985; Schoeder and Lockau, 1986; Malkin, 1986; Itoh et al., 1987), functions as A_1.

Phylloquinone in PS I particles can be almost completely extracted by treatment with diethyl ether or hexane-methanol without damaging other electron transfer components (Itoh et al., 1987; Biggins and Mathis, 1988). The extraction stops the rapid oxidation of A_0^- and enhances the charge recombination between P700$^+$ and A_0^- with a half-time of 30–50 ns (Ikegami et al., 1987; Biggins and Mathis, 1988). This produces the triplet state of P700 (Ikegami et al., 1987; Itoh et al., 1987) or re-excites P700 and results in delayed fluorescence (Itoh and Iwaki, 1988). Reconstitution of one molecule of phylloquinone per P700 fully recovers the A_1 function to mediate the reduction of F_A/F_B by A_0^- (Itoh et al., 1987; Biggins and Mathis, 1988; Ikegami et al., 1987; Iwaki and Itoh, 1989; Itoh and Iwaki, 1989a). The function of phylloquinone can also be replaced by artificial benzo-, naphtho-, and anthraquinones and the efficiency of quinone to function as A_1 mainly depends on redox potentials of quinones (Iwaki and Itoh, 1989, 1991b). The phylloquinone-binding site also binds inhibitors such as phenanthrolines, antimycine A, or mixothiazole which are known to inhibit quinone reactions in other electron transfer systems, as well as various quinones and quinonoid compounds (Itoh and Iwaki, 1989b). The quinone and its binding site were, therefore, proposed to be designated as the $Q_\emptyset$ and $Q_\emptyset$ site, respectively, in analogy to the Q_A or Q_B site of PS II (for discussion on binding site of herbicides at Q_B see Chapter by Van Renson, this volume) or of purple bacterial RCs. Redox potential of A_1 was estimated to be -830 mV from the quinone reconstitution study (Iwaki and Itoh, 1991a).

Strong-light irradiation of PS I compelx under highly reducing condition destroys A_1 (Inoue et al., 1989) presumably by overreduction of phylloquinone since the phylloquinone is known to be doubly reduced under such conditions (Sétif and Bottin, 1989). Irradiation with 350 nm ultraviolet light inhibited the extraction of phylloquinone and was assumed to have destroyed phylloquinone. However, this treatment depressed neither the so-called 'A_1^-' EPR signal (Ziegler et al. 1987) nor the flash-induced reduction of F_A/F_B (Palace et al., 1987; Biggins et al., 1990) and, thus, has been regarded as evidence against the idea that A_1 is phylloquinone. On the other hand, irradiation with 250 nm ultraviolet light results in destruction of A_1 (Iwaki and Itoh, unpublished data), Thus, the controversy is not yet fully resolved.

An EPR signal 'A_1' at g value of 2.0051 with a linewidth of 10.5 gauss, assigned to semiquinone radical of phylloquinone (Bonnerjea and Evans, 1982; Itoh et al., 1987; Mansfield et al., 1987; Smith et al., 1987), has also been a matter of controversy. The linewidth of EPR signal narrowed upon incubation in deuterated media (Mansfield et al., 1987). However, no line narrowing of A_1 EPR spectrum occurred when 65% of phylloquinone in PS I was deuterated, suggesting that the photoaccumulated EPR signal is not attributable to the phylloquinone radical (Barry et al., 1988). Thus, the controversy as to whether the EPR signal A_1 is a phylloquinone radical continues.

d. Tertiary electron acceptors, F_X, F_B and F_A

PS I complex contains 12 atoms of iron and zero valance sulphur (Høj and Møller, 1986; Golbeck et al., 1987) which are attributable to the three 4Fe-4S type, iron-sulphur centres F_X, F_B and F_A (Malkin, 1987; Cammack, 1988; Golbeck, 1988). F_X resides on the PSI-A and PSI-B polypeptides of RC complex (Sakurai and San Pietro, 1985; Golbeck and Cornelius, 1986; Høj et al., 1987; Golbeck et al., 1988a), and F_B and F_A on the PSI-C (Oh-oka et al., 1987; Hayashida et al., 1987). These iron-sulphur centres are difficult to be distinguished from each other by optical analysis because of their similar absorption spectra peaking at about 430 nm; these centres have been usually distinguished by their low temperature EPR signals (Malkin, 1987; Cammack, 1988; Golbeck, 1988) (Table 2). An electron acceptor named "P430", measured by flash spectrophotometry (Hiyama and Ke, 1971; Hiyama and Fork, 1980), is generally accepted as being F_A or F_B (Rutherford and Heathcote, 1985; Malkin, 1987; Andréasson and Vänngård, 1988).

(i) F_X (A_2)

F_X is characterized by an extremely low redox potential (about −730 mV) and by an unusual EPR signal (Malkin, 1987; Cammack, 1988; Golbeck, 1988) (Table 2). F_X is assumed to bridge the PSI-A and PSI-B RC subunits forming a [4Fe-4S] cluster as estimated from the studies by extended X-ray absorption fine structure (EXAFS) (McDermott et al., 1989), by Mössbauer (Petrouleas et al., 1989), or by comparison of amino-acid sequences with that of nitrogenase containing [4Fe-4S] cluster (Scheller et al., 1989a), although it had been earlier considered to be 2[2Fe-2S] Cluster (Golbeck et al., 1987; Bertrand et al., 1988). When PSI-C F_A/F_B protein is depleted, F_X shows the broadened EPR signal (Golbeck *et al.*, 1988a; Hoshina *et al.*, 1989; Parrett *et al.*, 1989) (see Fig. 3) and it is again narrowed on readdition of F_A/F_B protein (Golbeck *et al.*, 1988b; Parrett *et al.*, 1990). The structure of F_X is, thus, affected by the existence of PSI-C protein (or F_A/F_B clusters) which presumably binds nearby F_X.

(ii) F_A/F_B

The [4Fe-4S] clusters of F_A and F_B exist in close proximity and interact magnetically with each other. This conclusion was obtained from the observed changes in the EPR spectra of the F_A^-/F_B, F_A/F_B^-, or F_A^-/F_B^- state (Rutherford and Heathcote, 1985; Malkin, 1987 Andréasson and Vánngárd, 1988; Golbeck, 1988) (see Fig. 5). The EPR spectra change and show a more symmetrical nature, if the PSI-C protein is isolated from the PS I RC (Oh-oka *et al.* 1988b, 1991; Wynn and Malkin, 1988b). When the isolated spinach F_A/F_B protein was added to the cyanobacterial PS I RC complex, the EPR spectrum of the F_A/F_B became almost identical with that of the original PS I complex (Golbeck *et al.*, 1988b; Parrett *et al.*, 1990), indicating that the PS I RC complex can be reconstituted with the F_A/F_B protein from different organisms. The redox potentials of F_A (-530 mV) and F_B (-590 mV) are close to each other (Table 2).

e. Water-soluble redox components interacting with PS I RC: Plastocyanin and ferredoxin

P700 photoxidized *in situ* is reduced by a water-soluble copper protein, plastocyanin (or by water-soluble cytochrome c's in cyanobacteria and in some algae) within micro- to milliseconds. Plastocyanin binding to the net negatively charged lumen surface of the RC complex is affected by electrostatic interactions. At the high ionic strength or at the lower medium pH, the electrostatic repulsion between plastocyanin and PS I RC is decreased and results in an increase in the reaction rate with $P700^+$ (Tamura *et al.*, 1981). Plastocyanin is encoded by a nuclear gene, the sequence of which is available for *Silene pratensis* (Smeekens *et al.*, 1985a), spinach (Rother *et al.*, 1986), and *Arabidopsis thaliana* (Vorst *et al.*, 1988). X-ray crystallographic analysis shows that the copper of plastocyanin is ligated by residues His-37, Cys-84, His-87, and Met-92 (Colman *et al.*, 1978; Guss *et al.*, 1986)

At the reducing side of PS I RC (Stroma side) either F_A or F_B reduces a water-soluble protein ferredoxin. It then reduces $NADP^+$ or catalyzes the cyclic electron flow around PS I. Ferredoxin (Em = - 420 mV) is a 10 kD iron-sulphur protein which has one [2Fe-2S] cluster. The EPR spectrum shows g values of 2.05, 1.96, and 1.89 (Cammack, 1988). The nuclear gene coding for its apoprotein was sequenced in *Silene pratensis* (Smeekens *et al.*, 1985b) and pea (Dobres *et al.*, 1987).

IV. Electron Transfer in PS I

a. Forward electron transfer in PS I

After excitation of the isolated PS I complex by flash, the electron from P700 goes through A_0, A_1, F_X, and F_B/F_A and then goes back to $P700^+$ with a half-time of

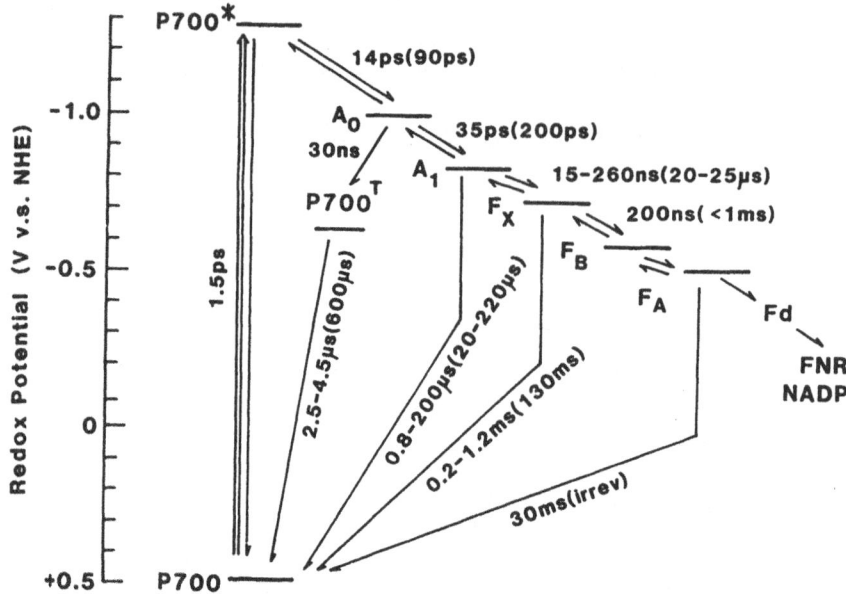

Figure 7. Electron Transfer in PS I. The components are placed at their approximate or presumed midpoint redox potential, E_m. Time represents a half-time measured at room temperature or at temperatures below 77K in the parentheses.

about 30 ms (Andréasson and Vänngård, 1988; Golbeck, 1988) if no electron acceptor such as ferredoxin or methyl viologen is added (Fig. 7). P700 reaches its excited state P700* within 1.5 ps (Shuvalov et al., 1986) after light absorption, and reduces A_0 within 10-14 ps (Frenton et al., 1979; Wasielewski et al., 1987) at room temperature and 90 ps at cryogenic temperature (1.5K) (Gillie et al., 1987). A_0^- is then reoxidized with a half-time of 35–40 ps by A_1 (phylloquinone) (Shuvalov et al., 1986; Fenton et al., 1979). The electron trnasfer from A_1^- to F_X occurs within 15–260 ns depending on preparations (Mathis and Sétif, 1988; Brettel, 1988; Bock et al., 1989).

The electron transfer among the iron-sulphur centres are proposed to be either in a linear $F_X \rightarrow F_B \rightarrow F_A$ or in a branched $F_X \rightarrow F_B$ or $F_X \rightarrow F_A$ path. The linear electron transport path has been generally accepted by most workers (Malkin, 1984; Inoue et al., 1986). Methyl viologen seems to accept electrons efficiently from F_B and with a lower affinity from F_A (Fujii et al., 1990).

b. Back reaction between P700+ and reduced electron acceptors

Rates for the charge recombination between P700+ and the various reduced acceptors have been studied in detail. The rates depend not only on temperature but also on the redox states or the presence of nearby components. F_X^- reduces P700+ with a half-time of 250 μs at room temperature when F_A and F_B are prereduced (Bottin et al., 1987) but with a $t_{1/2}$ of 1–1.2 ms if the F_A/F_B-protein is removed (Golbeck and Cornelius, 1986; Golbeck et al., 1988a; Hoshina et al., 1989, 1990). This is explained by the electrostatic interaction between charges on F_A^-/F_B^- and F_X^-. When forward electron transfer from A_1^- is blocked, the charge recombination between P700+ and A_1^- occurs with a half-time of 30 μs (Bottin et al., 1987). The radical pair was reported to last for 0.8 μs (Sétif and Bottin, 1989) or 250 μs (Brettel, 1989) when the iron-sulphur centres are prereduced.

When forward electron transfer from A_0^- is blocked, the radical pair ($P700^+$, A_0^-) lasts for 20–50 ns (120 ns at low temperature), as shown by flash absorption changes (Nuijs *et al.*, 1986; Ikegami *et al.*, 1987; Biggins and Mathis, 1988; Mathis *et al.*, 1988; Sétif and Bottin, 1989) and by delayed fluorescence from P700 (Itoh and Iwaki, 1988). Some centres return to the ground state of P700 via a triplet state of P700 (Ikegami *et al.*, 1987; Sétif and Bottin, 1989).

V. Modification of PS I RC: Interaction between Functional Components and Proteinaceous Environments inside the RC

In photosynthetic RCs, every elctron transfer step proceeds with a very high efficiency. This depends mainly on the mutual arrangements of prosthetic groups, their interaction with the RC proteins, and the specific nature of each component. The importance of the interaction with RC protein is exemplified by the multiple function of chlorophyll *a* as antenna, A_0 and P700 in PSI RC. A three-dimensional structure of purple bacterial RCs given by X-ray crystallography (Deisenhofer *et al.*, 1985; Allen *et al.*, 1987) now acts as a guide for the study of the function-structure relationship of photosynthetic RCs. To know the mechanisms, a powerful method is to modify the structure and see the resulting changes in function. This has been done mainly by two methods: the alteration of amino-acid residues by molecular biology and the alteration of prosthetic groups by the extraction/reconstitution method. The site-directed mutagenesis in PS I RC is only at its beginning stages, although it has been successfully adopted in PS II and purple bacterial RCs. On the other hand, as shown briefly in the later section, the exchange of prosthetic groups has been successfully done for quinones in PS I (Iwaki and Itoh, 1989, 1991b) as well as in purple bacterial RC (Gunner *et al.*, 1986). A common basis for the quinone function in these two types of RCs may soon emerge.

a. Exchange of quinone in PS I RC

Extraction of phylloquinone from PS I RC stops the rapid oxidation of A_0^- and enhances the charge recombination in nanosecond time range. Thus, in the measurement of P700 oxidation in microsecond time range, we can detect only a small amount of $P700^+$ (Fig. 8). Readdition of phylloquinone recovers the oxidation of A_0^- and enhances the amount of detectable $P700^+$. Various artificial quinones or quinonoid compounds stabilize $P700^+$ just as the intrinsic phylloquinone does (Fig. 8).

The results and conclusions of these experiments are summarized as follows (Iwaki and Itoh, 1989, 1991a, b; Itoh and Iwaki, 1991):

1) Various quinones almost fully substitute for the function of phylloquinone as the electron acceptor for A_0^- (Fig. 8).

2) Only the quinones which show the redox potential ($E_{1/2}$) electrochemically determined in apolar organic solvent dimethylformamide comparable to or more negative than that of phylloquinone, can reduce F_X and stabilize $P700^+$ for several tens of milliseconds. This indicates that these quinones exhibit redox midpoint potentials (E_m) more negative than that of F_X (–730 mV) *in situ* in the RC protein (Fig. 8).

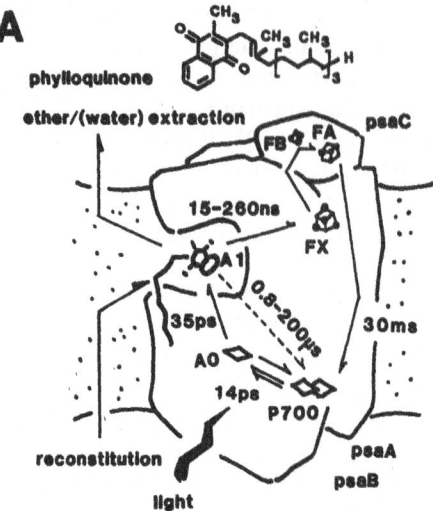

Photosystem I Reaction Center

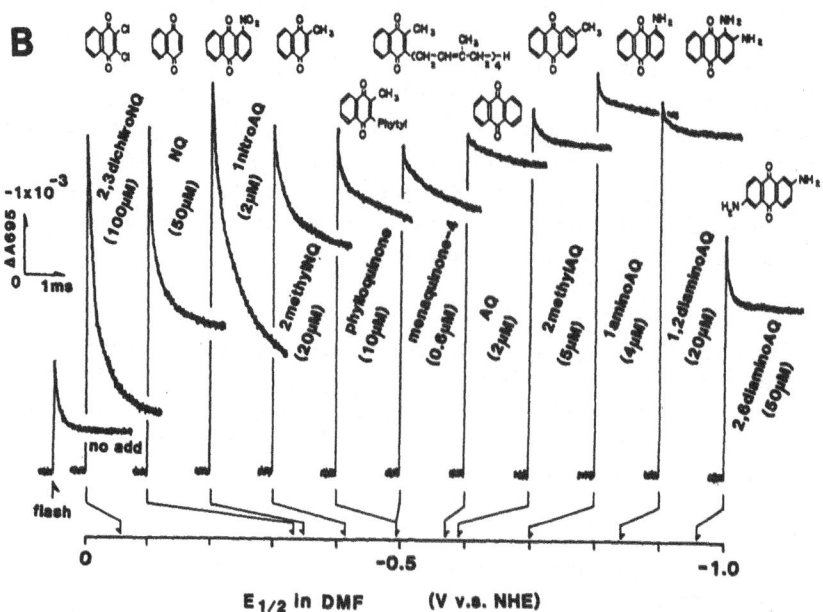

Figure 8. (A) Quinone extraction and reconstitution in PS I RC. (B) Flash-induced oxidation and following dark reduction of P700 in the quinone-reconstituted PS I RC. Phylloquinone (A_1) in PS I RC was extracted with diethyl ether and various quinones were introduced. Flash-excitation induces low extent of P700$^+$ in the phylloquinone-extracted PS I RC due to the faster charge recombination reaction. Quinones, whose redox potential in organic solvent are lower than that of phylloquinone, reduce F_A/F_B and give slower decay, while those with higher redox potential directly reduce P700$^+$ with a half-time of about 100 μs. Redox potentials of quinones polarographically measured *in vitro* in organic solvent dimethylformamide are indicated on the horizontal axis.

3) The higher potential quinones, such as dichloronaphthoquinone, do not reduce F_X but directly reduce P700⁺ with a faster 200 μs half-time. These quinones are estimated to exhibit *in situ* redox potentials more positive than that of F_X.

4) The quinones reconstituted to PS I RC, therefore, are assumed to shift their E_m to the negative by about 300 mV compared to those in dimethylformamide. The quinone which shows an $E_{1/2}$ of –500 mV in dimethylformamide was found to give the same E_m as that of F_X when reconstituted into PS I RC.

5) The intrinsic phylloquinone, whose $E_{1/2}$ in dimethylformamide is -600 mV, thus, is estimated to have an E_m value *in situ* in PS I RC of -830 mV, i.e., 100 mV more negative than E_m of F_X. The Em of PS I phylloquinone has been too low to be determined by other methods.

It is concluded that the quinone environments in the PS I RC destabilized semiquinone–while that in the Q_A site of purple bacterial (and perhaps in PS II) RC stabilizes it more than that in apolar organic solvent (Iwaki and Itoh, 1989, 1991a), since similar quinones show about 200 mV more positive redox potential than their $E_{1/2}$ in dimethylformaide in the primary electron acceptor (Q_A) site in the *Rhodobacter sphaeroides* RC (Gunner *et al.*, 1986). The PS I RC protein shifts the quinone redox potential to the negative while the purple bacterial or PS II RCs shift it to the positive. This comes from the difference in the protein structure in the quinone-binding sites in PS I RC and in purple bacterial or PS II RCs.

b. Structure of phylloquinone-binding site in PS I RC

The structure of quinone strongly affects its affinity to the PS I RC and provides information about the structure of the PS I Q_ϕ site. More hydrophobic quinones show higher affinity to the site (Iwaki and Itoh, 1989, 1991b). However, if the ring size of quinone is larger than that of anthraquinone, the affinity decreases, suggesting that the size of a quinone-binding pocket is as large as that of the methylated naphthoquinone ring of phylloquinone. Long hydrophobic tail of phylloquinone, on the other hand, is shown to be unnecessary for the function but significantly increases the affinity. The existence of a negative charge near the quinone Q_ϕ was also suggested from the specificity of affinity of various quinones. This charge is expected to destabilize the semiquinone–and to shift the redox potential of phylloquinone *in situ* in the binding site to be more negative. The reaction rates between A_0^- and the reconstituted quinones do not significantly vary depending on the molecular structure of quinones provided their redox potentials were appropriate. These results indicate that the RC protein fixes the quinone properly in its binding site and gives favourable reaction environments for the forward electron transfer from A_0^- to quinone. The ability of the PS I RC protein, demonstrated by the exchange experiments, is unique and cannot be replaced by the ever-known organic solvent.

VI. Conclusion and Future Studies

PS I provides a unique reaction system which is established on the highly organized combination of prosthetic groups and proteins. Roles of polypeptide subunits have not yet been fully underestood. Some of them may simply work to maintain the proper structure and others to facilitate interaction with water-soluble components, as is being shown by chemical modification or mutagenesis studies. It is interesting that PS I RCs in cyanobacteria and in higher plant chloroplasts show almost similar polypeptide composition, including the very small ones. This suggests that each polypeptide has a special function. Small polypeptides may also function in regulation of the interaction with other RCs or LHCs. Similar complex subunit composition has also been reported in PS II complex. However, the purple bacterial photosynthetic systems do not seem to have such complex polypeptide arrangement.

Nature has created highly efficient and organized photosynthetic RCs. By the modification or degradation of these complexes we are able to produce only less efficient systems. However, these artificial complexes may be just mimicking the old path of evolution in which less efficient photosynthetic systems have been lost. By using chlorophylls, quinones, iron, and proteins, photosynthetic organisms have created different types of highly efficient RCs with different energy input and output. To evolve these highly efficient systems, modification of protein, which changes the arrangement of prosthetic groups and the reaction environments, must have played an important role. Mutagenesis or exchange experiments seem to open a new way to investigate the path of evolution. PS I RC protein can be a new reaction system in which we can fix or change the geometry of functional molecules or modify solvent amino-acid. This can facilitate desired reactions by suppressing undesired reactions. Study of the excellent features of the PS I RC complex as well as that of other RCs will lead us to a deeper insight into the secrets of life.

Acknowledgements

The authors thank Ms. Masayo Iwaki for her discussion, critical reading of the manuscript and drawing the figures, and Drs. P. Mohanty and Govindjee for their helpful discussion and review of the manuscript. Thanks also are due to Mr. M. Kawamoto for drawing the figures.

VII. References

Allen JP, Feher G, Yeates TO, Komiya H and Rees DC (1987) Structure of the reaction center from *Rhodobacter sphaeroides* R-26: The protein subunits. *Proc Natl Acad Sci USA* **84**: 6162–6166

Andersen B, Koch B, Scheller HV, Okkels JS and Møller BL (1990) Nearest neighbour analysis of the photosystem I subunits in barley and their binding of ferredoxin. In: (M Baltsheffsky, ed.) *Current Research in Photosynthesis*, Vol II, Kluwer Acad Pub, Dordrecht, The Netherlands pp 671–674

Andersson B, Anderson JM and Ryrie IJ (1982) Transbilayer organization of the chlorophyll-proteins of spinach thylakoids. *Eur J Biochem* **123**: 465–472.

Andréasson L–E and Vänngård T (1988) Electron transport in photosystems I and II. *Annu Rev Plant Physiol Plant Mol Biol* **39**: 379–411

Astashkin AV, Dikanov SA and Tsvetkov Yu D (1988) The structure of the primary electron donors P865 and P700 of bacterial and plant photosynthesis based on magnetic resonance data. *Chem Phys Lett* **152**: 258–264

Barry BA, Bender CJ, McLntosh L, Ferguson-Miller S and Babcock GT (1988) Photoaccumulation in photosystem I does not produce a phylloquinone radical. *Israel J Chem* **28**: 129–132

Bassi R and Simpson D (1987) Chlorophyll-protein complexes of barley photosystem I. *Eur J Biochem* **163**: 221–230

Bengis C and Nelson N (1975) Purification and properties of the photosystem I reaction center from chloroplasts. *J Biol Chem* **250**: 2783–2788

Bengis C and Nelson N (1977) Subunit structure of chloroplast photosystem I reaction center. *J Biol Chem* **252**: 4564–4569

Bertrand P, Guigliarelli B, Gayda J–P, Sétif P and Mathis P (1988) An interpretation of the peculiar magnetic properties of center X in terms of a 2Fe-2S cluster. *Biochim Biophys Acta* **933**: 393–397

Biggins J and Mathis P (1988) Functional role of vitamin K_1 in photosystem I of the cyanobacterium *Synechocystis* 6803. *Biochemistry* **27**: 1494–1500

Biggins J, Tanguay NA and Frank HA (1990) Functional role of phylloquinone on the acceptor side of photosystem I In (M. Baltscheffsky, ed) *Current Research in Photosynthesis* Vol II, Kluwer Acad Pub, Dordrecht, pp 639–642

Blankenship RE and Fuller RC (1986) Membrane topology and photochemistry of the green photosynthetic bacterium *Chloroflexus aurantiacus*. In: (LA Staechelin and CJ Arntzen, eds) *Encyclopedia of Plant Physiology*, Vol 19, Photosynthesis III, Springer-Verlag, Berlin, pp 390–399

Bock CH, van del Est AJ, Brettel K and Stehlik D (1989) Nanosecond electron transfer kinetics in photosystem I as obtained from transient EPR at room temperature. *FEBS Lett* **247**: 91–96

Bonnerjea J and Evans MCW (1982) Identification of multiple components in the intermediary electron carrier complex of photosystem I. *FEBS Lett* **148**: 313–316

Bottin H, Sétif P and Mathis P (1987) Study of the PS I acceptor side by double and triple flash experiments. *Biochim Biophys Acta* **894**: 39–48

Breton J and Ikegami I (1989) Orientation of photosystem-I pigments: Low temperature linear dichroism spectroscopy of a highly-enriched P700 particles isolated from spinach. *Photosynth Res* **21**: 27–36

Brettel K (1988) Electron transfer from A_1^- to an iron-sulfur center with $t_{1/2}$=200 ns at room temperature in photosystem I Characterization by flash absorption spectroscopy. *FEBS Lett* **239**: 93–98

Brettel K (1989) New assignment for 250 μs kinetics in photosystem I: P-700^+ recombines with A_1^- (not F_X^-). *Biochim Biophys Acta* **976**: 246–249

Brettel K, Sétif P and Mathis P (1986) Flash-induced absorption changes in photosystem I at low temperature: evidence that the electron acceptor A_1 is vitamin K_1. *FEBS Lett* **203**: 220–224

Bruce BD and Malkin R (1988) Structural aspects of photosystem I from *Dunaliella salina*. *Plant Physiol* **88**: 1201–1206

Cammack R (1988) Electron paramagnetic resonance characterization of iron-sulfur proteins. *Methods in Enzymol* **167**: 427–436

Cantrell A and Bryant DA (1987) Molecular cloning and nucleotide sequence of the *psaA* and *psaB* genes of the cyanobacterium *Synechococcus* sp. PCC7002. *Plant Mol Biol* **9**: 453–468

Capaldi RA and Vanderkooi G (1972) The low polarity of many membrane proteins. *Proc Natl Acad Sci USA* **69**: 930–932

Colman PM, Freeman HC, Guss JM, Murata M, Norris VA, Ramshaw JAM and Venkatappa MP (1978) X-ray crystal structure analysis of plastocyanin at 2.7 Å resolution. *Nature* **272**: 319–324

Cushman, T, Hallick RB and Price CA (1988) The two genes for the P-700 chlorophyll *a* apoproteins on the *Euglena* Chloroplast genome contain multiple introns. *Curr Genet* **13**: 159–171

Deisenhofer J, Epp O, Miki K, Huber R and Michel H (1985) Structure of the protein subunits of the photosynthetic reaction centre of *Rhodopseudomonas viridis* at 3Å resolution. *Nature* **318**: 618–624

Dobres MS, Elliott RC, Watson, JC and Thompson WF (1987) A phytochrome regulated pea transcript encodes ferredoxin I. *Plant Mol Biol* **8**: 53–59

Dunn PPJ and Gray JC (1988) Localization and nucleotide sequence of the gene for the 8 kDa subunit of photosystem I in pea and wheat chloroplast DNA. *Plant Mol Biol* **11**: 311–319

Dunn PPJ, Packman LC, Pappin D and Gray JC (1988) N-terminal amino acid sequence analysis of the subunits of pea photosystem I. *FEBS Lett* **228**: 157–161.

Enami I, Ohta H and Katoh S (1986) Topographical studies on subunit polypeptides of the PS I reaction center complex in the thylakoid membranes of the thermophilic cyanobaterium *Synechococcus* sp. *Plant Cell Physiol* **27**: 1395–1405

Enami I, Ohta H and Miyaoka T (1987) Cross-linking studies on the membrane topography of photosystem I reaction center complex in *Synechococcus* sp. *Plant Cell Physiol* **28**: 101–111.

Evans MCW and Bredenkamp G (1990) The structure and function of the photosystem I reaction center. *Physiol Plant* **79**: 415–420

Fenton JM, Pellin MJ, Govindjee and Kaufmann KJ (1979) Primary photochemistry of the reaction center of photosystem I. *FEBS Lett* **100**: 1–4

Fish L, Kück U and Bogorad L (1985) Two partially homologous adjacent light-inducible maize chloroplast genes encoding polypeptides of the P700 chlorophyll *a*-protein complex of photosystem I. *J Biol Chem* **260**: 1413–1421

Ford RC, Picot D and Garavito RM (1987) Crystallization of the photosystem I reaction center. *EMBO J* **6**: 1581–1586

Franzén LG, Frank G, Zuber H and Rochaix JD (1989a) Isolation and characterization of cDNA clones encoding the 17.9 and 8.1 kDa subunits of photosystem I from *Chlamydomonas reinhardtii*. *Plant Mol Biol* **12**: 463–474

Franzén LG, Frank G, Zuber H and Rochaix JD (1989b) Isolation and characterization of cDNA clones encoding photosystem I subunits with molecular masess 11.0. 10.0 and 8.4 kDa from *Chlamydomonas reinhardtii*. *Mol Gen Genet* **219**: 137–144

Fujii T, Yokoyama E, Inoue K and Sakurai H (1990) The sites of electron donation of photosystem I to methyl viologen. *Biochim Biophys Acta* **1015**: 41–48

Gast P, Swarthoff T, Ebskamp FCR and Hoff AJ (1983) Evidence for a new early acceptor in photosystem I of plants. An ESR investigation of reaction center triplet yield and of the reduced intermediary acceptors. *Biochim Biophys Acta* **722**: 163–175

Gillie JK, Fearey BL, Hayes JM and Small GJ (1987) Persistent hole burning of the primary donor state of Photosystem I: Strong linear electron-phonon coupling. *Chem Phys Lett* **134**: 316–322

Girard J, Chua, NH, Bennoun P, Schmidt G and Delosme M (1980) Studies on mutants deficient in the photosystem I reaction centers in *Chlamydomonas reinhardtii*. *Curr Genet* **2**: 215–221

Golbeck JH (1988) Structure, function and organization of the photosystem I reaction center complex. *Biochim Biophys Acta* **895**: 167–204

Golbeck JH and Cornelius JM (1986) Photosystem I charge separation in the absence of centers A and B. 1 Optical characterization of center 'A$_2$' and evidence for its association with a 64-kDa peptide. *Biochim Biophys Acta* **849**: 16–24

Golbeck JH, McDermott AE, Jones WK and Kurtz DM (1987) Evidence for the existence of [2Fe-2S] as well as [4Fe-4S] clusters amoung F$_A$, F$_B$ and F$_X$. Implications for the structure of the photosystem I reaction center. *Biochim Biophys Acta* **891**: 94–98

Golbeck JH, Parrett KG, Mehari T, Jones KL and Brand JJ (1988a) Isolation of the intact photosystem I reaction center core containing P-700 and iron-sulfur center F$_X$. *FEBS Lett* **228**: 268–272

Golbeck, JH, Mehari T, Parrett KG and Ikegami, I (1988b) Reconstitution of the photosystem I complex from the P-700 and F$_X^-$ containing reaction center core protein and F$_A$/F$_B$ polypeptide. *FEBS Lett* **240**: 9–14

Gounaris K, Barber J and Harwood JL (1986) The thylakoid membranes of higher plants. *Biochem J* **237**: 313–326

Green BY (1988) The chlorophyll-protein complexes of higher plant photosynthetic membranes or just what green band is that? *Photosynth Res* **15**:3–32

Gunner MR, Robertson DE and Dutton PL (1986) Kinetic studies on the reaction center protein from *Rhodopseudomonas sphaeroides*: The temperature and free energy dependence of electron transfer between various quinones in the Q$_A$ site and the oxidized bacteriochlorophyll dimer. *J Phys Chem* **90**: 3783–3795

Guss JM, Harrowell PR, Murata M, Norris VA and Freeman HC (1986) Crystal structure analyses of reduced (Cu$^+$) popular plastocyanin at six pH values. *J Mol Biol* **192**: 361–387

Haworth P, Watson JL and Amtzen CJ (1983) The depletion, isolation and characterization of a light-harvesting complex which is specifically associated with photosystem I. *Biochim Biophys Acta* **724**: 151–158

Hayashida N, Matsubayashi T, Shinozaki K, Sugiura M, Inoue K and Hiyama T (1987) The gene for the 9 kDa polypeptide, a possible apoprotein for the iron-sulfur centers A and B of the photosystem I complex, in tobacco chloroplast DNA. *Curr Genet* **12**: 247–250

Hippler M, Ratajczak R and Haehnel W (1989) Identification of the plastocyanin binding subunit of photosystem I. *FEBS Lett* **250**: 280–284

Hiratsuka J, Shimada H, Whittier R, Ishibashi T, Sakamoto M, Mori M, Kondo C, Honji Y, Sun CR, Meng BY, Li YQ, Kanno A, Nishizawa, Y, Hirai A, Shinozaki K and Sugiura M (1989) The complete sequence of the rice (*Oryza sativa*) chloroplast genome: Intermolecular recombination between distinct tRNA genes accounts for a major plastid DNA inversion during the evolution of the cereals. *Mol Gen Genet* **217**: 185–194

Hiyama T and Ke B (1971) A further study of P430: A possible primary electron acceptor of photosystem I. *Arch Biochem Biophys* **147**: 99–108

Hiyama T and Fork DC (1980) Kinetic identification of component X as P430: A primary electron acceptor of photosystem I. *Arch Biochem Biophys* **199**: 488–496

Hiyama T, Watanabe T, Kobayashi M and Nakazato M (1987) Interaction of chlorophyll *a* with the 65 kDa subunit protein of photosystem I reaction center. *FEBS Lett* **214**: 97–100

Hiyama T, Yanai N, Takano Y, Ogiso H, Suzuki K and Terakado, K (1990) A photosystem 1 reaction center complex constituted only by two subunits (In: M. Baltcheffsky ed.) *Current Research in Photosynthesis*, Vol II, Kluwer Acad Pub, Dordrecht, The Netherlands, pp 587–590

Hoffman NE, Pichersky E, Malik VS, Ko K and Cashmore AR (1988) Isolation and sequence of a tomato cDNA clone encoding subunit II of the photosystem I reaction center. *Plant Mol Biol* **10**: 435–445

Høj PB and Møller BL (1986) The 110-kDa reaction center protein of photosystem I, P700-chlorophyll *a*-protein 1, is an iron-sulfur protein. *J Biol Chem* **261**: 14292–14300

Høj PB, Svendsen I, Scheller HV and Møller BL (1987) Identification of a chloroplast-encoded 9-kDa polypeptide as a 2[4Fe-4S] protein carrying centers A and B of photosystem I. *J Biol Chem* **262**: 12676–12684

Hoshina S and Itoh S (1987) Characterization of photosystem I chlorophyll-protein complexes reconstituted into phosphatidylcholine liposomes. *Plant Cell Physiol* **28**: 599–609

Hoshina S, Sue S, Kunishima N, Kamide K, Wada K and Itoh S (1989) Characterization and N-terminal sequence of a 5 kDa polypeptide in photosystem I core complex from spinach. *FEBS Lett* **258**: 305–308

Hoshina S, Sakurai R, Kunishima N, Wada K and Itoh S (1990) Selective destruction of iron-sulfur centers by heat/ethylene glycol treatment and isolation of photosystem I core complex. *Biochim Biophys Acta* **1015**: 61–68

Ikegami I and Itoh S (1986) Chlorophyll organization in P700-enriched particles isolated from spinach chloroplasts. CD and absorption spectroscopy. *Biochim Biophys Acta* **851**:75–85

Ikegami I and Itoh S (1988) Absorption spectroscopy of P-700-enriched particles isolated from spinach. Is P-700 a dimer or a monomer? *Biochim Biophys Acta* **934**: 39–46

Ikegami I and Katoh S (1975) Enrichment of photosystem I reaction center chlorophyll from spinach chloroplasts. *Biochim Biophys Acta* **376**: 588–592

Ikegami I and Katoh S (1989) Preparation and characterization of P700-enriched photosystem-I complexes from thermophilic cyanobacterium, *Synechococcus* sp. *Plant Cell Physiol* **30**: 175–182

Ikegami I and Ke B (1984) A 160-kilodalton photosystem-I reaction-center complex: Low-temperature absorption and EPR spectroscopy of the early electron acceptors. *Biochim Biophys Acta* **764**: 70–79

Ikegami I, Sétif P and Mathis P (1987) Absorption studies of photosystem I photochemistry in the absence of vitamin K-1. *Biochim Biophys Acta* **894**: 414–422

Ikeuchi M, Hirano A, Hiyama T and Inoue Y (1990) Polypeptide compostition of higher plant photosystem I complex: Identification of *psaI*, *psaJ* and *psaK* gene products. *FEBS Lett* **263**: 274–278

Inoue K, Sakurai H and Hiyama T (1986) Photoinactivation sites of photosystem I in isolated chloroplasts. *Plant Cell Physiol* **27**: 961–968

Inoue K, Fujii T, Yokoyama E, Matsuura K, Hiyama T and Sakurai H (1989) The photoinhibition site of photosystem I in isolated chloroplasts under extremely reducing condition. *Plant Cell Physiol* **30**: 65–71

Itoh S and Iwaki M (1988) Delayed fluorescence in photosystem I enhanced by phylloquinone (vitamin K-1) extraction with ether. *Biochim Biophys Acta* **934**: 32–38

Itoh S and Iwaki M (1989a) Vitamin K_1 (phylloquinone) restores the turnover of FeS centers in the ether-extracted spinach PS I particles. *FEBS Lett* **243**: 47–52

Itoh S and Iwaki M (1989b) New herbicide-binding site in the photosynthetic electron-transport chain. Competitive herbicide binding at the photosystem I Phylloquinone-(vitamin K_1)-binding site. *FEBS Lett* **250**: 441–447

Itoh S and Iwaki M (1991) Full replacement of the function of the secondary electron acceptor phylloquinone (= vitamin K_1) by non-quinone carbonyl compound in green plant photosystem I reaction center. *Biochemisty*. **30**: 5340–5346.

Itoh S, Iwaki M and Ikegami I (1987) Extraction of vitamin K-1 from photosystem I particles by treatment with diethyl ether and its effects on the A_1^- EPR signal and system I photochemistry. *Biochim Biophys Acta* **893**: 508–516

Iwaki M and Itoh S (1989) Electron transfer in spinach photosystem I reaction center containing benzo-, naphtho- and anthroquinones in place of phylloquinone. *FEBS Lett* **256**: 11–16

Iwaki M and Itoh S (1991a) Function of quinones and quinonoids in green plant photosystem I reaction center In: (J Bolton, ed) *Advances in Chemistry* Series No. 228, *Electron Transfer in Inorganic, Organic and Biological System*, American Chemical Society, pp. 163–178.

Iwaki M and Itoh S (1991b) Structure of the phylloquinone-binding (Q_ϕ) site in green plant photosystem I reaction center; The affinity of quinones and quinonoid compounds for the Q_ϕ site. *Biochemistry*, **30**: 5347–5352.

Ke B and Shuvalov VA (1987) Picosecond absorption spectroscopy in photosynthesis and primary electron transfer processes. In: (J Barber, ed) *The Light Reactions*, Elsevier, Amsterdam, pp 31–93

Kirsch W, Seyer P and Hermann RG (1986) Nucleotide sequence of the clustered genes for two P700 chlorophyll *a* apoproteins of the photosystem I reaction center and the ribosomal protein S14 of the spinach chloroplast chromosome. *Curr Genet* **10**: 843–855

Kobayashi M, Watanabe T, Struch A and Scheer H (1988) Mesochlorination of chlorophyll *a* in the course of pigment extraction. *FEBS Lett* **235**: 293–297

Koike H, Ikeuchi M, Hiyama T and Inoue Y (1989) Identification of photosystem I components from the cyanobacterium, *Synechococcus vulcanus* by N-terminal sequencing. *FEBS Lett* **253**: 257–263

Kok B (1959) Light induced absorption changes in photosynthetic organisms. II. A split-beam difference spectrophotometer. *Plant Physiol* **34**: 184–192

Küch U, Choquet Y, Schneider M, Dron M and Bennoun P (1987) Structural and transcription analysis of two homologous genes for the P700 chlorophyll *a*-apoproteins in *Chlamydomonas reinhardtii*: Evidence for *in vivo* trans-splicing. *EMBO J* **6**: 2185–2195.

Lagoutte B (1988) Cloning and sequencing of spinach cDNA clones encoding the 20 kDa PS I polypeptide. *FEBS Lett* **232**: 275–280.

Lagoutte B and Mathis P (1989) The photosystem I reaction center: Structure and photochemistry. *Photochem Photobiol* **49**: 833–844

Lam E, Ortiz W, Mayfield S and Malkin R (1984) Isolation and characterization of a light-harvesting chlorophyll *a/b* protein complex associated with photosystem I. *Plant Physiol* **74**: 650–655

Lehmbeck J, Rasmussen OF, Bookjans GB, Stummann BM and Hennigsen KW (1986) Sequence of two genes in pea chloroplast DNA coding for 84 and 82 kDa polypeptides of the photosystem I complex. *Plant Mol Biol* **7**: 3–10

Lundell, DJ, Glazer AN, Melis A and Malkin R (1985) Characterization of a cyanobacterial photosystem I complex. *J. Biol Chem* **260**: 646–654

Maeda H, Watanabe T, Kobayashi M and Ikegami I (1992) Presence of two chlorophyll *a* molecules at the core of photosystem I. *Biochim. Biophys. Acta* **1099**: 74–80

Malkin R (1984) Diazonium modification of photosystem I. A specific effect on iron-sulfur center B. *Biochim Biophys Acta* **764**: 63–69

Malkin R (1986) On the function of two vitamin K_1 molecules in the PS I electron acceptor complex. *FEBS Lett* **208**: 343–346

Malkin R (1987) Photosystem I. In: (J Barber, ed) Topics in Photosynthesis, Vol 8, *The Light Reactions* Elsevier, Amsterdam, pp 495–525

Mansfield RW and Evans MCW (1985) Optical difference spectrum of the electron acceptor A_0 in photosystem I. FEBS Lett **190**: 237–241

Mansfield RW and Evans MCW (1986) UV optical difference spectrum associated with the reduction of electron acceptor A_1 in photosystem I of higher plants. *FEBS Lett* **203**: 225–229

Mansfield RW, Nugent JHA and Evans MCW (1987) ESR characteristics of Photosystem I in deuterium oxide: Further evidence that electron acceptor A_1 is a quinone. *Biochim Biophys Acta* **894**: 515–523

Margulies MM (1989) Photosystem I core. *Plant Sci* **64**: 1–13

Mathis P (1990) Compared structure of plant and bacterial photosynthetic reaction centers: Evolutionary implications. *Biochim Biophys Acta* **1018**: 163–167

Mathis P and Rutherford AW (1987) The primary reactions of photosystems I and II of algae and higher plants In: (J. Amesz, ed.) *New Comprehensive Biochemistry*, Vol. 15, *Photosynthesis*. Elsevier, Amsterdam, pp. 63–96

Mathis P and Sétif P (1988) Kinetic studies on the function of A_1 in the photosystem I reaction center. *FEBS Lett* **237**: 65–68

Mathis P, Ikegami I and Sétif P (1988) Nanosecond flash studies for the absorption spectrum of the photosystem I primary acceptor A_0. *Photosynth Res* **16**: 203–210

McDermott AE, Yachandra VK, Guiles RD, Sauer K, Klein MP, Parrett KG and Golbeck JH (1989) EXAFS structural study of F_X, the low-potential Fe-S center in photosystem I. *Biochemisty* **28**: 8056–8059

Moënne-Loccoz P, Robert B, Ikegami I and Lutz M (1990) Structure of the primary electron donor in photosystem I: A resonance Raman study. *Biochemistry* **29**: 4740–4746

Møller BL, Scheller HV, Okkels JS, Koch B, Andersen B, Nielsen HL, Olsen I, Halkier BA and Høj PB (1990) Chloroplast encoded photosystem I polypeptides of barley. In: (M Baltscheffsky, ed) *Current Research in Photosynthesis*, Vol II, Kluwer Acad Pub, Dordrecht, pp 523–530

Mullet JE, Burke JJ and Arntzen CJ (1980) Chlorophyll proteins of photosystem I. *Plant Physiol* **65**: 814–822

Münch S, Ljungberg U, Steppuhn J, Schneiderbauer A, Nechushttai R, Beyreuther K and Herrmann RG (1988) Nucleotide sequences of cDNAs encoding the entire precursor polypeptides for subunits II and III of the photosystem I reaction center from spinach. *Curr Genet* **14**: 511–518

Nelson N (1987) Structure and function of protein complexes in the photosynthetic membrane. In: (J Amesz, ed.) *New Comprehensive Biochemistry*, Vol 15, photoynthesis, Elsevier, Amsterdam, pp. 213–231

Nelson N (1988) A veteran's look at the chloroplast H^+-ATPase and photosystem I reaction center. In: (JL Harwood and TJ Walton, eds.) *Plant Membranes—Structure, Assembly and Function. The Biochem Soc London*, pp. 159–167

Nitsch C, Braslavsky SE and Schatz GH (1988) Laser-induced optoacoustic calorimetry of primary processes in isolated photosystem I and photosystem II particles. *Biochim Biophys Acta* **934**: 201–212

Nuijs AM, Shuvalov VA, Van Gorkom HJ, Plijter JJ and Duysens LMN (1986) Picosecond absorbance difference spectroscopy on the primary reactions and the antenna-excited states in photosystem I particles. *Biochim Biophys Acta* **850**: 310–318

Oh-oka H, Takahashi Y, Wada K, Matsubara H, Ohyama K and Ozeki H (1987) The 8 kDa polypeptide in photosystem I is a probable candidate of an iron-sulfur center protein coded by the chloroplast gene *frxA*. *FEBS Lett* **218**: 52–54

Oh-oka H, Takahashi Y, Kuriyama K, Saeki K and Matsubara H (1988a) The protein responsible for center A/B in spinach photosystem I: Isolation with iron-sulfur center(s) and complete sequence analysis. *J Biochem* **103**: 962–968

Oh-oka H, Takahashi Y, Matsubara H and Itoh S (1988b) EPR studies of a 9 kDa polypeptide with an iron-sulfur cluster (s) isolated from photosystem I complex by n-butanol extraction. *FEBS Lett* **234**: 291–294

Oh-oka H, Takahashi Y and Matsubara M (1989) Topological considerations of the 9-kDa polypeptide which contains centers A and B, associated with the 14- and 19-kDa polypeptides in the photosystem I complex of spinach. *Plant Cell Physiol* **30**: 869–875

Oh-oka H, Itoh S, Saeki K, Takahashi Y and Matsubara H (1991) F_A/F_B protein from the spinach photosystem I complex: Isolation in a native state and some properties of the iron-sulfur clusters. *Plant Cell Physiol* **32**: 11–17

Ohyama K, Fukuzawa H, Kohchi T, Shirai H, Sano T, Sano S, Umesono K, Shiki Y, Takeuchi M, Chang Z, Aota S, Inokuchi H and Ozeki H (1986) Chloroplast gene organization deduced from complete sequence of liverwort *Marchantia polymorpha* chloroplast DNA. *Nature* **322**: 572–574

Okkels JS, Jepsen LB, Hønberg LS, Lehmbeck J, Scheller HV, Brandt P, Høyer-Hansen G, Stummann B, Henningsen KW, von Wettstein D and Møller BL (1988) A cDNA clone encoding a 10.8 kDa photosystem I polypeptide of barley. *FEBS Lett* **237**: 108–112

Okkels JS, Ascheller HV, Jepsen LB and Møller BL (1989) A cDNA clone encoding the precursor form 10.2 kDa photosystem I polypeptide of barley. *FEBS Lett* **250**: 575–579

Olson JM and Pierson BK (1987) Origin and evolution of photosynthetic reaction centers. *Orig Life* **17**: 419–430

O'Malley PJ and Babcock GT (1984) Electron nuclear double resonance evidence supporting a monomeric nature for $P700^+$ in spinach chloroplasts. *Proc Natl Acad Sci USA* **81**: 1098–1101

Ortiz W, Lam E, Chollar S, Munt D and Malkin R (1985) Topography of the protein complexes of the chloroplast thylakoid membrane. Studies of photosystem I using a chemical probe and proteolytic digestion. *Plant Physiol* **77**: 389–397

Ovchinnikov YuA, Abdulaev NG, Zolotarev AS, Shmuckler BE, Zargarov AA, Kutuzov MA, Telezhinskaya IN and Levina NB (1988a) Photosynthetic reaction centre of *Chloroflexus aurantiacus* I Primary structure of L-subunit. *FEBS Lett* **231**: 237–242

Ovchinnikov YuA, Abdulaev NG, Shmuckler BE, Zargarov AA, Kutuzov MA, Telezhinskaya IN, Levina NB and Zolotarev AS (1988b) Photosynthetic reaction centre of *Chloroflexus aurantiacus:* Primary structure of M-subunit. *FEBS Lett* **232**: 364–368

Palace GP, Franke JE and Warden JT (1987) Is phylloquinone an obligate electron carrier in photosystem I? *FEBS Lett* **215**: 58–62

Parrett KG, Mehari T, Warren PG and Golbeck JH (1989) Purification and properties of the intact P-700 and F_X-containing photosystem I core protein. *Biochim Biophys Acta* **973**: 324–332

Parrett KG, Mehari T and Golbeck JH (1990) Resolution and reconstitution of the cyanbacterial photosystem I complex. *Biochim Biophys Acta* **1015**: 341–352

Petrouleas V, Brand JJ, Parrett KG and Golbeck JH (1989) A Mössbauer analysis of the low-potential iron-sulfur center in photosystem I: Spectroscopic evidence that F_X is a [4Fe-4S] cluster. *Biochemistry* **28**: 8980–8983

Pichersky E, Hoffman NE, Bernatzky R, Piechulla B, Tanksley SD and Cashmore AR (1987) Molecular characterization and genetic mapping of DNA sequences encoding the Type I chlorophyll *a/b*-binding polypeptide of Photosystem I in *Lycopersicon esculentum* (tomato). *Plant Mol Biol* **9**: 205–216

Reilly P, Hulmes JD, Pan Y-CE and Nelson N (1988) Molecular cloning and sequencing of the *psaD* gene encoding subunit II of photosystem I from the cyanobacterium, *Synechocystis* sp. PCC6803. *J Biol Chem* **263**: 17658–17662

Robert B and Moënne-Loccoz P (1990) Is there a proteic substructure common to all photosynthetic reaction centers? In: (M Baltscheffsky, ed) *Current Research in Photosynthesis,* Vol I, Kluwer Acad Pub, Dordrecht, pp 65–68

Rögner M, Mühlenhoff U, Boekema EJ and Witt HT (1990) Mono-, di- and trimeric PS I reaction center complexes isolated from the thermophilic cyanobacterium *Synechococcus* sp. Size, shape and activity. *Biochim Biophys Acta* **1015**: 415–424

Rother C, Jansen T, Tyagi A, Tittgen J and Hermann RG (1986) Plastocyanin is encoded by an uninterrupted nuclear gene in spinach. *Curr Genet* **11**: 171–176

Rutherford AW and Mullet JE (1981) Reaction center triplet states in photosystem I and photosystem II. *Biochim Biophys Acta* **635**: 225–235

Rutherford AW and Heathcote P (1985) Primary photochemistry in photosystem-I. *Photosynth Res* **6**: 295–316

Ryrie IJ, Anderson JM and Glare T (1985) Transmembrane organization of individual polypeptides of photosystem I complex and the light-harvesting complex of photosystem II of thylakoid membranes. *Photobiochem Photobiophys* **9**: 145–157

Sakurai H and San Pietro A (1985) Association of Fe–S center(s) with the large subunit(s) of photosystem I particles. *J Biochem* **98**: 69–76

Schaffernicht H and Junge W (1981) Analysis of the complex band spectrum of P700 based on photoselection studies with photosystem I particles. *Photochem Photobiol* **34**: 223–232

Schantz R and Bogorad L (1988) Maize chloroplast genes *ndhD, ndhE,* and *psaC* sequences, transcripts and transcript pools. *Plant Mol Biol* **11**: 239–247

Scheller HV and Møller BL (1990) Photosystem I polypeptides. *Physiol Plant* **78**: 484–494

Scheller HV, Høj PB, Svendsen I and Møller BL (1988) Partial amino acid sequences of two nuclear-encoded photosystem I polypeptides from barley. *Biochim Biophys Acta* **933**: 501–505

Scheller HV, Svendsen I and Møller BL (1989a) Subunit composition of photosystem I and identification of center X as a [4Fe–4S] iron-sulfur cluster. *J Biol Chem* **264**: 6929–6934

Scheller HV, Okkels JS, Høj PB, Svendsen IB, Roepstorff P and Møller BL (1989b) The primary structure of a 4.0–kDa photosystem I polypeptide encoded by the chloroplast *psaI* gene. *J Biol Chem* **264**: 18402–18406

Scheller HV, Andersen B, Okkels JS, Svendsen IB and Møller BL (1990) Photosystem I in barley: Subunit PSI-F is not essential for the interaction with plastocyanin. In: (M Baltscheffsky, ed) *Current Research in Photosynthesis,* Vol II, Kluwer Acad Pub, Dordrecht, The Netherlands, pp. 679–682

Schoeder H-U and Lockau W (1986) Phylloquinone copurifies with the large subunit of photosystem I. *FEBS Lett* **199**: 23–27

Senge M, Dörnemann D and Senger H (1988) The chlorinated chlorophyll RCI, a preparation artefact. *FEBS Lett* **234**: 215–217

Sétif P and Bottin H (1989) Identification of electron-transfer reactions involving the acceptors A_1 of photosystem I at room temperature. *Biochemistry* **28**: 2689–2697

Sétif P, Mathis P and Vänngård T (1984) Photosystem I photochemistry at low temperature. Heterogeneity in pathways for electron transfer to the secondary acceptors and for recombination processes. *Biochim Biophys Acta* **767**: 404–414

Shinozaki K, Ohme M, Tanaka M, Wakasugi T, Hayashida N, Matsubayashi, T, Zaita, N, Chunwongse, J, Obokata, J, Yamaguchi-Shinozaki K, Ohto C, Torazawa K, Meng BY, Sugita M, Deno H, Kamogashira T, Yamada K, Kusuda J, Takaiwa F, Kato A, Tohdoh N, Shimada H and Sugiura M (1986) The complete nucleotide sequence of tobacco chloroplast genome: Its gene organization and expression. *EMBO J.* **5**: 2043–2049

Shuvalov VA, Nuijs AM, van Gorkom HJ, Smit HWJ and Duysens LNM (1986) Picosecond absorbance changes upon selective excitation of the primary electron donor P–700 in photosystem I. *Biochim Biophys Acta* **850**: 319–323

Smeekens S, de Groot M, van Binsbergen J and Weisbeck P (1985a) Sequence of the precursor of the chloroplast thylakoid lumen protein plastocyanin. *Nature* **317**: 456–458

Smeekens S, van Binsbergen J and Weisbeck P (1985b) The plant ferredoxin precursor: Nucleotide sequence of a full length cDNA clone. *Nucleic Acids Res* **13**: 3179–3194

Smith NS, Mansfield RW, Nugent JHA and Evans MCW (1987) Characterization of electron acceptors A_0 and A_1 in cyanobacterial photosystem I. *Biochim Biophys Acta* **892**: 331–334

Staehelin LA (1986) Chloroplast structure and supramolecular organization of photosynthetic membranes. In: (LA Staehelin and CJ Arntzen, eds) *Encyclopedia of Plant Physiology*, Vol 19, *Photosynthesis III*, Springer-Verlag, Berlin, pp 1–84

Steppuhn J, Hermans J, Nechushtai R, Ljungberg U, Thümmler F, Lottspeich F and Herrmann RG (1988) Nucleotide sequence of cDNA clones encoding the entire precursor polypeptides for subunits IV and V of the photosystem I reaction center from Spinach. *FEBS Lett* **237**: 218–224

Steppuhn J, Hermans J, Nechushtai R, Herrmann GS and Herrmann RG (1989) Nucleotide sequences of cDNA clones encoding the entire precursor polypeptide for subunit VI and of the plastome-encoded gene for subunit VII of the photosystem I reaction center from spinach. *Curr Genet* **16**: 99–108

Takahashi Y, Hirota K and Katoh S (1985) Multiple forms of P700-chlorophyll *a*-protein complex from *Synechococcus* sp.: The iron, quinone and carotenoid contents. *Photosynth Res* **6**: 183–192

Tamura N, Itoh S, Yamamoto Y and Nishimura M (1981) Electrostatic interaction between plastocyanin and P700 in the electron transfer reaction of photosystem I-enriched particles. *Plant Cell Physiol* **22**: 603–612

Thornber JP (1986) Biochemical characterization and structure of pigment-proteins of photosynthetic organisms. In: (LA Staehelin and CJ Arntzen, eds) *Encyclopedia of Plant Physiology*, Vol 19, *Photosynthesis III*, Springer-Verlag, Berlin, pp 98–142

Trebst A (1987) The three-dimensional structure of the herbicide binding niche on the reaction center polypeptides of photosystem II. *Z Naturforsch* **42c**: 742–750

Vacek K, Wong D and Govindjee (1977) Absorption and fluorescence properties of highly enriched reaction center particles of photosystem I and of artificial systems. *Photochem Photobiol* **26**: 269–276

Vorst O, Oosterhoff-Teertstra R, Vankan P, Smeekens S and Weisbeck P (1988) Plastocyanin of *Arabidopsis thaliana*; isolation and characterization of the gene and chloroplast import of the precursor protein. *Gene* **65**: 59–69

Warden JT and Golbeck JH (1987) Electron-spin resonance studies of the bound iron-sulfur centers in photosystem I. II Correlation of P-700 triplet production with urea/ferricyanide inactivation of the iron-sulfur clusters. *Biochim Biophys Acta* **891**: 286–292

Wasielewski MR, Fenton JM and Govindjee (1987) The rate of formation of $P700^+-A_0^-$ in photosystem I particles from spinach as measured by picosecond transient absorption spectroscopy. *Photosynth Res* **12**: 181–190

Webber AN and Malkin R (1990) Photosystem I reaction-centre proteins contain leucine zipper motifs. A proposed role in dimer formation. *FEBS Lett* **264**: 1–4

Witt I, Witt HT, Gerken S, Saenger W, Dekker JP and Rögner M (1987) Crystallization of reaction center I of photosynthesis. Low-concentration crystallization of photoactive protein complexes from the cyanobacterium *Synechococcus* sp. *FEBS Lett* **221**: 260–264

Wynn RM and Malkin R (1988a) Interaction of plastocyanin with photosystem I: A chemical cross-linking study of the polypeptide that binds plastocyanin. *Biochemistry* **27:** 5863–5869

Wynn RM and Malkin R (1988b) Characterization of an isolated chloroplast membrane Fe-S protein and its identification as the photosystem I Fe-S_A/Fe-S_B binding protein. *FEBS Lett* **229:** 293–297

Wynn RM and Malkin R (1990) The photosystem I 5.5 kDa subunit (the *psaK* gene product) An intrinsic subunit of the PSI reaction center complex. *FEBS Lett* **262:** 45–48

Zanetti G and Merati G (1987) Interaction between photosystem I and ferredoxin. Identification by chemical cross-linking of the polypeptide which binds ferredoxin. *Eur J Biochem* **169:** 143–146

Ziegler K, Lockau W and Nitschke W (1987) Bound electron acceptors of photosystem I. Evidence against the identity of redox center A_1 with phylloquinone. *FEBS Lett* **217:** 16–20

Zilber AL and Malkin R (1988) Ferredoxin cross-links to a 22 kDa subunit of photosystem I. *Plant Physiol* **88:** 810–814

Zuber H (1987) The structure of light-harvesting pigment-protein complexes. In: (J Barber, ed) *The Light Reactions*. Elsevier, Amsterdam, pp 197–259

3

Oxidation of Water to Molecular Oxygen

Govindjee and William J. Coleman

Department of Plant Biology
University of Illinois at Urbana-Champaign
265 Morrill Hall, 505 South Goodwin Avenue
Urbana, IL 61801-3707, USA

CONTENTS

84

ABSTRACT

Numerous papers have been published on the topic of water oxidation; however, the light-driven water splitting enzyme still remains one of the enigmas of photosynthesis. In this chapter, we have summarized the advances made on the oxidation of H_2O to O_2. The water-oxidizing complex includes, at least, the reaction center proteins D1 and D2 and a 33 kilodalton extrinsic protein; other proteins seem also to be involved. The 33 kDa extrinsic protein may turn out to be dispensable for the molecular mechanism of O_2 evolution. Chloride and calcium ions are also required, although their exact functions remain unknown. It is clear, however, that Mn undergoes dynamic changes as the oxygen clock moves from relaxed states S_0 to S_1 and S_1 to S_2. This is followed by conversion of S_2 to S_3 and S_3 to S_4 until O_2 is evolved. It is accepted that Mn is the charge accumulator, but it is considered likely now that at one of the steps, a histidine residue may act as a redox active ligand, and store the charge. It was generally believed that H^+ are released as 1, 0, 1, 2 during $S_0 \rightarrow S_1$, $S_1 \rightarrow S_2$, $S_2 \rightarrow S_3$, and $S_3 \rightarrow (S_4) \rightarrow S_0$ transitions. However, the currently accepted pattern is 1, 0.5, 1 and 1.5. The nature of the intermediates of water oxidation form H_2O to O_2 is still unknown. Although the recent knowledge about the 3–D crystal structure of the reaction center complex from purple photosynthetic bacteria has led to a more precise picture of a portion of the water oxidizing complex than known before, further understanding will come after this complex is crystallized and its 3–D structure known and after methods are evolved to trap and monitor transient intermediates in water oxidation.

I. Introduction

Molecular oxygen (O_2) was nearly absent when the earth's atmosphere was formed approximately 4.5 billion years ago (Rao et al., 1981; Wayne, 1988). Fossil evidence suggests that prokaryotes inhabited the earth as far back as 3.5 to 3.8 billion years ago. The appearance of significant levels of atmospheric O_2 approximately 2 billion years ago must have followed the rise of photosynthesis. It is considered highly likely that cyanobacterial photosynthesis may have been present much earlier (about 3.5 billion years ago) but O_2 did not accumulate in the atmosphere, either because various geochemical reactions were consuming it or because it was being used up by other aerobic forms of life that had also evolved (Veizer, 1988). Although it is conceivable that oxygen production was detrimental to many other organisms living at the time (since they were likely to have been anaerobic heterotrophs (Rao et al., 1981; Cavalier–Smith, 1987)), terrestrial organisms ultimately benefited from this new development in at least two respects. For example, the availability of O_2 as a terminal electron acceptor made it feasible to generate adenosine triphosphate (ATP) by respiration. Likewise, the introduction of large amounts of O_2 into the earth's atmosphere must have helped to create the ozone layer in the lower stratosphere, which blocks out short-wavelength solar UV radiation (Turco, 1985; Wayne, 1988).

For photosynthetic organisms, the evolution of oxygenic photosynthesis was a major biochemical breakthrough: it enabled them to exploit a virtually unlimited source of electrons by oxidizing water. Water is an extremely poor reductant, however, and thus a system to remove electrons from such a substrate (where the energy input for each step is restricted to a quantum of visible light) must be efficient at generating a strong oxidant and must provide a "charge-accumulating" complex to handle the four oxidizing equivalents needed to oxidize two H_2O molecules to dioxygen:

$$2 \ H_2O = O_2 + 4H^+ + 4e^-$$

The photosynthetic reactions that generate molecular oxygen are part of a highly-ordered series of photochemical events that are coupled to two electron-transfer mechanisms. The two mechanisms operate on opposite ends of a specialized pigment-protein complex, known as photosystem II (PS II) or the water-plastoquinol oxido-reductase, which is found within the cells of cyanobacteria, multicolored algae and higher green plants. One of these directly generates O_2 by removing four electrons and four protons from two water molecules. The splitting of water releases protons into the inner aqueous compartment of the thylakoid membrane. The other reduces plastoquinone (PQ) to plastoquinol (PQH_2), which is capable of shuttling two protons and two electrons. Another photosystem (PS I) transfers electrons from plastoquinol to nicotinamide dinucleotide phosphate ($NADP^+$), producing the NADPH required for the reduction of CO_2. On the other hand, ATP, also needed for CO_2 fixation, is synthesized by using the proton gradient generated by these mechanisms. Light absorption by chlorophylls (P680 and P700) at the heart of the two photosynthetic protein complexes (the reaction centers I and II) provides the energy to drive these two reactions (for details, see Fig. 1 and its legend). The molecular structure that accomplishes this feat is now being examined in detail.

For earlier reviews on photosystem II, see Govindjee (1980, 1982, 1984), Velthuys (1980, 1987), van Gorkom (1985), Diner (1986), Mathis (1986), Ort (1986) Andréasson and Vanngård (1988), Hansson and Wydrzynski (1990) Andersson and Styring (1991) and Vermaas and Ikeuchi (1991). The electron acceptor side of PS II has been discussed by Vermaas and Govindjee (1981), Crofts and Wraight (1983), Govindjee and Eaton-Rye (1986), and Govindjee and Wasielewski (1989), and the electron donor side by Ghanotakis and Yocum (1985), Govindjee et al. (1985), Renger and Govindjee (1985), Babcock (1987), Renger (1987a, b), Coleman and Govindjee (1987), Homann (1987), Brudvig et al. (1989), Rutherford (1989), Babcock et al. (1989) and Ghanotakis and Yocum (1990). Finally, the relation of the work on the reaction center of purple bacteria to photosystem II has been discussed by Michel and Deisenhofer (1988).

II. Proteins and Cofactors

The PS II complex is embedded within the lipid bilayer of the thylakoid membrane. There are three electron-transfer complexes in the membrane (Figs. 2 and 3): PS II (under discussion here), the cytochrome bf complex (which transfers electrons from PS II to PS I; see Cramer et al., 1991), and PS I (Golbeck and Bryant, 1991). The current picture of PS II (Fig. 2) suggests that in additon to the reaction center protein, composed of polypeptides D1 and D2 (molecular masses 39 kilodaltons each) and cytochrome b-559 (a 4 and 9 kilodalton heme-containing protein whose function is unknown), there are also at least three polypeptides of molecular mass 33, 24 and 17 kilodaltons that are bound to the inner surface of the thylakoid membrane and that contribute to oxygen production in plants. A number of other recently discovered polypeptides (with a molecular mass between 4 and 20 kilodaltons) are associated with PS II, but their functions are still uncharacterized. Other chlorophyll-containing pigment proteins (CP-43, molecular mass 52 kilodaltons; and CP-47, molecular mass 56 kilodaltons) help to transfer excitation energy into the reaction center. Several organic and inorganic ions (iron, manganese, calcium, bicarbonate and chloride) are also involved in catalyzing electron transfer, maintaining the protein structure, or

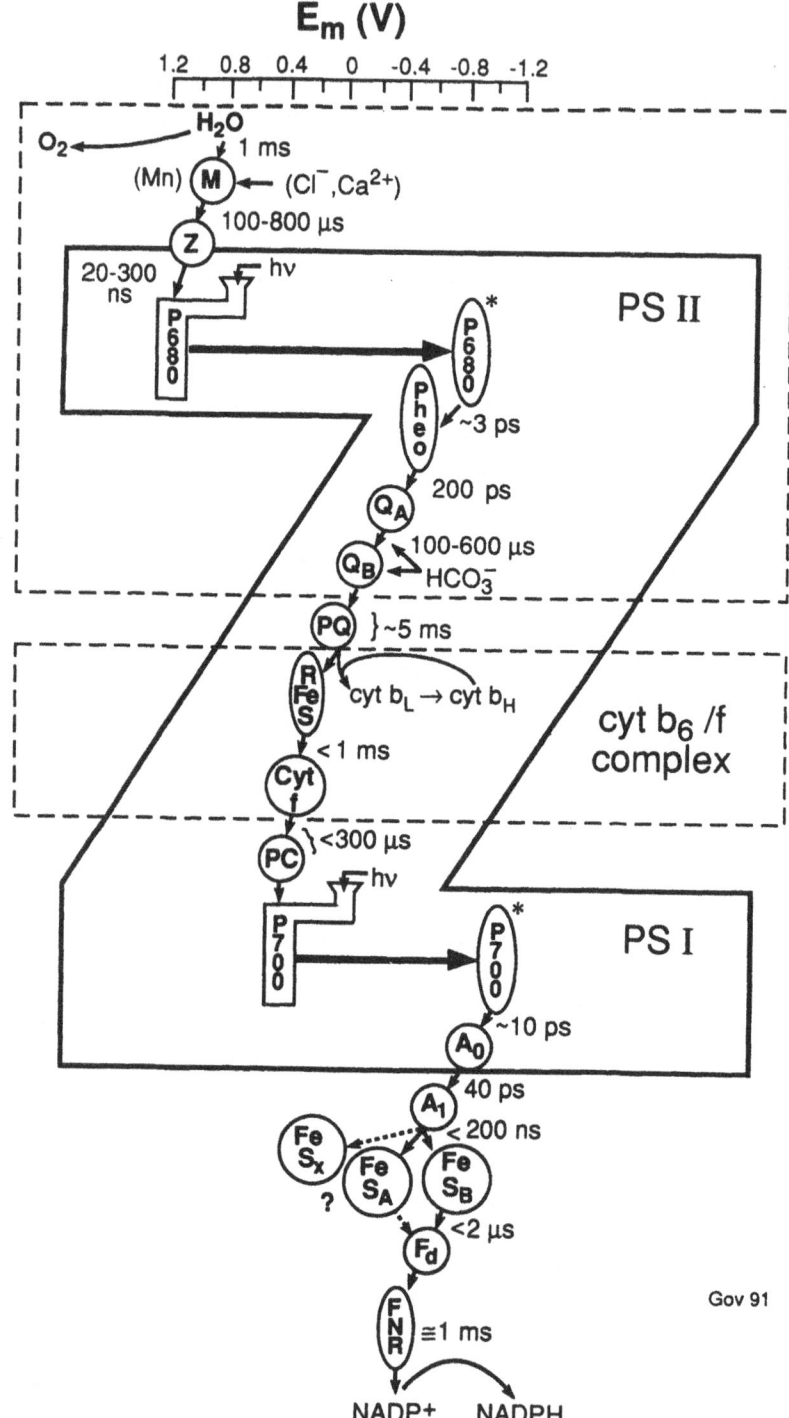

regulating activity. Since cyanobacteria do not contain the extrinsic 17 and 24 kilodalton proteins and since a *Synechocystis* mutant lacking extrinsic 33 kilodalton polypeptide (L. Sherman, B. Zilinskas and others, personal communication) can carry out O_2 evolution, these 3 extrinsic polypeptides cannot be considered an absolute requirement for the O_2 evolution mechanism. For other details, see Andersson and Styring (1991) and Ghanotakis and Yocum (1990).

Due to the inherent complexity of PS II, a number of advances in our understanding of how a charge separation is accomplished have come from simpler biological systems. The simplest chlorophyll-containing photosynthetic complexes that have been extensively studied as models for PS II occur in the purple non-sulfur bacteria (for reviews, see Feher *et al.*, 1989; Coleman and Youvan, 1990). This bacterial system also catalyzes the reaction that is intrinsic to all such complexes, that is, converting light energy into a transmembrane electrochemical potential gradient (generated by a charge separation and the pumping of H^+). Energy stored in this gradient is subsequently used by the complex known as the ATP synthase or coupling factor to synthesize ATP for cellular metabolism (Ort and Melandri, 1982; Jagendorf *et al.*, 1991).

Figure 1. (on facing page): The linear photosynthetic electron transfer pathway in plants and cyanobacteria: the Z Scheme. The energy from two separate photosystems (I and II) connected in series is used to oxidize water and reduce $NADP^+$. Photysystem (PS) II, whose reaction center core partly resembles that of the purple bacteria, oxidizes two H_2O molecules into $O_2 + 4H^+ + 4e^-$ and reduces plastoquinone (PQ) to PQH_2 in a $2e^-/2H^+$ reaction. Photysystem (PS) I, which resembles the reaction center of certain green sulfur bacteria, oxidizes the electron-carrying molecule plastocyanin (PC) and reduces $NADP^+$ to NADPH. The cytochrome b_6f complex functions analogously to its bacterial counterpart, the cytochrome bc_1 complex, but contains cytochrome f instead of cytochrome c_1. The PS II utilizes manganese (Mn), calcium (Ca^{2+}) and chloride (Cl^-) ions to split water and deliver the electrons to Z, which is a tyrosine residue on the D1 polypeptide. The other PS II reaction center components are similar in function to those of the purple bacteria but involve chlorophyll, pheophytin and plastoquinone instead of their bacterial analogs. One or two molecules of bicarbonate (HCO_3^-) also appear to bind to Fe and/or near the quinones, and facilitate electron/proton transfer from Q_A to $Q_B^{(-)}$. Apparently, HCO_3^- plays no role in purple photosynthetic bacteria. PS I has a pair of primary donor chlorophyll molecules whose long-wavelength absorption maximum is at 700 nm (P700). The light-generated excited state (P700*) donates an electron to another chlorophyll molecule (A_o), and the resulting cation ($P700^+$) is re-reduced by plastocyanin. From A_o, the electron continues to move in an energetically downhill process to a phylloquinone (A_1, a vitamin K_1 molecule), and then through a series of iron-sulfur centers (FeS_x, FeS_A, FeS_B and Fd or ferredoxin). At the end of the chain, a complex known as the ferredoxin-$NADP^+$ oxidoreductase (FNR) transfers two electrons and a proton to convert $NADP^+$ to NADPH. A proton gradient created between the outside of the membrane (the stroma) and the lumen is dissipated through the coupling factor complex (CF_o — CF_1) to produce ATP.

The relative ability of a compound to accept or donate an electron can be gauged by its midpoint redox potential (E_m). These values are also shown in the diagram. A compound with a higher (more positive) E_m is more oxidizing. A compound with a lower (more negative) E_m is more reducing. Thus, in plant and cyanobacterial photosynthesis, light energy is needed to remove electrons from H_2O in the highly positive H_2O/O_2 couple and to add electrons to $NADP^+$ in the highly negative $NADP^+/NADPH$ couple. Thermodynamic measurements indicate that the oxidation of two H_2O molecules into O_2, $4H^+$, and $4e^-$ should require an average oxidizing potential of about +0.81 volts for each electron removed at pH 7.0. Estimation of the Em, 7 (the midpoint redox potential at pH 7.0) of $P680/P680^+$ in the plant reaction center gives a value of about +1.1 to +1.2 volts. This value places an upper limit on the oxidizing potential available from each photon of red light absorbed by Photosystem II. The bacterial systems, in contrast, absorb longer-wavelength photons of lower energy.

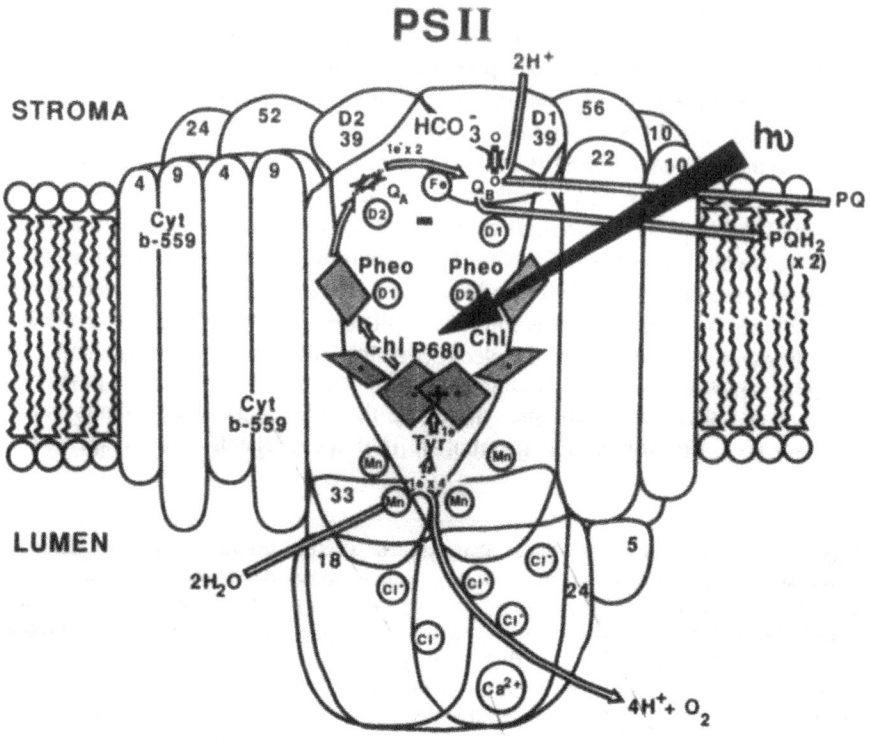

Figure 2. The photosystem II complex. The reaction center core consists of the D1 and D2 polypeptides (39 kD) which are closely associated in the membrane with cytochrome *b*559 (consisting of 4 and 9 kD subunits). The manganese responsible for catalyzing the water oxidation reactions is believed to bind to the D1 and D2 polypeptides. Numbers on the polypeptides refer to the actual or apparent molecular weights of the subunits. Polypeptides of molecular weight 52 (CP43) and 56 kD (CP47) are part of the chlorophyll-binding antenna complex. Other intrinsic polypeptides have an unknown function. Several extrinsic polypeptides (33, 24, 18 and 5 kD), forming part of the complex, appear to be involved in enhancing Ca^{2+} and Cl^- binding. The positions of the chromophores and the pathway of electron transfer within the reaction center are based largely on analogy to the purple bacterial system. In cyanobacteria, the 18 and 24 kD polypeptides on the lumen side are absent.

A major breakthrough in the field took place when the reaction center complex of the photosynthetic bacterium *Rhodopseudomonas viridis* was crystallized in 1982 (Michel, 1982). Elucidation of the three-dimensional structure by X-ray diffraction has made it possible to better understand how the light-driven charge separation is achieved (Deisenhofer and Michel, 1989). In idealized form, the reaction center mechanism involves five components: a primary donor chlorophyll pigment (P), capable of being converted to a reducing (i.e., electron-donating) excited state (P*) followed by oxidation to P^+, an electron-deficient ground state; a secondary electron donor (D) which can reduce P^+; a pheophytin electron acceptor (Pheo) which can accept an electron from P*; a primary quinone electron acceptor (Q_A), which is

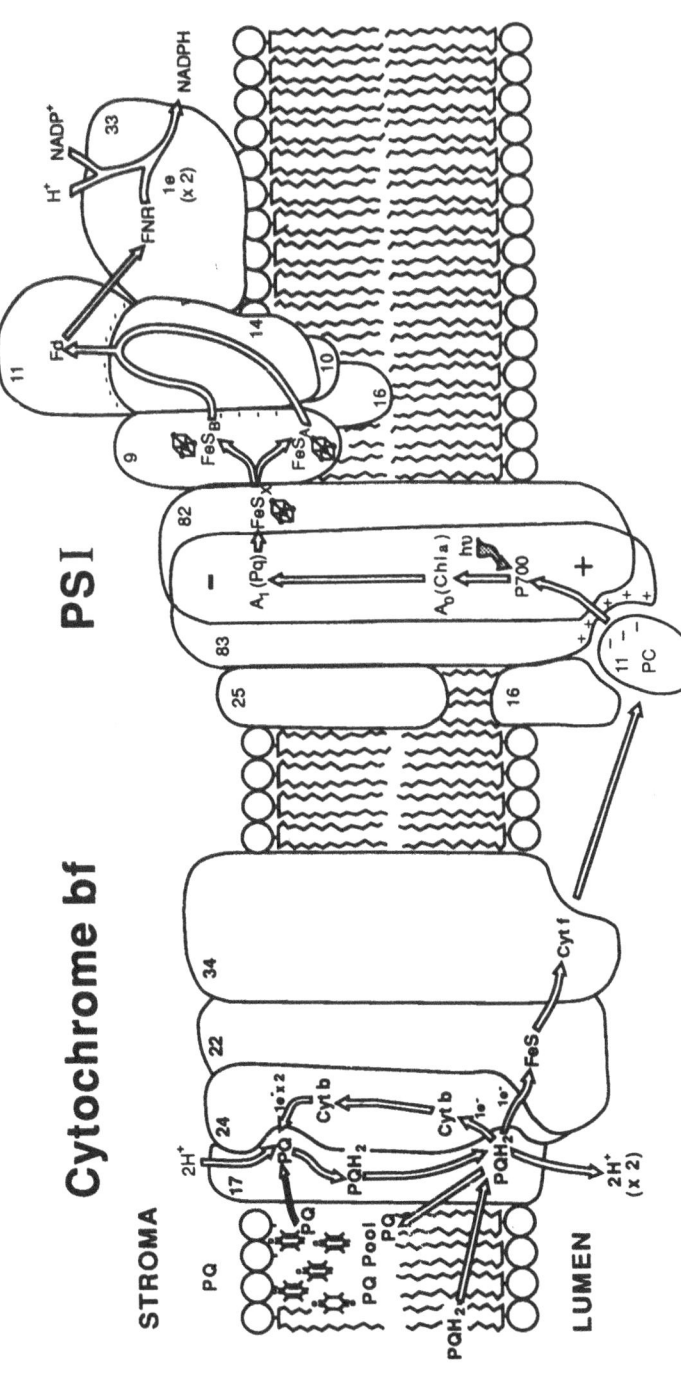

Figure 3. The cytochrome *bf* complex and photosystem I. Reactions in which plastoquinol is oxidized and plastocyanin is reduced are shown to occur in the cytochrome *bf* complex. The function of the photosystem I (PSI) complex is to oxidize the reduced plastocyanin and reduce NADP⁺. Soluble cytochrome *c* and flavodoxin can replace plastocyanin and ferredoxin in cyanobacteria and some red algae. Numbers on the polypeptides refer to the actual or apparent molecular weights of the subunits. For other details of electron transfer, see Fig. 1 and reviews by Cramer *et al.* (1991) and Golbeck and Bryant (1991).

tightly bound to the reaction center protein; and a secondary quinone acceptor (Q_B), which is tightly bound only in its singly-reduced state, and which is otherwise capable of diffusing between protein complexes in the membrane. A non-heme iron atom (Fe^{2+}) is situated between Q_A and Q_B, but does not directly participate in electron transport in bacteria. These components are anchored in a fixed arrangement within the scaffolding of the reaction center protein complex, such that together they span the thickness of the photosynthetic membrane. Absorption of a photon of light (hυ) by P (transforming it from P to P*) or transfer of an exciton to P initiates the charge separation, as the electron moves away from its positively-charged "hole".

$$D{\cdot}P{\cdot}Pheo{\cdot}Q_A{\cdot}Q_B \xrightarrow{\text{hυ}} D{\cdot}P^*{\cdot}Pheo{\cdot}Q_A{\cdot}Q_B \rightarrow$$

$$D{\cdot}P^+{\cdot}Pheo^-{\cdot}Q_A{\cdot}Q_B \longrightarrow D{\cdot}P^+{\cdot}Pheo{\cdot}Q_A^-{\cdot}Q_B \rightarrow$$

$$D^+{\cdot}P{\cdot}Pheo{\cdot}Q_A{\cdot}Q_B^-$$

Because under normal circumstances the component Q_B does not leave the complex until it has acquired two electrons (and, ultimately, two protons to form Q_BH_2), this member of the electron-accepting side of the complex must wait until the reaction center absorbs a second photon and generates another charge separation before it can diffuse to the next membrane-bound electron/proton transferring complex (known as the cytochrome *bf* complex in plants and cytochrome bc_1 in bacteria), which shuttles the protons to the inner aqueous phase of the thylakoid membrane.

In plants, the cation state of the primary donor, created as a result of the charge separation, is rapidly reduced by a donor known as "Z", which has recently been suggested or shown to consist of the phenolic group of tyrosine 161 on the D1 polypeptide (Barry and Babcock, 1987; Vermaas *et al.*, 1988; Debus *et al.*, 1988a, b; Metz *et al.*, 1989). The tyrosine radical generated by this reaction is believed to act as the acceptor for the electrons removed from water by the oxygen-evolving complex (Figs. 1 and 2).

III. Primary Photochemistry

The primary reactions of Photosystem II can be written as:

$$^1Chla + \text{hυ} \xrightarrow{1} {}^1Chla^* \text{ (light absorption; creation of excited state)}$$

$$^1Chla^* + P680 \xrightarrow{2} {}^1Chla + {}^1P680^* \text{ (creation of singlet excited state of P680)}$$

$$^1P680^* + Pheo \xrightarrow{3} {}^1P680^+ + Pheo^- \text{ (primary charge separation)}$$

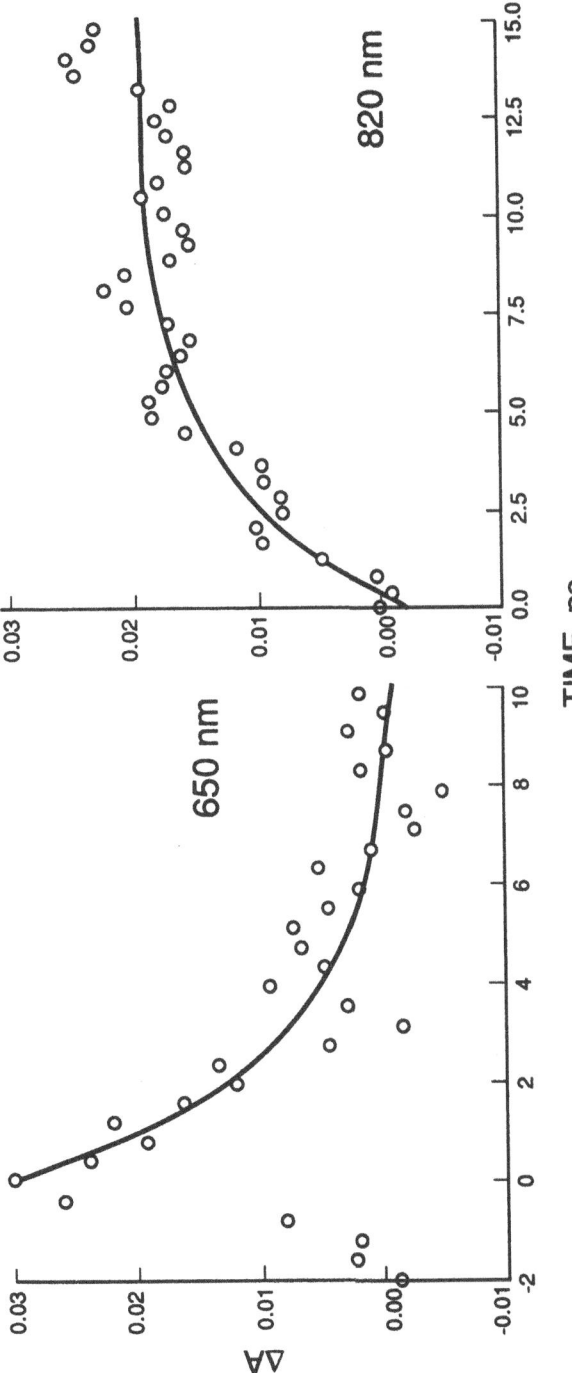

Figure 4. Primary photochemistry of photosystem II. Transient absorption changes at 650 nm (due to decay of ^{1}Chl$a$*) and at 820 nm (due to formation of P680$^+$) for photosystem II reaction centers following a 100 µJ, 500 fs laser flash at 610 nm. Left: absorption change at 650 nm; right: absorption change at 820 nm (after Wasielewski *et al.*, 1989).

$$^1P680^+ + Pheo^- \xrightarrow{4} P680 + Pheo \text{ (charge recombination, or loss of charges by}$$
$$\text{other mechanisms)}$$

$$^1P680^+ + Pheo^- \xrightarrow{5} {}^3P680 + Pheo \text{ (triplet state formation)}$$

Here, P680 is the reaction center chlorophyll a and pheophytin is the primary electron acceptor. For a general background on the primary events of photosynthesis, see Govindjee and Govindjee (1974).

The act of light absorption is the most rapid reaction and is estimated from the interaction of light with Chl a molecule. For example, Chl a goes into first singlet excited state by absorbing red light (680 nm). The 680 nm light oscillates with a frequency of 4.41×10^{14} cycles s^{-1}. Thus, one transition (or cycle) occurs in 2.5×10^{-15}s or 2.5 femtoseconds. This, then, is the approximate time of excitation (reaction 1). The excitation energy transfer in a reaction center must also be very rapid, considering the rise time for $^1P680^*$ production to be within the 500 fs instrument function (reaction 2). Wasielewski et al. (1989) have measured the lifetime of reaction 3 to be 3.0 ± 0.6 picoseconds in stable reaction center II preparations (to be discussed below). The lifetime of $P680^+$ — $Pheo^-$ (reaction 4) had been measured in reaction center II, prepared according to Nanba and Satoh (1987), to be 32 nanoseconds (ns) (see Takahashi et al., 1987). Hansson et al. (1988) have reported that the lifetime of $P680^+$ — $Pheo^-$ depends upon the number of antenna Chl a remaining attached to the reaction center. Okamura et al. (1987) have shown the formation of the triplet state of P680, and have established the radical pair nature of its precursor through its spin polarization characteristics. Nuijs et al. (1986a, b) have studied primary reactions, using a 35 ps laser, in photosystem II particles enriched in reaction centers (P680/80 Chl). This study, however, does not provide direct answers to the steps involved in reaction 3 above. We discuss below the direct measurement of the kinetics of charge separation in a stable reaction center of plant photosynthesis, using 500 fs laser flashes.

Wasielewski et al. (1989) provided the first direct measurement of charge separation in stabilized PS II reaction centers, at 4 C, with a 500 fs resolution. Using 610 nm exciting flashes, they observed the following: (1) an instrument-limited (0.5 $\pm$ 0.4 ps) decrease in absorbance at 670 nm and 485 nm, indicating the formation of the singlet excited state of P680 ($^1P680^*$); (2) an increase in absorbance at 820 nm (due to the formation of $P680^+$) with a time constant (1/e) of 3.0 ± 0.6 ps; (3) a decrease in absorbance at 670 nm and an increase in absorbance at 485 nm or at 538 nm reflecting the formation of $P680^+$ Pheophytin$^-$ with a time constant also of approximately 3 ps; and (4) a decrease in absorbance at 650 nm (at the isosbestic point for the $P680^+$ $Pheo^-$ — P680 Pheo changes), reflecting the decay of $^1P680^*$ with a time constant of 2.6 ± 0.6 ps that matches, within experimental error, the formation of $P680^+$ (Fig. 4). Furthermore, it was shown that when Pheo was prereduced by the addition of sodium dithionite, methylviologen and light, only the absorbance changes due to the formation of $^1P680^*$ were observed; the absorbance increase at 820 nm was eliminated, confirming that the observed changes are indeed due to the charge separation.

Since the back reaction between Pheo⁻ and P680⁺ has been measured to be 30 ns (Takahashi *et al.*, 1987; Danielius *et al.*, 1987) in Satoh and Nanba's D1, D2, cyt *b*-559 preparations, and is expected to be 2-4 ns in intact PS II, the forward electron flow from Pheo⁻ to Q_A is expected to be many orders of magnitude faster. Based on the measurements of the lifetime of chlorophyll *a* fluorescence of 180 to 300 ps from PS II in algae when the reaction centers are fully open, the reduction time of Q_A is suggested to be 300 ps (see Moya *et al.*, 1986; Holzwarth, 1987, for reviews on lifetime of fluorescence measurements). Our own (J. Fenton, N.S. Rao, E. Gratton, 1982, presented at the Midwest Photosynthesis Congress, Argonne, IL, USA) unpublished measurements of 200 ps in spinach thylakoids is in agreement with this conclusion. Since P680⁺ to P680 reaction occurs in 20 – 30 ns (P680⁺ is a quencher of Chl *a* fluorescence), the reduction time of Q_A to Q_A^- cannot be measured by fluorescence. However, the electron transfer from Pheo⁻ to Q_A can be measured by absorbance changes. Nuijs *et al.*, (1986a), from measurements with 35 ps resolution, concluded that the electron transfer from Pheo⁻ to Q_A occurs with a time constant of 270 ps in spinach PS II preparations enriched in P680 (1P680/80 Chl *a*). Eckert *et al.* (1988) have measured that Pheo⁻ transfer electrons to Q_A within 200 ps. On the other hand, Schatz *et al.* (1987), using *Synechococcus* PS II preparations enriched in P680 (also 1P680/80 Chl *a*), estimated that the electron transfer from Pheo⁻ to a subsequent quinone electron electron acceptor takes place in ~ 500 ps. Furthermore, the quantum yield of photoreduction of Q_A has been estimated to be 0.9 (see Thielen and van Gorkom, 1981). It is thus clear that no other reaction, such as the reduction of NADP⁺ by reduced Pheo⁻ (Arnon and Barber, 1990), can compete with the rapid and efficient reduction of Q_A by Pheo⁻.

IV. The Oxygen Clock

Removing four electrons from water by way of four separate photoacts presents a challenging chemical problem. Although a number of basic mechanisms are possible, some are better suited than others to the energetic constraints of the system (for details, see Renger 1987a, b, 1988; Renger and Govindjee, 1985; Brudvig *et al.*, 1989). Since the chemistry is confined to a biological milieu, the toxicity of certain intermediates, such as hydrogen peroxide, may impose additional restrictions on the reaction pathway. One of the objectives in examining the catalytic turnover of the oxygen-evolving complex has been to relate the kinetics of O_2-evolution to the primary charge-separating reactions. The classical data of Emerson and Arnold (1932a, b) suggested that the photochemical reactions leading to oxygen release occur within discrete clusters of Chl known as photosynthetic units. Years later, the identification of PS II as the oxygenic photosystem (containing P680 as the primary oxidant) made it possible to study the coupling between Chl oxidation and the oxidation of water. It is known that the component P680⁺ must eventually (by way of Z) extract an electron from water, but the rate at which each electron arrives has been found to vary in a periodic fashion. Measurements of the decay of P680⁺ following an excitation flash indicate the oxidized primary donor recovers an electron in darkness at different rates, depending on the number of flashes given to the system (van Best and Mathis, 1978; Brettel *et al.*, 1984; Sonneveld *et al.*, 1979; Schlodder *et ·al.*, 1985). For example, the halftime for the conversion of P680$^+$ to P680 is approximately 20 ns after the 1st and 5th flashes, but is much longer after the 2nd, 3rd and 4th flashes. The periodicity of four indicates that a cyclic reaction with four steps in involved in donating electrons to the reaction center.

The relationship between the periodicity of four in electron transfer to P680$^+$ and the production of oxygen from water becomes obvious when one compares the above data with the flash number dependence of O_2 release. Using a sensitive platinum electrode, Joliot *et al.* (1969) observed that the amount of O_2 evolved after a series of short, saturating flashes oscillates with a period of four. There is no O_2 evolution after the first flash and none (or very little) after the second flash, but a maximum amount after the third flash. Thereafter, the amplitude of the O_2 yield oscillates with a period of four until the differences gradually damp out.

Kok *et al.* (1970) explained these data by proposing (see Fig. 5) that the oxygen-evolving complexes act independently of one another, and that each is capable of existing in several transient oxidation states, known as the "S states" (for different

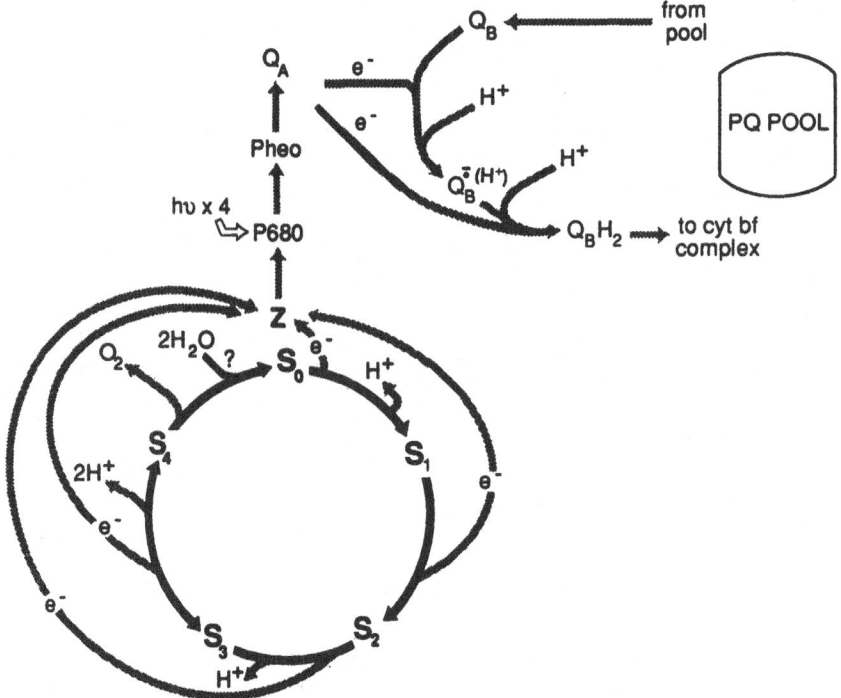

Figure 5. The four-step "oxygen clock". This mechanism delivers electrons one-by-one to the oxidized reaction center chlorophyll *a* molecule P680$^+$ as each incoming photon (hυ) oxidizes the primary donor P680. Each of the "S-state" intermediates represents an increasingly oxidized state of the oxygen-evolving complex. Protons (H$^+$) are shown to be released from the complex on the $S_0 \rightarrow S_1$ (1H$^+$), $S_2 \rightarrow S_3$ (1H$^+$) and $S_3 \rightarrow (S_4) \rightarrow S_0$ (2H$^+$) transitions. But, not only this release may be in non-integers, it may not even reflect the actual release from the "S" states, but from the protein. Molecular oxygen (O_2) is released only on the final transition. It is not yet known at what stage the two H_2O molecules bind or when the O—O bond is formed. On the other side of the photosystem, the two electron "quinone gate" removes reducing electrons in pairs (along with protons) for transport to the cytochrome *bf* complex. Here again, protons may first bind to the proteins.

models, see Mar and Govindjee, 1972). We call this the "oxygen clock" (see Govindjee and Coleman, 1990). The precise chemical nature of each state was, however, unknown. The need for a cycle having a periodicity of four arose from the assumption that it would take four positive charges, from four consecutive photoacts at P680, to oxidize two molecules of H_2O to yield a molecule of O_2. Advancement of each S-state during a series of flashes is limited by a dark relaxation step (i.e., an intervening S'-State), such that a finite time interval must elapse before the next incoming quantum can be utilized (Kok et al., 1970; Bouges–Bocquet, 1973; Joliot and Kok, 1975). Dark relaxation steps may reflect a rate limitation on PS II turnover due to the reoxidation of Q_A^- on the $S_0' \rightarrow S_1$, $S_1' \rightarrow S_2$ and $S_2' \rightarrow S_3$ transitions. However, the $S_3' \rightarrow S_4 \rightarrow S_0$ transition appears to be limited by reactions at the water-splitting site (Joliot et al., 1966; Bouges–Bocquet, 1973). A second intermediate state for the $S_2 \rightarrow S_3$ transition (S_2''), has also been proposed in order to explain the sigmoidal kinetics of this relaxation step (Bouges–Bocquet, 1973).

The other curious features of Joliot's results (the maximum on the third rather than the fourth flash and the damping-out of the oscillation with an increasing number of flashes) provided an insight into the relative oxidation sate of the "resting point" of the clock, as well as the dynamics of electron transfer between the oxygen-evolving complex and P680. Kok et al. (1970) explained these phenomena as follows: in darkness, there is a certain proportion of two S-states, labeled S_0 and S_1, wherein S_1 (which has one more oxidizing equivalent than S_0) outnumbers S_0 by approximately 3:1. In samples dark-adapted for hours, one starts only from S_1 (Vermaas et al., 1984). The predominant reaction after the first flash oxidizes S_1 to S_2. Likewise, the second flash converts S_2 to S_3 and the third flash converts S_3 to S_4. The very short-lived S_4 state (which has accumulated four positive charges from four successive reductions of P680$^+$ to P680) releases the O_2, which is quickly detected by the platinum/Ag/AgCl electrode, used in these studies. The S_4 state must then spontaneously return to the S_0 state, so that the cycle may start again. The assumption of a long-lived S_1 state which is stable in the dark helped to explain the puzzling observation of maximal O_2 yield on the third flash. The predominance of S_1 as the resting state occurs because in darkness the S_2 and S_3 states deactivate to S_1 (on the timescale of seconds) by losing their oxidizing equivalents (Forbush et al., 1971). The S_0 state is also gradually oxidized to S_1 in the dark through the slow reduction of oxidized "D", a tyrosine residue on the D2 protein (Styring and Rutherford, 1987).

The observed loss of phase coherence (i.e., damping) with increasing flash number was explained by postulating the contribution of two other factors. First, that a small probability exists for the failure of a given state S_n to advance to the S_{n+1} state (termed a "miss"; Kok et al., 1970; Forbush et al., 1971). This phenomenon could result from either closure of some of the phototraps (due to the presence of Q_A^-) or from a charge recombination within the system. Second, that a "double hit" occurs when the flash duration is just long enough to allow two photons to be absorbed by the same photosystem, permitting two turnovers of the S-states (Weiss and Sauer, 1970; Kok et al., 1970). This explains the small amount of O_2 often observed after flash 2. When these complications are taken into account, Joliot et al.'s data, including the damping of the oscillations of O_2 yield with increasing flash number, become readily understandable. After a large number of flashes, the S-states reach an equilibrium, such that the concentrations of S_0, S_1, S_2 and S_3 become equal and there is no longer any oscillation in the yield of O_2. The question of whether the value of the miss parameter (i.e., the probability) is identical or different for all of the S-state transitions has not been resolved (Delrieu, 1974; Wydrzynski, 1982).

V. The Manganese Complex as the Charge Accumulator

Discoveries of Joliot et al. (1969) and Kok et al. (1970) regarding the "oxygen clock" opened up the black box of oxygen evolution, but they still did not explain the physical make-up of the clock itself (the oxygen-evolving complex) or the chemistry of water in each of the S-states. Their experiments set off a long search for the chemical nature of the "charge accumulator" whose oxidation states are expressed by each of the numbered S-states. In the beginning, this elusive chemical entity was labeled with the letter "M", perhaps becasue it was assumed to be a metal atom (see discussion by Wydrzynski, 1982). Even before the "clock" was discovered, manganese (Mn) was suspected to comprise at least part of "M", because it has long been known that four Mn atoms per P680 are essential for O_2 evolution (e.g., see Yocum et al., 1981). Current models for the location of the four Mn atoms place them on the lumenal side of the D1–D2 RC complex (e.g., see Coleman and Govindjee, 1987). Several lines of evidence suggest that two of the Mn are in an environment that is different from that of the other two (e.g., see Kambara and Govindjee, 1985). The catalytic site for water-splitting is believed to consist of a minimum of two Mn atoms (e.g., see Haddy et al., 1989).

If one removes these Mn atoms, O_2 evolution is abolished, but the light-driven electron transfer from Z to $NADP^+$ remains intact, provided that artificial electron donors are supplied to the system (see Kok and Cheniae, 1966). These early studies established the role of Mn in O_2 evolution. Theoretically, Mn is a logical choice for the water-splitting catalyst, since it is capable of existing in several different oxidation states and is also known to catalyze reactions with H_2O_2 and O_2^- in other enzymes (Reed, 1986).

Many of the biologically relevant Mn complexes are paramagnetic, and a number of highly-sensitive measuring techniques have exploited this fact (Reed and Markham, 1984). Most notably, electron paramagnetic resonance (EPR) spectroscopy has been used to look at light-induced changes in the electronic structure of the Mn complex. Nuclear magnetic resonance (NMR) spectroscopy has also been used to examine the paramagnetic properties of the Mn in each observable S-state, by indirectly monitoring the ^1H_2O protons that are in contact with the metal atoms.

The question of the oxidation state of Mn has been approached from several different angles. Wydrzynski et al. (1975, 1978) pioneered the use of water proton (^1H) NMR to demonstrate dynamic changes in the oxidation state of Mn in chloroplast membranes. Srinivasan and Sharp (1986a, b) have unambiguously shown that changes in the S-states are related to redox changes in Mn. Tentatively, $S_0 \rightarrow S_1$ is assigned to a Mn(II) $\rightarrow$ Mn(III) transition, and $S_1 \rightarrow S_2$ to a Mn(III) $\rightarrow$ Mn(IV) transition; however, no redox change in Mn is attributed to the $S_2 \rightarrow S_3$ transition.

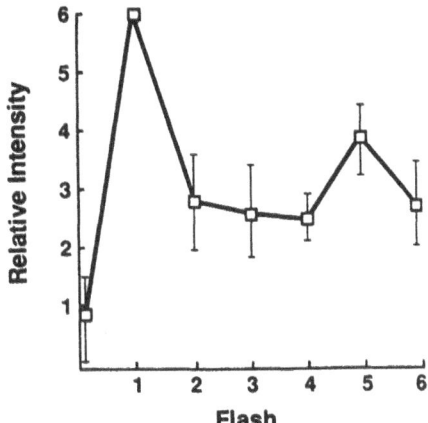

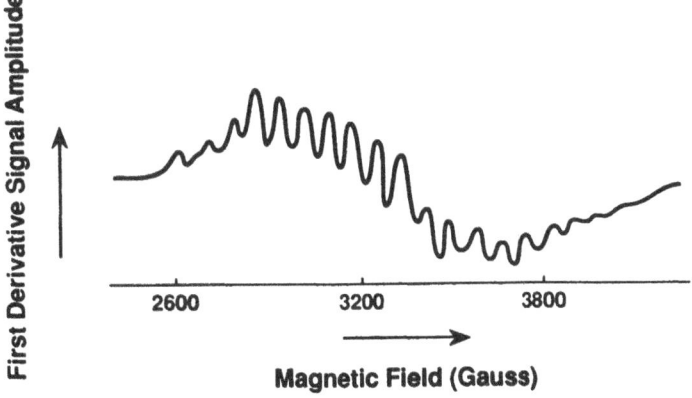

Figure 6. Electron paramagnetic resonance (EPR) spectroscopy of the oxygen clock frozen in the S_2-state. *Top* : Flash number dependence of the EPR signal; *bottom* : EPR signal at low temperature. The multiline signal is attributable to an oxidized manganese complex with unpaired electrons. The period-4 oscillation in the signal intensity versus flash number suggests that it is associated with the water-splitting reactions. Comparing the signal with those from synthetic Mn model complexes indicates that the oxygen-evolving system may contain a Mn(III) –Mn(IV) binuclear or Mn(III)$_3$ –Mn(IV) tetranuclear complex in the S_2-state (from Dismukes and Siderer, 1981).

Room-temperature EPR assays of Mn released by heating following a series of flashes have indicated a period of four oscillations in the estimated valence state of the Mn *in situ*, which suggests that the protein-bound Mn is present in both the Mn(II) and Mn(III) states, with Mn(IV) as a transient intermediate (Wydrzynski and Sauer, 1980). These results and conclusions require confirmation and extension.

Later, more direct methods were developed to observe the oxidation state of Mn by EPR at low temperature. Dismukes and Siderer (1981) reported the discovery of a 16–18 line Mn EPR signal in membranes that had been given one flash (which converts S_1 to S_2) immediately before being frozen to liquid helium temperatures. The amplitude of this signal oscillates with a period of four, indicating its involvement in the O_2 clock, but has a maximum on the first and fifth flashes, suggesting its identity with the S_2 state (Fig. 6). The signal also contains information about the electronic structure of the S_2–Mn complex. A comparison with signals obtained from model compounds suggests that the S_2 state may arise from a mixed-valence Mn(III)–Mn(IV) dimer or a Mn(III)$_3$–Mn(IV) tetramer, although other explanations cannot be excluded (see discussions by Dismukes, 1986, 1988).

An important conclusion regarding the possible association of water with the S_2–Mn complex has emerged in a related study. Isotope substitution experiments have shown that replacing $H_2{}^{16}O$ with the ^{17}O isotope broadens the low-temperature S_2 state multiline EPR signal, which suggests that an exchangeable oxygen atom is associated with the catalytic Mn during the formation of S_2 (Hansson et al., 1986). In another type of labeling experiment, $H_2{}^{16}O$ was replaced by $H_2{}^{18}O$ in samples which were poised in the S_3 state. The O_2 generated on a subsequent flash was found by mass spectroscopy to contain predominantly ^{16}O (Radmer and Ollinger, 1986). Although this experiment indicates that the bound O atoms within the catalytic site are exchangeable on the time scale of the experiment, they do not conclusively demonstrate that H_2O binds only at the level of the S_4 state. Isotope exchange experiments do indicate, however, that the Mn active site is in contact with the solvent during the S-state transitions.

An even more direct measurement of both the Mn oxidation state and the immediate physical environment of the metal has been obtained by X-ray absorption edge (XAE) and extended X-ray absorption fine structure (EXAFS) spectroscopy (see reviews by Sauer et al., 1988 and Sauer et al., 1991). EXAFS measurements have determined that there are several N or O ligands at a distance of about 1.9 Å from the Mn (Yachandra et al., 1986a; George et al., 1989). At least two Mn exist in the S_1 state as a binuclear complex, with an approximate metal-metal distance of 2.7 Å (Kirby et al., 1981). This distance is compatible with a μ_2–oxo or di–μ_2–oxo bridged structure (Kirby et al., 1981; Yachandra et al., 1986a; George et al., 1989; Guiles et al., 1990a), which has been observed in some Mn model complexes (Pecoraro, 1988). Bridges composed of μ_2–hydroxo linkages are also possible. The 2.7 Å separation remains constant in the S_0, S_1, S_2 and S_3 states (Guiles et al., 1987; Sauer et al., 1988). Polarized EXAFS measurements indicate that the 2.7 Å vector is oriented roughly perpendicular to the membrane normal (George et al., 1989). A 3.3 Å vector, oriented parallel to the membrane normal, has been attributed to a second Mn–Mn pair (see e.g., George et al., 1989). However, the low intensity of this peak does not preclude the possibility that one of the scatterers is a metal atom with a lower Z value than Mn. A peak at still greater distance (~ 4.3 Å) has also been observed, and could perhaps reflect a Mn–Ca interaction (Penner–Hahn et al., 1989).

Measurements of the X-ray absorption edge indicate that the energy increases when S_0 is converted to S_1 (Guiles et al. 1990a), and that an additional increase occurs when S_1 is converted to S_2 (Yachandra et al., 1987). These results suggest that S_2 is more oxidized than S_1, and that S_0 is more reduced than S_1; however, no further change in the position of the edge is discernible after the transition from S_2 to S_3 (Goodin et al., 1984; Sauer et al., 1988; Guiles et al., 1990b), suggesting that the oxidizing equivalent generated during this step is not stored on the Mn itself. This result raises the possibility that a redox-active component other than Mn stores the additional oxidizing equivalent in the S_3 state. Kambara and Govindjee (1985) had discussed the possibility of a redox active ligand. One candidate for this role is a histidine radical. The idea of a manganese-histidine cluster was first discussed in detail by Padhye *et al.* (1986) and has now been adopted by others (e.g., see Rutherford, 1989). An assignment of histidine residues in the Dl protein as possible ligands for manganese has been supported both theoretically (Coleman and Govindjee, 1987; Dismukes, 1988) and experimentally (*e.g.*, see Tamura *et al.*, 1989; Preston and Seibert, 1990; Ono and Inoue, 1991). Manganese may, however, be associated with other carboxylic group containing amino acids also (Coleman and Govindjee, 1987). For example, examination of site-directed D2 mutants of *Synechocystis* sp. PCC6803 suggests that Glutamic-69 of D2 may be one potential ligand to Mn involved in O_2 evolution (Vermaas *et al.*, 1990).

A third technique that has been used to study the chemical composition of the S-states (in addition to the magnetic resonance and X-ray measurements just described) is optical spectroscopy, since manganese complexes have unique absorption bands in the ultraviolet. Several laboratories have measured flash-induced ultraviolet absorption changes that are attributed to Mn (e.g., see Dekker *et al.*, 1984a, b, c; Lavergne, 1987; Renger and Hanssum, 1988). Recent evidence suggests that the $S_0 \rightarrow S_1$ transition converts an Mn(II) ion to Mn(III), but that subsequent transitions are all Mn(III) $\rightarrow$ Mn(IV) (Kretschmann *et al.*, 1988). It is clear from all of these studies that Mn undergoes dynamic changes, including changes in its oxidation state, during the S-state transitions, although the exact electronic state of Mn corresponding to each S-state remains to be discovered. It is important to note that Boussac *et al.* (1990) have observed absorbance changes, during the S_2 to S_3 conversion in Ca-depleted membranes, that are ascribable to redox changes in histidine (see above). Finally, Lavergne (1991) has suggested that this S_2 to S_3 transition in normal membranes may involve histidine or tryptophan.

Despite the many ingenious attempts to understand the role of Mn in removing electrons from water, the electrons by themselves are not the whole story. Clearly, protons must also be accounted for as products of the water oxidation reactions. Here, however, the question arises: are all four protons released at once, simultaneous with the release of O_2, or are they liberated with each S-state transition? This question was seemingly answered by careful measurements of proton release after a series of flashes, using a highly sensitive pH electrode or pH–sensitive dyes. Fowler (1977) and Saphon and Crofts (1977) demonstrated that four protons are released sequentially from the oxygen-evolving complex: one during $S_0 \rightarrow S_1$, none during $S_1 \rightarrow S_2$, one during $S_2 \rightarrow S_3$, and two during $S_3 \rightarrow S_4 \rightarrow S_0$. After making an important correction in the earlier work, Jahns *et al.* (1991) observed that H^+ release pattern is 1, 0.5, 1 and 1.5 (instead of 1, 0, 1 and 2, as mentioned above). These and other results (e.g., see Förster and Junge, 1985; Renger *et al.*, 1987) have important implications for the mechanism of the O_2 clock, although the interpretation depends on whether the protons originate directly from water or from some other source (e.g., the polypeptides). Sequential extraction of protons from bound water implies that the substrate ligand itself (H_2O) is undergoing chemical changes prior to the S_4 state. However, if proton release from H_2O occurs in concert with O_2 release, then this implies that no water oxidation occurs until the final S-state transition. A hybrid mechanism (stepwise proton release from water via the polypeptides) is also conceivable. It is too early to relate H^+ release pattern to water chemistry.

VI. Function of Chloride

The higher S-states accumulate some sort of net positive charge (particularly in S_2, since less H^+ release is observed during the $S_1 \rightarrow S_2$ transition). Stabilization of this positive charge may explain the essential role of chloride ions (Cl^-) in keeping the O_2 clock running. Izawa and co-workers (Izawa et al., 1969; Hind et al., 1969; Kelley and Izawa, 1978) demonstrated that Cl^- activates the oxygen-evolving complex. Using ^{35}Cl–NMR Critchley et al. (1982) and Baianu et al. (1984) showed that Cl^- associates and dissociates from the thylakoid membranes of halophytes freely and rapidly (with an exchange rate of greater than 3×10^4 s^{-1} at room temperature). These findings led to the proposal that Cl^- binding might be associated with the arrival of a positive charge on the oxygen-evolving complex from the oxidized reaction center $P680^+$ and that its release might coincide with the release of protons (Govindjee et al., 1983). ^{35}Cl–NMR experiments by Preston and Pace (1985) suggest that Cl^- may bind more tightly in the S_2 and S_3 states than in the S_0 and S_1 states, a finding which is consistent with the more positively charged character of the higher S-states. Homann et al. (1986) found that when Cl^- is absent, excitation with a single flash produces an abnormal S_2 state. EPR measurements also indicate that Cl^--depleted PS II membranes form an abnormal S_2-state, whose low-temperature spectrum has a single broad peak at $g = 4.1$ instead of the multiline signal centered at $g = 2$ (Ono et al., 1986). However, the $g = 4.1$ signal can be converted into the normal multiline signal by adding Cl^- (Ono et al., 1986). Thus, Cl^- might be required for the interconversion of the S_2-state intermediate into the final relaxed state.

Despite its apparent involvement in the water oxidation reactions, direct binding of chloride to the Mn in the lower S-states was considered doubtful since EXAFS data obtained by Yachandra et al. (1986b) did not indicate the presence of Cl^- within 2–3 Å of the Mn atoms. Several EPR studies of the S_2-state also failed to detect Cl^- within the coordination sphere of Mn (see review by Govindjee and Homann, 1989). However, M. Klein and co-workers (personal communication) have now observed signals that they attribute to Cl^- near Mn. Homann (1987) and Coleman and Govindjee (1987) have suggested that Cl^- most likely binds to positively-charged and other specific amino acids on the oxygen-evolving complex proteins. Within the last few years, ^{35}Cl–NMR has been used to observe Cl^- binding to PS II from spinach (see Coleman et al., 1987). These measurements suggest that a fairly large number of Cl^- ions bind to the oxygen-evolving complex. The bound ions appeared to be distributed between two major binding sites: one near the catalytic Mn, perhaps on the D1 and D2 polypeptides, and the other on the 33 kilodalton polypeptide. In contrast, Wydrzysnki et al. (1990a), also using ^{35}Cl–NMR measurements, conclude that only a single class of exchangeable chloride interaction sites occur in photosystem II.

The function of Cl^- may be to expedite the release of protons from water, and thus increase the efficiency of the water oxidation reactions, or to stabilize the Mn ions in the higher S-states. However, it is not yet clear whether the Cl^- effects, thus far studied, are specific chemical/catalytic effects of these ions or they reflect effects of the organization of the surface properties of the proteins, i.e., conformational changes (e.g., see Wydrzynski *et al.*, 1990b). Thus, the real role of Cl^- in photosystem II remains unknown.

Coleman (1990) has summarized the role of Cl^- in *in vitro* enzyme systems. Perhaps, we should dig deeper into this problem by comparing these systems with the *in vivo* system.

VII. Role of Calcium

Calcium appears to be intimately involved in the function of Cl^-, and seems to be required both for the oxidation of water and for the operation of the PS II reaction center, perhaps in a structural or regulatory capacity. Piccioni and Mauzerall (1976) first demonstrated a potential role for Ca^{2+} in O_2 evolution in cyanobacteria. Ca^{2+} can replace the function of the 17 and 24 kilodalton polypeptides in O_2 evolution in thylakoids from higher plants (see Ghanotakis *et al.*, 1984; Ono and Inoue, 1984, Boussac *et al.*, 1985; Homann, 1988). Calcium has been shown to play an important role in controlling other proteins by acting as an intracellular messenger (switching on and off their activity) and by maintaining their three-dimensional structure (Gerday *et al.*, 1988). However, the real role of Ca^{2+} in oxygen evolution remains to be established. Removal of calcium does result in a dramatic inhibition of O_2 evolution, and this effect is reversed upon the readdition of Ca^{2+} (e.g., see Murata and Miyao, 1985). Boussac and Rutherford (1988) have shown that NaCl-induced removal of Ca^{2+} results in the inhibition of the S_3 to the S_0 step, i.e., the step in which O_2 is released. Rutherford (1989) has discussed an interesting possibility in which Ca^{2+} is located close to the Mn cluster in the oxygen-evolving complex; he imagines that Ca^{2+} acts as a shuttle in bringing H_2O and Cl^- at the right moment, i.e., at the S_4 stage, to manganese involved in water oxidation. Furthermore, Sivaraja *et al.* (1989) suggest, on the basis of citric acid induced release of Ca^{2+}, that calcium serves the purpose of a "gate keeper" for the access of substrate water to the catalytic manganese. However, as yet, no direct evidence exists that proves any direct and specific role of Ca^{2+} in the O_2 evolution process. For reviews, see Homann (1990) and Yocum (1991).

In conclusion, although a great deal of fog has been lifted from water oxidation being a mysterious reaction, yet its mechanism has only begun to be understood at the molecular level. Although manganese clearly undergoes redox changes, and histidine appears to be involved at one of the steps, yet the roles of calcium and chloride are obscure. New methods need to be invented to monitor the intermediates of water oxidation. Also, there is much need to improve methods to prepare the simplest, yet stable, reaction center complexes that will retain not only quinones, but also manganese. Such simple preparations will help us obtain meaningful data on the molecular mechanism of O_2 evolution.

VIII. References

Andersson B and Styring S (1991) Photosystem II: Molecular organization, function, and acclimation. *Current Topics in Bioenergetics.* **16:** 1–81

Andréasson L–E and Vanngård T (1988) Electron transport in photosystem I and II. *Annu Rev Plant Physiol Plant Mol Biol* **39:** 379–411

Arnon DI and Barber J (1990) Photoreduction of $NADP^+$ by isolated reaction centers of photosystem II: Requirement for plastocyanin. *Proc Natl Acad Sci USA* **87:** 5930–5934

Babcock GT (1987) The photosynthetic oxygen-evolving process. In: (J Amesz, ed.) *Photosynthesis*, Elsevier Science Publishers BV Biomed Div, Amsterdam, pp 125–158

Babcock GT, Barry BA, Debus RJ, Hoganson CW, Atamian M, McIntosh L, Sithole I and Yocum CF (1989) Water oxidation in photosystem II: From radical chemistry to multielectron chemistry. *Biochemistry* **28:** 9557–9565

Baianu IC, Critchley C, Govindjee and Gutowsky HS (1984) NMR study of chloride ion interactions with thylakoid membranes. *Proc Natl Acad Sci USA* **81:** 3713–3717

Barry BA and Babcock GT (1987) Tyrosine radicals are involved in the photosynthetic oxygen-evolving system. *Proc Natl Acad Sci USA* **84:** 7099–7103

Bouges–Bocquet B (1973) Limiting steps in photosystem II and water decomposition in *Chlorella* and spinach chloroplasts. *Biochim Biophys Acta* **292:** 772–785

Boussac A and Rutherford AW (1988) Nature of the inhibition of the oxygen-evolving enzyme of photosystem II induced by NaCl washing and reversed by the addition of Ca^{2+} or Sr^+. *Biochemistry* **27:** 3476–3483

Boussac A, Zimmerman, J–L, Rutherford AW and Lavergne J (1990) Histidine oxidation in the oxygen-evolving photosystem II enzyme. *Nature* (London) **347:** 303–306

Boussac A, Maison–Peteri B, Etienne, A–L and Vernotte C (1985) Reactivation of oxygen evolution of NaCl-washed photosystem II particles by Ca^{2+} and/or the 24 kDa protein. *Biochim Biophys Acta* **808:** 231–234

Brettel K, Schlödder E and Witt HT (1984) Nanosecond reduction kinetics of photooxidized chlorophyll a_{II} (P680) in single flashes as a probe for the electron pathway, H^+ release and charge accumulation in the O_2-evolving complex. *Biochim Biophys Acta* **766:** 403–428

Brudvig GW, Beck WF and dePaula J (1989) Mechanism of photosynthetic water oxidation. *Annu Rev Biophys Biophys Chem* **18:** 25–46

Cavalier–Smith T (1987) The origin of cells: A symbiosis between genes, catalysts, and membranes. *Cold Spring Harbor Symp Quant Biol* **52:** 805–824

Coleman WJ (1990) Chloride binding proteins: Mechanistic implications for the oxygen-evolving complex of photosystem II. *Photosynth Res* **23:** 1–28

Coleman WJ and Govindjee (1987) A model for the mechanism of chloride activation of oxygen evolution in photosystem II. *Photosynth Res* **13:** 199–223

Coleman WJ and Youvan DC (1990) Spectroscopic analysis of genetically modified photosynthetic reaction centers. *Annu Rev Biophys Biophys Chem* **19:** 333–367

Coleman WJ, Govindjee and Gutowsky HS (1987) The location of the chloride binding sites in the oxygen evolving complex of spinach photosystem II. *Biochim Biophys Acta* **894:** 453–459

Cramer WA, Furbacher PN, Szczepaniak and Tae GS (1991) Electron transport between photosystem II and photosystem I. *Current Topics in Bioenergetics* **16:** 179–222

Critchley C, Baianu IC, Govindjee and Gutowsky HS (1982) The role of chloride in O_2 evolution by thylakoids from salt-tolerant higher plants. *Biochim Biophys Acta* **682:** 436–445

Crofts AR and Wraight C (1983) The electrochemical domain of photosynthesis. *Biochim Biophys Acta* **726:** 149–185

Danelius RV, Satoh K, van Kan PJM, Plijter JJ, Nuijs AM and van Gorkom HJ (1987) The primary reaction of photosystem II in the D1–D2-cytochrome *b-559* complex. *FEBS Lett* **213:** 241–244

Debus RJ, Barry BA, Babcock GT and McIntosh L (1988a) Site-directed mutagenesis identifies a tyrosine radical in the photosynthetic oxygen-evolving system. *Proc Natl Acad Sci USA* **85:** 427–430

Debus RJ, Barry BA, Sithole I, Babcock GT and McIntosh L (1988b) Directed mutagenesis indicates that the donor to P^+680 in Photosystem II is tyrosine-161 of the D1 polypeptide. *Biochemistry* **27:** 9071–9074

Deisenhofer J and Michel H (1988) The photosynthetic reaction center for the purple bacterium *Rhodopseudomonas viridis*. The Nobel lecture. *EMBO J* **8**: 2149–2170

Dekker JP, Plijter JJ, Ouwenhand L and Van Gorkom HJ (1984a) Kinetics of manganese redox transitions in the oxygen-evolving apparatus of photosynthesis. *Biochim Biophys Acta* **767**: 176–179

Dekker JP, Van Gorkom HJ, Brok M and Ouwenhand L (1984b) Optical characterization of photosystem II electron donors. *Biochim Biophys Acta* **764**: 301–309

Dekker JP, Van Gorkom HJ, Wensink J and Ouwenhand L (1984c) Absorbance difference spectra of the successive redox states of the oxygen-evolving apparatus of photosynthesis. *Biochim Biophys Acta* **767**: 1–9

Delrieu M-J (1974) Simple explanation of the misses in the cooperation of charges in photosynthetic O_2 evolution. *Photochem Photobiol* **20**: 441–454

Diner B (1986) The reaction center of photosystem II. *Encycl Plant Physiol*, New Series **19**: 422–436

Dismukes GC (1986) The metal centers of the photosynthetic oxygen-evolving complex. *Photochem Photobiol* **43**: 99–115

Dismukes GC (1988) The spectroscopically derived structure of the manganese site for photosynthetic water oxidation and a proposal for the protein binding sites for calcium and manganese. *Chemica Scripta* **28A**: 99–104

Dismukes GC and Siderer Y (1981) Intermediates of a polynuclear manganese center involved in photosynthetic oxidation of water. *Proc Natl Acad Sci USA* **78**: 274–278

Eckert HJ, Weise N, Bernarding J, Eichler HJ and Renger G (1988) Analysis of the electron transfer from Pheo⁻ to Q_A in PS II membrane fragments from spinach by time resolved 325 nm absorption changes in the picosecond domain. *FEBS Lett* **240**: 153–158

Emerson R and Arnold W (1932a) A separation of the reactions in photosynthesis by means of intermittent light. *J Gen Physiol* **15**: 391–420

Emerson R and Arnold W (1932b) The photochemical reaction in photosynthesis. *J Gen Physiol* **16**: 191–205

Feher G, Allen JP, Okamura MY and Reese DC (1989) Structure and function of bacterial photosynthetic reaction centres. *Nature* (London) **339**: 111–116

Forbush B, Kok B and McGloin M (1971) Cooperation of charges in photosynthetic O_2 evolution—II. Damping of flash yield oscillation, deactivation. *Photochem Photobiol* **14**: 307–321

Förster V and Junge W (1985) Stoichiometry and kinetics of proton release upon photosynthetic water oxidation. *Photochem Photobiol* **41**: 183–190

Fowler CF (1977) Proton evolution from photosystem II. Stoichiometry and mechanistic considerations. *Biochim Biophys Acta* **462**: 414–421

Gerday Ch., Bolis L and Gilles R, eds. (1988) Calcium and calcium binding proteins. Springer Verlag, New York

George GN, Prince RC and Cramer SP (1989) The manganese site of the photosynthetic water-splitting enzyme. *Science* **243**: 789–791

Ghanotakis DF and Yocum CF (1985) Polypeptides of photosystem II and their role in oxygen evolution. *Photosynth Res* **7**: 97–114

Ghanotakis DF and Yocum CF (1990) Photosystem II and the oxygen evolving complex. *Annu Rev Plant Physiol Plant Mol Biol* **41**: 255–276

Ghanotakis DF, Babcock GT and Yocum CF (1984) Calcium reconstitutes high levels of oxygen evolution in polypeptide depleted photosystem II preparations. *FEBS Lett* **167**: 127–130

Golbeck JH and Bryant D (1991) Photosystem I. *Current Topics in Bioenergetics* **16**: 83–177

Goodin DB, Yachandra VK, Britt RD, Sauer K and Klein MP (1984) The state of manganese in the photosynthetic apparatus. 3. Light-induced changes in the X-ray absorption (K-edge) energies of manganese in photosynthetic membranes. *Biochim Biophys Acta* **767**: 209–216

Govindjee (1980) The oxygen evolving system of photosynthesis. *Plant Biochem J* (India) Sircar Memorial Volume: 7–30

Govindjee, ed. (1982) *Photosynthesis: Energy Conversion by Plants and Bacteria*, Vol 1, Academic Press, New York

Govindjee (1984) Photosystem II: The oxygen evolving system of photosynthesis. In: (C. Sybesma, ed) *Advances in Photosynthesis Research*, Vol 1, Martinus Nijhoff/Dr. W. Junk Publishers, The Hague, pp 237–338

Govindjee and Coleman W (1990) How plants make oxygen. *Scientific American* **262**: 50–58

Govindjee and Eaton–Rye JJS (1986) Electron transfer through photosystem II acceptors: Interaction with anions. *Photosynth Res* **10**: 365–379

Govindjee and Govindjee R (1974) Primary events in photosynthesis. *Scientific American* **231**: 68–82

Govindjee and Homann PH (1989) Function of chloride in water oxidation in photosynthesis. In: (A Kotyk, ed) *Highlights of Modern Biochem.* Vol 1, VSP, Utrecht, pp 933–961

Govindjee and Wasielewski MR (1989) Photosystem II: From a femtosecond to a millisecond. In: (WR Briggs, ed) *Photosynthesis*, Alan R. Liss, Inc., New York, pp 71–103

Govindjee, Baianu IC, Critchley C, and Gutowsky HS (1983) Comments on the possible roles of bicarbonate and chloride ions in photosystem II. In: (Y Inoue, AR Crofts, Govindjee, N Murata, G Renger, and K Satoh, eds) *The Oxygen Evolving System of Photosynthesis.* Academic Press, Tokyo, pp 283–292

Govindjee, Kambara T and Coleman W (1985) The electron donor side of photosystem II: The oxygen evolving complex. *Photochem Photobiol* **42**: 187–210

Guiles RD, Yachandra VK, McDermott A, Cole J, Britt RD, Dexheimer SL, Sauer K and Klein MP (1987) Structural features of the manganese cluster in different state of the oxygen evolving complex to photosystem II: An X-ray absorption spectroscopy study. In: (J Biggins, ed) *Progress in Photosynthesis Research*, Vol 1, Martinus Nijhoff, Dordrecht, pp 561–564

Guiles RD, Yachandra VK, McDermott AE, Cole, JL, Dexheimer SL, Britt RD, Sauer K and Klein M (1990a) The S_0 state of photosystem II induced by hydroxylamine: Differences between the structure of the manganese complex in the S_0 and S_1 states determined by X-ray absorption spectroscopy. *Biochemistry* **29**: 486–496

Guiles RD, Zimmermann J–L, McDermott, AE, Yachandra VK, Cole JL, Dexheimer SL, Britt RD, Wieghardt K, Bossek U, Sauer K and Klein MP (1990b) The S_3 state of photosystem II: Differences between the structure of the manganese complex in the S_2 and S_3 states determined by X-ray absorption spectroscopy. *Biochemistry* **29**: 471–485

Haddy A, Aasa R and Andréasson L–E (1989) S-band EPR studies of the S_2-state multiline signal from the photosynthetic oxygen-evolving complex. *Biochemistry* **28**: 6954–6959

Hansson Ö and Wydrzynski T (1990) Current perceptions of photosystem II. *Photosynth Res* **23**: 131–162

Hansson Ö, Andréasson L–E and Vänngård T (1986) Oxygen from water is coordinated to manganese in the S_2 state of photosystem II. *FEBS Lett* **195**: 151–154

Hansson Ö, Duranton J and Mathis P (1988) Yield and lifetime of the primary radical pair in preparations of photosystem II with different antenna size. *Biochim Biophys Acta* **932**: 91–96

Hind G, Nakatani HY and Izawa S (1969) The role of chloride in photosynthesis. The chloride requirement of electron transport. *Biochim Biophys Acta* **172**: 277–289

Holzwarth AR (1987) Picosecond fluorescence spectroscopy and energy transfer in photosynthetic antenna pigments. *Topics Photosynth* **8**: 95–157

Homann PH (1987) The relations between the chloride, calcium and polypeptide requirements of photosynthetic water oxidation. *J. Bioenerg Biomembr* **19**: 105–123

Homann PH (1988) The chloride and calcium requirement of photosynthetic water oxidation: Effects of pH. *Biochim Biophys Acta* **934**: 1–13

Homann PH (1990) The role of calcium in photosynthetic water oxidation. In: (LJ Anghileri, ed) *The Role of Calcium in Biological Systems*, Vol. V, CRC Press, Inc, Boca Raton, FL, pp. 79–96

Homann PH, Gleiter H, Ono T and Inoue Y (1986) Storage of abnormal oxidants Σ_1, Σ_2 and Σ_3 in photosynthetic water oxidases inhibited by chloride removal. *Biochim Biophys Acta* **850**: 10–20

Izawa S, Heath RL and Hind G (1989) The role of the chloride ion in photosynthesis. III. The effect of artificial electron donors upon electron transport. *Biochim Biophys Acta* **180**: 388–398

Jagendorf AT, McCarty RE and Robertson D (1991) Coupling factor components: structure and function. In: (L Bogorad and IK Vasil, eds.) *The Photosynthetic Apparatus: Molecular Biology and Operation.* Academic Press, SanDiego, pp. 225–254

Jahns P, Lavergne J, Rappaport F and Junge W (1991) Stoichiometry of proton release during photosynthetic water oxidation: a reinterpretation of the responses of neutral red leads to a non-integer pattern. *Biochim Biophys Acta* **1057**: 313–319

Joliot P and Kok B (1975) Oxygen evolution in photosynthesis. In: (Govindjee, ed.) *Bioenergetics of Photosynthesis.* Academic Press, New York, pp 387–412

Joliot P, Hofnung M and Chabaud R (1966) Étude de l'émission d'oxygène par des algues soumises a un éclairement modulé sinusoidalement. *J Chim Phys* **63**: 1423–1441

Joliot P, Barbieri G and Chabaud R (1969) Un nouveau modèle des centres photochimique du système II. *Photochem Photobiol* **10**: 309–329

Kambara T and Govindjee (1985) Molecular mechanism of water oxidation in photosynthesis based on the functioning of manganese in two different environments. *Proc Natl Acad Sci USA* **82**: 6119–6123

Kelley PM and Izawa S (1978) The role of chloride ion in photosystem II. I Effects of chloride ion on photosystem II electron transport and on hydroxylamine inhibition. *Biochim Biophys Acta* **502**: 198–210

Kirby, JA, Robertson AS, Smith JP, Thompson AC, Cooper SR and Klein MP (1981) State of manganese in the photosynthetic apparatus. 1. Extended X-ray absorption fine structure studies on chloroplasts and di-µ-oxo-bridged dimanganese model compounds. *J Am Chem Soc* **103**: 5529–5537

Kok B and Cheniae GM (1966) Kinetics and intermediate steps of the oxygen evolving step in photosynthesis. In: (DR Sanadi, ed) *Current Topics in Bioenergetics*, Vol I, Academic Press, New York, pp 1–97

Kok B, Forbush B and McGloin M (1970) Cooperation of charges in photosynthetic oxygen evolution: I. A linear four step mechanism. *Photochem Photobiol* **11**: 457–475

Kretschmann H, Dekker JP, Saygin O and Witt HT (1988) An agreement on the quaternary oscillation of ultraviolet absorption changes accompanying the water splitting in isolated photosystem II complexes from the cyanobacterium *Synechococcus*. *Biochim Biophys Acta* **932**: 358–361

Lavergne J (1987) Optical difference spectra of the S-state transitions in the photosynthetic oxygen evolving complex. *Biochim Biophys Acta* **894**: 91–107

Lavergne J (1991) Improved UV-visible spectra of the S-transitions in the photosynthetic oxygen-evolving system. *Biochim Biophys Acta* **1060**: 175–188

Mar T and Govindjee (1972) Kinetic models of oxygen evolution in photosynthesis. *J Theoret Biol* **36**: 427–446

Mathis P (1986) Structural aspects of vectorial electron transfer in photosynthetic reaction centers. *Photosynth Res* **8**: 97–111

Metz JG, Nixon PJ, Rögner M, Brudvig GW and Diner BA (1989) Directed alteration of the D1 polypeptide of photosystem II: Evidence that tyrosine-161 is the redox component Z, connecting the oxygen-evolving complex to the primary electron donor, P680, *Biochemistry* **28**: 6960–6969

Michel H (1982) Three-dimensional crystals of a membrane protein complex. The photosynthetic reaction centre from *Rhodopseudomonas viridis. J Mol Biol* **158**: 567–572

Michel H and Deisenhofer J (1988) Relevance of the photosynthetic reaction center from purple bacteria to the structure of Photosystem II. *Biochemistry* **27**: 1–7

Moya I, Sebban P and Haehnel W (1986) Lifetime of excited states and quantum yield of chlorophyll *a* fluorescence *in vivo*. In: (Govindjee, J Amesz, and DC Fork, eds) *Light Emission by Plants and Bacteria*, Academic Press, Orlando, pp 161–190

Murata N and Miyao M (1985) Extrinsic membrane proteins in the photosynthetic oxygen-evolving complex. *Trends in Biochem Sciences (TIBS)* **10**: 122–124

Nanba O and Satoh K (1987) Isolation of a photosystem II reaction center consisting of D-1 and D-2 polypeptides and cytochrome *b*-559. *Proc Natl Acad Sci USA* **84**: 109–112

Nuijs AM, van Gorkom HJ, Plijter JJ and Duysens LNM (1986a) Primary charge separation and excitation of chlorophyll *a* in photosystem II particles from spinach as studied by picosecond absorption difference spectroscopy. *Biochim Biophys Acta* **848**: 167–175

Nuijs AM, Shuvalov VA, van Gorkom HJ, Plijter JJ and Duysens LNM (1986b) Picosecond absorbance difference spectroscopy on primary reactions and the antenna excited states in photosystem I particles. *Biochim Biophys Acta* **850**: 310–318

Okamura MY, Satoh K, Isaacson RA and Feher G (1987) Evidence of the primary charge separation in D_1D_2 complex of photosystem II from spinach: EPR of the triplet state. In: (J Biggins, ed) *Progress in Photosynthesis Research*, Vol 1, Martinus Nijhoff, The Hague, pp 379–391

Ono T and Inoue Y (1984) Ca^{2+}-dependent restoration of O_2 evolving activity in $CaCl_2$-washed PS II particles depleted of 33, 24 and 16 kDa polypeptides. *FEBS Lett* **168**: 281–286

Ono T and Inoue Y (1991) A possible role of redox-active histidine in the photoligation of manganese into a photosynthetic O_2-evolving enzyme. *Biochemistry* **30**: 6183–6188

Ono T, Zimmerman JL, Inoue Y and Rutherford AW (1986) EPR evidence for a modified S-state transition in chloride-depleted photosystem II. *Biochim Biophys Acta* **851**: 193–201

106

Ort D (1986) Energy transduction in oxygenic photosynthesis: An overview of structure and mechanism. *Encycl Plant Physiol*, New Series **19**: 143–196

Ort DR and Melandri BR (1982) Mechanism of ATP synthesis. In: (Govindjee, ed) *Photosynthesis: Energy Conversion by Plants and Bacteria*, Vol 1, Academic Press, New York, pp 537–587

Padhye S, Kambara T, Hendrickson DFN and Govindjee (1986) Manganese-histidine cluster as the functional center of the water oxidation complex in photosynthesis. *Photosynth Res* **9**: 103–112

Pecoraro VL (1988) Structural proposals for the manganese centers of the oxygen evolving complex. An inorganic chemist's perspective. *Photochem Photobiol* **48**: 249–264

Penner-Hahn JE, Fronko RM, Yocum CF, Betts SD and Bowlby NR (1989) X-ray absorption spectroscopy of the manganese sites in the photosynthetic oxygen evolving complex (Abstract No. 793). *Physiol Plant* **76**: A143

Piccioni R and Mauzerall D (1976) Increase effected by calcium ion in the rate of oxygen evolution from preparations of *Phormidium luridum*. *Biochim Biophys Acta* **423**: 605–609

Preston C and Pace RJ (1985) The S-state dependence of chloride binding to plant photosystem II. *Biochim Biophys Acta* **810**: 388–391

Preston C and Seibert M (1990) Partial identification of the high-affinity Mn-binding site in *Scenedesmus obliquus* photosystem II. In: (M Baltscheffsky, ed) *Current Research in Photosynthesis*, vol II, Kluwer Acad Publ, Dordrecht, The Netherlands pp 423-426

Radmer R and Ollinger D (1986) Do the higher oxidation states of the photosynthetic O_2-evolving system contain bound H_2O? *FEBS Lett* **195**: 285–289

Rao KK, Hall DO and Cammack R (1981) The photosynthetic apparatus. In: (H Gutfreund, ed) *Biochemical Evolution*, Cambridge University Press, Cambridge, pp 150–202

Reed GH (1986) Manganese: An overview of chemical properties. In: (VL Schramm and FC Wedler, eds.) *Manganese in Metabolism and Enzyme Function*, Academic Press, Orlando, pp 313–325

Reed GH and Markham GC (1984) EPR studies of Mn(II) complexes with enzymes and proteins. *Biol Magn Res* **6**: 73–142

Renger G (1987a) Mechanistic aspects of photosynthetic water cleavage. *Photosynthetica* **21**: 203–224

Renger G (1987b) Biological exploitation of solar energy by photosynthetic water splitting. *Angew Chemie* (Int Eng Ed) **26**: 643–660

Renger G (1988) On the mechanism of photosynthetic water oxidation to dioxygen. *Chemica Scripta* **28A**: 105–109

Renger G and Govindjee (1985) The mechanism of photosynthetic water oxidation. *Photosynth Res* **6**: 33–55

Renger G and Hanssum B (1988) Studies on the deconvolution of flash-induced absorption changes into the difference spectra of individual redox steps in the water-oxidizing enzyme system. *Photosynth Res* **16**: 243–259

Renger G, Wacker U and Völker M (1987) Studies on the protolytic reactions coupled with water cleavage in photosystem II membrane fragment from spinach. *Photosynth Res* **13**: 167–189

Rutherford AW (1989) Photosystem II, the water splitting enzyme. TIBS **14**: 227–232

Saphon S and Crofts AR (1977) Protolytic reactions in photosystem II: A new model for the release of protons accompanying the photooxidation of water. *Z Naturforsch* **32C**: 617–626

Sauer K, Guiles RD, McDermott AE, Cole JL, Yachandra VY, Zimmerman J-L, Klein MP, Dexheimer SL and Britt RD (1988) Spectroscopic studies of manganese involvement in photosynthetic oxygen evolution. *Chemica Scripta* **28A**: 87–91

Sauer K, Yachandra VK, Britt RD and Klein MP (1991) The photosynthetic water oxidation complex studied by EPR and X-ray absorption spectroscopy. In: (VL Pecoraro, ed) *Manganese Redox Enzymes*. VCH Publishers, N.Y., 47 pages, in press

Schatz GH, Brock H and Holzwarth AR (1987) Picosecond kinetics of fluorescence and absorbance changes in photosystem II particles excited at low photon density. *Proc Natl Acad Sci USA* **84**: 8414–8418

Schlödder E, Brettel K and Witt (1985) Relation between microsecond reduction kinetics of photooxidized chlorophyll a_{II} (P680) in photosystem II particles from *Synechococcus* sp. *Biochim Biophys Acta* **808**: 1232–131

Sivaraja M, Tso, J and Dismukes GC (1989) A calcium-specific site influences the structure and activity of the manganese cluster responsible for photosynthetic water oxidation. *Biochemistry* **28**: 9459–9464

Sonneveld A, Rademaker H and Duysens LNM (1979) Chlorophyll *a* fluorescence as a monitor of nanosecond reduction of the photooxidized primary donor P-680$^+$ of photosystem II. *Biochim Biophys acta* **548**: 536–551

Srinivasan AN and Sharp RR (1986a) Flash-induced enhancements in the proton NMR relaxation rate of photosystem II particles. *Biochim Biophys Acta* **850**: 211–217

Srinivasan AN and Sharp RR (1986b) Flash-induced enhancements in the proton NMR relaxation rate of photosystem II particles: Response to flash trains of 1-5 flashes. *Biochim Biophys Acta* **851**: 369–376

Styring S and Rutherford AW (1987) In the oxygen evolving complex of photosystem II the S_0 state is oxidized to the S_1 state by D^+ (Signal II$_{slow}$). *Biochemistry* **26**: 2401–2405

Takahashi Y, Hansson O, Mathis P and Satoh K (1987) Primary radical pair in the photosystem II reaction center. *Biochim Biophys Acta* **893**: 49–59

Tamura N, Ikeuchi M and Inoue Y (1989) Assignment of histidine residues in D1 protein as possible ligands for functional manganese in photosynthetic water-oxidizing complex. *Biochim Biophys Acta* **973**: 281–289

Thielen APGM and van Gorkom HJ (1981) Energy transfer and quantum yield in photosystem II. *Biochim Biophys Acta* **637**: 439–446

Turco RP (1985) The photochemistry of the stratosphere. In: (JS Levine, ed) *The photochemistry of Atmospheres*, Academic Press, Orlando, pp 77–128

van Best JA and Mathis P (1978) Kinetics of the oxidized primary electron donor of photosystem II in spinach chloroplasts and in *Chlorella* cells in the microsecond and nanosecond time ranges following flash excitation. *Biochim Biophys Acta* **503**: 178—188

van Gorkom HJ (1985) Electron transfer in photosystem II. *Photosynth Res* **6**: 97–112

Veizer J (1988) The earth and its life: Systems perspective. *Origins of Life Evol Bios* **18**: 13–39

Velthuys B (1980) Mechanisms of electron flow in photosystem II and towards photosystem I. *Annu Rev Plant Physiol* **31**: 545–567

Velthuys B (1987) Photosystem II reaction center. *Topics Photosynth* **8**: 341–378

Vermaas WFJ and Govindjee (1981) The acceptor side of photosystem II in photosynthesis. *Photochem Photobiol* **34**: 775–793

Vermaas WFJ and Ikeuchi M (1991) Photosystem II. In: (L Bogorad and IK Vasil, eds) *The Photosynthetic Apparatus: Molecular Biology and Operation*, Academic Press, San Diego, pp. 25–111

Vermaas WFJ, Renger G and Dohnt A (1984) The reduction of the oxygen evolving system in chloroplasts by thylakoid components. *Biochim Biophs Acta* **764**: 194–202

Vermaas WFJ, Rutherford AW, and Hansson Ö (1988) Site-directed mutagenesis in photosystem II of the cyanobacterium, *Synechocystis* sp. PCC 6803: The donor D is a tyrosine residue in the D2 protein. *Proc Natl Acad Sci USA* **85**: 8477–8481

Vermaas WFJ, Charité J and Shen G (1990) Glu-69 of the D2 protein in photosystem II is a potential ligand to Mn involved in photosynthetic oxygen evolution. *Biochemistry* **29**: 5325–5332

Wasielewski MR, Johnson DG, Seibert M and Govindjee (1989) Determination of the primary charge separation rate in isolated photosystem II reaction centers with 500 femtosecond time resolution. *Proc Natl Acad Sci USA* **86**: 524–528

Wayne RP (1988) *Principles and Applications of Photochemistry*. Oxford University Press, Oxford, pp 181–191

Weiss C and Sauer K (1970) Activation kinetics of photosynthetic oxygen evolution under 20–40 nanosecond laser flashes. *Photochem Photobiol* **11**: 495–501

Wydrzynski T (1982) Oxygen evolution in photosynthesis. In: (Govindjee, ed) *Photosynthesis: Energy Conversion by Plants and Bacteria*, Vol I, Academic Press, New York, pp 469–506

Wydrzynski T and Sauer K (1980) Periodic changes in the oxidation state of manganese in photosynthetic oxygen evolution upon illumination with flashes. *Biochim Biophys Acta* **589**: 56–70

Wydrzynski T, Zumbulyadis N, Schmidt PG and Govindjee (1975) Water proton relaxation as a monitor of membrane-bound manganese in spinach chloroplasts. *Biochim Biophys Acta* **408**: 349–354

Wydrzynski T, Marks SB, Schmidt PG, Govindjee and Gutowsky HS (1978) Nuclear magnetic relaxation by the manganese in aqueous suspensions of chloroplasts. *Biochemistry* **17**: 2155–2162

Wydrzynski T, Ångstrom J, Baumgart F, Renger G and Vanngård T (1990a) ^{35}Cl–NMR linewidth measurements of aqueous suspensions of photosystem II membrane fragments reveal only a simple hyperbolic dependence with chloride concentration. *Biochim Biophys Acta* **1018**: 55–60

Wydrzynski T, Baumgart F, MacMillan F and Renger G (1990b) Is there a direct chloride cofactor requirement in the oxygen-evolving reactions of photo-system II? *Photosynth Res* **25**: 59–72

Yachandra VK, Guiles RD, McDermott A, Britt RD, Dexheimer SL, Sauer K and Klein MP (1986a) The state of manganese in the photosynthetic apparatus. 4. Structure of the manganese complex in photosystem II studies using EXAFS spectroscopy. The S_1 state of the O_2-evolving photosystem II complex from spinach. *Biochim Biophys Acta* **850**: 324–332

Yachandra VK, Guiles RD, Sauer K and Klein MP (1986b) The state of Mn in the photosynthetic apparatus. 5. The chloride effect in photosynthetic oxygen evolution. Is halide coordinated to the EPR-active Mn in the oxygen evolving complex? Studies on the substructure of the low temperature multiline EPR signal. *Biochim Biophys Acta* **850**: 333–342

Yachandra VK, Guiles RD, McDermott AE, Cole JL, Britt DR, Dexheimer SL, Sauer K and Klein MP (1987) Comparison of the structure of the manganese complex in the S_1 and S_2 states of the photosynthetic O_2-evolving complex: An X-ray absorption spectroscopy study. *Biochemistry* **26**: 5974–5981

Yocum CF (1991) Calcium activation of photosynthetic water oxidation. *Biochim Biophys Acta* **1059**: 1–15

Yocum CF, Yerkes CT, Blankenship R, Sharp RR and Babcock GT (1981) Stoichiometry, inhibitor sensitivity, and organization of manganese associated with photosynthetic oxygen evolution. *Proc Natl Acad Sci USA* **78**: 7507–7511

4

Stoichiometry of Proton Uptake by Thylakoids During Electron Transport in Chloroplasts

Boris N. Ivanov

Institute of Soil Science and Photosynthesis
Russian Academy of Sciences, Pushchino,
Moscow Region, 142292, Russia

CONTENTS

ABBREVIATIONS

b_6^r	:	the heme group of cytochrome b_6, associated with site ''r'' quinone reducing site of bf complex
b_6^o	:	the heme group of cytochrome b_6, associated with site ''c'', quinoloxidizing site of bf complex;
DBMIB	:	2,5-dibromo-3-methyl-6-isopropyl-p-benzoquinone
DCCD	:	N,N'-dicyclohexylcarbodiimide
DCMU	:	3-(3',4'-dichlorophenyl)-1-1, dimethylurea
ETC	:	electron transport chain
FeCN	:	potassium hexacyanoferrate, potassium ferricyanide
MV	:	1,1'-dimethyl-4,4'-bipyridinium dichloride, methyl viologen
PQ, PQ^- and PQH_2	:	oxidized, semiquinone and quinol forms of plastoquinone
Q_B, Q_B^- and Q_BH_2	:	oxidized, semiquinone and reduced in protonation state forms of the secondary quinone acceptor of Photosystem II
PS I and PS II	:	Photosystem I and II
A_{515}	:	the light-induced absorption change at 515 nm
ΔH^+	:	transemembrane difference in electrochemical proton potential
$\Delta \psi$	:	transmembrane difference in electrical potential
b_6/f	:	cyt b_6/f complex

ABSTRACT

The results of measurements of proton-electron stoichiometry in chloroplast thylakoid membrane were examined. Various methods used for the measurement have been scrutinized. The stoichiometry determined by different methods, at different stages of proton accumulation by thylakoids has been analysed. Taking this point into account as well as the values of stoichiometry expected due to operation of Q-cycle, the results obtained by flash and continuous illumination have been examined. Special attention has been paid to the effect of potential difference across the thylakoid membrane on the estimation of stoichiometry. Based on the results of the stoichiometry measurements, and on the functioning of $b_6 f$ complex. It is proposed that the conditions facilitating the Q cycle operation depend on the (1) frequent excitation of photosystems and (2) on the extent of plastoquinone pool reduction. It is concluded on the basis of experimental data that under physiological conditions at saturating light intensity, the H^+/e^- ratio in photosynthetic electron transport chain of higher plants is 3. The significance of this stoichiometry for satisfying the energy requirements during photosynthetic carbon reduction is emphasized.

I. Introduction

In the early sixties, Jagendorf and Hind (1963) and Nenman and Jagendorf (1964) found that protons are absorbed by thylakoids upon illumination of chloroplast suspension. With the development of Mitchell's chemiosmotic theory (Mitchell, 1966) this absorption was considered to be one of the stages of energy transformation during photosynthesis, namely, the generation of ΔH^+ (transmembrane difference in electrochemical proton potential) across the thylakoid membrane during electron transfer through the photosynthetic ETC. The number of protons accumulated by thylakoids per electron transferred, the H^+/e^- ratio, is an important characteristic of this process. This constitutes the basis for drawing conclusions on the amount, localization, and mechanisms of operation of proton-electron coupling sites.

Several reviews discussing this problem have been published earlier (Schwartz, 1971; Ivanov and Muzafarov, 1974; Cox and Olsen, 1982; Ort, 1986; Cramer *et al.*, 1987). The present chapter discusses the peculiarities of different methods employed for determining 'proton-electron ratio' and analyses the possible causes of variability of true value of proton-electron stoichiometry in photosynthetic electron transport chain (ETC) of higher plants.

II. Main Methods for Determining Stoichiometry of Proton-electron Coupling and Analysis of Some Early Results

It is important to note that proton-electron ratio obtained by various methods describes either binding protons from the medium, or release of these ions inside thylakoids, or their stationary accumulation by thylakoids. To underline the differences, various conditions were used for experimental data obtained by different techniques. Besides the use of specified conditions for the method of determination of stoichiometry it was also indicated that the measured value may not be equal to real stoichiometry.

a. Stoichiometry-determination at the initial period of illumination of chloroplasts in continuous light; determination of $(V_{H^+})_a$ V_{e^-}

In a number of early investigations, the proton-electron ratio was estimated by the absorption of protons from the external medium during the first few seconds after light was switched on. The (V_{H^+}) a rate thus estimated was divided by the electron transport rate V_{e^-} to yield $(V_{H^+})_a/V_{e^-}$.

The $(V_{H^+})_a/V_{e^-}$ ratio obviously should be considered as stoichiometry of binding protons from the medium. In the pioneering studies of Karlish and Avron (1967, 1968), it was found that this ratio in the experiments using diquat and FeCN as electron acceptors decreased from 4–6 at pH 6.0 to 0.5–1.0 at pH 8.5. Later investigations considered these values to be erroneous (Karlish and Avron, 1971; Dilley, 1970).

Izawa and Hind (1967) took into account practically all the sources of errors and estimated H^+/e^- ratio to be 2. While analysing this study, as well as studies of Karlish and Avron (1967, 1968) and Dilley (1970), it should be noted that the protons released due to water splitting were not considered as those accumulated inside the thylakoids. It was suggested that these protons quickly became available for the measurement by the electrode. Therefore, when the $(V_{H^+})_a$ value was estimated in the experiments with FeCN (potassium hexacyanoferrate or potassium ferricyanide) as acceptor the rate of stationary acidification of suspension was added to the rate of its initial alkalization. With the $(V_{H^+})_a$ value calculated in this way, the $(V_{H^+})_a/V_{e^-}$ ratio in the experiments with FeCN was estimated to be 2.5. However, from the data it is seen that during the first three seconds of illumination the rate of proton efflux from thylakoids is very small. Taking into account the fact that the protons of water are released at the inner side of the thylakoid membrane, it is obvious that in this case there is no need for correction based on the assumption of "instant" appearance of these protons in the medium for estimating the value $(V_{H^+})_a$. The $(V_{H^+})_a/V_{e^-}$ ratio is 1.5 without this correction. The same value of stoichiometry of binding protons from medium in the experiments with FeCN was obtained in flashing light experiments almost after 11 years (Velthuys, 1978). This result indicated the possibility of uptake of three protons by thylakoids (see section IIIc). Unfortunately, from the data of Izawa and Hind, the H^+/e^- ratio was long considered to be 2.

b. Stoichiometry determination under flash illumination (H^+/e^- determination)

Under conditions of flash illumination, both the absorption of medium protons by thylakoids and the release of these ions inside thylakoids in response to a flash or a series of short flashes were measured. The value of proton-electron stoichiometry was calculated by estimating the number of protons, Δ_{H^+}, which left the medium or appeared inside thylakoids, and the number of transferred electrons, Δ_{e^-}.

Schliephake et al. (1968), using benzyl viologen, safranine T, indigo carmine as electron acceptors, observed that 2 H^+ are absorbed from the outer phase when 1 e^- is transferred from water. Since 1 H^+ is bound in response to the reduction of these acceptors, it was concluded that ETC components bind only 1 H^+ when 1 e^- is transferred. It is supported by the fact that the number of protons bound was decreased when DCMU (3-(3, 4-dichlorophenyl)-1, 1-dimethylurea) was used to disrupt electron transfer to photosystem I (PS I). Numerous subsequent determinations of H^+/e^- ratio under flash illumination supported these results and also showed that one proton is released inside the thylakoids due to water oxidation and one proton due to PQH_2 oxidation per electron passing through the chain (Junge and Auslander,

1974; Auslander and Junge, 1974; Graber and Witt, 1975; Hope and Morland, 1979; Hope and Mathews, 1983, 1984).

c. Stoichiometry determination under continuous illumination, $(V_{H^+})_e/V_{e^-}$

Under continuous illumination of chloroplasts a steady state is established when proton accumulation inside the thylakoids is compensated for the efflux of these ions into the medium. The method used by Schwartz (1968) for assessing an accumulation rate under these conditions is based on this assumption and with the implication that the proton efflux rate in the light should be equal to the initial rate of this efflux observed on switching off the light, $(V_{H^+})_e$. The ratio of proton accumulation rate to electron transport rate corresponds to the true H^+/e^- ratio since it characterizes not only the proton upake from the medium but also proton deposition during water oxidation. Consequently, the $(V_{H^+})/V_{e^-}$ ratio measured correctly is equal to the H^+/e^- ratio. In the experiments using electron accpetor, FeCN at pH 6.8 and NADP$^+$ at pH 7.4, Schwartz (1968) found that the $(V_{H^+})_a/V_{e^-}$ ratio was close to 2 and was independent of uncouplers being present at small concentrations in the chloroplast suspension. The latter proved the adequacy of the method.

Karlish and Avron (1971) in their experiments with diquat as electron acceptor found that $(V_{H^+})_e/V_{e^-}$ ratio decreased from 2 at pH 6.0 to 0.2 at pH 8.4. Valinomycin was observed to raise this ratio up to 4–5 and 0.8–1.0 at pH 6.0 and 8.4, respectively. The effect of valinomycin was explained on the basis of energy transfer inhibition by this antibiotic.

In the studies by Schroder et al. (1972), wherein FeCN was used as an acceptor, $(V_{H^+})_e/V_{e^-}$ ratio at pH 7.0 was close to 2. At pH 8.0, the ratio was slightly less than 1 but it reached 2 when valinomycin was added to the medium. In contrast to Karlish and Avron (1971), Schroder et al. (1972) suggested that valinomycin raised this ratio because of the enhancement of permeability of K^+ which results in suppression of a proton diffusion potential appearing across the membrane at the moment of switching off the light. In the absence of valinomycin, this potential slows down the proton efflux from thylakoids upon darkening and $(V_{H^+})_e$ turns out to be lower than the rate of the proton efflux in the light.

The effect of valinomycin on the value of the $(V_{H^+})_e/V_{e^-}$ ratio at pH 8.0 was studied by Ivanov and Muzafarov (1974), Chow and Hope (1977), and Ivanov et al. (1980b). However, it was observed that antibiotic concentration sufficient for membrane depolarization was lower approximately by one order of magnitude than the concentration in the experiments with FeCN as acceptor at which the $(V_{H^+})_e/V_{e^-}$ reaches 2 (Chow and Hope, 1977). These authors as well as Schroder et al. (1972) proposed that valinomycin raised the $(V_{H^+})_e/V_{e^-}$ ratio up to the value equal to the H^+/e^- ratio. According to their opinion, $(V_{H^+})_e$, in the absence of the antibiotic, is less than the proton efflux rate in the light due to the fact that only the efflux of those protons which have penetrated into lumen is registered after switching off the light. It was suggested that in light, protons released during oxidation of ETC components within the membrane enter not only into the lumen but also directly to the medium without entering intrathylakoid space. Chow and Hope (1977) ascribed the effect of valinomycin on the $(V_{H^+})_e/V_{e^-}$ ratio to the fact that this antibiotic is not only a K^+ permeability-inducing agent but also a lipophilic protonophore decreasing energy barriers at the membrane boundaries. Accordingly, proton efflux occurs from the membrane interior even after cessation of illumination.

In order to explain decrease in $(V_{H^+})_e/V_{e^-}$ ratio with the increase in pH, Ivanov *et al.* (1980a) suggested that protons released within the membrane in light can enter directly to the medium via CF_0-CF_1 channel-opening at pH > 7.2 (Portis *et al.*, 1975) and observes that after the light is switched off the protons were bound by buffering groups in the membrane. K^+-valinomycin complex probably neutralizes these groups since the thylakoid membranes lose protons upon addition of valinomycin to chloroplast suspension in the dark (Molothkovsky and Dzubenko 1972). It is essential to note that the magnitude of the valinomycin-chlorophyll ratio at which the $(V_{H^+})_e/V_{e^-}$ ratio reaches maximum in the experiments with FeCN is $3 \times 10^{-2}M$ (Schroder *et al.*, 1972; Chow and Hope, 1977; Ivanov *et al.*, 1980b), and this value coincides with the estimate of proton capacity ascribed to the intramembrane domains (Dilley *et al.*, 1987).

On the basis of the results described above it has been concluded that the H^+/e^- ratio is 2 for photosynthetic ETC of higher plants under continuous illumination. The $(V_{H^+})_e/V_{e^-}$ ratio measured in the experiments with low potential acceptors, however, differed from the value of 2 which was well reproduced in the experiments with FeCN. In studies of Chow and Hope (1977) and Ivanov *et al.* (1980b) it was found that with methylviologen as acceptor, this ratio only exceeded slightly the value of 1 at pH 8.0 at those valinomycin concentrations for which, the ratio was 2 with FeCN as acceptor.

III. Values of Stoichiometry Requiring the Participation of $b_6 f$ Complex in Proton Translocation

Stoichiometry of proton-electron coupling reported in the majority of studies carried out u ntil the mid-1980s coincided with one obtained by Witt's group in Berlin. However, even in the 1970s, some authors obtained the H^+/e^- ratios which did not correspond to the proposed Z scheme. In 1975, Mitchell suggested that a Q-cycle mechanism with participation of cytochromes of group b and quinone molecules operates in the bacterial, mitochondrial, and chloroplast ETCs, and in the last case this Q cycle provides the transmembrane translocation of 2 protons during one electron transfer between the photosystems (Mitchell, 1975, 1976). Later it was demonstrated that a protein complex consisting of cytochromes b_6 and f and the Rieske FeS centre participates in the Q cycle, and this complex is similar to cytochrome-FeS complexes of bacteria and mitochondria participating in the Q cycle (see Hauska *et al.*, 1983). The values of proton/electron stoichiometry in chloroplast exceeding those predicted by the old scheme were considered as an indication of the participation of $b_6 f$ complex.

a. Operation of Q cycle and expected values of proton-electron stoichiometry in chloroplast ETC

The mechanism of participation of $b_6 f$ complex in proton translocation is shown in Fig. 1. Figure 1 is a variant of Mitchell's Q cycle. Other mechanisms of membrane energization and proton translocation coupled with electron transfer via such cytochrome complexes of mitochondria, bacteria, and chloroplasts such as b-cycle

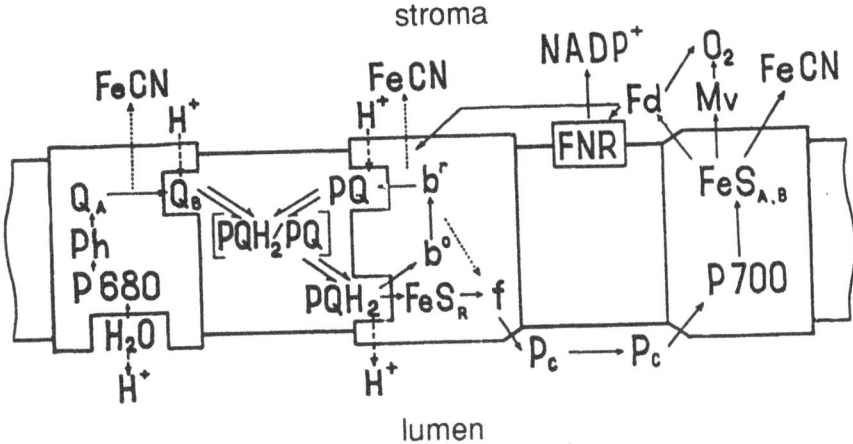

Figure 1. Schematic representation of electron transfer through chloroplast ETC including Q-cycling and of coupled proton accumulation by thylakoids.
P700 and P680, reaction center of PSI and PS2, respectively; Ph, pheophytin; Q_A, primary stable quinone acceptor of PS2; Q_B, secondary quinone acceptor of PS2, plastoquinone molecule in semiquinone from tightly bound with PS2 complex; f, chtochrome f; PQ and PQH_2, plastoquinone and plastoquinol; b_6^o and b_6^r, the heme groups of cytochrome b_6 associated with the site of PQH_2 oxidation and PQ reduction, respectively; FeS_R, the Rieske iron-sulfur centre; Pc, plastocyanine; $FeS_{A, B}$, two iron-sulfur acceptors of PSI; Fd, ferredoxin; FNR, ferredoxin-NADP reductase; $NADP^+$, odidized nicotinamide adenine dinucleotide phosphate; MV, methylviologen, FeCN, ferricyanide.

(Wikstrom and Krab, 1980), semiquinone cycle (Wikstrom and Krab, 1986) and their variant (Moss and Rich, 1987) have been suggested. The problem of stoichiometry will be discussed further on the basis of the modified Q cycle model (Crofts *et al.*, 1983) as it has been done in the majority of works discussed below. It should be noted that formation of two $Q_BH_2 = PQH_2\cdot$ molecules (Fig. 2) in PS II requires the stripping off of four electrons from water and is accompanied by the release of four protons into the inner phase of thylakoids. If the electron transfer via b_6-f complex is completed at the stage shown in Fig. 2b, the H^+/e^- ratio will be 4 or 2 on the basis of either electrons transferred to PS I or taken off from water. If at the donor side of the complex the PQH_2 molecule formed at its acceptor side is oxidized, i.e., the stage given in Fig. 2c is carried out, the H^+/e^- ratios will turn out to be 3.3 and 2.5, respectively. With increase in the number of electrons transferred via $b_6 f$ complex, both stoichiometries approach 3. To achieve the stationary level of proton uptake by thylakoids, about 100 electrons should be transferred to each PS I. The calculation shows that when the Q cycle operates in chloroplast ETC, the H^+/e^- ratio measured under stationary conditions must be very close to 3.

---

* PQ, PQ^-, and PQH_2 are oxidized, semiquinone, and quinol forms of plastoquinone. Q_B, Q_B^-, and Q_BH_2 are oxidized, semiquinone, and reduced in protonation state forms of the secondary quinone accpetor of PS II.

116

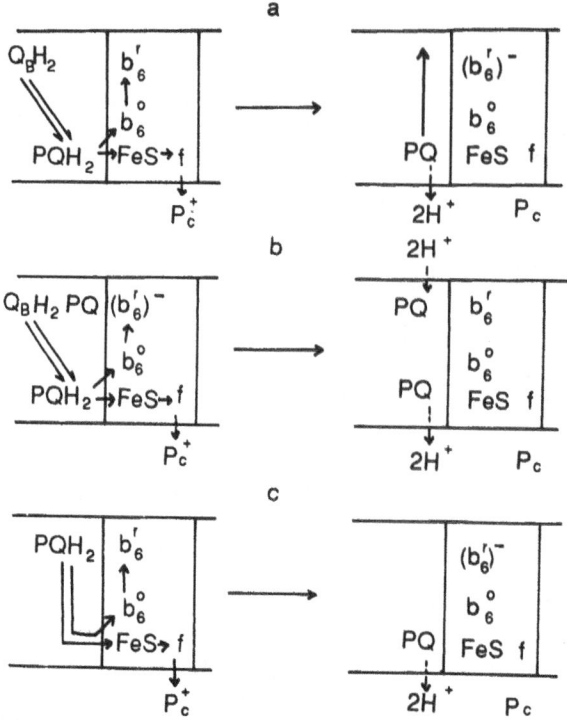

Figure 2. Stages of electron transfer and associated proton translocation catalysed by b_6/f complex depicted in terms of the modified Q-cycle model.

"a" and "b", subsequent oxidation of two PQH_2 molecules formed in PS2 by the complex; "c", oxidation of a PQH_2 molecule formed at the acceptor side of the complex. Designations are the same as in Fig. 1.

Let us predict the stoichiometry of protons binding from the medium. At acceptor side of PS II, proton binding occurs during formation of Q_B as well as of Q_BH_2 (Forster et al., 1981; Hope and Mathews, 1983; Robinson and Crofts, 1984). Though dependent on medium pH, stoichiometries of proton binding during semiquinone and quinol formation can differ from 1 (Hope and Mathews, 1983), at an average each electron participating in the formation of $Q_BH_2(PQH_2)$ induces binding of one proton from the medium.

For b_6-f complex the picture is more complicated. Localization of a single electron on cytochrome b_6 given in Fig. 2, i.e., the stabilization of $(b_6^o b_6^r)$ state, is not always likely to occur. According to the data of Velthuys (1979), Joliot and Joliot (1984a), Hope et al.(1987), and Rich (1988), prolonged localization of a single electron on cytochrome b_6, i.e., stabilization of $(b_6^o b_6^r)$ state (Fig. 2), occurs mostly under oxidizing conditions (when the plastoquinone pool is largely oxidized). Under reducing conditions created by the duroquinol addition into thylakoid suspension, there is a quick oxidation of cytochrome reduced in response to a flash; the $t_{1/2}$ for this oxidation is 20–40 msec (Joliot and Joliot, 1984b; Jones and Whitmarsh, 1985; Rich et al., 1987; Hope et al., 1987). However, it is not clear whether this time characterizes the oxidation of $(b_6^o b_6^r)$ and does not include the oxidation of the two-electron reduced intermediate $b_6^{o}b_6^{r}$ generated from double turnovers in some $b_6 f$ complexes. It was assumed (Joliot and Joliot, 1984a; Rich, 1988) that when only one electron was present on cytochrome b_6, protons were not bound from the medium. However, the data supporting the protonation of one-electron reduced complex in reducing conditions have been reported (Hope and Rich, 1989); the apparent pK_a of a such protonation was suggested to be 6.6. This magnitude of pK_a is close to that of pK_a for the reduced form of cytochrome b_6 i.e., 6.7, as reported by Furbacher et al. (1989). But the protonation of the once turnovered complex was observed also at pH values above pH 7.2 (Hope and Rich, 1989). Further research is required to determine whether this protonation is a result of plastosemiquinone formation.

Thus, the electron entering the fully oxidized cytochrome b_6 (stages a and c, Fig. 2) under certain conditions need not result in proton absorption by thylakoids. However, as in PS II, during PQH_2 formation at the r site of the complex, i.e., during the Q-cycle operation, each electron which has been transferred via cytochrome b_6 must induce proton binding from the medium. When chloroplasts are illuminated with a series of flashes, at an average one proton must be bound at the acceptor side of PS II and one proton at the acceptor side of $b_6 f$ complex, in response to each flash. Numerical value of the measured $\Delta H^+/\Delta e^-$ ratio may depend in this case on the method of determination of amount of electrons transferred. In a number of studies, the amount of protons bound from the medium after a series of flashes was divided by the amount of electrons taken off from water in response to the same number of flashes but during exclusive functioning of PS II (see Section IIIc). According to this procedure, for chloroplasts isolated from spinach grown at high irradiance which contained PS II, b_6-f, and PS I electron-transport complexes in stoichiometry of 1.6:1.0:1.0 (Chow and Hope, 1987), the $\Delta H^+/\Delta e^-$ ratio must be 2.25 with low potential acceptors and 1.62 with FeCN. However, there are conflicting views in the literature concerning the stoichiometry of the three electron-transport complexes stated above (Graan and Ort, 1986; Jursinic and Dennenberg, 1988; Chow et al., 1989).

The difficulties given above also arise during analysis of proton deposition inside the thylakoid space under conditions of flash illumination. Additionally, it is also necessary to take account of possible difference in the rate of evolution of protons originating from various sources. For instance, the slowest phase of acidification of intra thylakoid space after single-turnover flash was not attributed to proton transport events (Hope and Mathews, 1984). Rich (1988) has drawn attention to the fact that this phase could result from oxidation of the PQH_2 molecules formed at the acceptor side of b_6-f complex, i.e., at o site (stage c in Fig. 2).

b. $(V_{H^+})_e/Ve^-$ ratios close to 3

The first data on the fact that $(V_{H^+})_e/V_{e^-}$ ratio can be 3 were obtained in 1974–1977. Ivanov et al. (1974) in the experiments with FeCN as an acceptor found this value in cases of chloroplasts of plants grown at low irradiance. Ivanov and Akulova (1976) obtained the similar result when ATP was present in the suspension of isolated chloroplasts. The latter study was carried out also with FeCN as an acceptor and the hydrolysis of ATP was not the cause of the high value of $(V_{H^+})_e/V_{e^-}$ ratio.

Fowler and Kok (1976) and Rathenow and Rumberg (1980) in their experiments with FeCN as acceptor observed that the $(V_{H^+})_e/V_{e^-}$ ratio was increased from 2 to 3 with the decrease in actinic light intensity. Ivanov and Ovchinnikova (1984) confirmed these data and also showed that the $(V_{H^+})_e/V_{e^-}$ ratio approached 3 with the decrease of electron transport rate when DCMU was present in the suspension at low concentrations.

In earlier studies with low potential acceptors, the values of $(V_{H^+})_e/V_{e^-}$ ratio were usually less than in the experiments with FeCN. Ivanov et al. (1985) found that the presence of azide in the reaction mixture in the experiments with low potential acceptor could lead to a situation wherein $(V_H^+)_e$ was lower than the proton transport rate in the light. The $(V_{H^+})_e/V_{e^-}$ ratio measured in the absence of azide in the experiments with methylviologen as an acceptor was found to be 3 at saturating light intensity and 2 at low intensity. These experiments were carried out at pH 6.0 in the absence of valinomycin.

At pH 8.0, the $(V_{H^+})_e/V_{e^-}$ ratio was 1.2–1.5 at high light intensity but it increased to 3 upon addition of valinomycin or DCCD (N, N′-dicyclohexylcarbodiimide) to the reaction mixture (Ivanov et al., 1983).

The results described above indicate that under continuous illumination the $(V_{H^+})_e/V_{e^-}$ ratio in the experiments with FeCN and methylviologen as acceptors may be different. The difference in the values of the H^+/e^- ratio with these acceptors at high light intensity was confirmed in the work of Ivanov (1988). In parallel experiments it was shown that at pH 8.0 in the presence of ionophore valinomycin similar value of intra-thylakoid pH was maintained even when the reduction rates of these acceptors differed by 1.5, namely electron flow to methylviologen was at slower rate than to FeCN. Since membrane permeability in both the cases was the same, the observed difference in electron transport rates indicated that the number of H^+ released inside the thylakoids per one electron transferred from water to methylviologen was 1.5 times larger than the number of H^+ released per one electron transferred to FeCN.

c. High values of $\Delta H^+/\Delta e^-$ ratio

The $\Delta H^+/\Delta e^-$ ratios exceeding those obtained by Witt's group in Berlin were obtained for the first time by Fowler and Kok (1976). These experiments were carried out at pH 7.0 in the absence of valinomycin. The $\Delta H^+/\Delta e^-$ ratios were found to be 2 and 4 with FeCN and methylviologen as acceptors, respectively, suggesting that the protons released during water splitting are evolved inside the thylakoids. Taking into account that in the experiments with methylviologen the protons of the medium are bound during hydrogen peroxide formation, the authors concluded that the H^+/e^- ratios were 3 and 4 with FeCN and methylviologen, respectively.

The results of Fowler and Kok were initially questioned by Saphon and Crofts (1977). Indeed, in subsequent studies Velthuys (1978) found this ratio to be 1.5 with FeCN as acceptor. As mentioned earlier (see Section IIa), the $\Delta H^+/\Delta e^-$ ratio determined in the experiments with FeCN must be less than 2, even if cytochrome b_6/f complex binds protons from the medium in response to a flash.

In the work of Graan and Ort (1983), the proton number was determined in the experiments with methylviologen as an acceptor, and the electron number in the experiments with FeCN and dimethylquinone as an acceptor pair. It was found that the value of $\Delta H^+/\Delta e^-$ ratio increased from 1.7 to 2.6 at pH 8.1 upon addition of valinomycin to chloroplast suspension.

In similar experiments Hope et al., (1985) observed that when FeCN was used as an acceptor, the presence of valinomycin increased the $\Delta H^+/\Delta e^-$ ratio from 1.0 to 1.5–1.8. Additional proton uptake which resulted from valinomycin addition depended on flash frequency (half-maximal stimulation was observed at 3 Hz, the maximal one at 10–15 Hz). In contrast to the $(V_{H^+})_e/V_{e^-}$ ratio the maximal value of the $\Delta H^+/\Delta e^-$ ratio was obtained at valinomycin-to-chlorophyll ratio being around 10^{-3}. This indicates that the models of action of valinomycin for enhancement of the ratios are not completely identical. This is to be expected because the $\Delta H^+/\Delta e^-$ ratio unlike the $(V_H^+)_e/V_e^-$ ratio describes only binding of protons from the medium.

Hangarter et al. (1987) supported the data of Graan and Ort (1983). They have also shown that during electron flow from duroquinol to methylviologen in the presence of DCMU, i.e., through $b_6\text{-}f$ complex and PS I, the $\Delta H^+/\Delta e^-$ ratio is 1.7, whereas during electron flow from water to dimethylquinone in the presence of DBMIB (2, 5-dibromo-3-methyl-6-isopropyl-p-benzoquinone), i.e., electron flow through PS II only, this ratio is 1.0. These authors also used neutral red to study acidification of intra thylakoid space during electron flow through individual portions of ETC. It turned out that on an average, a two-fold proton deposition inside the thylakoid space was observed per electron transferred from duroquinol via $b_6\text{-}f$ complex to PS I as compared with the case when one electron was taken off from water. These data supported the early results of Velthuys (1980), who measured acidification of intra-thylakoid space in response to the single flash during complete non-cyclic flow from water to PS I.

The results of Hangarter et al. (1987) convincingly proved that the functioning of the $b_6 f$ complex could be coupled with the proton translocation into thylakoids. In this work, however, it was found that the $\Delta H^+/\Delta e^-$ ratio for the whole-chain electron transport to methylviologen decreased after 20–30 flashes from 2.6 to 1.6 in spite of the presence of valinomycin.

d. Proton translocation by $b_6\text{-}f$ complex incorporated into liposome membrane

The H^+/e^- ratio was measured during electron transfer from plastoquinol (Hurt et al., 1982) and duroquinol (Willms et al., 1987) to plastocyanino with participation of b_6/f complex incorporated into liposome membranes. In both the cases a value close to 2 was obtained. It indicates that the complex can translocate one additional proton across the membrane for each electron transferred through the complex. The investigation of functional activity of such incorporated complexes shows also that the process includes oxidation and reduction of quinols/quinones at two sites required by the Q cycle model and that these sites have no strict specificity to plastoquinol/-plastoquinone (Willms et al., 1988).

IV. Causes of the Observed Variability in Stoichiometry

a. Electrogenic function of b_6/f complex and the effect of valinomycin on $\Delta H^+/\Delta e^-$ ratio

It has been repeatedly suggested that "low" values of proton-electron stoichiometry result from inhibition of the electrogenic reaction, associated with $\Delta\psi$ of the cytochrome b_6-f complex, due to increase in $\Delta\psi$ under rapid flashes or in the strong light. The "disappearance" of a slow phase of the carotenoid (electrochromic) bandshift ΔA_{515}, the absorbance change at 515 nm (which is considered by the majority of authors as an increase in membrane potential $\Delta\psi$ due to transfer of an additional charge during electrogenic reactions occurring in cytochrome complex; see Jones and Whitmarsh (1985) in a sequence of rapid flashes was one of the evidences supporting this view (Bouges-Boiquet, 1977, 1981).

Bouges-Bocquet (1981) was one of the first who suggested that slow electrogenic reaction is inhibited under frequent flashes when ΔpH- and $\Delta\psi$-dependent free energy of the process including an electrogenic stage is not sufficient to provide charge transfer across the membrane against the membrane potential ($\Delta\psi$) which increases with a flash number. The study carried out by Hope and Matthews (1987) on isolated chloroplasts showed that during electron transport from water to methylviologen and FeCN the decrease of slow phase up to its complete elimination after 4 to 6 flashes given, at a frequency of 10 Hz correlated with the increase in the total signal ΔA_{515}. It was also found that at valinomycin concentrations, used in the experiments with FeCN as acceptor, in which an additional proton was bound in response to a flash (Hope et al., 1985), the total signal A_{515} was considerably decreased and slow phase was retained even after the sixth flash. These results agreed with Bouges-Bocquet's proposition stated above. Besides, the authors pointed out that they supported the suggestion that the increase of $\Delta H^+/\Delta e^-$ ratio in the presence of valinomycin resulted from the removal of restriction of Q cycle operation under $\Delta\psi$ reduction value (see Section IIIc).

Rich (1988) showed that the visible decrease or even the absence of slow phase of ΔA_{515} on the experimental curves can result from superposition of the full slow phase and of total A_{515} decay. The data of Van Kooten et al., 1983, similar to those of Hope and Matthews (1987), were very closely simulated; in the course of the simulation it was suggested that slow rise of the ΔA_{515} was present after every flash and that the decay of total ΔA_{515} was second order. To simulate the data obtained on algae (Bouges-Bocquet, 1981), it was sufficient to use a first-order decay. Finally, it has been shown that $\Delta\psi$ of 30–50 mv did not prevent the protonmotive action of b_6/f complex (Hope and Rich, 1989) .

In the experiments with chromatophores as in the experiments with chloroplasts proton absorption at the beginning of illumination with continuous light and in flashing light was stimulated upon addition of valinomycin into the reaction mixture (Petty et al., 1979). This stimulation is accounted for by both acceleration of electron transfer and by the decrease of in the rate of proton efflux back into the medium, as the $\Delta\psi$ is decreased in the presence of antibiotic valinomycin (Myatt and Jackson, 1986). Obviously, acceleration of electron transfer is caused by the acceleration of electrogenic reaction(s) when its (their) free energy is increased due to $\Delta\psi$ decrease.

In the experiments with chloroplasts the slowing down of onset of the slow phase with the increase in $\Delta\psi$ was found by Bouges-Bocquet (1981), whereas an acceleration of its onset in the presence of valinomycin was found by Hope and Matthews (1987).

Myatt and Jackson (1986) concluded that in the absence of valinomycin the slowing down of electron transfer with the increase of $\Delta\psi$ is due to the fact that not all the complexes can accomplish Q cycle within the intervals between rapid flashes, and this fact leads to the decrease in the average number of protons absorbed from the medium in response to a flash. The slowing down of the reaction of quinol oxidation by cytochrome complex under rapid flashes may also occur due to acidification of intra-thylakoid space. Taking into account the peculiarities of calculation of proton-electron stoichiometry in flashing light and the ratio between electron-transport complexes in chloroplasts (see Section IIIa) the proton binding by only a half of b_6/f complexes in response to a flash would yield in the experiments with methylviologen the $\Delta H^+/\Delta e^-$ ratio of 1.9. Thus, low values of the $\Delta H^+/\Delta e^-$ ratio indicating the absence of Q-cycle operation can be observed at rapid $\Delta H^+/\Delta e^-$ flashes due to the slowing down of transversal electron transfer. Consequently, there are insufficient grounds to confirm that under such conditions this pathway of electron transfer does not operate at all.

"High" values of the H^+/e^- ratio obtained by Velthuys (1978) and Fowler and Kok (1976) in the absence of valinomycin could be caused by excitation conditions, i.e., flash frequency of 2 and 1 Hz, respectively, and by peculiarities of the chloroplast preparation used.

b. Dependence of the measured stoichiometry on the flash frequency and the degree of plastoquinone pool reduction

In Section IIIa it has been pointed out that proton uptake with participation of b_6/f complex may not always occur when the first electron is transferred to cytochrome b_6. Slow oxidation of once-reduced cytochrome b_6 under oxidizing conditions $t_{1/2}$ is 1.2 sec (Joliot and Joliot, 1984a), 0.4 sec (Hope et al., 1987), 0.5 sec (Rich, 1988) is characterized as "non-physiological"; there are no precise data on the pathway of this oxidation though it was suggested that donors of PS I could be acceptors of electron in this case (Houchins and Hind, 1983). It was shown (Velthuys, 1979; Joliot and Joliot, 1984a) that only injection of the second electron to cytochrome b_6 leads to its rapid oxidation, $t_{1/2}$ is 20–30 sec (Dolan and Hind, 1974; Velthuys, 1979), which is accompanied, as it is suggested, by the formation of PQH_2, i.e., proton binding from the medium. Obviously, if the time between oxidations of two PQH_2 molecules by cytochrome complex is greater than the time of non-physiological oxidation of once-reduced cytochrome b_6; such a complex will not participate in proton uptake by thylakoids.

In the works of Schliephake et al. (1968), Junge and Auslander (1974), Auslander and Junge (1974), Graber and Witt (1975), Saphon and Crofts (1977), Hope and Morland (1979), and Hope and Matthews (1983, 1984), the flash frequency was 0.1 –0.5 Hz. Such intervals between flashes as compared with the time of $(b_6b_6)^-$ oxidation may account for the fact that in these studies the proton binding site in ETC "before" PS I was not observed except on the acceptor side of PS II. The given time $t_{1/2}$ of $(b_6b_6)^-$ oxidation also may account for the dependence of $\Delta H^+/\Delta e^-$ ratio on flash frequency which has been observed in the experiments with FeCN (Hope et al., 1985) (see Section IIIc).

The results obtained by Velthuys (1978) indicated that it is necessary to reduce PQ-pool for participation of $b_6 f$ complexes in proton binding from the medium; i.e., in the presence of DCMU and FeCN (the latter at a low concentration of 50 μM) proton binding was observed only in case of pre-illuminated chloroplasts.

The influence of a degree of quinone pool reduction on the operation of cytochrome complexes was demonstrated in the experiments with chromatophores of *Rhodopseudomonas sphaeroides* (Crofts et al., 1983). Using the data obtained on this bacterium, Robertson and Dutton (1988) proved that the dependence of participation of bc_1 complex in membrane energization upon the level of quinone pool reduction results from a statistical character of interaction of PQH_2 molecules with the complexes. The latter can lead to the fact that some complexes having once-reduced cytochrome b will not react with the second QH_2 molecule up to cytochrome oxidation in any "non-physiological" way.

The effect of plastoquinone pool reduction on the measured stoichiometry of proton-electron coupling must be exhibited most of all in chloroplasts of higher plants which are characterized by lateral asymmetry in the distribution of electron transport complexes in thylakoid membranes (Cox and Andersson, 1981; Andersson and Haehnel, 1982; Allred and Stachelin, 1986). Cytochrome complexes of stroma membranes probably will bind protons in response to every flash (at an average) only if the plastoquinone pool along all the thylakoid membranes is maintained in the reduced state.

The amount of PS I reaction centres, whose turnovers lead to $b_6 f$ complex turnovers, (see Chapter by Hoshina and Itoh, this volume) is not less than the amount of these complexes; however, the proton uptake by cytochrome complexes due to double turnovers after single flash should be very small under the low PQH_2 concentration. This also might be one of the reasons why proton uptake due to b_6 /f complex activity was not observed in the earlier studies.

In the previous section, we have discussed that incomplete turnover of some cytochrome complexes between the flashes can cause "low" values of $\Delta H^+/\Delta e^-$ ratio. Similar calculation can be also carried out for the present case of unfinished cycling with participation of some complexes. In this case it turns out that not only low values of $\Delta H^+/\Delta e^-$ but also the low values of H^+/e^- ratio (i.e., measured $(V_{H^+})_e/V_{e^-}$ ratio) are to be expected. The $(V_{H^+})_e/V_{e^-}$ ratio of 2 showing the absence of Q cycle was obtained by Ivanov et al., (1985) in the experiments with methylviologen under illumination of chloroplasts, adapted to some extent to the darkness, with a weak continuous light for 1–1.5 min. But the authors found that the $(V_{H^+})_e/V_{e^-}$ ratio was close to 3 in such weak light for the chloroplasts pre-illuminated with a strong light. After prolonged illumination of such pre-illuminated plastids with weak light, the $(V_{H^+})_e/V_{e^-}$ ratio in this light also became equal to 2. Plastoquinone pool reduction is one of the main consequences of pre-illumination with a strong light and the $(V_{H^+})_e/V_{e^-}$ ratio of 3 in the weak light after such pre-illumination is more likely due to this reduction. As it was pointed out by Hope and Mathews (1984), the illumination of thylakoids with 20–40 flashes at 10–20 Hz can lead to a more oxidized state of plastoquinone pool, for example, because the miss factor applies to the entry of electron into the pool (Kok *et al.*, 1970), but not to the flash-induced withdrawal of electrons via PS I. The decrease in $\Delta H^+/\Delta e^-$ (Hangarter *et al.*, 1987) and $(V_{H+})_e/V_{e^-}$ (Ivanov *et al.*, 1985) ratios in both the cases under relatively weak illumination may result from such a process.

c. The influence of the rate of electron flow under continuous illumination with strong light

When chloroplasts are illuminated with a strong light, the plastoquinone pool is reduced to a great extent. This seems to create the conditions for constant participation of all cytochrome complexes in proton uptake by thylakoids. But the operation of Q cycle in photosynthetic ETC of higher plants was discussed more often particularly in connection with the decrease of $(V_{H+})_e/V_{e^-}$ ratio from 3 to 2 by the increase of the intensity of continuous light (see Section IIIb). The absence of the third proton uptake in the strong light could not be explained by inhibition of electrogenic reactions in $b_6 f$ complex with the increase of $\Delta\psi$, since, for example rather high valinomycin concentration was used Rathenow and Rumberg (1980). Bendall (1982) and Rich (1984) have suggested that Q cycle is switched off at high light intensity due to the transfer of both the electrons from PQH_2 to cytochrome f which remains in an oxidized state under these conditions. The same explanation for $(V_H^+)_e/V_{e^-}$ of 2 at high light intensity was given recently by Schubert *et al.* (1990).

For a long time, it was not taken into consideration that the decrease of the $(V_H^+)_e/V_{e^-}$ ratio with the increase of light intensity was observed only when FeCN, a potent oxidizer capable of reacting with nearly all ETC components, was used as an acceptor. Rich (1988) showed that the steady-state acidification of chloroplast suspension under illumination in the presence of FeCN was observed even when DCMU was present. The rate of this acidification was slow in the weak light but it considerably increased in the strong light. The $(V_H^+)_e/V_{e^-}$ ratio found by division of $(V_H^+)_e$ by the difference of steady-state suspension acidification rates in the absence and the presence of DCMU turned out to be close to 3 at both low and high light intensities.

It is known that FeCN can accept electrons at different points of ETC. Depending on the relative rates of FeCN reduction at the sites depicted in Fig. 1 the H^+/e^- ratio could reach various values from 1 to 3. However, all the scientists who have measured the $(V_H^+)_e/V_{e^-}$ ratio with this acceptor (FeCN) in strong light obtained a value very slightly different from 2. In fact, this was an important observation which favours to the suggestion that the Q cycle is switched off under strong illumination.

It is unlikely that the H^+/e^- ratio of 2 is a consequence of the parallel reduction of FeCN at all the three possible points or only at the acceptor sides of photosystems. In both the cases the rates of electron transfer at the reduction sites must be the same. However, it was shown that in undestroyed, undamaged thylakoids, FeCN at concentrations up to 1 mM was very weakly reduced at the acceptor side of PS II (McCauley *et al.*, 1984).

Ivanov and Ovchinnikova (1984) suggested that in the strong light, FeCN was reduced not only at the acceptor side of PS I but also at the acceptor side of $b_6 f$ complex (see Chapter by Govindjee and Coleman, this volume). Moreover, the reduction in that second site was accompanied by a break in the Q cycle. In this case the transfer of each electron from water to FeCN must lead to accumulation of two protons inside thylakoids. Conformational changes of the thylakoid membrane occurring under the illumination can facilitate this process. It was shown (Horton and Cramer, 1974) that after chloroplast illumination with a strong light the rate of oxidation of cytochrome f, one of the complex components, by ferricyanide was increased.

V. Conclusions

The analysis of the available experimental data on stoichiometry of proton-electron coupling in chloroplasts has shown that there are no convincing proofs of the fact that under physiological conditions, i.e., under continuous illumination of high intensity, the ratio H^+/e^- is 2 in photosynthetic ETC of higher plants. This realization is critical for chloroplast energetics since this ratio determines the potential ratio of the "products of light stage of photosynthesis" ATP and NADPH, formed under non-cyclic electron transfer. If under optimal conditions the synthesis of one ATP molecule requires the efflux of three protons from thylakoids (Rathenow and Rumberg, 1980; Davenport and McCarty, 1981; Strotman and Lohse, 1988), then at the ratio H^+/e^- of 3 the ratio of ATP and NADPH must reach 2. In the majority of early works carried out on isolated chloroplasts, the value of P/e_2 ratio was obtained close to a unit and it was considered that this ratio could not exceed 1.33, since the ratio H^+/e^- wast accepted to be 2 (see rev. McCarty, 1980). Only in the works where special attention was paid to the preservation of the native structure of the lamellae system, the ratio P/e_2 in the experiments with methylviologen and ferredoxin as acceptors was obtained close to 2 without any corrections for electron transport uncoupled with ATP synthesis (Reeves and Hall, 1973; Robinson and Wiskich, 1976). In these same works, with FeCN as acceptor the ratio P/e_2 was 1.4–1.6. Robinson and Wiskich (1976) explained lower ratio in the experiments with this acceptor by the fact that even in the chloroplasts isolated thoroughly, a part of electrons is transferred to FeCN from the carriers located between the two photosystems (Section IVc). Thus, the data of these works correspond to the fact that the ratio H^+/e^- is equal to 3 under non-cyclic electron transport.

Obviously, all the requirements of the activity functioning chloroplasts in ATP production should not be limited by the requirements of CO_2 fixation cycle. This cycle would not be able to be in operate without parallel processes of metabolite transport and biosynthetic processes in chloroplast and cell. These processes require energy supply mainly in the form of ATP.

The available data indicate that cyclic electron transport with participation of PS I and b_6f complex may play a certain role in ATP synthesis in chloroplasts *in vivo* (Slovacek *et al.*, 1980). It is likely that the formation of ATP due to pseudocyclic electron transport also takes place (Shmeleva *et al.*, 1979; Furbank and Badger, 1983). However, taking into account a high rate of $NADPH^+$ reduction under physiological conditions it can be suggested that 0.5 ATP per 1 NADPH formed additionally to 1.5 ATP per 1 NADPH used in the Calvin cycle contributes considerably to the provision of energy for the processes accompanying this cycle.

Acknowledgements

I gratefully acknowledge Prof. Romanova's support of this work and deeply indebted to Prof. Krendeleva for many stimulating discussions. I wish to thank Prof. Ort and Drs. Rich, Hope, and Schubert for allowing me access to their unpublished and recently published information. Thanks are due to Nina Kljueva for the help in preparing the manuscript.

VI. References

Allred DR and Stachelin LA (1986) Spatial organization of the cytochrome b_6-f-complex within chloroplast thylakoid membranes. *Biochim Biophys Acta* **849:** 94–103

Anderson B and Haehnel W (1982) Location of photosystem I and photosystem II reaction centers in different thylakoid regions of stacked chloroplasts. *FEBS Lett* **146:** 13–17

Auslander W and Junge W (1974) The electric generation in photosynthesis of green plants. I. Vectorial and protolytic properties of the electron transport chain. *Biochim Biophys Acta* **333:** 59–70

Bendall DS (1982) Photosynthetic cytochromes of oxygenic organisms. *Biochim Biophys Acta* **683:** 119–151

Bouges-Bocquet B (1977) Cytochrome f and plastocyanin kinetics in *Chlorella pyrenoidosa* II. Reduction kinetics and electron field increase in the 10 ms range. *Biochim Biophys Acta* **462:** 371–379

Bouges-Bocquet B (1981) Factor regulating the slow electrogenic phase in green algae and higher plants. *Biochim Biophys Acta* **635:** 327–340

Chow WS and Hope AB (1977) Proton translocation, electron transport and photophosphorylation in isolated chloroplasts. *Aust J Plant Physiol* **4:** 647–665

Chow WS and Hope AB (1987) The stoichiometries of supramolecular complexes in thylakoid membranes from spinach chloroplasts. *Aust J Plant Physiol* **14:** 21–28

Chow WS, Hope AB and Anderson JM (1989) Oxygen per flash from leaf disks quantifies Photosystem II. *Biochim Biophys Acta* **973:** 105–108

Cox RP and Andersson B (1981) Lateral and transverse organization of cytochromes in the chloroplast thylakoid membrane. *Biochim Biophys Res Commun* **103:** 1326–1342

Cox RP and Olsen LF (1982) The organization of the electron-transport chain in the thylakoid membrane. In: (J Barber, ed) *Electron Transport and Photophosphorylation*, Elsevier Biomedical Press, Amsterdam, pp 49–79

Cramer WA, Black MT, Widger WR and Girvin ME (1987) Structure and function of photosynthetic cytochrome b-c_1 and b_6-f complexes. In: (J Barber, ed) *Topics in Photosynthesis*, Vol 8, *The Light Reactions*, Elsevier Science Press, Amsterdam, pp 447–493

Crofts AR, Meinhardt SW, Jones KR and Snozzi M (1983) The role of the quinone pool in the cyclic electron-transfer chain of *Rhodopseudomonas sphaeroides*. A modified Q-cycle mechanism. *Biochim Biophys Acta* **723:** 202–218

Davenport JW and McCarty RE (1981) Quantitative aspects of adenosine triphosphate-driven proton translocation in spinach chloroplasts thylakoid. *J Biol Chem* **256:** 8947–8954

Dilley RA (1970) The effect of various energy-conversion states of chloroplasts on proton and electron transport. *Arch Biochem Biophys* **137:** 270–283

Dilley RA, Theg SM and Beard WA (1987) Membrane-proton interactions in chloroplast bioenergetics: Localized proton domains. *Annu Rev Plant Physiol* **38:** 347–389

Dolan E and Hind G (1974) Kinetics of the reduction and oxidation of cytochromes b and f in isolated chloroplasts. *Biochim Biophys Acta* **357:** 380–385

Fowler CF and Kok B (1976) Determination of H^+/e^- ratios in chloroplasts with flashing light. *Biochim Biophys Acta* **423:** 510–523

Forster V, Hong Yu-Qun and Junge W (1981) Electron transfer and proton pumping under excitation of dark-adapted chloroplasts with flashing light. *Biochim Biophys Acta* **638:** 141–152

Furbacher PN, Girvin ME and Cramer WA (1989) On the question of inter-heme electron transfer in the chloroplast cytochrome b_6 in situ. *Biochemistry* **28:** 8990–8998

Furbank RT and Badger MR (1983) Oxygen exchange associated with electron transport and photophosphorylation in spinach thylakoids. *Biochim Biophys Acta* **723:** 400–409

Graan T and Ort DR (1983) Initial events in the regulation of electron transfer in chloroplasts. *J Biol Chem* **258:** 2831–2836

126

Graan T and Ort DR (1986) Detection of oxygen-evolving photosystem II centers inactive in plastoquinone reduction. *Biochim Biophys Acta* **852**: 320–330

Graber P and Witt HT (1975) Direct measurement of the protons pumped into the inner phase of the functional membrane of photosynthesis per electron transfer. *FEBS Lett* **59**: 184–189

Hangarter RP, Jones RW, Ort DR and Whitmarsh J (1987) Stoichiometries and energetics of proton translocation coupled to electron transport in chloroplasts. *Biochim Biophys Acta* **890**: 106–115

Hauska G, Hurt E, Gabellini N and Lockau W (1983) Comparative aspects of quinol-cytochrome c/plastocyanin oxidoreductases. *Biochim Biophys Acta* **726**: 95–133

Hope AB, Birch S and Matthews DB (1987) Further studies of proton translocation in chloroplasts after single-turnover flashes. IV Effects of cytochrome b_6 *f* complex inhibitors on proton uptake and cytochrome b_6 turnover. *Aust J Plant Physiol* **14**: 47–57

Hope AB, Handley L and Matthews DB (1985) Further studies of proton translocation in chloroplasts after single-turnover flashes. III. Conditions for the operation of an apparent Q-cycle in thylakoids. *Aust J Plant Physiol* **12**: 387–394

Hope AB and Matthews DB (1983) Further studies of proton translocation in chloroplasts after single-turnover flashes. I. Proton uptake. *Aust J Plant Physiol* **10**: 363–372

Hope AB and Matthews DB (1984) Further studies of proton translocation in chloroplasts after single-turnover flashes II. Proton deposition. *Aust J Plant Physiol* **11**: 267–276

Hope AB and Matthews DB (1987) The slow phase of the electrochromic shift in relation to the Q-cycle in thylakoids *Aust J Plant Physiol* **14**: 29–46

Hope AB and Morland A (1979) Proton translocation in isolated spinach chloroplasts after single-turnover actinic flashes. *Aust J Plant Physiol* **6**: 289–304

Hope AB and Rich PR (1989) Proton uptake by the chloroplast cytochrome of complex. *Biochim Biophys Acta* **975**: 96–103

Horton P and Cramer WA (1974) The accessibility of the chloroplast cytochromes *f* and *b*-559 to ferricyanide. *Biochim Biophys Acta* **368**: 348–360

Houchins JP and Hind G (1983) Kinetic evidence for involvement of two cytochrome *b*-563 hemes in photosynthetic electron transport. *Biochim Biophys Acta* **725**: 138–145

Hurt EC, Hauska G and Shahak Y (1982) Electrogenic proton translocation by the chloroplast cytochrome b_6 *f* complex reconstituted into phospholipid vesicles. *FEBS Lett* **149**: 211–216

Ivanov BN (1988) The effect of valinomycin on proton transport in chloroplast thylakoid membranes. *Biologicheskiye Membrani (USSR)* **5**: 628–634

Ivanov BN and Akulova EA (1976) The effect of ATP and of phosphorylation conditions on the H^+/e^- ratio and proton transport is isolated chloroplast. *Dokladi Akad Nauk USSR* **277**: 999–1002

Ivanov BN and Muzafarov EN (1974) Determination of link between proton uptake and electron transport in isolated chloroplasts. In: (EA Akulova, ed) *Methodi Issled. Fotosint Transporta Electrona*. Akad Nauk USSR, Pushchino, pp 69–83

Ivanov BN and Ovchinnikova VI (1984) Changes of stoichiometry of proton-electron coupling in photosynthetic electron transport chain. Biofyzika USSR **29**: 770–774

Ivanov BN, Ovchinnikova VI and Shmeleva VL (1983) The effects of valinomycin and dicyclohexylcarbodiimide on measured stoichiometry of coupling between proton and electron trasnports in chloroplasts. Biokhimiya USSR **48**: 1954–1964

Inanov BN, Povalyaeva TV and Akulova EA (1980a) The H^+/e^- ratio in isolated chloroplasts. I. The effects of media pH and of acceptor on value of the H^+/e^- ratio during electron transfer from water to ferricyanide and methylviologen. Biofizika USSR **1980**: 183–184

Ivanov BN, Povalyaeva TV and Akulova EA (1980b) The H^+/e^- ratio in isolated chloroplasts. II The effect of valinomycin and gramicidin on value of the H^+/e^- ratio. *Biofizika (USSR)* **1980**: 184–185

Ivanov BN, Shmeleva VL, Muzafarov EN and Akulova EA (1974) The P/2e and H^+/e^- ratios in chloroplasts of plants grown at different light intensities. *Dokladi Akad Nauk USSR* **217**: 971–974

Ivanov BN, Shmeleva VL and Ovchinnikova VI (1985) Stoichiometry of proton uptake in isolated pea chloroplasts under different light intensities. *J Bioenerg Biomembr* **17**: 239–249

Izawa S and Hind G (1967) The kinetics of the pH rise in illuminated chloroplasts suspensions. *Biochim Biophys Acta* **143**: 377–390

Jagendorf AT and Hind G (1963) Studies on the mechanism of photophosphorylation. In: *Photosynthetic Mechanism of Green Plants*, Publ 1145 NAS-NRC, Washington, DC, pp 599–610

Joliot P and Joliot A (1984a) Electron transfer between the two photosystems. I Flash excitation under oxidizing conditions. *Biochim Biophys Acta* **765:** 210–218

Joliot P and Joliot A (1984b) Electron transfer between the two photosystems. II Equilibrium constants. *Biochim Biophys Acta* **765:** 219–226

Jones RW and Whitmarsh J (1985) Origin of the electrogenic reaction in the chloroplast cytochrome *b/f* complex. *Photobiochem Photobiophys* **9:** 119–127

Junge W and Auslander W (1974) The electric generation in photosynthesis of green plants. I Vectorial and protolytic properties of the electron transport chain. *Biochim Biophys Acta* **333:** 59–70

Jursinic PA and Denneberg RJ (1988) Enhanced oxygen yields caused by double turnovers of Photosystems II induced by dichlorobenzoquinone. *Biochim Biophys Acta* **934:** 177–185

Karlish SJD and Avron M (1967) Relevance of proton uptake induced by light to the mechanism of energy coupling in photophosphoryltion. *Nature* **216:** 1107–1109

Karlish SJD and Avron M (1968) Analysis of light induced proton uptake in isolated chloroplasts. *Biochim Biophys Acta* **153:** 878–888

Karlish SJD and Avron M (1971) Energy transfer inhibition and ion movements in isolated chloroplasts. *Eur J Biochem* **20:** 51–57

Kok B, Forbush B and McGloin MP (1970) Cooperation of charges in photosynthetic O_2 evolution. I A linear four step mechanism. *Photochem Photobiol* **11:** 457–475

McCarty RE (1980) Photosynthetic phosphorylation by chloroplasts of higher plants. In: (KC Smith, ed) *Photochemical Photobiological Revs*. Vol 5, Pergaman Press. New York, pp 1–47

McCauley SW, Taylor SE, Dennenberg RJ and Melis A (1984) Measurement of the relative electron-transport capacity of photosystem I and photosystem II in spinach chloroplasts. *Biochim Biophys Acta* **765:** 186–195

Mitchell P (1966) Chemiosmotic coupling in oxidative and photosynthetic phosphorylation. *Biol Rev* **41:** 445–502

Mitchell P (1975) Protonmotive redox mechanism of the cytochrome b-c₁ complex in the respiratory chain: protonmotive ubiquinone cycle. *FEBS Lett* **56:** 1–6

Mitchell P (1976) Possible molecular mechanisms of the protonmotive function of cytochrome systems. *J Theor Biol* **62:** 327–367

Molotkovsky Yu G and Dzubenko VS (1972) Inhibition of photophosphorylation in chloroplasts by valinomycin. *Dokladi Akad Nauk SSSR* **204:** 1272–1275

Moss DA and Rich PR (1987) The effect of pre-reduction of cytochrome *b*-563 on the electron-transfer reactions of the cytochrome *bf* complex in higher plant chloroplasts. *Biochim Biophys Acta* **894:** 189–197

Myatt JF and Jackson JB (1986) "Backlash" and the coupling between electron transport and proton translocation in bacterial chromatophores. *Biochim Biophys Acta* **848:** 212–223

Neumann J and Jagendorf AT (1964) Light-induced pH changes related to phosphorylation by chloroplasts. *Arch Biochim Biophys* **107:** 109–119

Ort DR (1986) Energy transduction in oxygenic photosynthesis: An overview of structure and mechanism. In: (LA Staehelin and CJ Arntzen, eds) *Encyclopedia of Plant Physiology*, New Series, Vol 19, *Photosynthesis III, Photosynthetic Membranes and Light Harvesting Systems*, Springer-Verlag, Berlin, pp 96–143

Petty K, Jackson JB and Dutton PL (1979) Factors controlling the binding of two protons per electron transferred through the ubiquinone and cytochrome *b/c₂* segment of *Rhodopseudomonas sphaeroides* chromatophores. *Biochim Biophys Acta* **546:** 17–42

Portis AR, Magnusson RP and McCarty RE (1975) Conformational changes in coupling factor 1 may control the rate of electron flow in spinach chloroplasts. *Biochem Biophys Res Commun* **64:** 877–884

Rathenow M and Rumberg B (1980) Stoichiometry of proton translocation during photosynthesis. *Ber Bunsenges Phys Chem* **84:** 1059–1062

Reeves SG and Hall DO (1973) The stoichiometry (ATR/2e ratio) of non-cyclic photophoshorylation in isolated spinach chloroplasts. *Biochim Biophys Acta* **314:** 66–78

Rich PR (1984) Electron and proton transfers through quinones and cytochrome *bc* complexes. *Biochim Biophys Acta* **768:** 53–79

Rich PR (1988) A critical examination of the supposed variable proton stoichiometry of the chloroplast cytochrome *bf* complex. *Biochim Biophys Acta* **932**: 33–42

Rich PR, Heathcote P and Moss DA (1987) Kinetic studies of electron transfer in a hybrid system constructed from the cytochrome *bf*-complex and Photosystem I. *Biochim Biophys Acta* **892**: 138–151

Robertson DE and Dutton PL (1988) The nature and magnitude of the charge-separation reactions of ubiquinol cytochrome oxidoreductase. *Biochim Biophys Acta* **935**: 273–291

Robinson HH and Crofts AR (1984) Kinetics of proton uptake and oxidation-reduction reactions of the quinone acceptor complex of PS II from pea chloroplasts. In: (C Syhesema, ed.) *Advanced Photosynthesis Research*, Vol 1, Martinus Nijhoff/Dr Junk Publishers, The Hague, pp 477–480

Robinson SP and Wiskich JT (1976) Factor affecting the ADP/O ratio in isolated chloroplasts. *Biochim Biophys Acta* **440**: 131–146

Saphon S and Crofts AR (1977) The H^+/e^- ratio in chloroplasts is 2. Possible errors in its determination. *Z Naturforsch* **32b**: 810–816

Schliephake W, Junge W and Witt HT. (1968) Correlation between field formation, proton translocation, and the light reactions in photosynthesis. *Z Naturforsch* **23b**: 1571–1578

Schroder H, Muhle H and Rumberg B (1972) Relationship between ion transport phenomena and phosphorylation in chloroplasts. In: (G Forty, M Avron and A Melandri, eds.) *Proc 2nd Intern Congr Photosynth Res*, Dr W. Jung, NV The Hague, pp 919–930

Schubert K, Liese F and Rumberg B (1990) Analysis of the variability of the H^+/e stoichiometry in spinach chloroplasts. In: (M Baltsheffsy, ed) *Current Research in Photosynthesis*, Vol 3, Kluwer Acad Publ, Dordrecht, The Netherlands, pp 279–282

Schwartz M (1968) Light induced proton gradient links electron transport and phosphorylation. *Nature* **219**: 915–919

Schwartz M (1971) The relation of ion transport to phosphorylation. *Annu Rev Plant Physiol* **22**: 469–487

Shmeleva VL, Ivanov BN and Bityukova LV (1979) Light-dependent oxygen consumption by isolated pea chloroplasts during $NADP^+$ photoreduction. *Biokhimiya (USSR)* **44**: 911–916

Slovacek RE, Crowther D and Hind G (1980) Relative activities of linear and cyclic electron flows during chloroplasts CO_2-fixation. *Biochim Biophys Acta* **592**: 496–505

Strotmann H and Lohse D (1988) Determination of the H^+/ ATP ratio of the H^+ transport-coupled reversible chloroplast ATPase reaction by equilibrium studies. *FEBS Lett* **229**: 308–312

Van Kooten O, Gloudemans AGM and Vredenberg WJ (1983) On the slow component of P-515 and the flash-induced reduction of cytochrome *b*-563 in chloroplast membranes. *Photobiochem Photobiophys* **6**: 9–14

Velthuys BR (1978) A third site of proton translocation in green plant photosynthetic electron transport. *Proc Natl Acad Sci USA* **75**: 6031–6032

Velthuys BR (1979) Electron flow through plastoquinone and cytochrome b_6 and *f* in chloroplasts. *Proc Natl Acad Sci USA* **76**: 2765–2769

Velthuys BR (1980) Electron and proton transfer events in chloroplasts during a short series of flashes. *FEBS Lett* **115**: 167–170

Wikstrom M and Krab K (1980) Respiration-linked H^+ translocation in mitochondria: Stoichiometry and mechanisms. *Curr Top Bioenerg* **10**: 51–101

Wikstrom M and Krab (1986) The semiquinone cycle. A hypothesis of electron transfer and proton translocation in cytochrome *bc*-type complexes. *J Bioenerg Biomembr* **18**: 181–193

Willms I, Malkin R and Chain RK (1987) Oxidation-reduction reactions of cytochrome b_6 in a liposome-incorporated cytochrome b_6-*f* complex. *Arch Biochem Biophys* **258**: 248–258

Willms I, Malkin R and Chain RK (1988) Quinone interaction with the chloroplast cytochrome b_6-*f* complex. *Arch Biochem Biophys* **264**: 36–44

COORDINATION AND REGULATION

5

Regulation of the 32 kD-D1 Photosystem II Reaction Center Protein

Sudhir K. Sopory[1], Maria Lucia Ghirardi[2], Bruce M. Greenberg[3], Tedd Elich[2], Marvin Edelman[4] and Autar K. Mattoo[2, 5]

[1]School of Life Sciences, Jawaharlal Nehru University, New Delhi, India;
[2]Beltsville Agricultural Research Center, Plant Molecular Biology Laboratory, USDA, Beltsville, MD, USA;
[3]Department of Biology, The University of Waterloo, Waterloo, Canada;
[4]Department of Plant Genetics, The Weizmann Institute of Science, Rehovot, Israel;
[5]Address correspondence to this author at Plant Molecular Biology Laboratory, ARS, US Department of Agriculture, Building 006, BARC-West, Beltsville, Maryland 20705–2350, USA.

CONTENTS

ABBREVIATIONS

PS I	:	Photosystem I;
PS II	:	Photosystem II;
Chlorophyll	:	Chl;
DCMU	:	3-(3, 4-Dichlorophenyl)-1, 1-dimethylurea;
SDS PAGE	:	Sodium dodecylsulfate-polyacrylamide gel electrophoresis.

132

ABSTRACT

The 32 kD-D1 protein is a chloroplast-encoded gene product that, in association with D2 and cytochrome b_{559}, constitutes the photosystem II (PS II) reaction center. The protein has additional characteristics that have attracted considerable attention. Among these are its rapid light-dependent degradation and its being the target site for many PS II herbicides. We review the current model for the organization and function of the PS II reaction center components in analogy with the bacterial reaction center. The different steps in the life cycle of the 32 kD-D1 protein including its post-translational modifications are also described. Finally, we address the question of light regulation of the protein at the transcriptional, post-transcriptional, and post-translational levels. The various models proposed for the structure and function of the 32 kD-D1 protein are being tested by means of cyanobacterial and *Chlamydomonas* transformation systems. Some recent results on the bioengineering of the herbicide-binding sites of the protein are also summarized.

I. Introduction

Photosynthesis is the process used by higher plants, algae, and some bacteria for the conversion of solar energy into chemical form. In higher plants, photosynthesis takes place in the chloroplasts, organelles that contain a set of highly organized vesicle-forming membranes known as "thylakoids". The thylakoids are differentiated into appressed lamellae (grana) and unappressed or stroma-exposed lamellae. This structural differentiation is accompanied by lateral segregation of the multi-subunit protein complexes along the membrane.

Photosynthesis in higher plants is mediated by two chlorophyll-containing protein complexes, PS I and PS II. Each photosystem consists of a reaction center core that binds the chromophores responsible for initial charge separation and stabilization, a tightly bound chlorophyll (Chl) *a*-containing core complex, and an accessory set of Chl*a/b*-containing polypeptides, which harvest the light and transfer it to the reaction center (Glazer and Melis, 1987).

PS II particles, free of the Chl*a/b* light-harvesting antenna, contain polypeptides of 47 kD, 43 kD, 32 kD (D1) and 30 kD (D2), the two subunits of cytochrome b_{559} (9 and 4 kD) and, in some cases, subunits of 34, 23, and 17 kD (Satoh *et al.*, 1983; Tang and Satoh, 1985; Ikeuchi *et al.*, 1985) that constitute the oxygen-evolving complex. Earlier studies had suggested the 47 kD and 43 kD polypeptides as the reaction center core of PS II, based on fluorescence emission and light-induced absorption characteristics (Cam and Green, 1983; Nakatani *et al.*, 1984). However, with the solution of the crystal structure of bacterial photosynthetic reaction center (Deisenhofer *et al.*, 1985; Deisenhofer and Michel, 1989), predictions were made that the PS II reaction center was located on the D1/D2 polypeptides, in analogy with the L and M subunits of the bacterial reaction center (Trebst, 1986; Barber, 1987). These predictions were fulfilled by the isolation of a physiologically active PS II reaction center particle from spinach (Nanba and Satoh, 1987) comprising D1, D2, and cytochrome b_{559}.

Although a functional PS II would require a stable association of the constituent proteins for electron transfer, D1 is known to undergo a high rate of turnover in the light (Edelman and Reisfeld, 1978; Mattoo, *et al.*, 1984) while both D2 and cytochrome b_{559} are stable as are also the bacterial L and M subunits. The fast turnover rate of D1 has been proposed to have a role in the regulation of a number of thylakoid functions (Mattoo and Edelman, 1985), mainly as a way to protect the thylakoid from excess irradiation. Another intriguing consequence of the fast turnover rate of D1 would be the constant assembly and disassembly of the PS II reaction center (Ghirardi *et al.*, 1990), which could involve intramembrane cycling of the PS II polypeptides between grana and stroma lamellae under physiological growth conditions (Ghirardi *et al.*, 1990) and under photoinhibitory conditions (Adir *et al.*, 1990; Guenther and Melis, 1990).

The 32 kD-D1 protein has attracted a lot of interest also because it is the site of action of the PS II-specific herbicides atrazine (Pfister *et al.*, 1981) and diuron (Mattoo *et al.*, 1981, 1986).

The importance of the light-induced turnover of D1 and its function as a component of the PS II reaction center and as the target site for PS II herbicides have combined to put it in the forefront of photosynthesis research. It has now been possible with the information available to build some models of the protein's life cycle in the thylakoids, of the molecular mechanisms of its function as a reaction center protein component, and of the herbicide-binding niche on the protein. Recent developments in genetic transformation of cyanobacterial and *Chlamydomonas* chloroplasts provide us with the tools necessary to test hypotheses posed by the models and to answer questions regarding mechanisms and signals involved in the dynamics of this membrane protein.

II. Photosynthetic Reaction Centers

A photosystem is an organized structure composed of pigments and proteins, which catalyzes the conversion of light energy to chemical energy. Within a photosystem, photosynthetic electron transport is triggered by a chlorophyll (Chl)-protein complex known as the "reaction center". The key property of a reaction center is its ability to stabilize the photoinduced charge separation between a primary donor (usually a special pair or a dimer of Chls) and an electron acceptor. Our knowledge of photosynthetic reaction centers was revolutionized when the crystal structure of the reaction center from the anaerobic bacterium *Rhodopseudomonas viridis* was solved (Deisenhofer, *et al.*, 1985; Deisenhofer and Michel, 1989), for which the Nobel Prize in Chemistry in 1988 was awarded to H. Michel, J. Deisenhofer, and R. Huber. The bacterial reaction center is built on a scaffold of four polypeptide subunits: L (21–24 kD), M (24–28 kD), H (28–35 kD), and a cytochrome (Deisenhofer *et al.*, 1985; Allen *et al.*,1988; Deisenhofer and Michel, 1989). The cofactors necessary for electron transport through the bacterial reaction center and associated with the L and M subunits are four bacteriochlorophyll *a* (BChl*a*) molecules, two bacteriopheophytins (BPheo), two quinones (Q_A and Q_B), a non-heme iron, and a carotenoid (Fig. 1). Two of the BChl*a* molecules form a dimer, which is the special pair (P960).

The DNA sequence of the genes for the L and M subunits revealed that the proteins are highly homologous to each other (Youvan *et al.*, 1984). Data from the crystallographic studies showed that L and M have similar secondary structures (Deisenhofer *et al.*, 1985;, Allen *et al.*, 1988; Deisenhofer and Michel, 1989), with each subunit containing five membrane-spanning helices and together forming a heterodimer (Fig. 1). The helices of the proteins and the associated cofactors are organized with two-fold rotational symmetry about an axis perpendicular to the plane of the membrane.

134

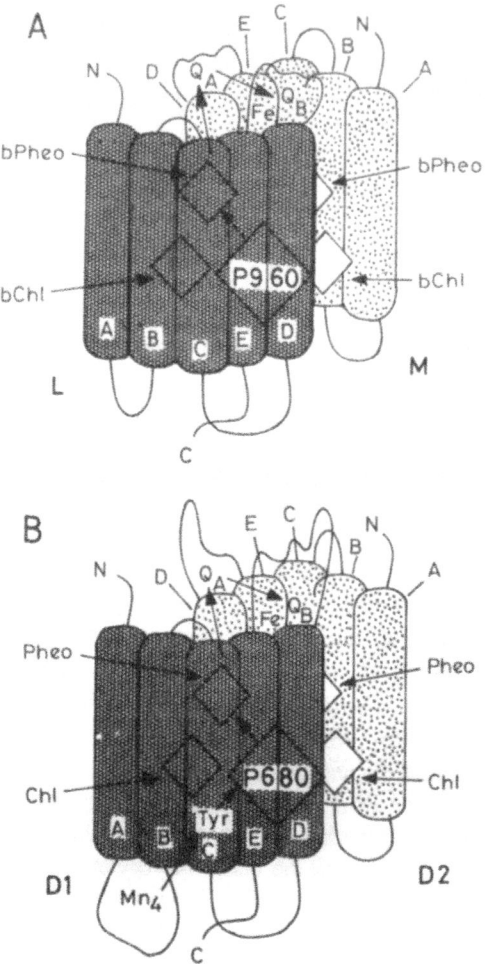

Figure 1. Structural comparison of the purple bacteria (A) and PS II (B) reaction centres. Each protein subunit (L, M, D1, and D2) and five membrane-spanning helices labeled A through E. Both reaction centers contain the special pair Chls (P960 and P680), two bPheos, two accessory bChls, Q_A, Q_B, and a non-heme iron. The helices and cofactors are arranged with two-fold rotational symmetry about an axis perpendicular to the membrane. The preferred pathway of electron transport through each complex is indicated. (Taken from Greenberg, 1990.)

When P960 absorbs a photon it is elevated to an excited state (P960*) and passes an electron to BPheo, the primary electron acceptor (Fig. 1), forming $bChl^+$ and $BPheo^-$ (see Deisenhofer and Michel, 1989). This reaction occurs in the picosecond time frame. The electron on $bPheo^-$ quickly (in 200 ps) migrates to the secondary electron acceptor, Q_A, and finally to Q_B (within about 100 µs). The rapid spatial isolation of the electron away from the P960 stabilizes the $BChl^+/Q_B^-$ charge separation. This provides sufficient time for the reduction of $P960^+$ by the cytochrome associated with the reaction center. Cytochrome reduction of $P960^+$ is much faster (nanoseconds) than the rate of charge recombination between P960$^+$ and Q$_B$ (hundreds of microseconds).

Q$_B$ must receive two electrons before it can diffuse out of the reaction center as reduced quinone (quinol, QH$_2$). However, P960 only has the capacity to pass one electron at a time with each photon input. Therefore, Q$_B$ must be reduced in two steps, and for this reason is referred to as a "two-electron gate". The first electron received by Q$_B$ results in a semiquinone anion radical (Q$_B^-$), which must be stabilized by the reaction center until a second electron arrives. When doubly reduced, Q$_B^{-2}$ is protonated to form quinol it dissociates from the reaction center and becomes a part of the quinone pool.

The PS II reaction center and the *R. viridis* reaction center are so exceptionally similar that it was possible to propose a structure and mechanism for PS II (Fig. 1; see also Trebst, 1986; Barber, 1987; Michel and Deisenhofer, 1988; Mattoo *et al.*, 1989). The structural similarity of the bacterial and the PS II reaction centers (Youvan *et al.*, 1984) was based on homology of the primary amino-acid sequences between the L and M subunits and their counterparts in PS II, the 32 kD-D1 and 30 kD-D2 polypeptides. The homology is most notable in the functional Chl and quinone-binding domains (see Michel and Deisenhofer, 1988). Extension of the crystal structure of the *R. viridis* reaction center to modeling PS II revealed that the D1 and D2 subunits can be folded in a manner analogous to the L and M polypeptides (Fig. 1; see also Trebst, 1986; Barber, 1987). The putative composition of the PS II reaction center was confirmed when a minimal and functional PS II particle, which contained only D1, D2, cytochrome b$_{559}$, four Chl*a*s, and two Pheos was isolated (Nanba and Satoh, 1987; Marder *et al.*, 1987). This preparation was found to be active in primary photoinduced charge separation between P680$^+$ and Pheo$^-$ (Nanba and Satoh, 1987; Seibert *et al.*, 1988). It was previously established that the 32 kD-D1 protein binds Q$_B$ (Vermaas *et al.*, 1983). Thus, the D1 and D2 proteins are predicted to be the apoproteins of the PS II reaction center, each forming five membrane-spanning helices (Trebst, 1986; Barber, 1987; Michel and Deisenhofer, 1988). Furthermore, the two proteins are proposed to bind all the cofactors necessary for electron transport through the reaction center (i.e., the special pair of Chl*a* molecules [P680], two accessory Chl*a*s, two Pheos, two quinones, and a non-heme iron).

The mechanism of electron transport through the PS II reaction center is proposed to be similar to the *R. viridis* reaction center (Fig. 1; see also Barber, 1987; Mattoo *et al.*, 1989). The time frames of the redox reactions in PS II are very close to the rates in the bacterial system (cf. Deisenhofer and Michel, 1989; Rutherford, 1989). In PS II, P680 is a dimer of Chl*a* molecules that is an energy trap because it absorbs at a longer wavelength (680 nm) than the antenna Chls (630–670 nm). When P680 absorbs the energy of a photon its redox potential is elevated to a level that allows it to reduce the Pheo bound to the D1 protein, forming P680$^+$/Pheo$^-$. The electron is then rapidly passed through the reaction center from Pheo to Q$_A$, and finally to Q$_B$, spatially separating the electron across the membrane from P680$^+$. Concomitantly, P680$^+$ is reduced by an electron that ultimately comes from water (see below).

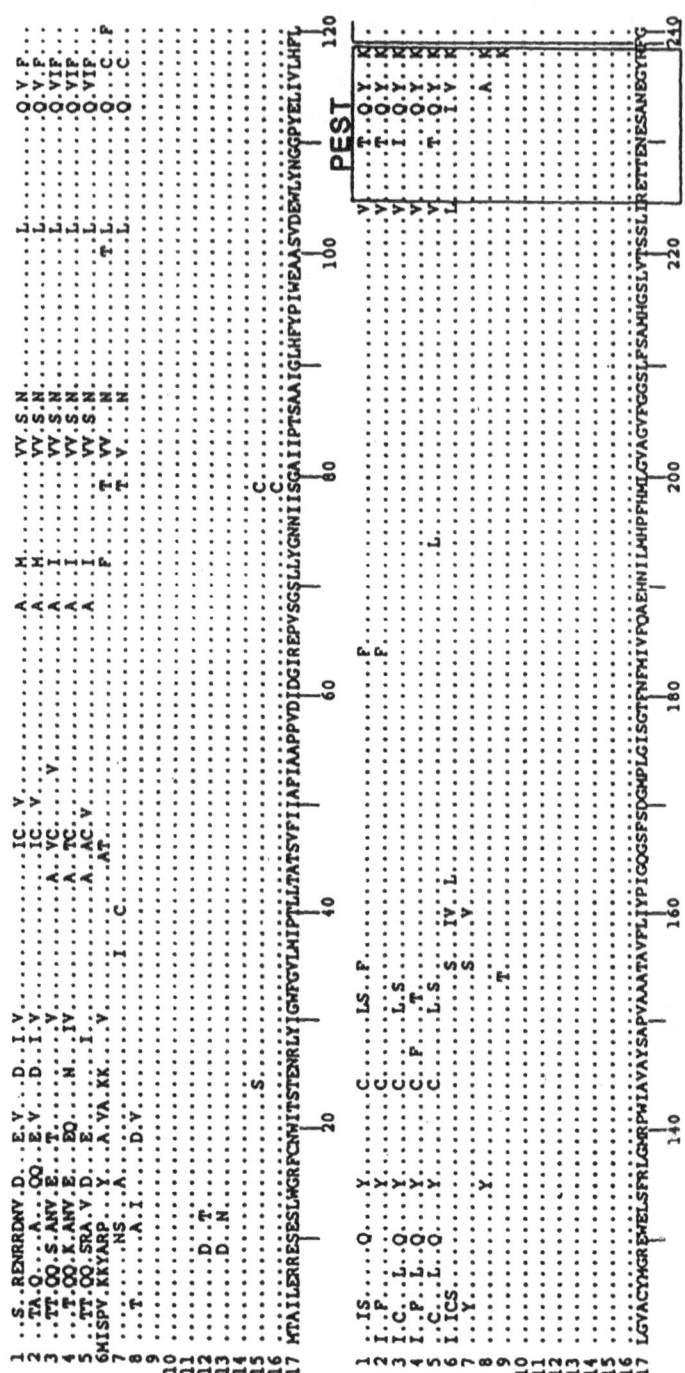

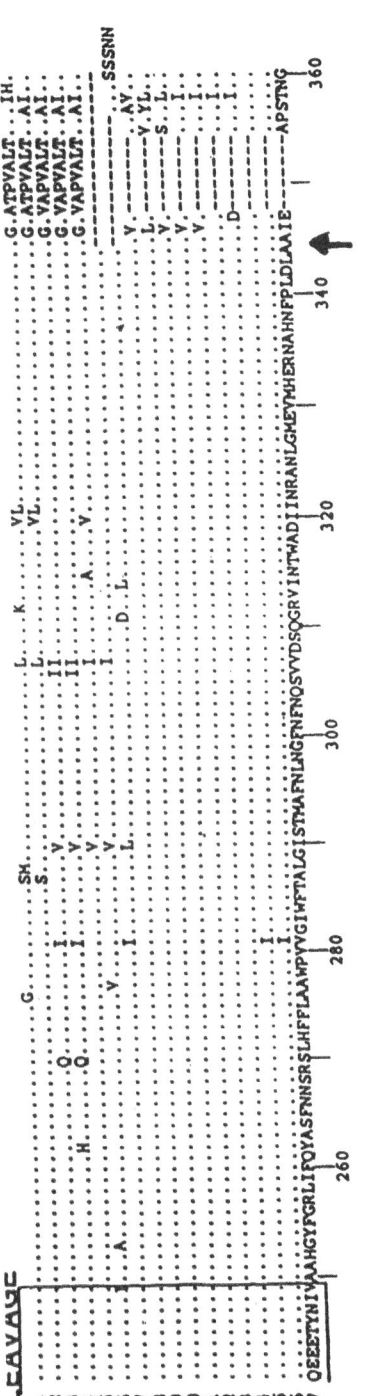

Figure 2. Comparison of deduced amino-acid sequences from *psbA* genes. The pest-like region and putative cleavage domain, which is in the area of the protein that also binds Q$_B$, are demarked. Sequences: [1] *Anacystis nidulans* copy I, and [2] II/III (Golden *et al.*, 1986). [3] *Anabaena* 7120 copy I and [4] copy II (Curtis and Haselkorn, 1984). [5] *Fremyella displosiphon* (Mulligan *et al.*, 1984). [6] *Euglena gracilis* (Karabin *et al.*, 1984). [7] *Chlamydomonas reinhardtii* (Erickson *et al.*, 1984) [8] *Marchantia polymorphia* (Ohyama *et al.*, 1986). [9] *Zea mays* (Larrinua and McLaughlin, 1987). [10] *Sinapis alba* (Link and Langridge, 1984). [11] *Brassica napus* (Reith and Strauss, 1987). [12] *Medica sativa* (Aldrich *et al.*, 1986a). [13] *Pisum sativum* (Oishi *et al.*, 1984). [14] *Glycine max* (Spielman [15] *Petunia hybrida* (Aldrich *et al.*, 1986b). [16] *Amaranthus hybridus* (Hirschberg and McIntosh, 1983). [17] *Solanum nigrum* (Goloubinoff and Stutz, 1983). [15] *Petunia hybrida* (Aldrich *et al.*, 1986b). [16] *Amaranthus hybridus* (Hirschberg and McIntosh, 1983). [17] *Solanum nigrum* (Goloubinoff *et al.*, 1984), *Spinacia oleracea*, *Nicotiana debneyi* (Zurawski *et al.*, 1982), and *Nicotiana tabacum* (Sugita and Sugiura, 1984). The arrow indicates the site of processing of the 33.5 kD precursor. Adapted from Mattoo *et al.* (1988).

Q_B has been known for some time to be a two-electron gate in PS II (Bouges-Bocquet, 1973; Velthuys and Amesz, 1974; Rich and Moss, 1987) which is analogous to the bacterial reaction center. When the first electron reaches Q_B, a semiquinone anion radical is formed. After Q_B is reduced to quinol by a second electron, it dissociates from the reaction center, blends in with the plastoquinone pool, and migrates to the cytochrome b_6/f complex (Rich and Moss, 1987).

The path of P680$^+$ reduction is different from that of P960$^+$. A redox active amino-acid residue in the 32 kD-D1 protein, tyr161, is responsible for donating an electron to P680$^+$ (Debus et al., 1988), replacing the cytochrome function in the bacterial reaction center. When P680$^+$ oxidizes tyr161, a tyr$^+$ radical is formed. This tyr$^+$ radical is then reduced by an electron that is split from water. Tyr161 in the 32 kD-D1 protein is referred to as Z, and has a symmetrical counterpart in the D2 protein (Tyr160, referred to as D) that is mechanistically silent.

The electron hole in the tyr$^+$ (Z$^+$) radical is filled by the oxygen-evolving complex, which is a Mn$_4$ cluster. The Mn complex is bound to the lumenal side of the D1/D2 heterodimer and is stabilized by three extrinsic polypeptides (Rutherford, 1989). The consecutive release of four electrons is required to form one di-oxygen molecule from two water molecules. The Mn cluster has a four-electron capacity, which is represented by five different oxidation states (S-states, S_0 to S_4). This allows the Mn complex to mediate the transfer of one electron at a time from the two water molecules to tyr$^+$. In the complex, S_0 is fully reduced and S_4 is fully oxidized. A one-photon input at P680 is required for extraction of the electron from tyr161, and tyr$^+$ re-reduction advances the S-state clock one click until it reaches S_4. The fully oxidized Mn$_4$ cluster is returned from S_4 to S_0 in a radiationless transition by releasing one O_2 molecule and accepting four electrons from two water molecules (see Chapter by Govindjee and Coleman, this volume).

III. The psbA Gene

The 32 kD-D1 protein is coded by the *psbA* gene, which in eukaryotic photosynthetic organisms is located on the chloroplast genome (Zurawski et al., 1982). The *psbA* genes from several higher plant species, algae, and cyanobacteria have been sequenced (Fig. 2). The deduced amino-acid sequence of the *psbA* gene predicted the hydrophobic nature of the protein (Zurawski et al., 1982) with five membrane-spanning helices (Trebst, 1986; Barber, 1987). Across evolution this gene shows a remarkable degree of sequence homology at the nucleotide level, as well as at the amino-acid level. For example, the *psbA* genes from higher plants share over 90% nucleotide sequence homology (Fig. 2), while the deduced amino-acid sequences of some plants (e.g., tobacco, *Solanum nigrum,* and spinach) are 100% homologous (Zurawski et al., 1982; Goloubinoff et al., 1984). Moreover, the sequence homology between higher plant and cyanobacterial *psbA* genes at the amino-acid level is around 85% (Golden et al., 1986).

The location of the *psbA* gene in the chloroplast genome of several plant species is highly conserved. In most cases, the gene is in the large single copy region just outside one of the inverted repeats (Whitfeld and Bottomley, 1983; Palmer, 1985). The orientation of the gene is such that transcription proceeds towards the inverted repeat. Notable exceptions to this organization are found in *Chlamydomonas* where the gene is localized in the inverted repeat (Erickson *et al.*, 1984), and in *Euglena* and peas, which do not have inverted repeats (Karabin *et al.*, 1984; Oishi *et al.*, 1984). Also, there are introns in the *Chlamydomonas* and *Euglena psb*A genes (Erickson *et al.*, 1984; Karabin *et al.*, 1984), while the higher plant *psb*A genes are devoid of introns (Zurawski *et al.*, 1982; Palmer, 1985).

Transcription of the *psb*A gene is under the control of a bacteria-like promoter. The gene contains a regulatory sequence at –10/–35 region (relative to the start of transcription) that is homologous to *Escherichia coli* promoters (Zurawski *et al.*, 1982; Gruissem and Zurawski, 1985). Also, similar to bacteria, the promoter is 50 to 100 bp 5′ to the initiation codon. The *psb*A promoter is one of the strongest chloroplast promoters (Gruissem and Zurawski, 1985), and specific mutations lower its strength. The rate of transcription of the *psb*A gene is under developmental and light control (Deng and Gruissem, 1987), although there is a basal level of transcription even in non-photosynthetic tissues.

IV. Metabolic Life History of the 32 kD-D1 Protein

a) Synthesis and maturation of the precursor

Translation of the *psb*A mRNA is regulated by light and the transcript is quite stable (Fromm *et al.*, 1985). Translation initiation could begin at one of two methionine codons (met-1 or met-37). If translation begins at met-1, a product of about 39 kD is predictable (Zurawski *et al.*, 1982), while initiation at met-37 would yield a protein of approximately 35 kD (Hirschberg and McIntosh, 1983). The precursor of 32 kD-D1, synthesized *in vivo* (Reisfeld *et al.*, 1982) and *in organello* (Grebanier *et al.*, 1978; Ellis and Barraclough, 1978; Hoffman-Falk *et al.*, 1982) has a mobility of 33.5 to 34.5 kD in sodium dodecyl sulfate (SDS) polyacrylamide gels, and thus it could be assumed that translation is initiated at met-37. However, it has been shown conclusively that translation begins at met-1 (Eyal *et al.*, 1987; Michel *et al.*, 1988). Therefore, the 32 kD-D1 protein runs with anomalous mobility in SDS polyacrylamide gel electrophoresis (PAGE), as is common for some hydrophobic proteins.

The 33.5–34 kD precursor of the 32 kD-D1 protein is initially inserted into the unappressed stroma lamellae (Mattoo and Edelman, 1987; Mattoo *et al.*, 1988) and likely assembled there into the PS II reaction center (Adir *et al.*, 1990). The exciting finding that the 33.5–34.5 kD precursor is processed *in vivo* to its mature size by a C-terminal cleavage, in contrast to the frequently encountered N-terminal processing of proteins, was first reported by Edelman and co-workers (Marder *et al.*, 1984). The determination of the site of processing of D1 was based on the distribution of individual amino-acid abundance in proteolytic fragments of the 32 kD-D1 protein and its precursor. Given the deduced amino-acid sequence of the protein and an established peptide map for the proteases employed, computer analysis comparing the predicted labeling patterns with those derived experimentally indicated that processing occurred at the C-terminus of the protein. This finding was later directly confirmed by protein sequencing data, which demonstrated that the processed protein had one amino acid removed at the N-terminus (Michel *et al.*, 1988), and alanine 344 formed the C-terminus of the mature protein (Takahashi *et al.*, 1988). The processing enzyme has been characterized from spinach chloroplasts (Inagaki *et al.*, 1989). It has a molecular weight of 34,000 and seems loosely associated with thylakoids, presumably with the unappressed lamellae.

Insight into the function of the processing event came from the finding that mutants that do not process the 32 kD-D1 protein lack oxygen evolution although reaction center assembly and primary photochemistry are unaffected (Diner *et al.*, 1988; Taylor *et al.*, 1988). This result suggests that cleavage of the C-terminal fragment is necessary for the assembly or stability of the oxygen-evolving complex of PS II. The question then arises as to why this fragment is translated in the first place. Possibilities include that it is required for membrane integration or for initial reaction center assembly.

b) Membrane translocation and degradation of the protein

The fate of the newly synthesized 32 kD-D1 protein was studied by pulse-chase experiments in conjunction with subcellular fractionation of chloroplast membranes (Mattoo and Edelman, 1987). Such studies showed that the 32 kD-D1 protein translocates from stroma lamellae to the grana partitions, in a post-translational event that results in the accumulation of the protein in the grana stacks. Translocation of the 32 kD-D1 protein to the grana stacks occurs after the protein is processed to the mature form in stroma lamellae. Very little is known about the signals responsible for intra-thylakoid protein translocation. The process is known to be stimulated by but not entirely dependent on light (F.E. Callahan and S.K Sopory, unpublished). It is conceivable that a light-induced modification causes changes in surface charge that energetically favors integration of the protein in the more hydrophobic grana environment. It is noteworthy that an increase in surface charge due to phosphorylation of light harvesting complex (LHC) II has been proposed to induce the intrathylakoid movement of the LHC polypeptide from grana to stroma lamellae (Haworth *et al.*, 1982; Horton, 1983; Larsson *et al.*, 1983). In contrast to what happens to LHC II, one has to invoke a decrease in surface charge (or increase in hydrophobicity) in order to explain the movement of the 32 kD-D1 protein in the opposite direction, from stroma lamellae to grana.

One of the most characteristic features of the 32 kD-D1 protein in the grana is its light-induced rapid turnover (Reisfeld *et al.*, 1978a, b; Mattoo *et al.*, 1984). Its steady state distribution in grana and stroma lamellae has been found to be 85 and 15, respectively (Wettern, 1986; Staehelin, 1986; Callahan *et al.*, 1989), which may reflect the differences observed in the time spent by the protein in the two membrane regions (9–18 minutes in stroma lamellae versus 6–12 hours in grana stacks at 30 μE m^{-2} s^{-1}) (Callahan *et al.*, 1987). The 32 kD-D1 protein is synthesized at rates comparable to the most abundant chloroplast protein ribulose-1, 5-*bis*phosphate carboxylase (RuBPcase)/oxygenase; yet, due to its fast turnover the 32 kD-D1 protein remains a low-abundance thylakoid component. Its half-life is dependent on the light intensity (Mattoo *et al.*, 1984; Greenberg *et al.*, 1987) and light quality (Gaba *et al.*, 1987; Greenberg *et al.*, 1989a, b) at which the plant is grown, in contrast to the analogous L subunit of the photosynthetic bacterial reaction center and other thylakoid membrane polypeptides that are relatively stable. The relative instability of the 32 kD-D1 protein has been linked to the presence of an alpha-helix destabilizing region between helices D and E (Figs. 1B & 2; Greenberg *et al.*, 1987), which is common to many rapidly degraded eukaryotic proteins (Rogers *et al.*, 1986) and may signal polypeptide cleavage in its vicinity. This region (PEST-like sequence) in the 32 kD-D1 protein is enriched in glutamate, serine, and threonine residues and is bordered by arg225 and arg238, two positively charged amino-acids. It is interesting that PEST regions are absent from the bacterial non-oxygenic L subunit (Mattoo *et al.*, 1988; Mattoo *et al.*, 1989) and from the rest of the chloroplast genome (Greenberg *et al.*, 1987).

Degradation of the 32 kD-D1 protein is blocked by the herbicides DCMU and atrazine (Mattoo *et al.*, 1981, Mattoo *et al.*, 1984), which block electron transfer from Q_A to Q_B by displacing Q_B from its binding site on D1 (Vermaas *et al.*, 1983; Trebst, 1987). Light-induced turnover of D1 has been correlated with occupancy of the Q_B site (Critchley, 1988; Ohad *et al.*, 1990). Conditions under which the site is unoccupied by Q_B, such as DCMU (3-(3, 4-Dichlorophenyl)-1, 1-dimethylurea) or atrazine treatment, or darkness, lead to an inhibition of D1 turnover.

Mutations that confer herbicide tolerance have been mapped to at least six positions (phe211, val219, ala251, phe255, ser264, leu275) in the *psbA* gene (Hirschberg and McIntosh, 1983; Goloubinoff *et al.*, 1984; Erickson *et al.*, 1985; Hirschberg *et al.*, 1987; Trebst, 1987; Johanningmeier *et al.*, 1987; Ajlani *et al.*, 1989). The area of the protein where the mutation occurs is phylogenetically conserved and is inclusive of the loop between helices D and E, which contains the Q_B binding site (Fig. 1). The most commonly found mutations in higher plants are ser264 to thr and ser264 to ala conversions (Hirschberg and McIntosh, 1983; Goloubinoff *et al.*, 1984; Trebst, 1987). This mutation confers atrazine tolerance, but leaves the plants sensitive to DCMU. However, a mutation of ser264 to gly in algae and cyanobacteria confers DCMU resistance (Erickson *et al.*, 1985; Hirschberg *et al.*, 1987; Ajlani *et al.*, 1989), while providing little atrazine tolerance. The mutations at the other positions provide differential herbicide tolerance traits, and double mutations (Hirschberg *et al.*, 1987; Ajlani *et al.*, 1989) can confer higher degrees of herbicide tolerance.

At least two photosensitizers have been implicated in the 32 kD-D1 protein degradation (see Section Vc) (Greenberg *et al.*, 1989a). At all light conditions where degradation of the 32 kD-D1 protein occurs, the same primary degradation fragment is observed suggesting a common proteolytic pathway (Greenberg *et al.*, 1989b). A primary degradation product of the 32 kD-D1 protein is a membrane-bound 23.5 kD fragment, which is thought to originate from cleavage in the region between arg238 and ile248 (Greenberg *et al.*, 1987). The cleavage site is located next to the PEST-like region, in the hydrophilic loop between helices D and E. This region contains the binding site for Q_B and DCMU, and is phylogenetically conserved in higher plants, algae, and cyanobacteria (Fig. 2; Mattoo *et al.*, 1988). The steady-state level of the 23.5 kD fragment was estimated to be one-fourth that of the 32 kD-D1 protein, implying that its turnover rate is four times faster.

Prior to its degradation, the 32 kD-D1 protein is modified. In pulse-chase experiments this novel form, identified on SDS-PAGE as a 32.5 kD band, appears solely in the grana after degradation of the 32 kD-D1 protein has commenced (Ghirardi *et al.*, 1990; Callahan *et al.*, 1990). This modified form, 32*, is structurally similar to the 32 kD-D1 protein, as shown by partial proteolysis and antibody labeling.

Formation of 32* is blocked by inhibitors of 32 kD protein degradation such as DCMU, Dinoseb, and propylgallate. The physiological steady-state ratio of 32* to 32 is about 1:4 in several higher plants (Callahan et al., 1990). 32* might represent a damaged form of the 32 kD-D1 protein, which is marked for proteolytic degradation.

Light-induced inactivation of PS II (photoinhibition) is detected at light intensities at which photosynthesis is saturated. Photoinhibition of photosynthesis in vivo results in both the loss of PS II activity and increase in the rate of turnover of the 32 kD-D1 protein, suggesting an interrelationship between the two phenomena (Ohad et al., 1984). Evidence (Ohad et al., 1988, 1990) has been presented for the existence of an irreversibly modified form of the 32 kD-D1 protein in Chlamydomonas, which follows a reversible photoinactivation of the PS II reaction center. It is not clear whether this modification is related to the physiological, light-induced degradation of the 32 kD-D1 protein, or whether it is a distinct phenomenon that occurs only under photoinhibitory conditions.

The fast turnover rate of the 32 kDa protein has been suggested to be involved in the regulation of thylakoid processes (Mattoo et al., 1981; Mattoo and Edelman, 1985). Indeed, the 32 kD-D1 protein has been designated as a "light meter" in the thylakoids. It has been suggested that its fast turnover protects the other photosynthetic components from the deleterious effect of excessive irradiation (Ellis, 1981), which could be achieved by quick replacement of the damaged protein in a "damage-repair" cycle (Guenther and Melis, 1990; Ghirardi et al., 1990; Adir et al., 1990). It is also possible that changes in the chlorophyll antenna size of the photosystems, which are sensitive to light intensity, are somehow linked to the rate of turnover of the 32 kD-D1 protein.

c) Post-translational modifications

The 32 kD-D1 protein is presently known to undergo as many as five post-translational modifications: (1) C-terminal processing; (2) covalent palmitoylation; (3) removal of the initiating methionine residue; (4) N-acetylation of the resulting N-terminal threonine residue (i.e., threonine-2 from the deduced amino-acid sequence); and (5) O-phosphorylation of the N-acetyl threonine. Processing of newly synthesized 32 kD-D1 protein was the first of the above listed modifications to be discovered, and has been discussed above.

Incubation of Spirodela plants, an aquatic duckweed, with ^{3}H-palmitate led to the rapid radiolabeling of two thylakoid proteins, 32 kD-D1 protein and LHCP (Mattoo and Edelman, 1987). Integrity of the label was demonstrated by chromatographic co-migration of the extracted, hydrolyzed label with palmitate standards. The insensitivity of the label to treatment with NH_2OH suggested an ether or amide linkage. Fractionation of labeled thylakoids demonstrated that the palmityolated 32 kD-D1 protein was located in the grana partitions. The palmityolation occurs post-translationally, is light-stimulated, and is apparently transient in nature. This modification has been speculated to be important in PS II assembly and/or protein translocation; however, there is no experimental evidence supporting or refuting either role.

Phosphorylation of thylakoid proteins has been intensively studied since 1978, mostly in regard to phosphorylation of LHC II. There appear to be at least 8 to 12 thylakoid proteins that are phosphorylated, apparently all on threonine residues (Michel and Bennett, 1987). The kinase(s) responsible is regulated by the redox state of the plastoquinone pool (Haworth *et al.*, 1982).

While considerable evidence has been amassed that indicates phosphorylation of LHC II as the mechanism behind state 1-state 2 transitions (Haworth *et al.*, 1982), the role of phosphorylation in the other thylakoid proteins is unknown and, for the most part, unaddressed. In fact, unambiguous identification of the other phosphorylated proteins and characterization of their phosphorylated sites had been lacking for quite some time. In this regard, although phosphorylation of 32 kD-D1 protein had been suggested as early as 1982 (Owens and Ohad, 1982), it was not until recently that this modification was rigorously established and mapped.

Michel and Bennett (1987) reported that PS II core particle isolated from ^{32}P-labeled chloroplasts contains four phosphoproteins. Tryptic digestion of these phosphorylated PS II cores resulted in the release of four phosphopeptides which were subsequently purified by Fe^{+3}-chelate affinity chromatography and reversed-phase HPLC. Sequencing by Edman degradation identified one peptide as the N-terminus of the *psb*H gene product missing the initiating methionine and containing a phosphothreonine as the second amino acid from the N-terminus (Fig. 3). The other three peptides were resistant to this procedure and were subsequently sequenced by tandem mass spectrometry, which revealed them to be the N-termini of the 32 kD-D1, D2 and CPa-2 proteins of PS II (Michel *et al.*, 1988). Each peptide contained an N-acetyl-O-phosphothreonine residue. In the case of 32 kD-D1 and D2 proteins, this residue corresponds to threonine-2 of the deduced amino-acid sequence, while in the case of CPa-2 it corresponds to threonine-15 (Fig. 3.). The fact that all of

D1:   N-Acetyl-Thr-Ala-Ile-Leu-Glu-Arg
              (P)
D2:   N-Acetyl-Thr-Ile-Ala-Val-Gly-Lys
              (P)
CPa-2: N-Acetyl-Thr-Leu-Phe-Asn-Gly-Thr-Leu-Thr-Leu-Ala-Gly-Arg
              (P)
psbH: NH$_2$-Ala-Thr-Gln-Thr-Val-Glu-Ser-Ser-Arg
              (P)

Figure 3. Amino-acid sequence of the N-terminal tryptic phosphopeptides from PS II core proteins. Positions of phosphorylation are indicated (P). After Michel and Bennett (1987) and Michel *et al.* (1988).

these proteins contain a phosphorylation site on or near their N-terminus suggests a common role for this modification on PS II core polypeptides. This role has not yet been directly addressed experimentally.

Michel *et al.* (1988) also demonstrated that the N-terminal methionine residue, where translation of 32 kD-D1 protein is initiated, is removed and the resulting N-terminal threonine acetylated. Whether or not these events occur in concert is unknown. Indeed, whether these events are truly post-translational, rather than co-translational, is also unknown. N-terminal amino-acid residues in other proteins have been implicated as determinants of protein stability (Bachmair *et al.*, 1986), and it has been suggested that N-terminal acetylation prevents digestion by aminopeptidases (Michel *et al.*, 1988).

V. Photoregulation of 32 kD-D1 Protein Metabolism

Light is one of the major factors that regulates the turnover of the 32 kD-D1 protein. In a number of nuclear-coded proteins, light affects the process of transcription. For the 32 kD-D1 protein, however, light regulates mRNA translation and protein degradation. The effects and the possible mechanisms of action of light at various stages of 32 kD-D1 protein turnover are described in the following sections.

a) Regulation of *psb* A gene transcription

Of the large number of proteins that are encoded by the chloroplast genome (see Marder and Barber, 1989) the two that exist as major products of *in vitro* and *in organello* protein synthesis are the large subunit of RuBPcase and the 32 kD-D1 protein (Ellis, 1981). As described in Section III, the *psb*A gene is located near the inverted repeat (Bedbrook *et al.*, 1978). Most of the genes for PS II proteins are part of polycistronic transcription units, except the *psb*A gene (Westhoff and Herrmann, 1988).

The expression of *psb*A gene has been studied both at the transcriptional and at the translational levels under various developmental states and light regimes. Bedbrook *et al.* (1978) showed that the *psb*A transcript level was higher in mature chloroplasts than in etioplasts. Evidence for the light stimulation of the 32 kD-D1 protein mRNA has accumulated in a number of plant species. Link (1982) measured the steady-state transcript levels in dark- and light-grown mustard seedlings using cloned DNA probes. The level of 1.22 kb mRNA for 32 kD-D1 protein was almost undetectable in dark-grown seedlings but became a major constituent of plastid mRNA pool in light-grown seedlings. Subsequent kinetic studies showed that light initiates the stimulation of *psb*A transcript only 48 hours after germination; the amount continued to increase in light but in dark-grown seedlings it leveled off and even declined after 72 hours. By 96 hours, about a five-fold increase in light was obtained (Hughes *et al.*, 1987). Similar results showing light stimulation of *psb*A transcript have been obtained with spinach (Herrmann *et al.*, 1985; Deng and Gruissem, 1987), barley (Kreuz *et al.*, 1986; Mullet and Klein, 1987; Klein and Mullet, 1987), peas and mung beans (Thompson *et al.*, 1983), tobacco (Richter, 1984), *Spirodela* (Reisfeld *et al.*, 1978b), and maize (Altman *et al.*, 1984).

The abundance of 32 kD-D1 protein mRNA indicates that the gene is probably triggered on by a strong promoter. A comparison of the *psb*A gene promoter with that of the *rbc*L gene showed strong homologies between them but not with other genes. Besides "–35 region" and "Pribnow box", three consensus sequences were observed (Sugita and Sugiura, 1984). A recent study on the *psb*A gene promoter of wheat has shown a strong resemblance to the sequences in other dicot species (Hanley-Bowdoin and Chua, 1988).

It is known that the nuclear genes triggered by light contain specific light-regulatory elements (LRE) like GT boxes, GC boxes, and AT boxes at the 5′ upstream region (Kuhlemeir *et al.*, 1987). Two groups of LREs could be distinguished. The first one, the CCTTATCAT motif (Grab and Stuber, 1987), was deciphered by computer analysis (similarities), and the second one, which corresponds to nuclear protein factor GT 1 (Green *et al.*, 1987, 1988), was revealed by experimental data. In fact, homology search by computer revealed the presence of sequences comprising box II′, II″ and III in the tobacco *psb*A gene (Victor and Los,' 1990). However, experimental evidence in favor of the role of LRE-like elements upstream of *psb*A gene is still awaited.

From these results, two points emerge regarding the *psb*A gene. One, it is not part of a polycistronic transcriptional unit, and two, it is regulated by light and contains LRE-like elements. The nature of the photoreceptor that modulates the level of the *psb*A transcript has been addressed in several studies.

The experiments of Link (1982) and Hughes and Link (1988) suggested that the receptor-pigment involved in stimulation of the *psb*A gene transcription was phytochrome. Northern blot analysis was used to detect changes in the level of the *psb*A transcript in dark-grown seedlings given different light treatments. The level of the *psb*A transcript, which was very low in dark-grown seedlings, increased appreciably upon irradiation with red light. The red light effect was reversed by far-red light, strongly indicating phytochrome action. A similar criterion was used to probe for phytochrome regulation of *psb*A transcript levels in peas (Thompson *et al.*, 1983) and maize (Zhu *et al.*, 1985).

It should be noted, however, that for both tobacco cell cultures (Richter, 1984) and *Spirodela* (Gressel, 1978) evidence has been presented showing a large enhancement in the *psb*A transcript levels by blue light, rather than by red light. These results suggest an involvement of cryptochrome (blue light receptor) in this response. Blue light was not effective in *Sinapis* (Hughes and Link, 1988). The reason for this difference can be attributed to the method of growing plants, and it is also possible that a high irradiance response is operative. In fact, irradiations with ruby-red light suggest that high irradiance response may be involved in modulating the level of the transcript in that organism (Hughes and Link, 1988).

Hughes and Link (1988) found that DCMU did not affect *psb*A transcript level but norflurazon completely inhibited light-induced accumulation of the *psb*A transcript. Norflurazon is known to retard carotenoid formation leading to photooxidation of chlorophyll. These data, therefore, suggested that photoconversion of protochlorophyllide to chlorophyll and its stabilization in thylakoids might be an important signal for the increase in the *psb*A transcript (Mullet *et al.*, 1990). In fact, phytochrome does not seem to be the predominant photoreceptor for *psb*A gene expression and its effect may be indirect (Hughes and Link, 1988).

Fromm *et al.* (1985) demonstrated that light-mediated increase in *psb*A transcript is the result of developmental changes related to conversion of etioplast to chloroplast. In their experiments on fully green *Spirodela,* they found that even though the 32 kD-D1 protein level varied according to the light conditions, there was no difference in the accumulation of 32 kD-D1 mRN.A. These data suggested that, in a fully developed and mature chloroplast, light-dark transitions do not change the mRNA levels and, therefore, regulation of *psb*A gene expression may well be at the post-transcriptional/translational level. It is possible that during the early phase of plastid development the organelle machinery is not competent to handle the RNA processing/translation. Thus, in the event of increased transcription of *psb*A, its mRNA level accumulates, and once the chloroplast is fully developed, the level is maintained at a steady state without undergoing major changes with light-dark cycles (Thompson, 1988). Such a view is also reflected in the studies on tomato where circadian rhythms in gene transcript levels were observed. Northern blots performed with RNA obtained at six different time points in a 24-hour period showed only slight alterations in *psb*A mRNA level but considerable fluctuations were seen in LHCP II mRNA (Piechulla, 1988). It seems, therefore, that light-mediated increase in the transcription of some genes during early stages may be more a function of chloroplast development than a direct effect of light. The differences could be due to the ability of a specific mRNA to process or recruit itself in the translational machinery.

Most of the work described above addressed light regulation of *psb*A transcript level only in higher plants. However, there are reports on light regulation of *psb*A transcription in the cyanobacterium *Synechocystis* (S6803). In this organism, *psb*A transcripts were not detected in dark-grown cultures (Mohamed and Jansson, 1990). In *Synechococcus*, two *psb*A genes exist. The expression of *psb*A-1 was found to be higher than that of *psb*A-2 (Golden *et al.*, 1986). The *psb*A-1 gene was expressed at low light intensity (< 50 µE m⁻²s⁻¹), whereas *psb*A-II expression was barely detectable in samples grown in light intensities below 100 µE m⁻²s⁻¹ and was abundant at high light intensities. The northern blot analysis showed that the changes in the abundance of *psb*A transcript occurred within a few minutes of transfer to different light environment (Golden *et al.*, 1990).

b) Post-transcriptional regulation

The experiments of Fromm *et al.* (1985) showed that the level of 32 kD-D1 protein is regulated post-transcriptionally since the steady-state level of *psb*A transcript did not change during light-dark transitions whereas the synthesis of the protein required light. Further support for the post-transcriptional regulation of *psb*A transcript abundance has come from other laboratories.

Deng *et al.* (1987) found that the increase in *psb*A transcript level did not correlate with an increase in its transcription. Mullet and Klein (1987) studied the relationship between changes in RNA levels and transcription during plastid development in barley seedlings. They found that the transcription rates declined as a function of plant age resulting in decreased levels of *rbc*L and *psa*A-*psb*A specific mRNAs.

In the light of the above data, two questions need to be addressed. One, how is mRNA stability achieved? Two, how does light promote the recruitment of mRNA in the protein-synthesizing machinery?

It has been suggested that 3′ inverted repeats help in mRNA stabilization. RNA molecules possessing 3′ inverted repeats were found to be substantially more stable than those lacking them (Stern and Gruissem, 1987). However, it is not clear how the differential accumulation would be achieved through inverted repeats. It is also unlikely that some exonucleases could accomplish this, since RNA processing generates the same 3′ ends in dark-grown and green plants. It is probable that some ''factors'' are required to preferentially protect a certain set of mRNA from rapid decay and/or render others more susceptible. It was found that chemically synthesized RNAs containing inverted repeats associate with proteins present in processing extract and that such associations may occur with the regions constituting inverted repeats (Stern and Gruissem, 1987). Whether these RNA-binding proteins are developmentally or environmentally regulated is unknown.

Illumination of dark-grown plants causes an increase in membrane-bound and stromal polysomes while the *psb*A transcript shows only a moderate redistribution (Klein *et al.*, 1988). It was hypothesized that the *psb*A transcript is always associated with polysomes (*i.e.*, there was no inhibition of translation initiation) but the protein could not be synthesized due to a block in chain elongation. The block could be overcome by chlorophyll as is known also for chlorophyll *a* apoprotein of PS I. Recent work (Mullet *et al.*, 1990) supports the notion that chlorophyll is required for accumulation of the full-length 32 kD-D1 transcript: pulse-labeled plastids of dark-grown barley plants show an accumulation of radioactivity in low MW translation intermediates of 15 and 23 kD polypeptides which, upon illumination, chase into the full-length 32 kD-D1 protein.

c) Photoreceptors in protein degradation

As has been discussed in the previous sections, the 32 kD-D1 protein undergoes light-dependent degradation. The exact mechanism of this photodegradation is not well understood. Recent studies have addressed possible photoreceptors involved in the process. Greenberg *et al.* (1989a) quantified the degradation of 32 kD-D1 protein over a broad spectral range. It was found that maximum degradation rates per photon flux occurred in the ultraviolet region. In the far red and visible light the shape of the action spectrum for the degradation rate was coincident with absorption spectrum of the blue photosynthetic pigments, chlorophyll and carotenoids. In order to check whether chlorophyll was the only photoreceptor responsible for degradation of the protein, the action spectrum was expressed as a function of the quantum yield of chlorophyll at each wavelength. A straight line was obtained between 447 and 731 nm, indicating that indeed chlorophyll is the only photoreceptor in the visible region of the spectrum. However, the quantum yield of the rate of degradation did not follow the flat response in the UV-C, UV-B, UV-A, and violet regions (254 nm to 429 nm), thus indicating that another absorbing species acts as a photosensitizer (Greenberg *et al.*, 1989a) in these spectral regions.

The above evidence for two different photoreceptors, one in the ultraviolet and the second in the visible, was complemented by the synergistic effect on the 32 kD-D1 protein degradation rate by mixing visible and ultraviolet light. The rate of degradation of 32 kD-D1 protein in plants illuminated with a mixture of visible and ultraviolet light was much greater than would have been expected from using visible or ultraviolet light independently at the same final fluence rate. These results are evidence for more than one photoreceptor contributing to the process (Attridge *et al.*, 1984). Further, when the bulk chlorophyll was reduced by growing plants in intermittent light/dark cycles, it was found (Greenberg *et al.*, 1989b) that in these chlorophyll-depleted plants the degradation rate spectrum coincided with the absorption spectrum of plastoquinone (in its various redox states), most comparable with the plastosemiquinone. Thus, the removal of bulk chlorophyll apparently unmasked a photosensitizer in the ultraviolet region.

148

Degradation of the 32 kD-D1 protein has been proposed to reflect excessive electron flow through PS II (Kyle, 1985), which is sensed by the accumulation of quinone radicals at the Q_B site. However, the analogous Q_B-containing L subunit of photosynthetic bacterial reaction center does not exhibit fast light-induced turnover. Fast turnover of this protein seems to be a characteristic of organisms catalyzing oxygenic photosynthesis. On the basis of these observations, Greenberg et al. (1990) proposed a model for the light-induced degradation of the 32 kD-D1 protein (Fig. 4).

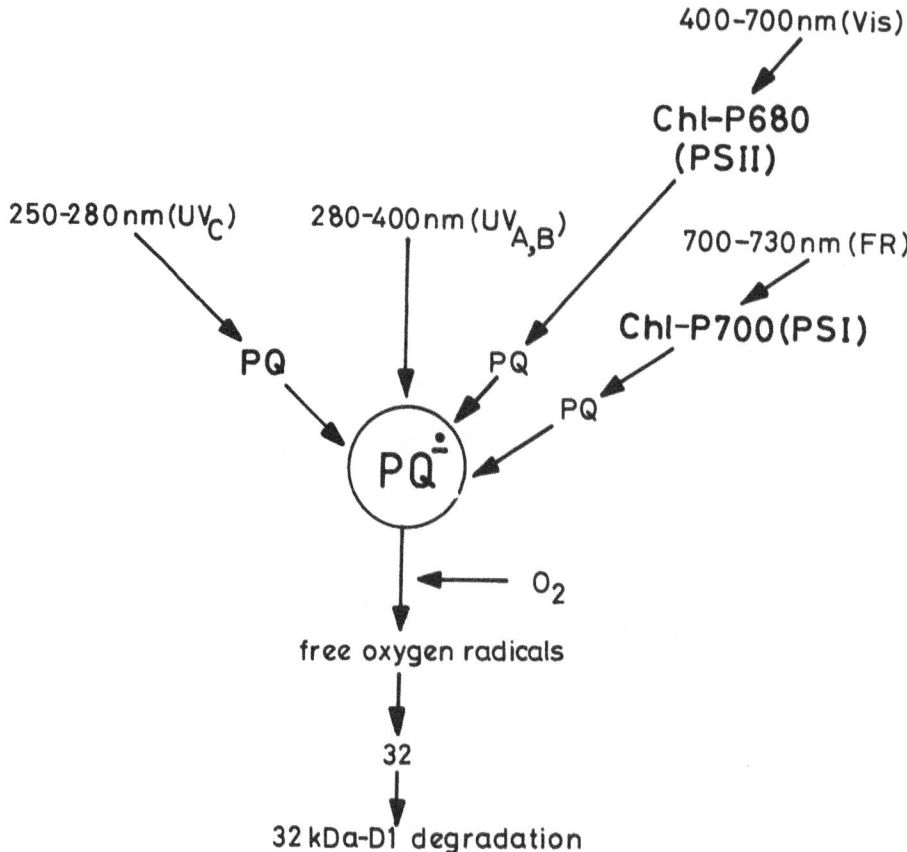

Figure 4. A model of the light-induced degradation of the 32 kD-D1 protein in the thylakoids. The absorption of light by two types of photoreceptors (shadowed) results in the formation of plastosemiquinone anion radical [PQ⁻] (see text). In UV-C (200–280 nm), the anion radical is formed by photoconversion of PQ (Creed et al., 1983); UV-A and UV-B (280–400 nm) irradiations excite PQ⁻ directly (Amesz, 1977). In visible light, the radical is transiently formed at the Q_B site during linear electron flow through PS II (Crofts and Wraight, 1983); in the far-red it is formed during PSI-mediated cyclic electron flow (Crofts and Wraight, 1983). The transient PQ⁻ species may react with O_2, producing free oxygen radicals that attack the 32 kD-D1 protein, marking it for subsequent proteolytic degradation. (Adapted from Greenberg et al., 1990)

According to this model, photodamage to the 32 kD-D1 protein occurs whenever a long-lived plastosemiquinone anion is generated (either by electron transfer from the reaction center to Q_B or by ultraviolet-mediated photoconversion of plastoquinone into a radical) in an oxygen-enriched environment. This species can readily react with oxygen, resulting in the formation of O^{-2} (superoxide free radicals). These radicals may interact with each other or may reduce Fe present in the PS II complex from Fe^{+3} to Fe^{+2}, causing the formation of secondary free radicals like hydroxyl radicals, which in turn may attack the 32 kD-D1 protein. Thus, the degradation of 32 kD-D1 protein may be a free-radical-mediated process. This idea is supported by the work of Sopory *et al.* (1990) showing that degradation of the 32 kD-D1 protein is inhibited by free radical scavengers such as uric acid and propylgallate, which do not affect PS II electron flow. As mentioned in Section IVb, a modified form of the 32 kD-D1 protein has recently been reported (Ghirardi *et al.*, 1990; Callahan *et al.*, 1990) in the thylakoids of *Spirodela* and other higher plants as an intermediate in the degradation of the protein. In agreement with the scheme described above, the modified form is absent from the thylakoids of plants treated with propylgallate. Under these conditions, the 32 kD-D1 protein does not turn over.

VI. Bioengineering of Functional Domains

DNA transformation and site-directed mutagenesis have been instrumental in forwarding our understanding of functional domains of the 32 kD-D1 protein. The major biological systems used for this purpose have been the cyanobacteria. The PS II reaction center of these oxygenic photosynthetic bacteria is similar to that of higher plants, in both structure and function.

Golden and Haselkorn (1985) exploited the transformability of *Synechococcus* (*Anacystis nidulans* R2). They first selected for a *Synechococcus* mutant resistant to the herbicide diuron. They next cloned a fragment from the *psb*A gene which, based on previous findings (Hirschberg and McIntosh, 1983; Goloubinoff *et al.*, 1984), was thought to contain within it the site for herbicide binding. The cloned fragment was then used to transform wild-type cells to diuron resistance. This experiment once and for all proved that mutation within *psb*A is responsible for that phenotype.

Horovitz *et al.* (1989) capitalized on the *Synechococcus* system to study interactions between two amino acids on the 32 kD-D1 polypeptide backbone involved in herbicide binding. To carry out their study, Horovitz *et al.* transformed cells by mutated *psb*A genes using site-directed mutagenesis, an oligonucleotide-mediated technique in which nucleotides are added, deleted, or substituted in a segment of DNA whose sequence is known (Zoller and Smith, 1982; Sambrook *et al.*, 1989). A "thermodynamic cycle" that consisted of two parallel and identical substitutions at residues 264 and 255 was prepared and the mutants analyzed for herbicide-binding strengths. Strict additivity for the two positions was demonstrated with a whole series of herbicides, indicating a lack of interactions between these two amino-acid residues.

As powerful as the *Synechococcus* system is, it has a major drawback to obligate phototrophy. Many PS II mutations are simply lethal to this organism. More recent studies have thus turned to transformable *Synechocystis* 6803, an organism that can grow photoheterotrophically on glucose, a trait which is crucial for maintenance of certain PS II defective mutants. This cyanobacterium has a relatively small genome size (5,000 kbp; Herdman *et al.*, 1979) and a short generation time (14–24 hours; Williams, 1988). In addition, the *Synechocystis* genes coding for the 32 kD-D1 (Jensson *et al.*, 1988) and D2 (Williams and Chisholm, 1987) polypeptides have been identified and mutations in these genes result in specific alterations in PS II functions. Site-directed mutations can be generated *in vitro* at a chosen site of the *psb*A gene and the mutated gene can then be used to replace the existing copy by homologous recombination, an efficient means of transformation in *Synechocystis* 6793 (Dzelzkalns and Bogorad, 1988).

Debus *et al.* (1988) used the *Synechocystis* system to pinpoint a functional site on the 32 kD-D1 protein involved in electron transfer. PS II contains two redox-active tyrosines. One of these, Z, reduces the reaction center chlorophyll, P680, and transfers the oxidizing equivalent to the oxygen-evolving complex. The second, D, has a long-lived free radical state of unknown function. Using site-directed mutagenesis, Debus and colleagues changed tyr161 of the D1 protein to Phe. The resulting mutant assembled PS II but was unable to grow photosynthetically and exhibited particularly altered fluorescence properties. These alterations showed that forward electron transfer to $P680^+$ was disrupted in the mutant.

The above studies reveal the power of bioengineering as a basic research tool for pinpointing functional and structural motifs in the 32 kD-D1 protein. Presently, several laboratories are investigating aspects such as the cleavage site, the processing site, and the relationship of herbicide binding to linear electron transport by means of site-directed mutagenesis. The recent development of DNA transformation technique for *Chlamydomonas* using high velocity microprojectiles coated with DNA to bombard the cells (Boynton *et al.*, 1988) promises to expand bioengineering of the 32 kD-D1 protein to eukaryotic cells.

VII. References

Adir N, Shochat S and Ohad I (1990) Light-dependent D1 protein synthesis and translocation is regulated by Reaction Center II. *J Biol Chem* **265**: 12563–12568

Ajlani G, Kirilovsky D, Picaud M and Asteir C (1989) Molecular analysis of psbA mutations responsible for various herbicide resistance phenotypes in *Synechocystis* 6714. *Plant Mol Biol* **13**: 469–479

Aldrich J, Cherney B and Merlin E (1986a) Sequence of the chloroplast-encoded psbA gene for the QB polypeptide of alfalfa. *Nucl Acids Res* **14**: 9537

Aldrich J, Cherney B, Merlin E, Christopherson LA and Williams C (1986b) Sequence of the chloroplast-encoded psbA gene for the QB polypeptide of petunia. *Nucl Acids Res* **14**: 9536

Allen JP, Feher G, Yeates TO, Komiya H and Rees DC (1988) Structure of the reaction center from *Rhodobacter sphaeroides* R-26: Protein-cofactor (quinones and Fe^{2+}) interactions. *Proc Natl Acad Sci USA* **85**: 8487–8491

Altman A, Cohen BN, Weissbach M and Brot N (1984) Transcriptional activity of isolated maize chloroplasts. *Arch Biochem Biophys* **235**: 26–33

Amesz J (1977) Plastoquinone. In: (A Trebst and M Avron, eds.,) *Encyclopedia of Plant Physiology*, Vol 5, Springer-Verlag, Berlin, pp 238–246

Attridge T, Black M and Gaba V (1984) Photocontrol of hypocotyl elongation in light-grown *Cucumis sativus*. *Planta* **162**: 422–426

Bachmair A, Finley D and Varshavsky A (1986) *In vivo* half-life of a protein is a function of its amino-terminal residue. *Science* **234**: 179–186

Barber J (1987) Photosynthetic reaction centres: A common link. *TIBS* **12**: 321–326

Bedbrook JR, Link G, Coen DM, Bogorad L and Rich A (1978) Maize plastid gene expressed during photoregulated development. *Proc Natl Acad Sci USA* **75**: 3060–3064

Bouges-Bocquet B(1973) Electron transfer between the two photosystems in spinach chloroplasts. *Biochim Biophys Acta* **314**: 250–256

Boynton JE, Gillham NW, Harris EH, Hosler JP, Johnson AM, Jones AR, Randolph-Anderson BL, Robertson D, Klein TM, Shark KB and Sanford JC (1988) Chloroplast transformation in *Chlamydomonas* with high velocity microprojectiles. *Science* **240**: 117–120

Callahan FE, Edelman M and Mattoo AK (1987) Post-translational acylation and intra-thylakoid translocation of specific chloroplast proteins. In: (J Biggins, ed) *Progress in Photosynthesis Research*, Vol 3, Nijhoff, the Hague, pp 799–802

Callahan FE, Wergin WP, Nelson N, Edelman M and Mattoo AK (1989) Distribution of thylakoid proteins between stromal and granal lamellae in *Spirodela*. *Plant Physiol* **91**: 629–635

Callahan FE, Ghirardi ML, Sopory SK, Mehta AM, Edelman M and Mattoo AK (1990) A novel metabolic form of the 32 kDa-D1 protein in the grana-localized reaction center of Photosystem II. *J Biol Chem* **265**: 15357–15360

Cam EL and Green BR (1983) Relationship between the two minor chlorophyll a-protein complexes and the photosystem II reaction centre. *Biochim Biophys Acta* **724**: 291–293

Creed D, Hales BJ and Porter G (1983) Photochemistry of the plastoquinones. *Proc R Soc London Series A* **334**: 505–521

Critchley C (1988) The chloroplast thylakoid membrane system is a molecular conveyor belt. *Photosynth Res* **19**: 265–276

Crofts T and Wraight CA (1983) The electrochemical domain of photosynthesis. *Biochim Biophys Acta* **726**: 149–185

Curtis S and Haselkorn R (1984) Isolation, sequence and expression of two members of the 32 kd thylakoid membrane protein gene family from the cyanobacterium *Anabaena* 7120. *Plant Mol Biol* **3**: 249–258

Debus RJ, Barry BA, Babcock GT and McIntosh L (1988) Site-directed mutagenesis identifies a tyrosine radical involved in the photosynthetic oxygen-evolving system. *Proc Natl Acad Sci USA* **85**: 427–430

Debus RJ, Barry BA, Sithole I, Babcock GT and McIntosh L (1989) Directed mutagenesis indicates that the donor to P680$^+$ in photosystem II is TYR-161 of the D1 polypeptide. *Biochemistry* **27**: 9071–9074

Deisenhofer J and Michel H (1989) The photosynthetic reaction centre from the purple bacterium *Rhodopseudomonas viridis*. *EMBO J* **8**: 2149–2170

Deisenhofer J, Epp O, Miki K, Huber R and Michel H (1985) Structure of the protein subunits in the photosynthetic reaction centre of *Rhodopseudomonas viridis* at 3A resolution. *Nature* **318**: 618–624

Deng X-W and Gruissem W (1987) Control of plastid gene expression during development: The limited role of transcriptional regulation. *Cell* **49**: 379–387

Deng X-W, Stern DB, Tonkyn JC and Gruissem W (1987) Plastid run on transcription. Application to determine the transcriptional regulation of plastid genes *J Biol Chem* **262**: 9641–9648

Diner BA, Ries DF, Cohen BN and Metz JG (1988) COOH-terminal processing of polypeptide D1 of the Photosystem II reaction center of *Scenedesmus obliquus* is necessary for the assembly of the oxygen-evolving complex. *J Biol Chem* **263**: 8972–8980

Dzelzkalns VA and Bogorad L (1988) Molecular analysis of a mutant defective in photosyntheitc oxygen evolution and isolation of a complementing clone by a novel screening procedure. *EMBO J* **7**: 333–338

Edelman M and Reisfeld A (1978) Characterization, translation and control of the 32,000 Dalton chloroplast membrane protein in *Spirodela*. In: (G Akoyunoglou and JH Argyroudi-Akoyunoglou, eds) *Chloroplast Development*, Elsevier/North Holland, Amsterdam, pp 641–652

Ellis RJ (1981) Chloroplast proteins: Synthesis, transport and assembly. *Annu Rev Plant Physiol* **32**: 111–137

Ellis RJ and Barraclough R (1978) Synthesis and transport of chloroplast proteins inside and outside the cell. In: (G. Akoyunoglou *et al.*, eds.) *Chloroplast Development*, Elsevier/North Holland, Amsterdam, pp 185–194

Erickson JM, Rahire M and Rochaix J-D (1984) *Chlamydomonas reinhardtii* gene for the 32,000 molecular weight protein of Photosystem II contains four introns and is located entirely within the chloroplast inverted repeat. *EMBO J* **3**: 2753–2762

Erickson JM, Rahire M, Rochaix J-D and Mets L (1985) Herbicide resistance and cross-resistance: Changes at three distinct sites in the herbicide-binding protein. *Science* **228**: 204–207

Eyal Y, Goloubinoff P and Edelman M (1987) The amino terminal region delimited by Met1 and Met37 is an integral part of the 32 kDa herbicide binding protein. *Plant Mol Biol* **8**: 337–343

Fromm H, Devic M, Fluhr R and Edelman M (1985) Control of psbA gene expression in mature *Spirodela* chloroplasts. Light regulation of 32-kd protein synthesis is independent of transcript level. *EMBO J* **4**: 291–295

Gaba V, Marder JB, Greenberg B, Mattoo AK and Edelman M (1987) Degradation of the 32 kD herbicide binding protein in far red light. *Plant Physiol* **84**: 348–352

Ghirardi ML, Callahan FE, Sopory SK, Elich TD, Edelman M and Mattoo AK (1990) Cycling of the Photosystem II reaction center core between grana and stroma lamellae. In: (M Baltscheffsky, ed) *Current Research in Photosynthesis*, Vol **1**, Kluwer Acad Publ, Dordrecht, The Netherlands, pp 733–738

Glazer AN and Melis A (1987) Photochemical reaction centers: Structure, organization, and function. *Annu Rev Plant Physiol* **38**: 11–45

Golden SS and Haselkorn R (1985) Mutation to herbicide resistance maps within the psbA gene of *Anacystis nidulans* R2. *Science* **229**: 1104–1107

Golden SS, Brusslan J and Haselkorn R (1986) Expression of a family of psbA genes encoding a photosystem II polypeptide in the cyanobacterium, *Anacystis nidulans* R2. *EMBO J* **5**: 2789–2798

Golden SS, Schaefer MR, Bustos S, Nolty MS and Cho DS (1990) Regulation of cyanobacterial gene families encoding proteins of Photosystem II. In: (M Baltscheffsky, ed) *Current Research in Photosynthesis*, Vol. **3**, Kluwer Acad Publ, Dordrecht, The Netherlands, 440–481

Goloubinoff P, Edelman M and Hallick RB (1984) Chloroplast-coded atrazine resistance in *Solanum nigrum* biotypes are isogenic except for a single condon change. *Nucl Acids Res* **24**: 9489–9496

Grab V and Stuber K (1987) Discrimination of phytochrome dependent light inducible from non-light inducible plant genes. Prediction of a common light-responsive element (LRE) in phytochrome dependent light inducible plant genes. *Nucleic Acid Res* **15**: 9957–9973

Grebanier AE, Coen DM, Rich A and Bogorad L (1978) Membrane proteins synthesized but not processed by isolated maize chloroplasts. *J Cell Biol* **78**: 734–746

Green PJ, Kay SA and Chua N-H (1987) Sequence-specific interactions of a pea nuclear factor with light-responsive elements upstream of the rbcs-3A gene. *EMBO J* **6**: 2543–2549

Green PJ, Young MM, Cuozzo M, Kano-Mirakami Y, Silverstasn P and Chua N-H (1988) Binding site requirements for pea nuclear protein factor GT-1 correlate with sequences required for light-dependent transcriptional activation of the rbcs-3A gene. *EMBO J* **7**: 4035–4044

Greenberg BM (1990) Photosynthesis. In: (YH Hui, ed) *The Encyclopedia of Food Science and Technology, Wiley-Interscience*, New York: in press.

Greenberg BM, Gaba V, Mattoo AK and Edelman M (1987) Identification of a primary *in vivo* degradation product of the rapidly-turning-over 32 kd protein of photosystem II. *EMBO J* **6**: 2865–2869

Greenberg BM, Gaba V, Canaani O, Malkin S, Mattoo AK and Edelman M (1989a) Separate photosensitizers mediate degradation of the 32-kDa photosystem II reaction center protein in the visible and UV spectral regions. *Proc Natl Acad Sci USA* **86**: 6617–6620

Greenberg BM, Gaba V, Mattoo AK and Edelman M (1989b) Degradation of the 32 kDA Photosystem II reaction center protein in UV, visible and far-red light occurs through a common 23.5 kDa intermediate. *Z. Naturforschung* **44c**: 450–452

Greenberg BM, Sopory S, Gaba V, Mattoo AK and Edelman M (1990) Photoregulation of protein turnover in the PSII reaction center. In: (M Baltscheffsky, ed) *Current Research in Photosynthesis*, Vol **1**, Kluwer Acad Publ, Dordrecht, The Netherlands, pp 209–216

Gressel J (1978) Light requirement for the enhanced synthesis of a plastid mRNA during *Spirodela* greening. *Photochem Photobiol* **27**: 167–169

Gruissem W and Zurawski G (1985) Analysis of promoter regions for the spinach chloroplast rbcL, atpB and psbA genes. *EMBO J* **4**: 3375–3383

Guenther JE and Melis A (1990) The physiological significance of photosystem II heterogeneity in chloroplasts. *Photosynth Res* **23**: 105–109

Hanley-Bowdoin L and Chua N-H (1988) Transcription of the wheat chloroplast gene that encodes the 32 kd polypeptide. *Plant Mol Biol* **10**: 303–310

Haworth P, Kyle DJ, Horton P and Arntzen CJ (1982) Chloroplast membrane protein phosphorylation. *Photochem Photobiol* **36**: 743–748

Herdman M, Janvier M, Rippka R and Stanier RY (1979) Genome size of cyanobacteria. *J Gen Microbiol* **111**: 73–85

Herrmann RG, Westhoff P, Alt J, Tittgen J and Nelson N (1985) In: (LV VlotenDoting, GSP Groot and PC Hall, eds.) *Molecular Form and Function of the Plant Genome*, Plenum Publ Corp, New York, pp 233–256

Hirschberg J and McIntosh L (1983) Molecular basis of herbicide resistance in *Amaranthus hybridus*. *Science* **222**: 1346–1349

Hirschberg J, Ohad N, Pecker I and Rahat A (1987) Isolation and characterization of herbicide resistant mutants in the cyanobacterium *Synechococcus* R2. *Z. Naturforsch* **42c**: 758–761

Hoffman-Falk H, Mattoo AK, Marder JB, Edelman M and Ellis RJ (1982) General occurrence and structural similarity of the rapidly-synthesized, 32,000-dalton protein of the chloroplast membrane. *J Biol Chem* **257**: 4583–4587

Horovitz A, Ohad N and Hirschberg J (1989) Predicted effects on herbicide binding of amino acid substitutions in the D1 protein of photosystem II. *FEBS Lett* **243**: 161–164

Horton P (1983) Control of chloroplast electron transport by phosphorylation of thylakoid proteins. *FEBS Lett* **152**: 47–52

Hughes JE and Link G (1988) Photoregulation of psbA transcript level in mustard cotyledons. In: (Govindjee, HJ Bohnert, W Bottemslay, DA Bryant, JE Mullet, WL Ogren, H Pakarasi and CR Sommerville, eds) *Molecular Biology of Photosynthesis*, Kluwer Acad Publ, Dordrecht, pp 423–439

Hughes JE, Neuhaus H and Link G (1987) Transcript levels of two adjacent chloroplast genes during mustard (*Sinapis alba* L.) seedling development are under temporal and light control. *Plant Mol Biol* **9**: 355–363

Ikeuchi M, Yuasa Y and Inoue Y (1985) Simple and discrete isolation of an O_2-evolving PS II reaction center complex retaining Mn and the extrinsic 33 kDa protein. *FEBS Lett* **185**: 316–332

Inagaki N, Fujita S and Satoh K (1989) Solubilization and partial purification of a thylakoidal enzyme of spinach involved in the processing of D1 protein. *FEBS Lett* **246**: 218–222

Jansson, C, Debus RJ, Osiewacz HD, Gurevitz M, and McIntosh L (1987) Construction of an obligate photoheterotrophic mutant of the cyanobacterium, *Synechocystis* 6803. Inactivation of the psbA gene family. *Plant Physiol* **85**: 1021-1025

Johanningmeier U, Bodner U and Wildner GF (1987) A new mutation in the gene coding for the herbicide-binding protein in *Chlamydomonas*. *FEBS Lett* **211**: 221–224

Karabin, GD, Farley M and Hallick RB (1984) Chloroplast gene for M_r 32000 polypeptide of photosystem II in *Euglena gracilis* is interrupted by four introns with conserved boundary sequences. *Nucl Acids Res* **12**: 5801–5812

Klein RR and Mullet JE (1987) Control of gene expression during higher plant chloroplast biogenesis. Protein synthesis and transcript levels of psbA, psaA-psaB, and rbcL in dark-grown and illuminated barley seedlings. *J Biol Chem* **262**: 4341–4398

Klein RR, Mason MS and Mullet JE (1988) Light regulated translation of chloroplast proteins. I Transcripts of psaA-psaB, psbA and rbcL are associated with polysomes in dark-grown and illuminated barley seedlings. *J Cell Biol* **106**: 289–301

Kreuz K, Dehesh K and Apel K (1986) The light-dependent accumulation of the P700 chlorophyll a protein of the photosystem I reaction centre. *Eur J Biochem* **159**: 459–467

Kuhlemeir C, Green PJ and Chua N-H (1987) Regulation of gene expression in higher plants. *Annu Rev Plant Physiol* **38**: 221–257

Kyle DJ (1985) The 32000 Dalton Q_B protein of Photosystem II. *Photochem Photobiol* **41**: 107–126

Larrinua IM and McLaughlin WE (1987) A gene cluster in the *Z. mays* plastid genome is homologous to part of the S10 operon of *E. coli*. In: (J. Biggins, ed) *Progress in Photosynthesis Research*, Vol 4, Nijhoff, The Hague, pp 649–652

Larsson UK, Jergil B and Andersson B (1983) Changes in the lateral distribution of the light-harvesting chlorophyll-a/b-protein complex induced by its phosphorylation. *Eur J Biochem* **136**: 25–29

Link G (1982) Phytochrome control of plastid mRNA in mustard (*Sinapis alba* L.) *Planta* **154**: 81–86

Link G and Langridge U (1984) Structure of the chloroplast gene for the precursor of the Mr 32,000 photosystem II protein from mustard. *Nucl Acids Res* **12**: 945–958

Marder JB and Barber J (1989) The molecular anatomy and function of thylakoid proteins. *Plant Cell and Environ* **12**: 595–614

Marder JB, Goloubinoff P and Edelman M (1984) Molecular architecture of the rapidly metabolized 32-kilodalton protein of photosystem II. *J Biol Chem* **259**: 3900–3908

Marder JB, Chapman DJ, Telfer A, Nixon PJ and Barber J (1987) Identification of psbA and psbD gene products, D1 and D2, as proteins of a putative reaction center of photosystem II. *Plant Mol Biol* **9**: 325–333

Mattoo AK and Edelman M (1985) Photoregulation and metabolism of a thylakoidal herbicide-receptor protein. In: (JB St. John, E Berlin and PC Jackson, eds) *Frontiers of Membrane Research in Agriculture*, Rowman and Allanheld, Totowa, pp 23–24

Mattoo AK and Edelman M (1987) Intramembrane translocation and posttranslational palmitoylation of the chloroplast 32-kDa herbicide-binding protein. *Proc Natl Acad Sci USA* **84**: 1497–1501

Mattoo AK, Pick U, Hoffman-Falk H and Edelman M (1981) The rapidly-metabolized 32,000-dalton polypeptide of the chloroplast is the "proteinaceous shield" regulating photosystem II electron transport and mediating diuron sensitivity. *Proc Natl Acad Sci USA* **78**: 1572–1576

Mattoo AK, Hoffman-Falk H, Marder JB, and Edelman M (1984) Regulation of protein metabolism: coupling of photosynthetic electron transport to *in vivo* degradation of the rapidly metabolized 32-kilodalton protein of the chloroplast membranes. *Proc Natl Acad Sci USA* **81**: 1380–1384

Mattoo AK, Marder JB, Gaba V and Edelman M (1986) Control of 32 kDa thylakoid protein degradation as a consequence of herbicide binding to its receptor. In: (G Akoyunoglou, ed) *Regulation of Chloroplast Differentiation*, Alan R. Liss, Inc, pp 607–613

Mattoo AK, Callahan FE, Greenberg BM, Goloubinoff P and Edelman M (1988) Molecular dynamics of the 32,000-dalton photosystem II herbicide-binding protein. In: (PA Hedin, JJ Mann and RM Hollingworth eds) *Biotechnology for Crop Protection*, ACS Symposium Series, Washington, DC, **379**: 248–357

Mattoo AK, Marder JB and Edelman M (1989) Dynamics of the Photosystem II Reaction Center. *Cell* **56**: 241–246

Michel HP and Bennett J (1987) Identification of the phosphorylation site of an 8.3 kDA protein from Photosystem II of spinach. *FEBS Lett* **212**: 103–108

Michel H and Deisenhofer J (1988) Relevance of the photosynthetic reaction center from purple bacteria to the structure of photosystem II. *Biochemistry* **27**: 1–7

Michel H, Hunt DF, Shabanowitz J and Bennett J (1988) Tandem mass spectrometry reveals that the three photosystem II proteins of spinach chloroplast contain N-acetyl-O-phosphothreonine *J Biol Chem* **263**: 1123–1130

Mohamed A and Jensson C (1990) Transcriptional light regulation of psbA gene expression in *Synechocystis 6803*. In: (M Baltscheffsky, ed) *Current Research in Photosynthesis*, Vol 3, Kluwer Acad Publ Dordrecht, The Netherlands 565–568

Mullet JE and RR Klein (1987) Transcription and RNA stability are important determinants of higher plant chloroplast RNA levels. *EMBO J* **6**: 1571–1579

Mullet JE Klein PG and Klein RR (1990) Chlorophyll regulates accumulation of the plastid-encoded chlorophyll-apoproteins CP43 and D1 by increasing apoprotein stability. *Proc Nat Acad Sci USA* **87**: 4038–4042

Mulligan B, Schulte N, Chen L and Bogorad L (1984) Nucleotide sequence of a multiple-copy gene for the B protein of photosystem II of a cyanobacterium. *Proc Natl Acad Sci USA* **81**: 2693–2697

Nakatani HY, Ke B, Dolan E and Arntzen CJ (1984) Identity of the Photosystem II reaction center polypeptide. *Biochim Biophys Acta* **765**: 347–352

Nanba O and Satoh K (1987) Isolation of a Photosystem II reaction center consisting of D-1 and D-2 polypeptides and cytochrome b-559. *Proc Natl Acad Sci USA* **84**: 109–112

Ohad I, Kyle DJ and Arntzen CJ (1984) Membrane protein damage and repair: Removal and replacement of inactivated 32-Kilodalton polypeptide in chloroplast membranes. *J Cell Biol* **99**: 481–485

Ohad I, Koike H, Shochat S and Inoue Y (1988) Changes in the properties of reaction center II during the initial stage of photoinhibition as revealed by thermoluminescence measurements. *Biochim Biophys Acta* **933**: 288–298

Ohad I, Adir N, Koike H, Kyle DJ, and Inoue Y (1990) Mechanism of photoinhibition *in vivo*. A reversible light-induced conformational change of reaction center II is related to an irreversible modification of the D1 protein. *J Biol Chem* **265**: 1972–1979

Ohyama K, Fukuzawa H, Kohchi T, Shirai H, Sano T, Sano S, Umesono, Shiki Y, Takeuchi M, Chang K, Aota A-I, Inokuchi H and Ozeki H (1986) Chloroplast gene organization deduced from the complete sequence of liverwort *Marchantia polymorpha* chloroplast DNA. *Nature* **322**: 572–574

Oishi KK, Shapiro DR and Tewari KK (1984) Sequence organization of a pea chloroplast gene coding for a 34,500-dalton protein. *Molec Cell Biol* **4**: 2556–2563

Owens GC and Ohad I (1982) Phosphorylation of *Chlamydomonas reinhardii* chloroplast membrane proteins *in vivo* and *in vitro*. *J Cell Biol* **93**: 712–718

Palmer JD (1985) Comparative organization of chloroplast genomes. *Annu Rev Genet* **19**: 325–354

Pfister K, Steinback KE, Gardner G and Arntzen CJ (1981) Photoaffinity labeling of an herbicide receptor protein in chloroplast membranes. *Proc Nalt Acad Sci USA* **78**: 981–985

Piechulla B (1988) Plastid and nuclear mRNA fluctuations in tomato leaves—Diurnal and circadian rhythms during extended dark and light periods. *Plant Mol Biol* **11**: 345–353

Reisfeld A, Jakob KM and Edelman M (1978a) Characterization of the 32000 dalton chloroplast membrane protein. II. The molecular weight of chloroplast messenger RNAs translating the precursor to P-32000 and full-size RUBP carboxylase large subunit. In: (G Akoyunolglou *et al.*, eds) *Chloroplast Development*, Elsevier, North Holland, pp 669–674

Reisfeld A, Gressel J, Jakob KM and Edelman M (1978b) Characterization of the 32,000 dalton membrane protein. I. Early synthesis during photoinduced plastid of *Spirodela*. *Photochem Photobiol* **27**: 161–165

Reisfeld A, Mattoo AK and Edelman M (1982) Processing of a chloroplast-translated membrane protein *in vivo*. *Eur J Biochem* **124**: 125–129

Reith M and Strauss NA (1987) Nucleotide sequence of the chloroplast gene responsible for triazine resistance in *Canola*. *Theor Appl Genet* **73**: 357–363

Rich P and Moss DA (1987) The reactions of quinones in higher plants. In: (J Barber, ed) *The Light Reactions*, Elsevier, Amsterdam, pp 421–445

Richter G (1984) Blue light control of the level of two plastid mRNAs in cultured plants cells. *Plant Mol Biol* **3**: 271–276

Rodermal SR and Bogorad L (1985) Maize plastid photogenes: Mapping and photoregulation of transcript levels during light-induced development. *J Cell Biol* **100**: 463–476

Rogers S, Wells R and Rechsteiner M (1986) Amino acid sequences common to rapidly degraded proteins: The PEST hypothesis. *Science* **234**: 364–368

Rutherford AW (1989) Photosystem II, the water splitting enzyme. *TIBS* **14**: 227–232

Sambrook J, Fritsch EF and Maniatis T (1989) Oligonucleotide mediated mutagenesis. In: *Molecular Cloning* Vol 2, Cold Spring Harbor Laboratory Press, New York, pp 15.51–15.80

Satoh K, Nakatani HY, Steinback KE, Watson J and Arntzen CJ (1983) Polypeptide composition of a photosystem II core complex. Presence of a herbicide-binding protein. *Biochim Biophys Acta* **724**: 142–150

Seibert M, Picorel R, Rubin AB and Connolly JS (1988) Spectral, photo-physical and stability properties of the isolated PS II reaction center. *Plant Physiol* **87**: 303–306

Shinozaki K, Ohme M, Tanaka M, Wakasugi T, Hayoshide N, Matsubayushi T, Zaita N, Chunwongse J, Obokata J, Yamoguchi-Shinozaki K, Ohto C, Torazawa K, Meng By, Sugita M, Deno H, Kamogashia T, Yawoda K, Kusuda J, Takaiwa F, Kato A, Tohsloh N, Shimada H and Sugiura M (1986) The complete nucleotide sequence of the tobacco chloroplast genome: Its organization and expression. *EMBO J* **5**: 2043–2050

Sopory SK, Greenberg BM, Mehta RA, Edelman M and Mattoo AK (1990) Free radical scavengers inhibit light-dependent degradation of the 32 kDa Photosystem II reaction center protein. *Z Naturforschung* **45c**: 412–417

Spielmann A and Stutz E (1983) Nucleotide sequence of soybean chloroplast DNA regions which contain the psbA and trnH genes and cover the ends of the large single copy region and one end of the inverted repeats. *Nucl Acids Res* **11**: 7157–7167

Staehelin LA (1986) Chloroplast structure and supramolecular organization of photosynthetic membranes. In: (A Pirson and MH Zimmerman, eds) *Encyclopedia of Plant Physiology* New Series Vol **19**, Springer-Verlag, Berlin pp 1–84

Stern DB and Gruissem W (1987) Control of plastid gene expression: 3′ inverted repeats act as mRNA processing and stabilizing elements, but do not terminate transcription. *Cell* **51**: 1145–1157

Sugita M and Sugiura M (1984) Nucleotide sequence and transcription of the gene for the 32,000 dalton thylakoid membrane protein from *Nicotiana tabacum*. *Mol Gen Genet* **195**: 308–318

Takahashi M, Shiraishi T and Asada K (1988) COOH-terminal residues of D1 and the 44 kDa CPa-2 at spinach Photosystem II core complex. *FEBS Lett* **240**: 6–8

Tang X-S and Satoh K (1985) The oxygen-evolving photosystem II core complex. *FEBS Lett* **179**: 60–64

Taylor MAQ, Packer JCL and Bowyer JR (1988) Processing of the D1 polypeptide of the photosystem II reaction center and photoactivation of a low fluorescence mutant (LF-1) of *Scenedesums obliquus*. *FEBS Lett* **237**: 229–233

Thompson WF (1988) Photoregulation. Diverse gene responses in greening seedlings. *Plant Cell Environ* **11**: 319–328

Thompson WF, Everett M, Polans NO, Jorgensen RA and Palmer JD (1983) Phytochrome control of RNA levels of developing pea and mung-bean leaves. *Planta* **158**: 487–500

Trebst A (1986) The topology of the plastoquinone and herbicide binding peptides of photosystem II in the thylakoid membrane. *Z Naturforsch* **41c**: 240–245

Trebst A (1987) The three-dimensional structure of the herbicide binding niche on the reaction center polypeptides of photosystem II. *Z. Naturforsch* **42c**: 742–750

Velthuys BR and Amesz J (1984) Charge accumulation of the reducing side of system 2 photosynthesis. *Biochim Biopys Acta* **333**: 85–94

Vermaas WFJ, Arntzen CJ, Gu LQ and Yu CA (1983) Interactions of herbicides and azidoquinones at a photosystem II binding site in the thylakoid membrane. *Biochim Biophys Acta* **723**: 266–275

Victor ES and Los DA (1990) Transcription of some chloroplast genes could be under phytochrome control: A computer prediction and analysis of light-responsive sequences. In: (M Baltscheffsky, ed) *Current Research in Photosynthesis*, Vol 3, Kluwer Acad Publ, Dordrecht, The Netherlands pp 593–596

Westhoff P and Herrmann RG (1988) Complex RNA maturation in chloroplasts. *Eur J Biochem* **171**: 551–564

Wettern M (1986) Localization of 32000 dalton chloroplast protein pools in thylakoids: Signficance in atrazine binding. *Plant Sci* **43**: 173–177

Wettern M and Ohad I (1984) Light-induced turnover of thylakoid polypeptides in *Chlamydomonas reinhardtii*. *Israel J Bot* **33**: 256–263

Whitfeld PR and Bottomley W (1983) Organization and structure of chloroplast genes. *Annu Rev of Plant Physiol* **34**: 279–310

Williams JGK (1988) Construction of specific mutations in the PS II photosynthetic reaction center by genetic engineering methods in the cyanobacterium, *Synechocystis* 6803. *Methods Enzymol* **167**: 766–779

Williams JK and Chisholm DA (1987) Molecular sequence of both genes from the cyanobacterium *Synechocystis* 6803. In: (J Biggins, ed) *Progress in Photosynthetic Research*, Vol 4, Nijhoff, The Hague, 809–812

Youvan DC, Bylina EJ, Alberti M, Begusch H and Hearst JE (1984) Nucleotide and deduced polypeptide sequences of the photosynthetic reaction-center, B870 antenna, and flanking polypeptides from *R. capsulata*. *Cell* **37**: 949–957

Zhu YS, Kung SD and Bogorad L (1985) Phytochrome control of levels of mRNA complementary to plastid and nuclear genes of maize. *Plant Physiol* **79**: 371–376

Zoller MJ and Smith M (1982) Oligonucleotide-directed mutagenesis using M13-derived vectors: an efficient and general procedure for the production of point mutations in any fragment of DNA. *Nucl Acids Res* **10**: 6487–6500

Zurawski G, Bonhert H-J, Whitfeld PR and Bottomley W (1982) Nucleotide sequence of the gene for the 32,000-M_r thylakoid membrane protein from *Spinacia oleracea* and *Nicotiana debneyi* predicts a totally conserved primary translation product of M_r 38950. *Proc Natl Acad Sci USA* **79**: 7699–7703

6

Regulation of Electron Transport at the Acceptor Side of Photosystem II by Herbicides, Bicarbonate and Formate

Jack J.S. van Rensen

Laboratory of Plant Physiological Research
Agricultural University,
Arboretumlaan 4, 6703 BD Wageningen
The Netherlands

CONTENTS

ABBREVIATIONS

Atrazine	:	2-chloro-4-ethylamino-6-isopropylamino-1, 3, 5-triazine;
Bromoxyni	:	3, 5-dibromo-4-hydroxy-benzonitril;
DCMU	:	3-(3, 4-dichlorophenyl)-1, 1-dimethylurea;
DBMIB	:	dibromothymoquinone;
Dinoseb	:	2, 4-dinitro-6-s-butyl-phenol;
DNOC	:	4, 6-dinitro-o-cresol;
Ioxynil	:	3, 5-diiodo-4-hydroxy-benzonitril;
kd	:	apparent dissociation constant;
L	:	low molecular weight subunit;
M	:	medium molecular weight subunit;
Metribuzin	:	4-amino-6-isopropyl-3-methylthio-1, 2, 4-triazine-5-one;
NO	:	nitric oxide;
PS I	:	photosystem I;
PS II	:	photosystem II;
Q_A	:	primary quinone electron acceptor of PS II;
Q_B	:	secondary quinone electron acceptor of PS II;
Simeton	:	2-methoxy-4, 6 bis (ethylamino)-1, 3, 5-triazine

158

ABSTRACT

The photosystem II reaction center can be considered as a water-plastoquinone oxido-reductase. Using four photons it transfers four electrons from two molecules of water to plastoquinone producing molecular oxygen and two molecules of doubly reduced plastoquinone. Our understanding of the structure and function of this complex has greatly increased during the recent years. The basis of the reaction center of photosystem II is formed by the D1 and D2 proteins, both having a molecular mass of about 32 kDa. The D1 protein contains not only the binding site for the physiological electron carrier Q_B, but also the binding sites for several classes of herbicides and for bicarbonate and formate.

Both the diuron-type and the phenol-type herbicides act by replacing the physiological electron carrier Q_B from its binding site at the D1 protein. Because the herbicides cannot be reduced, the electron flow is interrupted between the primary electron acceptor of photosystem II Q_A, and the plastoquinone pool. There appears a relation between the residence time of a herbicide at the D1 protein and its activity as an inhibitor of electron flow.

Incubation of isolated chloroplasts with formate, while flushing them with nitrogen gas, results in full inhibition of electron flow activity, which can be restored by addition of bicarbonate. This antagonistic action of formate and bicarbonate is located at the D1 protein and affects electron flow between Q_A and the plastoquinone pool. The advances in this field should encourage future work on the mechanism of the action of formate and bicarbonate at the molecular level as well as on their action *in vivo*.

The study of triazine-resistance in weeds and herbicide-resistance in algae and photosynthetic bacteria has resulted in the recognition of a common binding niche for Q_B, herbicides, formate and bicarbonate at the D1 protein including the hydrophobic transmembrane helices IV and V and the parallel helix connecting these on the matrix side of the D1 protein.

I. Introduction

The acceptor side of photosystem (PS) II is a fascinating topic because of the various different aspects involved and the remarkable progress recently made. PS II is itself at the basis of the primary plant production. It converts light energy into chemical energy: the light-induced charge separation results in positive and negative charges. The positive charges are used for the splitting of water and release of oxygen (see chapter by Govindjee and Coleman, this volume). The negative charges at the acceptor side lead to electron flow toward photosystem I (PS I); this electron flow is coupled with ATP formation. After another charge separation in PS I $NADP^+$ is reduced. NADPH, H^+ and ATP are used for the reduction of CO_2.

It was very exciting when, a few years ago, Deisenhofer, Michel and Huber succeeded in crystallizing the membrane-bound reaction center protein of the photosynthetic bacterium *Rhodopseudomonas viridis* and subseqently determined its structure (see Michel *et al.*, 1986). This was such an important achievement that they were awarded the Nobel prize in chemistry in 1988. This structure enabled to predict the structure of the green plant PS II (Trebst, 1986).

Since their introduction between 1950 and 1960 many herbicides, especially the triazines and ureas, have been used as tools in the study of the mechanism of photosynthesis. They appear to act specifically at the PS II acceptor side. This is also the site of the antagonistic action of formate and bicarbonate on electron transport.

In 1970, the first incident of triazine resistance was found in the U.S.A. Since then this resistance has spread all over the world. The study of this phenomenon has greatly improved our understanding of PS II. The investigations of induced resistance against several herbicides in algae has provided the basis for the topology of the herbicide and the Q_B-binding site.

II. The Photosynthetic Electron Transport Pathway

The process of photosynthesis takes place within the chloroplast. The chloroplast contains the stroma in which the grana are embedded. The light energy conversion processes are located in the grana, while CO_2-reduction occurs in the stroma. Grana consist of stacks of thylakoids, i.e., vesicle-like structures having an internal space surrounded by a membrane. The grana are interconnected by unappressed stroma thylakoids. The thylakoid membrane contains the electron and proton translocating components (Fig. 1).

Photosynthesis starts with the absorption of light energy by the chlorophylls of the two photosystems (PS II and PS I). The excitation energy is transferred to the reaction centers: P680 in PS II, and P700 in PS I. P680 and P700 are specialized chlorophyll a molecules which are able to accomplish a charge separation, resulting in $P680^+$ and $Pheo^-$ in light reaction II, and $P700^+$ and Chl^- in light reaction I (Pheo is pheophytin; Chl is a bound chlorophyll monomer; see Wasielewski et $al.$ 1987, 1989). The charge separations are followed by electron and proton translocating reactions.

The electron-hole of $P680^+$ is filled up via various steps, indicated in Fig. 1 as S, by an electron which is ultimately derived from water. The "splitting" of water yields not only electrons, but also oxygen and protons. From $Pheo^-$, the electron is transported to the first stable quinone electron acceptor Q_A and then to the secondary quinone electron acceptor Q_B. Because of its microenvironment, Q_A acts as a one-electron carrier, it can be reduced as far as the semiquinone state. Q_B acts as a two-electron gate. It accumulates two electrons, becomes protonated with two protons from the stroma and exchanges with the plastoquinone pool (PQ and PQH_2, see Crofts and Wraight, 1983). From PQH_2, electrons are transferred to the cytochrome

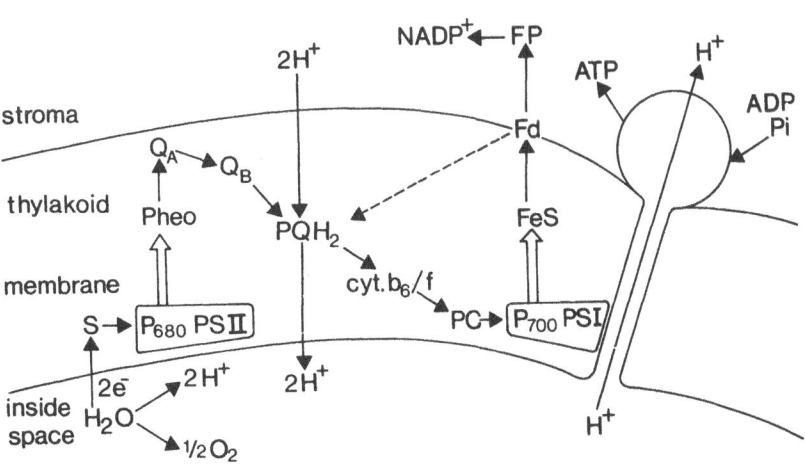

Fig. 1. Electron and proton transport pathways in photosynthetic electron flow. (For details see text)

b$_6$/cytochrome f complex and two protons are liberated into the internal lumen of the thylakoid. Plastocyanin (PC) is the primary electron donor to P700$^+$. From Chl$^-$ the electron is transferred to FeS centres X, A and B and then to Fd (soluble ferredoxin), and *via* a ferredoxin-NADP-reductase (a flavoprotein, FP) to NADP$^+$. Under certain conditions a cyclic electron flow is possible by transfer of electrons from reduced Fd to the plastoquinone pool. The protons which have been accumulated in the internal space of the thylakoid can flow back to the stroma through the ATPase, which phosphorylates ADP to ATP. A more detailed review on this energy transduction process has been written by Ort (1986).

The thylakoid membrane contains four major intrinsic polypeptide complexes: the PS II reaction centre, the PS I reaction centre, the cytochrome b$_6$/f complex and the ADP-phosphorylating complex (ATPase). For a general review on the structure and function of PS I and PS II, we refer to Andréasson and Vänngård (1988) (*see* Chapter by Hoshina and Itoh, for PS I). The current perceptions of PS II structure and function were recently reviewed by Hansson and Wydrzynski (1990). The structure and function of the Cytochrome b$_6$/f complex were described by Hauska (1985) and O'Keefe (1988); those of the chloroplast coupling factor (ATPase) by Haraux (1986).

On special interest for this chapter is the PS II reaction centre, which can be considered as a water-plastoquinone oxido-reductase; using four photons it transfers four electrons from two molecules of water to plastoquinone producing molecular oxygen and two molecules of doubly reduced plastoquinone. The five most important polypeptide subunits of the PS II reaction centre have molecular weights of 47, 43, 33, 32 and 10 kDa, respectively. A model of the organization of PS II is presented in Fig. 2. The two largest proteins are the chlorophyll-binding proteins CP47 an CP43. They are coded by the chloroplast *psb*B and *psb*C genes, respectively; they function as light-harvesting antenna and may play a role in the assembly of the PS II complex (Carpenter and Vermaas, 1989). The smallest complex capable of the primary charge separation upon illumination consists of the 33, 32 and 10 kDa proteins (Nanba and Satoh, 1987; Satoh, 1988). These are the D2, D1 proteins and cytochrome b559, respectively; they are coded by the chloroplast genes *psb*D, *psb*A and *psb*E, respectively. The *psb*A gene was sequenced by Zurawski *et al.* (1982). The PS II core complex contains in addition several low-molecular-mass components, e.g., the 10 kDa phosphoprotein and the 5, 4.8, 4 and 2 kDa polypeptides, coded by the *psb*H, *psb*L, *psb*I, *psb*F and *psb*K genes, respectively (*see* Ikeuchi and Inoue, 1988; Ikeuchi *et al.*, 1989).

Our current conception about the many aspects of PS II activity involving protein and cofactor interactions was greatly stimulated by the crystallization and X-ray diffraction analysis of the reaction centre from *Rhodopseudomonas viridis* (Deisenhofer *et al.*, 1984; Deisenhofer *et al.*, 1985). It was realized that there are significant structural and functional homologies between the D1 and D2 proteins in PS II and the L (low molecular weight) and the M (medium molecular weight) subunit of this photosynthetic purple bacterium (Trebst, 1986; Barber, 1987). The D2 protein contains the binding site for Q$_A$ while the D1 protein binds Q$_B$.

The 32 kDa protein was described first by Ellis (1977) as "peak D" on the basis of electrophoretic patterns. It is one of the most studied plant proteins (review: Kyle,

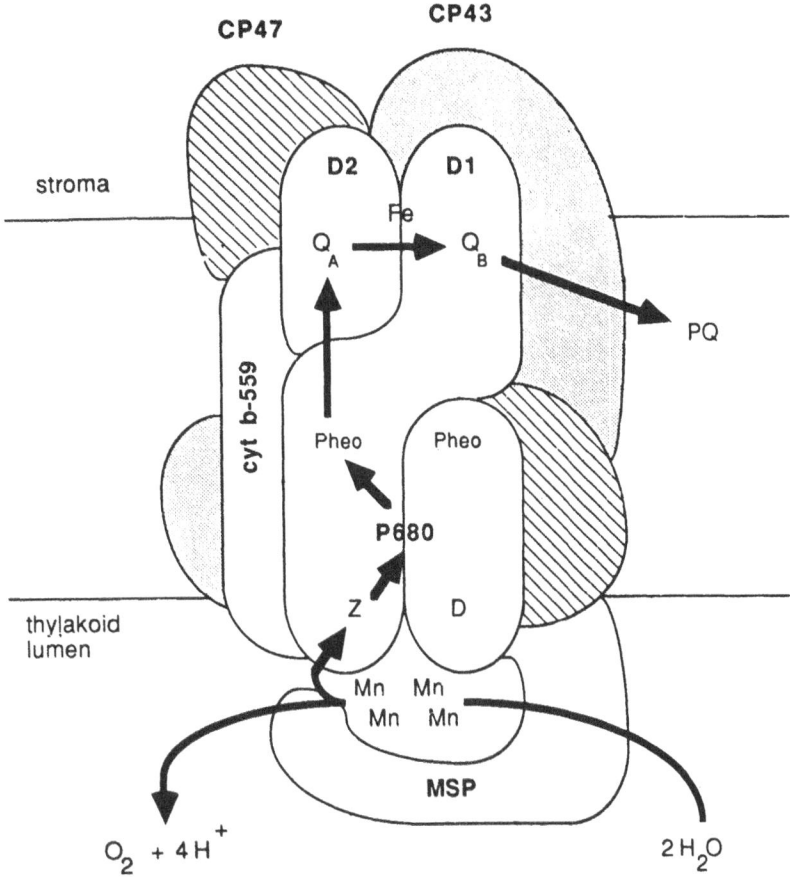

Fig. 2 Simplified two-dimensional model of the PS II core complex in the thylakoid membrane (*from:* Carpenter and Vermaas, 1989, with permission).

1985). It is now usually indicated as the D1 protein (Chua and Gillham, 1977). In earlier literature it was referred to by several names: according to its size, the 32 kDa protein; because it binds PS II herbicides, the herbicide binding protein (Pfister and Arntzen, 1979); Q_B-protein because it contains the binding site for Q_B and since it has a high turnover rate, the rapidly turning over protein.

The hydrophobicity plot of the D1 protein was initially thought to indicate that there were seven membrane spanning helices (Rao *et al.*, 1983), but later data suggested only five helices (Trebst, 1986). A scheme of the protein is presented in Fig. 3. The binding sites for Q_B, herbicides, bicarbonate and formate are suggested to be located between the fourth and the fifth helix. The amino acid residues forming the binding niche of Q_B and the herbicides are known (Trebst, 1987), but an exact steric model of the shape of the pocket has been hard to establish.

The functional components of PS II are very similar to those of the bacterial photosystem, in particular in the functional events on the acceptor side that involve Q_A, Q_B and the non-heme iron that is located between the two quinones (Ikegami and Katoh, 1973; Diner *et al.*, 1990). The charge separation between P680 and

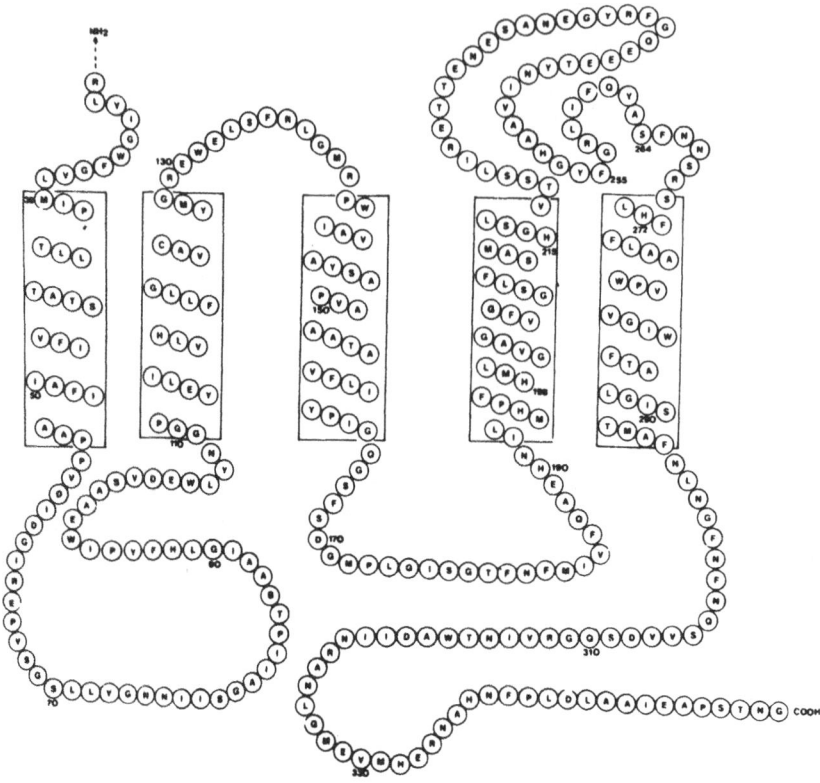

Fig. 3. Folding of the amino acid sequence of the D1 protein of PS II in five hydrophobic helices spanning the membrane (*from:* Trebst, 1986, with permission).

pheophytin has a half-time of 3 ± 0.6 ps (Wasielewski *et al.*, 1989). The electron may then be passed on to the first stable acceptor plastoquinone, Q_A, in 250-300 ps, or recombine with the hole on the primary donor P680 in 2-30 ns, depending on whether Q_A is reduced or absent, to form the ground or triplet states of P680. Electron transfer to Q_B takes place in about 300 μs. After protonation, the doubly reduced $Q_B H_2$ may then exchange with plastoquinone in the pool.

III. Inhibition of Electron Flow by Herbicides

Weed control using organic chemicals commenced in 1932, when 4, 6-dinitro-$\underline{O}$-cresol (DNOC) was first used as a weed-controlling agent. The phenoxyacetic acids such as 2, 4-D followed in the 1940s. Chemical weed control was widely accepted when the ureas (1951), the triazines (1955) and the bipyridiniums (1960) became available. The latter three groups of herbicides act through their inhibitions on the photosynthetic process. The bipyridiniums accept electrons at a site closely beyond PS I; the ureas and triazines effectively block electron transport at the level of the PS II acceptor site; *see* reviews by Pfister and Urbach (1983); Sandmann and Böger (1986); Renger (1986) and Van Rensen (1989).

Research on the action of herbicides inhibiting photosynthesis has yielded much detailed information about their mechanisms of action. Furthermore, our understanding of the photosynthetic process has been greatly enhanced by the use of these chemicals as specific inhibitors. This chapter highlights important events of the research on PS II herbicides and surveys some recent developments.

i) Action of diuron-type herbicides

The most important inhibitors of this class are the urea and triazine herbicides. Both groups contain large numbers of active chemicals with a common chemical structure of a sp^2 hybrid bound to N, O, or =CH and attached to a lipophillic substituent (Trebst and Draber, 1979). The site and mode of action are the same for all the members of these groups, but the differences in activity are caused by the various lipophillic side chains. In photosynthesis research, the urea 3-(3, 4-dichlorophenyl)-1, 1-dimethylurea (diuron=DCMU) and the triazine, 2-chloro-4-ethylamino-6-isopropylamino-1, 3, 5-triazine (atrazine) are the best known.

Wessels and Van der Veen (1956) were the first to show that diuron inhibited the Hill reaction in isolated chloroplasts. The stimulation of fluorescence by diuron was explained by Duysens and Sweers (1963) by assuming that this inhibitor prevented the reoxidation of Q_A^-. By studying various parts of the electron transport pathway the site of inhibition of these herbicides was located between Q_A and the PQ pool. PS I-dependent electron transport and cyclic electron flow was inhibited only at very high concentrations (Van Rensen, 1969).

The study of the mode of action of these herbicides was greatly stimulated by the introduction of a new technique by Tischer and Strotmann (1977). After a radiolabeled herbicide had been bound to isolated chloroplasts, other (non-radiolabeled) herbicides were added to see if they could replace the labeled compound. By this replacement technique they showed that phenylureas, triazines, triazinones, pyridazinones and biscarbamates compete for the same binding site. Moreover, the relative concentration of specific binding sites was found to be one per 300-500 chlorophyll molecules, i.e., about one per electron transfer chain.

A characteristic of these herbicidal inhibitors is the reversibility of their effects. Van Rensen and Van Steekelenburg (1965) found that the inhibition of oxygen evolution in algae by diuron and by 2-methoxy-4, 6-bis (ethylamino)-1, 3, 5,-triazine (simeton) could be removed easily by washing. Izawa and Good (1965) showed that diuron was reversibly bound to chloroplasts. This implied that only weak bonds were involved in the interaction of these herbicides and the receptor molecule in the thylakoid membrane. Tischer and Strotmann (1979) measured the ΔH for binding of 4-amino-6-isopropyl-3-methylthio-1, 2, 4-triazine-5-one (metribuzin) and found it to be -50 kJ/mol. According to Shipman (1981) this binding energy is much too small for covalent binding in a protein, and it is also not consistent with hydrogen bonding. He suggested that polar components of the herbicide bind *via* coulombic interactions at or near a highly polar protein site, probably a protein salt bridge or the terminus of an α-helix on the D1 protein.

The proteinaceous nature of the binding site of these herbicides was confirmed by experiments with the water-soluble enzyme trypsin. Regitz and Ohad (1975) and

Renger (1976) demonstrated that diuron-sensitivity of the Hill reaction was removed by proper treatment of chloroplasts with trypsin. These observations were extended by Böger and Kunert (1979) and Van Rensen and Kramer (1979) to the triazine and phenolic herbicides. Trypsin treatment lowers the binding affinity of these herbicides (Trebst, 1979; Steinback *et al.*, 1981). At the same time, Q_A becomes accessible to ferricyanide. Therefore, Renger (1976) postulated the presence of a protein shield above the quinone acceptors. He assumed that this protein acted as an allosteric regulator for the electron transport between Q_A and the PQ pool. Moreover, this protein was assumed to contain the binding sites for diuron-type herbicides. This shielding protein was later identified as the "rapidly turning over" D1 protein of PS II (Mattoo *et al.*, 1981). Essential for this identification was the photo-affinity labelling technique: radioactive azidoatrazine was found to bind covalently to a 32 kDa protein upon irradiation (Pfister *et al.*, 1981a). After this finding this protein was named the herbicide binding protein.

ii) Action of phenol-type herbicides and other herbicidal inhibitors

The phenol-type herbicides include DNOC, 2, 4-dinitro-6-s-butylphenol (dinoseb), 3, 5-dibromo-4-hydroxybenzonitril (bromoxynil) and 3, 5-diiodo-4-hydroxy-benzonitril (ioxynil). Because of its similarity to 2, 4-dinitrophenol DNOC was long considered to act only by uncoupling of oxidative phosphorylation. However, in 1964, Kerr and Wain reported that it also inhibited the Hill reaction. Van Rensen *et al.* (1977) and van Rensen and Hobé (1979) demonstrated that at low concentrations DNOC inhibited photosynthetic electron transport at the same site as diuron, while at high concentration PSI-dependent electron transport was uncoupled. Because of this dual effect of inhibition and uncoupling, Moreland (1980) classified these phenolic herbicides as inhibitory uncouplers.

Trebst (1987) has classified two families of PS II herbicides: the serine 264 (SER) family and the histidine 215 (HIS) family. The SER family contains the diuron-type herbicides, as described in the preceding section; the phenol-type herbicides belong to the HIS family. Although inhibiting electron flow at the same site, the HIS herbicides do not have the common essential substructure of herbicides of the SER family; furthermore, the interaction with the receptor site in the thylakoid membrane is also different (Van Rensen *et al.*, 1978). Herbicides of the SER group have hydrogen bridges to the nitrogen backbone of the amino acid sequence of the D1 protein and to serine 264; those of the HIS group have the hydroxyl group pushed away from serine 264 towards histidine 215 of the D1 protein (Draber *et al.*, 1989).

Trebst and Draber (1979) studied structure-activity relationships of a large number of halogenated nitro- and dinitrophenols and found that the activity of ureas and triazines was related to lipophilicity and electronic parameters, whereas, with the phenol-type herbicides, activity was more related to steric parameters. These herbicides have a lag-phase for the inhibition of PS II activity, which disappears after preillumination or mild trypsin digestion, indicating hindered accessiblity to the binding site (Böger and Kunert, 1979; Thiel and Böger, 1986). While diuron-type herbicides bind predominantly or even exclusively to the D1 protein, the phenol-type herbicides bind to additional polypeptides in PS II besides the D1 protein

(Oettmeier *et al.*, 1980). Furthermore, weeds resistant against triazine herbicides are still susceptible towards phenol-type herbicides, sometimes even more than the wild-type (Pfister and Arntzen, 1979; Jansen *et al.*, 1986).

A large number of phenol analogues have been described as potent inhibitors of PS II; they include benzoquinones (Oettmeier *et al.*, 1978), naphthoquinones (Pfister *et al.*, 1981b), anthraquinones (Oettmeier *et al.*, 1988), pyrones (Kawamura *et al.*, 1980), pyridones (Trebst *et al.*, 1985); hydroxyquinolines (Draber *et al.*, 1989), cyanoacrylates (Phillips and Huppatz, 1984; Huppatz *et al.*, 1990), phloroglucinol derivatives (Yoneyama *et al.*, 1990), and the antibiotics stigmatellin and aurachin (Oettmeier *et al.*, 1990).

iii) Interaction of herbicides with Q_B

Interaction of diuron-type herbicides with plastoquinone was first proposed by Van Rensen (1969, 1971). In 1974, Velthuys and Amesz suggested that diuron alters the midpoint potential of Q_A making it difficult to reduce Q_B. It is now widely accepted that the mechanism of action is a displacement of Q_B from its binding site at the D1 protein. This was independently and simultaneously proposed by Velthuys (1981) for the PS II complex, and by Wraight (1981) for the reaction centre of purple photosynthetic bacteria. This binding of the herbicides appears to be competitive with plastoquinone; the rate of release of the inhibitor from the site is many times slower than the rate of release for plastoquinone (Vermaas *et al.*, 1984). This implies that the inhibitor stays rather long at the binding site on the D1 protein instead of Q_B. The inhibitory herbicides cannot be reduced by Q_A and electron transfer beyond this point is thereby prevented.

iv) Kinetics of herbicide binding

As is already implied, the Q_B-quinone is not permanently bound to the D1 protein. In its oxidized and reduced state, Q_B and Q_BH_2, it is only weakly bound and can easily leave its binding site; the semiquinone Q_B^- is more tightly bound. Owing to the equilibrium between Q_A and Q_B, an electron shared between the two molecules spends about 95% of the time on Q_B and the remainder of the time on Q_A. Also herbicides bind and release at their binding site on the D1 protein. These exchange parameters can be measured using a method, initiated by Vermaas *et al.* (1984) and adapted by Naber (1989). It is based on the flash-induced oxygen evolution patterns of isolated broken chloroplasts, which are measured in the absence and in the presence of herbicides. The exchange parameters are obtained by fitting experimental data to those calculated with a kinetic model. This model is derived from the following equations.

$$Sn.Q_A.Q_B + I \underset{E_3}{\overset{E_1}{\rightleftarrows}} Sn.Q_A.I + Q_B$$

$$\overset{E_2}{Sn.(Q_A.Q_B)^- + I \rightleftarrows Sn.(Q_A.I)^- + Q_B}$$
$$\underset{E_4}{}$$

In these equations, Sn (where $n = 0, 1, 2, 3$) represents the redox state of the oxygen evolving complex. In the presence of slowly exchanging herbicides, having residence times on the D1 protein of the same of order of magnitude as the duration of the flash train or longer, the oscillation is hardly damped compared to the control. In this case only the amplitude of the signal is diminished. However, when the herbicides exchange is occurring with the same or higher frequency than the firing of the flashes, the damping of the oscillation is considerably stronger. This is caused by the fact that then reaction centres are blocked for a certain time span, and start making turnovers at the moment the herbicide is displaced by a PQ-molecule. Thus, centres can get out of phase with each other, and produce O_2 at different flashes. By comparing flash patterns obtained with different flash frequencies and herbicide concentrations, the exchange parameters E_1 to E_4 can be calculated.

Using this method, Naber (1989) found that for the herbicides atrazine, diuron and ioxynil the E_1-parameters are much higher than those for E_2. This means that these herbicides can bind much easier to the D1 protein when Q_B is in its oxidized form than when it is in its semiquinone form. This result is consistent with the fact that Q_B^- has a much higher affinity for its binding environment than Q_B and Q_BH_2.

The dissociation rates E_3 and E_4 can be used to calculate the time that a herbicide stays at its binding site on the D1 protein. This residence time equals the inverse of the parameters $E_3 + E_4$. In Table 1 the residence times are presented for a few herbicides in a triazine-susceptible and resistant biotype of *Chenopodium album*. Compared with the residence time of Q_B at the D1 protein, which is about 20 ms, those of the herbicides are at least 10-fold higher. The only exception is the value of atrazine in the resistant biotype which is close to that of Q_B; this explains the resistance. Comparing the ratios of the resistant (R) over susceptible (S) values of the residence times with those of the I_{50} (concentration yielding 50% inhibition), it appears that the residence time is inversibly related with I_{50}. This means that the inhibition of a herbicide is stronger when the time it stays on the D1 protein is longer.

Table 1. Comparison of residence times of herbicides at the D1 protein and their activity in a triazine-susceptible and resistant plant

	C. album susceptible (S)	C. album resistant (R)	R/S ratio	I_{50} R/S
atrazine	6.7	0.058	.0087	290
DCMU	390	500	1.3	1.6
ioxynil	0.25	4	16.0	0.1

Residence time in seconds; I_{50}, herbicide concentration in μM which inhibits 50% electron flow.

A detailed study of the exchange parameters of several herbicides led Naber (1989) to the conclusion that the "on" kinetics of a compound to a binding environment are determined principally by the accessibility of the niche to the compound. This is determined by the chemical structure of the herbicide, especially its molecular dimensions, charges and hydrophobicity. The differences in activity between herbicides are largely due to variations in the release kinetics. A stationary binding, resulting in a significant electron transport inhibition, requires a strict molecular shape.

IV. The Antagonistic Action of Bicarbonate and Anions Like Formate on Electron Flow

It is generally known that CO_2 is required for photosynthesis. Carbon dioxide is fixed by ribulose 1, 5-bisphosphate carboxylase and further reduced to carbohydrate. However, CO_2 appears also to be involved in photosynthetic electron transport. In 1958, Warburg and Krippahl discovered that CO_2 accelerates the production of oxygen upon illumination of isolated chloroplasts in the presence of an electron acceptor such as ferricyanide. This phenomenon was confirmed by many workers, but there was little agreement as to the conditions necessary for observing the dependence of Hill reaction on CO_2 and on the significance of such dependence. In 1973, Stemler and Govindjee described a method whereby reproducible and large increases in the rate of the Hill reaction in isolated chloroplasts could be observed upon addition of bicarbonate to "CO_2-depleted" samples. The method depends on depletion of the chloroplasts from CO_2 by flushing the suspension in the dark with nitrogen gas, while the chloroplasts are suspended in a medium of pH between 5.5 and 6.0, and containing a high anion concentration; usually the formate ion is used. The resulting electron transport rate is then measured at a pH of usually 6.5. It is considered likely that CO_2 is the diffusing species, while bicarbonate is the binding species. Because the stimulation is evoked by the addition of a bicarbonate solution to CO_2-depleted chloroplasts, the phenomenon is usually named the bicarbonate-effect. There are several reviews available on the bicarbonate-effect. For the early literature see Govindjee and Van Rensen (1978); and for the recent literature see Van Rensen and Snel (1985) and Blubaugh and Govindjee (1988).

The bicarbonate-effect is illustrated in Fig. 4. The electron transport rate is recorded as oxygen evolution. The Hill reaction rate is very low after 30 minutes of "CO_2 depletion" (and is usually close to zero after about 60 minutes depletion); it is fully recovered after incubation during 60–90 seconds in the dark with $NaHCO_3$. It is our experience that it is very easy to inactivate the chloroplasts by the CO_2 depletion procedure. Therefore, we suggest that one should always check if the recovery by addition of bicarbonate has taken place.

i) Action of bicarbonate at the molecular level

Although Stemler (1982) has advocated a role of CO_2 on the donor side of PS II, most researchers favour the idea that bicarbonate has a major effect on the acceptor side of PS II. The observations in favour of a role of bicarbonate at this site are as follows:

168

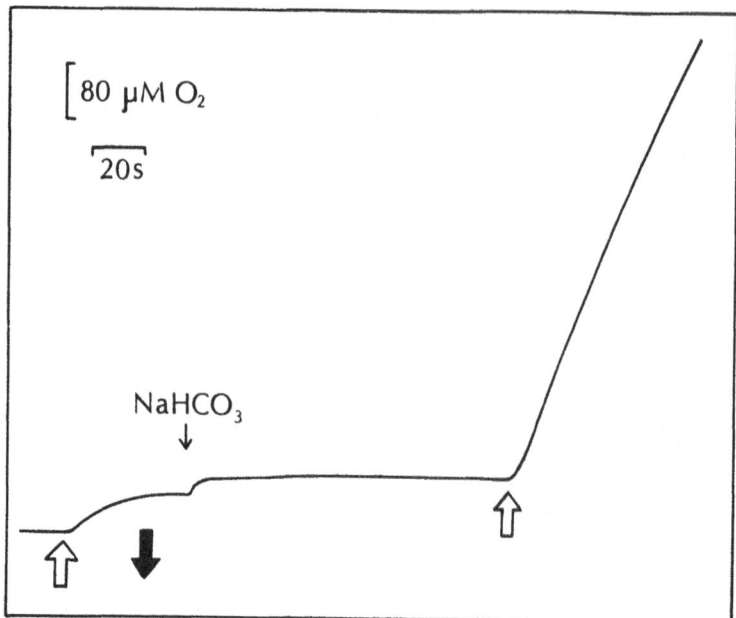

Fig. 4. Recording of the Hill reaction of chloroplasts depleted of CO_2 during 30 minutes, before and after the addition of 10 mM $NaHCO_3$. Arrow pointing upwards, light on; arrow downwards, light off.

The first indication for an effect of bicarbonate on the reducing side of PS II was obtained by Wydrzynski and Govindjee (1975), who measured chlorophyll *a* fluorescence induction kinetics in chloroplasts after CO_2 depletion. The variable chlorophyll *a* fluorescence monitors the redox state of Q_A; Q_A is a quencher of fluorescence, whereas Q_A^- is not. Therefore, a rapid accumulation of Q_A^- due to an inhibition of electron transport beyond Q_A is easily detected by fluoresence induction measurements. CO_2 depletion causes a fast increase in the variable fluorescence yield, similar but not identical to that observed in normal chloroplasts in the presence of the herbicide DCMU (Wydrzynski and Govindjee, 1975). As noticed earlier, DCMU is known to block the reoxidation of Q_A^- by the secondary quinone acceptor Q_B.

The decay of the chlorophyll *a* fluorescence yield after a saturating flash monitors the reoxidation of Q_A^-. The half-time of the Q_A^- reoxidation is increased upon CO_2 depletion, indicating again the inhibition of Q_A^- reoxidation. Detailed information about the effect of CO_2 depletion was obtained by measuring the decay of the chlorophyll *a* fluorescence yield after various numbers of short saturating flashes. By this techinque the following events are monitored. After odd number of flashes:

$Q_A Q_B \xrightarrow{h\nu} Q_A^- Q_B \longrightarrow Q_A Q_B^-$. After even number of flashes:

$Q_A Q_B^- \xrightarrow{h\nu} Q_A^- Q_B^- \longrightarrow Q_A Q_B^{2-}$; Q_B^{2-} becomes protonated and exchanges with the plastoquinone pool:

$Q_A Q_B^{2-} + 2H^+ + PQ \longrightarrow Q_A Q_B + PQH_2$. Govindjee *et al.* (1976) found no differences in the decay of the fluorescence yield after various number of flashes in control and in CO_2-depleted chloroplasts, to which bicarbonate was added. In CO_2-depleted chloroplasts, however, they found little effect on the decay after one or two flashes, but a very large slowing down of the decay after three or more flashes. Robinson et al. (1984) showed that in CO_2-depleted chloroplasts the Q_A^--decay after one flash is about five-fold slower than in control chloroplasts, but after three or more flashes it is 36-fold slower. Although the absolute values of the rates of Q_A^--decay in this type of experiment depend on the conditions of the experiment, it now appears that there is a smaller inhibition of CO_2 depletion on the Q_A^- reoxidation by Q_B, a larger inhibition of this reoxidation by Q_B^- and a much larger inhibition of the protonation of Q_B^{2-} and/or exchange of Q_BH_2 with the plastoquinone pool. In CO_2-depleted chloroplasts three electrons can be stored leading to $Q_A^-Q_B^{2-}$. (For a detailed study, see Eaton-Rye and Govindjee, 1988a).

The site of inhibition of CO_2–depletion was also determined by studying its effect on various parts of the electron transport chain (Khanna et al., 1977). There was no effect of CO_2 depletion on electron transport from reduced diaminodurene to methyl viologen, indicating the absence of an effect on PS I-dependent electron transport. A large bicarbonate-effect was demonstrated on the electron flow from water to oxidized diaminodurene in the presence of dibromothymoquinone (DBMIB), indicating an effect before the plastoquinone pool. Since the electron flow from water to silicomolybdate in the presence of DCMU was not affected by CO_2 depletion, it was concluded that the bicarbonate-effect was located between Q_A and the PQ pool (Khanna et al., 1981). The same conclusion was drawn from the absence of a bicarbonate-effect on electron transport in trypsin-treated chloroplasts in which ferricyanide accepts electrons directly at the Q_A site (Van Rensen and Vermaas, 1981).

The localization of the bicarbonate-effect between Q_A and the plastoquinone pool was further concluded from the interaction of bicarbonate (or formate) with PS II inhibiting herbicides. Van Rensen and co-workers (Van Rensen and Vermaas, 1981; Khanna et al., 1981; Van Rensen, 1982; Vermaas et al., 1982; Snel and Van Rensen, 1983) studied the interaction of bicarbonate and herbicides through their effects on electron transport in isolated chloroplasts. By adding different concentrations of bicarbonate to CO_2-depleted chloroplasts various rates of restoration of the Hill reaction were obtained. It was demonstrated that the system thylakoid membrane versus bicarbonate has Michaelis-Menten kinetics and that it can be treated like a system enzyme versus substrate. From double reciprocal plots of the rate of the Hill reaction as a function of the bicarbonate concentration (an example is shown in Fig. 5) the apparent dissociation constant (Kd) of the thylakoid-bicarbonate complex could be calculated. When 100 mM formate is present in the reaction medium, the apparent Kd appears to be about 1 mM bicarbonate. The Kd for bicarbonate depends on the presence of both formate and of herbicides. In the presence of low concentrations of formate the apparent Kd decreases, approaching 80 μM $NaHCO_3$ in the absence of formate (Snel and Van Rensen, 1984). In the presence of urea, triazine or phenol-type herbicides the Kd for bicarbonate increases by at least two-fold (Fig. 5). This means that these herbicides decrease the apparent affinity of the thylakoid membrane for bicarbonate.

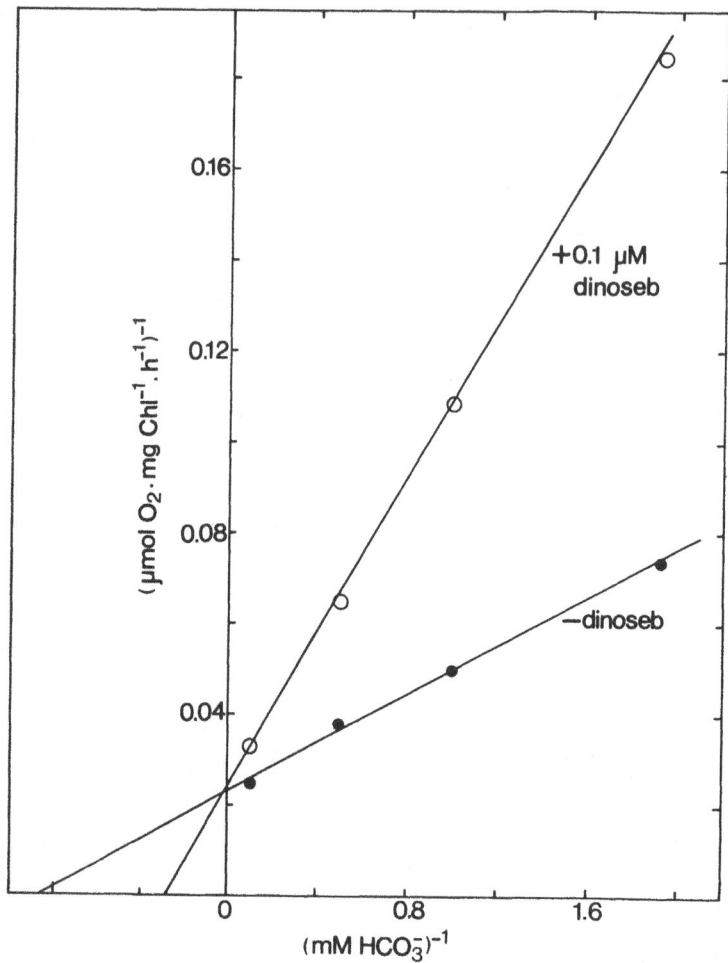

Fig. 5. Double reciprocal plot of the rate of the Hill reaction of CO_2-depleted chloroplasts as a function of the added bicarbonate concentration in the absence and in the presence of the herbicide dinoseb.

The above arguments lead strongly to the suggestion that the binding site of bicarbonate is located on the D1 protein of photosystem II. Michel and Deisenhofer (1988) compared the primary structure of the L and M polypeptides of the bacterial reaction centres with the D1 and D2 polypeptides of PS II and suggested that glutamate in bacteria is replaced by bicarbonate in PS II as a ligand to the non-heme iron. It was demonstrated by Shopes *et al.* (1989) that the bicarbonate-effect is absent in photosynthetic bacteria. Van Rensen *et al.* (1988) showed that the kinetics of bicarbonate binding to thylakoids are influenced by the redox state of this non-heme iron. In addition to the binding site at the non-heme iron, Blubaugh and Govindjee (1988) suggested a second binding site in the D1 protein at arginine-257 and histidine-252; the bicarbonate ions at both sites binding cooperatively. Diner and Petrouleas (1990) demonstrated that NO binds to the non-heme iron of PS II and slows down, like formate, electron flow between Q_A and Q_B. This slowing is completely reversed by the addition of bicarbonate, indicating that NO, like formate, displaces bicarbonate from the reaction centre. These data also argue in favour of bicarbonate as a ligand to the non-heme iron. More recently, J. Cao, W. Vermaas and Govindjee (1989, personal communication) have obtained evidence that not only the D1 protein, but also the D2 protein may be involved in the bicarbonate-effect as two arginine mutants (R 233 and R 251) of the D2 protein show a ten-fold increased sensitivity to formate.

There have been early suggestions that bicarbonate and formate may interfere with protonation reactions near Q_B (Govindjee and Van Rensen, 1978). It is possible that H_2CO_3 is involved in the protonation of Q_B^{2-} or its proteinaceous environment, since the pK_a of $(CO_2 + H_2O)$ is 6.4 at 25°C. $CO_2/HCO_3^-/CO_3^{2-}$ could serve as a proton shuttle between Q_B and the external aqueous phase. Formate is not able to function in such a way, because the pK_a of formate is 3.8. Recently, evidence for such a function was obtained by Van Rensen et al., (1988) and Eaton-Rye and Govindjee (1988b). Blubaugh and Govindjee (1988) discussed a mechanism in which the bicarbonate at arginine-257 of the D1 protein protonates a histidine (most likely histidine-252) in D1, as part of a mechanism for protonation of Q_B^-.

ii) The bicarbonate-effect *in vivo*

The current knowledge of the bicarbonate-effect is almost exclusively based on experiments carried out with isolated thylakoid membranes. The observations of a bicarbonate-effect *in vivo* is difficult to distinguish, due to the obvious requirement for CO_2 in the Calvin cycle.

Garab et al. (1983) demonstrated effects of bicarbonate on the energization of the thylakoid membrane in leaves. Mende and Wiessner (1985) studied the bicarbonate-effect in intact cells of the green alga *Chlamydobotrys stellata*, and concluded that both sides of PS II are affected in the absence of bicarbonate. Chlorophyll fluorescence experiments in leaves under conditions of very low photosynthesis and measurements of electron flow, i.e., independent of carbon assimilation (leaf discs infiltrated with methyl viologen, acting as a terminal electron acceptor in stead of CO_2) have suggested that CO_2 (bicarbonate) modifies the redox state of the quinone electron acceptors of PS *in vivo* independently of carbon assimilation and thereby acts as a cofactor for efficient PS II electron flow in the leaf (Ireland et al., 1987). Thermoluminescence of leaves provided evidence that CO_2 facilitates the flow of electrons from the reduced Q_A towards PS I (Garab et al., 1988). El-Shintinawy and Govindjee (1990) infiltrated spinach leaf discs with a CO_2-depleting solution (containing formate) and studied the effect on oxygen evolution and the simultaneous quenching of the variable fluorescence. From their results they suggested two sites of action of bicarbonate: one between pheophytin and Q_A; the other (the most dramatic effect) beyond Q_A.

Stemler (1989) reported that formate does not inhibit electron flow by removing bound bicarbonate, but it does so by binding to empty sites. This conclusion was based on the observation of the absence of a formate-induced release of CO_2 in maize thylakoids at pH 6.0 monitored by a mass spectrometer. In contrast, Govindjee et al. (1991) have shown, using a sensitive membrane inlet mass spectrometer and a sensitive differential CO_2 infrared gas analyser, that formate treatment releases micromolar quantities of CO_2 from spinach and pea thylakoids at pH 6.5: both the CO_2 release and the bicarbonate-reversible inhibition of electron flow are pH dependent and occur within minutes of formate treatment. These results are consistent with the hypothesis that native bound bicarbonate is required for electron flow in thylakoids.

V. Resistance to Herbicides

A weed is resistant when it survives and grows normally at the usually effective dose of a herbicide. Resistance of a plant to a herbicide is most often due to conditions which, via various mechanisms, prevent the herbicide reaching its site of action in the plant. Some of these conditions are limited uptake of the herbicide, limited transport, or metabolic detoxification. A rare case of resistance is an alteration of the receptor site for the herbicide. This has happened in relation to the triazine herbicide family. There are only very few reports of resistance to other PS II herbicides. Important information about herbicide resistance can be found in the book, edited by LeBaron and Gressel (1982).

i) Triazine-resistance as found in weeds

In the last decade resistance against triazine herbicides has developed in many weeds and in many countries over the world. This resistance is specific for all triazine herbicides; the concentration of triazine herbicide causing 50% inhibition of the Hill reaction in isolated chloroplasts from resistant biotype is often about 1000 times higher than in chloroplasts from the sensitive plants. There is little difference in sensitivity for urea herbicides, like diuron, while resistant biotypes are somewhat more sensitive to phenol-type herbicides. The resistance is caused by lack of binding of triazine herbicides (Pfister and Arntzen, 1979) and could be correlated with an alteration of the D1 protein. After it was found that the chloroplast *psb*A gene codes for the D1 protein, Hirschberg *et al.* (1984) demonstrated a change from adenine in the susceptible to guanine in the resistant biotype. This leads to a substitution of serine for glycine at position 264 within the D1 protein. Up to now, many triazine-resistant plants have been investigated; in all such cases triazine resistance is correlated with this same mutation (Naber, 1989).

The alteration of the D1 protein does not only result in triazine resistance, but also in a change in the equilibrium of the $Q_A^- Q_B$ to $Q_A Q_B^-$ reaction, originally assumed to be due to a large (ten-fold) decrease in the rate of electron flow from Q_A^- to Q_B (Pfister and Arntzen, 1979; Vermaas and Arntzen, 1983). The impaired rate of electron transport between Q_A and Q_B was suggested to be related with a significantly lowered rate of light- and CO_2-saturated photosynthesis, growth and productivity at the whole plant level (McCloskey and Holt, 1990 and references cited). However, an influence of impaired rate of electron flow at the reducing side of PS II on the overall electron flow may be questioned. The lower rate of electron transport from Q_A^- to Q_B in the resistant biotype is still about ten times faster than the oxidation of reduced plastoquinone. The latter reaction remains the rate limiting step having a half-time of about 20 ms. Both Ort *et al.* (1983) and Jansen *et al.* (1986)

measured a lower rate of electron flow between Q_A and plastoquinone in resistant chloroplasts; the whole chain electron transport activity was, however, not significantly different in the two biotypes.

Jansen and Pfister (1990) measured the rate of Q_A^- oxidation in 28 different plant species and found an average half-time of 314 ± 46 μs. In five different triazine-resistant plants this average was 946 ± 100 μs. This three-fold decrease in rate is much less than previously reported. These authors suggested that the changed kinetics of the Q_A/Q_B reaction does not simply decrease primary photosynthetic efficiency and biomass productivity via a direct effect on photosynthetic electron flow. They speculated that the mutation in the D1 protein affects also some other functional aspects of the PS II complex important for regulation photosynthesis and biomass production, e.g., the turnover of the D1 protein. Evidence for such a speculation was reported by Hart and Stemler (1990). They compared triazine-resistant and susceptible *Brassica napus* plants, grown under low photon flux density conditions and then exposed to high light intensity. Measuring photon yield, F_v/F_m (variable chlorophyll *a* fluorescence/maximum chlorophyll *a* fluorescence) and light-saturated O_2 evolution they found no differences when plants were grown at low light intensity. After exposure to high light intensity they found differences between resistant and susceptible biotypes that led to the conclusion that resistant plants are more sensitive to photoinhibition. Comparable results were obtained by van Rensen and De Vos (1990, unpublished results) using resistant and susceptible plants of *Chenopodium album*. Due to a lower rate of electron flow between Q_A and Q_B, the time that an electron remains at Q_A is longer in the resistant plant; this effect is of increasing importance at higher light intensities. When Q_A is in its reduced state, no charge separation is possible in PS II, leading to increased photoinhibition. In this way, productivity in resistant plants may become lower.

ii) Herbicide-resistance induced in algae, cyanobacteria and photosynthetic bacteria

Triazine-resistance initiated the study of the molecular biology of the D1 protein. Many mutants of algae and bacteria resistant to various herbicides have been obtained by growing them at a lower concentration of a herbicide for selection pressure, or by site directed mutagenesis (Carpenter and Vermaas, 1989). The organisms most often used for these studies are the green alga *Chlamydomonas reinhardtii*, the cyanobacteria *Synechococcus* sp. and *Synechocystis* sp. PCC 6714, and the purple bacteria *Rhodopseudomonas viridis*, *Rhodobacter* (*Rb.*) *sphaeroides* and *Rb. capsulatus*. The matter was recently reviewed by Trebst (1991).

In triazine- or triazinone-resistant green algae and cyanobacteria, the serine 264 change is the most prominent and most easily obtained mutation. Due to the different codon usage for serine in higher plants (AGT) *versus* algae (TCT), serine is altered to an alanine in algae and cyanobacteria instead of serine to glycine in higher plants. There are now mutants resistant to several herbicides: atrazine, diuron, metribuzin, cyanoacrylate, terbutryn, ioxynil and bromacil. These mutations are all located at or between the fourth and the fifth helix of the D1 protein on the following amino acids: phe 211, val 219, ala 251, phe 255, gly 256, ser 264, asp 266 and leu 275.

Interestingly, in spite of the high conservation of the D1 protein through the plant kingdom, it appears that several amino acids can be changed without complete loss of function: all mutants grow photoautotrophically. In many mutants the functioning of the acceptor side of PS II has been studied, with variable results. A recent study illustrating this point is that of Etienne *et al.* (1990). They investigated electron transport properties of PS II in five herbicide-resistant mutants of *Synechocystis*, one of *Synechococcus*, one of *Chlamydomonas* and one of *Chenopodium album*. Two mutants had an almost unimpaired PS II electron transfer. In most cases, the initial phase of the electron transfer from Q_A^- to Q_B was unaltered, but the electron transport equilibrium between the two acceptors was displaced. In the *Chlamydomonas* mutant examined, the electron flow from Q_A^- to Q_B was slowed down.

VI. The Architecture of the Herbicide and Q_B-Binding Site

When the X-ray structure of the reaction centre of *R. viridis* had become available (Deisenhofer *et al.*, 1984), the homology of the L and D1 protein allowed Trebst (1986) to propose a new folding of the D1 protein. The folding model was based on amino acids conserved at essential positions in both the L/M and D1/D2 subunits, and on the then known amino acid changes in herbicide-resistant plants. These changes were found in a common binding niche. The original model was then improved using information from herbicide-resistant photosynthetic bacteria (Trebst, 1987). Further amino acid changes in herbicide-resistant algae and photosynthetic bacteria were recently accommodated in this binding niche (Trebst, 1991).

The two-dimensional folding of the herbicide and Q_B binding niche in Fig. 6 indicates most amino acid mutations identified till now. They are all between amino acids 211 and 275. This area includes the hydrophobic transmembrane helices IV and V and the parallel helix connecting these on the matrix side of the D1 protein. The helices IV and V are connected to each other by an Fe atom bridged via two histidines. Via the Fe the D1 protein is also connected to the D2 protein by two further histidine bridges (Trebst, 1991). This area is the Q_B site where a plastoquinone molecule is bound during electron transport. As yet, there is no long range effect of the known mutations on the folding. The altered amino acids have no major contribution to the Q_B binding, even the hydrogen bridge of the OH of serine 264 is not essential for Q_B binding. This is different for those amino acids identified by photoaffinity labelling (Dostatni *et al.*, 1988; Oettmeier *et al.*, 1989). These amino acids could well be those that cannot be mutated without loss of Q_B function; they fit into the folding pattern of Fig. 6. Furthermore, they added to an important development in understanding further features in the D1 protein structure.

As already described in Section III, two families of PS II herbicides may be distinguished. The SER family including the urea and triazine herbicides; and the HIS family including the phenol-type herbicides. Dostatni *et al.* (1988) were able by direct sequencing to demonstrate that tyrosine 235 and tyrosine 254 (Y in Fig. 6) were targets for azido-monuron. Oettmeier *et al.* (1989) identified valine 249 as one amino acid involved in binding of azido-ioxynil (a cyanophenol herbicide). They suggested that the 2-position of ioxynil points towards this amino acid and that ioxynil

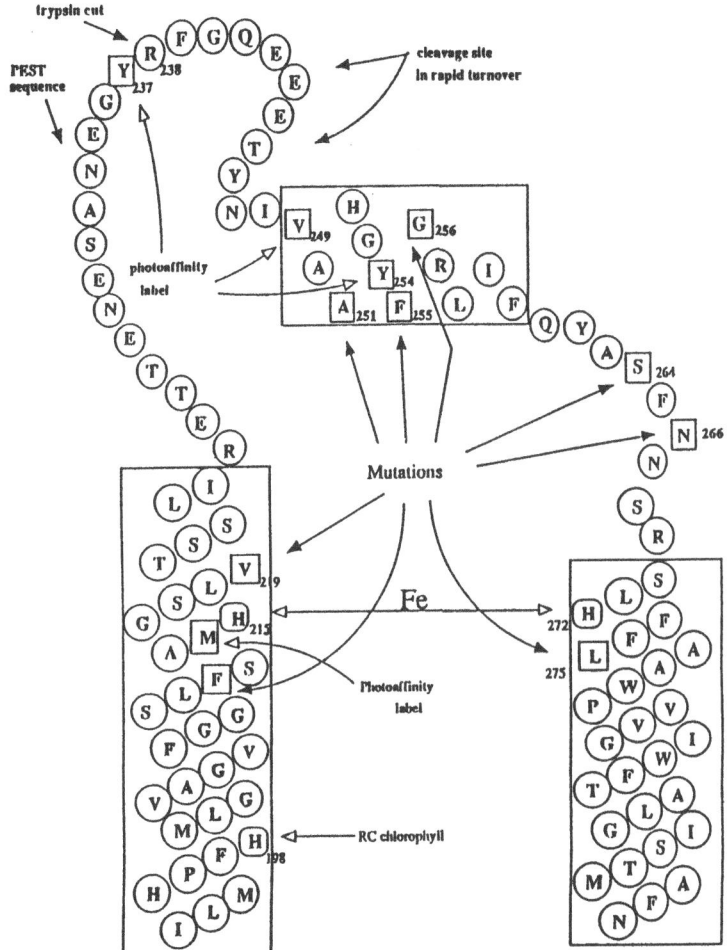

Figure 6. Model of the folding of the amino acid sequence in the binding niche of the quinone and herbicide binding site on the D1 protein of PS II. Indicated are two transmembrane helices (IV and V) and the parallel helix on the matrix side of the thylakoid membrane (*from*: Trebst, 1991, with permission).

in its binding niche is sandwiched between valine 249 and asparagine 266 (N in Fig. 6). This places the ioxynil binding site in a different environment of the D1 protein as compared to the SER family herbicides. As one likely orientation of azido-atrazine is that the alkyl side chains of atrazine are oriented towards the stretched amino acid sequence which contains the serine 264; and the azido-substituent is oriented towards methionine 214 on helix IV. Terbutryn is proposed to bind to asparagine 266 (N in Fig. 6) and serine 264.

In conclusion, already much is known about the amino acids involved in Q_B and herbicide binding. More details about the tertiary structure of the binding niche will be revealed by studying herbicide-resistant mutants and molecular modelling of herbicides.

VII. References

Andréasson L-E and Vänngård T (1988) Electron transport in photosystems I and II. *Ann Rev Plant Physiol Plant Mol Biol* **39**: 379–411

Barber J (1987) Photosynthetic reaction centers: a common link. *TIBS* **12**: 321–326

Blubaugh DJ and Govindjee (1988) The molecular mechanism of the bicarbonate effect at the plastoquinone reductase site of photosynthesis. *Photosynthesis Res* **19**: 85–128

Böger P and Kunert K-J (1979) Differential effects of herbicides upon trypsin-treated chloroplasts. *Z Naturforsch* **34C**: 1015–1020

Carpenter SD and Vermaas WFJ (1989) Direct mutagenesis to probe the structure and function of photosystem II *Physiol Plant* **77**: 436–443

Chua NH and Gilham NW (1977) The site of synthesis of the principal thylakoid membrane polypeptides in *Chlamydomonas reinhardtii*. *J Cell Biol* **77**: 441–452

Crofts AR and Wraight CA (1983) The electrochemical domain of photosynthesis. *Biochim Biophys Acta* **726**: 149–185

Deisenhofer J, Epp O, Miki K, Huber R and Michel H (1984) X-ray structure analysis of a membrane protein complex. Electron density map at 3 Å resolution and a model of the chromatophores of the photosynthetic reaction center from *Rhodopseudomonas viridis*. *J Mol Biol* **180**: 385–398

Deisenhofer J, Epp O, Miki K, Huber R and Michel H (1985) X-ray structure analysis at 3 Å resolution of a membrane protein complex: Folding of the protein subunits in the photosynthetic reaction center from *Rps. viridis*. *Nature* **318**: 618–624

Diner BA and Petrouleas V (1990) Formation by NO of nitrosyl adducts of redox components of the photosystem II reaction center. II Evidence that HCO_3^-/ CO_2 binds to the acceptor-side non-heme iron. *Biochim Biophys Acta* **1015**: 141–149

Diner BA, Petrouleas V and Wendoloski JJ (1990) The iron-quinone electron-acceptor complex of photosystem II. *Physiol Plant* **1**: 423–436

Dostatni R, Meyer HE and Oettmeier W (1988) Mapping of two tyrosine residues involved in the quinone-(Q_B) binding site of the D-1 reaction center polypeptide of photosystem II. *FEBS Lett* **239**: 207–210

Draber W, Pittel B and Trebst A (1989) Modeling of photosystem II inhibitors of the herbicide-binding protein. In: (PS Magee *et al.*, eds.) *ACS Symposium Series No 413* "Probing Bioactive Mechanisms". American Chemical Society pp 215–228

Duysens LNM and Sweers HE (1963) Mechanism of two photochemical reactions in algae as studied by means of fluorescence. In: (*Jpn Soc Plant Physiol*, eds) *Studies on Microalgae and Photosynthetic Bacteria*, University of Tokyo Press, Tokyo pp 353–372

Eaton-Rye JJ and Govindjee (1988a) Electron transfer through the quinone acceptor complex of photosystem II in bicarbonate-depleted spinach thylakoid membranes as a function of actinic flash number and frequency. *Biochim Biophys Acta* **935**: 237–247

Eaton-Rye JJ and Govindjee (1988b) Electron transfer through the quinone acceptor complex of photosystem II after one or two actinic flashes in bicarbonate-depleted spinach thylakoid membranes. *Biochim Biophys Acta* **935**: 248–257

El-Shintinawy F and Govindjee (1990) Bicarbonate effects in leaf discs from spinach. *Photosynthesis Res* **24**: 189–200

Ellis RJ (1977) Protein synthesis by isolated chloroplasts. *Biochim Biophys Acta* **463**: 185–218

Etienne A-L, Ducruet J-M, Ajlani G, and Vernotte C, (1990) Comparative studies on electron transfer in photosystem II of herbicide-resistant mutants from different organisms. *Biochim Biophys Acta* **1015**: 435–440

Garab G, Sanchez Burgos AA, Zimányi L and Faludi-Dániel A (1983) Effect of CO_2 on the energization of thylakoids in leaves of higher plants. *FEBS Lett* **154**: 323–327

Garab G, Rozsa Z and Govindjee (1988) Carbon dioxide affects charge accumulation in leaves. *Naturwissensch* **75**: 517–519

Govindjee and Van Rensen JJS (1978) Bicarbonate effects on the electron flow in isolated broken chloroplasts. *Biochim Biophys Acta* **505**: 183–213

Govindjee, Pulles MJP, Govindjee R, van Gorkom HJ and Duysens LNM (1976) Inhibition of the reoxidation of the secondary electron acceptor of photosystem II by bicarbonate depletion. *Biochim Biophys Acta* **449**: 602–605

Gōvindjee, Weger HG, Turpin DH, Van Rensen JJS, De Vos OJ and Snel JFH (1991) Formate releases carbon dioxide/bicarbonate from thylakoid membranes. Naturwissensch **78**: 168–170

Hansson Ö and Wydrzynski T (1990). Current perceptions of photosystem II. *Photosynthesis Res* **23**: 131–162

Haraux F (1986) Integrated functioning of the chloroplast coupling factor. *Biochemie* **68**: 435–449

Hart J and Stemler A (1990) Increased sensitivity to photoinhibition in triazine-resistant *Brassica napus* L. Plant Physiol **93**(1)S: 116

Hauska G (1985) Organization and function of cytochrome b_6/bc_1 complexes. In: (KE Steinback *et al.*, eds) *Molecular Biology of the Photosynthetic Apparatus*, Cold Spring Harbor Laboratory pp 79–87

Hirschberg J, Bleeker A, Kyle DJ, McIntosh L and Arntzen CJ (1984) The molecular basis of triazine-herbicide resistance in higher-plant chloroplasts. *Z Naturforsch* **39c**: 412–420

Huppatz JL, McFadden HG and McCaffery LM (1990) Cyanoacrylate inhibitors as probes for the nature of the photosystem II herbicide binding site. *Z Naturforsch* **45c**: 336–342

Ikegami I and Katoh S (1973) Studies on chlorophyll fluorescence in chloroplasts. II Effect of ferricyanide on the induction of fluorescence in the presence of 3-(3, 4-dichlorophenyl)-1, 1-dimethylurea. *Plant Cell Physiol* **14**: 829–836

Ikeuchi M and Inoue Y (1988) A new 4.8-kDa polypeptide intrinsic to the PS II reaction center, as revealed by modified SDS-PAGE with improved resolution of low-molecular-weight proteins. *Plant Cell Physiol* **29**: 1233–1239

Ikeuchi M, Koike H and Inoue Y (1989) Identification of psbI and psbL gene products in cyanobacterial photosystem II reaction center preparation. *FEBS Lett* **251**: 155–160

Ireland CR, Baker NR and Long SP (1987) Evidence for a physiological role of CO_2 in the regulation of photosynthetic electron transport in intact leaves. *Biochim Biophys Acta* **893**: 434–443

Izawa S and Good NE (1965) The number of sites sensitive to 3-(3, 4-dichlorophenyl)-1, 1-dimethylurea, 3-(4-chlorophenyl)-1,1-dimethylurea and 2-chloro-4-(2-propylamino)-6-ethylamino-s-triazine in isolated chloroplasts. *Biochim Biophys Acta* **102**: 20–38

Jansen MAK and Pfister K (1990) Conserved kinetics at the reducing side of reaction-center II in photosynthetic organisms; changed kinetics in triazine-resistant weeds. *Z Naturforsch* **45c**: 441–445

Jansen MAK, Hobé JH, Wesselius JC and van Rensen JJS (1986) Comparison of photosynthetic activity and growth performance in triazine-resistant and susceptible biotypes of *Chenopodium album*. *Physiol Vég* **24**: 475–484

Kawamura M, Yoshida S, Takahashi N and Fujita Y (1980) Pyrone derivatives: effective inhibitors of photosynthetic electron flow. *Plant Cell Physiol* **21**: 745–753

Kerr MW and Wain RL (1964) Inhibition of the ferricyanide Hill reaction of isolated bean leaf chloroplasts by 3, 5-diiodo-4-hydroxybenzonitril (ioxynil) and related compounds. *Ann Appl Biol* **54**: 447–450

Khanna R, Govindjee and Wydrzynski T (1977) Site of bicarbonate efect in Hill reaction. Evidence from the use of artificial electron acceptors and donors. *Biochim Biophys Acta* **462**: 208–214

Khanna R, Pfister K, Keresztes A, van Rensen JJS and Govindjee (1981) Evidence for a close spatial location of the binding sites for CO_2 and for photosystem II inhibitors. *Biochim Biophys Acta* **634**: 105–116

Kyle DJ (1985) The 32000 dalton Q_B protein of photosystem II. *Photochem Photobiol* **41**: 107–116

LeBaron HM and Gressel J (1982) *Herbicide Resistance in Plants*, John Wiley and Sons, New York

Mattoo AK, Pick U, Hoffmann-Falk H, and Edelman M (1981) The rapidly metabolized 32,000-dalton polypeptide of the chloroplast is the 'proteinaceous shield' regulating photosystem II electron transport and mediating diuron herbicide sensitivity. *Proc Natl Acad Sci USA* **78**: 1572–1576

McCloskey WB and Holt JS (1990) Triazine resistance in *Scenecio vulgaris* parental and nearly isonuclear backcrossed biotypes is correlated with reduced productivity. *Plant Physiol* **92**: 954–962

Mende D and Wiessner W (1985) Bicarbonate *in vivo* requirement of photosystem II in the green alga *Chlamydobotrys stellata*. *J Plant Physiol* **118**: 259–266

Michel H and Deisenhofer J (1988) Relevance of the photosynthetic reaction center from purple bacteria to the structure of photosystem II. *Biochemistry* **27**: 1–7

Michel H, Epp O and Deisenhofer J (1986) Pigment-protein interactions in the photosynthetic reaction center from *Rhodopseudomonas viridis*. *EMBO J* **5**: 2445–2451

Moreland DE (1980) Mechanisms of action of herbicides. *Ann Rev Plant Physiol* **31**: 597–638

178

Naber JD (1989) Molecular aspects of herbicide binding in chloroplasts. PhD thesis, Agricultural University Wageningen pp 1–116

Nanba O and Satoh K (1987) Isolation of a photosystem II reaction center consisting of D-1 and D-2 polypeptides and cytochrome b_{559}. *Proc Natl Acad Sci USA* **84:** 109–112

O'Keefe D (1988) Structure and function of the chloroplast cytochrome bf complex. *Photosynthesis Res* **17:** 189–216

Oettmeier W, Reimer S and K Link (1978) Quantitative structure-activity relationship of substituted benzoquinones as inhibitors of photosynthetic electron transport. *Z Naturforsch* **33c:** 695–703

Oettmeier W, Masson K and Johanningmeier U (1980) Photoaffinity labeling of the photosystem II herbicide binding protein. *FEBS Lett* **118:** 267–270

Oettmeier W, Masson K and Donner A (1988) Anthraquinone inhibitors of photosystem II electron transport. *FEBS Lett* **231:** 259–262

Oettmeier W, Masson K, Höhfeld J, Meyer HE, Pfister K and Fischer H-P (1990) (^{125}I) Azido-ioxynil labels val 249 of the photosystem II D-1 reaction center protein. *Z Naturforsch* **44c:** 444–449

Oettmeier W, Dostatni R, Majewski C, Höfle G, Fecker T, Kunze B and Reichenbach H (1990) The aurachins, naturally occurring inhibitors of photosynthetic electron flow through photosystem II and the cytochrome b_6/f-complex. *Z Naturforsch* **45c:** 322–328

Ort DR (1986) Energy transduction in oxygenic photosynthesis; an overview of structure and mechanism. In: (LA Staehelin and CJ Arntzen, eds) *Encyclopedia of Plant Physiology, N.S., Photosynthesis III*, Springer-Verlag, Berlin pp 143–196

Ort DR, Ahrens WH, Martin B and Stoller EW (1983) Comparison of photosynthetic performance in triazine resistant and susceptible biotypes of *Amaranthus hybridus*. *Plant Physiol* **72:** 925–930

Pfister K and Arntzen CJ (1979) The mode of action of photosystem II-specific inhibitors in herbicide-resistant weed biotypes. *Z Naturforsch* **34c:** 996–1009

Pfister K and Urbach W (1983) Effects of biocides and growth regulators; physiological basis. In: (OL Lange *et al.*, eds) *Encyclopedia of Plant Physiology, N.S., 12 D, Plant Ecology, IV* Springer-Verlag, Berlin pp 329–391

Pfister K, Steinback KE and Arntzen CJ (1981a) Photoaffinity labeling of a herbicide receptor protein in chloroplast membranes. *Proc Natl Acad Sci USA* **78:** 981–985

Pfister K, Lichtenthaler HK, Burger G, Musso H and Zahn M (1981b) The inhibiton of photosynthetic light reactions by halogenated naphthoquinones. *Z Naturforsch* **36c:** 645–655

Phillips J and Huppatz JL (1984) Cyanoacrylate inhibitors of photosynthetic electron transport. Nature of the interaction with the receptor site. *Z Naturforsch* **39c:** 335–337

Rao JKM, Hargrove PA and Argos P (1983) Will the seven helix bundle be a common structure for integral membrane proteins? *FEBS Lett* **156:** 165–169

Regitz G and Ohad I (1975) Changes in the protein organization in developing thylakoids of *Chlamydomonas reinhardtii* Y-1 as shown by sensitivity to trypsin. In: (M Avron, ed) *Proc IIIrd Int Congress Photosynthesis*, Elsevier, Amsterdam pp 1615–1625

Renger G (1976) Studies on the structural and functional organization of system II of photosynthesis; the use of trypsin as a structurally selective inhibitor at the outer surface of the thylakoid membrane. *Biochim Biophys Acta* **440:** 287–300

Renger G (1986) Herbicide interaction with photosystem II, recent developments. *Physiol Vég* **24:** 509–521

Robinson HH, Eaton-Rye JJ, van Rensen JJS and Govindjee (1984) The effects of bicarbonate depletion and formate incubation on the kinetics of oxidation-reduction reactions of the photosystem II quinone acceptor complex. *Z. Naturforsch* **39c:** 382–385

Sandmann G and Böger P (1986) Sites of herbicide inhibition at the photosynthetic apparatus. In: (LA Staehelin and CJ Arntzen, eds) *Encyclopaedia of Plant Physiol, N.S. Photosynthesis III*, Springer-Verlag, Berlin pp 595–602

Satoh K (1988) Reality of P-680 chlorophyll protein-Identification of the site of primary photochemistry in oxygenic photosynthesis. *Physiol Plant* **72:** 209–212

Shipman LL (1981) Theoretical study of the binding site and mode of action for photosystem II herbicides. *J Theor Biol* **90:** 422–427

Shopes RH, Blubaugh DJ, Wraight CA and Govindjee (1989) Absence of a bicarbonate-depletion effect in electron transfer between quinones in chromatophores and reaction centers of *Rhodobacter sphaeroides*. *Biochim Biophys Acta* **974:** 114–118

Snel JFH and Van Rensen JJS (1983) Kinetics of the reactivation of the Hill reaction in CO_2-depleted chloroplasts by addition of bicarbonate in the absence and in the presence of herbicides. *Physiol Plant* **57**: 422–427

Snel JFH and Van Rensen JJS (1984) Reevaluation of the role of bicarbonate and formate in the regulation of photosynthetic electron flow in broken chloroplasts. *Plant Physiol* **75**: 146–150

Steinback KE, Pfister K and Arntzen CJ (1981) Trypsinmediated removal of herbicide binding sites within the photosystem II complex. *Z Naturforsch* **36c**: 98–108

Stemler A (1982) The functional role of bicarbonate in photosynthetic light reaction II. In: (Govindjee, ed) *Photosynthesis Vol. II, Development, Carbon Metabolism and Plant Productivity*, Academic Press, New York pp 513–539

Stemler A (1989) Absence of a formate-induced release of bicarbonate from photosystem II. *Plant Physiol* **91**: 287–290

Stemler A and Govindjee (1973) Bicarbonate ion as a critical factor in photosynthetic oxygen evolution. *Plant Physiol* **52**: 119–123

Thiel A and Böger P (1986) Binding of ioxynil to photosynthetic membranes. *Pest Biochem Physiol* **25**: 270–278

Tischer W and Strotmann H (1977) Relationship between inhibitor binding and inhibition of photosynthetic electron transport. *Biochim Biophys Acta* **460**: 113–125

Tischer W and Strotmann H (1979) Some properties of the DCMU-binding site in chloroplasts. *Z. Naturforsch* **34c**: 992–995

Trebst A (1979) Inhibition of photosynthetic electron flow by phenol and diphenylether herbicides in control and trypsin-treated chloroplasts. *Z Naturforsch* **34c**: 986–991

Trebst A (1986) The topology of the plastoquinone and herbicide binding peptides of photosystem II in the thylakoid membrane. *Z Naturforsch* **41c**: 240–245

Trebst A (1987) The three-dimensional structure of the herbicide binding niche on the reaction center polypeptides of photosystem II. *Z. Naturforsch* **42c**: 742–750

Trebst A (1991) The molecular basis of resistance of photosystem II herbicides. In: (JC Caseley *et al.*, eds) Butterworths-Heinemann Ltd, Oxford pp. 145-164

Trebst A and Draber W (1979) Structure activity correlations of recent herbicides in photosynthetic reactions. In: (H Geissbühler, ed) *Advances in Pesticide Science*, part 2. Pergamon Press, Oxford pp 223–234

Trebst A, Depka B, Ridley SM and Hawkins AF (1985) Inhibition of photosynthetic electron transport by halogenated 4-hydroxy-pyridines. *Z Naturforsch* **40c**: 391–399

Van Rensen JJS (1969) Polyphosphate formation in *Scenedesmus* in relation to photosynthesis. In: (H Metzner, ed) *Progress in Photosynthesis Research, Vol. III*. H Laupp jr, Tübingen pp 1769–1776

Van Rensen JJS (1971) Action of some herbicides in photosynthesis of *Scenedesmus*, as studied by their effects on oxygen evolution and cyclic photophosphorylation. Meded Landbouwhogeschool Wageningen **71(9)**: 1-80

Van Rensen JJS (1982) Molecular mechanisms of herbicide action near photosystem II. *Physiol Plant* **54**: 515–521

Van Rensen JJS (1983) Interaction of photosystem II herbicides with bicarbonate and formate in their effects on photosynthetic electron flow. *Z Naturforsch* **39c**: 374–377

Van Rensen JJS (1989) Herbicides interacting with photosystem II. In: (AD Dodge, ed) *Herbicides and Plant Metabolism*, Cambridge University Press, Cambridge pp 21–36

Van Rensen JJS and Hobé JH (1979) Mechanism of action of the herbicide 4, 6-dinitro-*o*-cresol. *Z Naturforsch* **13c**: 1021–1023

Van Rensen JJS and Kramer HJM (1979) Short-circuit electron transport insensitive to diuron-type herbicides induced by treatment of isolated chloroplasts with trypsin. *Plant Sci Lett* **17**: 21–27

Van Rensen JJS and Snel JFH (1985) Regulation of photosynthetic electron transport by bicarbonate, formate and herbicides in isolated broken and intact chloroplasts. *Photosynthesis Res* **6**: 231–246

Van Rensen JJS and Van Steekelenburg PA (1965) The effect of the herbicides simetone and DCMU on photosynthesis. Meded Landbouwhogeschool Wageningen **65(13)**: 1–8

Van Rensen JJS and Vermaas WFJ (1981) Action of bicarbonate and photosystem II inhibiting herbicides on electron transport in pea grana and in thylakoids of a blue-green alga. *Physiol Plant* **51**: 106–110

Van Rensen JJS, Tonk WJM and De Bruijn SM (1988) Involvement of bicarbonate in the protonation of the secondary quinone electron acceptor of photosystem II via the non-heme iron of the quinone-iron acceptor complex. *FEBS Lett* **226:** 347–351

Van Rensen JJS, Van der Vet W and Van Vliet WPA (1977) Inhibition and uncoupling of electron transport in isolated chloroplasts by the herbicide 4, 6-dinitro-*o*-cresol. *Photochem Photobiol* **25:** 579–583

Van Rensen JJS, Wong D and Govindjee (1978) Characterization of the inhibition of photosynthetic electron transport in pea chloroplasts by the herbicide 4, 6-dinitro-*o*-cresol by comparative studies with 3-(3, 4-dichlorophenyl)-1, 1-dimethylurea. *Z Naturforsch* **33c:** 413–420

Velthuys BR (1981) Electron dependent competition between plastoquinone and inhibitors for binding to photosystem II. *FEBS Lett* **126:** 277–281

Velthuys BR and Amesz J (1974) Charge accumulation at the reducing side of system 2 of photosynthesis. *Biochim Biophys Acta* **333:** 85–94

Vermaas WFJ and Arntzen CJ (1983) Synthetic quinones influencing herbicide binding and photosystem II electron transport; the effects of triazine-resistance on quinone binding properties in thylakoid membranes. *Biochim Biophys Acta* **725:** 483–491

Vermaas WFJ, Dohnt G and Renger G (1984) Binding and release kinetics of inhibitors of Q_A^- oxidation in thylakoid membranes. *Biochim Biophys Acta* **765:** 74–83

Vermaas WFJ, Van Rensen JJS and Govindjee (1982) The interaction between bicarbonate and the herbicide ioxynil in the thylakoid membrane and the effects of amino acid modification on bicarbonate action. *Biochim Biophys Acta* **681:** 242–247

Warburg O and Krippahl G (1958) Hill-Reaktionen. *Z Naturforsch* **13b:** 509–514

Wasielewski MR, Fenton JM and Govindjee (1987) The rate of formation of $P700^+-A_0^-$ in photosystem I particles from spinach as measured by picosecond transient absorption spectroscopy. *Photosynthesis Res* **12:** 181–190

Wasielewski MR, Johnson DG, Seibert M and Govindjee (1989) Determination of the primary charge separation rate in isolated photosystem II reaction centers with 500-fs time resolution. *Proc Natl Acad Sci USA.* **86:** 524–528

Wessels JSC and Van der Veen R (1956) The action of some derivatives of phenylurethan and of 3-phenyl-1, 1-dimethylurea on the Hill reaction. *Biochim Biophys Acta* **19:** 548–549

Wraight CA (1981) Oxidation-reduction physical chemistry of the acceptor quinone complex in bacterial photosynthetic reaction centers; evidence for a new model of herbicide activity. *Isr J Chem* **21:** 348–354

Wydrzynski T and Govindjee (1975) A new site of bicarbonate effect in photosystem II of photosynthesis; evidence from chlorophyll fluorescence transients in spinach chloroplasts. *Biochim Biophys Acta* **387:** 403–408

Yoneyama K, Konnai M, Honda I, Yoshida S, Takahashi N, Koike H and Inoue Y (1990) Phloroglucinol derivatives as potent photosystem II inhibitors. *Z Naturforsch* **45c:** 317–321

Zurawski G, Bohnert HJ, Whitfield PR and Bottomley W (1982) Nucleotide sequence of the gene for the 32,000-M thylakoid membrane protein from *Spinacia oleraceae* and *Nicotiana debneyi* predicts a totally conserved primary translation product of M_r 38,950. *Proc Natl Acad Sci USA* **79:** 7699–7703

7

Photosynthesis and Herbicides: Effects of Pyridazinones on Chloroplast Function and Biogenesis

N.V. Karapetyan

A.N. Bakh Institute of Biochemistry
Russian Academy of Sciences
Moscow 117071 Russia

CONTENTS

ABBREVIATIONS

CP_a	: Pigment-protein complex of Photosystem II
CP1	: Pigment-protein complex of Photosystem I
DCMU	: dichlorophenyldimethyl urea
F_0	: initial fluorescence
ΔF	: variable fluorescence
LHC	: light-harvesting complex
m-RNA	: messenger RNA
PS I	: Photosystem
PS II	: Photosystem II; PAGE, polyacrylamide gel electrophoresis
Q_A (Q_A)	: primary stable electron acceptor of Photosystem II in oxidized (reduced) state
SAN 6706	: 4-chloro-5-(dimethylamino)-α, α, α,-(trifluoro-m-tolyl)-3(2H)-pyridazinone
SAN 9785	: 4-chloro-5-(dimethylamino)-2-phenyl-3 (2H)-pyridazinone
SAN 9789	: 4-chloro-5-(methylamino)-2-α, α, α-(trifluoro-m-tolyl)-3 (2H)-pyridazinone
SDS	: sodium dodecyl sulphate

182

ABSTRACT

Pyridazinones (norflurazon, metflurazon, SAN 9785) show multiple action on photosynthesis. The mode of action depends not only on herbicide molecule but also on the stage of chloroplast development and the conditions of treatment. Acting on green leaves or mature chloroplasts, these herbicides mainly inhibit the photosystem II activity by blocking the electron efflux from Q_A^- like diuron. SAN 9785 is a more efficient inhibitor than metflurazon and norflurazon. Treatment of greening leaves by pyridazinones causes bleaching of chlorophylls and carotenoids and gradual inactivation of photosystem II. Effect of pyridazinones on unicellular algae is similar to that on greening leaf. The action of norflurazon and metflurazon is maximum when they completely block the biosynthesis of carotenoids. Carotenoidless seedlings show no photosystem II activity but the photosystem I remains active. These seedlings are very sensitive to light: photodestruction of chloroplasts takes place even at low light intesity and is accompanied by chlorophyll bleaching, changes in the fatty acids of lipids, loss of CPa (pigment-protein complex of photosystem II) and light harvesting complex polypeptides and 70S ribosomes. The biosynthesis of CPa polypeptides is not blocked by norflurazon but carotenoids are necessary for their assembly. Thus, pyridazinones can elucidate the structural role of carotenoids in the biogenesis and stabilization of active photosystem II and in regulation of stoichiometry of the photosystems in plants.

I. Introduction

Herbicides are chemical compounds used to kill weeds. Blocking the various processes in plants (photosynthesis, respiration, the biosynthesis of chlorophylls, carotenoids, fatty or amino acids etc.), these compounds destroy the metabolism of plants and thereby bring about the death of plants. Most herbicides block photosynthesis, especially the primary processes of photosynthesis. They mostly inhibit electron transport in chloroplasts at the acceptor side of photosystem II (PS II) or photosystem I (PS I). Inhibition of electron flow stops photosynthesis; light energy absorbed by chlorophylls cannot be used in photosynthesis and causes destruction of chloroplasts. Another large group of herbicides inhibit photosynthesis by changing cell metabolism. Inhibition of biosynthesis of chlorophylls and carotenoids destroys the structure of pigment-protein complexes or chloroplasts membrane. It causes gradual destruction of photosynthetic apparatus and bleaching of chlorophyll, and the subsequent destruction of pigment-protein complexes, membranes, ribosomes, and other plant parts (Feierabend and Schubert, 1978; Fedtke, 1982; Feierabend, 1984; Sandmann and Böger, 1989; Goldfeld and Karapetyan, 1989).

Chlorophyll bleaching, as a result of changes in metabolism, is observed as a common effect of different types of herbicides, like bipyridiliums, diphenyl ethers, photoherbicides (photodynamic herbicides), and pyridazinones. As efficient artificial electron acceptors of PS I, bipyridiliums block the electron transport to $NADP^+$, reduced herbicides are oxidized by oxygen, and the activated oxygen destroys the photosynthetic apparatus, causing chlorophyll bleaching (Moreland, 1980; Merzlyak, 1989). The mechanisms of bleaching induced by diphenyl ethers or photoherbicides are not different: treatment of plants by δ-aminolevulinic acid or by diphenyl ethers stimulates the accumulation of porphyrins, which induces photodestruction of chloroplast components (Matringe and Scalla, 1987).

The effect of pyridazinones is complex and in this chapter the mechanism of action of pyridazinone herbicides is reviewed. The various effects of pyridazinones have been discussed earlier (see Eder, 1979; Fedtke, 1982; Bose and Mannan, 1988; Goldfeld and Karapetyan, 1989; Sandmann and Böger, 1989). We focus here on the action of pyridazinones, mainly norflurazon and its derivatives, on the activity and biosynthesis of PS II. These herbicides bring about a decrease in chlorophyll content as a result of their photodestruction. They have multiple effect on photosynthetic apparatus:

1) inhibition of the photosynthetic electron transport on the reducing side of PS II;
2) under certain conditions, inhibition of the biosynthesis of carotenoids; and
3) changes in the fatty acid composition of lipids.

The types of herbicidal effect of pyridazinones and their efficiency are determined by:

1) substitutions on the herbicide molecule;
2) herbicide concentration in active site which depends on the condition of treatment (feeding via leaves or roots);
3) duration of treatment;
4) intensity of illumination;
5) resistance of the plant to the herbicides; and
6) the degree of development of chloroplasts (etiolated, greening, or green leaves).

II. Action of Pyridazinones on Mature Chloroplasts

Pyridazinones, at high concentrations (about 100 µM), block photosynthetic oxygen evolution, variable fluorescence, and photosynthetic CO_2 assimilation (Hilton et al., 1969; Kunert and Böger, 1979; Karapetyan et al., 1981; Grumbach, 1982; Mannan and Bose, 1986; Laskay and Lehoczki, 1986). It was shown that pyridazinones are bound to the same 32 kD protein as diuron (Tischer and Strotmann, 1977), which causes inhibition of electron flow on the acceptor side of PS II (Pfister and Arntzen, 1979). The possible effect of these herbicides on the donor side of PS II of *Chlorella* was suggested on the basis of measurements of thermoluminescence and delayed fluorescence (Herczeg et al., 1979).

The detailed investigation of pyridazinone action on the fluorescence induction of isolated chloroplasts has shown that these herbicides block mainly electron transport on the acceptor side of PS II (Karapetyan et al., 1981). This was based on the similarity of the variable fluorescence kinetics of chloroplasts treated with pyridazinones or diuron (DCMU), measured using double beam fluorimeter (Karapetyan and Klimov, 1971). The addition of 100 µM pyridazinones caused the enhancement of fluorescence yield which indicates the inhibition of reoxidation of Q_A^-, the PS II stable primary electron acceptor (Fig.1). At high light intensity, the maximal level of fluorescence of chloroplasts treated with herbicides was about the same as that of control (untreated) chloroplasts. Illumination by light 1 (light primarily absorbed by PS I) after light 2 (light primarily absorbed by PS II) was turned off did not enhance the fluorescence lowering as in the case of control chloroplasts, but brought about an increase of fluorescence yield due to an accumulation of additional Q_A^-. Artificial electron donors like $MnCl_2$ or NH_2OH did not change the rate of electron transport of chloroplasts in the presence of pyridazinones (Mannan and Bose, 1985b). If pyridazinones were to inhibit electron transport on the donor side of PS II (Herczeg et al., 1979), the intensity of variable fluorescence would be lower and the fluorescence rise would be delayed. This was observed on the addition of norflurazon to

184

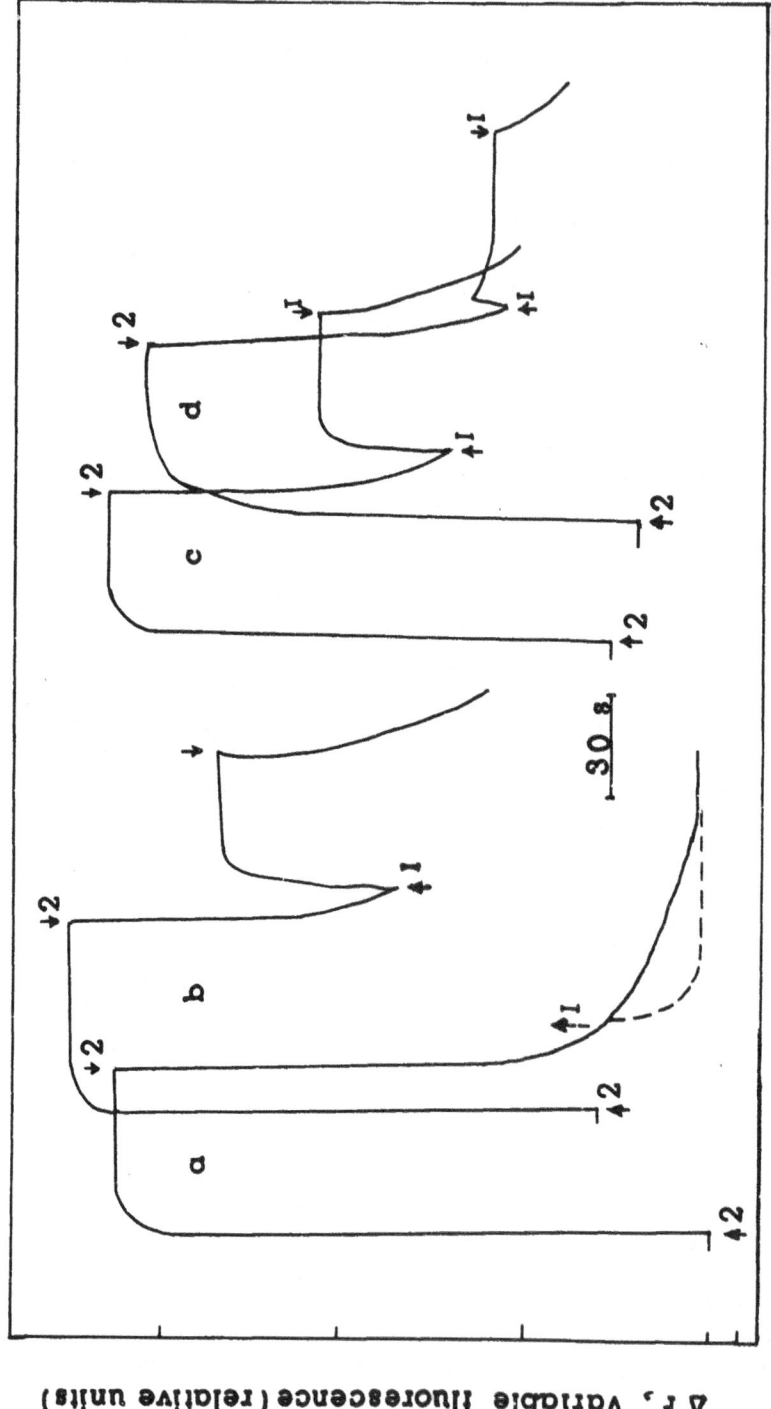

Time , sec

ΔF, Variable fluorescence (relative units)

Figure 1. The kinetics of variable fluorescence (ΔF) at 20°C of pea chloroplasts: a—without addition; b—with 10 μM diuron; c—with 100 μM SAN 9785; d—with 100 μM norflurazon. ↑1, ↑2 light on and ↓1, ↓2 light off of light 1 ($\lambda > 710$ nm) or light 2 ($640 < \lambda > 700$ nm). F_0 is the initial fluorescence.

chloroplasts; the maximal fluorescence level was lower than that in the presence of SAN 9785 (4-chloro-5-(dimethylamino)-2-phenyl-3(2H)-pyridazinone) or diuron, and rise in fluorescence increase was delayed and light 2–light 1 shift caused an increase in fluorescence followed by slow lowering in fluorescence yield (Fig. 1). This may indicate that norflurazon acts at both the acceptor and the donor side of PS II electron tansport chain and the low efficiency of norflurazon as inhibitor of electron efflux from PS II as compared to SAN 9785 and metflurazon. Even at a high concentration, norflurazon does not block Q_A^- oxidation completely, and light 1 activates Q_A^- reoxidation (Karapetyan et $al.$, 1981)

Pyridazinones, even at high concentrations, do not completely inhibit photosynthetic oxygen evolution. The photosynthetic electron transport chain in the presence of high concentrations of pyridazinones also indicates the high yield of delayed fluorescence in isolated chloroplasts in comparison to that of diuron (Karapetyan et $al.$, 1981). It is interesting to note that the yield of delayed fluorescence of chloroplasts treated with SAN 9785 (the most efficient inhibitor of electron transport among the pyridazinones) was lower than that for norflurazon or metflurazon.

The maximal level of variable fluorescence of $Chlorella$ cells in the presence of norflurazon even at strong illumination is lower than that for metflurazon or SAN 9785, which also indicates a lower efficiency of norflurazon as an inhibitor of PS II (Karapetyan et $al.$, 1981). This is not due to lack of penetration of norflurazon because the same is observed with isolated chloroplasts (Ridley and Ridley, 1979; Mannan and Bose, 1985a). About ten times higher concentrations of pyridazinones as compared to diuron are required for maximal inhibition, which could be due to non-specific binding of pyridazinones with proteins (Mannan and Bose , 1985b; Laskay and Lehoczki, 1986).

Pyridazinones only partly block the light-induced ΔpH development of chloroplasts and norflurazon is the least efficient (Karapetyan et $al.$, 1981); the inhibitory effect of pyridazinones on ΔpH was more pronounced at non-saturating intensities of light. These data show that the effect of pyridazinones on photosynthetic electron transport is similar but not identical to that of diuron, although these herbicides bind to the same PS II protein, Q_B (for detailed discussion see chapter by Jack van Rensen, this volume; Pfister and Arntzen, 1979).

The prolonged (24–72 hr) treatment of green leaves by pyridazinones even at strong illumination (10,000 lux) did not significantly change the pigment apparatus (Table 1). The content of chlorophylls (especially chlorophyll b) increased, which caused decrease in chlorophyll a/b ratio from 3:1 for control to 2:1 for treated leaves, but chlorophyll to carotenoid ratio showed negligible change (Rakhimberdieva et $al.$, 1982). This long-term treatment of leaves by pyridazinones caused only minor changes in the state of chlorophylls: a small blue shift of 740 nm band and the enhancement of 685 nm band were observed in low temperature fluorescence spectrum. However, marked changes were found in the kinetics of fluorescence induction. The long-term treatment of green leaves by these herbicides had the same effect as the addition of diuron: the enhancement of maximal level of variable fluorescence was observed, followed by a slow decline. Similar data were obtained at high concentrations (200 µM) of pyridazinones but at a low light intensity

Table 1. Pigment composition of green or greening barley leaves illuminated 72 hr with 30 μM pyridazinones

Sample	Chlorophylls			Carotenoids	Chl/Car
	a	b	a/b		
green					
control	535	174	3.1	202	3.5
+ metflurazon	615	297	2.1	232	3.9
+ SAN 9785	514	240	2.1	216	3.5
+ norflurazon	496	229	2.2	199	3.6
greening					
control	478	172	2.8	149	4.4
+ metflurazon	243	72	3.4	54	5.8
+ SAN 9785	409	144	2.8	124	4.4
+ norflurazon	289	85	3.4	65	5.7

(Laskay and Lehoczki, 1986). The incubation of detached leaves with herbicides during 72 hr of greening did not change either the content of chlorophylls and carotenoids or the ratio of linolenate to linoleate. However, it changed the state of chlorophylls, which caused an enhancement in initial fluorescence (F_0) yield. It is interesting to note that chloroplasts isolated from greening leaves treated for 72 hr with pyridazinones displayed high activity of Hill reaction (Laskay and Lehoczki, 1986). Probably, these herbicides are washed out from bound sites during chloroplast isolation because the binding constants of pyridazinones are of magnitude less than that for diuron (Tischer and Strotmann, 1977). So, pyridazinones did not change the structure of the already developed photosynthetic apparatus of green leaves (and algae). They only act as electron transport inhibitors, and at high intensity of illumination they induce either a change in the state of chlorophylls or cause a small photobleaching of pigments.

III. Effect of Pyridazinones on the Biogenesis of Chloroplasts

The long-term treatment by pyridazinones of developing chloroplasts (etiolated or greening leaves and cultivated alga, as well as seedlings that are grown from seeds incubated with herbicides) causes significant changes in the structure of photosynthetic apparatus. If 1–5 μM concentrations of pyridazinones are lethal for alga, then higher concentrations of herbicides are required for plants because of their slow penetration in leaves and roots. Detached etiolated leaves of barley, grown without pyridazinones, were incubated first with various herbicides for 24 hr in the dark, then illuminated by intense light. The formation of photosynthetic apparatus—biosynthesis of pigments, appearance of pigment-protein complexes, changes in the lipids and in the ultrastructure of chloroplasts etc.—were monitored. These parameters were dependent on the intensity of illumination and concentration of the herbicides.

Leaves greening in the presence of 30 μM norflurazon at 10,000 lux differed from control ones after merely 3 hr of treatment (Rakhimberdieva *et al.*, 1982). Both samples contained PS II reaction centres, appearing as low temperature fluorescence induction, but the state of chlorophylls was altered as indicated by the high yield of F_0 and the changes in 77K emission spectra. These changes were more pronounced after 6 hr greening. Besides, in the kinetics of variable fluorescence of treated leaves, the slow fluorescence transient maximum was absent, and the fluorescence yield change after actinic light was switched off was lower than the F_0 level.

Maximal changes were observed in pigment content after 72 hr illumination of leaves in the presence of norflurazon. The content of carotenoids and chlorophylls decreased by 50 per cent, there was a change to the chlorophyll *a/b* ratio (Table 1), fluorescence band at 696 nm disappeared, and a blue shift from 740 nm to 728 nm was observed (Rakhimberdieva *et al.*, 1982). The yield of F_0 in treated leaves was sharply enhanced, and the kinetics of variable fluorescence differed as compared to control leaves (Fig. 2). The initial increase was followed by a slow decrease in leaves incubated with norflurazon or metflurazon. The intensity of fluorescence after actinic light was turned off decreased to a lower level than the F_0 level, which was slowly reversible in the dark. The addition of diuron strongly delayed the light-induced decrease of fluorescence. However, the decrease of variable fluorescence to lower level after light was switched off was independent of diuron, which blocks the electron flow between photosystems. Thus, this phenomenon is not connected with electron transport. It reflects changes in the pigment-protein complexes of thylakoids. The light-induced fluorescence decrease was also observed for algae grown with norflurazon (Karapetyan *et al.*, 1983, 1985). These changes did not occur in leaves treated with SAN 9785 (Rakhimberdieva *et al.*, 1982).

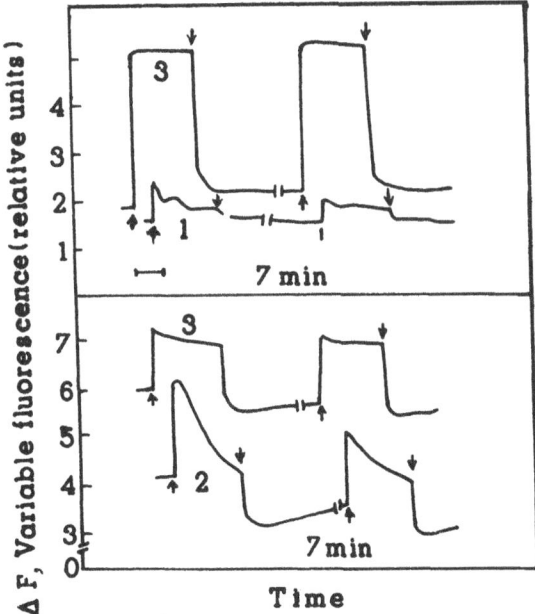

Figure 2. The kinetics of variable fluorescence of greening leaves illuminated for 72 hr with 30 μM norflurazon: control without (1) and with 10 μM diuron (3); norflurazon-treated leaves without (2) or with 10 μM diuron (3).

Similar results were obtained in experiments wherein detached etiolated barley seedlings were treated with high concentrations of pyridazinones at low light intensity (Mannan and Bose, 1986). It was shown that norflurazon had low inhibitory effect on the biosynthesis of carotenoids at various stages on greening. The synthesis of carotenoids and chlorophylls in treated leaves continued for about 12 hr. Then the rate of their accumulation decreased and after 48 hr α- and β-carotenes disappeared, whereas the content of xanthophylls remained high. The total content of carotenoids was about 2.5 times lower and the content of general lipids was about 3 times higher than in the control leaves, although the ratio of saturated to unsaturated fatty acids was unchanged (Laskay et al., 1983a). Thus, the composition of fatty acids in chloroplasts of leaves allowed to green in the presence of norflurazon is significantly different from that of seedlings grown without norflurazon (Bolychevtseva et al., 1988).

The photosynthetic CO_2 assimilation for leaves greened for 72 hr with 200 μM norflurazon was 50 times less although there were only insignificant changes in the pigments (Laskay et al., 1985). The variable fluorescence of these leaves decreased sharply and the slow induction phase transient maximum disappeared in kinetics. The yield of F_0 in treated leaves constituted about 98 per cent of maximal fluorescence (Rakhimberdieva et al., 1982). Thylakoids of treated leaves do not contain CPa (photosystem II complex) and LHC (light harvesting complex) oligomers. Only CP1 (photosystem I complex) and LHC monomers were found (Laskay et al., 1985).

SAN 9785 (200 μM) partly blocked photosynthetic CO_2 assimilation, variable fluorescence, and its kinetics, although control and treated leaves had the same pigment-protein complexes including CPa (Laskay et al., 1986).

IV. Action of Pyridazinones on Algae

The cultivation of *Scenedesmus* with 1μM norflurazon or metflurazon for 48 hr caused decrease in carotenoid content and bleaching of chlorophylls, accompanied by a decrease of photosynthetic oxygen evolution (Kunert and Böger, 1979). In the case of heterotrophic growth of alga or of alga bubbled with nitrogen +0.4 per cent CO_2 in the presence of 10 μM norflurazon, the decrease in carotene amount was minimal and while the xanthophyll content increases, the amount of chlorophyll did not change. Thus, norflurazon induces only a small decrease in carotene content in algae as a result of inhibition of carotene biosynthesis. A significant decrease in carotenoids during the cultivation of algae in air plus CO_2 is due to their photodestruction (Kunert and Böger, 1979).

In *Scenedesmus* and *Bumilleriopsis*, grown with 1-5 μM norflurazon in light at 30 W m^{-2}, chlorophyll bleaching was accompanied by decrease of PS II activity as measured by variable fluorescence (Karapetyan et al., 1983). With a 10 per cent chlorophyll bleaching, the yield of variable fluorescence decreased, and it disappeared with 50 per cent bleaching of chlorophylls. The PS II reaction centres also decreased with 30 per cent chlorophyll loss as is evident from 77K fluorescence transient and it disappeared when 60 per cent of chlorophylls were bleached (Karapetyan et al., 1985). The intensity of F_0 did not change during initial bleaching, but it was enhanced sharply when the alga lost PS II activity. In contrast to *Scenedesmus*,

Bumilleriopsis cells were more resistant to norflurazon; the variable fluorescence of these algae remaining unchanged even at 60 per cent level of chlorophyll bleaching (Karapetyan *et al.*, 1985).

It is important to note that the illumination by intense light of algae that have lost 50–70 per cent chlorophyll caused a decrease in fluorescence yield which was slowly reversible in the dark. Since this light-induced lowering of fluorescence yield was observed even in the presence of diuron, it was not due to PS II. The PS I was more stable to photodestruction but even its activity disappeared at 80–90 per cent bleaching of chlorophylls.

The loss of chlorophylls caused significant changes in the pigment system: blue shifts (from 686 to 682 nm and from 720 to 704 nm) were found in 77K fluorescence spectra, the band at 696 nm decreased gradually (Fig. 3). The presence of 682 nm

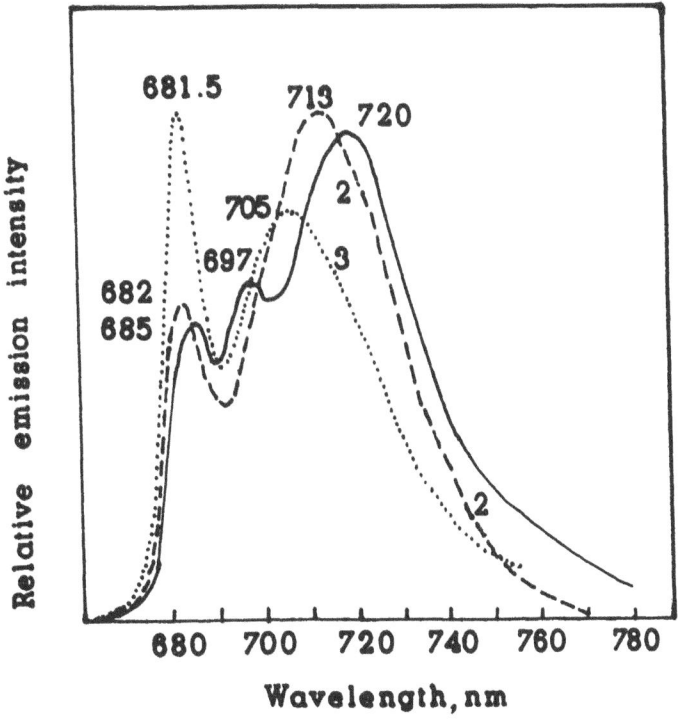

Figure 3. 77K fluorescence emission spectra of alga cells S. *acutus* grown without (1) or with 1 μM norflurazone (2, 3); chlorophyll bleaching is 70% (2) or 85% (3). Exciting beam 430 nm.

band in the fluorescence spectrum of algae with inactive PS II indicates the LHC stability of this alga to norflurazon. The carotenoids in the alga were also not destroyed; this determines the stability of LHC. The 720 to 704 nm shift is due to the damage of the peripheral antenna complex of PS I. The addition of low concentration of norflurazon (30 μM) caused stimulation of photosynthetic apparatus of *Chlorella*: an increase in chloroplast size, amount of photosynthetic membranes, and pigments per cell was observed (Herczeg *et al.*, 1980). The composition of pigments did not change but the ratio of saturated to unsaturated fatty acids increased. The rate of photosynthetic oxygen evolution on chlorophyll basis remained constant on treatment by metflurazon or SAN 9785, but decreased in the case of norflurazon.

The kinetics of variable fluorescence of alga grown for 24 hr with and without pyridazinones differed insignificantly. Remarkable changes in kinetics and the yield of variable fluorescence were found after 72 hr algal cultivation although the algae were able to evolve oxygen (Lehoczki *et al.*, 1982a, b). The addition of 10 μM diuron to these algae caused an enhancement in variable fluorescence, suggesting that the electron transport in acceptor part of PS II was not inhibited.

Inactivation of PS II in alga during the growth with norflurazon could be the result of not only destruction of reaction centres, but also suppression of biosynthesis of CPa-polypeptides. For this reason, norflurazon action on the expression of genes coding the polypeptides of thylakoids of red alga *Cyanidium caldarium* grown in dark or under illumination was studied. The polypeptide compositions of membranes of algae grown in the dark with or without norflurazon were identical, i.e., at these conditions all the genes responsible for coding of these polypeptides were transcribed (Yurina *et al.*, 1989). Although the major polypeptide profile of thylakoids from alga grown with norflurazon in light was similar to that of control, the synthesis of eight polypeptides was inhibited and that of three was stimulated. Thus, photodestruction of chloroplasts in norflurazon-grown alga in light has an effect on the expression of genes coding the thylakoid polypeptides (Linden *et al.*, 1990).

The photodestruction of chloroplasts because of carotenoid deficiency also blocks a number of cytoplasmic messenger RNA (mRNA) coding chloroplasts proteins (Mayfield and Taylor, 1984; Taylor, 1989). The inhibition of the accumulation of LHC mRNA is due to the low rate of transcription of the corresponding gene (Batschauer *et al.*, 1986). This inhibition could be a result of damage of a signal from the chloroplasts to the nucleus which is necessary for the expression of nuclear genes coding chloroplast proteins. (For discussion on polypeptide composition see chapter by Tyagi *et al.*, this volume).

The resistance of the alga to pyridazinones could be regulated. The growth of *Chlorella* with cerulenin (an antibiotic which blocks the biosynthesis of fatty acids) did not cause changes in the kinetics of the variable fluorescence, but the addition of diuron or pyridazinones to these cells did not increase the fluorescence yield (Lehoczki *et al.*, 1982b). This increase was observed for control cells indicating the inhibition of PS II acceptor side. Thus, PS II of alga grown with cerulenin is insensitive to these herbicides. It is quite probable that changes in chloroplast lipids induced by cerulenin alter the PS II resistance to inhibitors as a result of changes of 32 kD protein necessary for diuron binding. The decreased affinity of norflurazon for binding to phytoene desaturase was another possible reason for resistance against this herbicide (Linden *et al.*, 1990; Chamovitz *et al.*, 1990).

Action of high concentrations of norflurazon on greening etiolated leaves was similar to the action of low concentration of this herbicide on alga: in both cases the amount of carotenoids changed insignificantly, the chlorophyll did not bleach completely even at strong illumination. The PS II inactivation occurred at 50–60 per cent level of chlorophyll bleaching, whereas PS I still remained active. Thus, alga could be used as a good model to study the mechanisms of norflurazon effects on the development of photosynthetic apparatus in leaves.

V. Photosynthetic Apparatus of Seedlings Deficient in Carotenoids as a Result of Pyridazinone Action

The action of pyridazinones on photosynthetic apparatus is maximum when seeds are germinated and the seedlings are raised in the presence of 100 µM norflurazon. About 99 per cent inhibition of carotenoids biosynthesis takes place under these conditions (Lehoczki et al., 1982a). Therefore, these seedlings could only be grown at very low light intensity in order to prevent chlorophyll bleaching. Upon binding to desaturase membrane complex, the pyridazinones inhibit the transformation of phytoene and phytofluene of colour carotenoids (Sandmann and Böger, 1989). This desaturase is the most norflurazon-sensitive enzyme in carotenoid biosynthesis.

For seedlings grown with 100 µM norflurazon the content of protochlorophyllide was similar to or even higher that that of control seedlings (Klockare et al., 1981; Bolychevtseva et al., 1987). Although the amount of carotenoids in treated seedlings was only 0.5 per cent of control (Axelsson et al., 1982), the amount of polyunsaturated fatty acids in mono- and digalactosyldiacylglycerol was about the same (Sandelius et al., 1981). Thus, the decrease in chlorophyll content in norflurazon-grown seedlings is not due to the inhibition of chlorophyll biosynthesis. The chlorophyll bleaching in carotenoid-deficient seedlings was accompanied by a decrease in the content of linoleic acid in galactolipids, which was not observed in leaves containing carotenoids (Sandelius et al., 1981). Thus, carotenoids protect galactolipids from photodestruction.

Barley seedlings grown with 100 µM norflurazon at 30 lux light intensity mainly contain chlorophyll a; the amount of chlorophyll b is very low and carotenoids are absent (Table 2). The pigment content of these seedlings is different from that of

Table 2. Characteristics of chloroplasts from barley seedlings grown with 100 µM norflurazon at various intensities of continuous light

	10 lux		30 lux		100 lux	
	control	+ norfl	control	+ norfl	control	+ norfl
Chl a (µg/g w.w.)	295	126.5	343.4	33.4	677	4.1
Chl b (µg/g w.w.)	35.7	5.2	51.5	1.5	131.1	0.8
Carotenoids (µg/g w.w.)	97.5	5.2	118.3	3.3	194.1	1.6
$C_{18:2}$ (mol %)	16.0	20.9	13.9	28.5	13.8	36.5
$C_{18:3}$ (mol %)	64.0	63.0	70.1	55.3	68.0	44.7
Malonaldehyde (nmol/g w.w.)	11.2	14.7	12.5	30.0	12.2	0.8
70S/80S + 80S (%)	42.4	43.5	40.9	42.6	38.8	—
Chl/P700	210	100	230	50	—	—
$\Delta F/F_0$ (20°C)	2.7	—	3.2	—	—	—

control (Lehoczki et al., 1982a; Bolychevtseva et al., 1987). This is indicated by a blue shift in the low temperature fluorescence spectrum: from 687 to 685 and from 740 to 720 nm (Fig. 4). The band at 696 nm is absent and the intensity of emission band at 685 nm is low. The latter band appears broadened because of the appearance of shoulder at 673 nm (Öquist et al., 1980; Lebedev et al., 1988). The proportional

192

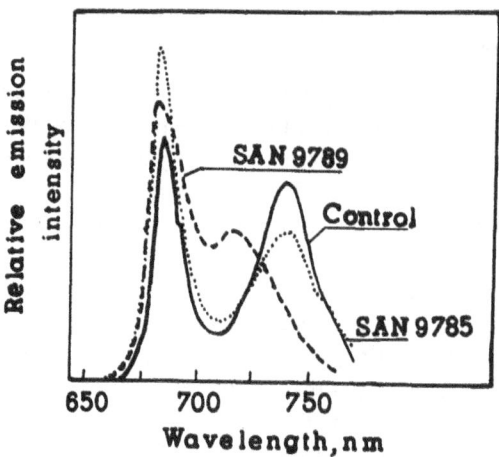

Figure. 4. 77K fluorescence emission spectra of barley seedlings grown without or with 100 μM SAN 9785 or norflurazon.

decrease in fluorescence intensity at 685 nm and 696 nm indicates the absence of CPa emission. The similarity of excitation spectra for fluorescence bands at 685 nm and 720 nm for treated seedlings suggests that both emission bands are due to the same pigment-protein complex. The 436 nm band in these excitation spectra indicates that both emission bands belong to PS I chlorophyll *a*. The carotenoid bands at 491, 497, and 510 nm and pheophytin band at 540 nm are not found in the excitation spectra of emissions at 685 and 720 nm (Fig. 5).

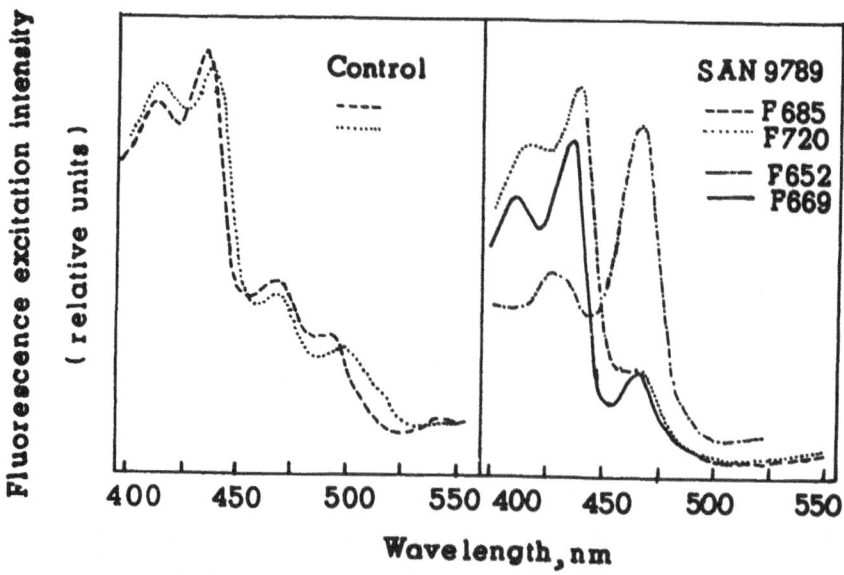

Figure 5. 77K fluorescence excitation spectra of control (left) and norflurazon (SAN 9789) treated (right) barley seedlings.

The absence of variable fluorescence at room temperature and at 77K, of fluorescence band at 696 nm and pheophytin band in fluorescence excitation spectrum in norflurazon-treated seedlings indicates that these seedlings have no active PS II. At the same time, PS I is still active: light-induced difference spectrum of isolated chloroplasts from those seedlings shows negative band at 696 nm due to P700[+] (Bolychevtseva et al., 1988). However, PS I antenna is changed because of absence of emission from peripheral antenna that is indicated by shortwave shift from 740 to 720 nm.

Leaves of control barley seedlings grown at 10–30 lux have normal chloroplasts with developed grana system. But the chloroplasts in norflurazon-grown seedlings contain only stromal lamellae; granae are absent, although the chloroplasts have the same size and form as control (Bolychevtseva et al., 1988). Electron microscopy studies show that chloroplasts of seedlings grown with norflurazon at 10 or 30 lux contain starch grains and ribosomes indicating that these chloroplasts are active in protein and carbohydrate synthesis. As was shown earlier (Karapetyan et al., 1987), norflurazon-treated leaves grown at 10 or 30 lux contain both 70S and 80S ribosomes, and their ratio is similar to that of control seedlings (Table 2). On the other hand, seedlings grown at 100 lux exhibit only 80S ribosomes and no 70S ribosomes; furthermore, the amount of chloroplast rRNA in plastids of these seedlings displays a sharp decline. The loss of 70S ribosomes in rice seedlings grown with 20 μM norflurazon at high intensity has been shown earlier (Feierabend and Schubert, 1978; Feierabend et al., 1984). The destruction of 70S ribosomes is accompanied by the loss in activity of some enzymes that are synthesized on 80S ribosomes, as well as by the inhibition of activity of peroxisome enzymes.

The polypeptide composition of thylakoids of seedlings grown in low light (10 or 30 lux) without norflurazon is typical of normal chloroplasts. The SDS-PAGE (sodium dodecyl sulphate-polyacrylamide electrophoresis gel) analysis shows polypeptides of CPI (67, 19–18 kD), CPa (52, 45, 33–30 kD) and LHC (29, 26, 25–24 kD) as well as coupling factor (58, 56 kD). Electrophoretic profile of thylakoid polypeptides of seedlings grown at the same low light intensities but with norflurazon is different: the amount of 52, 45, 33–30 kD (CPa) and 26–24 kD (LHC) is decreased (Karapetyan et al., 1987; Bolychevtseva et al., 1988). However, the polypeptide composition of the chloroplast-soluble fraction of control and norflurazon-treated seedlings is similar: about 50 components are found with very intense bands 53 and 14, 4 kD of ribulose bisphosphate carboxylase (RuBPcase).

The results indicated that the absence of PS II activity in norflurazon-treated seedlings could be connected not only with chlorophyll bleaching but also with destruction of CPa polypeptides. The lack of PS II function is quite obvious from chloroplast ultrastructure which shows the plastids of treated mutant, while norflurazon-treated wild-type plants showed a large decrease in LHC mRNA.

The scavengers of singlet oxygen or hydroxyl did not prevent chlorophyll bleaching at intense illumination of seedlings grown with metflurazon (Feierabend, 1984), which indicates contribution of superoxide ion or chlorophyll triplets. In norflurazon-treated seedlings, the concentration of chlorophyll triplets is increased (Krasnovsky et al., 1980), which could increase the content of free radical intermediates. Long chain carotenoids quench singlet oxygen more efficiently than phytoene, which accumulates in norflurazon-treated seedlings. The latter does not protect against photodestruction. Therefore, structural and functional changes in these seedlings are the result of photodestruction processes which take place in carotenoid-deficient leaves even at low intensities: at 10 lux, bleaching of pigments; at 30 lux, damage of lipids; and at 100 lux, destruction of protein-synthesizing apparatus (Karapetyan et al., 1991).

Different types of changes are observed in photosynthetic apparatus of seedlings grown with SAN 9785. At low concentration of this herbicide the changes in the content of chlorophylls and lipids are insignificant, but in treated leaves the ratio of linolenic to linoleic acids decreased and the ratio of emissions at 695 to 743 nm increased, e.g., the emission emanating from PS II was enhanced (Mannan and Bose, 1985b; Bose and Mannan, 1988). Photosynthetic CO_2 assimilation of these leaves was low, whereas the yield of variable fluorescence was enhanced as a result of inhibition of Q_A^- in seedlings which do not have grana. Thus, the carotenoids are necessary not only for photoprotection, but also for structural stability (Karapetyan et al., 1991 (also see chapter by Sharma and Hall, this volume)). Carotenoids and chlorophyll b are localized mainly in LHC. The low amount of LHC could be connected not only with their photodestruction but also with the suppression of its biosynthesis. However, the similar content of LHC mRNA in control and treated seedlings grown at low light intensity shows that translation is affected (Batschauer et al., 1986).

Thus, photodestruction in seedlings grown with norflurazon takes place even at very low light intensity (30 lux or less) as a result of carotenoid deficiency. Bleaching of chlorophylls and destruction of photosynthetic apparatus is accompanied by a decrease in linolenate and an increase in linoleate and other fatty acids (Table 2). Selective damage of unsaturated fatty acids is typical for photooxidative destruction (Bolychevtseva et al., 1987). Another indication on photooxidative processes in chloroplasts is the accumulation of malonaldehyde (Table 2). A lower amount of malonaldehyde in seedlings grown with norflurazon at 100 lux as compared to 30 lux is in agreement with lowered chlorophyll content which sensitized photodestruction and malonaldehyde formation.

The evidence for photooxidative damage induced by the chlorophyll was obtained from the experiments with chlorophyll-deficient mutant treated with norflurazon (Mayfield and Taylor, 1984). Plants grown in dim light had high levels of LHC mRNA, as did norflurazon-treated wild-type plants. Transferring the plants to high light had no effect on LHC mRNA in the norflurazon-treated reoxidation. The content of CP1 of treated leaves decreased and that of CPa for control and treated seedlings was similar. Probably lipid environment is important for electron transport: this conclusion is confirmed by data on activation of electron transport in acceptor side of PS II in chloroplasts where the ratio of linolenate to linoleate was changed by electrolysis without change in carotenoid content (Horvath et al., 1986). In seedlings in high light intensity grown with 100–500 µM SAN 9785, chlorophyll a/b ratio decreased, while the xanthophyll/carotene ratio, emissions at 685 and 740 nm, as well as the ratio of CPa to CP1 increased suggesting an enrichment of PS II in the seedlings grown in the presence of this pyridazinone (Mannan and Bose, 1985b).

The reason behind the inactivation or disappearance of PS II is not clear. Reaction centres of both photosystems contain β-carotene. However, if carotenoid-deficient

PS I could be synthesized and kept active, it is not clear why the carotenoid-deficient PS II cannot be synthesized and kept active.

As has been shown, photooxidative destruction took place even at very low light intensities (10 lux or less). It did not, however, damage 70S ribosomes, synthesis of RuBPcase, and PS I polypeptides. In carotenoidless seedlings, PS II is very sensitive to light. So if photodestruction could be excluded by growing seedlings in flashing light (2.5 msec flash, 12 min dark), it is possible to preserve PS II. It was shown that flash-illuminated seedlings grown with 100 μM norflurazon contained even more chlorophyll than control seedlings (Table 3), but this is not enough to form PS II. Moreover, the CPa polypeptides are synthesized in treated seedlings: electrophoretic profile of both types of seedlings is similar (Karapetyan *et al.*, 1991). Norflurazon does not inhibit the biosynthesis of CPa apoproteins, which are incapable of assembling into the membrane without β-carotene. Thus, β-carotene is the important structural component for the biosynthesis and stability of active PS II (Karapetyan *et al.*, 1991).

Table 3. Pigments and photosystem activity of barley seedlings grown with 100 μM norflurazon at flash light

Sample	Chl*a*	Chl*b*	Carotenoids	Chl/P700	$\Delta F/F_0$
Control	51.8	4.1	103	150	0.5
+ norflurazon	59.3	5.3	3	60	—

Pigments content in μg/g (w/w).

VI. Summary

A comparative study of the effects of SAN 9785 and norflurazon (or metflurazon) on photosynthetic apparatus indicates that though these are very similar molecules, they differ in their mechanism of action. The use of pyridazinones provides a convenient approach to studying the structure of photosynthetic apparatus, the stoichiometry and modifications of photosystems of higher plants and algae, as well as changes in fatty acid composition of chloroplast lipids. While SAN 9785 causes the increase of PS II with respect to PS I, norflurazon and metflurazon cause the decrease of PS II. Under certain conditions these herbicides bring about total inactivation or disappearance of PS II. The changes in the amount of pigment-protein complexes, e.g. the increase of CPa content induced by SAN 9785, must be connected with the effect of herbicides on the protein-synthesizing apparatus. The increase of CPa content induced by SAN 9785 and the decrease of CPa content induced by norflurazon treatment could perhaps be the result of the action of these herbicides on the same site in the processes responsible for acceleration or delay of biosynthesis of CPa polypeptides. The knowledge of the mechanism of genetic regulation of biosynthesis of photosystem proteins would allow us to regulate the activity of photosynthetic apparatus.

Acknowledgements: Supported in part by Indo-USSR long term project (Biology).

VII. References

Axelsson L, Dahlin C and Ryberg H (1982) The function of carotenoid during chloroplast development. V Correlation between carotenoid content, ultrastructure and chlorophyll b to chlorophyll a ratio. Physiol Plant 55: 111–116

196

Batschauer A, Mosinger E, Kreuz K, Dorr I and Apel K (1986) The implication of plastid derived factor in the transcriptional control of nuclear genes encoding the LHPC. *Eur J Biochem* **154**: 625–634

Bolychevtseva YuV, Chivkunova OB, Merzlyak MN and Karapetyan NV (1987) Effect of norflurazon on chlorophyll, fatty acid and lipid peroxidation products content in barley seedlings grown under different illumination conditions. *Biochemistry* (USSR) **52**: 160–167

Bolychevtseva YuV, Turishcheva MS, Bezsmertnaya IN and Karapetyan NV (1988) Polypeptide composition and structure of chloroplasts of barley grown in the presence of norflurazon under weak illumination. *Biochemistry* (USSR) **53**: 677–686

Bose S and Mannan RM (1988) Effect of substituted pyridazinones on the structure and functioning of photosynthetic apparatus. In: (Randhir Singh and SK Sawhney, eds) *Advances in Frontier Areas of Plant Biochemistry*, Prentice-Hall of India, New Delhi, pp 19–31

Chamovitz D, Pecker I, Sandmann G, Böger P and Hirschberg J (1990) Cloning a gene coding for norflurazon resistance in cyanobacteria. *Z Naturforsch* **45c**: 482–486

Dahlin C (1988) Correlation between pigment composition and apoproteins of the LHCP II in wheat. *Physiol Plant* **74**: 342–348

Eder FA (1979) Pyridazinones, their influence on the biosynthesis of carotenoids and the metabolism of lipids in plants (survey of literature). *Z Naturforsch* **34c**: 1052–1054

Fedtke C (1982) *Biochemistry and Physiology of Herbicide Action*, Springer-Verlag, Berlin, Heidelberg, New York, pp 1–201

Feierabend J (1984) Comparison of the action of bleaching herbicides. *Z Naturforsch* **39c**: 450–454

Feierabend J and Schubert B (1978) Comparative investigation of the action of several chlorosis-inducing herbicides on the biogenesis of chloroplasts and leaf microbodies. *Plant Physiol* **61**: 1017–1022

Goldfeld MG and Karapetyan NV (1989) Physico-chemical Basis of Herbicide Action. *VINITI Moscow Serie Biochemistry*, Vol 30, pp 1–144

Grumbach KH (1982) Herbicides which inhibit electron transport or produce chlorosis and their effect on chloroplast development in radish seedlings. I Chlorophyll a fluorescence transients and photosystem 2 activity. *Z Naturforsch* **37c**: 268–275

Herczeg T, Lehoczki E and Szalay L (1979) The prompt effects of pyridazinone herbicides on the primary processes of photosynthesis. *FEBS Lett* **108**: 226–228

Herczeg T, Lehoczki E, Rojik G, Vass I, Farkas T and Szalay L (1980) Stimulatory effects of pyridazinone herbicides on *Chlorella, Plant Sci Lett* **19**: 285–294

Hilton Jl, Scharen AL, StJohn JB, Moreland DE and Norris KH (1969) Modes of action of pyridazinone herbicides. *Weed Sci* **17**: 541–545

Horvath G, Droppa M, Szito T, Mustardy LA, Horvath LI and Vigh L (1986) Homogenous catalytic hydrogenation of lipids in the photosynthetic membrane: Effects on membrane structure and photosynthetic activity. *Biochim Biophys Acta* **849**: 325–336

Karapetyan NV and Klimov VV (1971) A device for measurement of the kinetics of the light-induced changes in fluorescence yield of photosynthetic organisms. *Plant Physiol* (USSR) **18**: 223–228

Karapetyan NV Rakhimberdieva MG, Lehoczki E and Krasnovsky AA (1981) Action of pyridazinone herbicides on photosynthetic electron transport of chloroplasts and *Chlorella*. Biochemistry (USSR) **46**: 2082–2088

Karapetyan NV, Strasser R and Böger P (1983) Variable fluorescence and fluorescence spectra of algae after herbicide-induced pigment bleaching. *Z Naturforsch* **38c**: 556–562

Karapetyan NV, Strasser R and Böger P (1985) Changes in photosystem 2 during alga growth in the presence of herbicides inducing chlorophyll bleaching. *Plant Physiol* (USSR) **32**: 70–78

Karapetyan NV, Bolychevtseva Yu V, Turishcheva MS and Bezsmertnaya IN (1987) Changes in photosystem activity and structure of chloroplasts in barley seedlings grown with norflurazon in dim light. *Proc Indian Natl Sci Acad* **B54**: 369–372

Karapetyan NV, Bolychevtseva YuV and Rakhimberdieva MG (1991) The necessity of carotenoids for the assembly of active photosystem 2 reaction centers. In: (RH Duglas, J Mohn and G Ronto, eds) *Light in Biology and Medicine*, Vol 2, Plenum Press, New York, pp. 45–54

Klockare B, Axelsson L, Ryberg H, Sandelius AS and Widell FO (1981) The function of carotenoids during chloroplast development. I Effects of the herbicide SAN 9789 on chlorophyll synthesis and plastid ultrastructure. In: (G Akoyunoglou, ed) *Photosynthesis V Chloroplast Development*, Balaban, Philadelphia, pp 277–284

Krasnovsky AA Jr, Kovalev YuV and Lehoczki E (1980) Luminescence analysis of chlorophyll triplet state in barley leaves with blocked carotenoid biosynthesis. *Dokl AN SSSR* **256**: 726–730

Kunert KJ and Böger P (1979) Influence of bleaching herbicides on chlorophyll and carotenoids. *Z Naturforsch* **34c**: 1047–1051

Laskay G and Lehoczki E (1986) Photosynthetic properties of green barley leaves after treatment with pyridazinone herbicides: comparison with the effects of diuron. *J Exp Bot* **37**: 1558–1567

Laskay G, Lehoczki E, Maroti I and Szalay L (1983a) Effects of pyridazinone herbicides during chloroplast development in detached barley leaves. I Effects on pigment accumulation and fluorescence properties. *Z Naturforsch* **38c**: 736–740

Laskay G; Farkas T, Lehoczki E and Gulya K (1983b) Effects of pyridazinone herbicides during chloroplast development in detached barley leaves. II Effects on lipid content, fatty acid composition and ultrastructure of chloroplast. *Z Naturforsch* **38c**: 741–747

Laskay G, Lehoczki E, Dobi AL and Szalay L (1985) Effects of pyridazinone herbicides during chloroplast development in detached barley leaves. III Effects of SAN 6706 on photosynthetic activity and chlorophyll-protein complexes. *Z Naturforsch* **41c**: 585–590

Laskay G, Lehoczki E, Dobi AL and Szalay L (1986) Photosynthetic characteristics of detached barley leaves during greening in the presence of SAN 9785. *Planta* **169**: 123–129

Lebedev NN, Pakshina EV, Bolychevtseva YuV and Karapetyan NV (1988) Fluorescence characterization of chlorophyll-proteins in barley seedlings grown with the herbicide norflurazon under low irradiance. *Photosynthetica* **22**: 371–376

Lehoczki E, Rakhimberdieva MG and Karapetyan NV (1982a) Effect of blocking of carotenoid synthesis with pyridazinones on photosystem 2 formation in barley leaves. *Plant Physiol* (USSR) **29**: 682–686

Lehoczki E, Rakhimberdieva MG and Karapetyan NV (1982b) Specificity of the photosystem 2 functioning in *Chlorella* grown in the presence of cerulenin or pyridazinone herbicides. *Appl Biochem Microbiol* (USSR) **18**: 405–410

Linden H, Sandmann G, Chamovitz D, Hirschenberg J and Böger P (1990) Biochemical characterization of *Synechococcus* mutants selected against the bleaching herbicide norflurazon. *Pest Biochem Physiol* **36**: 46–51

Mannan RM and Bose S (1985a) Inhibition of photosynthetic electron transport in wheat chloroplast thylakoids by the herbicide BASF 13-33. *Indian J Biochem Biophys* **22**: 171–183

Mannan RM and Bose S (1985b) BASF 13-338 induced changes in the structure and functioning of the photosynthetic apparatus in wheat seedlings. *Photochem Photobiol* **41**: 63–72

Mannan RM and Bose S (1986) Changes in net CO_2 assimilation in wheat seedlings grown in presence of BASF 13-338. *Indian J Biochem Biophys* **23**: 114–118

Matringe M and Scalla R (1987) Photoreceptors and respiratory electron flow involvement in the activity of acifluorfen-methyl and LS 82556 of non-chlorophyllous soybean cells. *Pest Biochem Physiol* **27**: 267–274

Mayfield SP and Taylor WC (1984) Carotenoid-deficient maize seedlings fail to accumulate LHCP mRNA. *Eur J Biochem* **144**: 79–84

Merzlyak MN (1989) Activated oxygen and oxidative processes in the membrane of plant cell. *VINITY Moscow ser Plant Physiol*, Vol 6, pp 1–164

Moreland DE (1980) Mechanisms of action of herbicides. *Ann Rev Plant Physiol* **31**: 597–638

Öquist G, Samuelsson G and Bishop NI (1980) On the role of β-carotene in the reaction center chlorophyll a antennae of photosystem 1. *Physiol Plant* **50**:63–70

Pfister K and Arntzen CJ (1979) The mode of action of photosystem 2 specific inhibitors in herbicide-resistant weed biotypes. *Z Naturforsch* **34c**: 996–1009

Rakhimberdieva MG, Lehoczki E, Karapetyan NV and Krasnovsky AA (1982) Effect of pyridazinones and cerulenin on biosynthesis and functional state of photosystem 2 in barley leaves. *Biochemistry (USSR)* **47**: 637–646

Ridley SM and Ridley J (1979) Interaction of chloroplasts with inhibitors. Location of carotenoid synthesis and inhibition during chloroplast development. *Plant Physiol* **63**: 392–398

Sandelius AS, Axelsson L, Klockare B, Ryberg H and Widell KO (1981) The function of carotenoids during chloroplast development. IV Protection of galactolipids from photodecomposition sensitized by early forms of chlorophyll. In: (G Akoyunoglou, ed) *Photosynthesis V Chloroplast Development*, Balaban, Philadelphia, pp 305–309

Sandmann G and Böger P (1989) Inhibition of carotenoid biosynthesis by herbicides. In: (P Böger and G Sandmann, eds) *Target Sites of Herbicide Action*, CRC Press, Boca Raton, Florida, pp 25–44

Taylor WC (1989) Regulatory interactions between nuclear and plastid genomes. *Annu Rev Plant Physiol Plant Mol Biol* **40**: 211–233

Tischer W and H Strotmann (1977) Relationship between inhibitor binding by chloroplasts and inhibition of photosynthetic electron transport. *Biochim Biophys Acta* **460**: 113–125

Vijayan P and Bose S (1991) Inhibition of the cation-induced revercible changes in recitation energy distribution in thylakoids of BASF 13.338 gram per seedling. In: (SK Sinha, PV Sane, SC Bhargava and PK Agarwal, eds) *Proceedings of International Congress of Plant Physiology*, Society of Plant Physiology and Biochemistry, pp 610–620

Yurina NP, Karakashev GV, Odintsova MS and Karapetyan NV (1989) Norflurazone effect on synthesis and composition of thylakoid polypeptides in unicellular alga, *Cyanidium caldarium. Dokl AN SSSR* **306**: 739–742

8

Interactions between Electron Transport and Carbon Assimilation in Leaves: Coordination of Activities and Control

Christine H. Foyer

Laboratoire du Métabolisme, INRA
78026 Versailles, Cédex
France

CONTENTS

ABBREVIATIONS

ϕPS I	:	index of the relative quantum efficiency for electron transport by PS I
ϕPS II	:	index of the relative quantum efficiency for electron transport by PS II
DHAP	:	dihydroxyacetone phosphate
E4P	:	erythrose 4-phosphate
Fd	:	ferredoxin
F6P	:	fructose 6-phosphate
FBP	:	fructose-1-6 bisphosphate
FBPase	:	fructose-1-6-bisphosphatase
GAP	:	glyceraldehyde phosphate
GBP	:	glycerate-1, 3 bisphosphate
GlP	:	glucose 1-phosphate
G6P	:	glucose 6-phosphate
GSH	:	reduced glutathione
GSSG	:	oxidised glutathione
JpsI	:	product of ϕPSI and incident irradiance
JpsII	:	product of ϕPSII and incident irradiance
NADP-MDH	:	NADP-malate dehydrogenase
P700$^\bullet$	:	the fraction of the P700 pool that remains unoxidised
P700$^+$	:	the fraction of the P700 pool that is oxidised
Pi	:	inorganic phosphate
PGA	:	3-phosphoglycerate

PQ	:	plastquinone
PS I	:	photosystem I
PS II	:	photosystem II
Q_A	:	the primary stable electron acceptor of PS II
qQ	:	coefficient for the photochemical quenching of chlorophyll *a* fluorescence
qNP	:	coefficient for non-photochemical quenching of chlorophyll *a* fluorescence
R5P	:	ribose 5-phosphate
RuBP	:	ribulose-1, 5-bisphosphate
RuBPcase	:	ribulose-1, 5-bisphosphate carboxylase/oxygenase
S7P	:	sedoheptulose 7-phosphate
SBP	:	sedoheptulose-1, 7-bisphosphate
Xu5P	:	xylose 5-phosphate

ABSTRACT

Concepts of control within photosynthetic systems have generally been discussed in terms of the independent regulation of two discrete, spatially separated reaction sequences. These are the electron transport processes and the carbon reduction cycle, occurring in the thylakoid membrane and the stroma, respectively. The former comprise two photosystems operating sequentially to achieve light-driven reduction of $NADP^+$ with concomitant production of a proton gradient. This is used to generate ATP. NADPH and ATP thus produced are consumed during the assimilation and reduction of CO_2 to the level of sugar phosphate. The producer-consumer relationship between electron transport and CO_2 assimilation ensures their tight coupling by virtue of the cycling of intermediates. However, *in vivo* regulation is complicated by the necessity to reconcile the conflicting requirements of the thylakoid reactions and the stromal enzymes.

Several regulatory mechanisms are involved in the modulation of the activity state of key enzymes of the carbon reduction cycle so as to match their activity to the availability of the products of electron flow. High levels of ATP and NADPH are needed to drive high rates CO_2 reduction but high rates of electron transport are difficult to maintain when the electron acceptor NADP and the substrate for photophosphorylation ADP are not freely available. Measurements of (NADPH)/(NADP) ratios and phosphorylation potentials (ATP)/(ADP) + (Pi) *in vivo* give much lower values than would be predicted from *in vitro* measurements with isolated thylakoids. In addition, these parameters are surprisingly stable *in vivo* over a wide range of conditions. The molecular mechanisms whereby electron transport is restrained when ADP and NADP are in short supply are not fully understood. Under these conditions the quantum efficiency of photosystem II is down-regulated and thermodynamic constraints exert a restraining control on the rate of electron flow. The mechanisms that serve to decrease the quantum efficiency of photosystem II also facilitate the coordinate and harmless conversion of light energy directly to heat.

Photoinhibition is a further mechanism that causes a restriction of electron transport. It is exerted under conditions of excessive irradiation and results in a stable down-regulation of photosystem II function.

Much progress has been made in recent years in understanding the relative contributions made by each regulatory process but our insights are far from complete. The most important question to address, however, is why such tight regulation exists. The answers to this question are complex but undoubtedly such precise regulation confers a physiological advantage. In short, co-regulation serves to prevent deleterious effects that would otherwise occur. Precise coordination of reaction rates prevents continuous oscillations in metabolite flux, allows optimization of resources, and yet confers a degree of flexibility that is essential for the avoidance of the detrimental effects of light in a hazardous and constantly changing environment. The most destructive of these are the mechanisms that produce toxic derivatives of oxygen which are an inevitable consequence of the operation of the electron transport chain in an aerobic environment.

I. Introduction

In terms of metabolic regulation, photosynthesis may be considered to consist of at least two distinct sets of reactions that must pull together if the overall process is to remain efficient and stable. All components are essential for photosynthesis to proceed and all must take some part in control. In the context of the chloroplast, we may conveniently divide the individual photosynthetic processes into (1) the membrane reactions that convert light energy into chemical energy, stored temporarily as ATP and NADPH, and (2) the stromal reactions that use this ATP and NADPH to drive the reduction of CO_2 to the level of sugar phosphate. However, in any consideration of the regulation of photosynthesis one must remember that the photosynthetic activity of the chloroplast is integrated with the metabolic processes of the plant cell. Assimilate is exported in strict stoichiometric exchange for phosphate liberated by cytosolic reactions, predominantly sucrose synthesis. The degree of coordination observed between thylakoid and stromal reactions is remarkable. Precise coupling between CO_2 assimilation and sucrose synthesis also allows parallel responses to changes in environmental factors and metabolic limitations. Photosynthesis proceeds in a dynamic and fluctuating environment and it is essential that there is precise co-ordination that allows adjustments in the fluxes both within and between individual reaction sequences so that the system adapts to the new conditions. A key feature of photosynthesis is the degree of coupling and interdependence that exists between the component processes.

Various interactions occur between the photochemical processes and the assimilation reactions. Both participate in the turnover of the ATP and NADPH pools. In addition the light reactions greatly alter the stromal environment in terms of pH, ionic status, and oxidation-reduction potential. All of these exert influence over the activation state of the carbon reduction pathway through modulation of component enzymes. A principal aim of metabolic regulation within the chloroplast is to ensure provision of ATP and NADPH, at the correct stoichiometry, to drive carbon assimilation. Accordingly, the rate of turnover of NADPH and ATP by the carbon reduction pathway has feedback effects on the transthylakoid ΔpH and the redox state of the the electron transport system. These, in turn, control primary photochemistry in (PS II), the electron transfer rate between the photosystem and the activities of membrane-bound enzymes such as the ATP synthase, ferredoxin-NADP reductase, and the thylakoid protein kinase.

The photosynthetic electron transport system is linked inextricably to carbon assimilation via the regeneration and use of ATP and NADPH. Electron transport will operate most efficiently when the supply of NADP, ADP, and inorganic phosphate (Pi) is non-limiting. The composite parameter of the redox $((NADPH_2)/(NADP))$ and phosphorylation potentials $((ATP)/(ADP)+(Pi))$, termed "assimilatory power" or "assimilatory force" (Arnon et al., 1985), will then be low. Assimilatory force is kept low when the reduction and assimilation of CO_2 by the carbon reduction cycle is rapid. This can only occur when the light-modulated enzymes of the cycle are appropriately activated. In all plants so far examined, under steady-state conditions, the assimilatory force is kept relatively low and constant even when irradiance is saturating. This precise co-regulation of electron transport and the carbon reduction cycle is reflected by the constancy of the ATP and NADPH levels despite large variations in incident irradiance and electron flux. The mechanisms that regulate the thylakoid electron transport processes and photophosphorylation coordinate the synthesis of ATP and NADPH with the rate at which these metabolites are used in metabolism. It has previously been suggested that the in vivo capacity for electron transport can exceed that for CO_2 fixation in air at saturating irradiance by a factor of two or more (von Caemmerer and Farquhar, 1981; Leegood et al., 1985; Stitt, 1986). Thus, the phosphorylating electron transport chain can be restrained considerably in high light even under optimal conditions. It is pertinent to consider the mechanisms that exert regulation within the electron transport system.

Electron transport generates the proton motive force that drives ATP synthesis. The proton motive force across the thylakoid membrane consists almost entirely of ΔpH owing to counter-ion movement which negates most of the contribution from the electrical potential difference. This ΔpH is generated by proton release into the thylakoid lumen and accompanying alkalization of the stroma. Acidification of the thylakoid lumen occurs as a result of proton release from water oxidation and the oxidation of plastoquinol by cytochrome b_6/f complex (for discussion on stoichiemetry of protein uptake, ΔpH, see chapter by Ivanov, this volume). As the proton motive force develops it progressively constrains electron transport.

It has always been assumed that as light intensity increases the *in vivo* metabolic capacity for ATP consumption will fall below the capacity for ATP production and that the resulting excessive ΔpH would act to restrict electron transport. *In vitro* experiments with isolated thylakoids and intact chloroplasts have shown that the rate of plastoquinol oxidation decreases with decreasing lumen pH. This has been accepted as a fundamental feature of the restriction of electron transport (Weis *et al.*, 1987). The relevance of this mechanism *in vivo* is not, however, completely resolved as will be discussed further below. More recently it has become clear that ΔpH also has direct effects on the quantum efficiency of PS II (Weis *et al.*, 1987; Horton, 1989; Foyer *et al.*, 1990).

In the short term the quantum yield of photochemistry of PS II is regulated by the non-photochemical quenching of excitation energy induced by a high transthylakoid ΔpH. This major regulatory response allows harmless dissipation of excess excitation energy as heat in the PS II antenna (Horton, 1989). It is of fundamental importance in the protection of the photosynthetic processes against photodestruction. In the short term, the photosynthetic apparatus cannot attenuate the absorption of light and the energy therein cannot be stored except through effective photosynthesis. Excess excitation energy must be dissipated to prevent over reduction of the electron transport chain. The imposition of ΔpH tiggers this rapidly reversible type of down-regulation of PS II.

The rate-limiting step in electron transport resides between the photosystems at the level of plastoquinol oxidation. It has long been recognized that there is a decrease in the rate of plastoquinol oxidation by the chytochrome b_6/f complex as the lumen pH decreases. In pea leaves, for example, this kinetic limitation begins to cause a decrease in the efficiencies of the photosystems at 100 μmol m^{-2} s^{-1} irradiance. We have argued previously (Foyer *et al.*, 1990) on the basis of the evidence from intact systems that the degree of control of electron flow *in vivo* in terms of the limitation of plastoquinol oxidation is relatively constant and does not play a major role in attenuation of electron flux except under extreme conditions. The imposition of this type of control over photosynthetic electron flow has no obvious advantages and could, in fact, be detrimental to the photochemical processes since it would favour overreduction of PS II and oxidation of PS I. Indeed, measurements on whole leaves have indicated that PS II is less reduced than would be expected for a given oxidation state of PS I and also that PS I becomes even more oxidized with increasing irradiance (Weis *et al.*, 1987; Harbinson *et al.*, 1990; Foyer *et al.*, 1990).

Measurements of the kinetics of reduction of oxidized P700, (the reaction centre chlorophyll of PS I, P700$^+$) in leaves in air have shown that the $t_{1/2}$ for P700$^+$ reduction does not vary with increasing irradiance (Harbinson and Hedley, 1989). This suggests that the resistance for electron flow between PS I and PS II is not subject to large changes in response to increasing irradiance. This observation is not consistent with the original model for photosynthetic control where an increasing restriction of the rate of electron flow through the electron transport chain is brought about by a decreasing intra-thylakoid pH which decreases the rate of reaction of plastoquinol with the Reiske iron-sulphur centre of the cytochrome b_6/f complex (Bendall, 1982).

Direct measurements of the bulk lumen pH with increasing irradiance have shown that this parameter is remarkably constant above the low irradiance range (Heldt *et al.*, 1973). There is, however, a system for the attenuation of non-cyclic electron flow by direct, stable, long-term down-regulation of PS II. It is called "photoinhibition" and it occurs under extreme conditions of irradiance or stress. This type of down-regulation replaces, at least in part, regulation exerted by the transthylakoid ΔpH gradient at the level of plastoquinol oxidation and also by the short-term non-photochemical quenching processes.

Many of the measurements contained in this article were obtained by the combined application of biochemical and biophysical techniques. These have greatly aided our investigative and analytical ability. Chlorophyll *a* fluorescence quenching analysis and the *in vivo* measurement of PS I by 820 nm absorption changes have been particularly helpful. The fluorescence yield from chlorophylls associated with PS II is much greater than that emanating from PS I chlorophylls at room temperature. Therefore, room temperature fluorescence measurements are related solely to the efficiency of the PS II reaction centre and the factors that modify its efficiency (Krause and Weis, 1991). The fluorescence yield *in vivo* is lowered by the two fundamentally different mechanisms following the onset of electron transport and carbon assimilation. The fluorescence yield is controlled first by the rate of reoxidation of the first stable electron acceptor to PS II (Q). This type of quenching is essentially linked to photochemistry and is thus called "photochemical quenching" and denoted as qQ or qP. The second type of quenching mechanism is related to the energization of the membrane and its associated processes and is called "non-photochemical quenching" (qNP). While photochemical quenching is caused by charge separation at PS II reaction centres, non-photochemical quenching may be due to a number of non-radiative de-excitation processes in PS II. Much of the non-photochemical quenching is linked to the formation of the transthylakoid ΔpH and the internal acidification of the thylakoid lumen that occurs in the light and is thus called "energy-dependent" quenching or qE.

The quantum efficiency of PS II (ΦPS II) can be calculated from chlorophyll *a* fluorescence measurements (Genty *et al.*, 1989, 1990) and that of PS I (ϕPS I) is obtained from the light-induced absorbance changes at 820 nm (Harbinson and Woodward, 1987; Harbinson and Hedley, 1989; Weis *et al.*, 1987). Such non-invasive probes have been used successfully to give *in vivo* measurments of photochemical efficiency and flux rates rates in intact leaves. Furthermore, these measurements can be effectively coupled to the analysis of metabolic status in terms of concurrent enzyme activities and metabolite levels via freeze-clamping techniques (Fig. 1) in order to give a concerted analysis of electron transport and metabolism (Harbinson *et al.*, 1990; Harbinson and Foyer, 1991). In the following discussion the carbon reduction cycle will be referred to as the Benson-Calvin cycle.

II. Regulation with Respect to Irradiance

In a series of elegant experiments, Genty, Harbinson, Briantais and Baker (Genty *et al.*, 1989, 1990; Harbinson *et al.*, 1989) clearly demonstrated that in crop plants such as pea and barley there is tight coupling between the quantum efficiencies of the photosystems, and also that, when photorespiration is eliminated, there is a good

Figure. 1a. A leaf chamber for the simultaneous measurements of electron transport and carbon metabolism.

The whole chamber. This design of leaf chamber allows the simultaneous measurement of CO_2 fixation, water vapour release, chlorophyll fluorescence, and light-induced absorbance changes at 820 nm on intact, attached leaves. In addition, the leaf sample being measured can be rapidly frozen by a chilled metal block which is forced up by a pneumatic piston through a disposable, gas-tight lower window. The metal block with the frozen leaf sample attached is stopped by the upper window, the bulk of which is a thin plastic film. Further liquid nitrogen is added either manually, via the funnel or using a cryogenic solenoid-operated valve to ensure that the leaf sample remains frozen while being retrieved. The fibre optic bundles carry the measuring beams necessary for the measurement of chlorophyll fluorescence and absorbance changes, and the actinic light used to drive photosynthesis. Photodiodes with optical filters are situated above and below the leaf to detect the measuring beams. The one situated below the leaf is fixed on a drop arm which is released by a pneumatic piston to allow free access to the leaf by the chilled metal block used for freezing. The solid construction of the chamber is necessary to allow measurement of the absorbance changes and to withstand the shock of the piston forcing block from the lower half against the upper half.

correlation between the efficiency of each photosystem with quantum yield of CO_2 assimilation (Baker, 1991; Genty *et al.*, 1990) and O_2 evolution (Seaton and Walker, 1990). Although the coupling between these parameters is remarkably close, the system is dynamic allowing a rapid and concerted response over a large range of irradiances (see chapter by Critchley and Russell, this volume). With increasing irradiance there is initially an increase in the flux of electrons through the photosystem, but the increase

Figure 1b. Dis-assembled chamber. This view shows the partly dis-assembled chamber with a pea leaf lying across the lower gas exchange chamber. The foamed siliconed rubber sealing ring used to produce a non-damaging gas-tight seal against the upper and lower surfaces of the leaf can be clearly seen on the underside of the upper position of the chamber. A small area of the leaf is enclosed to ensure that the chlorophyll fluorescence and absorbance changes from all the photosynthesizing leaf area can be measured. The electrical connections to the photodiode situated below the leaf can be seen at the bottom of the picture.

This chamber was designed by Jeremy Harbinson and constructed in the Mechanical Workshops of the John Innes Institute, Norwich UK.

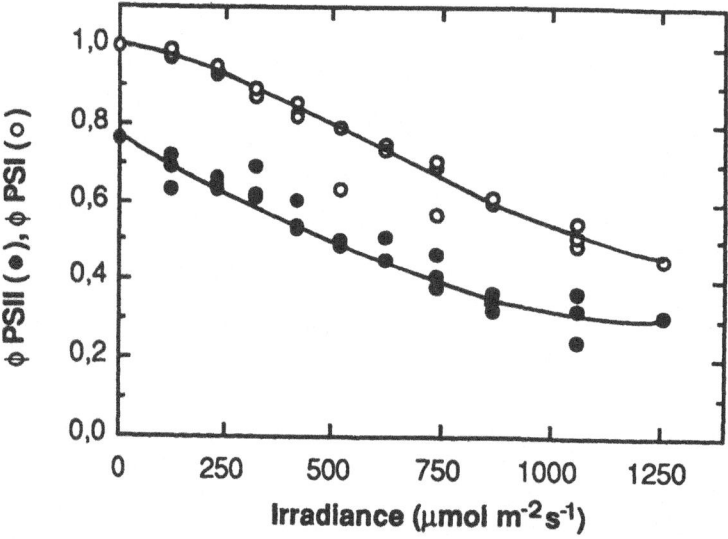

Figure 2. The relationships between ϕPS I, ϕPS II. and incident irradiance measured in intact pea leaves in air. Data taken from Harbinson *et al.* (1990).

in irradiance is accompanied by a progressive decrease in the photochemical efficiency of both photosystems (Fig. 2). As a result of this decreasing photochemical efficiency, the index of electron flux through each of the photosystems can be obtained using the product of the quantum efficiency of either photosystem and irradiance, i.e., JPS II and JPS I (Harbinson *et al.*, 1990; Harbinson and Foyer 1991). This tends to saturate with increasing irradiance. The activation states of light-modulated Benson-Calvin cycle enzymes are increased to accompany the increase in flux through the photosystems (Fig. 3). Measurements of the activity of NADP-malate dehydrogenase

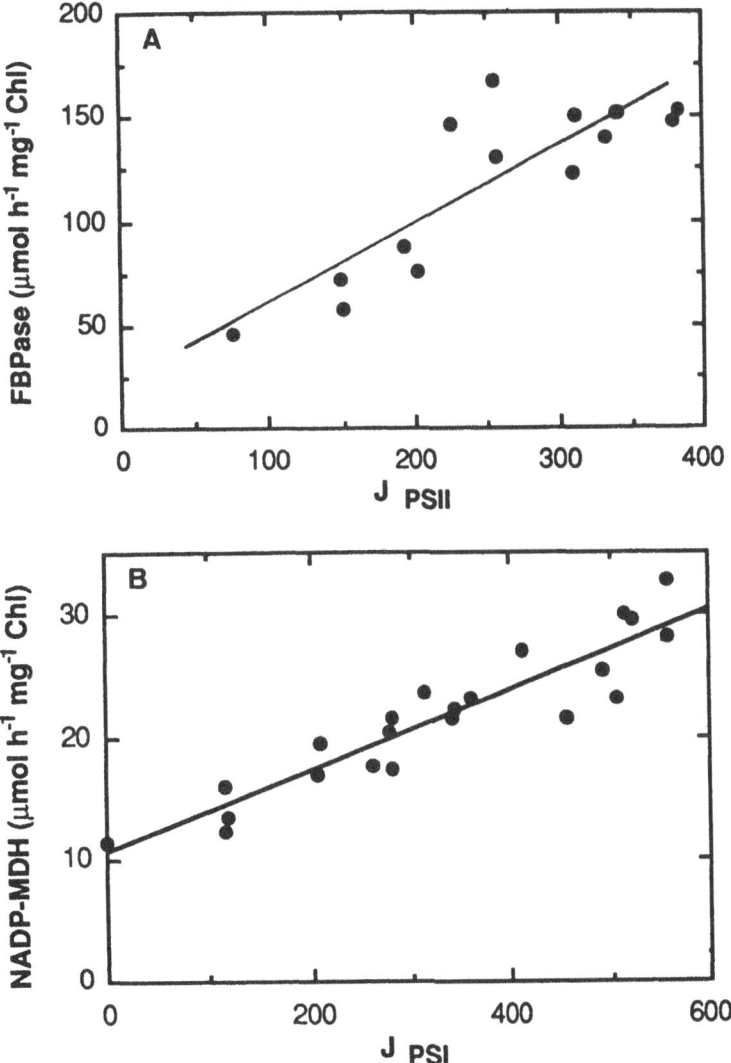

Figure 3A. The relationship between fructose-1,6-*bis*phosphatase activity and the flux of electrons through PS II (JPSII), as measured by the product of ϕPS II and incident irradiance in pea leaves exposed to varying irradiance in air. **B.** The relationship of the activity of NADP-MDH and the flux of electron through PS I (JPSI) measured as in (A).

208

NADPMDH) suggest that as the flux of electrons through the photosystems increases, the redox state of the stroma also increases. Even so, the stroma is kept in a relatively oxidized state (Harbinson et al., 1990).

These measurements were made under steady-state conditions and it appears that in these circumstances the activity of the Benson-Calvin cycle can effectively increase in proportion to increasing irradiance such that very small changes in the stromal NADPH/NADP ratios occur. This is indeed the measured response *in vivo* where NADPH levels remain relatively unchanged during the daily period. In fact, NADPH/NADP ratios can actually decrease since net NADP synthesis occurs in the light (Servaites et al., 1991; Fredeen et al., 1990). In most steady-state by conditions, there is not a detectable restriction of PS I activity by the availability of NADP as an electron acceptor nor any change in the capacity for electron flow between the photosystems. The relative capacities of the thylakoids and the stroma are co-ordinated such that electron transport is predominantly non-cyclic in air when irradiance is varied. The situation is somewhat different, however, when the CO_2 supply is decreased to the compensation point. In this situation the relationship between ϕPS I and ϕPS II becomes increasingly curvilinear (Fig. 4B) in sharp contrast to the

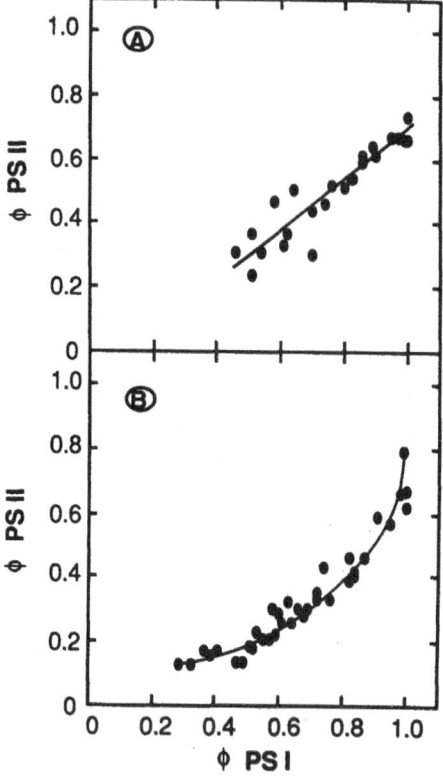

Figure 4. The relationships between ϕPS I and ϕPS II measured in pea leaves in air (A) and in CO_2-free air (B) when exposed to a series of increasing irradiances. Data in (A) taken from Harbinson et al. (1990) and data in (B) taken from Harbinson and Foyer (1991).

approximately linear relationships between ϕPS I and ϕPS II found in air (Fig. 4A Harbinson and Foyer, 1991). Such a non-linear relationship may be explained in terms of the operation of cyclic electron flow around PS I. The consistent depression of ϕPS II relative to ϕPS I with increasing irradiance suggests that the amount of cyclic electron flow around PS I relative to non-cyclic electron flow increases with irradiance. The quantum efficiency of PS II is much more restricted than that of PS I. The stroma remains oxidized at the CO_2 compensation point but there is a partial breakdown in the co-regulation of electron transport and stromal metabolism. The efficient regulation of both non-cyclic and cyclic electron flow in these circumstances limits the increase in the degree of reduction of stromal components.

III. The Induction Processes of Photosynthesis

When dark-adapted leaves, leaf protoplasts, or isolated chloroplasts are illuminated after a period of darkness, CO_2 assimilation begins only slowly after a lag phase or induction period, following which an acceleration to a steady-state rate is achieved (Fig. 5). The induction phenomenon is an intrinsic feature of the photosynthetic process (Osterhout and Hass, 1919) and is attributed to two features arising from the regulated capacity of the Benson-Calvin cycle (Fig. 6) and its feedback control of electron transport (Prinsley and Leegood, 1986). First, the activation of component enzymes is co-ordinated with, and responds to, the light-driven operation of the thylakoid

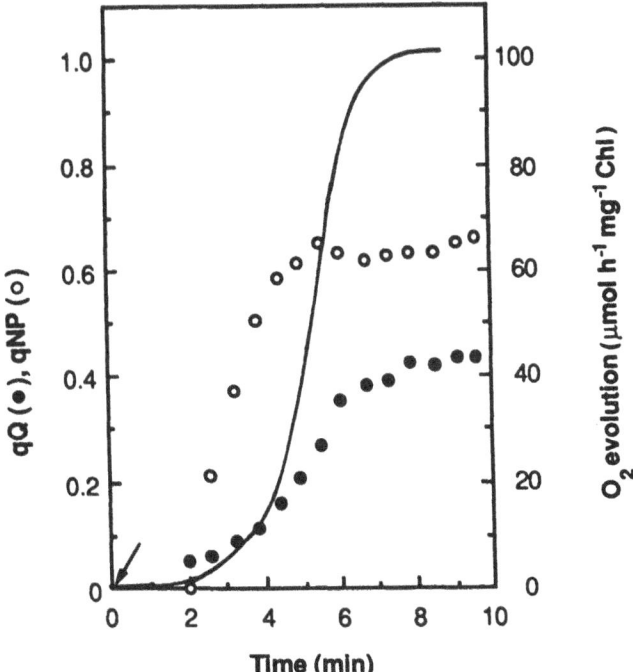

Figure 5. Induction of photosynthesis in isolated intact spinach chloroplasts. O_2 evolution (solid lines) was measured in the oxygen electrode with simultaneous measurement of photochemical quenching, qQ (●) and non-photochemical quenching, qNP (O), of chlorophyll a fluorescence.

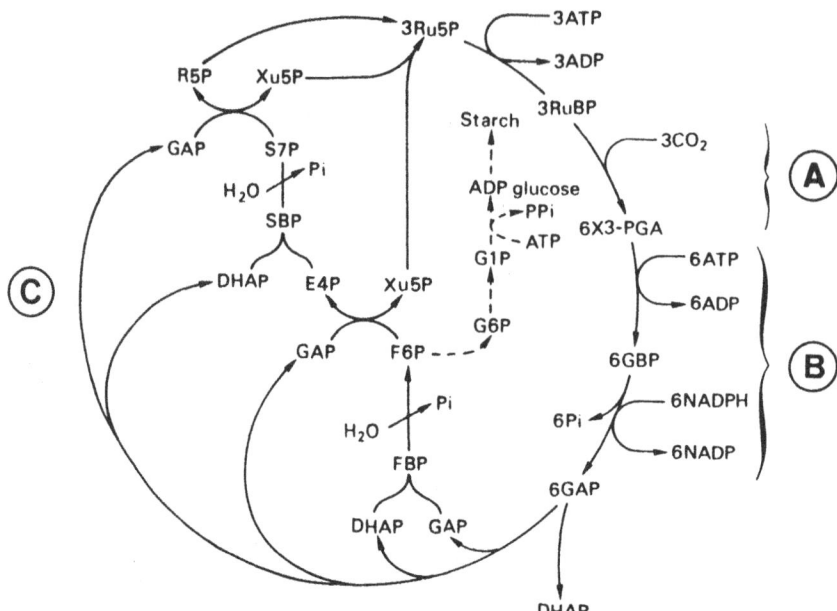

Figure 6. Diagrammatic representation of the autocatalytic Benson-Calvin cycle showing the phases of carboxylation (A), reduction (B), and regeneration of substrate (C).

membrane reactions (Laing *et al.*, 1981; Edwards and Walker, 1983; Cseke and Buchanan, 1986). Second, the Benson-Calvin cycle is autocatalytic in that it produces its own substrate, ribulose-1, 5-*bis*phosphate (RuBP). Significant increases in the component metabolite pools are required in order to facilitate rapid flux through the cycle (Robinson and Walker, 1980). Co-regulation of the rate of the Benson-Calvin cycle with the on-going rate of electron transport is essential and is achieved via several mechanisms. First, assimilatory force drives the Benson-Calvin cycle and rapid consumption of ATP and NADPH allows electron transport to proceed at near maximum capacity (Heber *et al.*, 1986). In this regard the conversion of phosphoglycerate to triose phosphate in the reductive phase of the cycle consumes two-thirds of the ATP and all of the NADPH used in the cycle. This unfavourable reaction is freely reversible and is driven towards triose phosphate production by high phosphoglycerate contents and ATP/ADP ratios. Second, the light-dependent regulation of several of the enzymes of the Benson-Calvin cycle allows carbon assimilation to proceed at rates commensurate with the on-going rate of electron transport. A furthur dimension is added to the regulation of the Benson-Calvin cycle when modulation by component metabolites is considered. For example, a build-up of 3-phosphoglycerate will cause a feedback inhibition of ribulose-1,5- *bis*phosphate carboxylase (RuBPcase) when the regeneration of RuBP is slow and RuBP levels fall (Furbank *et al.*, 1987; Servaites *et al.*, 1991).

Recycling of inorganic phosphate by carbohydrate biosynthesis provides a furthur means of feedback control (Foyer, 1989; Stitt, 1986) and this is considered to give rise to the very long induction phase that can last up to an hour or more after the initial rapid induction period is complete. The orthophosphate content of the stroma and cytosol is particularly important in the co-ordination of the rate of photosynthesis with sucrose and starch synthesis. Since the chloroplast is largely dependent on sucrose synthesis for the Pi required for the production of ATP by the thylakoid ATP synthase, it is pertinent to ask the question whether the Pi concentration ever becomes limiting for ATP synthesis. A direct answer to this question is difficult to obtain since accurate measurements of the unbound mobile Pi pool of the cytoplasmic compartment of the leaf mespohyll cannot be made. However, there are undoubtedly conditions, for example low temperatures, where the rate of carbon assimilation, which uses Pi in the form of ATP, exceeds that of sucrose and starch synthesis which recycle Pi. In these circumstances, Pi availability will limit ATP synthesis and stimulate proton leakage through the ATP synthase (Evron and Avron, 1990).

The induction phase of photosynthesis may be viewed as the opening of a series of electron gates that form control points regulating electron flow from the thylakoid membranes to electron acceptors in the chloroplast stroma. For example, ferredoxin-NADP reductase has been considered a "gate" because it requires reductive activation in the light (Satoh, 1981). Many components of the system must be activated to attain a rate consistent with the on-going electron flux. The low activities of enzymes result in very poor PS II quantum efficiency and an absence of oxidized P700. In the initial 10–20 seconds following the onset of illumination the quantum efficiency of PS II (ϕPS II) falls or remains constant, P700 is largely unoxidized and the NADP pool is relatively oxidized. These results are consistent with a limitation after P700 but before NADP during this period presumably at the level of NADP-ferredoxin oxidoreductase. However, limitation at this level is short-lived and the NADP pool then becomes extensively reduced for up to 60 seconds after the onset of illumination. Reduction of the NADP pool is followed by the increase in the activation state of the thiol-modulated enzymes such as fructose-1, 6-bisphosphatase (Fig. 7). The thioredoxin pool is fully reduced within the first minute of irradiance, the remaining time required for enzyme activation lies in the hysteretic response of the enzymes in the activation process.

Five enzymes of the Benson-Calvin cycle require activation in the light; of these, fructose 1, 6-bisphosphatase, sedoheptulose 1, 7-bisphosphatase, phosphoribulokinase and NADP-dependent glyceraldehyde dehydrogenase are activated by the thioredoxin system (Cseke and Buchanan, 1986) and thus respond to the flux of electrons from PS I, as is clearly evidenced when irradiance is varied (Fig. 3). All of these enzymes exist in virtually inactive forms in the dark. NADP-glyceraldehyde phosphate dehydrogenase activation is rather different from the other three enzymes since it appears to activate fully even in low light. When spinach leaves are illuminated following a short period of darkness the 3-phosphoglycerate pool initially increases and then falls sharply (Fig. 8), suggesting that 3-phosphoglycerate is turned over rapidly following the onset of illumination. Further, the synthesis of triose phosphate (Fig. 6) and its subsequent export to the cytosol occur within seconds of illumination. Triose phosphate export and its conversion to phosphoglycerate occur in the earliest stages of induction and, thus, supply both ATP and reducing power for biosynthetic processes in the cytosol such as sucrose synthesis and nitrate assimilation. These observations would suggest that the rate of the NADP-glyceraldehyde phosphate

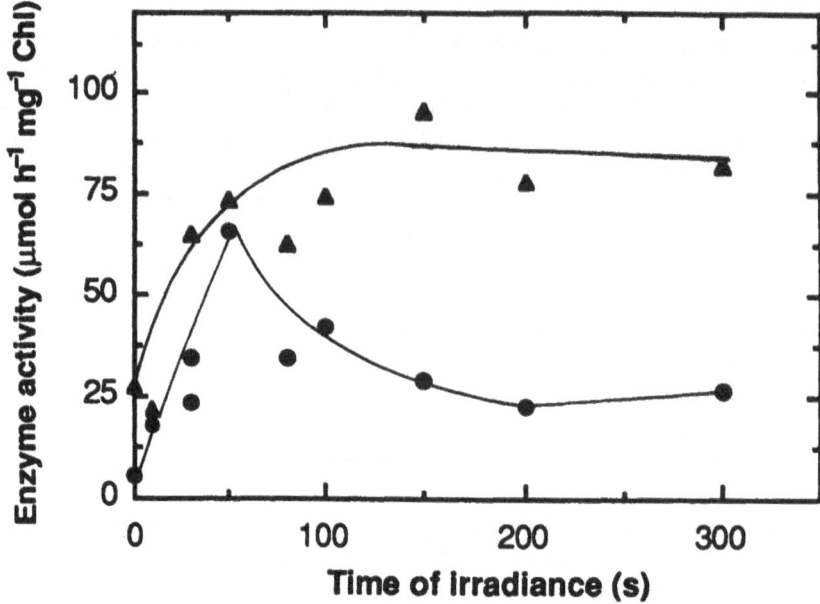

Figure 7. The activation of fructose-1, 6-*bis*phosphatase (▲) and NADP-MDH (●) during the induction phase of photosynthesis as measured in pea leaves in air using the apparatus shown in Fig. 1. Each point is the activity determined after pooling six separate leaf discs in which metabolism was stopped at the times indicated.

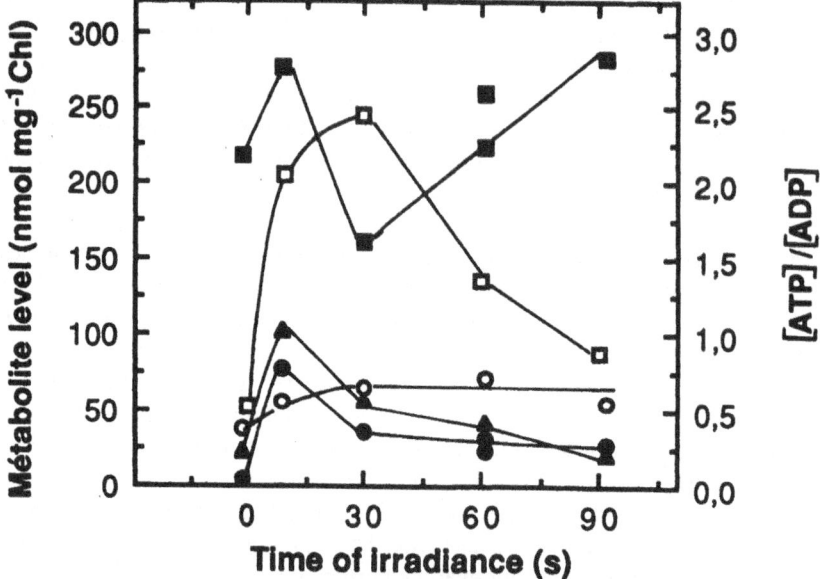

Figure 8. Metabolite changes measured during the induction phase of photosynthesis of spinach leaf discs in air supplemented with 5 per cent CO_2. Each point is data taken from pooled leaf disc samples (four in each case) which had been pretreated for 5 minutes under the above conditions and 500 μmol $m^{-2} s^{-1}$ irradiance. They were then given 1 minute dark. Samples were freeze-clamped upon subsequent illumination at the times given. ATP/ADP (□) fructose-1, 6-*bis*phosphate (■), RuBP (O), triose phosphate (▲).

dehydrogenase reaction provides a significant limitation only during the first seconds of induction. However, the regulation of NADP-glyceraldehyde phosphate dehyrogenase is highly complex and poorly understood (Leegood, 1990). It exists in a number of aggregated forms. The enzyme shows hysteretic properties, its activation is sensitive to NADPH, ATP, and Pi, and the enzyme protein has been shown to be phosphorylated (Leegood, 1990; Guitton and Mache, 1987). The two enzymes, phosphoglycerate kinase and NADP-glyceraldehyde phosphate dehydrogenase, form an enzyme couple with negative co-operativity. However, the negative co-operativity is reduced upon illumination and there is apparently channeling of glycerate-1, 3-*bis*phosphate between the two enzymes (Macioszek and Anderson, 1987; Macioszek *et al.*, 1990)

It is evident that the activation of the thiol-modulated enzymes of the regenerative phase of the Benson-Calvin cycle (Fig. 6) which activate in 1–2 minutes after the onset of irradiance form a major gate or limitation of the assimilation rate early in induction. For example, fructose-1, 6-*bis*phosphatase activity increases progressively in an approximately hyperbolic fashion with time of irradiance during the induction period (Fig. 7) and broadly parallels the changes in efficiencies of both photosystems. Previously it was considered that light activation of RuBPcase could not be an important factor contributing to the induction photosynthesis (Walker, 1981) but recent evidence has suggested that some activation of RuBPcase is required after a period of darkness or an increase in irradiance and that the activation process is slow, requiring several minutes (Seemann *et al.*, 1988; Woodrow and Mott, 1989). It is clear that in response to illumination RuBPcase has unique mechanisms for the regulation of activation state. These include the inhibitor carboxyarabinitol-1-phosphate (CA1P) and the activase protein (Berry *et al.*, 1987; Gutteridge *et al.*, 1986; Portis, 1990). RuBPcase activase facilitates activation of RuBPcase by removing inhibitors such as CA1P, and also RuBP, bound to inactive sites on the enzyme (Portis, 1990).

There are, therefore, two distinct phases in the regulation or limitation of electron flow in leaves during photosynthetic induction. In the first phase the limitation of electron transport is after PS I. During this time, the resistance for electron flow between the photosystems, though high, is not limiting. During the second phase in which electron transport continues to increase the limitation lies between the two photosystems. The increasing rate of electron transport is determined by the decreasing resistance to electron flow between the two photosystems. These changes occur in parallel with increases in the rate of CO_2 fixation (Harbinson and Hedley, 1989).

When P700 oxidation is limited on the acceptor side quantitative information about PS I efficiency cannot be obtained via the absorption change at 820 nm. Nonetheless, and increasing $\Delta A820$ in parallel with the overall increase in øPS II implies that, although PS I is initially completely limited on its acceptor side, with time this limitation is relieved. It has been demonstrated that during the increase in the $P700^+$ pool, there are only small changes in the $t_{1/2}$ for $P700^+$ reduction. Thus, the changes in øPS II, which is also the quantum yield for non-cyclic electron flow, are not the result of changes in the degree of restriction of electron flow between the two photosystems as measured by the $t_{1/2}$ for $P700^+$ reduction (Harbinson and Hedley, 1989), but rather the availability of PS I electron acceptors.

As the turnover of the Calvin cycle increases there is increased demand for NADPH and ATP. This results in increased regeneration of NADP and ADP and a relief of the restriction of electron transport on the acceptor side. Pseudocyclic electron flow is high during the induction phase (Marsho *et al.*, 1979). This coupled process uses reducing power (Foyer *et al.*, 1984; Neubauer and Schreiber, 1989) and yet provides ATP (Egneus *et al.*, 1975) (see Section VII). The processes of cyclic and pseudocyclic electron flow serve to generate ATP when non-cyclic electron flow is limited by the availability of NADP. The initial phase of over-reduction may be necessary, however, to force the activation of the enzymes of the Benson-Calvin cycle whose electrochemical potentials for reduction are not entirely favourable. Once the thiol-modulated enzymes of the Benson-Calvin cycle become active the NADPH/NADP ratio falls to a level very close to the dark level.

IV. NADP-malate Dehydrogenase as a Physiological Indicator of the Physiological Redox State of the Stroma

A major uncertainty has persisted for many years as to whether direct measurements of NADPH/NADP ratios give quantitative information about the intra-chloroplast redox ratios. Measured pyridine nucleotide (NADP, NADPH) levels in leaves have consistently shown that the stromal NADP system is largely oxidized (Fredeen *et al.*, 1990; Takahama *et al.*, 1981) and indicate that under steady-state conditions an NADPH/NADP ratio of 2 or 3 is the maximum value that can be sustained. This result was initially unexpected but may be rationalized in terms of the necessity to minimize the production of oxygen-free radicals via the reaction between reduced ferredoxin and dioxygen. Reduced ferredoxin, and possibly other components of the electron transport chain, will reduce O_2 to O_2^-. Oxygen, thus, serves as an electron acceptor and although this is a coupled process that may supply some of the ATP needed to drive the CO_2 reduction cycle it also generates toxic products, O_2^- and H_2O_2, that can instigate harmful cascade reactions generating even more toxic compounds. In spite of this it is evident that reducing power is required to drive CO_2 fixation, nitrite reduction, enzyme activation, and many other biosynthetic pathways. Consequently, it would be most useful to find an accurate measure of reducing power within the stroma in order to determine exactly how the oxidation-reduction balance is maintained.

Using *in vitro* systems consisting of ruptured chloroplasts in the presence of NADP and ferredoxin, light-dependent reduction of NADP goes virtually to completion, with over 95 per cent of the available NADP subject to reduction by light. Clearly there is a significant difference between the measured reduction state of the pyridine nucleotide pool *in vivo* and values that are obtained *in vitro*. Differences between measured pyridine nucleotide ratios in leaf extracts and calculated potentials can be explained in terms of metabolite binding to proteins affecting the levels of free and thermodynamically-active pyridine nucleotides *in situ*.

The measurement of photosynthetically active metabolites in leaves, protoplasts, and chloroplasts has added considerably to the understanding of the regulation of photosynthesis. However, the interpretation of such data often involves the assumption that measured amounts of metabolites reflect the free concentrations available to photosynthetic enzymes and processes. This assumption may not be valid in the case of the chloroplast stroma because the stroma has a very high protein concentration (measured values extrapolate to 18 mg soluble protein mg^{-1} chlorophyll for 100 per cent intact chloroplasts). Up to 90 per cent of the stromal protein is RuBPcase, which has an active site concentration of at least 4 mM. Calculations based on the Kd values for various stromal metabolites that bind to the RuBP binding site of RuBPcase have suggested that a considerable proportion of certain stromal metabolites such as pyridine nucleotides might be bound to this enzyme *in vivo* (Ashton, 1982). Indeed, the binding of stromal metabolites to RuBPcase has been found to have a considerable inhibitory action on the reactions for which these metabolites are substrates (Furbank *et al.*, 1987).

For the above reasons measured values of NADPH and NADP are of qualitative but not of quantitative significance and must be regarded with caution. According to the suggestion of Scheibe and Stitt (1988), NADP-MDH may be used as a physiological indicator of the redox state of the stroma. Indeed, NADP-MDH may provide a more accurate physiological method of measurement of the relevant reduction state of the stroma than direct measurement of the NADP/NADPH pool. NADP-MDH is localized exclusively in the chloroplast stroma. It is subject to light-activation by modulation of the thiol groups of the enzyme and is modulated by the NADPH/NADP ratio (Crawford *et al.*, 1989; Miginiac-Maslow *et al.*, 1990; Scheibe, 1987). The relationship between the measured redox state of the NADPH/NADP system and NADP-MDH during photosynthetic induction in pea leaves is hyperbolic. The activities of NADP-MDH in steady-state conditions in leaves are low (Harbinson *et al.*, 1990; Harbinson and Foyer, 1991). These observations suggest that measurements of leaf NADP-MDH activity in most situations are an accurate reflection of the overall stromal redox state.

V. Regulation in a Dynamic Light Environment

Regulation of both the light reactions and photosynthetic carbon assimilation must be extremely dynamic in order to take advantage of the highly variable light regimes within plant canopies. In natural environments the quantity of light received by the leaves is rarely constant even at the top of a canopy where leaf movement in the wind, intermittent shading by nearby leaves, and even the degree of cloud cover result in a highly dynamic light environment. In the depths of the leaf canopy the situation becomes even more extreme and the efficient utilization of brief exposures to high intensity light becomes increasingly important. Sunflecks reaching forest understorey plants can contribute 40 to 70 per cent of their daily light absorption and hence carbon gain (Pearcy, 1990). It is evident from the work of Pearcy and his colleagues (Kirschbaum and Pearcy 1988a, b; Pearcy, 1990) that the ability to use sunflecks efficiently depends on the induction processes of the photosynthetic machinery (Edwards and Walker, 1983) (see chapter by Critchely and Russell, this volume). In order to use transient increases in irradiance it appears to be necessary to briefly decouple electron transport from CO$_2$ assimilation (Kirschbaum and Pearcy, 1988b). The occurrence of a lightfleck induces a burst of O$_2$ evolution as a result of increased non-cyclic electron transport. This promotes reduction of phosphorylglycerate to triose phosphate in the reductive phase of the Benson-Calvin cycle (Fig. 6). This is then used to regenerate RuBP, which only requires ATP, in the regenerative phase of the cycle. Leaves exposed to frequent sunflecks showed complete induction of photosynthesis. Sunflecks can be used efficiently because of post-sunfleck CO_2 assimilation occurring after the higher light exposure has ended. Post-lightfleck CO_2 fixation occurs because of the capacity to rapidly build up an assimilary force in terms of high-energy metabolites, such as phosphoglycerate and triose phosphate, that can be used after the lightfleck that passed. Carbon gain during short lightflecks can be twice that predicted from steady-state photosynthetic rates (Pearcy, 1990). The induction state achieved during one lightfleck carries over into the next and enzyme activity can actually continue to increase in between successive lightflecks. The requirement for the induction of photosynthesis depends largely on the light-dependent activation state of the enzymes of the Benson-Calvin cycle and also stomatal opening.

VI. Protection of the Photosynthetic Processes

When irradiance is low the attainment of maximum quantum yield requires that PS II and PS I are excited at appropriate rates. This is achieved, at least in part, by reversible phosphorylation of a mobile population of the light harvesting chlorophyll *a*/*b* binding protein that adjusts the relative cross-sections of the photosystems (Horton, 1983, 1989). However, at light levels approaching saturation, and beyond, the photosynthetic apparatus must alter its regulatory position in favour of energy dissipation and decreased quantum efficiencies in order to maintain optimum redox poise and prevent overreduction and photodestruction. Prolonged exposure to light levels where excitation energy exceeds the capacity for carbon assimilation can lead to photoinhibition. The term "photoinhibition" is defined as a reduction in photosynthetic activity by excessive light and is evidenced by a light-induced loss of quantum efficiency and a change in chlorophyll fluorescence characteristics (Powles, 1984). Photoinhibition has become more or less synonymous with photoinhibitory damage to the PS II reaction centre but also comprises photoprotective processes which serve to dissipate excitation energy as heat before it reaches the PS II reaction centre. Thus, measured levels of photoinhibition may be the combined result of both processes occurring simultaneously. The molecular mechanism(s) of non-radiative energy dissipation remain to be elucidated. Various mechanisms have been suggested including the operation of the futile cycle of electrons around PS II (Schreiber and Neubauer, 1987). According to the model of photosynthetic energy dissipation proposed by Butler and co-workers (Kitajima and Butler, 1975) the overall yield of chloropyll *a* fluorescence is a function of the rate constants for fluorescence (k_F), non-radiative energy dissipation in the pigment bed (k_D), exciton transfer from PS II to PS I (k_T), and the photochemical activity of PS II (k_p). Photoprotective dissipation of excess excitation energy via an increase in the rate constant for non-radiative dissipation has been correlated with the appearance of the carotenoid zeaxanthin in the thylakoid membranes. A linear relationship between the rate constant for radiationless energy dissipation in the antenna chlorophylls (k_D), as calculated from chlorophyll *a* fluorescence analysis, and the leaf content of zeaxanthin has been consistently demonstrated (Demmig-Adams, 1990) (see chapter by Sharma and Hall, this volume).

Leaves of shade-grown plants are far more susceptible to photoinhibition than leaves of plants grown in high light. Low-light-grown plants exposed to sudden

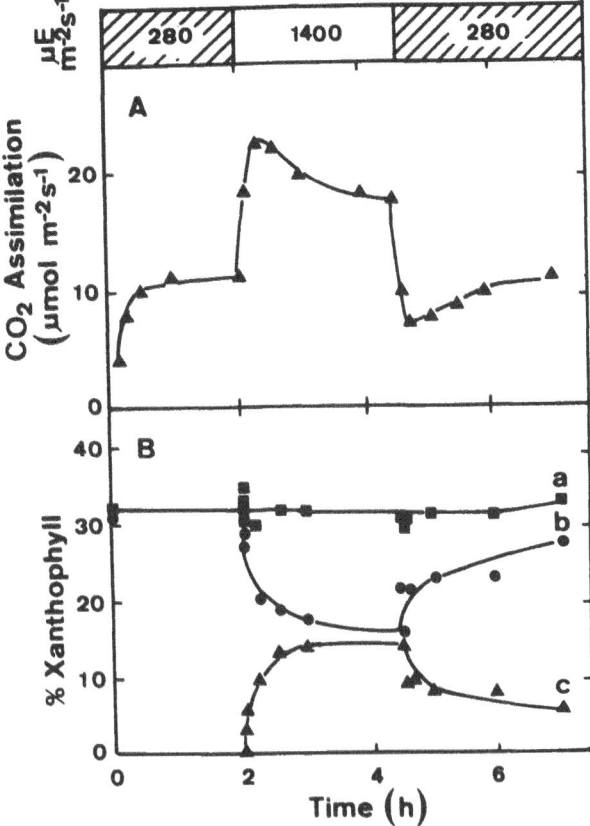

Figure. 9. CO_2 assimilation (A) and turnover of the xanthophyll cycle components (B) in intact barley leaves response to sudden transitions in irradiance from the growth irradiance of 280 µmol m^{-2} S^{-1} to a photoinhibitory irradiance (14.00 µmol m^{-1} S^{-1}) and subsequently a return to 280 µmol m^{2-} S^{-1}. The xanthophyll components are expressed as a percentage of the total xanthophyll pool. Violoxanthin plus antheraxanthin plus zeaxanthin (a), violoxanthin (b), zeaxanthin plus antheraxanthin (c).

transition to high light show rapid conversion of violaxanthin to zeaxanthin (Fig. 9). Upon the transition from the growth irradiance (280 µmol m^{-2} s^{-1}) to a photoinhibitory irradiance (1400 µmol m^{-2} s^{-1}) the CO_2 assimilation rate of barley leaves doubles within minutes (Fig. 9) but the high irradiance rapidly causes a decrease in quantum efficiency. This is because the capacity to make effective use of absorbed light energy in excess of that normally experienced is low. Following exposure of low-light-grown plants to high light the activities of NADP-MDH and fructose-1, 6- *bis*phosphatase rise to near maximum values. Such observations suggest that the transition to high light causes the stroma to become highly reduced. Highly reducing conditions predispose the photosynthetic membranes to photoinhibition. As photoinhibitory loss of thylakoid function occurs, there is a progressive decrease in the capacity for non-cyclic electron flow. This is evidenced by a decrease in the activity of thiol-modulated enzymes. Thus, photoinhibition results in oxidation of the stroma and in this situation there is a strong competition for a limited supply of reducing power (Dujardyn and Foyer 1989; Foyer *et al.*, 1989; Miginiac-Maslow *et al.*, 1990).

Thus, as a result of photoinhibitory loss of PS II function ϕPS II becomes limiting and the stroma becomes relatively oxidized.

The scavengers for active oxygen species are lower in plants grown in low light than in high light. Shade plants have about four times less of the xanthophyll cycle components than sun plants. Such a decreased capacity in the protective systems results in increased susceptibility to active oxygen species and free radicals. Le Gouallec and Cornic (1988) observed an enhancement of photoinhibition of CO_2 assimilation at 20 per cent O_2 compared to 1 per cent O_2. Thus, the environment in which leaves develop can have a profound effect on their susceptibility to photoinhibition. More frequently, however, the immediate environmental conditions to which leaves are exposed is most important in determining susceptibility to photoinhibition, for example, when plants normally acclimated to high light are exposed to environmental stresses such as low temperature.

The formation of zeaxanthin occurs via a complex reaction system known as the xanthophyll or violaxanthin cycle (Yamamoto, 1979). This cycle consists of the light-dependent de-epoxidation of violaxanthin, via antheraxanthin, to zeaxanthin as in reaction (1) and light-independent epoxidation of zeaxanthin via antheraxanthin to violaxanthin as in reaction (2).

$$\text{Violaxanthin} \xrightarrow{\text{2H (Ascorbate)}} \text{Antheraxanthin} \xrightarrow{\text{2H (Ascorbate)}} \text{Zeaxanthin} \quad (1)$$

$$\text{Zeaxanthin} \xrightarrow{\text{NADPH} + \text{H}^+ + \text{O}^2} \text{Antheraxanthin} \xrightarrow{\text{NADPH} + \text{H}^+ + \text{O}^2} \text{Violaxanthin} (2)$$

The violaxanthin cycle is considered to be located close to PS II. However, violaxanthin depoxidase is present in etiolated leaves and the xanthophyll cycle operates well in the absence of a functional PS II if a transthylakoid ΔpH is established (Pfündel and Strasser, 1988). The violaxanthin de-epoxidase requires ascorbate as an essential cofactor, has an acidic (pH 5.0) pH optimum, and is inhibited by dithiotreitol.

If excess excitation energy exceeds the capacity of the dissipative pathways, photodamage to the PS II reaction centre and the inactivation of primary photochemistry results. Turnover of a central PS II core polypeptide, the herbicide-binding 32 kD (DI protein) polypeptide, is enhanced (Ohad et al., 1990) (see chapter by Sopory et al. this volume). Recovery from this type of photodamage can be prevented by the addition of inhibitors of 70S ribosome function, such as chloramphenicol, that prevent synthesis of the D1 polypeptide. Net damage to PS II occurs when the rate of destruction of the D1 protein exceeds its rate of replacement in the membrane. Repair occurs only slowly at low temperatures or when the leaves are deficient in nitrogen. However, this type of photodamage may also be viewed as a protective device that serves to avoid or prevent overreduction when the energy dissipative systems are overwhelmed or ineffective. This may be necessary at low temperatures where zeaxanthin synthesis is much reduced or prevented.

VII. The Complex Role of Oxygen: Protection and Destruction

The relationship between photosynthesis and oxygen is complex because oxygen has multiple roles. First, it has a protective function in the dissipation of energy through the processes of photorespiration and pseudocyclic electron flow. Oxygen has been considered a rather ineffective electron acceptor (Egneus *et al.*, 1975). However, the capacity for superoxide generation as a result of oxygen reduction by the electron transport chain may be higher than 20 $\mu mol\ h^{-1}\ mg^{-1}$ chlorophyll. Electron transport to oxygen poises the electron carriers of the cyclic pathway while remaining a coupled process leading to ATP synthesis. It is suggested that this reaction makes an important contribution to the establishment of the overall $ATP/2e^-$ ratio of 1.5 required for CO_2 assimilation. Non-cyclic electron flow is believed to provide an $ATP/2e^-$ of 1.33 and thus pseudocyclic electron phosphorylation could produce additional ATP. Superoxide, the product of pseudocyclic electron flow, is rapidly converted to H_2O_2 by the action of superoxide dismutase and this H_2O_2 is the substrate for the stromal ascorbate peroxidase, the first enzyme of the ascorbate-glutathione cycle (Fig. 10). For every two electrons that are used in the formation of H_2O_2 a further two are required for its metabolism via the ascorbate-glutathione cycle. This H_2O_2-scavenging system is highly efficient (Anderson *et al.*, 1983). Chlorophyll *a* fluorescence quenching is observed upon addition of H_2O_2 to isolated intact chloroplasts. The quenching observed is characteristic of a Hill reagent (Neubauer and Schreiber, 1989). Hence, H_2O_2 production and scavenging could serve as a valve reaction preventing photoinhibitory damage. It is important to note the role of ascorbate in dissipative and protective processes. Ascorbate is an outstanding antioxidant in plant tissues. Its function in H_2O_2 detoxification is a necessity since chloroplasts do not contain catalase and photosynthesis is particularly sensitive to low (10 μM) concentrations of H_2O_2 because the thiol-modulated enzymes of the Benson-Calvin cycle are readily oxidized and thus inactivated. Ascorbate has also an important secondary antioxidant function in the formation and regeneration of membrane-bound quenchers such as α-tocopheol and zeaxanthin.

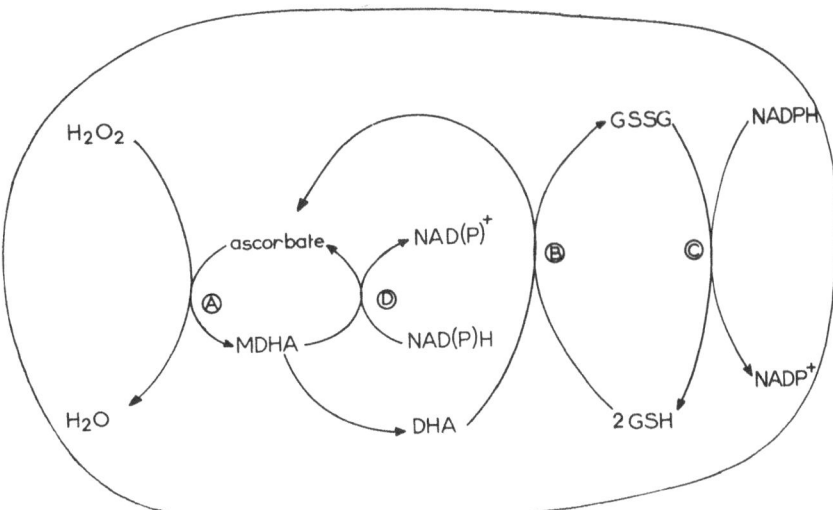

Figure 10. Diagrammatic representation of the ascorbate-glutathione cycle of the chloroplast with component enzymes ascorbate peroxidase (A), dehydroascorbate (DHA) reductase (B), glutathione reductase (C), and mono-dehydroascorbate (MDHA) reductase (D).

Since superoxide formation leads to the generation of hydrogen peroxide and other reactive radicals, there is the possibility that if scavenging of active oxygen species is not always efficient, then detrimental reactions will occur and loss of function will result. The protective effects of O_2 might be counterbalanced, causing loss of CO_2 fixation and photodamage. O_2 has been shown to play an important role in photoinhibition in isolated thylakoids, membrane fragments, and PS II particles. No turnover of Q_B-binding protein takes place in the absence of oxygen (Richter *et al.*, 1990; Nedbal *et al.*, 1990) and anaerobiosis largely prevents photoinactivation of purified D1/D2/cyt b_{559} complexes (McTavish *et al.*, 1989). Oxygen is thought to accelerate photoinactivation in purified PS II complexes via the production of radicals or singlet oxygen (Telfer *et al.*, 1990). The partial elimination of the oxygen effects in the presence of superoxide dismutase and catalase tends to support this conclusion. The detrimental effects of oxygen are less apparent *in vivo* than in *in vitro* systems presumably because of the enormous benefit of energy utilization by photorespiratory carbon metabolism. However, photoinhibition has been shown to be promoted by O_2 under circumstances where energy turnover by photosynthetic carbon assimilation and photorespiration are low, for example at low temperatures.

VIII. Conclusions

The presence of the energy-generating thylakoid membranes in the same intracellular compartment as the Benson-Calvin allows a tight coregulation of the two processes. Effective regulation is based on a metabolic compromise that allows the efficient function of both processes simultaneously. The precision of this regulation is evidenced by the remarkable degree of stability of the phosphorylation and redox potentials. These are kept rather constant in order to allow efficient non-cyclic electron transport and prevent the overreduction of the electron transport system that leads to photoinhibitory damage. The ATP/ADP and NADPH/NADP ratios are kept relatively constant by virtue of (1) the sensitive adjustments of the quantum efficiencies of PS II and PS I, (2) the provision of cyclic and pseudocyclic phosphorylation in situations when CO_2 assimilation capacity is decreased relative to irradiance, and (3) the adjustments of the activation states of certain Benson-Calvin enzymes and also their modulation by component metabolite levels. Our understanding of the action of mechanisms involved in photosynthetic control has advanced significantly following the introduction of techniques for studying the efficiencies of the photosystems *in situ* in intact leaves.

The quantum efficiencies of both photosystems decline with increasing irradiance. Measurements of P700$^+$ reduction have suggested that there is little change in the restriction of electron flow between PS I and PS II as irradiance is increased. The measured relationships between the quantum efficiencies of PS I and II found in pea leaves and other species subjected to a range of irradiances imply a predominant role for non-cyclic electron flow in most circumstances. However, the relationships between the quantum efficiencies of PS I and II in limiting CO_2 and also when the CO_2 assimilation capacity is low are consistent with the operation of cyclic electron flow in these circumstances.

It is evident that the supply of reducing equivalents to the stroma by thylakoid electron transport chain is strictly regulated. Fluxes of reducing equivalents pass through many pathways in the stroma. The high NADPH/NADP ratios produced during the induction phase of photosynthesis are not encountered in steady-state conditions. The rate of activation of thiol-modulated Benson-Calvin cycle enzymes such as fructose-1,6-*bis*phosphatase is rather low. The marked changes in the redox state of the stroma occurring during the induction of photosynthesis are consistent with the requirements of reductive activation of the Benson-Calvin cycle enzymes. Following transitions to high irradiance, such as occurs during sunflecks, the rate of electron transport is restrained by the activation state of the Benson-Calvin cycle enzymes. In this regard, it is important to note that nearly all our knowledge of the regulation of leaf photosynthesis is extrapolated from steady-state measurements in constant light.

The protective effects of O_2 as an energy sink in both photorespiration and pseudocyclic electron flow are essential in preventing overreduction. Univalent reduction of oxygen to superoxide and its subsequent dismutation to H_2O_2 and O_2 are an inevitable consequence of photosynthetic electron flow in an oxygen environment. Efficient scavenging of H_2O_2 is essential because very low H_2O_2 concentrations cause severe inhibition of Benson-Calvin cycle enzymes. H_2O_2 metabolism in the stroma is a highly efficient process and may serve as a valve reaction protecting against photodamage. Further experiments are required in order to determine whether H_2O_2 production and reduction has a sufficient flux *in vivo* to have significance in ATP synthesis and also to determine the importance of pseudocyclic electron flow in the regulation that exists between electron transport and carbon assimilation.

Acknowledgements

I am deeply indebted to my friend and colleague Jeremy Harbinson, who has been my collaborator in much of the work described in this manuscript, for discussions regarding the nature of photosynthetic regulation. I wish to thank Graham Noctor for critical reading of the manuscript and discussion. I gratefully acknowledge the research collaboration with Bob Furbank in the production of the data in Figure 8 and with Marie Dujardyn and Yves Lemoine in the experiments contained in Figure 9. "If I have seen further it is because I stood on the shoulders of giants."

IX. References

Anderson JW, Foyer CH and Walker DA (1983) Light-dependent reduction of hydrogen peroxide by intact spinach chloroplasts. *Biochim Biophys Acta* **724**: 69–74

Arnon DI, Whatley FR and Allen MB (1958) Assimilatory power in photosynthesis. *Science* **127**: 1026–1034

Ashton AR (1982) A role for ribulose-1, 5 bisphosphate carboxylase as a metabolic buffer. *FEBS Lett* **145**: 1–6

Baker NR (1991) A possible role for photosystem II in environmental perturbations of photosynthesis. *Physiol Plant* **81**: 563–570

Bendall DS (1982) Photosynthetic cytochromes of oxygenic organisms. *Biochim Biophys Acta* **683**: 119–157

Berry JA, Lorimer GH, Pierce J, Seemann J, Meek J and Freas S (1987) Isolation, identification and synthesis of 2-carboxyarabinitol-1-phosphate, a diurnal regulator of ribulose bisphosphate carboxylase activity. *Proc Natl Acad Sci USA* **84**: 734–738

Cseke C and Buchanan BB (1986) Regulation of the formation and utilisation of photosynthate in leaves. *Biochim Biophys Acta* **853**: 43–63

Crawford NA, Droux M, Kosower NS and Buchanan BB (1989) Evidence for function of the ferredoxin/thioredoxin system in the reductive activation of target enzymes of isolated intact chloroplasts. *Arch Biochem Biophys* **271**: 223–239

Demmig-Adams B (1990) Carotenoids and photoprotection in plants: A role for the xanthophyll zeaxanthin. *Biochim Biophys Acta* **1020**: 1–24

Dujardyn M and Foyer CH (1989) Limitation of CO_2 assimilation and regulation of Benson-Calvin cycle activity in barley leaves in response to changes in irradiance, photoinhibition and recovery. *Plant Physiol* **91**: 1562–1568

Edwards G and Walker DA (1983) Induction. In: *C3, C4: mechanisms and Cellular and Environmental Regulation of Photosynthesis*, Blackwell Science Publications, Oxford, London, pp 156–200

Egneus H, Heber U, Matthiesen U and Kirk MR (1975) Reduction of oxygen by the electron transport chain of chloroplasts during assimilation of carbon dioxide. *Biochim Biophys Acta* **408**: 252–268

Evron Y and Avron M (1990) Characterization of an alkaline pH- dependent proton "slip", in the ATP synthase of lettuce thylakoids. *Biochim Biophys Acta* **109**: 115–120

Foyer CH (1989) The role of orthophosphate in photosynthetic control: Studies using phosphorus-31 nuclear magnetic resonance. *Plant Physiol (Life Science Advances)* **8**: 81–89

Foyer CH, Anderson J and Walker DA (1984) Light dependent reduction of hydrogen peroxide via the ascorbate-glutathione cycle in intact spinach chloroplasts. In: (C Sybesma, ed) *Advances in Photosynthesis Research*, Vol III Martinus Nijhoff/Dr W Junk Publishers, pp 689–692

Foyer CH, Dujardyn M and Lemoine Y (1989) Responses of photosynthesis and the xanthophyll and ascorbate-glutathione cycles to changes in irradiance, photoinhibition and recovery. *Plant Physiol Biochem* **27**: 751–760

Foyer CH, Furbank R, Harbinson J and Horton P (1990) The mechanisms contributing to photosynthetic control of electron transport by carbon assimilation in leaves. *Photosynth Res* **25**: 83–100

Fredeen AL, Raab T, Rao IM and Terry N (1990) Effects of phosphorus nutrition on photosynthesis in *Glycine max* (L.) Merr. *Planta* **181**: 399–405

Furbank RT, Foyer CH and Walker DH (1987) Interactions between ribulose-1, 5-bisphosphate carboxylase and stromal metabolites. III Corroboration of the role of this enzyme as a metabolite buffer. *Biochim Biophys Acta* **894**: 165–173

Genty B, Briantais JM and Baker NR (1989) The relationship between the quantum yield of photosynthetic electron transport and photochemical quenching of chlorophyll fluorescence. *Biochim Biophys Acta* **990**: 87–92

Genty B, Harbinson J and Baker NR (1990) Relative quantum efficiencies of the two photosystems of leaves in photorespiratory and non-photorespiratory conditions. *Plant Physiol Biochem* **28**: 1–10

Guitton C and Mache R (1987) Photophosphorylation *in vitro* of the large subunit of the ribulose 1,5-bisphosphate carboxylase and of the glyceraldehyde 3-phosphate dehydrogenase. *Eur J Biochem* **166**: 249–254

Gutteridge S, Parry MAJ, Burton S, Keys AJ, Mudd A, Feeny J, Servaites JC and Pierce J (1986) A nocturnal inhibitor of carboxylation in leaves. *Nature* **324**: 274–276

Harbinson J and Foyer CH (1991) Relationships between the efficiencies of photosystems I and II and stromal redox state in CO_2-free air: Evidence for cyclic electron flow *in vivo*. *Plant Physiol* **91**: 41–49

Harbinson J, Genty B and Baker NR (1989) Relationship between the quantum efficiencies of photosystems I and II in pea leaves. *Plant Physiol* **90**: 1029–1034

Harbinson J, Genty B and Foyer CH (1990) The relationship between photosynthetic electron transport and stromal enzyme activity in pea leaves: Towards an understanding of the nature of photosynthetic control, *Plant Physiol* **94**: 545–553

Harbinson J and Hedley CL (1989) The kinetics of p-700$^+$ reduction in leaves: A novel *in situ* probe of thylakoid functioning. *Plant Cell Environ* **12**: 357–369

Harbinson J and Woodward FI (1987) The use of light induced absorbance changes at 820 nm to monitor the oxidation state of P-700 leaves. *Plant Cell Environ* **9**: 131–140

Heber U, Neimanis S, Dietz KJ and Vill J (1986) Assimilatory power as a driving force in photosynthesis. *Biochim Biophys Acta* **852**: 144–155

Heldt HW, Werden K, Milovancev M and Geller G (1973) Alkalisation of the chloroplast stroma caused by light-dependent proton flux into the thylakoid space. *Biochim Biophys Acta* **314**: 224–241

Horton P (1983) Control of electron transport by the thylakoid protein kinase. *FEBS Lett* **152**: 47–52

Horton P (1989) Interactions between electron transport and carbon assimilation: Regulation of light-harvesting and photochemistry. In: (WR Briggs, ed) *Photosynthesis, Plant Biology Series* Vol 8, Alan R Liss Inc, New York, pp 393–406

Kirschbaum MUF and Pearcy RW (1988a) Gas exchange analysis of the fast phase of photosynthetic induction in *Alocasia macrorrhiza*. *Plant Physiol* 87: 818–821

Kirschbaum MUF and Pearcy RW (1988b) Concurrent measurements of O_2 and CO_2 exchange during lightflecks in *Alocasia macrorrhiza* (L.) G. Don. *Planta* 174: 527–533

Kitajima M and Butler WL (1975) Quenching of chlorophyll fluorescence and primary photochemistry in chloroplasts by dibromothymoquinone. *Biochim Biophys Acta* 376: 105–115

Krause GH and Weis E (1991) Chlorophyll fluorescence and photosynthesis: The basics. *Annu Rev Plant Physiol Plant Mol Biol* 42: 313–349

Laing WA, Stitt M and Heldt HW (1981) Control of CO_2 fixation. Changes in the activity of ribulose phosphate kinase and fructose and sedoheptulose bisphosphatase in chloroplasts. *Biochim Biophys Acta* 637: 348–359

Leegood RC (1990) Enzymes of the Calvin cycle. In: (PJ Lea, ed) *Methods in Plant Biochemistry*, Vol 3, Academic Press, London, pp 15–37

Leegood RC, Walker DA and Foyer CH (1985) Regulation of the Benson-Calvin cycle. In: (J Barber and NR Baker, eds) *Photosynthetic Mechanisms and the Environment*. Elsevier Science Publishers, Amsterdam, New York, pp 191–258

Le Gouallec JL and Cornic G (1988) Photoinhibition of photosynthesis in *Elatostema repens*. *Plant Physiol Biochem* 26: 705–712

Macioszek J and Anderson LE (1987) Changing kinetic properties of the two-enzyme phosphoglycerate kinase/NADP-linked glyceraldehyde-3-phosphate dehydrogenase couple from pea chloroplasts during photosynthetic induction. *Biochim Biophys Acta* 892: 185–190

Macioszek J, Anderson JB and Anderson LE (1990) Isolation of chloroplastic phosphoglycerate kinase. Kinetics of the two-enzyme phosphoglycerate kinase/glyceraldehyde 3-phosphate dehydrogenase couple. *Plant Physiol* 94: 291–296

Marsho TV, Behrens PN and Radmer KJ (1979) Photosynthetic oxygen reduction in isolated intact chloroplasts and cells from spinach. *Plant Physiol* 64: 656–659

McTavish H, Picorel R and Seibert M (1989) Stabilisation of isolated photosystem II reaction center complex in the dark and in the light using polyethylene glycol and an oxygen-scrubbing system. *Plant Physiol* 89: 452–456

Miginiac-Maslow M, Decottignies P, Jacquot JP and Gadal P (1990) Regulation of corn leaf NADP-malate dehydrogenase light activation by the photosynthetic electron flow. Effect of photoinhibition studied in a reconstituted system. *Biochim Biophys Acta* 1017 273–279

Nedbal L, Masojidek J, Komenda J, Prasil O and Setlik I (1990) Three types of photosystem II photoinactivation 2. Slow processes. *Photosynth Res* 24: 89–97

Neubauer C and Schreiber U (1989) Photochemical and non-photochemical quenching of chlorophyll fluorescence induced by hydrogen peroxide. *Z. Naturforsch* 44c: 262–270

Ohad I, Adir N, Kioke H and Kyle D (1990) Mechanism of photoinhibition *in vivo*. A reversible light-induced conformational change of reaction center II is related to an irreversible modification of the D1 protein. *J Biol Chem* 265: 1972–1979

Osterhout WJV and Hass ARC (1919) On the dynamics of photosynthesis. *J Gen Physiol* 1: 1–16

Pearcy RW (1990) Sunflecks and photosynthesis in plant canopies. *Annu Rev Plant Physiol Plant Mol Biol* 41: 421–453

Pfündel E and Strasser RJ (1988) Violaxanthin de-epoxidase in etiolated leaves. *Photosynth Res* 15: 67–73

Portis AR Jr (1990) Rubisco activase. *Biochim Biophys Acta* 1015: 15–28

Powles SB (1984) Photoinhibition of photosynthesis induced by visible light. *Annu Rev Plant Physiol* 35: 15–44

Prinsley RT and Leegood RC (1986) Factors affecting photosynthetic induction in spinach leaves. *Biochim Biophys Acta* 849: 244–253

Richter M, Rühle W and Wild A (1990) Studies on the mechanism of photosystem II photoinhibition. A two-step degradation of D_1-protein. *Photosynth Res* 24: 237–243

Robinson SP and Walker DA (1980) The significance of light activation of enzymes during the induction phase of photosynthesis in isolated chloroplasts. *Arch Biochem Biophys* 202: 617–623

224

Satoh K (1981) Fluorescence induction and activity of ferredoxin-NADP reductase in *Bryopsis* chloroplasts. *Biochim Biophys Acta* **638:** 327–333

Scheibe R (1987) NADP$^+$ malate dehydrogenase in C3 plants. Regulation and role of a light-activated enzyme. *Physiol Plant* **71:** 393–400

Scheibe R and Stitt M (1988) Comparison of NADP-malate dehydrogenase activation, Q_A reduction and O_2 evolution in spinach leaves. *Plant Physiol Biochem* **26:** 473–482

Schreiber U and Neubauer C. (1987) The polyphasic rise of chlorophyll fluorescence upon onset of strong continuous illumination: II Partial control by the photosystem II donor side and possible ways of interpretation. *Z Naturforsch* **42c:** 1255–1264

Seemann JR, Kirschbaum MUF, Sharkey TD and Pearcy RW (1988) Regulation of ribulose 1,5-bisphosphate carboxylase activity in *Alocasia macrorrhiza* in response to step changes in irradiance. *Plant Physiol* **88:** 148–152

Seaton GGR and Walker DA (1990) Chlorophyll fluorescence as a measure of photosynthetic carbon assimilation. *Proc R Soc Lond B* **242:** 29–35

Servaites JC, Shieh WJ and Geiger DR (1991) Regulation of photosynthetic carbon reduction cycle by ribulose bisphosphate and phosphoglyceric acid. *Plant Physiol* **97:** 1115–1121

Stitt M (1986) Limitation of photosynthesis by carbon metabolism. I. Evidence for excess electron transport capacity in leaves carrying out photosynthesis in saturating light and CO_2. *Plant Physiol* **81:** 1115–1122

Takahama U, Shimizu-Takahama M and Heber U (1981) The redox state of the NADP system in illuminated chloroplasts. *Biochim Biophys Acta* **637:** 530–539

Telfer A, He W-Z and Barber J (1990) Spectral resolution of more than one chlorophyll electron donor in the isolated photosystem II reaction center complex. *Biochim Biophys Acta* **101:** 143–151

Von Caemmerer S and Farquhar GD (1981) Some relationships between the biochemistry of photosynthesis and the gas exchange of leaves. *Planta* **153:** 376–387

Walker DA (1981) Photosynthetic induction. In: (G Akoyunoglou, ed) *Proceedings of the 5th International Congress on Photosynthesis*, Vol IV, Balaban International Sciences Series, Philadelphia, pp 189–202

Weis E, Ball JR and Berry J (1987) Photosynthetic control of electron transport in leaves of *Phaseolus vulgaris*. Evidence for regulation of photosystem II by the proton gradient. In: (J Biggins, ed) *Progress in Photosynthesis Research*, Vol 2, Martinus Nijhoff Publ, Dordrecht, pp 553–556

Woodrow IE and Motte KA (1989) Rate limitation of non-steady state photosynthesis by ribulose 1,5-bisphosphate carboxylase in spinach. *Aust J Plant Physiol* **16:** 487–500

Yamamoto HY (1979) Biochemistry of the violaxanthin cycle in higher plants. *Pure and Appl Chem* **51:** 639–648

9

Leaf Senescence-induced Alterations in Structure and Function of Higher Plant Chloroplasts

Anil Grover* and Prasanna Mohanty

School of Life Sciences,
Jawaharlal Nehru University
New Delhi 110067, India

CONTENTS

ABBREVIATIONS

Chl	:	Chlorophyll
Cyt	:	Cytochrome
DAD	:	Diaminodurene
DBMIB	:	2, 5-Dibromo-3-methyl-6-isopropyl-p-benzoquinone
DCPIP	:	2, 6-Dichlorophenol indophenol
DPC	:	1, 5-Diphenyl carbazide
FeCN	:	Ferricyanide

* Present Address: Department of Plant Molecular Biology, University of Delhi South Campus, New Delhi 110021, India

MA	:	Methylamine
MV	:	Methylviologen
PC	:	Plastocyanin
PD	:	Phenylenedimine
PD_{ox}	:	oxidized p-Phenylenediamine
PMS	:	Phenazine methosulphate
PQ	:	Plastoquinone
PS I	:	Photosystem I
PS II	:	Photosystem II
RuBPcase	:	Ribulose-1, 5-*bis*phosphate carboxylase
SDS-PAGE	:	Sodium dodecyl sulphate-polyacrylamide gel electrophoresis
TMPD	:	N, N, N´, N´-Tetramethyl-p-phenylenediamine

ABSTRACT

Leaf senescence constitutes a crucial aspect of plant life. In this chapter, we highlight various possible reactions involved in the decline of photochemical activity of chloroplasts during leaf senescence. Loss of photosynthetic pigments is the most conspicuous event during this phase. Electron transport activities catalysed by photosystem II and photosystem I (PS II and PS I) also decline during senescence. The progress of senescence appears to affect not only the oxygen-evolving complex but also the $Q_A \rightleftarrows Q_B$ site of PS II and cytochrome b_6/f complex in the electron transport chain. As a combined result of all these perturbations, an imbalance between PS II and PS I catalysed activities is induced, which eventually limits photosynthesis in senescing leaves. The studies on molecular aspects of leaf senescence focus on identification of specific changes in gene expression associated with the onset and progress of senescence phas   A critical review of the current state of knowledge pertaining to senescence-induced changes in thyla iid structure and functional capacities of the photosynthetic membranes is presented. It points to specit directions for future investigation to get a better understanding of the mechanism(s) involved in regulati energy transfer and energy conservation in the senescing chloroplasts.

I. Introduction

The process of senescence is an integral phase in the development cycle of a plant and it is shown in varied ways. For example, the initiation of embryo growth brings about cessation of cotyledonary activities and the formation of new leaves results in the degeneration of older ones. The development of reproductive organs such as pod and ear is the signal for senescence of the leaves or the whole plant in most annuals. In view of the fact that the manipulation of leaf senescence may possibly have significant agronomic implications, considerable emphasis has been placed on physiological and biochemical characterization of the leaf senescence process. However, studies on leaf senescence have largely been made at the whole-plant and whole-organ levels. It is necessary to relate the senescence-associated changes occurring at the whole-plant and whole-leaf levels to those taking place in the individual organelles of a cell for better understanding of the causalty of the mechanism of senescence. We shall focus here on leaf senescence-associated changes in chloroplasts in relation to photosynthetic light reaction. Our interest in taking chloroplast senescence as a case study is two-fold: (1) structure and function of chloroplasts determine the leaf photosynthetic efficiency, which has an important role in determining, ultimately, the crop yield (see Amthor and McCree, 1990); and (2) chloroplasts, among all the cell organelles, seem to be most sensitive to leaf senescence (see Biswal and Biswal, 1988). In this chapter, the term "chloroplast senescence" has been distinguished from "chloroplast aging". Senescence of chloroplasts mainly refers to leaf senescence-associated *in vivo* changes. Chloroplast aging, on the other hand, indicates time-dependent alterations in chloroplast under *in vitro* conditions.

Several studies have shown that ribulose-1, 5-*bis*phosphate carboxylase (RuBPcase), the major chloroplastic protein, is circumstantially related to the remobilization of nitrogen from the leaves to the growing seeds and to leaf senescence (Huffaker and Miller, 1978; Abrol *et al.*, 1984; Grover *et al.*, 1985; Grover and Sinha, 1988). However, we shall not discuss the details of RuBPcase as we wish to confine ourselves in this chapter to leaf senescence-induced alterations in the structure and function of higher plant chloroplasts *vis-à-vis* the efficiency of the light reaction of

228

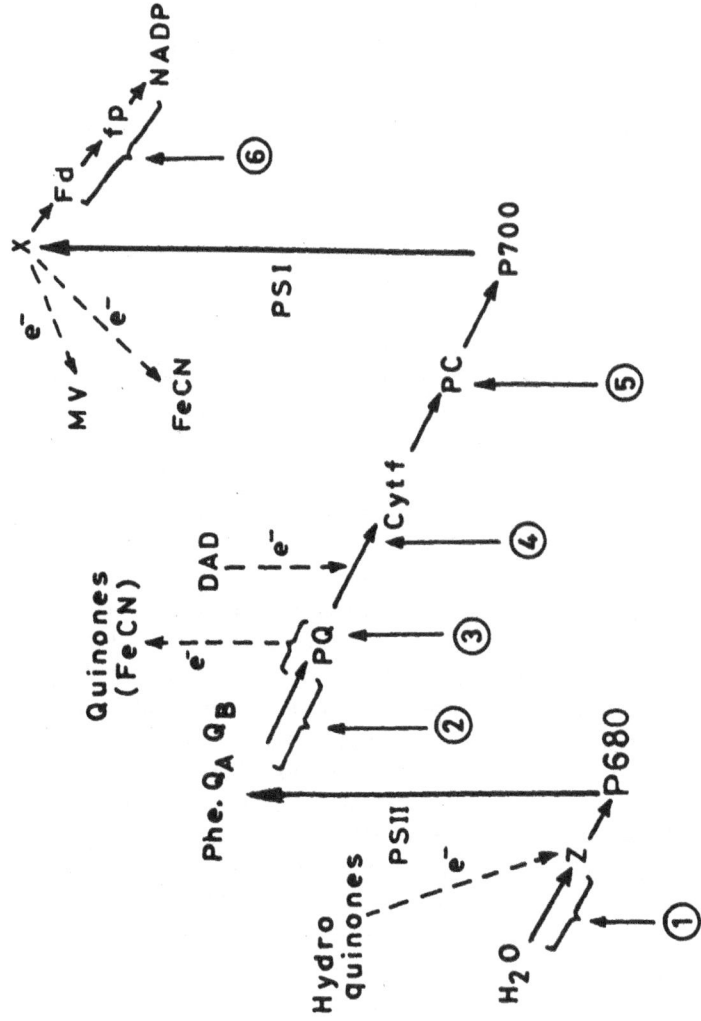

Figure 1. The Z scheme of photosynthetic electron transport. Note the sites of inhibition (numbered arrows) and main sites of electron donation and acceptance by artificial reductants and oxidants (broken arrows). Q_A: primary electron acceptor for PS II; Q_B: secondary stable quinone electron acceptor; P680: reaction centre chlorophyll of photosystem II; P700: reaction centre chlorophyll of photosystem I; Phe: pheophytin; Z: tyrosine amino-acid residue at 161 position of D1 polypeptide; MV: methylviologen; FeCN: ferricyanide; DAD: diaminodurene; 1: hydroxylamine, ammonia, tris; 2: 3-(3,4-dichlorophyenyl)-1,1-dimethylurea (DCMU); 3: dibromothymoquinone (DBMIB); 4: 1-ethyl-3-(3-dimethylaminopropyl)-carbodiimide (EDAC); 5: polycations, HgCl$_2$, KCN; 6: disalicylidenpropanediamine (DSPD), pyrophosphate. For further details, refer to the text and also see Govindjee and Coleman (Chapter 3).

photosynthesis. For fundamental details on RuBPcase, see Andrews and Lorimer (1987) and for the light reaction of photosynthesis, see Govindjee and Whitemarsh (1982), Packham and Barber (1987), and Govindjee and Coleman (see Chapter 3). Details concerning the metabolic and regulatory events of leaf senescence have been extensively discussed in several reviews (Thimann, 1980; Thomas and Stoddart, 1980; Sestak, 1982; Grover and Sinha, 1988; Nooden and Guiamet, 1989). Critical aspects of chloroplast senescence have been reviewed earlier by Woolhouse (1982, 1984) and Gepstein (1988).

In order to appreciate and assess the nature of alterations in the photochemical activity of chloroplasts, we begin with highlights of the current model of photosynthetic electron transport chain, the "Z" scheme. A comparison will be made here of the various senescing systems which have been exploited for studying chloroplast senescence. This will be followed by analysis of data on compositional changes in photosynthetic pigments and structural disorganization of chloroplast lamellar system. How the macro- and micro-changes in chloroplast structure influence its active functioning will then be presented, followed by a brief discussion on the molecular events accompanying chloroplast senescence.

II. Current Model of Photosynthetic Electron Transport Chain

The most widely accepted scheme of photosynthetic electron transfer, the Z scheme, postulates operation, in series, of two specific redox reactions initiated by two specific reaction centres associated with two separate light harvesting assemblies. These two systems are known as photosystem II and photosystem I (see Duysens, 1989; Govindjee and Coleman, Chapter 3). The reaction centre chlorophylls (Chls) in PS II and PS I have been designated as P680 and P700, respectively. The components of the electron transport chain include plastoquinone (PQ), which connects PS II, via cytochrome (Cyt) b_6/f, with PS I. Ferredoxin (Fd) is an intermediate preceding nicotinamide adenine dinucleotide phosphate (NADP+). The ferredoxin-NADP+ reductase (fp, a flavoprotein) is required for NADP+ reduction by PS I (Fig. 1). The details are discussed in the legend of Fig. 1. For different aspects of the Z scheme, the reader is referred to recent reviews by Barber and Marder (1986), Andreasson and Vanngard (1988), Anderson and Thomson (1989), and Hansson and Wydrzynski (1990).

A current understanding of the sites of various electron transport components in thylakoid membranes has been illustrated in Fig. 2. The typical internal structure of a higher plant chloroplast contains areas of appressed lamellas known as grana, connected by single unappressed stroma lamellas. Most of the thylakoid proteins, consisting of more than 50 different polypeptides (Westhoff et al., 1988), are organized into the following six intrinsic membrane-spanning complexes: PS I reaction centre, PS II reaction centre, light harvesting complex associated with PS I (LHC I), light harvesting complex associated with PS II (LHC II), cytochrome b_6/f complex, and chloroplast coupling factor. Further details are described in the legend of Fig 2. In association with a limited number of extrinsic proteins (e.g., a 33 kD PS II protein and a copper protein plastocyanin), these complexes carry out light harvesting and

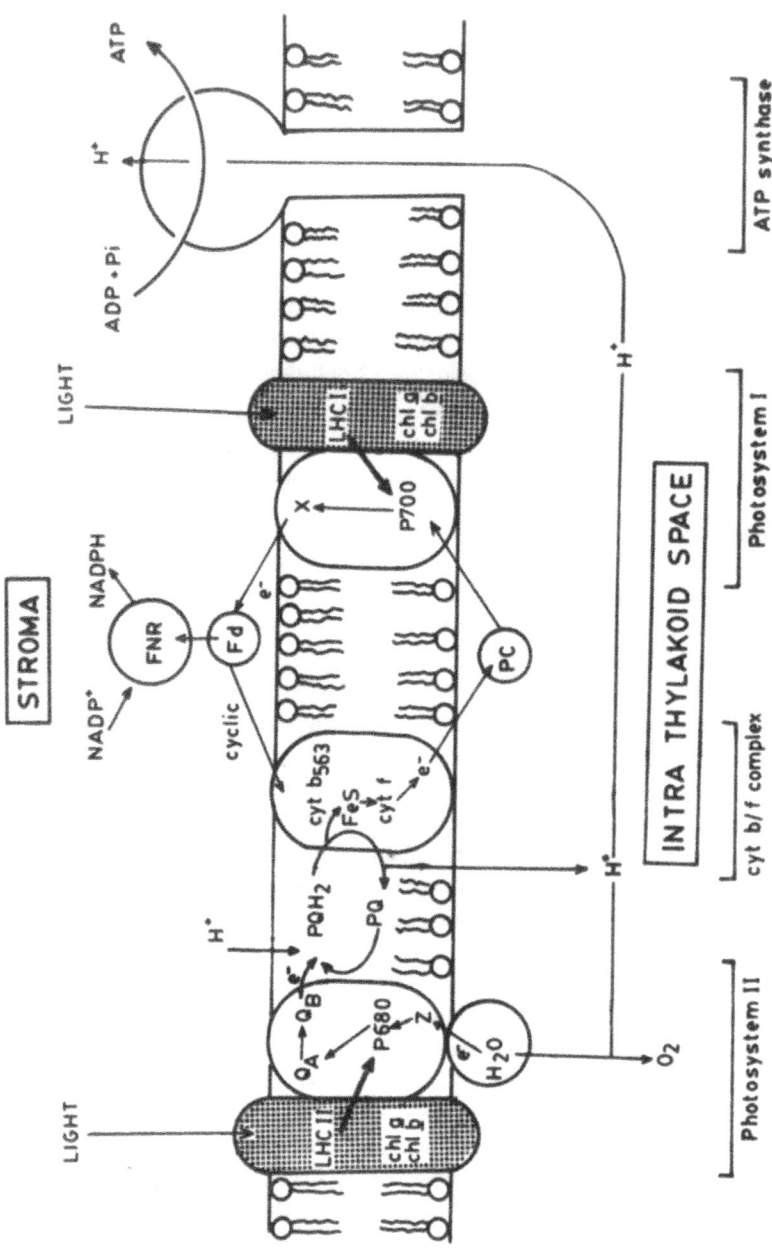

Figure 2. The topographical distribution of electron carriers in thylakoid membrane. Most of the thylakoid proteins are organized into intrinsic membrane-spanning complexes such as photosystems I and II, light harvesting complexes (LHC) I and II, the cytochrome b/f complex and ATP synthase complex. For more details on polypeptide components so far identified for various complexes, see Simpson (1990). PQ and PQH_2: oxidized and reduced plastoquinone; FNR: ferredoxin NADP reductase. Refer to Fig. 1 for other abbreviations used. These complexes carry out light harvesting, electron transport, and ATP generation.

its distribution, electron transport adenosine triphosphate (ATP) generation, and other related activities.

The active PS II complex is localized in the granal region of the higher plant chloroplasts (see Sane, 1977). It mediates non-cyclic electron transport from water to PQ. PS II contains Chl a, pheophytin, β-carotene, P680, Mn^{++}, Cl^-, Ca^{++}, quinone electron acceptors known as Q_A and Q_B, and several intrinsic and a few extrinsic polypeptides, including the 33 kD protein of the O_2-evolving complex mentioned above (see Peter and Thornber, 1988; Simpson, 1990). The Cyt b_6/f complex is present in the both grana and stroma lamellae (see Willey and Gray, 1988). This complex accepts electrons from the PQ pool and reduces PC. Electron flow from PS II occurs via the Cyt b_6/f complex and PC to PS I, which is present mostly in stroma, although some are found in the end membranes of the grana thylakoids. It contains Chl a, P700, β-carotene, and several intrinsic and extrinsic polypeptides (see Andreasson and Vanngard, 1988; Reilly and Nelson, 1988; Lagoutte and Mathis, 1989; Bassi *et al.*, 1990; Simpson, 1990; Hoshina and Itoh, Chapter 2). Light harvesting complex II and internal antenna of PS II are associated with PS II (see Chitinis and Thornber, 1989) while LHC I and internal antenna of PS I are associated with PS I (see Bassi *et al.*, 1990).

Although the path of electron transport has been deciphered by kinetic absorption measurements to a large extent (see Govindjee and Coleman, Chapter 3, and Hoshina and Itoh, Chapter 2), some of the knowledge on chloroplast functioning has been derived from the use of artificial electron donors and acceptors. The sites where commonly used electron donors, acceptors, and inhibitors act are shown in Fig. 1 (also see Izawa, 1980).

III. Comparison of Various Senescing Systems

Studies on chloroplast senescence employ one of the following three experimental conditions: (1) attached leaves senescing naturally and/or coupled to induction treatments such as incubation in the dark or hormonal application; (2) detached leaves, incubated in light or dark, with or without hormones; and (3) chloroplasts isolated from normal leaves induced to senesce under *in vitro* conditions. Chloroplast senescence under in *in vivo* or *in situ* conditions (chloroplasts inside attached or detached leaves) provide cellular reflection of the leaf senescence process. On the other hand, aging induced by experimental treatments in isolated chloroplasts (*in vitro* system) gives detailed information on the biochemical aspects of the chloroplasts membranes and other macromolecules, but obviously ignores intracellular considerations. Therefore, *in vitro* aging of isolated chloroplasts is not comparable to leaf senescence. Data on *in vitro* chloroplast aging has been excluded in this chapter except in a few instances where such studies reflect results comparable to leaf senescence.

Before proceeding further, one should also consider whether the phenomenology of senescence is the same or different in attached and detached leaves. It is important to appreciate the concept of leaf senescence in order to consider this question. Earlier studies regarded senescence as a process of deteriorative events which precede the death of a mature cell (Beevers, 1976). According to this view, attached leaves that are close to the completion of their life span and detached leaves which represent induced termination of life activities might be assumed to follow similar physiology. However, awareness has grown that senescence is a physiologically programmed process which makes possible remobilization of minerals from older to younger plant parts (Stoddart and Thomas, 1982). Regarding leaf senescence as a set of reactions merely related to the death of the system has, therefore, been disputed. Based on these developments, it is perhaps logical to regard senescence in attached and detached leaves to be different in many respects.

Direct comparison of attached and detached leaf senescence has been carried out in a few instances. The amino acids accumulate continuously in detached senescing oat leaves while the content of amino acids declines after an early increase in attached oat leaves (Thimann et al., 1974). Peroxidase activity increases with senescence in attached rice leaves (Kar and Mishra, 1976), but remains unchanged or decreases in detached barley (Srivastava et al., 1983) and pigeon pea (Grover and Sinha, 1985) leaves. Attached and detached barley leaves have been shown to differ with respect to endoproteinases activity profiles and electrophorograms (Miller and Huffaker, 1985). Differences in derivatives of Chl breakdown in attached and detached leaves have been reported (Maunders et al., 1983). The attached and detached wheat leaves have been found to follow nearly the same pattern in senescence-associated decline in absorbance of Chls at various wavelengths (Grover et al., 1986b).

Relatively few comparative studies have been carried out on the nature and sequence of alterations of thylakoid components in attached and detached leaves (Mishra and Biswal, 1982). Because most papers in the published literature do not clearly compare the characteristics of chloroplast senescence in attached and detached leaves, we shall attempt to clarify this issue by indicating whether leaves were attached or detached while comparing the data in this chapter.

IV. Analysis of Photosynthetic Pigments

The alterations in chloroplast constituents during senescence are of both quantitative and qualitative nature. These include loss of photosynthetic pigments, disorganization of thylakoid lamellas and stromal components, and decline in activities of the photochemical reactions. The most conspicuous feature of chloroplast senescence is the changes in the composition of the photosynthetic pigments (Sestak, 1977, 1978, 1982). In fact, except for the findings of Thomas and Stoddart (1975), who examined the mutants of *Festuca pratensis* and established that Chl loss and other senescence parameters are not strictly related to each other, Chl degradation has been considered synonymous with the syndrome of leaf senescence (Thimann, 1980). In many studies, Chl loss (or the resulting leaf yellowing) has been used as a sole criterion for analysis of the effect of various physiological and environmental variables on leaf senescence (Lindoo and Nooden, 1977; Nooden and Nooden, 1985).

During leaf senescence, marked differences in degradation rates of Chl a, Chl b, and carotenoids have been documented (Biswal and Mohanty, 1976a; Grover et al., 1986a; Kura-Hotta et al., 1987). Since the ratio of Chl a to Chl b decreases during senescence (Biswal and Mohanty, 1976a; Dhindsa et al., 1981) Chl a appears, in general, more susceptible to degradation than Chl b. Carotenoids are, however, relatively more stable than Chl during senescence (Biswal and Mohanty, 1976a). The carotenoid to total Chl ratio was found to increase from 0.16 to 0.5 during four days of dark incubation of detached wheat leaves (Grover *et al.*, 1986a). The differential rates of Chl and carotenoid losses could result, if we speculate, that some of the degradation product of Chl are used in the formation of carotenoids. The higher retention of carotenoids as compared to total Chl could have a physiological significance by way of protecting the leftover Chl molecules from damage against excess light and against free oxygen radicals (for further details on this topic, see Young and Britton, 1990; Winston, 1990). In the event, when more light is absorbed by a leaf than can be used in photosynthesis, carotenoids are thought to play a crucial role towards photoprotection of Chl (Demmig-Adams *et al.*, 1989; Sharma and Hall, Chapter 19).

It appears that various spectral forms of Chl *a* do not degrade to a similar extent in the primary wheat leaves. The far red spectral absorbing forms, which include the ones absorbing at 692, 700, and 708 nm, are shown to be extremely sensitive to senescence (Grover *et al.*, 1986b). Their loss is one of the most selective indicators of leaf senescence (Grover *et al.*, 1986b). In an analogous experiment on *in vitro* chloroplast (or thylakoid) aging, relatively higher sensitivity of 678 and 708 nm spectral forms of Chl *a* has been shown (Brody, 1983). The above experiments raise the possibility that Chl-protein complexes harbouring these long wavelength forms of Chl *a* are relatively more susceptible to degradation during leaf senescence.

The exact mechanism of Chl degradation is not yet clear. However, both photooxidation and enzymatic metabolism are thought to be important in accomplishing this event. Since leaves senesce faster in dark than in light (Thimann, 1980), enzymatic degradation appears to have a more significant role in regulating Chl concentrations. In this respect, the pathway of *in vitro* Chl bleaching implicates the involvement of chlorophyllases, lipoxygenases, and peroxidases. Pure chlorophyllase extracted from soluble and membrane-bound fractions have been shown to possess reasonably high specific activity towards Chl degradation (Kuroki *et al.*, 1981). Lipoxygenase-mediated polyunsaturated fatty acid oxidation and production of free radicals can potentially react with Chl and oxidize it (Klein *et al.*, 1984). The third possibility involves peroxidase which can bleach Chl in the presence of certain phenolics (Kato and Shimizu, 1985). It is still not clear which of these three pathways predominates under *in vivo* conditions.

Using *Phaseolus vulgaris* and *Hordeum vulgare* leaves, Maunders *et al.* (1983) obtained evidence to show that the initial step in Chl degradation pathway involves hydroxylation at C-10 position. According to Woolhouse (1984), the product of this hydroxylation reaction is not a stable degradation product but an intermediate which is further metabolized. Certain polar fluorescent compounds have also been implicated as catabolites of Chl but their chemical nature is not yet fully revealed (Duggelin *et al.*, 1988). In addition to unravelling the chemistry of Chl degradation, it is important to understand how the lipid and protein moieties surrounding the pigments in the thylakoid membrane influence their turnover. It is possible that the differential degradation of various spectral forms of Chl *a* as noted earlier (Brody, 1983; Grover *et al.*, 1986b) could be due to established differences in lipid environment of PS II and PS I core complexes (Siegenthaler and Rawyler, 1986; Murata *et al.*, 1990).

Understanding the mechanistic details of Chl degradation is currently a major challenge to plant biochemists and physiologists, and a lot of further work needs to be done on this topic (for earlier attempts, see Amir-Shapira et al., 1987; Matile et al., 1989; Llewellyn et al., 1990a, b; Ronning et al., 1991).

V. Structural Alterations of Chloroplasts

Details on alterations of chloroplast lamellar structure during leaf senescence have mainly come from three independent lines of research: ultrastructural examinations, compositional analysis of lipids, and absorption and fluorescence spectral studies.

a. *Ultrastructural examinations*

Since 1960, several workers compared chloroplast ultrastructure in mature and senescent leaves (see Stoddart and Thomas, 1982). Dodge (1970) observed that in senescing *Betula* leaves, chloroplasts are usually among the first cell organelles to show changes in structure. This fact, since then, has been confirmed in several species (see Biswal and Biswal, 1988; Gepstein, 1988). Chloroplasts of senescing leaves, in general, show reduced volume and their shape markedly changes from nearly oval to spherical. There is an initial increase in the number of chloroplast lamellas and the size of the grana. In rice leaves, the number of thylakoids per granum increases from 2 to 15 in young leaves to 20 to 30 in aged leaves (Hashimoto et al., 1989). Granal and stromal thylakoids also show change in their plane of orientation with advancing senescene (Hashimoto et al., 1989).

In the later part of senescence, the stromal thylakoids appear increasingly segmented and, as a result, pockets of granal remnants consisting of disorganized stacks are observed (Bricker and Newman. 1982). In the final stage of senescence, dense lipid-soluble globules have been observed in plastids (Paharia, 1986; Reddy, 1986; Hashimoto et al., 1989). These globules occupy much of the interior space of the plastids in senescent chloroplasts (Fig. 3). It is believed that these globules are possibly formed from the lipids after disintegration of granal and stromal lamellas (Burke et al., 1984). Analysis of lipid composition of isolated plastoglobuli during senescence of *Fagus sylvatica* leaves revealed that carotenoid esters predominated in the final stage of senescence (Tevini and Steinmuller, 1985).

The outer envelope of the chloroplast appears to maintain its integrity till the later stages of senescence, probably in order to allow the accomplishment of degradation of chloroplast constituents in a controlled manner (Woolhouse, 1984). The eventual degradation of the chloroplast envelope can be regarded as a sign of death of the chloroplast.

b. *Compositional analysis of lipids*

The senescing thylakoid membranes undergo several qualitative changes in their lipid composition. Evidence in recent years suggests that changes in lipid composition have a decisive effect on senescence-associated degradation of other components such as Chls (discussed above) and some membrane proteins (Harwood et al., 1982; Hilditch et al., 1989). Both the chloroplast envelope and the thylakoid membranes are characterized by a low phospholipid content and by the presence of glycolipids (Joyard et al., 1989). The major lipid compounds are galactolipids which contain

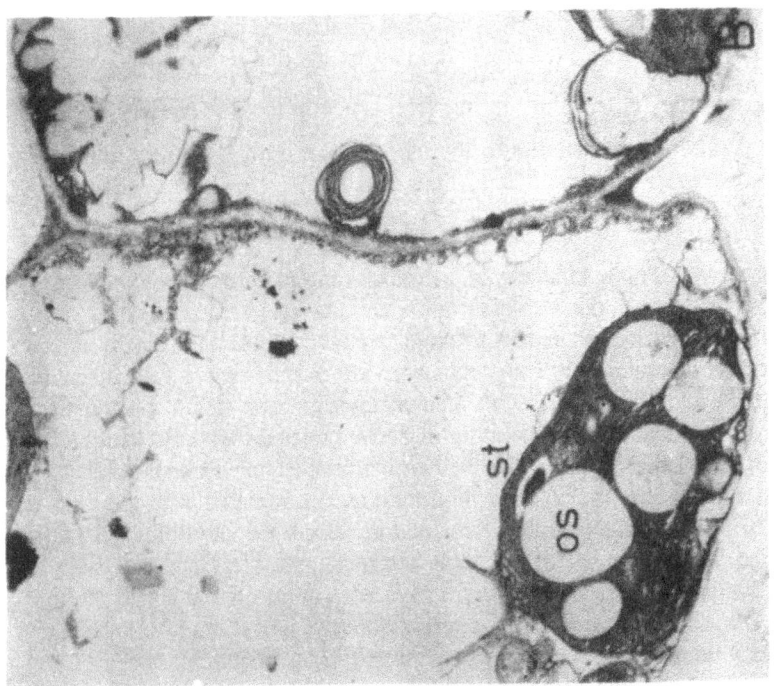

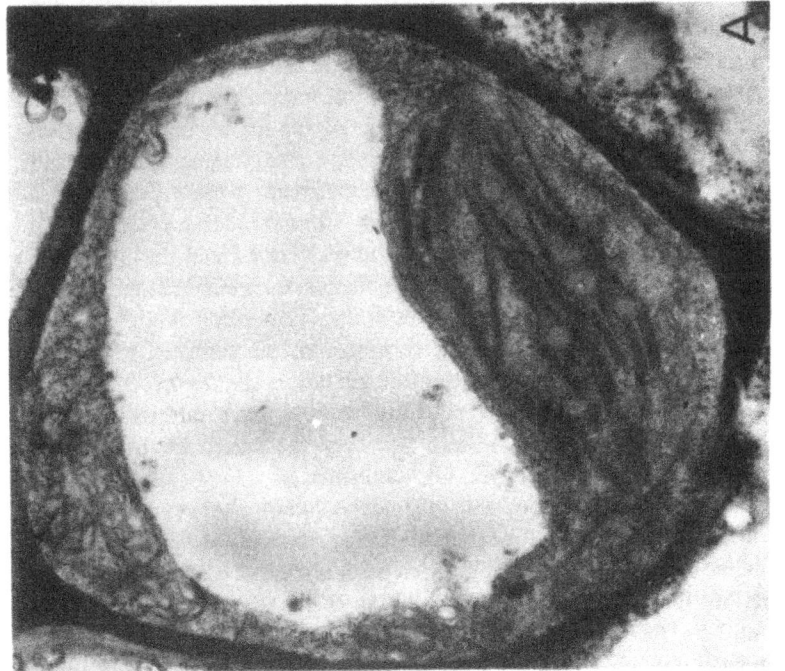

Figure 3. The ultrastructural changes in chloroplasts of *Vigna unguiculata* mesophyll cells. A. Mesophyll cells from control (non-senescent) plants. See the starch-filled chloroplasts (-X 20,000) with few and small osmiophilic globuli. Note the abundance of cytoplasm with mitochondria. B. Mesophyll cells from senescent plants. See the increase in osmiophilic globuli. The size of the starch grain is also reduced (-X 8,000). g: grana; os: osmiophilic granules; st: starch granules (PV Reddy and SK Sinha, 1986, unpublished).

one or two galactose molecules attached to the glycerol backbone corresponding, respectively, to monogalactosyl diacylglycerol, and digalactosyl diacylglycerol (Joyard et al., 1989). Detailed work on alteration in lipid composition in senescing chloroplasts is available mainly on bean leaves. The galactolipids are observed to decrease preferentially, over the other lipids, during senescence in these leaves (Novitskaya et al., 1977). Within the galactolipids, the monogalactosyl diglycerides appear to degrade at a faster rate than the others (Fong and Heath, 1977).

The membranes of senescing bean leaves are also found to have a high degree of leakiness to electrolytes. This increased electrolyte leakiness is temporally associated with rise in the level of malondialdehyde, a product of lipid peroxidation, indicating that membrane leakiness is due to higher lipid peroxidation (Venkatarayappa et al., 1984). Phase transition temperature for thylakoid membrane lipids is drastically altered during senescence: it is below $-30°C$ in young bean leaves and is about $30°C$ in mature leaves (McKersie and Thompson, 1978). Venkatarayappa et al. (1984) also reported that the lipid phase transition temperature of partly purified thylakoid membranes from senescing bean leaves is well above their growth temperature. From the preceding discussion, it appears that the onset of degeneration is correlated with the time at which the chloroplast lipid transition temperature enters the physiological range (McKersie and Thompson, 1978). In some instances, the increased activities of superoxide dismutase and catalase enzymes have been implicated in enhanced lipid peroxidation and cumulative membrane deterioration (Dhindsa et al., 1981). Changes in lipid composition could possibly cause general loss of membrane integrity and stimulate intracellular decompartmentalization.

A somewhat detailed insight into the influence of lipid environment and coordination among various reactions of chloroplast senescence was obtained using non-yellowing mutants of Festuca pratensis. This mutant has a unique characteristic: the mutant leaves (cultivar Bf 993) retain Chl during senescence but lose soluble proteins at a rate similar to that of the control yellowing leaves (cultivar Rossa) (Thomas and Stoddart, 1975). The mutant genotype also retains a number of chlorophyll proteins and other hydrophobic low molecular weight polypeptides (Thomas, 1982; Hilditch 1986). The light harvesting Chl a/b binding protein (LHC II), the reaction centre protein D1, and Cyt b_6/f underwent a rapid decline in the tissues of wild type during senescence; in contrast, the mutant retained high amounts of these proteins throughout senescence (Davies et al., 1988; Hilditch et al., 1989). However, the 33 kD extrinsic protein of the oxygen-evolving complex was found to be equally labile in both the mutant and the wild genotypes (Hilditch et al., 1989). It has also been shown that the non-yellowing mutants have unusually stable galactolipid bilayers (Harwood et al., 1982). However, the precise primary lesion of the non-yellowing mutation is, as yet, not identified.

The formation of lipid peroxides and other concomitant changes in the lipid composition are relatively better understood during induced aging of isolated chloroplasts. It has been found that light induces peroxidation of lipids during aging of chloroplasts but this is not a cause of inhibition of the photochemical activity (Hoshina et al., 1975). The inhibition of photochemical activity in isolated chloroplasts has been correlated, among other factors, with changes in lecithin to lysolecithin, liberation of free unsaturated fatty acids, and accumulation of free radicals (Hoshina *et al.*, 1975; Kochritz *et al.*, 1984; DuPont and Siegenthaler, 1986).

c. *Absorption and fluorescence spectral studies*

The physical state of the membranes—particularly that of the membrane lipids, strongly influences the absorbance and fluorescence spectral profiles of the Chls (Hoshina *et al.*, 1983). Therefore, senescence-associated changes in the characteristics of absorption and fluorescence spectra can provide significant clues to the structural and functional changes in thylakoid membranes (for a background on fluorescence of photosynthetic systems, see Govindjee *et al.*, 1986). Sestak (1972) showed that the red band of the absorption spectrum of chloroplasts from old leaves shifted by 1 nm, at 77K, to longer wavelengths than in chloroplasts isolated from young leaves of *Zea mays*. It was suggested that either the differential rates of decline of various forms of Chl *a* or the rearrangement of chloroplast internal structure during senescence caused red shift in the absorption spectrum. At room temperature, however, similar red shift of 4 nm was observed in thylakoids isolated from detached senescing barley leaves (Biswal and Mohanty, 1976b). No change in the position of the red absorption band was noticed in detached senescing wheat leaves (Grover *et al.*, 1986a). Thus, the shift of the red absorption band in senescing leaves does not seem to be universal phenomenon. We have also noted that senescence causes a considerable rise in light scattering by isolated thylakoids (Grover *et al.*, 1986a). The increased scattering could be ascribed to changes in size, orientation, and/or number of thylakoids per granum in senescing chloroplasts.

Further evidence for the changes in membrane structure during senescence has come through Chl *a* fluorescence yield and emission data. Fragata (1975) reported that aging of isolated chloroplast decreases the intensity of Chl *a* fluorescence. Dark stress-induced senescence in detached barley leaves caused a sharp decline in the total fluorescence intensity with time (Biswal *et al.*, 1979). In the latter study, decline in ratio of Chl *a* fluorescence intensity at 685 (F 685) and 735 (F 735) nm was also noted and was thought to be due to loss of variable fluorescence at F 685 nm originating from PS II. This in turn could result from loss in O_2-evolving capacity due to damage to the electron transport carriers such as the primary acceptor of PS II and disorganization of PS II pigments. Recently, Gruszecki *et al.* (1991) analysed the time course of fluorescence changes and *in vivo* photochemical activities using photoacoustic spectroscopy in pea leaves. Decrease in photosynthetic energy storage capacity due to onset of senescence paralleled the loss in energy transfer from carotenoids to Chl *a*, suggesting that energy transfer mechanism in PS II is impaired prior to PS I.

Analysis of fluorescence transients is being intensively used as a probe for photochemical potential of intact leaves and isolated chloroplasts in a variety of stress conditions (see Hetherington and Oquist, 1988), but so far, this technique has not been fully exploited for leaf senescence. We note that detailed analysis of the changes in Chl *a* fluorescence yield and emission feature are required to gain insight into the early events of foliar senescence.

In summary, both the lamellar and stromal components of chloroplast in senescing leaves have been shown to be extensively disorganized with respect to their structural integrity. The disorganization of structure must have a profound influence on photochemical activity (to be discussed below). The hierarchy of these events needs to be deciphered in order to allow us to both understand and delay senescence.

VI. Alterations in Photochemical Activity

It has been established that the thylakoid photochemical activities limit photosynthesis during senescence (Harding *et al.*, 1990). Decline in photochemical activity of chloroplasts during leaf senescence is observed whether it is expressed on unit fresh weight or unit Chl basis. Therefore, loss of Chl as well as loss of components of electron transport chain and alterations due to disorganization of lameller system might limit electron transport activity in senescing leaves.

a. *Rate-limiting steps in electron transport chain*

Electron transport under *in vivo* conditions remains coupled to the phosphorylation process. The uncouplers such as methylamine (MA), gramicidin (GD), and NH_4Cl disconnect the phosphorylation process from electron transport activity. Measurements of electron transport activities during senescence have been made in both coupled (or basal) and uncoupled states of chloroplasts. The rate of coupled, noncyclic electron transport from water to methyl viologen (MV), expressed on Chl basis, for chloroplasts from old *Phaséolus* leaves was only 20% of that found for chloroplasts from younger leaves (Jenkins and Woolhouse, 1981a). Similarly, drastic reduction in $H_2O \rightarrow MV$ electron transport activity was reported for several other senescing systems (McRae *et al.*, 1985; Sabat *et al.* 1985; Grover *et al.*, 1986a).

Theoretically, factors such as a change in the characteristics of reaction centres, a preferential loss of one of the two photosystems, and/or loss of some components of electron transport chain can potentially result in reduction of the whole chain ($H_2O \rightarrow MV$) electron transport activity (see Fig. 1). Possible alterations in number of antenna pigments and the reaction centre complexes have been addressed in some plant systems. For instance, Jenkins *et al.* (1981b) argued that the reduction in size of the light harvesting antenna due to loss of Chl would alter the light saturation kinetics of chloroplast activity. Since similar light saturation curves for electron transport activities were obtained employing chloroplasts isolated from mature and senescent *Phaseolus* leaves, it indicated that the antenna size associated with each reaction centre did not change (Jenkins *et al.*, 1981b). During foliar senescence of rice seedlings, Kura-Hotta *et al.* (1987) showed a more rapid loss of PS II reaction centre complex than LHC II. This preferential breakdown of PS II reaction centre can impose an imbalance between PS II and Cyt b_6/f activities. More detailed investigation is, however, necessary to characterize the antenna sizes and the reaction centres of PS II and PS I during leaf senescence.

It has been shown that the rates of electron transport (in saturating photon flux density conditions) through PS I [2, 6-dichlorophenol indophenol H_2 ($DCPIPH_2$) $\rightarrow$ MV or N, N, N′, N′-tetramethyl-p-phenylenediamine (TMPD) $\rightarrow$ MV] and PS II [$H_2O \rightarrow$ ferricyanide, in presence of oxidized-p-phenylenediamine (PD_{ox})] declined by approximately 25 and 33% respectively, during foliar senescence in *Phaseolus* (Jenkins and Woolhouse, 1981b). In contrast, $H_2O \rightarrow MV$ activity declined by nearly 80% in the same period (Jenkins and Woolhouse, 1981a). Based on these observations, it was suggested that the impairment of electron flow between the two photosystems limited the non-cyclic electron transport in chloroplasts from older leaves to a large extent (see Figs. 1, 2). Employing the same species, McRae *et al.* (1985) have also shown that rate of non-cyclic electron transport ($H_2O \rightarrow MV$) declines to a greater degree than the activities of either of the photosystems.

Several attempts have been made to unravel the rate-limiting step between the photosystems responsible for decline in non-cyclic electron transport ($H_2O \rightarrow MV$) activity. Data from kinetics of chloroplast fluorescence emission analysis indicated that the flow of electrons out of the PQ pool, rather than the transfer of electrons into the pool through PS II, limits electron transport in attached senescing *Phaseolus* leaves (Jenkins *et al.*, 1981a). In a study on thylakoids isolated from barley leaves of different ontogenetic stages, Holloway *et al.* (1983) found that the rate of electron flow from H_2O to MV or FeCN diminishes significantly during senescence. They suggested that PQ limits electron transport activity prior to leaf maturity, whereas during senescence, the rate limiting step was on the reducing side of PQ, which is in accordance with the findings of Jenkins *et al.* (1981a). Holloway *et al.* (1983) further carried out a quantitative analysis of various electron transport components. It was concluded that the rate-limiting step in $H_2O \rightarrow$ ferricyanide (FeCN) (electron acceptance at PS I side) reaction in senescing barley leaves was probably the transfer of electrons from the PQ to Cyt b_6/f complex. In chloroplasts from senescing cotyledons of *Cucurbita maxima*, a relatively faster degradation of Cyt f is indicated by the fact that Chl/Cyt f ratio of 400, found in young leaves, increased to 1200 in old leaves (Harnischfeger, 1973). A rather indirect evidence regarding senescence-associated decline in PC level was obtained employing KCN, which is a specific antagonist to PC. In detached senescing *Beta vulgaris* leaves, the sensitivity of $DCPIPH_2 \rightarrow MV$ reaction towards KCN decreased markedly, which could be due to reduction in the pool size of PC (Sabat *et al.*, 1985). Considering this reduction appreciable, it is possible that PC co-limits non-cyclic electron transport activity in senescing leaves.

The photosystems mediating photosynthetic electron transport are spatially separated in thylakoid membranes. Accordingly, the whole chain electron transport activity is achieved by lateral diffusion of mobile electron carriers, specifically PQ and PC (Haehnel, 1984). This raises a possibility of the contention that the rate at which PQ diffuses between PS II and Cyt f may limit the rate of photosynthetic electron transport. For this to happen, the lipid bilayer fluidity of thylakoid should decline concomitant to the progress of senescence. The results of McRae *et al.* (1985) with senescing primary bean leaves have shown that the lipid bilayer fluidity is not affected during senescence, thereby suggesting that changes in photosynthetic electron transport activity are not due to alterations in thylakoid fluidity.

b. *Preferential loss of photosystem*

The question regarding which of the two photosystems (PS II or PS I) is more sensitive to damage during leaf senescence has not been unequivocally answered. As mentioned earlier, during senescence of attached *Phaseolus* leaves, the saturation rates of electron transport through PS I (DCPIPH$_2$ → MV or TMPD → MV) and PS II [H$_2$O → PD$_{ox}$ (→ FeCN)] decline by approximately 25 and 33%, respectively (Jenkins and Woolhouse, 1981b). In senescing detached *Beta vulgaris* leaves, the saturating H$_2$O → FeCN activity of uncoupled chloroplasts was found to be only 10% of maximal activity after six days of leaf senescence, In contrast, saturation rates for DCPIPH$_2$ → MV showed a rise during senescence (Sabat *et al.*, 1985).

A transient rise in PS I activity was also observed in thylakoids isolated from detached wheat leaves (Grover *et al.*, 1986a). However, in the latter work, when the electron transport activity was computed on a unit fresh weight instead of a unit Chl basis, DCPIPH$_2$ → MV activity showed a continuous decline (Grover *et al.*, 1986a). To explain the transient rise in PS I activity (and to a lesser extent in PS II also) it was suggested that senescence induces uncoupling of electron transport and thereby enhances the electron transport rates. However, there remains the distinct possibility of enhancement of uncoupled PS I activity during leaf senescence. Increased activity of PS I has been observed with reduced DCPIP as donor (SC Sabat, A Grover and P Mohanty, unpublished), although the nature of this enhancement in PS I activity is unknown.

In contrast to the above studies (Jenkins and Woolhouse, 1981b; Sabat *et al.*, 1985; Grover *et al.*, 1986a), Bricker and Newman (1982) reported 18% higher loss of PS I (TMPD → tetrazolium blue) as compared to PS II (H$_2$O → DCPIP) electron transport activities during senescence of cotyledonary chloroplasts in soybean. In the study of Jenkins *et al.* (1981b), though PS II electron transport activity in chloroplasts from older leaves declined faster than the corresponding PS I activity by marginally 8%, older leaves contained a smaller proportion of Chl associated with PS I. The preceding discussion suggests that no final conclusion has as yet emerged regarding the relative sensitivity of the electron transport activities of the two photosystems.

The above-mentioned assessments of the relative sensitivity of PS II and PS I have been made on the basis of activity profiles of individual partial reactions. Since different sets of donors and acceptors are used for measuring PS II (in which case water is generally used as an electron donor) and PS I (for which usually exogenous donors are used) electron transport activities, data obtained by comparing activity profiles may not always give the true representation. Hence, what one really needs to do is to attempt a combined study of proteins associated specifically with the reaction complexes of PS II and PS I and related photochemical activity, at both, low and high light intensities.

In chloroplasts isolated from mature and senescent soybean cotyledons, Bricker and Newman (1982) analysed chloroplast-proteins using sodium dodecyl sulphate-polyacrylamide gel electrophoresis (SDS-PAGE). Ten Chl-protein bands were resolved in their system: three of these were P700 Chl *a* protein complexes associated with the reaction centre of PS I, five were light harvesting Chl *a/b* protein complexes associated with antenna of PS II, and two others corresponded to PS II reaction centre protein complexes. The amount of P700 Chl *a* protein complexes decreased more rapidly during senescence than the amount of Chl protein complexes associated with PS II. Electron transport activity of PS I declined more rapidly than that of PS II in their study. Thus, the electron transport activity profiles and the *per se* loss of reaction complex proteins were correlated.

Ben-David *et al.* (1983) have used antibodies against the individual subunits of protein complexes in the thylakoid membranes to follow the amounts of these polypeptides during foliar senescence in *Phaseolus vulgaris*. No change was found in the amount of polypeptides of PS I reaction centre and the chloroplast coupling factor. A significant decrease in the amount of different components of the Cyt b_6/f complex was observed. This study concluded that Cyt b_6/f is a possible candidate for the limitation in the electron transport rate in senescing leaves, which is in accordance to the findings of Holloway *et al.* (1983) mentioned earlier.

Roberts *et al.* (1987) have shown that the rate at which various thylakoid proteins are synthesized is differentially affected during senescence of primary bean leaves. They noted that synthesis of a 32 kD herbicide-binding protein (known as the D1 protein, see Sopory *et al.*, Chapter 5) continues throughout senescence, whereas the formation of α and β subunits of ATPase, the 68 kD PS I reaction centre polypeptide, Cyt b_6/f complex, and the structural apoprotein of the LHC II declines. The western blot analysis indicated that the Cyt b_6/f depleted at a relatively faster rate during senescence.

c. *Sequential loss in PS II complex*

A major advance in recent years has been the development of the concept that degradation within the PS II complex proceeds in a well-ordered and sequential manner. This conclusion has been derived using electron donors Mn^{++} and 1,5-diphenyl carbazide (DPC), which feed electrons to PS II. Biswal and Mohanty (1976a) observed that Mn^{++} ions stop supporting DCPIP reduction after four days in detached senescing barley leaves kept in darkness; however, DPC sustained Hill reaction activity till the seventh day. It was suggested that, since these two exogenous donors fed electrons at different sites between H_2O and the PS II reaction centre, senescence caused a sequential disruption in the function of these sites. Similar to these observations, dark stress-induced loss in DCPIP photoreduction was restored fully by DPC, but only partly by Mn^{++} ions in attached *Zea mays* leaves (Choudhury and Biswal, 1979). However, the precise positions where these two exogenous electron donors feed electrons to PS II complex is not fully settled (Hsu *et al.*, 1987). We suggest that further critical analysis be made with isolated PS II particles from senescing leaves, as it may provide a better understanding of this problem. Nonetheless, since the addition of exogenous electron donors before the PS II reaction centre restores the DCPIP-Hill activity reaction to a certain extent, the O_2-evolving system could be more labile during leaf senescence. The basis of the sensitivity of the O_2-evolving system can potentially arise from enzymes involved in water oxidation which are being actively investigated in many laboratories (Renger and Govindjee, 1985).

In summary, loss in photochemical efficiency during senescence on a unit leaf area or fresh weight basis occurs due to: (1) extensive losses of photosynthetic pigments; (2) specific alterations in electron transport components; and (3) specific loss of one of the reaction centres. The different behaviour in the loss of photochemical activities in dark- and light-incubated senescing leaves, noted by several workers, could be attributed to the susceptibility of senescing tissues to photoinhibition and/or light-controlled differential gene expression changes (Powles, 1984; Kasemir *et al.*, 1988). As far as photochemical activity on a unit Chl basis is concerned, it seems that the rate-limiting step lies somewhere either at the Cyt b_6/f or at the PC level and not within the photosystem complexes. Some major sites which appear to be relatively sensitive to senescence are depicted in Fig. 4. There may be some species-

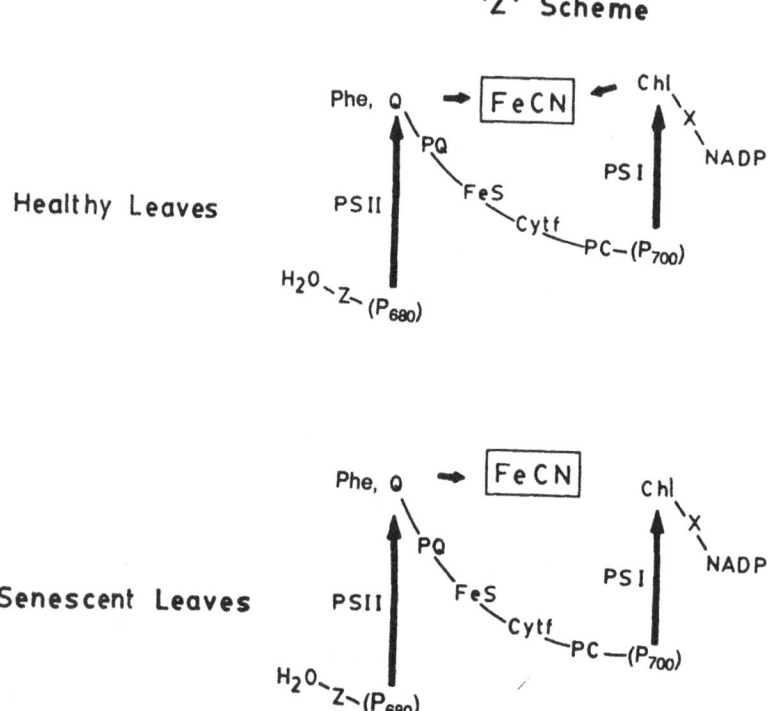

Figure 4. Senescence induced alterations in ferricyanide acceptor site (for details see Sabat *et al.* 1989a)

OEC

$$H_2O \rightarrow [Mn\ cluster] \rightarrow Z \rightarrow P_{680} \rightarrow Phe \rightarrow Q_A \rightarrow Q_B \rightarrow PQ \rightarrow Cytb_6/f \rightarrow PC \rightarrow$$

PS II

$$P_{700} \rightarrow A_0 \rightarrow A_1 \rightarrow Fex \rightarrow (Fe_A Fe_B) \rightarrow Fd \rightarrow FNR \rightarrow NADP$$

PS I

Figure 5. Senescence-triggered alterations at selective sites in the electron transport chain. ⇑ indicates locations of alterations. Electron transport components are same as in Fig. 1. Additional electron transport components shown in this figure are: OEC: oxygen-evolving complex; A_0: possibly a chlorophyll molecule; A_1: a phylloquinone or vitamin K_1; Fex, Fe_A and Fe_B: a series of Fe-S centres.

specificity in the location of rate-limiting steps, but the scarcity of data prevents any generalization at present. Interestingly, the biochemical evidence presented above suggests that PS II is more sensitive than PS I while electron micrographs have revealed that granum (which has mostly PS II complex) integrity is maintained for longer than stromal (which has both PS II and PS I complexes) components. The apparent paradox between the ultrastructural and biochemical studies, obviously, needs further attention.

VII. Change in pH Optimum of PS II Activity

Optimum pH of PS II electron transport activity in chloroplasts from senescing detached beet spinach and attached primary wheat leaves showed a distinctive shift towards acidic side (Sabat *et al.*, 1985, 1989a). This shift in pH optimum was observed in basal as well as methylamine-induced uncoupled electron transport activity. Interestingly, pH optimum of FeCN-Hill activity in the presence of PD_{ox} did not change. Based on the differential results obtained in the presence or absence of PD_{ox}, we have argued that senescence involves a change in the electron acceptance site of FeCN (Sabat *et al.*, 1985, 1989a, b; Mohanty *et al.*, 1986). We suggest that FeCN, which accepts electrons from both PS II and PS I in healthy leaves (Govindjee and Bazzaz, 1967), accepts electrons from only PS II in chloroplasts of senescing leaves (Fig. 5).

The observed change in this characteristic of FeCN acceptance site may possibly arise due to inefficient transfer of electrons from the reducing side of PS II towards the reaction centre of PS I. The decrease in the electron transfer between the two photosystems possibly results either from the decline in the pool size of the various electron transport carriers including PQ, PC, and cytochromes (Ben-David *et al.*, 1983; Sabat *et al.*, 1985, 1989a, Roberts *et al.*, 1987) or due to loss of PS II centres (Kura-Hotta *et al.*, 1987).

Earlier, an acid shift in the pH optimum of FeCN reduction was observed in senescing *Cucurbita maxima* cotyledons (Harnischfeger, 1974) and during *in vitro* chloroplast aging (Siegenthaler and Rawyler, 1977). Both the earlier and the most recent pH shift experiments show that the measurement of photochemical rates at any one constant pH may lead to misleading conclusions. For a critical study on this topic, see Harnischfeger (1974).

In summary, it can be stated that experiments indicating change in FeCN acceptance site again point to the fact that electron transport carriers seem to rate limit non-cyclic electron transport activity and, as a result, the two photosystems appear to be inefficiently coupled with respect to energy transfer in senescing leaves.

VIII. Change in Photophosphorylation Efficiency

Two types of photophosphorylation reactions, namely cyclic and non-cyclic, can be accomplished in isolated chloroplasts (see Trebst, 1974). Although cyclic photophosphorylation has not been unequivocally proven to exist *in vivo*, it appears to be associated with PS I only, while the non-cyclic reaction is known to exist *in vivo* and is coupled to the reduction of $NADP^+$ or artificial electron acceptors MV or FeCN and involves both photosystem.

Efficiency of photophosphorylation markedly declines during the phase of leaf senescence. Cyclic ATP synthesis per unit Chl was shown to decline continuously from the time of full leaf expansion to leaf maturity in *Populus deltoides* (Hernandez-Gil and Schaedle, 1973). The loss of FeCN-supported non-cyclic photophosphorylation has been found to be somewhat greater than phenazine methosulphate (PMS) catalysed PS I supported cyclic photophosphorylation during barley leaf senescence (Biswal and Mohanty, 1978). A limiting factor involved in decline of non-cyclic ATP synthesis can possibly arise from preferential inactivation of the O_2-evolving system associated with PS II.

Reduction in photophosphorylation efficiency can also result from the concomitant decrease in ATP synthase or H^+-ATPase activity as found in several senescing systems (Scherings and Kuiper, 1974; Biswal and Mohanty, 1978). Camp *et al.* (1982) demonstrated a loss of coupling factor during wheat leaf senescence. The synthesis rate of α and β subunits of ATPase has been found to decline in the senescing bean leaves (Roberts *et al.*, 1987). However, these findings are in contrast to the observations of Ben-David *et al.* (1983), who found no change in the amount of polypeptides of coupling factor during senescence of oat and barley leaves. The loss in Cyt b_6/f complex during leaf senescence was not accompanied by any loss in coupling factor in this study. It is possible that coupling factor may lose its latent ATPase activity (Biswal and Mohanty, 1978) without *per se* loss of the protein. The loss in coupling capacities during leaf senescence needs further investigation.

IX. Change in Chloroplast Number

We now discuss what happens to the chloroplast number per cell during senescence. The question that needs to be addressed is whether all chloroplasts of a cell senesce at the same uniform rate or whether there is a population of chloroplasts which is relatively more sensitive and gets affected to a greater extent. Two hypotheses worth considering are: (1) chloroplasts senesce through lysosomal interaction with the vacuole so that senescence would be characterized by a decline in chloroplast number per mesophyll cell; and (2) chloroplast number per mesophyll cell remains constant, at least until general cellular lysis occurs.

The first hypothesis has been supported by the independent analyses of a few workers. In senescing attached primary wheat leaves, Wittenbach *et al.* (1982) noticed, on per protoplast basis, a close correlation between the decline in chloroplast number and loss of total Chl and soluble proteins. Lamppa *et al.* (1980) observed a marked decline in chloroplast number, on mesophyll cell basis, during senescence of pea leaves. Camp *et al.* (1982), using naturally senescing flag and second leaf of wheat, observed that electron transport rates in isolated thylakoids prepared from these leaves began to decline at mid-senescence when activities were expressed on a leaf area basis, whereas on a Chl basis, the rates declined only at the later stages of senescence. This suggested that chloroplasts remaining in the leaf were photochemically active until late senescence; hence, the loss of photosynthetic activity during senescence was related to loss of whole chloroplasts as opposed to general degradation of the entire chloroplast population. Kura-Hotta *et al.* (1987) also suggested that the loss of photosynthesis in senescing leaves of rice seedlings is mainly caused by a decrease in the number of whole chloroplasts. This conclusion was based on the finding that various thylakoid proteins including reaction centre complexes, LHC I and LHC II, disappear more or less in parallel during senescence.

The alternative hypothesis emphasizing that senescence involves temporal degradation of components, such as Chl and thylakoid proteins, within the chloroplast has gained more support. With dark-incubated and naturally senescing leaves of barley seedlings, Martinoia et al. (1983) observed that chloroplast number decreased to a much lesser extent than did the contents of Chl, proteins, and RuBPcase. Mae et al. (1984) reported much faster degradation of Chl content than chloroplast number in naturally senescing primary wheat leaves. Using attached and detached primary wheat leaves, Wardley et al. (1984) concluded that constituents of the chloroplast are most likely degraded in situ. Our data also pointed out that the loss of Chl on per unit fresh weight and on a unit chloroplast number basis remain parallel in both naturally and dark-induced senescing primary wheat leaves (Grover et al., 1987). This again suggests that there is a uniform extent of loss in Chl in all the chloroplasts. This fact has been further confirmed in senescing leaves of rice seedlings (Hashimoto et al., 1989) as well as in senescing leaves of different soybean genotypes (Ford and Shibles, 1988).

To account for the in situ degradation of Chl, thylakoid proteins, RuBPcase, etc., chloroplasts must possess their own set of catabolic enzymes. There should either be an increased synthesis of these enzymes or their specific activities should rise during the senescence phase.

X. Senescence-specific Proteases

Do chloroplasts possess the enzymes needed for the gradual and controlled degradation of their constituents? A general rise in activities of lipases, nucleases, and proteases has been documented in senescing leaves of several species (Frith and Dalling, 1980; Peoples and Dalling, 1988; Vierskov and Thimann, 1988). Several senescence-induced proteases have the ability to hydrolyse chloroplast proteins, particularly RuBPcase (Peoples et al., 1980; Miller and Huffaker, 1982; Peoples and Dalling, 1988). The localization of such proteases inside the chloroplast had remained a subject of controversy for long. However, many experiments indicate that chloroplasts possess proteolytic enzymes (Waters et al., 1982; Dalling et al., 1983; Wardley et al., 1984; Grover and Sinha, 1988).

Studies on protein degradation during chloroplast senescence have so far concentrated on loss of stromal proteins, particularly RuBPcase (for the obvious reason that RuBPcase is the single most abundant source of nitrogen for reallocation). It remains to be followed how the chloroplast membrane proteins are affected during leaf senescence. The herbicide-binding Q_B protein (the D1 protein) of chloroplasts was shown to have a high turnover rate in healthy as well as senescent leaves (Roberts et al., 1987). Liu and Jagendorf (1986) isolated one endopeptidase and three aminopeptidases from intact chloroplasts of pea seedlings, which could account for the rapid metabolism of D1 protein. We are unaware of any report that shows enzymes leading to rapid degradation of Cyt b_6/f protein complex, which is currently thought to rate limit the photochemical activities in senescing leaves (see Section VI). The mechanism of degradation of electron transport components and other constituents of thylakoids, in healthy as well as senescent tissues, merit attention.

XI. Nuclear Control of Chloroplast Senescence

Whether chloroplast senescence is under the control of its own DNA or nuclear DNA may not be yet certain; however, most data favour the nuclear control. In *Elodea densa,* mesophyll cell protoplasts with nuclei showed all features of chloroplast senescence, while those without nuclei maintained green chloroplasts for a longer period (Yoshida, 1961). This fact has been further supported by studies using protein synthesis inhibitors. In one such study, Makovetzki and Goldschmidt (1976) showed that cycloheximide (which inhibits cytoplasmic protein synthesis) delays loss of Chl from detached *Anacharis canadensis* leaflets while chloramphenicol (which inhibits protein synthesis in chloroplasts) accelerated the loss.

A further support for the nuclear control of chloroplast senescence is based on the non-yellowing *Festuca pratensis* mutants. In these mutants, Chl is retained while the protein level declines in the usual way during senescence (Hilditch *et al.,* 1989). This mutation has been observed to follow Mendelian segregation, indicating that the loci controlling Chl degradation are under nuclear control. In addition, it has been observed that isolated chloroplasts under *in vitro* conditions senesce at a relatively much slower rate (Choe and Thimann, 1977), suggesting that some component(s) of the cell from the outside of the chloroplast might have a role in its senescence. During the course of chloroplast development, the cooperation between the nuclear and plastidic genome has been studied intensively and it is generally accepted that the plastidic development is controlled by the genetic information encoded in the nucleus to a large extent (Westhoff *et al.,* 1988; Tyagi *et al.,* 1989). Recently, it has also been shown that plastids are involved in controlling extraplastidic events (see Oelmuller, 1990). The nuclear-plastid interactions during chloroplast senescence have not been given much attention.

XII. Chloroplast Senescence: Molecular Events

Change in gene expression is fundamental to the plant development cycle. The biochemical and physiological manifestation of the leaf senescence process must also be under genetic control. However, several questions, in this context, remain to be answered: What gene expression changes accompany the senescence syndrome? Which genes govern the initial triggering of the process of senescence? What is the nature of the product (s) synthesized by senescence-related genes? The answers to such questions are crucial for our understanding of the molecular basis of leaf senescence. With the current upsurge in the advancement of molecular techniques of gene isolation and cloning, RNA isolation, *in vitro* and *in vivo* protein labelling, and their separation by electrophoresis, it is becoming increasingly possible to address oneself to the fundamental aspects of chloroplast and leaf senescence.

An attempt to isolate senescence-associated new mRNA was first made by Watanbe and Imaseki (1982) using *in vitro* translation coupled to one-dimensional SDS-PAGE analysis. This work was done on senescent wheat leaves. Kawakami and Watanbe (1988a, b) employed two-dimensional protein separation method, which provides a better resolution (O'Farrell, 1975), for resolving polypeptides related to senescence of radish cotyledons. Malik (1987) used the same two-dimensional protein separation technique for examining polypeptides associated with oat leaf senescence. All the above studies indicated that several novel polypeptides are specifically expressed during the senescence phase. However, no further characterization of the newly synthesized polypeptides was attempted in these analyses.

Since chloroplasts senesce at a relatively faster rate, it is possible that the early gene expression changes important for induction of leaf senescence are associated with the chloroplasts. Recently, several attempts have been made in this direction. Martin and Sabater (1989) isolated chloroplast mRNA from senescent and kinetin-treated barley leaves and translated them *in vitro*. Their results indicate that isolated intact chloroplasts from senescent barley leaves specifically synthesized polypeptides of 80, 60, 58, 50, 48, and 35 kD molecular weights. Guera *et al.* (1989) studied the distribution of polypeptides synthesized by leaf chloroplasts between the sub-chloroplast fractions such as stroma, chloroplast envelope, and thylakoids. Senescence-related newly synthesized polypeptides were localized in each of these fractions.

The functional role of polypeptides synthesized specifically during senescence phase is not yet known. It is possible that many of the biochemical alterations described in the earlier sections of this chapter may be related to the polypeptides induced in chloroplasts at the time of senescence. However, this needs to be established through additional research. Information on the identification of genes which govern the synthesis of novel polypeptides at the senescence stage is also scarce. In recent years, the refinements in recombinant DNA techniques and nucleotide sequence analyses have led to an explosion of information on the chloroplast molecular biology. The complete chloroplast genome sequence of *Nicotiana tabacum* (Shinozaki *et al.*, 1986), *Marchantia polymorpha* (Ohyama *et al.*, 1986), and *Oryza sativa* (Hiratsuka *et al.*, 1989) has been determined. (see Chapter 1) In addition, methods to isolate intact plastids which are active in protein synthesis have been fairly well standardized (see Gnanam *et al.*, 1988). It is hoped that eventually such an analysis will provide clues for an understanding of the mechanistic details regarding temporal relationship among losses of various chloroplast proteins during senescence of leaves. Furthermore, the detailed molecular analysis may provide clues as to how the expression of chloroplast genes is regulated in the senescing leaves. In general, the chloroplast gene expression has been shown to be controlled at the level of transcription (Lerbs *et al.*, 1984; Mullet, 1988) or translation (Monroy *et al.*, 1987; Westhoff *et al.*, 1988). Synthesis of various thylakoid proteins, during chloroplast biogenesis, has been found to be largely controlled at translational level (Westhoff *et al.*, 1988). Though controlled expression has been noted with respect to several genes in senescing chloroplasts (Garcia *et al.*, 1983; Kasemir *et al.*, 1988; Jayabaskaran *et al.*, 1990), the mechanism responsible for this has not been studied in detail.

XIII. Concluding Remarks

When mature leaves (for instance, leaves supporting developing pods in legumes and flag leaf in cereals) senesce, nitrogen liberated from hydrolysis of RuBPcase is mobilized to developing tissues such as seeds and grains (Grover et al., 1985; Huffaker and Miller, 1987; Grover and Sinha, 1988). Does leaf senescence-associated loss of photochemical activity of chloroplasts have any role in remobilization of nitrogen, carbon, and other elements? It is important to ask this question because proteins involved in the light reactions of photosynthesis represent about one-quarter of leaf protein nitrogen in a C_3 sun leaf (Evans and Seemann, 1989). According to Thomas (1977), nitrogen released from Chl could supplement mobilizable nitrogen from leaf proteins to the extent of about 10%. However, the implications of nitrogen mobilization from Chl for further growth of the plant are obscure (Llewellyn et al. 1990a). Furthermore, it is hard to predict this contribution because chloroplast senescence with respect to the photochemical apparatus has mainly been analysed using primary leaves. Efforts should be made to compare the details obtained with primary leaves with the functioning at the reproductive nodes of the crop plants. In leaves supporting reproductive structures, several studies have shown that water stress and sink size, among other factors, can significantly alter the senescence pattern (see Amthor and McCree, 1990). Senescence in mature leaves under constrained environments may, therefore, show different programming. Thus, understanding of the mechanism of senescence and its control would certainly be important with agriculture point of view.

An important consideration worth rigorous analysis is to look for reaction(s) of chloroplasts which can provide reliable markers for plant senescence, as it has been shown that chloroplasts reflect the earliest changes associated with leaf senescence. In this regard, alterations in the physical properties of chlorophylls and carotenoids as reflected by the absorption and fluorescence spectral profiles could be of key importance. This search should take into account that the marker parameter has to be easy and routinely analysable so that breeders can use it as a criterion for selecting progeny with delayed senescence traits.

Acknowledgements

The work discussed in this chapter was supported by the PL 480 ICAR-USDA project under contract No. FG-IN-575; IN-SEA-170 and FG-IN-679, IN-ARS 402. Grateful thanks are due to Prof. Patrick Breen, Oregon State University, Corvallis, USA for many valuable suggestions. We thank Dr. S.C. Sabat for his collaboration and Prof. Govindjee, University of Illinois, USA, and Prof. S.K. Sinha, Indian Agricultural Research Institute, New Delhi, for their comments on the manuscript.

XIV. References

Abrol YP, Kumar PA and Nair TVR (1984) Nilrate uptake, its assimilation and grain nitrogen accumulation Adv Cereal Sci Tech 6: 1–48

Amir-Shapira D, Goldschmidt EE and Altmann A (1987) Chlorophyll catabolism in senescing plant tissues: In vivo breakdown intermediates suggest different degradation pathways for Citrus fruit and parsley leaves. Proc Natl Acad Sci USA 84: 1901–1905

Amthor JS and McCree KJ (1990) Carbon balance of stressed plants: A conceptual model for integrating research results. In: (RG Alscher and JR Cumming, eds) Stress Responses in Plants: Adaptation and Acclimation Mechanisms, Wiley-Liss, A John Wiley and Sons, Inc, pp 1–15

Anderson JM and Thomson WW (1989) Dynamic molecular organization of the plant thylakoid membrane. In: (WR Briggs, ed) Photosynthesis, Alan R Liss, Inc, New York, pp 161–182

Andreasson LE and Vanngard T (1988) Electron transport in photosystems I and II. *Annu Rev Plant Physiol Mol Biol* **39**: 379–411

Andrews TJ and Lorimer GH (1987) Rubisco: Structure, mechanisms, and prospects for improvement. In: (MD Hatch and NK Boardman, eds) *The Biochemistry of Plants—A Comprehensive Treatise,* Vol 10, Academic Press, pp 131–218

Barber J and Marder JB (1986) Photosynthesis and the application of molecular genetics. In: (GE Russell, ed) *Biotechnology and Genetic Engineering Reviews,* Vol 4, Intercept Ltd, Ponteland, Newcastle upon Tyne, UK, pp 355–404

Bassi R, Rigoni F and Giacometti GM (1990) Chlorophyll binding proteins with antenna function in higher plants and green algae. *Photochem Photobiol* **52**: 1187–1206

Beevers L (1976) Senescence. In: (J Bonner and JE Varner, eds.) *Plant Biochemistry,* Academic Press, New York, pp 771–794

Ben-David H, Nelson N and Gepstein S (1983) Differential changes in the amount of protein complexes in the chloroplast membrane during senescence of oat and bean leaves. *Plant Physiol* **73**: 507–510

Biswal UC and Biswal B (1988) Ultrastructural modifications and biochemical changes during senescence of chloroplasts. *Int Rev Cytol* **113**: 271–321

Biswal UC and Mohanty P (1976a) Aging induced changes in photosynthetic electron transport of detached barley leaves. *Plant and Cell Physiol* **17**: 323–331

Biswal UC and Mohanty P (1976b) Dark stress induced senescence of detached barley leaves: Alteration in the absorption characteristic and photochemical activity of the chloroplasts isolated from senescing leaves. *Plant Sci Lett* **7**: 371–379

Biswal UC and Mohanty P (1978) Changes in the ability of photophosphorylation and activities of surface-bound adenosine triphosphatase and ribulose diphosphate carboxylase of chloroplasts isolated from the barley leaves senescing in darkness. *Physiol Plant* **44**: 127–133

Biswal UC, Singhal GS and Mohanty P (1979) Dark stress induced senescence of barley leaves: Changes in chlorophyll a fluorescence of isolated chloroplasts. *Indian J Exp Biol* **17**: 262–264

Bricker TM and Newman DW (1982) Changes in the chlorophyll-proteins and electron transport activities of soyabean (*Glycine max* L. cv. Wayne) cotyledon chloroplasts during senescence. *Photosynthetica* **16**: 239–244

Brody SS (1983) Spectral changes in chloroplasts during aging. *Photochem Photobiol* **37**: 585–586

Burke JJ, Kalt-Torres W, Swafford JR, Burton JW and Wilson RF (1984) Studies on genetic male-sterile soybeans. III The initiation of monocarpic senescence. *Plant Physiol* **75**: 1058–1063

Camp PJ, Huber SC, Burke JJ and Moreland DE (1982) Biochemical changes that occur during senescence of wheat leaves. I Basis for reduction of photosynthesis. *Plant Physiol* **70**: 1641–1646

Chitinis PR and Thornber JP (1988) The major light-harvesting complex of photosystem II: Aspects of its molecular and cell biology. In: (Govindjee, HJ Bohnert, DA Bryant, JE Mullet, WL Ogren, H Pakrasi and CR Somerville, eds) *Molecular Biology of Photosynthesis,* Kluwer Acad Publ Dordrecht, The Netherlands, pp 259–281

Choe HT and Thimann KV (1977) The retention of photosynthetic activity by senescing chloroplasts of oat leaves. *Planta* **135**: 101–107

Choudhury NK and Biswal UC (1979) Changes in photoelectron transport of chloroplast isolated from dark stressed leaves of maize seedlings. *Experientia* **35**: 1036–1037

Cuello J, Quiles MJ and Sabater B (1984) Role of protein synthesis in the regulation of senescence in detached barley leaves. *Physiol Plant* **57**: 260–266

Dalling MJ, Tang AB and Huffaker RC (1983) Evidence for the existence of peptide hydrolase activity associated with chloroplasts isolated from barley mesophyll protoplasts. *Z. Pflanzenphysiol* **111**: 311–318

Davies TGE, Thomas H and Rogers LJ (1988) Catabolism of cytochrome f in a senescence mutant of *Festuca pratensis. Biochem Soc Trans* **16**: 1054

Demmig-Adams B, Winter K, Kruger A and Czygan FC (1989) Light stress and photoprotection related to carotenoid zeaxanthin in higher plants. In: (WR Briggs, ed) *Photosynthesis,* Alan R Liss, Inc, New York, pp 375–391

Dhindsa RS, Dhindsa PP and Thorpe TA (1981) Leaf senescence: Correlated with increased levels of membrane permeability and lipid peroxidation, and decreased levels of superoxide dismutase and catalase. *J Exp Bot* **32**: 93–101

Dodge JD (1970) Changes in chloroplast fine structure during the autumnal senescence of *Betula* leaves. *Ann Bot* **34**: 817–824

Duggelin T, Bortlik K, Gut H, Matile P and Thomas H (1988) Leaf senescence in *Festuca pratensis:* Accumulation of lipofuscin-like compounds. *Physiol Plant* **74**: 131–136

DuPont J and Siegenthaler PA (1986) A parallel study of pigment bleaching and cytochrome breakdown during aging of thylakoid membranes. *Plant Cell Physiol* **27**: 437–484

Duysens LNM (1989) The discovery of the two photosynthetic systems: A personal account. *Photosynth Res* **21**: 61–79

Evans JR and Seemann JR (1989) The allocation of protein nitrogen in the photosynthetic apparatus: Costs, consequences, and control. In: (WR Briggs, ed.) *Photosynthesis*, Alan R Liss, Inc, New York, pp 183–205

Fong F and Heath RL (1977) Age dependent changes in phospholipids and galactolipids in primary bean (*Phaseolus vulgaris*) leaves. *Phytochemistry* **16**: 215–217

Ford DM and Shibles R (1988) Photosynthesis and other traits in relation to chloroplast number during soyabean senescence. *Plant Physiol* **86**: 108–111

Fragata M (1975) Effects of aging on chlorophyll fluorescence and photosystem II electron transport in isolated chloroplasts. *Can J Bot* **53**: 2842–2845

Frith GJT and Dalling MJ (1980) The role of peptide hydrolases in leaf senescence. In (KV Thimann, ed) *Leaf Senescence*, CRC Press, Florida, pp 117–130

Garcia S, Martin M and Sabater B (1983) Protein synthesis by chloroplasts during the senescence of barley leaves. *Physiol Plant* **57**: 260–266

Gepstein S (1988) Photosynthesis. In: (LD Nooden and AC Leopold, eds) *Senescence and Aging in plants*, Academic Press, New York, pp 85–104

Gnanam A, Subbaiah CC and Mannan M (1988) Protein synthesis by isolated chloroplasts. In: (Govindjee, HJ Bohnert, W Bottomley, DA Bryant, JE Mullet, WL Ogren, H Pakrasi, CR Somerville, eds) *Molecular Biology of Photosynthesis*, Kluwer Academic Publishers, The Netherlands, pp 777–800

Govindjee, Amesz J and Fork DC (1986) (eds) *Light Emission by Plants and Bacteria*, Academic Press, Orlando, Florida

Govindjee and Bazzaz M (1967) On the Emerson enhancement effect in ferricyanide Hill reaction in chloroplast fragments. *Photochem Photobiol* **6**: 885–894

Govindjee and Whitemarsh J (1982) Introduction to photosynthesis: Energy conversion by plants and bacteria. In: (Govindjee, ed) *Photosynthesis*, Vol. 1. *Energy Conversion by Plants and Bacteria*, Academic Press, pp 1–16

Grover A, Koundal KR and Sinha SK (1985) Senescence of attached leaves: Regulation by developing pods. *Physiol Plant* **63**: 87–92

Grover A, Sabat SC and Mohanty P (1986a) Effect of temperature on photosynthetic activities of senescing detached wheat leaves. *Plant Cell Physiol* **27**: 117–126

Grover A, Sabat SC and Mohanty P (1986b) Relative sensitivity of various spectral forms of photosynthetic pigments to leaf senescence in wheat (*Triticum aestivum* L). *Photosynth Res* **10**: 223–229

Grover A, Sabat SC and Mohanty P (1987) Does the loss of leaf chlorophyll during senescence arise due to loss in chloroplast number or chlorophyll content. *Biochem Physiol Pflanzen* **192**: 481–484

Grover A and Sinha SK (1985) Senescence of detached leaves in pigeon pea and chick pea: Regulation by developing pods: *Physiol Plant* **65**: 503–507

Grover A and Sinha SK (1988) Reproductive sink induced senescence of leaves in crop plants. In: (R Singh and SK Sawhney, eds) *Recent Advances in Frontier Areas of Plant Biochemistry*, Prentice Hall of India, India, pp 59–81

Gruszecki WI, Veeranjaneyulu K, Zelent B and Leblanc RM (1991) Energy transfer process during senescence: Fluorescence and photoacoustic studies of intact pea leaves. *Biochim Biophys Acta* **1056**: 173–180

Guera A, Martin M and Sabater B (1989) Subchloroplast localization of polypeptides synthesized by chloroplasts during senescence. *Physiol Plant* **75**: 382–388

Haehnel W (1984) Photosynthetic electron transport in higher plants. *Annu Rev Plant Physiol* **35**: 659–693

Harding SA, Guikema JA and Paulsen GM (1990) Photosynthetic decline from high temperature stress during maturation of wheat. I. Interaction with senescence processes. *Plant Physiol* **92**: 648–653

Hansson O and Wydrzynski T (1990) Current perception of photosystem II. *Photosynth Res* **23**: 131–162

Harnischfeger G (1973) Chloroplast degradation in ageing cotyledons of pumpkins. *J Exp Bot* **24**: 1236–1246

Harnischfeger G (1974) Studies on chloroplast degradation *in vivo*. III Effect of aging on Hill activity of plastids from *Cucurbita* cotyledons. *Z Pflanzenphysiol* **71**: 308–312

Harwood JL, Jones AVHM and Thomas H (1982) Leaf senescence in non-yellowing mutant of *Festuca pratensis*. III Total acyl lipids of leaf tissue during senescence. *Planta* **156**: 152–157

Hashimoto H, Kura-Hotta M, Katoh S (1989) Changes in protein content and in the structure and number of chloroplasts during leaf senescence in rice seedlings. *Plant Cell Physiol* **30**: 707–715

Hernandez-Gil R and Schaedle M (1973) Functional and structural changes in senescing *Populus deltoides* (Bartr.) chloroplasts. *Plant Physiol* **51**: 245–249

Hetherington SE and Oquist G (1988) A comparison of chlorophyll fluorescence measurements, post-chilling growth and visible symptoms of injury in *Zea mays*. *Physiol Plant* **72**: 241–247

Hilditch P (1986) Immunological quantification of the chlorophyll a/b binding protein in senescing leaves of *Festuca pratensis* Huds. *Plant Sci* **45**: 95–99

Hilditch PI, Thomas H, Thomas BJ, and Rogers LJ (1989) Leaf senescence in a non-yellowing mutant of *Festuca pratensis*: Proteins of photosystem II. *Planta* **177**: 265–272

Hiratsuka J, Shimada H, Whittier R, Ishibushi T, Sakamoto M, Mori M, Kondo C, Honji Y, Sun CR, Meng BY, Li YQ, Kanno A, Nishizawa Y, Hirai A, Shinozaki K and Sugiura M (1989) The complete sequence of the rice (*Oryza sativa*) chloroplast genome: Intermolecular recombination between distinct tRNA genes accounts for a major plastid DNA inversion during the evolution of cereals. *Mol Gen Genet* **217**: 185–194

Holloway PJ, Maclean DJ and Scott KJ (1983) Rate limiting steps of electron transport in chloroplasts during ontogeny and senescence of barley. *Plant Physiol* **72**: 795–801

Hoshina S, Kazi T and Nishida K (1975) Photoswelling and light-inactivation of isolated chloroplast. I Change in lipid content in light-aged chloroplasts. *Plant Cell Physiol* **16**: 465–474

Hoshina S, Mohanty P and Fork DC (1983) Temperature dependent changes in absorption and fluorescence properties of the cyanobacterium *Anacystis nidulans*. *Photosynth Res* **5**: 347–360

Hsu B-D, Lee J-Y and Pan R-L (1987) The high-affinity binding site for manganese on the oxidizing side of photosystem II. *Biochim Biophys Acta* **890**: 89–96

Huffaker RC and Miller BL (1978) Reutilization of ribulose bisphosphate carboxylase. In: (HW Siegelman and G Hind, eds) *Brookhaven Symposium on Photosynthetic Carbon Assimilation*, Plenum Publishing Corporation, New York, pp 139–152

Izawa S (1980) Acceptors and donors for chloroplast electron transport. *Meth Enzymol* **69**: 413–433

Jayabaskaran C, Kuntz M, Guillemant P and Weil J-H (1990) Variations in the levels of chloroplast tRNAs and aminoacyl tRNA synthetases in senescing leaves of *Phaseolus vulgaris*. *Plant Physiol* **92**: 136–140

Jenkins GI, Baker NR, Bradbury M and Woolhouse H (1981a) Photosynthetic electron transport during senescence of the primary leaves of *Phaseolus vulgaris* L. III Kinetics of chlorophyll fluorescence emission from intact leaves. *J Exp Bot* **31**: 999–1008

Jenkins GI, Baker NR and Woolhouse H (1981b) Changes in chlorophyll content and organization during senescence of the primary leaves of *Phaseolus vulgaris* L. in relation to photosynthetic electron transport. *J Exp Bot* **32**: 1009–1020

Jenkins GI and Woolhouse HW (1981a) Photosynthetic electron transport during senescence of the primary leaves of *Phaseolus vulgaris* L. I Non-cyclic transport. *J Exp Bot* **32**: 467–478

Jenkins GI and Woolhouse HW (1981b) Photosynthetic electron transport during senescence of the primary leaves of *Phaseolus vulgaris* L. II The activity of photosystems one and two, and a note on the site of reduction of ferricyanide. *J Exp Bot* **32**: 989–997

Joyard J, Block MA, Alban C and Douce R (1989) The plastid envelope membranes and their contribution to plastid membrane biogenesis. In: (WR Briggs, ed) *Photosynthesis*, Alan R Liss, Inc, New York, pp 331–345

Kar M and Mishra D (1976) Catalase, peroxidase and polyphenoloxidase activities during rice leaf senescence. *Plant Physiol* **57**: 315–319

Kasemir H, Rosemann D and Oelmuller R (1988) Changes in ribulose-1, 5-bisphosphate carboxylase and its translatable small subunit mRNA levels during senescence of mustard (*Sinapis alba*) cotyledons. *Physiol Plant* **73**: 257–264

Kato M and Shimizu S (1985) Chlorophyll metabolism in higher plants. VI Involvement of peroxidase in chlorophyll degradation. *Plant Cell Physiol* **26**: 1291–1301

252

Kawakami N and Watanbe A (1988a) Change in gene expression in radish cotyledons during dark-induced senescence. *Plant Cell Physiol* **29**: 33–42

Kawakami N and Watanbe A (1988b) Effects of light illumination on the population of translatable mRNA in radish cotyledons during dark-induced senescence. *Plant Cell Physiol* **29**: 347–353

Klein BP, Grossman S, King D, Cohen BS and Pinsky A (1984) Pigment bleaching, carbonyl production and antioxidant effects during the anaerobic lipooxygenase reaction. *Biochim Biophys Acta* **793**: 72–79

Kochritz A, Heike B and Hoffman P (1984) Linolenic acid-induced inhibition of photosynthetic electron transport in chloroplasts isolated from wheat seedlings (*Triticum aestivum L.*) of different age. *Biochem Physiol Pflanzen* **179**: 219–225

Kura-Hotta M, Satoh K and Katoh S (1987) Relationship between photosynthesis and chlorophyll content during leaf senescence of rice seedlings. *Plant Cell Physiol* **28**: 1321–1329

Kuroki M, Shioi Y and Sasa T (1981) Purification and properties of soluble chlorophyllase from the leaf sprouts. *Plant Cell Physiol* **22**: 717–725

Lagoutte B and Mathis P (1989) The photosystem I reaction centre: Structure and photochemistry. *Photochem Photobiol* **49**: 833–844

Lamppa GK, Elliot LV and Bendich AJ (1980) Changes in chloroplast number during pea leaf development. An analysis of a protoplast population. *Planta* **148**: 437–443

Lerbs S, Lerbs W, Klyanchko Ml, Romanko EG, Kulaeva ON, Wollgiehn R and Parthier B (1984) Gene expression in cytokinin-and light-mediated plastogenesis of *Cucurbita* cotyledons: ribulose-1, 5-bisphosphate carboxylase oxygenase. *Planta* **162**: 289–298

Lindoo SJ and Nooden LD (1977) Studies on the behaviour of the senescence signal in Anoka soyabeans. *Plant Physiol* **59**: 1136–1140

Liu XU and Jagendorf AT (1986) Neutral peptidases in the stroma of pea chloroplasts. *Plant Physiol* **81**: 603–608

Llellwyn CA, Fauzi R, Mantoura C and Brereton RG (1990a) Products of chlorophyll photodegradation-1. Detection and separation. *Photochem Photobiol* **52**: 1037–1041

Llellwyn CA, Fauzi R, Mantoura C and Brereton RG (1990b) Products of chlorophyll photodegradation. 2. Structural identification. *Photochem Photobiol* **52**: 1043–1047

Mae T, Kai N, Makino A and Ohira K (1984) Relation between ribulose bisphosphate carboxylase content and chloroplast number in naturally senescing primary leaves of wheat. *Plant Cell Physiol* **25**: 333–336

Makovetzki S and Goldschmidt EE (1976) A requirement for cytoplasmic protein synthesis during chloroplast senescence in the aquatic plant *Anacharis canadensis*. *Plant Cell Physiol* **17**: 859–862

Malik NSA (1987) Senescence in oat leaves: Changes in translatable mRNAs. *Physiol Plant* **70**: 438–446

Martin M and Sabater B (1989) Translational control of chloroplast protein synthesis during senescence of barley leaves. *Physiol Plant* **75**: 374–381

Martinoia E, Heck U, Dalling MJ and Matile Ph (1983) Changes in chloroplast number and chloroplast constituents in senescing barley leaves. *Biochem Physiol Pflanzen* **178**: 147–155

Matile P, Duggelin T, Schellenberg M, Rentsch D, Bortlik K, Peisker C and Thomas H (1989) How and why is chlorophyll broken down in senescent leaves. *Plant Physiol Biochem.* **27**: 595–604

Maunders MJ, Brown SB and Woolhouse HW (1983) The appearance of chlorophyll derivatives in senescing tissue. *Phytochem* **22**: 2443–2446

McKersie BD and Thompson JE (1978) Phase behaviour of chloroplast and microsomal membranes during leaf senescence. *Plant Physiol* **61**: 639–643

McRae DG, Chambers JA and Thompson JE (1985) Senescence related changes in photosynthetic electron transport are not due to alterations in thylakoid fluidity. *Biochim Biophys Acta* **810**: 200–208

Miller BL and Huffaker RC (1982) Hydrolysis of ribulose-1, 5-bisphosphate carboxylase by endoproteinases from senescing barley leaves. *Plant Physiol* **69**: 58–62

Miller BL and Huffaker RC (1985) Differential induction of endoproteinases during senescence of attached and detached barley leaves. *Plant Physiol* **78**: 442–446

Mishra AN and Biswal UC (1982) Differential changes in the electron transport properties of thylakoid membranes during aging of attached and detached leaves and of isolated chloroplasts. *Plant Cell Environ* **5**: 27–30

Mohanty P, Sabat SC and Grover A (1986) Age induced alterations in electron transport chain in wheat (*Triticum asstivum L.*) leaf chloroplasts. In: (C Rajamanickam, ed) *Current Trends in Life Sciences, XIII. Biomembranes: Structure, Biogenesis and Transport*, Today and Tomorrow Printers and Publishers, New Delhi, pp 23–26

Monroy AF, McCarthy SA and Schwartzbach SD (1987) Evidence for translational regulation of chloroplast and mitochondrial biogenesis in *Euglena. Plant Sci* **51**: 61–76

Mullet JE (1988) Chloroplast development and gene expression. *Annu Rev Plant Physiol* **39**: 475–502

Murata N, Higashi S-I and Fujimura Y (1990) Glycerolipids in various preparations of photosystem II from spinach chloroplasts. *Biochim Biophys Acta* **1019**: 261–268

Nooden LD and Guiamet JJ (1989) Regulation of assimilation and senescence by the fruit in monocarpic plants. *Physiol Plant* **77**: 267–274

Nooden LD and Nooden SM (1985) Effects of morphactin and other auxin transport inhibitors on soyabean senescence and pod development. *Plant Physiol* **78**: 263–266

Novitskaya GV, Rutskaya A and Malotkovsky YG (1977) Changes in lipid composition and the activity of bean chloroplast membranes with aging. *Fiziol Rast* **24**: 35–43

Oelmuller R (1990) Photoxidative destruction of chloroplasts and its effect on nuclear gene expression and extraplastidic enzyme levels. *Photochem Photobiol* **49**: 229–239

O'Farrell PH (1975) High resolution two-dimensional electrophoresis of proteins. *J Biol Chem* **250**: 4007–4021

Ohyama K, Fukuzawa H, Kohchi T, Shirai H, Sano T, Sano S, Umesono K, Shiki Y, Takeuchi M, Chang Z, Aota S, Inokuchi H and Ozeki H (1986) Chloroplast gene organisation deduced from complete sequence of liverwort *Marchantia polymorpha* chloroplast DNA. *Nature* **322**: 572–574

Packham NK and Barber J (1987) Structural and functional comparison of anoxygenic and oxygenic organisms. In: (J Barber, ed) *Topics in Photosynthesis*, Vol 8. *The Light Reactions*, Elsevier Science Publishers BV, The Netherlands, pp 1–30

Paharia S (1986) Response of *Vigna radiata* Wilczek to water stress and sink size. PhD Thesis, Post Graduate School, Indian Agricultural Research Institute, New Delhi

Peoples MB, Beilharz VC, Waters SP, Simpson RJ and Dalling MJ (1980) Nitrogen redistribution during grain growth II. Chloroplast senescence and the degradation of ribulose-1, 5-bisphosphate carboxylase. *Planta* **149**: 241–251

Peoples MB and Dalling MJ (1988) The interplay between proteolysis and amino acid metabolism during senescence and nitrogen reallocation. In: (LD Nooden and AC Leopold, eds) *Senescence and Aging in Plants*, Academic Press, New York, pp 181–217

Peter GF and Thornber JP (1988) The antenna components of photosystem II with emphasis on the major pigment-protein, LHC IIb. In: (H Scheer and S Schneider, eds) *Photosynthetic Light-harvesting Systems—Organization and Function*, Walter de Gruyter and Co, Berlin, pp 175–186

Powles SB (1984) Photoinhibition of photosynthesis induced by visible light. *Annu Rev Plant Physiol* **35**: 15–44

Reddy PV (1986) Combined effect of water stress and reproductive sink on physiological and biochemical changes in cowpea. PhD Thesis, Post Graduate School, Indian Agricultural Research Institute, New Delhi, India

Reilly P and Nelson N (1988) Photosystem I complex. In: (Govindjee, HJ Bohnert, W Bottomley, DA Bryant, JE Mullet, WL Ogren, H Pakrasi and CR Somerville, eds) *Molecular Biology of Photosynthesis*, Kluwer Acad Publ, Dordrecht, The Netherlands, pp 485–496

Renger G and Govindjee (1985) The mechanism of photosynthetic water oxidation. *Photosynth Res* **6**: 33–55

Roberts DR, Thompson JE, Dumbroff EB, Gepstein S and Mattoo AK (1987) Differential changes in the synthesis and steady state levels of thylakoid proteins during bean leaf senescence. *Plant Mol Biol* **9**: 343–353

Ronning CM, Bouwkamp JC and Solomas T (1991) Observations on the senescence of a mutant non-yellowing genotype of *Phaseolus vulgaris* L. *J Exp Bot* **42**: 235–241

Sabat SC, Grover A and Mohanty P (1985) Alterations in the characteristics of photosystem II and photosystem I catalyzed electron transports in chloroplasts isolated from senescing detached beet-spinach (*Beta vulgaris*) leaves. *Indian J Exp Biol* **23**: 711–719

Sabat SC, Grover A and Mohanty P (1989a) Senescence induced alterations in the electron transport chain in wheat (*Triticum aestivum* L.) leaf chloroplasts. *J. Photochem. Photobiol.* **3**: 175–183

Sabat SC, Grover A and Mohanty P (1989b) Selective alterations in photosynthetic pigment characteristics and photoelectron transport during senescence of wheat leaves. In: (GS Singhal, J Barber, RA Dilley, Govindjee, R Haselkorn and P Mohanty, eds) *Photosynthesis—Molecular Biology and Bioenergetics*. Narosa Publishing House, pp 343–351

254

Sane PV (1977) The topography of the thylakoid membrane of the chloroplast. In: (A Trebst and M Avron, eds) *Encyclopaedia of Plant Physiology*, Vol 5, Springer-Verlag, Berlin, pp 522–542

Scherings G and Kuiper PJC (1974) Effect of age on adenosine triphosphatase activity from tobacco chloroplasts. *Physiol Plant* **32**: 182–184

Sestak Z (1972) Leaf age and the shape of the red absorption band of maize chloroplasts. *Photosynthetica* **6**: 75–79

Sestak Z (1977) Photosynthetic characteristics during ontogenesis of leaves. 2 Photosystems, components of electron transport chain and photophosphorylation. *Photosynthetica* **11**: 449–474

Sestak Z (1978) Photosynthetic characteristics during ontogenesis of leaves 3 Carotenoids. *Photosynthetica* **12**: 89–109

Sestak Z (1982) Leaf ontogeny and photosynthesis. In: (CB Johnson, ed) *Physiological Processes Limiting Plant Productivity*, Butterworth, pp 147–158

Shinozaki K, Ohme M, Tanaka M, Wakasugi T, Hayashida N, Matsubayashi T, Zaita N, Chunwongse J, Obokata J, Yamaguchi-Shinozaki K, Ohto C, Torozawa K, Meng BY, Sugita M, Deno H, Kamogashira T, Yamada K, Kusuda J, Takaiwa F, Kato A, Tohdoh N, Shimada H and Sugiura M (1986) The complete nucleotide sequence of the tobacco chloroplast genome. *EMBO J* **5**: 2043–2049

Siegenthaler PA and Rawyler A (1977) Aging of the photosynthetic apparatus. V Change in pH dependence of electron transport and relationships to endogenous free fatty acids. *Plant Sci Lett* **9**: 265–273

Siegenthaler PA and Rawyler A (1986) Acyl lipids in thylakoid membranes: Distribution and involvement in photosynthetic functions. In: (LA Staehelin and CJ Arntzen, eds) *Photosynthesis III. Photosynthetic Membranes and Light Harvesting Systems. Encyclopaedia of Plant* Physiology, Vol 19, Springer-Verlag, Berlin, pp 693–705

Simpson DJ (1990) The structure of photosystem I and II. In: (M Baltscheffsky, ed) *Current Research in Photosynthesis*, Vol II, Kluwer Acad Publ, Dordrecht, The Netherlands, pp 725–732

Srivastava SK, Vashi DJ and Naik BI (1983) Control of senescence by polyamines and guanidines in young and mature barley leaves. *Phytochem* **22**: 2151–2154

Stoddart JL and Thomas H (1982) Leaf senescence. In: (D Boulter and B Parthier, eds) *Encyclopaedia of Plant Physiology*, Vol 14 A, Springer-Verlag, Berlin, pp 592–636

Tevini M and Steinmuller D (1985) Composition and function of plastoglobuli. II Lipid composition of leaves and plastoglobuli during beech leaf senescence. *Planta* **163**: 91–96

Thimanh KV (1980) The senescence of leaves. In: (KV Thimann, ed) *Senescence in Plants*, CRC Press, Florida, pp 85–115

Thimann KV, Tetley RR and Thanh TV (1974) The metabolism of oat leaves during senescence II. Senescence in leaves attached to the plant. *Plant Physiol* **54**: 859–862

Thomas H (1977) Ultrastructure, polypeptide composition and photochemical activity of chloroplasts during foliar senescence of a non-yellowing mutant genotype of *Festuca pratensis* Huds. *Planta* **137**: 53–60

Thomas H (1982) Leaf senescence in a non-yellowing mutant of *Festuca pratensis*. I. Chloroplast membrane polypeptides. *Planta* **154**: 212–218

Thomas H and Stoddart JL (1975) Separation of chlorophyll degradation from other senescence processes in leaves of mutant genotype of a meadow fescue (*Festuca pratensis* L.). *Plant Physiol* **56**: 438–441

Thomas H and Stoddart JL (1980) Leaf senescence. *Annu Rev Plant Physiol* **31**: 83–111

Trebst A (1974) Energy conservation in photosynthetic electron transport of chloroplasts. *Annu Rev Plant Physiol* **25**: 423–458

Tyagi AK, Kelkar NY, Kapoor S and Maheshwari SC (1989) Genes of the photosynthetic apparatus of higher plants—Structure, expression and strategies for their engineering. In: (GS Singhal, J Barber, RA Dilley, Govindjee, P Haselkorn and P Mohanty, eds) *Photosynthesis—Molecular Biology and Bioenergetics*, Narosa Publishing House, New Delhi, pp 3–20

Veierskov B and Thimann KV (1988) The control of protein breakdown and synthesis in the senescence of oat leaves. *Physiol Plant* **72**: 257–264

Venkatarayappa T, Fletcher RA and Thompson JE (1984) Retardation and reversal of senescence in bean leaves by benzyladenine and decapitation. *Plant Cell Physiol* **25**: 407–418

Wardley TM, Bhalla PL and Dalling MJ (1984) Changes in the number and composition of chloroplasts during senescence of mesophyll cells of attached and detached primary leaves of wheat (*Triticum aestivum* L.). *Plant Physiol* **75**: 421–424

Watanbe A and Imaseki H (1982) Changes in translatable mRNA in senescing wheat leaves. *Plant Cell Physiol* **23**: 489–497

Waters SP, Noble ER and Dalling MJ (1982) Intracellular localization of peptide hydrolases in wheat (*Triticum aestivum* L.) leaves. *Plant Physiol* **69**: 575–579

Westhoff P, Grune H, Schrubar, H, Oswald A, Streubel M, Ljungberg U and Herrman RG (1988) Mechanisms of plastid and nuclear gene expression during thylakoid membrane biogenesis in higher plants. In: (H Scheer and S Scheider, eds) *Photosynthetic Light-harvesting Systems—Organization and Function*, Walter de Gruyter and Co Berlin, pp 261–276

Willey DL and Gray JC (1988) Synthesis and assembly of the cytochrome b-f complex in higher plants. In: (Govindjee, HJ Bohnert, W Bottomley, DA Bryant, JE Mullet, WL Ogren, H Pakrasi and CR Somerville, eds.) *Molecular Biology of Photosynthesis*, Kluwer Academic Publishers, The Netherlands, pp 497–516

Winston GW (1990) Physiochemical basis for free radical formation in cells: Production and defenses. In: (RG Alscher and JR Cumming, eds) *Stress Responses in Plants: Adaptation and Acclimation Mechanisms*, Wiley-Liss, A John Wiley and Sons Inc, pp 57–86

Wittenbach VA, Lin W and Hebert RB (1982) Vacuolar localization of proteases and degradation of chloroplasts in mesophyll protoplasts from senescing primary wheat leaves. *Plant Physiol* **69**: 98–102

Woolhouse HW (1982) Leaf senescence. In: (H Smith and D Grierson, eds) *Molecular Biology of Plant Development*, Vol 18, Botanical Monograph, Blackwell Scientific Publications Oxford, pp 256–281

Woolhouse HW (1984) The biochemistry and regulation of senescence in chloroplasts. *Can J Bot* **62**: 2934–2942

Yoshida Y (1961) Nuclear control of chloroplast activity in *Elodea* leaf cells. *Protoplasma* **54**: 476–492

Young A and Britton G (1990) Carotenoids and stress. In: (RG Alscher and JR Cumming, eds) Stress Responses in Plants: Adaptation and Acclimation Mechanisms, Wiley-Liss, A John Wiley and Sons, Inc, pp 87–112

10

Maximizing Light Interception: The Role of Chloroplast Disposition in Cells and Thylakoid Membrane Architecture

Christa Critchley and A. Wendy Russell

Department of Botany,
The University of Queensland
QLD 4072, Australia

CONTENTS

ABBREVIATIONS

PPFD	: photosynthetic photon flux density;
PS II	: photosystem II;
PS I	: photosystem I;
LHC II	: light harvesting chlorophyll a/b binding protein complex of PS II;
PQ	: plastoquinone;
Q_B	: secondary electron acceptor of PS II;
PC	: plastocyanin;
cyt b_6 /f	: cytochrome b_6 /f complex.

258

ABSTRACT

We discuss the various strategies adopted by higher plants to maximize light interception. It may be possible to increase productivity through better use of incident light by the leaves. Interception of light can be influenced in different ways at four levels of organization in the plant. The first level of organization is that of the canopy, where plant habit and canopy structure in terms of leaf orientation towards the incoming light, space occupied by leaves, and shading are most important. The second level is concerned with leaf structure and anatomy, tissue layering, and chloroplast disposition in cell layers. Demonstrated light gradients in terms of light quantity as well as quality play a very important role here and suggest that self-shading occurs at this level also. In individual cells, chloroplasts may or may not be different with regard to light interception and self-shading. Good data are not yet available, but circumstantial evidence from some studies points to the possibility of self-shading by chloroplasts in individual cells. The fourth level of adaptation to differing light quantity and quality is at the thylakoid membrane level. Virtually nothing is known about their vertical heterogeneity. We suggest that there is a distinct possibility for differences in surface distribution of photosystems between successive layers of thylakoids in a grana stack. It will require more sophisticated methods of analysis and experimental approaches to prove the occurrence of self-shading and its relevance to quantum efficiency of photosynthesis at this molecular level.

I. Introduction

"All biomass production depends on photosynthesis", reads the opening sentence of an article by Lloyd Evans in a book about photosynthesis and productivity (Evans 1975). Yet, plant productivity in terms of biomass production is *not* directly correlated with the rate of photosynthesis measured on the basis of leaf area, and neither is crop production in terms of harvested plant product yield. Not even the carboxylation-efficient C_4 plants perform better at the level of short-term crop growth rate than C_3 species (Gifford, 1974). The contention that it may be possible to increase crop yield by increasing photosynthetic capacity and rate can no longer be supported (see also Workshop Program and Abstracts. The Genetics and Physiology of Photosynthesis and Crop Yield. Cambridge, UK, 20–24 July 1987). Moreover, Evans (1975) considered dark respiration, photosynthate partitioning, and growth potential to be much more important determinants of crop (and generally plant) productivity than photosynthetic capacity per se. Improved fertilizer treatment, better pest and disease resistance, and increases in harvest index have provided the most significant enhancements in crop yield (Beadle *et al.*, 1985).

The relationship between total dry matter production and intercepted total solar radiation is somewhat more promising. Studies on agricultural systems such as maize and soybean crops show that production rates at least during the early season are a linear function of photosynthetic photon flux density (PPFD) interception (Loomis and Gerakis, 1975). The story becomes more complex, however, at later stages of crop growth where the foliage density increases (Loomis and Gerakis, 1975).

Light interception is dependent on the disposition of the photosynthetic tissue relative to the incoming radiation so that plant habit as well as leaf orientation and morphology may be more significant determinants of crop productivity than rate of photosynthesis on a leaf area basis. In this paper, we discuss the factors that may be important in terms of light interception by leaves at the cellular and subcellular level. Plant habit, leaf orientation, and canopy structure are significant ecophysiological factors. It is no less important to understand tissue-specific factors such as cellular disposition of chloroplasts, suborganellar architecture of the photosynthetic membranes, and their regulation to complete our understanding of light interception in photosynthetic energy conversion. The topics dealt with are: (1) distribution and disposition of chloroplasts in cells and tissues and their contribution to effective photon interception, (2) structural and functional properties of individual chloroplast, (3) three-dimensional structure of the thylakoid membrane network in higher plants (including many crop plants) and its relevance to photon absorption, excitation energy distribution, electron flow, and energy conversion, and (4) chloroplast and thylakoid membrane responses to a rapidly and constantly changing light environment, i.e., the nature of short-term dynamic changes in chloroplasts.

A distinction is made between light (full spectrum of solar irradiance impinging on the earth's surface) and PPFD (that part of the full spectrum that is available for absorption by the photosynthetic pigments, namely the wavelengths between 400 and 700 nm).

II. Chloroplasts in Cells and Tissues

Trees, and other plant species with large and extensive canopies, have layers of leaves, often quite easily recognizable by their different colouring, with what has become known as sun or shade properties. In a recent study on Australian mangroves it was shown that about 85 per cent of the PPFD incident at the top of the canopy was attenuated within the first metre and that several photosynthetic characteristics such as light saturation, maximal photosynthetic rate, and light compensation point changed in the leaves in accordance with this decrease in irradiance (Carter *et al.*, 1990). Canopies, whether formed by individuals or plant communities, are also characterized by the way in which the leaves occupy space to maximize PPFD interception. The location of leaves on stems and branches and their orientation with respect to the direction of the incoming irradiance will tend to maximize PPFD interception by the total available green surface area. Nevertheless there is apparently room for compromise, evidenced by the plant's investment in leaves that are shaded by others, or in leaves with an orientation perpendicular rather than horizontal to the direction of the incoming irradiance. The trade-off is between maximizing photon interception and maintaining effective ability to use those photons with the appropriate degree of flexibility. Flexibility is required because in most plants irradiance reaching any part of a leaf or a canopy changes constantly through changes in cloudiness and/or wind-induced leaf movements. Excellent examples of the kind of compromise for flexibility that most plants seem to be able to make are provided by physiological responses to sunflecks by rainforest and other species (Pearcy, 1990).

It is important to note that few of the plant species studied to date have shown themselves to be completely incapable of adaptation to changes in photon irradiance levels. It is in fact true to say that except for very few specialists at either extreme, i.e., bright midsummer sunlight (2000 μmol quanta^{-1} m^{-2} sec^{-1}) and deep green shade (5–10 μmol quanta^{-1} m^{-2} sec^{-1}), most unstressed plants can adapt to a large range in irradiance quantity. Nutritional factors (e.g. P, N) and water supply must be adequate. The concept of genetically determined shade plants, or shade ecotypes, was certainly questioned by the studies on *Solanum dulcamara* by Ferrar and Osmond (1986) and on *Alocasia macrorrhiza* by Chow *et al.* (1988b). These studies and others (Anderson and Osmond, 1987) showed that some species, considered to be shade ecotypes, are nothing more than N-starved plants unable to build or maintain a photosynthetic apparatus suited for full sunlight irradiances.

At the whole plant level, chloroplasts experience diurnal, seasonal, and geographical variations in irradiance. We also tend to overlook the fact that up until the last few centuries, most of the planet was covered in forest and that it was in such an environment that the majority of higher plants evolved. From the top of the canopy to the forest floor we find extremes of light, in terms of both quantity and quality (plant-shaded areas are enriched in longer wavelength red and green light); these extremes are susceptible to long-term or very short-term changes (Pearcy *et al.*, 1985). At the level of the leaf there exist less obvious but demonstrated light gradients (Terashima and Saeki, 1983, 1985; Vogelmann *et al.*, 1989). We can theoretically, if not yet experimentally, extend these observations to the chloroplast and even the grana level (S.C. Brown, unpublished, has calculated that red light is not likely to penetrate a grana stack further than five layers), and begin to appreciate that many levels of heterogeneity of structure and function have arisen, providing the flexibility to respond to rapid changes.

In a series of pioneering studies in the mid-1980s, Terashima and co-workers showed that in broad-leaf species such as *Camellia japonica* L., *Spinacia oleracea* L., and *Glycine max* L., chloroplasts in different tissues of the leaf had distinctly different photosynthetic and morphological characteristics. These characteristics and the differences between them are similar to those found in sun and shade plants in general. The cells closest to the upper epidermis, the palisade parenchyma, contain sun-type chloroplasts and those further towards the lower epidermis, in the spongy mesophyll, shade-type chloroplasts (Terashima and Inoue, 1984, 1985a, b). Whatley (1979) had earlier shown the same phenomenon to occur in her study of plastid development in different tissues of *Phaseolus vulgaris* L.

More recently, Long and co-workers (1989) and Postl and Bolhàr-Nordenkampf (1990) presented evidence that the same is true even for a narrow-leaf and monocotyledonous species such as *Zea mays* L. These authors showed that characteristically different light response curves for leaves were obtained when adaxial or abaxial or both sides of the leaf were illuminated and their gas exchange rates measured (see Chapter by Long, this volume). Leverenz (1988) suggested that mutual shading of chloroplasts in leaves not only decreased the amount of absorbed light, but also altered the photosynthetic light response curve. He showed that during actual measurements of photosynthetic light curves significant alterations took place in the experimental material. These changes were mainly due to irradiance since temperature, CO_2 concentration, or species had no discernible effect on the nature of the light response curve. Leverenz's experiments also demonstrated that, at least in the species he examined, the minimum time for a reacclimation to a new irradiance was 24 hours (Leverenz, 1988). Chloroplast movements in cells and/or alterations in chloroplast ultrastructure could be responsible for these changes. It would be very interesting to make gas exchange measurements similar to those done by Postl and Bolhàr-Nordenkampf (1990) with maize on either side of the leaves of *Alocasia macrorrhiza*, which shows significant chloroplast movement when exposed to high PPFD (Chow *et al.*, 1988b). Terashima and co-workers (1986) have indeed shown that with appropriate changes in illumination conditions, morphological and biochemical characteristics of chloroplasts could be altered over a period of several days. It is not known, however, whether chloroplast movement in response to light played a role in the species they used. Terashima and Saeki (1983, 1985) and Vogelmann and co-workers (1989) showed that steep gradients in light environment exist within a leaf. Evidence for differential light acclimation in leaf tissues was also seen earlier in fluorescence studies by Powles and Björkman (1982).

It seems, therefore, that layers of chloroplasts within a leaf adjust their photosynthetic properties in a manner that is consistent with the availability of PPFD and in an analogous fashion to whole plants that live in a canopy. Leverenz (1988), in fact, suggests that canopies should be considered as "a collection of independently functioning photosynthetic cells rather than as a collection of leaves."

III. Properties of Individual Chloroplasts

Given the substantial flexibility and adaptability that can be observed in tissue and whole leaf photosynthetic systems to varying light, questions must be asked about the mechanism that underlies this flexibility. For example, do chloroplasts within a single cell shade each other? Terashima and co-workers (1986) suggested that individual cells of certain tissues contain chloroplasts with very similar properties. While this may have been the case for those properties that were measured, any more subtle differences may not have been expressed in those parameters or may not have been measurable at all. Populations of different chloroplasts from individual cells or cell layers may be distinguishable by flow cytometric methods (Ashcroft *et al.*, 1986, Xu *et al.*, 1990), but these methods have not yet been sufficiently exploited. In macroalgae, it has been suggested that number and arrangement of chloroplasts in fronds can lead to a degree of self-shading that is effective in preventing photoinhibition in these algae (Sealey *et al.*, 1990). Furthermore, the resolution with which it is possible to produce starch pictures on leaves (Walker, 1976; Edwards and Walker, 1983) also seems to indicate heterogeneous light environments within a cell. On the other hand it is clearly evident in some species that a response to light by the chloroplasts leads to specific positioning of the chloroplasts within the cells. For example, in *Alocasia* irradiance with high light leads to the chloroplasts being lined up along the anticlinal walls of both palisade and mesophyll cells, whereas under low light conditions the chloroplasts accumulate along the periclinal walls of the cells (Chow *et al.*, 1988b). In some species a significant response is the folding of leaves as in *Oxalis oregona* and *Macroptilium atropurpureum* (Ludlow and Björkman, 1984). Much more work is required on well-chosen species to advance our knowledge of the regulation of irradiance interception at the cellular level.

Do grana lamellae shade themselves? If the answer is yes, which is likely, why do we assume that layers of thylakoids in grana stacks are more or less vertically homogeneous? What would the surface distribution of complexes look like in the top thylakoid of a granum compared to the 15th or 35th into the stack? Do the top and third granal membranes in a high PPFD-adapted chloroplast have the same lateral distribution of electron transport complexes and photosystems?

While we believe that the recent proposal by Albertsson and co-workers (1990) for a concentric, circular domain organization of photosystems in the thylakoid membrane is consistent with evidence of heterogeneity in PS II and PS I and constitutes a significant advance, we favour the idea of arrays or domains of interlocking geometry in which all electron transport components, including different forms of PS II and PS I, are spaced such that photon absorption and electron transfer are maximal for a particular irradiance quantity and quality impinging on that lateral domain. Vertically through the stack more irradiance, should it become available, and irradiance of different quality can be absorbed. The concept of concentric and vertically homogeneous grana domains will be hard to reconcile with the existence or irregularly shaped stroma lamellae. In fact, if grana thylakoids are self-shading, there must be vertical heterogeneity in the composition and spatial distribution of electron transfer complexes and photosystems. Currently available techniques for fractionation of thylakoid membranes are not capable of probing and revealing such subtle differences, let alone achieving technically the separation vertically of appressed layers of grana thylakoids. Closer inspection of freeze-fracture pictures of thylakoids with a view to detecting heterogeneity rather than looking for homogeneity may provide some evidence for vertical heterogeneity.

Why do PS II and PS I have an uneven distribution of chlorophyll when quantum yields of NADP reduction and the operation of the Z scheme demand that it be even? The quantum yield measurements that are possible these days show near ideal values. They appear to be accurate to a fault when considering the presence of 25–40 per cent inactive PS II centres (Graan and Ort, 1986; Ort and Whitmarsh, 1990). Due to the smaller antenna size of the inactive PS II centres and possibly their comparative obscurity in the grana stacks, their effect on quantum yield would be expected to be relatively small, certainly less than 25–40 per cent. Do the active photosystems in fact *have* different amounts of chlorophyll associated with them? Calculations of the chlorophyll to reaction centre ratios assume that all chlorophyll is associated at any one time with one or the other of the two photosystems. Why is this so when one of the first chlorophyll protein complexes to be isolated and characterized was the light harvesting chlorophyll *a/b* binding protein complex of PS II (LHC II) and when it was shown that it can exist within the thylakoid membrane as a detached unit (Andersson and Anderson, 1985), perhaps a trimer (Kühbrandt, 1990)?

Experimental evidence coupled with the assumptions referred to above suggests that in sun leaves the chlorophyll distribution is 60 per cent in PS II compared with 40 per cent in PS I (Anderson and Osmond, 1987). In shade-adapted species, this imbalance is even more marked. Shade chloroplasts also have more extensive grana stacks and the degree of stacking may be related to the relative imbalance of chlorophyll between the two photosystems. It may also mean that within a large grana stack a sizeable proportion of LHC II is functionally unattached to any photosystem and that a large proportion of this unattached LHC II is found in grana stacks. If there is more LHC II than PS II reaction centre complexes then it may also be possible that this "extra" LHC II mainly functions in stacking and not in light harvesting. Moreover, if it is buried in relatively deeply shaded thylakoids in the large grana stacks of shade-adapted leaves it would not absorb a lot of light and may be associated with 25–40 per cent of the PS II centres that are non-functional (Graan and Ort, 1986). This would be particularly relevant for shade leaves although not for the reasons commonly quoted, i.e., an increase in light harvesting capacity. Instead it would serve to provide the flexibility required for the exploitation of sunflecks.

Following the discovery that the light reactions of photosynthesis involved two photoreactions, sensitized by different wavelengths of light (Emerson *et al.*, 1957; Eley and Myers, 1967), an apparently paradoxical phenomenon was observed. This was that the quantum yield was independent of wavelength within the range 580–685 nm and remained fairly constant in this range (Emerson and Lewis, 1943) despite the apparent wavelength-dependence of the two photoreactions. This suggested that there was a mechanism regulating the distribution of energy between the two photosystems (Myers, 1971; Butler, 1978).

This constancy of quantum yield has been shown to extend not only over a range of wavelengths but over a wide range of species and light environments. It appears that the photosynthetic system is so finely tuned and well-regulated that any non-stressed plant, operating in an environment to which it has adapted, will work at close to maximum efficiency under limiting light. The fact that three-dimensional arrangements of the thylakoid membrane play a role in the regulation of energy distribution is demonstrated by the variations in the extent of grana stacking in sun-adapted and shade-adapted species, both of which have high quantum efficiencies despite adaptation to light of very different quantity and quality (Lichtenthaler *et al.*, 1981; Glick *et al.*, 1985).

Light collected by pigment molecules is transferred as excitation energy via resonance to other pigments and finally to the reaction centre chlorophylls of the two photosystems (Sauer, 1975). It was suggested that PS II centres could communicate via energy transfer and that they could also transfer excitation energy to the lower energy centre of PS I (Joliot *et al.*, 1968). This PS II to PS I excitation energy transfer was termed "spillover" (Myers and Graham, 1963) and was based largely on interpretation of fluorescence data. Bonaventura and Myers (1969) later observed and measured a phenomenon based on fluorescence induction data which they termed the "state I to state II transition". This transition was made in plants adjusting from a period of darkness to steady-state photosynthesis, and in plants adapted to light preferred by PS I (light I) when illuminated by PS II (light II). There appeared to be a slow change (measured in minutes) leading to a change in the amount of excitation energy delivered to each photosystem so that under light I, more energy was delivered to PS II, but as the system adjusted to light I, this proportion decreased in favour of PS I. Although this was taken as evidence of a light-dependent mechanism for regulaiton of the energy distribution between the two photosystems (Bonaventura and Myers, 1969), it was not discussed in terms of dynamic responses of the thylakoid membrane to irradiance changes.

In the 1960s, Izawa and Good (1966a, b) demonstrated that cations were involved in membrane stacking, Murata (1969) reported that Mg^{2+} exerted some control over the distribution of energy between PS I and II, Homann (1969) recorded an increase in the fluorescence of chloroplasts following addition of cations, and Murakami and co-workers (1975) noticed an efflux of cations in response to proton uptake. Obviously, the cationic environment of the thylakoid membrane was under the control of light via photosynthetic electron transport, but none of these observations have so far been satisfactorily interpreted or explained.

Many other biophysical and biochemical phenomena associated with thylakoid structure and function have been recorded, but as yet we have no sensible integrated model. The very first experimental and conceptual advance was the discovery of "heterogeneity". Park and Sane (1971) first suggested the lateral differentiation of the thylakoid membrane based on studies of the distribution of particles between appressed (stacked) and non-appressed regions. They suggested that the stroma lamellae contain only PS I and that the grana contain both photosystems. Some support was provided for this suggestion when Staehelin (1976) showed that low-salt-induced unstacking lead to randomization of freeze-fracture particles along the membrane. These results contributed to the formulation of Barber's model, which suggested LHC II to be involved not only in state I to state II transitions, but also in stacking and unstacking of the membranes (Barber, 1980).

From the time of this early work until the present, fractionation techniques have been central to studies of differentiation between grana and stroma membranes. It is apparent that mild mechanical or detergent fragmentation of thylakoid membranes initially disrupts stroma lamellae which can then be separated from grana fractions via centrifugation. The properties of the fractions depend on the method of fractionation and separation. A very important adaptation of this technique was developed by Andersson and Akerlund (1978). It involves mechanical fragmentation in a solution of specific cation composition so that appression is maintained. Non-appressed and appressed membranes then form right-side-out and inside-out vesicles, respectively. For the first time it could be demonstrated that grana fractions were derived exclusively from appressed regions (Andersson et al., 1985).

Miller and Staehelin (1976) isolated the coupling factor from stroma-exposed membrane fractions and reported its exclusive location in the non-appressed regions of the chloroplast membrane. This had been suspected as steric constraints would seem to disallow this bulky group residence within the closely appressed grana. Jennings and co-workers (1979) established a similar location for NADP-reductase. Andersson and Anderson (1980) provided and collated extensive evidence for different locations of the two photosystems and published the now widely-accepted theory of lateral heterogeneity of chloroplast membranes. The theory states that the coupling factor, NADP-reductase, and PS I reside exclusively or preferentially in the stroma-exposed lamellae, the appressed regions being enriched in PS II and LHC II. The only complex that appears not to be exclusively located in either region is the cytochrome b_6/f complex, which may be evenly distributed in both regions or in some domain between them (Melis et al., 1986; Morrissey et al., 1986; Hinshaw and Miller, 1989).

This and other evidence brings into questions many of the assumptions about the interactions between PS II and PS I. Not only do the two photosystems appear to be located in different membrane regions, but extensive studies show that the stoichiometries of active centres operating at any one time may vary and deviate from the required 1:1 ratio (Melis and Harvey, 1981). Lee and Whitmarsh (1989), however, provide convincing evidence and arguments for "a relatively fixed photosynthetic apparatus that is able to perform efficiently under a very wide range of light intensities". Unfortunately they do not offer an explanation as to neither *how* that efficient performance is achieved nor how the very rapid turnover of one of the PS II reaction centre proteins, D1 can be accounted for. Infact it calls into question the concept of a relatively fixed photosynthetic apparatus. It appears that the two photosystems do not carry out the Z scheme of electron transport originally proposed by Hill and Bendall (1960) as a connected unit (Anderson, 1981) (for discussion on Z scheme see Chapter by Govindjee and Coleman, this volume). Alternatively, it is possible that the Z scheme does not operate stoichiometrically and that coordination of the two photosystems is achieved in some other way. The cytochrome b_6/f complexes as the intermediary electron transport units are crucial in this regard. In fact the location of the complexes, with regard to lateral distribution as well as association with the two photosystems, is not clear, to say the least. Evidence has been presented which indicates that a significant proportion of the cytochrome b_6/f complex is located in the grana margins (Anderson et al., 1990). Blackwell and Whitmarsh (1990) recently provided some experimental evidence that plastoquinone may not work as a long-range electron shuttle. We may, therefore, require a new model for electron transport in order to make progress with explanations of the more recent experimental evidence.

Lateral heterogeneity creates a dilemma in the whole understanding of electron transport. If the links of the electron transport chain are physically separated, how do they communicate? How do the electrons pass along it? If lateral heterogeneity is genuine, there must be in operation something other than the relatively straightforward linear electron transfer processes suggested by the Z scheme.

The acceptance of the concept of lateral differentiation of protein complexes also refocussed attention on spillover (Anderson, 1981). The spillover theory relied on the two photosystems in close proximity being supplied by common light harvesting chlorophyll-protein complexes or a pigment-protein complex which could somehow partition energy to either system (Myers and Graham, 1963). Lateral heterogeneity as currently understood effectively rules out spillover between PS I and PS II, yet there has been no adequate explanation for the data that first indicated it despite the arguments of Anderson and Andersson (1988).

Even before the lateral heterogeneity of PS I and PS II was established, Melis and Homann (1976) found heterogeneity in PS II centres. It appeared that two distinct populations of PS II existed, having different light absorption activities and different locations within the membrane. The α centres were located in the grana and the β centres resided with PS I in the stroma-exposed regions. It was also established that the physical distinction between the two types of PS II was the size of the LHC associated with them (Sundby et al., 1986) and a deficiency of the plastoquinone pool associated with these β centres was noted. This led to the suggestion that the centres were related in a two-step developmental process whereby β centres, lacking peripheral LHC II, were converted to active α centres by addition of LHC II and a functional plastoquinone pool. This was followed by grana formation induced by the interaction of LHC II complexes (Abadia et al., 1985). The interpretation was supported by Larsson and co-workers (1987), who showed that at least two populations of LHC II existed, consistent with an inner and a peripheral component. This inner component was described as integral of PS II, whereas the peripheral component was seen to become dissociated under various conditions (e.g. phosphorylation). It has been suggested that the conversion from PS II α to PS II β is triggered by LHC II phosphorylation (Timmerhaus and Weis, 1990).

Graan and Ort (1986) discovered inactive PS II centres. Chylla and Whitmarsh (1989) confirmed their existence. They determined that inactive centres were slow to reoxidize Q_A. They also found that there was heterogeneity among PS II with respect to reduction of Q_B, the secondary acceptor of PS II. For some time it was not clear whether these two populations and the α and β centres were the same, as there was evidence that a population of inactive (non-Q_B-reducing) centres resided within the grana. Not only has the existence of this population been questioned (Chow et al., 1988a), but to date we have no clear indication yet as to how and to what extent all the variously defined PS IIs are identical. What seems clear is that although the ratio of *total* PS II to PS I may be 1:1, the ratio of *active* PS II to PS I may vary considerably with growth irradiance intensity, quality, and perhaps other parameters.

Other work began to uncover various levels of heterogeneity among thylakoid membrane complexes. Andreasson and co-workers (1988) made an important advance when they discovered heterogeneity in PS I centres, finding a small but active population of PS I α centres in grana fractions. Levels of heterogeneity were also found in PS II α centres (Albertsson and Yu, 1988) and LHC II (Bassi et al., 1990).

Recent work especially by Albertsson and co-workers (1990) based on new fractionation studies presents a remarkable new model of the lateral arrangement of complexes within grana lamellae. Perhaps its most striking and at the same time disconcerting feature is a band of low-activity PS I complexes around the outside of the grana stack. It was originally considered that PS I was completely excluded from the stacks and this new model represents a significant departure from the principles enunciated in the theory of lateral heterogeneity. While emphasizing that theirs is only a static model, the authors make no attempt to reconcile it with the dynamic nature of the system or spillover.

IV. Three-dimensional Structure of Green Membranes and Effectiveness of Light Interception

Inasmuch as the thylakoid membrane is responsible for providing the energy and oxygen required to maintain our biosphere, it is arguably the most important membrane on the planet. It is also an extremely unusual membrane, being made up of a unique cocktail of lipids, some of which are found in no other system, and an unusually high proportion of protein (Murphy, 1985). This exotic composition is expressed in a unique three-dimensional structure. The thylakoid membrane forms an intricate network of membrane lamellae, of which some are exposed to the stroma and others lie tightly appressed with one another in stacked grana.

The existence of grana has been established for nearly 60 years (Weier, 1938) but both their three-dimensional structural form in terms of the relationships between their composite thylakoids and a satisfactory explanation of their function with respect to light-capture, electron transport, and energy transduction continue to elude researches (Miller and Lyon, 1985). In this section we briefly cover the history and recent advances of research into the three-dimensional structure of higher plant thylakoid membranes and comment on the connection between this structure and the various functions of this membrane, as this link is essential to advancing our understanding of this important membrane system and to devising sensible approaches to research.

Higher plant chloroplasts were recognized as early as the 1670s by Antoni Van Leeuwenhoek as green granules within the plant cell. Improvements in microscopy brought them firmly into focus and onto the research agenda in the late 19th and early 20th century (Weier, 1938). Grana were first described by Meyer (1883) and originally visualized by Metzner (1937) with a fluorescence microscope. Intrinsic chlorophyll fluorescence has been used as a central parameter in investigations of chloroplast structure and function for more than half a century.

Light microscopic techniques provided little information about the complex internal structure of the thylakoid membrane and there was much debate about fixation techniques and whether they yielded native structures (Weier, 1938). The advent of electron microscopy in the 1940s revealed much more about this intricate system and the concept of grana as stacks of lamellar discs with intergranal lamellae stretching between them was established. Since that time, largely due to the clarity and consistency of electron micrographs, it has been taken for granted that these pictures are representative of the native state of the thylakoid system. This may not be an entirely valid assumption (Wildman *et al.*, 1980; van Spronsen *et al.*, 1989).

Interpretation of early electron micrographs led most investigators to believe that the lamellae of the grana formed closed, separate discs. Menke (1962) named these discs "thylakoids", from a Greek word meaning sack-like, and the name has stuck, despite later indications of its inappropriateness (Heslop-Harrison, 1963). Menke believed the thylakoids were closed vesicles embedded singly or in stacks in the stroma and that larger sheets of closed double lamellae traversed the chloroplasts between them, representing the intergranal lamellae (Menke, 1962). Weier (1961) suggested that the grana were connected at regular intervals by a system of branching channels. Weier and Thomson (1962) studied many electron micrographs and from these constructed three-dimensional diagrams of intergranal connections explaining how they could account for the two-dimensional electron micrographic pictures. Heslop-Harrison (1963) suggested open fret connections linking thylakoids from neighbouring grana via a complex intergranal fretwork. His original model indicated connections within a plane. Later study of Weier and Thomson's fretwork connections in relation to inter- and intrathylakoid spaces indicated that the intrathylakoid space was continuous throughout the entire system. This led to Heslop-Harrison's famous postulate "that the entire lamellar system of the chloroplast including all the grana, constitutes a single enormously complex, membrane-bounded cavity, separate and distinct from the stroma" (Heslop-Harrison, 1963). This concept of continuity of the thylakoid membrane system was supported by work on the development and morphogenesis of the thylakoid network. Von Wettstein (1967) had already hinted at a continuous lamellar system when he described the stages of chloroplast development.

Wehrmeyer (1963) looked also at the structure in terms of development and stressed that grana and stroma lamellae were part of the same membrane system. He believed that the grana were the active sites of photosynthesis, while the stroma lamellae were involved with the formation and connection of grana. His work emphasized that not only was the system internally continuous, it was also generated by the continuous extension of an existing network. Not only is there considerable agreement that the thylakoid membrane is continuous (Anderson, 1975; Gunning and

Steer, 1975), but apart from the requirements of Mitchell's chemiosmotic hypothesis, a number of earlier and more recent observations support this claim. For example, if the outer chloroplast envelope is carefully disrupted under controlled osmotic conditions, stroma-less thylakoid membrane systems can be isolated with a high degree of intactness. The lamellae of grana may move apart somewhat but the membranes are never seen to separate (Weier et al., 1967). However, in low salt concentrations, thylakoids lose their grana completely by literally unstacking, forming what appear in electron micrographs to be concentric rings of double lamellar sheets. Under certain conditions this change can be completely reversed and the native configuration regained (Izawa and Good 1966a, b). Schönknecht and co-workers (1990) recently calculated that the thylakoid lumen constitutes a single electric unit.

Wehrmeyer (1964) followed up his development work with a study of the structure of the mature system, derived probabilities of fretwork connections arising from different sides of granal lamellae, and concluded that there were, most commonly, two exits leading to stroma lamellae, from each thylakoid of a granum. Paolillo and Falk (1966) conducted a similar study and their conclusion was that the intergranal lamellae wound around the grana in helices, rather like those illustrated by Wehrmeyer (1964) and Heslop-Harrison (1963). Paolillo's work culminated in a complex model involving a highly regular pattern of right-handed helices (Paolillo, 1970). He and his co-workers had arrived at the ultimate in ordered three-dimensional thylakoid structure.

Paolillo's model has been accepted and reinterpreted by several workers and remains the most comprehensive model to date. Brangeon and Mustardy (1979) reiterated and supported the helical arrangement of Paolillo. It may, however, not quite represent the flexible and changeable structure that seems to be operating.

Wildman and co-workers (1980) returned to the light microscope and presented a model in which the grana were arranged like a string of beads. They saw grana as integral, disc-shaped structures arranged spirally around stroma lamellae. Paolillo and Rubin (1980) came out in opposition to their model and suggested that the grana were extremely irregularly arranged and were rarely seen as regular columns or discs. Interestingly, quite recent studies using confocal scanning laser microscopy, a new technique which makes use of chlorophyll fluorescence from the chloroplast, seem to contradict Paolillo's model with regard to the regular spiral arrangement of the stroma lamellae and support the light microscopic observations of early workers (Menke, 1962; Wildman et al., 1980). They show the grana as being very regular indeed and do not appear to correlate with electron micrographs in this regard (Van Spronsen et al., 1989). These somewhat inconsistent observations remain unexplained and emphasize how incomplete our knowledge in this subject really is. The most likely model of three-dimensional thylakoid structure is that suggested by Clowes and Juniper (1968).

V. The Nature of Rapid Responses to Irradiance Change in Chloroplasts

Grana stacking is the most intriguing feature of the thylakoid membrane structure and its function is not understood (Miller and Lyon, 1985). In vitro experiments suggested that stacked thylakoids absorb less light than unstacked ones (Jennings and Zucchelli, 1985). One suggestion as to why thylakoids stack was recently advanced by Terashima and Takenaka (1990), who propose that extensive grana stacking, particularly in shade-adapted species, is important in post-illumination ATP formation, presumably by "holding" the ΔpH established in high light. While we do not believe that this is the sole purpose of grana stacking, it is certainly very likely that it is part of the explanation. The compromise is between short-term costs of additional photosynthetic machinery that may not be working most of the time, to be traded off against the benefit of flexibility and capacity to respond positively to brief, irregular fluctuations in light, i.e., a significant long-term benefit for overall growth and survival of the plant.

Why thylakoid membranes stack and how they do it has in our opinion not yet been resolved satisfactorily. It is, however, a crucial question because it will only be possible to answer questions about photosynthetic efficiency and effectiveness of light interception if we can understand what lies behind the apparent compromise solution, i.e., self-shading. Is that compromise in higher plant thylakoid stacking analogous to the way in which a canopy is structured such as those we find in tropical rainforests? Differences in light quality impinging on successive layers of individual grana stacks may play a similarly important role in the structural and functional characteristics of the lamellae as they do in a canopy. Chow and co-workers (1990) have indeed shown that adaptation to a particular irradiance quality leads to significantly reduced quantum efficiency at any other irradiance. This makes sense if quantum efficiency is regulated by changes to the pigment architecture or some other feature in the various individual LHC II complexes as well as by reversible associations of these LHC II complexes with the reaction centres. If this is so then we must look for reasons for the compromise to lie in gains in flexibility and variety perhaps. Just as rainforest canopies are highly flexible and adaptable structures, so are thylakoid membranes. Wind and gap formation through death or destruction of individuals requires shaded species to adapt to new light conditions very rapidly. In fact, recent work on sunflecks indicates that such rapid exploitation of high irradiances is indeed possible (Pearcy, 1990). The reason for this must lie in the dynamic nature of the thylakoid membrane structure and the flexibility of its response to changing light conditions. In the context of heterogeneity of photosynthetic complexes themselves as well as the heterogeneity of their lateral and perhaps vertical distribution, it is most important to consider the heterogeneity of the irradiance micro-environment because the three may be causally linked.

Melis and co-workers have extensively studied the stoichiometry of the two photosystems in different species and more particularly under different light conditions (Melis et al., 1985). Their work is supported by a very large body of literature on the molecular differences between the thylakoids of sun and shade plants (Anderson and Osmond, 1987). Their repeated conclusion is that the ratio of PS II to PS I is not the invariant 1 as predicted by the Z scheme of electron transport and maintained by Whitmarsh and Ort (1984), but that it varies between species and in response to different light qualities and quantities. In 1985, Melis and co-workers (1985) published a hypothesis suggesting that the photosynthetic apparatus is a highly adaptable, dynamic system which can alter its structure and composition in line with a changing light environment. They suggested that ATP and NADPH are the signals which monitor these environmental changes. The concentrations of these molecules and their ratio regulate biochemical changes involving both biosynthesis and degradation processes (possibly via protein phosphorylation) to readjust the levels of the two photosystem complexes.

It has been suggested that adjustments to stoichiometries of functional centres is a regulated phenomenon essential to the attainment of optimal photosynthetic efficiency (Chow *et al.*, 1990). The quantum yields of plants adapted to PS I preferred light (far-red-enriched) when placed in light preferable to PS II and vice versa were compared with quantum yields of the same plants under the light to which they had adapted. Quantum yields of non-adapted plants under both light qualities were significantly lower.

These stoichiometric changes are reminiscent of state I to state II transitions but operate on quite a different time scale, having an apparent half-time of about 20 hours. They are testimony to the extreme adaptability and flexibility of the photosynthetic membrane system and appear to underlie extremely complex regulatory mechanisms which may well be connected.

Many people have referred to the dynamic nature of the membrane, but have invariably done so from within the framework of a static model. "Dynamic" can refer to changes in energy transfer or phosphorylation states or photosystem stoichiometries. The term has also been used to refer to actual movement, most commonly the lateral movement or "migration", particularly of LHC II from the grana to the stroma lamellae, but also, recently, the movement or dynamic rearrangement of the membrane itself.

The idea that components cycle between the grana and stroma lamellae was first introduced by Critchley (1988). This idea was held by other people in different forms and was supported by data involving the distribution of the polypeptides of the different complexes between the two regions. Attention focussed, in particular, on the rapidly turning over D1 protein of PS II (Mattoo and Edelman, 1985). This protein is central to current studies on photoinhibition (the reduction of quantum yield of photosynthesis by excess light) from which several workers now suggest that membrane stacking is a mechanism for the protection of the plant from the damaging effects of excessive irradiances (Maenpaa *et al.*, 1987; Aro *et al.*, 1990; Hundal *et al.*, 1990). Perhaps these workers envisage the upper grana layers acting as sunglasses for those buried deeper inside the stack, reducing the effective irradiance level and changing light quality impinging on those centres. High PPFD light is also considered to produce damaging species which degrade components of the system (Richter *et al.*, 1990a, 1990b). Guenther and Melis (1990) and Ohad and co-workders (1990) interpreted the suggested cycling as a repair cycle by which damaged peptides were removed and replaced via the stroma. The degradation of D1, however, is an enzymatic cleavage and appears to be highly regulated even under non-photoinhibitory conditions (Challahan *et al.*, 1990; Virgin *et al.*, 1990a, b).

Critchley (1988, 1990) suggested that in conjunction with LHC II phosphorylation, the metabolism of D1 is a regulatory trigger which effects changes in the architecture of the membrane system, alternately exposing and shading regions of the membrane in response to varying light conditions. These changes also involve heterogeneous populations of the photosystems including a pool of non-functional PS II (as noted by Graan and Ort, 1986) or PS II that is not in use (Critchley, 1988). This hypothesis is supported by the work of Pearcy (1990) on sunflecks which indicates that shade-adapted species can make use of brief exposures to flashes of high irradiance with very high efficiency, suggesting an elaborate and dynamic system which can effect instantaneous increases in its light-capturing and electron transfer capacity. The important concept in this theory is the dynamic nature of the thylakoid membrane, both structurally and functionally, and the relative flexibility conferred by membrane stacking.

Most of the researchers who looked at the membrane in three dimensions were still coming to terms with the continuity and flexibility of the system they were studying. It is hard to conceive of a huge, enclosed sack folding and rearranging into such a complex and intricate structure as the thylakoid membrane system. It is even more difficult to imagine it unfolding and refolding or unstacking and restacking within minutes in response to irradiance changes. Unfortunately, it appears that when scientists became aware of the level of complexity they were dealing with, they put three-dimensional thylakoid structure in the "too hard" basket. Discovering new levels of complexity and heterogeneity we are beginning to realize how limited was and still is, our understanding of the function and regulation of the membrane. It is perhaps necessary, then, that the last 20 years have been devoted to elucidation of what thylakoid membranes do and how they do it. Rapidly gaining insights into the manner and function of grana stacking, the membrane's most characteristic and intriguing feature, we may soon be able to take the conceptual step into the third dimension and possibly even the conceptual leap into the fourth dimension (time) and shed some light on the true nature of this most complex and very important biological system.

VI. References

Abadia J, Glick RE, Taylor SE, Terry N and Melis A (1985) Photochemical apparatus organization in the chloroplasts of two *Beta vulgaris* genotypes. *Plant Physiol* **79**: 872–878

Albertsson PÅ and Yu SG (1988) Heterogeneity among photosystem II (alpha). Isolation of thylakoid membrane vesicles with different functional antennae size of photosystem II (alhpa). *Biochim Biophys Acta* **936**: 215–221

Albertsson PÅ and Andreasson E, Svensson P (1990) The domain organization of the plant thylakoid membrane. *FEBS Lett* **273**: 36–40

Anderson JM (1975) The molecular organization of chloroplast thylakoids. *Biochim Biophys Acta* **416**: 191–235

Anderson JM (1981) Consequences of spatial separation of photosystem 1 and 2 in thylakoid membranes of higher plant chloroplasts. *FEBS Lett* **124**: 1–10

Anderson, JM and Andersson B (1988) The dynamic photosynthetic membrane and regulation of solar energy conversion. *TIBS* **13**: 351–355

Anderson JM, Goodchild DJ and Thomson WW (1990) The granal margins of plant thylakoid membranes: An important non-appressed domain. In: (M. Baltscheffsky, ed) *Current Research in Photosynthesis*, Vol II, Kluwer Acad Publ, Dordrecht, The Netherlands, pp 803–808

Anderson JM and Osmond CB (1987) Shade-sun responses: Compromises between acclimation and photoinhibition. In: (DJ Kyle, CB Osmond and CJ Arntzen, eds) *Photoinhibition*. Elsevier Science Publishers, Amsterdam, pp 1–38

Andersson B and Akerlund HE (1978) Inside-out membrane vesicles isolated from spinach thylakoids. *Biochim Biophys Acta* **503**: 462–472

Andersson B and Anderson JM (1980) Lateral heterogeneity in the distribution of chlorophyll-protein complexes of the thylakoid membranes of spinach chloroplasts. *Biochim Biophys Acta* **595**: 427–440

Andersson B and Anderson JM (1985) The chloroplast thylakoid membrane: Isolation, subfractionation and purification of its supramolecular complexes. In: (HF Linskens and JF Jackson, eds) *Modern Method of Plant Analysis*, New Series, Vol 1, Springer Verlag, Heidelberg, pp 231–258

Anderson B, Sundby C, Akerlund HE and Albertsson PA (1985) Inside-out thylakoid vesicles. An important tool for the characterization of the photosynthetic membrane. *Physiol Plant* **65**: 322–330

Andreasson E, Svensson P, Weibull C and Albertsson PA (1988) Separation and characterization of stroma and grana membranes—evidence for heterogeneity in antenna size of both photosystem I and photosystem II. *Biochim Biophys Acta* **939**: 339–350

Aro EM, Tyystjarvi E and Nurmi A (1990) Effects of light and temperature on PS II heterogeneity. In: (M Baltscheffsky, ed) *Current Research in Photosynthesis*, Vol II, Kluwer Acad Publ, Dordrecht, The Netherlands, pp 439–442

Aschroft RG, Preston C, Cleland RE and Critchley C (1986) Flow cytometry of isolated chloroplasts and thylakoids. *Photobiochem Photobiophys* **13**: 1–14

Barber J (1980) An explanation for the relationship between salt-induced thylakoid stacking and the chlorophyll fluorescence changes associated with changes in spillover of energy from photosystem II to photosystem I. *FEBS Lett* **118**: 1–10

Bassi R, Rigoni F and Giacometti GM (1990) Chlorophyll binding proteins with antenna function in higher plants and green algae. *Photochem Photobiol* **52**: 1187–1206

Beadle CL, Long SP, Imbamba SK, Hall DO and Olembo RJ (1985) Photosynthesis in relation to plant production in terrestrial environments. Natural resources and the environment series, Vol 18, United Nations Environment Programme, Tycooly Publishing Ltd, Oxford, p. 132.

Blackwell MF and Whitmarsh J (1990) Effect of integral membrane proteins on the lateral mobility of plastoquinone in phosphatidylcholine proteoliposomes. *Biophys J* **58**: 1259–1271

Bonaventura C and Myers J (1969) Fluorescence and oxygen evolution from *Chlorella pyrenoidosa*. *Biochim Biophys Acta* **189**: 366–383

Brangeon J and Mustardy L (1979) The ontogenetic assembly of intra-chloroplastic lamellae viewed in 3-dimension. *Biol Cell* **36**: 71–80

Butler WL (1978) Energy distribution in the photochemical apparatus of photosynthesis. *Annu Rev Plant Physiol* **29**: 345–378

Callanhan, FE, Ghirardi ML, Sopory SK, Mehta AM, Edelman M and Mattoo AK (1990) A novel metabolic form of the 32kDa-D1 protein in the grana-localized raction centre of photosystem II. *J Biol Chem* **265**: 15357–15360

Carter DR, Cheeseman JM, Clough BF, Lovelock C, Sim RG and Ong JE (1990) Photosynthetic characteristics of the mangrove, *Bruguiera parviflora* (Roxb.) Wright & Arn., under natural conditions. In: (M Baltscheffsky, ed) *Current Research in Photosynthesis*, Vol IV, Kluwer Academic Publishers, Dordrecht, The Netherlands, pp 859–862

Chow WS, Anderson JM and Hope AB (1988a) Variable stoichiometries of photosystem II to photosystem I reaction centres. *Photosynth Res* **17**: 277–281

Chow WS, Quian L, Goodchild DJ and Anderson Jm (1988b) Photosynthetic acclimation of *Alocasia macrorrhiza* (L.) G. Don to growth irradiance: Structure, function and composition of chloroplasts. *Aust J Plant Physiol* **12**: 107–122

Chow WS, Melis A and Anderson JM (1990) Adjustments of photosystem stoichiometry in chloroplasts improve the quantum efficiency of photosynthesis. *Proc Natl Acad Sci USA* **87**: 7502–7506

Chylla RA and Whitmarsh J (1989) Inactive photosystem II complexes in leaves. Turnover rate and quantitation. *Plant Physiol* **90**: 765–772

Clowes FAL and Juniper BE (1968) *Plant Cells*. Blackwell Scientific Publications, Oxford, p 137

Critchley C (1988) The chloroplast thylakoid membrane is a molecular conveyor belt. *Photosynth Res* **19**: 265–276

Critchley C (1990) The role of LHC II phosphorylation and D1 turnover in the structure and function of photosystem II. In: (M Baltscheffsky, ed) *Current Research in Photosynthesis*, Vol II, Kluwer Acad Publ, Dordrecht, The Netherlands, pp 899–902

Edwards G and Walker DA (1983) C_3, C_4: *Mechanisms, and Cellular and Environmental Regulation, of Photosynthesis*. Blackwell Scientific Publications, Oxford, p 205

Eley JH and Myers J (1967) Enhancement of photosynthesis by alternated light beams and a kinetic model. *Plant Physiol* **42**: 598–607

Emerson R and Lewis CM (1943) The dependence of the quantum yield of *Chlorella* photosynthesis on wavelength of light. *Amer J Bot* **30**: 165–178

Emerson R, Chalmers R and Cederstand C (1957) Some factors influencing the long-wave limit of photosynthesis. *Proc Natl Acad Sci USA* **43**: 133–143

Evans LT (1975) Beyond photosynthesis—the role of respiration, translocation and growth potential in determining productivity. In: (JP Cooper, ed) *Photosynthesis and Productivity in Different Environments*. Cambridge University Press, Cambridge, pp 501–507

Ferrar PJ and Osmond CB (1986) Nitrogen supply as a factor influencing photoinhibition and photosynthetic acclimation after transfer of shade grown *Solanum dulcamara* to bright light. *Planta* **168**: 563–570

Gifford RM (1974) A comparison of potential photosynthesis, productivity and yield of plant species with differing photosynthetic metabolism. *Aust J Plant Physiol* **1**: 107–117

Glick RE, McCaulay SW and Melis A (1985) Effect of light quality on chloroplast membrane organization and function in pea. *Planta* **164**: 487–494

Graan T and Ort DR (1986) Detection of oxygen-evolving photosystem II centers inactive in plastoquinone reduction. *Biochim Biophys Acta* **852**: 320–330

Guenther JE and Melis A (1990) The physiological significance of photosystem II heterogeneity in chloroplasts. *Photosynth Res* **23**: 320–330

Gunning BES and Steer MW (1975) *Ultrastructure and the Biology of Plant Cells*. Edward Arnold, London, pp. 100–104

Heslop-Harrison J (1963) Structure and morphogenesis of lamellar systems in grana-containing chloroplasts. I Membrane structure and lamellar architecture. *Planta* **60**: 243–260

Hill R and Bendall F (1960) Function of the two cytochrome components in chloroplasts: A working hypothesis. *Nature* **186**: 136–137

Hinshaw JE and Miller KR (1989) Localization of light-harvesting complex II to the occluded surfaces of the photosynthetic membranes. *J Cell Biol* **109**: 1725–1731

Homann PH (1969) Cation effects on the fluorescence of isolated chloroplasts. *Plant Physiol* **44**: 932–936

Hundal T, Aro EM, Carlberg I and Andersson B (1990) Restortation of light induced photosystem II inhibition without *de novo* protein synthesis. *FEBS Lett* **267**: 203–206

Izawa S and Good NE (1966a) Effects of salts and electron transport on the conformation of isolated chloroplasts. I Light scattering and volume changes. *Plant Physiol* **41**: 533–543

Izawa S and Good NE (1966b) Effects of salts and electron transport on the conformation of isolated chloroplasts. II Electron microscopy. *Plant Physiol* **41**: 544–552

Jennings RC and Zucchelli G (1985) The influence of membrane stacking on light absorption by chloroplasts. *Photobiochem Photobiophys* **9**: 215–221

Jennings RC, Garlaschi FM, Gerola PD and Forti G (1979) Partition zone penetration by chymotrypsin, and the localization of the chloroplast flavoprotein and photosystem II. *Biochim Biophys Acta* **546**: 207–219

Joliot P, Joliot A and Kok B (1968) Analysis of the interactions between the two photosystems in isolated chloroplasts. *Biochim Biophys Acta* **153**: 635–652

Kühlbrandt W (1990) Structure of the light-harvesting chlorophyll a/b protein complex by high-resolution electron crystallography. In: (M Baltscheffsky, ed) *Current Research in Photosynthesis*, Vol II, Kluwer Acad Publ, Dordrecht, The Netherlands pp 217–222

Larsson UK, Anderson JM and Andersson B (1987) Variations in the relative content of the peripheral and inner light-harvesting chlorophyll a/b-protein complex (LHC II) subpopulations during thylakoid light adaptation and development. *Biochim Biophys Acta* **894**: 69–75

Lee WJ and Whitmarsh J (1989) Photosynthetic apparatus of pea thylakoid membranes. Response to growth light intensity. *Plant Physiol* **89**: 932–940

Leverenz JW (1988) The effects of illumination sequence, CO_2 concentration, temperature and acclimation on the convexity of the photosynthetic light response curve. *Physiol Plant* **74**: 332–341

Lichtenthaler HK, Buschmann C, Doll M, Fietz HJ, Bach T, Kozel U, Meier S and Rahmsdorf U (1981) Photosynthetic activity, chloroplast ultrastructure and leaf characteristics of high-light and low-light plants and of sun and shade leaves. *Photosynth Res* **2**: 115–141

Long SP, Farage PK, Bolhàr-Nordenkampf and Rohrhofer U (1989) Separating the contribution of the upper and lower mesophyll to photosynthesis in *Zea mays* L. leaves. *Planta* **177**: 207–216

Loomis RS and Gerakis PA (1975) Productivity of agricultural ecosystems. In: (JP Cooper, ed) *Photosynthesis and Productivity in Different Environments.* Cambridge University Press, Cambridge, pp 145–172

Ludlow MM and Björkman MM (1984) Paraheliotropic leaf movement in Siratro as a protective mechanism against drought-induced damage to primary photosynthetic reactions: Damage by excessive light and heat. *Planta* **161:** 505–518

Maenpaa P, Andersson B and Sundby C (1987) Differences in the sensitivity to photoinhibition between PS II in the appressed and non-appressed thylakoid regions. *FEBS Lett* **215:** 343–350

Mattoo AK and Edelman M (1985) Photoregulation and metabolism of a thylakoidal herbicide-receptor protein. In: (JB St. John, E Berlin and PC Jackson, eds) *Frontiers of Membrane Research in Agriculture.* Rowman & Allanheld, Ottowa, pp 23–34

Melis A and Harvey GW (1981) Regulation of photosystem stoichiometry, chlorophyll a and chlorophyll b content and relation to chloroplast ultrastructure. *Biochim Biophys Acta* **637:** 138–145

Melis A and Homann PH (1976) Heterogeneity of the photochemical centres in system II of chloroplasts. *Photochem Photobiol* **23:** 343–350

Melis A, Svensson P and Albertsson PÅ (1986) The domain organization of the chloroplast thylakoid membrane. Localization of photosystem I and of the cytochrome b$_e$/f complex. *Biochim Biophys Acta* **850:** 402–412

Melis A, Manodori A, Glick RE, Ghiradi ML, McAuley SW and Neale PJ (1985) The mechanism of photosynthetic membrane adaptation to environmental stress conditions. A hypothesis on the role of electron transport capacity and of the ATP/NADPH pool in the regulation of thylakoid membrane organization and function. *Physiol Vég* **23:** 757–765

Menke W (1962) Structure and chemistry of plastids. *Annu Rev Plant Physiol* **13:** 27–44

Metzner P (1937) Über den Bau der Chloroplasten. *Ber Deut Bot Ges* **55:** 16

Meyer PA (1883) *Das Chlorophyllkorn in chemischer, morphologischer und biologischer Beziehung.* A Felix, Leipzig

Miller KR and Lyon MK 1985. Do we really know why chloroplast membranes stack? *TIBS* **6:** 219–222

Miller, KR and Staehelin LA (1976) Analysis of the thylakoid outer surface. Coupling factor is limited to unstacked membrane regions. *J Cell Biol* **68:** 30–47

Morrissey, PJ, McAuley SW and Melis A (1986) Differential detergent-solubilization of integral thylakoid complexes in spinach chloroplasts. Localization of photosystem II, cytochrome b$_6$/f complex and photosystem I. *Eur J Biochem* **160:** 389–393

Murakami S, Torres-Pereira J and Packer L (1975) Structure of the chloroplast membrane—Relation to energy coupling and ion transport. In: (Govindjee, ed) *Bioenergetics of Photosynthesis.* Academic Press, New York, pp 555–618

Murata N (1969) Control of excitation transfer in photosynthesis. II Magnesium ion-dependent distribution of excitation energy between two pigment systems in spinach chloroplats. *Biochim Biophys Acta* **189:** 171–181

Murphy DJ (1985) The molecular organization of the photosynthetic membranes of higher plants. *Biochim Biophys Acta* **864:** 33–94

Myers K (1971) Enhancement studies in photosynthesis. *Annu Rev Plant Physiol* **22:** 289–312

Myers J and Graham JR (1963) Enhancement in *Chlorella. Plant Physiol* **38:** 105–116

Ohad, I Adir N, Koike H, Kyle DJ and Inoue Y (1990) Mechanism of photoinhibition *in vivo*. A reversible light-induced conformational change of reaction centres II is related to an irreversible modification of the D1 protein. *J Biol Chem* **265:** 1972–1979

Ort, DR and Whitmarsh J (1990) Inactive photosystem II centers: A resolution of discrepancies in photosystem II quantitation? *Photosynth Res* **23:** 101–104

Paolillo DJ Jr (1970) The three-dimensional arrangement of integral lamellae in chloroplasts. *J Cell Sci* **6:** 243–255

Paolillo DJ Jr and Falk RH (1966) The ultrastructure of grana in mesophyll plastids of *Zea mays. Amer J Bot* **53:** 173–180

Paolillo DJ Jr and Rubin G (1980) Reconstructions of the grana fretwork system of a chloroplast. *Amer J Bot* **67:** 575–584

Park RB and Sane PV (1971) Distribution of function and structure in chloroplast lamellae. *Annu Rev Plant Physiol* **22:** 395–430

Pearcy RW (1990) Sunflecks and photosynthesis in plant canopies. *Annu Rev Plant Physiol Plant Mol Biol* **41**: 421–453

Pearcy RW, Osteryoung K and Calkin HW (1985) Photosynthetic responses to dynamic light environments by Hawaiian trees. Time course of CO_2 uptake and carbon gain during sunflecks. *Plant Physiol* **79**: 896–902

Postl W and Bolhàr-Nordenkampf HR (1990) The light response of CO_2 concentration separated for the upper and lower side of a maize leaf. In: (M Baltscheffsky, ed) *Current Research in Photosynthesis*, Vol IV, Kluwer Acad Publ, Dordrecht, The Netherlands pp 31–34

Powles SB and Björkman O (1982) Photoinhibition of photosynthesis: Effect of chlorophyll fluorescence at 77K in intact leaves and in chloroplast membranes of *Nerium oleander*. *Planta* **156**: 97–107

Richter M, Rühle W and Wild A (1990a) Studies on the mechanism of photosystem !I photoinhibition. I. A two-step degradation of D1-protein. *Photosynth Res* **24**: 229–235

Richter M, Rühle W and Wild A (1990b) Studies on the mechanism of photosystem II photoinhibition. II. The involvement of toxic oxygen species. *Photosynth Res* **24**: 237–243

Sauer K (1975) Primary events and the trapping of energy. In: (Govindjee, ed) *Bioenergetics of Photosynthesis*. Academic Press, New York, pp 116–182

Schönknecht G, Althoff G and Junge W (1990) The electric unit size of thylakoid membranes. *FEBS Lett* **277**: 65–68

Sealey RV, Williams ML and Cobb AH (1990) Adaptation of *Codium fragile* (Suringar) Hariot fronds to photosynthesis at varying flux density. In: (M Baltscheffsky, ed) *Current Research in Photosynthesis.*, Vol II, Kluwer Acad Publ, Dordrecht, The Netherlands pp 455–458

Staehelin LA (1976) Reversible particle movements associated with unstacking and restacking of chloroplast membranes *in vivo*. *J Cell Biol* **71**: 136–158

Sundby C, Melis A, Maenpaa P and Andersson B (1986) Temperature-dependent changes in the antenna size of photosystem II. Reversible conversion of photosystem IIα to photosystem IIβ. *Biochim Biophys Acta* **851**: 475–483

Terashima I and Inoue Y (1984) Comparative photosynthetic properties of palisade tissue chloroplasts and spongy tissue chloroplasts of *Camellia japonica* L.: Functional adjustment of the photosynthetic apparatus to light environment within a leaf. *Plant Cell Physiol* **25**: 555–563

Terashima I and Inoue Y (1985a) Palisade tissue chloroplasts and spongy tissue chloroplasts in spinach: Biochemical and ultrastructural differences. *Plant Cell Physiol* **26**: 63–75

Terashima I and Inoue Y (1985b) Vertical gradient in photosynthetic properties of spinach chloroplasts dependent on intra-leaf light environment. *Plant Cell Physiol* **26**: 781–785

Terashima I and Saeki T (1983) Light environment within a leaf. I. Optical properties of paradermal sections of *Camellia* leaves with special reference to differences in the optical properties of palisade and spongy tissues. *Plant Cell Physiol* **24**: 1493–1501

Terashima I and Saeki T (1985) A new model for leaf photosynthesis incorporating the gradients of light environment and of photosynthetic properties of chloroplasts within a leaf. *Ann Bot* **56**: 489–499

Terashima I, Sakaguchi S and Hara N (1986) Intra-leaf and intracellular gradients in chloroplast ultrastructure of dorsiventral leaves illuminated from the adaxial or abaxial side during their development. *Plant Cell Physiol* **27**: 1023–1031

Terashima I and Takenaka A (1990) Factors determining light response characteristics of leaf photosynthesis. In: (M Baltscheffsky, ed) *Current Research in Photosynthesis.*, Vol IV, Kluwer Acad Publ, Dordrecht, The Netherlands pp 299–306

Timmerhaus M and Weis E (1990) Regulation of photosynthesis: α– to β–conversion of photosystem II and thylakoid protein phosphorylation. In: (M Baltscheffsky, ed) *Current Research in Photosynthesis.* Vol II, Kluwer Acad Publ, Dordrecht, The Netherlands pp 771–774

van Spronsen EA, Sarafis V, Brakenhoff J, van der Voort HTM and Nanninga N (1989) Three-dimensional structure of living chloroplasts as visualized by confocal scanning laser microscopy. *Protoplasma* **148**: 8–14

Virgin I, Ghanotakis DG and Andersson B (1990a) Light-induced D1-protein degradation in isolated photosystem II core complexes. *FEBS Lett* **269**: 45–48

Virgin I, Hundal T, Styring S and Andersson B (1990b) Consequences of light-induced D1-protein degradation on thylakoid membrane organization. In: (M Baltscheffsky, ed) *Current Research in Photosynthesis*, Vol II, Kluwer Acad Publ, Dordrecht, The Netherlands pp 423–426

Vogelmann TC, Bornman JF and Josserand S (1989) Photosynthetic light gradients and spectral régime within leaves of *Medicago sativa*. *Phil Trans R Soc Lond B* **323**: 411–421

von Wettstein D (1967) The formation of plastid structures. *Brookhaven Symp Biol* **11**: 138–159

Walker DA (1976) Plastids and intracellular transport. In: (CR Stocking and U Heber, eds) *Transport in Plants*, Vol III, Springer Verlag, Berlin, pp 85–136

Wehrmeyer W (1963) Über Membranbildungsprozesse im Chloroplasten. I. Zur Morphogenese der Granamembranen. *Planta* **59**: 280–295

Wehrmeyer W (1964) Zru Klärung der strukturellen Variabilität der Chloroplastengrana des Spinats in Profil und Aufsicht. *Planta* **62**: 280–295

Weier TE (1938) The structure of the chloroplast. *Bot Rev* **IV**: 497–529

Weier TE (1961) The ultramicrostructure of starch-free chloroplasts of fully expanded leaves of *Nicotiana rustica*. *Amer J Bot* **48**: 615–629

Weier TE and Thomson WW (1962) The grana of starch-free chloroplasts of *Nicotiana rustica*. *J Cell Biol* **13**: 89–108

Weier TE, Stocking CR and Shumway LK (1967) The photosynthetic apparatus in chloroplasts of higher plants. *Brookhaven Symp Biol* **19**: 353–374

Whatley JM (1979) Plastid development in the primary leaf of *Phaseolus vulgaris*: Variations between different types of cell. *New Phytol* **82**: 1–10

Whitmarsh J and Ort DR (1984); Stoichiometries of transport complexes in spinach chloroplasts. *Arch Biochem Biophys* **231**: 378–389

Wildman SG, Jope CA and Atchison BA (1980) Light microscope analysis of the three-dimensional structure of higher plant chloroplasts. Position of starch grains and probable spiral arrangement of stroma lamellae and grana. *Bot Gaz* **141**: 24–36

Xu C, Auger J and Govindjee (1990) Chlorophyll a fluorescence measurements of isolated spinach thylakoids obtained by using single-laser-based flow cytometry. *Cytometry* **11**: 349–358

CARBON ASSIMILATION AND PARTITIONING

11

Structure, Function and Regulation of Ribulose 1, 5-*Bis*phosphate Carboxylase in Higher Plants

R.C. Sachar, Daman Saluja and P. Murali

Biochemistry and Molecular Biology Laboratory
Department of Botany
University of Delhi
Delhi 110 007, India

CONTENTS

ABBREVIATIONS

ACO_2	:	activator CO_2
ADP	:	adenosine 5′-diphosphate
AMP	:	adenosine 5′-monophosphate
AMP-PCP	:	γ, β-methylene adenosine 5′-triphosphate
ATP	:	adenosine 5′-triphosphate
BP	:	binding protein
CA	:	2-carboxy D-arabinitol
CA1P	:	2-carboxy D-arabinitol 1-phosphate
CABP	:	2-carboxy-keto arabinitol-1, 5-*bis*phosphate

cAMP	:	cyclic adenosine 3′, 5′-monophosphate
CAT	:	chloramphenicol acetyl transferase
Chl. PK_1	:	chloroplast protein kinase 1
cpn	:	chaperonin
CTP	:	cytosine 5′-triphosphate
EGTA	:	Ethylene glycol *bis*-tetra acetic acid
GTP	:	guanosine 5′-triphosphate
K_{act}	:	activation constant
kDa	:	kilo Dalton
L, LS, LSU	:	large subunit
LSUBP	:	large subunit binding protein
LSU BP Complex	:	large subunit binding protein complex
NAD^+	:	nicotinamide adenine dinucleotide, oxidized
NADPH	:	nicotinamide adenine dinucleotide, reduced
NADP	:	nicotinamide adenine dinucleotide phosphate
prCAB	:	precursor chlorophyll A/B binding protein
prSS	:	precursor small subunit
rbc L	:	ribulose bisphosphate carboxylase large subunit
rbc S	:	ribulose bisphosphate carboxylase small subunit
RuBPcase	:	Ribulose-1, 5-*bis*phosphate carboxylase/oxygenase
RuBP	:	Ribulose-1, 5-*bis*phosphate
S, SS, SSU	:	small subunit
SDS-PAG	:	sodium dodecyl sulfate-polyacrylamide gel.

ABSTRACT

Ribulose-1, 5-*bis*phosphate carboxylase/oxygenase (E.4.1.1.39), often referred to as RuBPcase, has been extensively studied both at the structural and at the functional level becasue of its paramount role in the dark fixation of CO_2 during photosynthesis. RuBPcase has been purified to electrophoretic homogeneity from a wide variety of plants. It is a multimeric enzyme (MW 550,000) and comprises eight large subunits (MW 55,000) and eight small subunits (MW 12,000–16,000) in higher plants. The carboxylase and oxygenase activities of the enzyme reside in its large subunit, while its small subunit has been assigned a regulatory role in the catalytic activity. The large subunit of RuBPcase is encoded in the chloroplast and translated *in situ*. The small subunit is encoded in the nucleus and synthesized in the cytoplasm as a precursor polypeptide with a "transit peptide" at the N´-terminus. The transit peptide is responsible for the transport of the small subunit into the chloroplast. The precursor small subunit is processed to maturity in the chloroplast by a stromal enzyme and is then assembled with the large subunit to form the holoenzyme. A high molecular weight binding protein (29 S) is associated with the large subunit and is crucial for the assembly of the large subunits. The transport and assembly of subunits into holoenzyme are energy-dependent processes.

RuBPcase is regulated at the transcriptional and post-translational level. Transcription of large and small subunit genes are controlled by light and developmentally. The catalytic activity of RuBPcase is modulated by inhibitors and activators. Carbamylation of a lysine residue at the active site results in the activation of the enzyme. RuBPcase activase has been shown to mediate the light-mediated activation of RuBPcase. The substrate RuBP inhibits the activity of decarbamylated RuBPcase *in vivo*. Carboxy arabinitol-1-phosphate (CA1P) plays a vital role in the dark and light regulation of RuBPcase in *Phaseolus*, *Nicotiana*, *Panicum*, and *Solanum*. However, this mechanism does not operate in pea, spinach, *Cicer*, *Triticum* and *Zea mays*.

RuBPcase is a phosphoprotein in moss, spinach, pea, and *Cicer*. *In vivo* labelling of the enzyme with [^{32}P] has revealed that phosphorylation occurs at the small subunit. The *in vitro* dephosphorylation of the enzyme with alkaline phosphatase dissociated the small subunit from the large subunit octameric core. Concomitantly, there was a significant decrease in the catalytic activity of RuBPcase. Thus, phosphorylation seems to be necessary for the optimum biological activity of RuBPcase.

I. Introduction

Ribulose-1, 5-*bis*phosphate carboxylase/oxygenase (RuBPcase: E.C.4.1.1.39) catalyses the carboxylation of ribulose-1, 5-*bis*phosphate (RuBP), the primary step in dark reaction of photosynthesis. The pathway of conversion of CO_2 into sugars was mapped by Bassham and Calvin (1957). Since then, RuBPcase has been studied extensively because of its crucial role in fixation of atmospheric CO_2 into the pentose sugar, RuBP. On an annual basis, this enzyme fixes about 10^{11} tons of CO_2 and is the most abundant protein in the world (Ellis, 1979).

RuBPcase has a complex structure with a molecular mass of 550,000. There are eight large and eight small subunits in the holoenzyme. The large subunit (LSU) has a molecular weight of 55,000, while the molecular weight of the small subunit (SSU) varies from 12,000 to 16,000. Catalytic activity is ascribed to LSU. The SSU seems to play a regulatory role. The LSU is coded by the chloroplast genome and synthesized *in situ*. On the other hand, SSU is transcribed in the nucleus and synthesized on cytoplasmic ribosomes as a precursor polypeptide. The precursor of SSU is then transported into the chloroplast, where it is processed to maturity and assembled into the holoenzyme. However, the precise role of SSU in the regulation of the enzyme is not clearly understood. The transport of SSU into chloroplast and its subsequent assembly with LSU to form a holoenzyme is quite intricate. The assembly process involves the association of a high molecular weight protein complex and nucleoside triphosphate (ATP).

In some plant systems, RuBPcase has been shown to be a phosphoprotein. However, there is no unanimity of opinion about the site of phosphorylation. Phosphorylation has been reported for both the subunits of RuBPcase.

Photorespiration is oxidation of RuBP under high light and temperature releasing CO_2. This is catalysed by the oxygenase activity associated with RuBPcase. The oxygenation of RuBP necessarily decreases the photosynthetic efficiency and hence plant productivity.

RuBPcase does indeed play a vital role in plant productivity. Manipulation of this enzyme for better photosynthetic rates is one possibility for enhancing crop yield or plant productivity. Thus, researchers are probing various facets of this unique enzyme to get a better insight into the structure-function relationship, so as to engineer this enzyme at the genetic level.

The existing literature on the enzyme has been extensively reviewed (Dean et al., 1989; Jensen and Bahr, 1977; Lorimer, 1981a; Lubben et al., 1988; Manzara and Gruissem, 1988; Miziorko and Lorimer, 1983; Tobin and Silverthorne, 1985). The present review is aimed at projecting our current understanding about the structure, function, and metabolic regulation of RuBPcase in higher plants.

II. Genomic Organization

a. *Organization and expression of RuBPcase genes*

RuBPcase comprises of eight large (MW 55,000) and eight small (MW 14,000) subunits. The LSU is coded by the chloroplast genome (Chan and Wildman, 1972; Coen et al., 1977) and its mRNA is translated on chloroplast ribosomes (Blair and Ellis, 1973). The SSU is, however, encoded by the nuclear genome (Berry-Lowe et al., 1982; Coruzzi et al., 1983; Dunsmuir et al., 1983) and its mRNA is translated on cytosolic ribosomes as a precursor polypeptide. The precursor of SSU is transported into the chloroplast, where it is processed and assembled into the holoenzyme (Kawashima and Wildman, 1972; Highfield and Ellis, 1978). The RuBP carboxylase LSU (rbcL) gene is continuous (~1.4 kB), exists as a single copy, and encodes for a polypeptide of 475 amino acids. The sequence is highly conserved in many widely different species (see Manzara and Gruissem, 1988). Based on the sequence data of the coding regions of the rbcL gene, Ritland and Clegg (1987) have proposed evolutionary relationships in higher plants.

The SSU of RuBPcase is encoded as a multigene family in the nuclear genome of higher plants (soybean: Berry-Lowe et al., 1982; pea: Coruzzi et al. 1983; *Petunia*: Dunsmuir et al., 1983; tomato: Suguita et al., 1987; duckweed: Wimpee et al., 1983). In all the higher plants studied so far, the mature SSU coding sequences are highly conserved. Dean and co-workers (1989) have proposed that the conserved hexadecapeptide (residues between amino acids 61 and 76) in the SSU may be important structurally and/or functionally. The six aromatic amino acids in this region could be crucial for the binding of the SSU to the octameric SSU core. Although variations do exist in the structure of the LSU gene, many of the nucleotide substitutions are silent and are located in the transit peptide region. The entire complement of the RuBP carboxylase SSU (rbcS) gene has been isolated and sequenced in *Petunia* and tomato (Dean *et al.*, 1985; 1987; Suguita *et al.*, 1987). There are two loci for SSU containing a single gene in tomato and *Petunia*. The third locus in tomato consists of three genes, whereas in *Petunia* there are six genes (Manzara and Gruissem, 1988). In pea, five rbcS genes are clustered at a single locus (Polans *et al.*, 1985).

In dicots investigated so far, the rbcS gene has at least two introns that are present at a definite position (Berry-Lowe *et al.*, 1982; Cashmore, 1983). In *Petunia*, tomato, potato, and tobacco, there is an additional intron downstream of the second intron (Dean *et al.*, 1989). In *Lemna gibba*, the rbcS gene lacks the first intron (Wimpee *et al.*, 1983), while in wheat and corn the second intron is missing (Broglie *et al.*, 1983; Lebrun *et al.*, 1987).

The *in vivo* expression patterns of rbcS and rbcL genes have been studied in tomato (Suguita and Gruissem, 1987), maize (Nelson *et al.*, 1984; Sheen and Bogorad, 1985; 1986), amaranth (Berry *et al.*, 1985), mustard (Oelmuller *et al.*, 1986) and pea (Sasaki *et al.*, 1981, 1984; Thompson *et al.*, 1983). In pea, *Petunia* and tomato, the expression of rbcS gene is highest in leaf tissue compared to other tissues. Variation between the highly expressed rbcS gene to the least expressed gene is 25-fold in *Petunia*, 10-fold in pea and 14-fold in tomato (Manzara and Gruissem, 1988).

b. *Regulation of RuBPcase genes*

Expression of rbcS genes is known to be tissue specific and also regulated by light and cytokinin. In addition to the differential and organ-specific gene expression, the rbcS genes are also developmentally regulated in maize (Nelson *et al.*, 1984), etiolated seedlings of amaranth (Berry *et al.*, 1985) and tomato (Piechulla *et al.*, 1986). Light and dark experiments with matured leaves of pea revealed that the lag in the induction of the two genes of SSU during greening was a case of developmental regulation (Kuhlemier *et al.*, 1987a).

Tobin and Klein (1975) showed that light increases the levels of translatable messenger RNA (mRNA) in *Lemna gibba*. The relative abundance of mRNA in light-grown *Lemna* could be ascribed to phytochrome-mediated increased transcription (Silverthorne and Tobin, 1984; Tobin, 1981). Increased levels of translatable poly (A) mRNA of SSU and LSU in light-grown pea plants, in comparison to etiolated plants, provided convincing evidence for the role of light in the control of nuclear and chloroplast gene expression (Bedbrook *et al.*, 1980; Sasaki *et al.*, 1981, 1984; Smith and Ellis, 1981; Shinozaki and Suguira, 1982; Thompson *et al.*, 1983). The increased levels of SSU mRNA in light-grown pea seedlings have been attributed to an increased rate of transcription in light (Galaghar and Ellis, 1982). The light-mediated accumulation of SSU mRNA in seven-day-old etiolated seedlings of mung bean is a consequence of the development of chloroplast rather than the direct response of phytochrome action (Coruzzi *et al.*, 1984). Nevertheless, other workers have claimed that phytochrome can have a direct role in SSU gene expression (Fourcroy *et al.*, 1985; Wehmeyer *et al.*, 1990).

Several common sequences have been located upstream of the rbcS gene in tomato, *Petunia* (Dean *et al.*, 1985), pea (Green *et al.*, 1987) and *Nicotiana* (Poulsen *et al.*,

1986) that may have a regulatory role in RuBPcase gene expression. A sequence of ~ 100 bases upstream of the *Nicotiana* rbcS gene has been identified which is responsibe for the light-mediated and organ-specific expression of a chimeric gene (Poulsen *et al.*, 1986). A similar region of ~ 250 bases from the upstream region of pea rbcS gene could direct light-regulated and organ-specific expression of chloramphenicol acetyl transferase (CAT) gene in transgenic *Petunia* plants (Inamine *et al.*, 1985). Kuhlemeier *et al.* (1987b) identified a region of about 58 base pairs in the 5´ upstream region of the pea rbcS gene that functions as a light-responsive element. Analysis of this region resulted in the identification of a nuclear factor GT-1 which binds to a specific sequence in the upstream region. Presumably, this factor interacts with another factor which alters its activity under light and dark conditions. At present, the precise molecular mechanism of rbcS gene expression is not well understood.

The regulation of rbcL gene expression by light and chloroplast development has been studied in many higher plants (Berry *et al.*, 1985, 1986; Inamine *et al.*, 1985; Kuhlemeier *et al.*, 1987b; Rodermel and Bogorad, 1985; Sasaki *et al.*, 1987; Tobin and Silverthorne, 1985). However, transcriptional control of rbcL gene expression is not well understood. According to Gruissem and Zurawski (1985a) and Deng and Gruissem (1987), it appears that the transcriptional control of chloroplast gene expression is dictated by the promotor strength and the expression of the individual genes is regulated at the post-transcriptional level. The rbcL promotor region is well defined in spinach and maize (Gruissem and Zurawski, 1985b; Hanley-Bowdoin and Chua, 1987). Two consensus sequences cpt1 and cpt2 have been identified in the rbcL promotor region and shown to be of functional significance in chloroplast gene expression (Gruissem and Zurawski, 1985b). In rye leaves, Winter and Feierabend (1990) have shown a stoichiometric expression of large and small subunits of RuBPcase.

In suspension cultures of *Petunia hybrida*, cytokinin seems to regulate the rbcS gene expression (Charles *et al.*, 1985). Cytokinin treatment resulted in the accumulation of SSU mRNA and the level of this induction is further modulated by light.

Although the precise factors which control RuBPcase gene expression in the nucleus and chloroplast are not well understood, some aspects of gene regulation have been elucidated. There is a strong indication that the light-mediated regulation of rbcS gene expression is primarily at the transcriptional level. The nature of the chemical signal generated by light that eventually culminates in the induction of SSU and LSU mRNA is not known. RuBPcase gene expression undoubtedly provides a good model system for understanding the control mechanism of nuclear and chloroplast gene expression in higher plants.

c. *Transport and processing of the SSU*

The SSU, like most other chloroplast proteins, is synthesized as a large precursor with an extra sequence of 57 amino acids at the amino-terminus and is designated as the transit peptide (Chua and Schmidt, 1978). The transit peptide is required for the transport of precursor polypeptides into the chloroplast (Anderson and Smith,

1986; Mishkind *et al.*, 1985). Partial or complete proteolytic cleavage of the transit peptide from precursor small subunit (prSS) blocked its transport into the chloroplast (Mishkind *et al.*, 1985). A mature SSU synthesized *in vitro* that lacked the transit peptide failed to be imported into the chloroplast (Anderson and Smith, 1986). Further, a foreign protein with the transit sequence of SSU of RuBPcase could be successfully imported into the chloroplast (Cashmore *et al.*, 1985; Lubben and Keegstra, 1986; Van den Broeck *et al.*, 1985; Wasmann *et al.*, 1986). The amino-acid sequence of the transit peptide of SSU of RuBPcase exhibited a remarkable degree of homology in 22 plant species (Keegstra *et al.*, 1989). Furthermore, amino-acid homology has also been observed in the transit peptides of different chloroplast proteins (Karlin-Neuman and Tobin, 1986). The structure-function relationship of the transit peptide has been studied by generating and analysing deletions in the transit peptide (Ostrem *et al.*, 1989; Reiss *et al.*, 1987; Wasmann *et al.*, 1988). While large deletions abolished protein transport, small deletions did not impair protein transport, but did result in the improper processing of the precursor polypeptide (Ostrem *et al.*, 1989). Abnormal prSS, synthesized *in vitro* with amino-acid analogs, failed to be transported into isolated chloroplasts (Robinson and Ellis, 1985). Thus, the structure-function relationship of the transit peptide at the primary structure level is not well understood (Lubben *et al.*, 1988).

The entire mechanism of protein transport can be classified into four steps: (1) the binding of the precursor protein on the chloroplast outer envelope membrane; (2) translocation of the bound precursor across the envelope membrane; (3) proteolytic processing of the precursor into a mature polypeptide; and (4) localization of the mature polypeptide into its respective compartment. The literature on the transport of proteins into chloroplasts has been extensively reviewed (Chua and Schmidt, 1979; della-Cioppa *et al.*, 1987; Ellis, 1981; Lubben *et al.*, 1988; Schmidt and Mishkind, 1986).

Precursor binding studies have been performed using complex mixtures of translation products of poly(A) mRNA (Grossman *et al.*, 1980; Pfisterer *et al.*, 1982) and also with purified precursors (Cline *et al.*, 1985; Friedman and Keegstra, 1989; Olsen *et al.*, 1989). Binding assays performed under cold conditions (0–4°C) inhibited the transport of bound precursor across the membrane (Friedman and Keegstra, 1989; Grossman *et al.*, 1980). However, when import assay was performed at 25°C, the bound precursors could be transported into the chloroplasts. Binding of the precursor to the chloroplast envelope is an ATP-dependent process. Olsen and co-workers (1989) demonstrated that very low concentrations of ATP (50–100 µM) enhance the high binding affinity of many chloroplastic precursors.

Cline and co-workers (1985) and Friedman and Keegstra (1989) demonstrated the involvement of membrane protein(s) in the binding of prSS to the chloroplast outer envelope membrane. No direct evidence, however, has been provided. Pretreatment of intact chloroplasts with thermolysin abolished the binding of prSS and precursor chlorophyll *a/b* binding protein (prCAB) significantly (Cline *et al.*, 1984). This indicated that some protein component of the outer envelope membrane was crucial for the high-affinity binding of the precursor. Cornwell and Keegstra (1987) have identified a chloroplast outer envelope surface protein (66 kD) associated with the binding of prSS by employing photoactivatable cross-linking reagents. Pain and co-workers (1988) identified a 30 kD envelope membrane protein as the receptor for prSS using anti-idiotypic antibodies. The anti-idiotypic antibodies blocked the import of prSS into the chlororplasts. Kaderbhai and co-workers (1988) have also located a 30 kD protein in the inner envelope membrane by cross-linking studies. The authors identified this protein as the phosphate translocator of the inner envelope and suggested that the phosphate translocator has a function in protein transport. Although various groups have identified different proteins as the receptor for prSS, it is quite possible that the receptor has more than one polypeptide subunit as the various approaches have revealed this complexity (Keegstra et al., 1989).

The transport of proteins across biological membranes is a step requiring energy. Experiments with inhibitors and ionophores revealed that ATP is necessary for protein transport across chloroplast envelope. Cline and co-workers (1985) supported this finding by demonstrating that nigericin-induced inhibition of light-driven import of proteins could be overcome by the exogenous addition of ATP. Moreover, the requirement of ATP could not be replaced by any other nucleoside triphosphate or analogs of ATP (Flugge and Hinz, 1986; Pain and Blobel, 1987). Furthermore, based on ionophoric studies, Flugge and Hinz (1986) and Pain and Blobel (1987) suggested that ATP is the sole energy source for the translocation of prSS into the chloroplast and there is no involvement of any proton motive force. Although Olsen and co-workers (1989) observed that other nucleotides (cytosine 5′-triphosphate and guanosine 5'-triphosphate—CTP and GTP) could facilitate binding of prSS (though only by 50 per cent), CTP and GTP could not drive protein translocation. Both these reactions seem to require ATP inside the chloroplast rather than outside it (Olsen et al., 1989; Theg et al., 1989). Import of precursors was possible even in the absence of exogenously added ATP, provided ATP was present in the stroma (Theg et al., 1989). Conclusive evidence for the precise site of ATP utilization during protein translocation is still not available (Keegstra et al., 1989).

Different hypotheses have been put forward to explain the role of ATP in protein translocation into chloroplasts. One view is that ATP is required for the unfolding of the precursor polypeptide before its transportation into the mitochondria (Eilers and Schatz, 1988). However, experimental evidence does not support this hypothesis for the transport of polypeptides into chloroplasts (Olsen et al., 1989; Theg et al., 1989). Other workers have claimed that the role of ATP in protein transport into the chlororplast is through the phosphorylation of the precursor or components of the transport apparatus (Hinz and Flugge, 1988; Pain and Blobel, 1987; Soll and Buchanan, 1983). There is considerable evidence for the phosphorylation of the outer and inner envelope membrane proteins (Hinz and Flugge, 1988; Pain and Blobel, 1987). A number of envelope-bound protein kinases have been identified and characterized (Soll, 1988; Soll and Buchanan, 1983; Soll et al., 1988). However, none of these kinases possess the same inhibitor or substrate specificity for the protein translocation reaction.

The prSS is processed to its mature form in the stroma (Smith and Ellis, 1979). A protease has been purified that can cleave the precursors of both stromal and thylakoid proteins (Robinson and Ellis, 1984a). The enzyme was specific to cytoplasmically synthesized precursors of chloroplast proteins. The processing of the prSS (20 kD) to its mature form (14 kD) takes place in two steps (Robinson and

Ellis, 1984b). The precursor is initially cleaved to form an intermediate of 18 kD which is cleaved further to form the mature SSU.

d. *Large subunit binding protein*

The newly synthesized LSU is associated with a high molecular weight protein (29S). This has been referred to as RuBPcase large subunit binding protein (LSUBP) (Barraclough and Ellis, 1980; Ellis, 1977). The LSUBP has been purified from pea and has a molecular weight of 720,000. The SDS-PAG (sodium dodecyl sulphate-polyacrylamide gel) analysis of the binding protein revealed two subunits of molecular weights 60,000 and 61,000. The binding protein is a heteromer of $\alpha_6\beta_6$ composition. The α and β subunits of the binding protein are encoded in the nucleus and synthesized as precursor proteins of identical molecular weight (Lennox and Ellis, 1986; Musgrove *et al.*, 1987). The LSU and the binding protein are immunologically distinct. However, antiserum raised against binding protein does cross-react with SSU to a certain extent. The SSU of *Pseudomonas oxalaticus* was also immunoprecipitated by the binding protein antibodies. It has been envisaged that binding protein and SSU bind to LSU at a common site and hence possess similar domains which are recognized by the polyclonal antibodies raised against binding protein (Hemmingsen and Ellis, 1986). The antibodies raised against LSUBP from pea also cross-react with binding protein from other higher plants and to a high molecular weight protein from *E. coli* (Hemmingsen *et al.*, 1988). The LSUBP has been related to the groEL protein from *E. coli*, a heat shock protein involved in the assembly of multimeric proteins (Goloubinoff *et al.*, 1989a). The binding protein in plants is now designated as chloroplast cpn60.

e. *Assembly of subunits*

The newly synthesized LSU is initially complexed with a high molecular weight protein before it is assembled with SSU into the holoenzyme. Illumination of chloroplasts for 30 to 60 minutes resulted in the release of LSU from the LSU-BP complex (Barraclough and Ellis, 1980). In isolated pea chloroplasts, the newly synthesized LSU monomer exists in association with a subunit of binding protein forming a heterodimer (7S) which was considered to represent the subunit pools for the assembly of RuBPcase (Roy *et al.*, 1979). In isolated pea chloroplasts, it is well resolved that the newly synthesized LSU is present as 7S and 29S complexes (Roy *et al.*, 1982). Illumination of *in vivo* or *in vitro* labelled chloroplasts resulted in a decline in the radioactivity from the 29S complex, with a concomitant increase in radioactivity in the holoenzyme (18S) of RuBPcase (Roy *et al.*, 1982). Since the relative levels of radioactivity in the 7S and 29S complexes remained unaltered, it was considered that both these complexes contribute LSU for the assembly of the enzyme. The formation of 7S and 29S complexes cannot be attributed to the presence of low levels of SSU because, even under *in vivo* conditions wherein SSU is synthesized continuously, the 7S and 29S complexes are labelled prior to the incorporation of the labelled LSU into the holoenzyme. It is not explicit which of the two complexes is synthesized first during illumination of chloroplasts and whether an exchange of LSU occurs between the two complexes. The lag phase (~ 30 minutes) which precedes the incorporation of LSU into holoenzyme in isolated chloroplasts must, therefore, be a normal feature of assembly (Roy *et al.*, 1982).

In isolated pea chloroplasts, when the assembly of [^{35}S]-labelled LSU into holoenzyme was monitored as a function of time in light and dark, it was observed that labelled LSU assembled into holoenzyme only in light (Bloom *et al.*, 1983). In dark, the [^{35}S]-labelled LSU remained bound to the binding protein complex and hence failed to assemble into the holoenzyme. However, it is not known whether light is required for the dissociation of the newly synthesized LSU from the binding protein complex. The role of light could be to provide ATP through photophosphorylation, since ATP could substitute for light in the uptake of precursor polypeptide by chloroplasts (Grossman *et al.*, 1980). In pulse chase experiments, ATP promotes the assembly of large and small subunits of RuBPcase into the holoenzyme in the absence of light (Bloom *et al.*, 1983). The requirement of Mg^{2+} and ATP hydrolysis has been shown to be necessary for the dissociation of the 29S complex; analogs of ATP and other nucleotides failed to dissociate the complex (Milos and Roy, 1984). However, hydrolysis of ATP was not mandatory for the assembly, since analogs of ATP or GTP could efficiently stimulate the assembly of LSU and SSU into holoenzyme (Bloom *et al.*, 1983; Milos and Roy, 1984).

Milos and Roy (1984) assigned two functions to the LSUBP: (1) LSU-BP complex (7S) represents an obligatory intermediate in the assembly of $L_8 S_8$ and alternatively, (2) the binding protein could be a non-obligatory storage site. The 29S complex dissociates in the presence of Mg^{2+}-ATP and the thus dissociated LSU now sediments at 7S (Cannon *et al.*, 1986). The LSU from such 7S complexes could assemble into RuBPcase even in the absence of ATP at 24°C, thereby indicating that 7S LSU are assembly competent (Cannon *et al.*, 1986). It is further envisaged that the 29S complex does not directly serve as a donor of LSU to SSU. It is, however, not clear whether SSU binds to the 7S complex directly or LSU is first released from the 7S complex before assembly (Cannon *et al.*, 1986; Milos and Roy, 1984).

The conditions under which LSU is synthesized in the chloroplast also affect the assembly of LSU. Chloroplasts suspended in low concentration of sorbitol (0.188 M) showed increased levels of assembly (three-fold) than at high concentration of sorbitol (0.375 M) (Roy *et al.*, 1988a). This is due to the fact that 29S complex tends to dissociate at low concentration of sorbitol. Thus, the increased assembly of RuBPcase at low sorbitol concentration could be ascribed to the relative abundance of 7S complexes (Chaudhari and Roy, 1989; Roy *et al.*, 1988a). The extent of aggregation of LSU-BP complex (29S) is dependent on protein concentration and the complex also dissociates extensively into binding protein monomers and 7S complexes in the presence of light and ATP (Roy *et al.*, 1988b). Apparently, binding protein interacts with LSU during synthesis or shortly thereafter and facilitates the correct folding of the LSU polypeptide chain (Goloubinof *et al.*, 1989a).

More recently, attempts have been made to clone the genes of LSU and SSU of higher plants in *E. coli* using expression vectors (Bradley *et al.*, 1986; Gatenby *et al.*, 1985, 1987; van der Vies *et al.*, 1986). The genes for LSU and SSU were expressed in *E. coli* but failed to assemble into a biologically active holoenzyme (Bradley *et al.*, 1986; Gatenby *et al.*, 1987). In contrast, the genes of cyanobacterial RuBPcase led to the formation of an active enzyme when expressed in *E. coli* (Bradley *et al.*, 1986; Gatenby *et al.*, 1985; van der Vies *et al.*, 1986). Conceivably, the LSU of higher plants does not assume a competent assembly conformation when expressed in *E. coli*. The reason could be that *E. coli* lacks the assembly binding protein. The chaperonin groEL of *E. coli* shares extensive amino-acid and structural homology with LSUBP (Hemmingsen *et al.*, 1988; Pushkin *et al.*, 1982). Assembly of prokaryotic dimeric (L_2) and hexadecameric ($L_8 S_8$) RuBPcase in *E. coli* was achieved by expressing RuBPcase genes from *Anacystis nidulans* (Goloubinof *et al.*, 1989b). The cyanobacterial RuBPcase expressed in *E. coli* did exhibit enzyme activity which is sensitive to 2-carboxy-keto arabinitol-1, 5-*bis*phosphate (CABP), a specific inhibitor of RuBPcase. Since the L_8 core proteins were produced even in the absence of SSU, these studies indicated that groEL proteins are required only for the assembly of LSU.

The SSU isolated from cyanobacteria has only 40 per cent homology with the higher plants (Bloom *et al.*, 1983). In higher plants, a highly conserved sequence of 16 amino acids located at the N′-terminus was identified in the SSU (Mazur and Chui, 1985). Wasmann and co-workers (1989) deleted and replaced this conserved region of the SSU gene from pea with that from *Anacystis nidulans* SSU gene. Precursor SSU synthesized *in vitro* from the chimeric gene was incubated with isolated intact chloroplasts of pea to assay import and assembly of SSU into the holoenzyme. The fusion protein of SSU lacking 16 amino acids was imported into chloroplasts and processed to maturity, but failed to assemble with the LSU of pea. On the contrary, the SSU of *A. nidulans* carrying the 16 amino acids from pea SSU was found to assemble with the LSU of pea. Thus, it was proposed that this conserved region of SSU polypeptide is involved in the assembly of LSU and SSU in higher plants (Wasmann *et al.*, 1989). The SSU of *Chlamydomonas reinhardtii* also contains a similar sequence of amino acids at a position analogous to the higher plant sequence, but bears little homology with the higher plant sequence (Goldschmidt-Clermont and Rahiri, 1986). The prSS of *Chlamydomonas* is processed by pea chloroplasts but failed to assemble with pea LSU (Goldschmidt-Clermont and Rahiri, 1986; Mishkind *et al.*, 1985). Thus, it appears that, although higher plants and green algae show similar import processes, the mechanism of assembly of RuBPcase in higher plants is different. The assembly domain could be crucial for interacting with LSU-BP and/or LSU, for the efficient assembly of holoenzyme in higher plants. For understanding the precise requirement for assembly of LSU and SSU into holoenzyme, it is extremely necessary to develop an *in vitro* assembly system. Then it will be possible to study the association of phosphorylated and dephosphorylated SSU with LSU. This approach will elucidate the precise physiological significance of phosphorylation in the association of LSU and SSU. Figure 1 depicts the assembly of RuBPcase in higher plants.

III. Physicochemical Properties

a. *Subunit structure*

RuBPcase has been purified to electrophoretic homogeneity from many higher plants, algae and photosynthetic bacteria (Lorimer, 1981a). There is a good deal of similarity in the subunit structure of the enzyme both in higher plants and in algae (Anderson *et al.*, 1968; Kawashima and Wildman, 1970; Kieras and Hasselkorn, 1968;

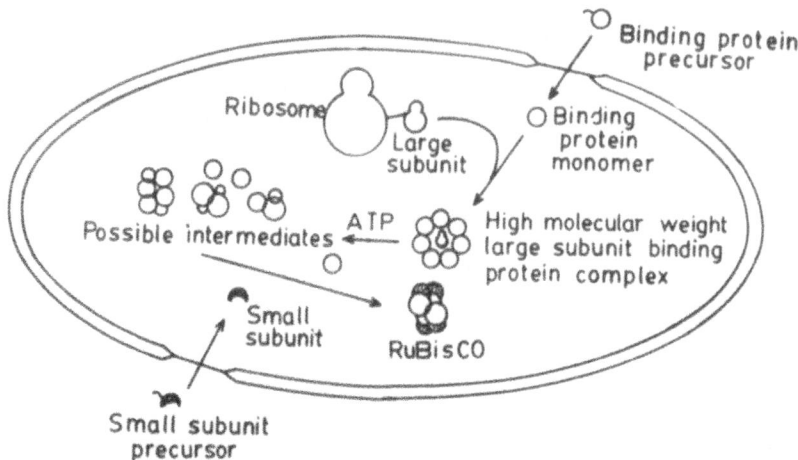

Figure 1. Schematic representation of the assembly of higher plant RuBPcase (after Roy, 1989).

Siegel *et al.*, 1972). The holoenzyme comprises eight LSU (MW 51,000–58,000) and eight SSU (MW 12,000–18,000). However, in bacterial systems, there is considerable variation with regard to the molecular size and subunit structure of RuBPcase. In algae and higher plants, the molecular mass of the holoenzyme is about 560,000, whereas in bacterial systems it varies from 120,000 to 430,000. Variation in the subunit composition and molecular size observed in bacterial RuBPcase may not reflect the true subunit composition of the enzyme (Tabita and McFadden, 1974). Since kinetic properties of RuBPcase are different in bacteria and higher plants, it is likely that there are differences in structure of RuBPcase in prokaryotes and eukaryotes (Andrews and Abel, 1981; Badger, 1980; Gordon *et al.*, 1980). Careful studies are necessary to elucidate the subunit structure of RuBPcase in bacteria and blue green algae.

b. *Amino-acid composition and sequence of subunits*

A comparison of tryptic peptide analysis, amino-acid sequence, and immunological properties revealed that LSU and SSU are non-identical even in the same species. The LSUs of different species also exhibited variation in their amino-acid sequences. The genes coding for LSU and SSU have been cloned and the sequence of amino acids has been deduced from their respective codons (Dron *et al.*, 1983; McIntosh *et al.*, 1980; Zurawski *et al.*, 1981).

There is more than 85 per cent homology in the amino-acid sequences of LSU from evolutionarily most distinct eukaryotes (*Chlamydomonas* versus maize). Out of the total of 475 amino acids, the residues between 169 and 220 and 321 and 340 are fully conserved. The region between 169 and 192 contains only neutral or basic amino acids. Insertion of acidic amino acids in this region resulted in the loss of catalytic activity of RuBPcase. Substitution of the glycine residue by asparagine at position 174 brings about inactivation of the enzyme (Dron *et al.*, 1983). The N′-terminal of the mature LSU is alanine in wheat, barley, and rye. However, the nucleotide sequence of LSU gene revealed that the initiation codon was located 14 amino-acid residues before the N′-terminal alanine. The post-translational cleavage of the 14 amino acids at the N′-terminus converts the LSU into its mature form (Langridge, 1981).

A lysine residue has been shown to play a pivotal role both at the catalytic and at the activation site. Affinity labelling studies have revealed that in all higher eukaryotes, a lysine residue at positions 175/176 and 335 are crucial for its catalytic function, whereas the lysine residue at position 201 is implicated in the activation of enzyme by CO_2 (Hartman et al., 1978; Lorimer, 1981b; Lorimer and Miziorko, 1980). There is considerable variation in the amino-acid composition of LSU isolated from *Rhodospirillum rubrum* as compared to higher plants (Robinson et al., 1980). There is only 28 per cent homology of amino-acid sequences of LSU between *R. rubrum* and higher plants. Although a lysine residue does interact with pyridoxal phosphate, the lysine residue in LSU of *R. rubrum* is not present at position 176. Furthermore, the peptide glycine-arginine-proline sequence is not homologously arranged in *R. rubrum* as in higher plants (Robinson et al., 1980). However, like the eukaryotic enzyme, *R. rubrum* enzyme possesses both carboxylase and oxygenase activities (McFadden, 1974; Ryan et al., 1974) and is activated by CO_2 and Mg^{2+} (Christeller and Laing, 1978; O'Leary et al., 1979; Whitman et al., 1979).

The amino-acid sequence of the SSU is much more conserved than that of LSU. The mature SSU in spinach, pea, and soybean has about 70 per cent homology (Takabe and Akazawa, 1975). The amino-acid residues in the regions 3–5, 11–21, 38–45, 49–75, and 109–117 show much more homology than in other regions of the polypeptide chain. However, there is no homology in the transit peptide sequence in higher plants and *Chlamydomonas*.

c) *Quaternary structure*

Based on electron microscopic study and X-ray diffraction analysis of crystalline RuBPcase, a bilayer model structure has been proposed for this enzyme in tobacco. According to this model, each layer consists of four large and four small subunits with four-fold axis of symmetry (Eisenberg et al., 1978). Electron microscopic studies suggested an eclipsed arrangement of the layers of LSU. The top view of the enzyme revealed a square structure with a central cylindrical hole of about 20 Å diameter that runs through the enzyme. Although the precise position of the SSU in the enzyme is not well defined, it has been speculated that it may be in the same plane as the LSU at the periphery (Tabita and McFadden, 1974).

Roy and co-workers (1978) examined the structure of RuBPcase, using cross-linking reagents. Since the two small subunits could be cross-linked using cross-linking reagents, it is considered that SSU is grouped in dimers. Each LSU also interacts with SSU and there is consistent association between LS_1 and SS_1. This approach suggested that all the eight SSUs cannot be in the same plane around the central core of LSU as proposed by Eisenberg and co-workers (1978). Based on these observations and using tilted-specimen electron microscopy and image processing of spinach RuBPcase, Barcena and Shaw (1985) suggested that SSUs are arranged equatorially around the core of eight LSUs in such a manner that the holoenzyme appears to be octagonal. The enzyme from *Alcaligenes eutrophus* is an oligomer and comprises eight large and eight small subunits (Bowien *et al.*, 1980). Electron microscopy of negatively stained samples of the enzyme , isolated from *A. eutrophus*, revealed that the molecule has dimensions of about $13 \times 13 \times 10.5$ nM and exhibited a four-fold axis of symmetry. However, X-ray diffraction studies of crystallized RuBPcase from *A. eutrophus* revealed that the enzyme is non-globular and forms two central layers. Later, studies carried out by Donnelly *et al.*, (1984), using small angle neutron scattering analysis, suggested that the enzyme in sol state is almost a hollow sphere and further favoured a two-layer model of RuBPcase from *A. eutrophus*. The differences in the quaternary structure could also be due to different conditions used for the growth of crystals. This is quite tenable, since a conformational change has been observed in the enzyme upon binding to bicarbonate and divalent cations (Grebanier *et al.*, 1978; Tomimatsu and Donovan 1981). The presence of carbonate and divalent cations also alters the reactivity of the enzyme towards cross-linking reagents (Grebanier *et al.*, 1978) and reagents that bind to RuBP binding site (Hartman *et al.*, 1978). Further studies are necessary for a vivid picture of the quaternary structure of RuBPcase in prokaryotes and eukaryotes.

d) *Function of subunits*

Both carboxylation and oxygenation activities have been shown to reside in the LSU of *rubrum*. The functional enzyme in *R. rubrum* consists of a dimer of LSUs as compared to eukaryotic RuBPcase, which is a heteromer ($L_8 S_8$). The absence of SSU does not affect the activation of the enzyme by CO_2 and metal ion. This suggested that activation site is also confined to the LSU (Christeller and Laing, 1978; O'Leary *et al.*, 1979; Whitman *et al.*, 1979). More recently, it has been demonstrated that the activator CO_2 binds to the LSU at epsilon amino group of lysine residues in plants having $L_8 S_8$ composition of RuBPcase (Lorimer, 1981a, Lorimer and Miziorko, 1980). Similarly, the binding of the metal ion (cobalt) to amino acids of the LSU of RuBPcase has been shown in spinach (Miziorko *et al.*, 1982). Genetic and biochemical studies conclusively proved that the catalytic site is present on the large subunit. A point mutation in LSU gene of *Chlamydomonas* resulted in catalytically inactive enzyme. Later, a revertant mutant was isolated with catalytically active RuBPcase, thereby confirming the catalytic role of the LSU (Dron *et al.*, 1983; Spreitzer and Mets, 1980; Spreitzer *et al.*, 1982).

The fact that the octamer of LSU, free of SSU, did possess 25 to 30 per cent of the catalytic activity further supported the view that the catalytic site is confined to LSU. The binding of the substrate RuBP to LSU has also been demonstrated by active site-directed inhibitors. These inhibitors bind irreversibly to the catalytic site on the LSU, thereby inhibiting the binding of RuBP (Hartman *et al.*, 1978). Badger and Lorimer (1981) presented kinetic and physical evidence to show that the effectors such as NADH and 6-phosphogluconate also exert their effect by interacting at the catalytic site located on the LSU.

Another function ascribed to the LSU of RuBPcase is the oxygenase activity. The LSU from different plants exhibit about 80 per cent homology in amino acids.

However, some of the amino acids at the active sites are different in LSU isolated from tobacco, maize and wheat. Parry and co-workers (1987) observed a difference in oxygenase activity in these plants. They further suggested that the extent of partitioning between the carboxylation and oxygenation reaction in different plants could be ascribed to the difference in the sequence of amino acids at the active site. However, at present, it is difficult to rule out the possibility that SSU may also play an important role in partitioning between carboxylase and oxygenase reactions.

Based on analogy of other heteromeric enzymes, it was considered that the SSUs in RuBPcase could have a regulatory role. However, unequivocal evidence for the regulatory role of SSU remains to be demonstrated. Circumstantial evidence suggests that the SSU regulates both the carboxylase and oxygenase reactions. Nevertheless, both these reactions occur in the absence of SSU in higher plants and in $R.$ $rubrum$. However, in higher plants the enzyme exhibits low activity in the absence of SSU. The K_m of the enzyme for CO_2 is much higher in $R.$ $rubrum$ (K_m 1.0 mM). The affinity of the higher plant enzyme for CO_2 (20 µM) was about 10-fold higher than that of $R.$ $rubrum$ (300 µM) (Christeller and Laing, 1978; Yeoh et $al.$, 1980, 1981). Based on these observations, it has been speculated that the binding of SSU to LSU could be responsible for an increased affinity and low K_m for CO_2 in higher plants. This argument, however, does not account for the relatively high K_m observed for the $L_8 S_8$ enzyme isolated from $Anabaena$ (K_m 293 µM) and $Synechococcus$ (K_m 250 µM) (Andrews and Abel, 1981; Badger, 1980).

Gibson and Tabita (1977) reported two molecular forms of RuBPcase from $R.$ $sphaeroides$. Form I has both LSU and SSU, whereas form II has only LSU. Further studies indicated a higher rate of activation of form I enzyme by HCO_3^- and Mg^{2+}. The authors ascribed it to the presence of SSU. LSU from the two forms are, however, not identical as revealed by immunological studies (Gibson and Tabita, 1977). Thus, difference in the rate of activation of the two molecular forms of the enzyme could be ascribed to the intrinsic property of LSU and not to the mere presence or absence of SSU (Miziorko and Lorimer, 1983). RuBPcase, isolated from $Synechococcus$, is activated as rapidly in the absence of SSU as in its presence (Andrews and Ballment, 1984). The authors concluded that the role of SSU in enzyme activation could be due to the stabilizing of the E-CO_2-Mg^{2+} complex, or perhaps by decreasing deactivation. Furthermore, the LSU has been shown to be catalytically incompetent in the absence of SSU. The activity of purified RuBPcase from $Synechococcus$ is proportional to the amount of SSU in the preparation and complete recovery in activity could be achieved in $vitro$ by the addition of SSU (Andrews and Ballment, 1984).

Treatment of purified RuBPcase from moss and spinach with alkaline phosphatase resulted in the dissociation of the SSU (dimer) from the catalytic LSU (Kaul et $al.$, 1986). The removal of the SSU did not affect the association of the LSU. However, the octameric LSU exhibited a dramatic loss (60 to 70 per cent) in enzyme activity and also had a lag phase (Kaul et $al.$, 1986). Treatment of $Chromatium$ RuBPcase with alkali also resulted in the dissociation of SSU with a concomitant loss (85 per cent) in enzyme activity (Takabe and Akazawa, 1973). The enzyme also exhibited altered pH optimum (pH 8.3) in the presence of Mg^{2+}. The authors consider that SSU could provide a conformational stabilization to the LSU in the holoenzyme.

The small subunits also modulate the effect of effectors on RuBPcase activity. The activity of RuBPcase from *R. rubrum* is stimulated by effectors such as NADH, fructose 6-phosphate, fructose 1, 6-*bis*phosphate, and 6-phosphogluconate (Whitman *et al.*, 1979). Since the enzyme lacks SSU, this site is present on the LSU (Gibson and Tabita, 1977; Tabita and McFadden, 1972). Incharoensakdi and co-workers (1986) have demonstrated that the presence of SSU enhances the extent of activation by CO_2 and Mg^{2+}, both in *Aphanotheca* and in *Chromatium*. The SSU also stimulates catalysis suggesting that the SSU interacts with both the activation and the catalytic sites.

From the foregoing observations, it is evident that the LSU possesses the sites of catalysis and activation. The fact that absence of SSU does result in the loss of enzyme activity suggests that SSU has some regulatory role. However, the molecular mechanism by which SSU stimulates the enzyme activity is far from clear.

IV. Activation of the Enzyme

a) *Molecular mechanism of activation*

Earlier, Pon and co-workers (1963) reported that CO_2 stimulates the activity of RuBPcase *in vitro*. The rate of CO_2 fixation *in vitro* by RuBPcase was, however, much lower than that observed *in vivo*. Lorimer and co-workers (1976) and Badger and Lorimer (1976) observed that CO_2 serves not only as a substrate for RuBPcase but also as an activator for carboxylation and oxygenation reactions. The activator CO_2 forms a ternary complex with the enzyme and metal ion. The kinetic data suggested that activation of RuBPcase involves an ordered addition of CO_2 and Mg^{2+}. Since the activation of RuBPcase is enhanced by the addition of Mg^{2+}, it was considered that the metal ion stabilizes the complex formation between the enzyme and CO_2. Kinetic turnover experiments using the reaction intermediate analog, CABP, revealed that the activator CO_2 is different from the substrate CO_2. It is only the substrate CO_2 that is fixed during the carboxylation reaction (Lorimer, 1979). Furthermore, CO_2, Mg^{2+} and CABP form a very stable complex with RuBPcase (Miziorko, 1979). The K_d of RuBPcase for CABP is very low (10^{-11} M) (Pierce *et al.*, 1980).

Enzyme (inactive) + CO_2 $\xrightarrow{\text{slow}}$ Enzyme – CO_2 + Mg^{2+} $\xrightarrow{\text{fast}}$

Enzyme – CO_2 – Mg^{2+} (active)

It was also observed that the activation of RuBPcase is brought about by CO_2 and not by HCO_3^-. Furthermore, the CO_2-mediated activation was dependent on pH of the medium (Lorimer *et al.*, 1976). The enzyme exhibits an increased affinity for metal ion after the CO_2 activator site is occupied (Miziorko and Mildvan, 1974). The activator CO_2 interacts with the e´-amino group of a lysyl residue (Lorimer *et al.*, 1976). Lorimer and Miziorko (1980) using spinach RuBPcase presented a more conclusive evidence for the formation of carbamate by $^{14}CO_2$. The activator $^{14}CO_2$ was fixed to the enzyme by treatment with diazomethan. The methylated enzyme-activator $^{14}CO_2$ did not undergo any exchange with large excess of unlabelled CO_2

and remained bound to enzyme even on treatment with protease and SDS and on fractionation by HPLC. The diazomethan fixes the CO_2 by esterification to NE-methoxy carbonyl lysine (a stable structure), since carbamate formation requires free CO_2 and uncharged amino group. This explained the necessity of free CO_2 and pH dependence for the activation reaction.

The carbamate formation occurs with the e´-amino group of lysine 208 on the large subunit (Lorimer, 1985) in spinach and lysine 191 in the *R. rubrum* enzyme (Donnelly *et al.*, 1983). The labile carbamate is stabilized by Mg^{2+} ion (Hall *et al.*, 1981; Miziorko, 1979; Pierce *et al.*, 1980). The reactive lysine residue has been identified by methylation of carbamate in the isolated quaternary structure of the enzyme-CO_2-Mg-CABP complex. It is, however, not clear whether the activation is a result of (1) conversion of a positively charged side chain of lysine to a negatively charged group of carbamate, or (2) induction of a conformational change or even (3) introduction of a ligand for direct coordination of Mg^{2+} ion.

Studies with site-directed mutagenesis, involving replacement of lysine residue with a glutamine residue, resulted in a loss of tight coupling of Mg^{2+} with the enzyme (Estelle *et al.*, 1985). Model building by Anderson and co-workers (1989) suggested that not only the charge (carbamate) but also the length of the lysine side chain is important for the interaction with Mg^{2+}. Glutamine is too short to form proper metal binding site instead of carbamylated lysine.

b) *RuBPcase activase*

Activation of RuBPcase is a prerequisite for its optimal catalytic activity and hence for photosynthetic carbon assimilation. Although regulation of the enzyme in presence of light has been reported from leaf tissue, intact isolated chloroplasts, and reconstituted chloroplast system, the physiological significance of light activation remained equivocal until a mutant of *Arabidopsis thaliana* was isolated which had low RuBPcase activity *in vivo* compared to the wild type (Somerville *et al.*, 1982). Somerville and co-workers (1982) showed that this mutant was defective for RuBPcase activation *in vivo* and required high levels of CO_2 for growth of the mutant. Under non-photorespiratory conditions, the rate of CO_2 fixation in the mutant was saturated several hundred times higher than under normal conditions. However, the rate of photorespiration in the mutant was identical to that of the wild type even at highest CO_2 concentrations. The identical kinetic properties (V_{max} and K_m for CO_2 and RuBP) and the isoelectric patterns of RuBPcase from the mutant and wild-type plants conclusively established that low photosynthetic rate in the mutant was not due to a defective enzyme. Moreover, the enzyme in isolated chloroplasts and protoplasts of the mutant could not be activated even under high levels of CO_2. Thus, it became evident that the low activity of RuBpcase in the mutant was due to lack of its activation. It could be ascribed to the presence of an inhibitor, or the absence of an activator, either of which could be a stromal metabolite or a structural component of the chloroplast.

Analysis of stromal extracts from the wild type and mutant form of *thaliana* on two-dimensional gels revealed that the mutant lacked two stromal polypeptides of 47 and 50 kD (Salvucci *et al.*, 1985). The requirement for a specific protein for light activation of RuBPcase is based on genetic analysis of the mutant which exhibited a typical Mendelian ratio and co-segregation of the 50 kD polypeptide (Salvucci *et al* 1985; Somerville *et al.* 1982). Stromal extracts from spinach and wild type of *thaliana* activated RuBPcase activity in the mutant *in vitro,* in a reconstituted system consisting of thylakoid membranes, stromal fraction and RuBPcase (Salvucci *et al.,* 1985). In contrast, stromal extracts from the mutant failed to bring about any light-mediated activation of RuBPcase from spinach and wild-type *A. thaliana.*

Illumination of isolated chloroplasts (Lorimer *et al.*, 1983) and leaves (Somerville *et al.*, 1982) results in an increased activation state of RuBPcase *in vivo*. Increased pH and Mg^{2+} concentration in illuminated chloroplasts is responsible for light-mediated activation of RuBPcase (Jensen and Bahr, 1977; Portis *et al.*, 1977; Purczeld *et al.*, 1978). This is inconsistent with the fact that RuBPcase activation saturates at a higher light intensity than that required for stromal alkalinization (Heber *et al.*, 1982). The differences in the activation states of RuBPcase in leaves under saturating and subsaturating light conditions (Machler and Nosberger, 1980), together with the identification of a mutant of *A. thaliana* that lacks RuBPcase activase, indicates that RuBPcase activation *in vivo* is not a spontaneous process as observed for the isolated enzyme.

Portis and co-workers (1977) and Purczeld and co-workers (1978) have shown that fructose *bis*phosphatase and sedoheptulose *bis*phosphatase activation states are affected by Mg^{2+} concentration and stromal pH, but are not affected by the mutation for RuBPcase activase in the mutant. Hence, in the mutant of *A. thaliana*, it is unlikely that the activation state of RuBPcase is affected by changes in Mg^{2+} concentration and pH. The suggestion that accumulation of high levels of RuBP in the mutant of *A. thaliana* might prevent the activation of RuBPcase *in vivo* (Salvucci *et al.*, 1986) is no longer favoured, as light activation could not be restored in leaves of this mutant by inhibiting the accumulation of RuBP.

Activation of RuBPcase by a partly purified RuBPcase activase has been demonstrated in a reconstituted system (Portis *et al.*, 1986). In this reconstituted system, RuBPcase activation increased by illumination and was dependent on the concentration of thylakoid membranes, RuBPcase, and RuBPcase activase. However, the activation state remained low in the dark. A large molar excess of RuBPcase activase was necessary for the optimal activation in the reconstituted system, the precise reason for which is not understood. In dark, the activation state of RuBPcase has been shown to be a function of CO_2 concentration under low levels of RuBP (Salvucci *et al.*, 1986; Schnyder *et al.*, 1984). In light, however, RuBP concentration is high and is in excess of the active site concentration (Molt *et al.*, 1984; Perchorowitz and Jensen, 1983). The spontaneous CO_2/Mg^{2+} mechanism (Lorimer and Miziorko, 1980; Lorimer *et al.*, 1976) for *in vitro* activation of RuBPcase does not hold true, since the CO_2 requirement for full activation greatly exceeds the CO_2 concentration available inside the illuminated chloroplasts. On the contrary, the K_{act} of CO_2 observed in the reconstituted system was found to be 4 µM, which is well within the physiological range of 10 µM found inside the chloroplast (Portis *et al.*, 1986). In the absence of RuBP, however, the K_{act} of CO_2 was about 23 µM which is in concurrence with the values observed for *in vitro* RuBPcase activation (Lorimer *et al.*, 1976; Perchorowitz and Jensen, 1983). Thus, the CO_2 response of RuBPcase in the reconstituted system can account for RuBPcase activation *in vivo* (Portis *et al.*, 1986).

The strong inhibitory nature of RuBP (Jordan and Chollet, 1983) on RuBPcase and high levels of RuBP (optimum of 34 mM) observed in the reconstituted system of Portis and co-workers (1986) provides the clue for the mechanism *in vivo* regulation of RuBPcase activation under high levels of RuBP in light. Despite the high levels of RuBP (3–12 mM) during steady-state photosynthesis in several species (Molt *et al.*, 1984; Perchorowitz and Jensen, 1983) the activase system can maintain RuBPcase in the active state. On the cessation of light, the high levels of RuBP inactivates the enzyme and the inactivation proceeds with equal facility in the presence or absence of activase (Portis *et al.*, 1986). Portis and co-workers (1986) proposed a functional model for the light and dark regulation of activation states of RuBPcase *in vivo*. In light, there is a dynamic equilibrium between the active and inactive RuBPcase. RuBPcase activase system brings about activation of RuBPcase, while the higher levels of substrate deactivate the enzyme. With the decrease in light intensity, the activation stimulus reduces and RuBPcase is deactivated by the high levels of RuBP. Although this model appears to be feasible, the precise molecular mechanism of activation/inactivation of RuBPcase remains to be determined.

It is evident that an initial increase in RuBPcase activity is directly proportional to the RuBPcase activase concentration (Robinson *et al.*, 1988). However, the rate of activation is also dependent on RuBPcase concentration. This indicated that the activation process involves a second order reaction dependent on the concentration of both RuBPcase and activase. A correlation has been drawn between *in vivo* decline in RuBPcase activation and the tightly bound RuBP (Brooks and Portis, 1988), suggesting that the enzyme-RuBP complex is formed *in vivo*. Carbamylation of the enzyme has been associated with *in vivo* activation of the enzyme-RuBP complex by RuBPcase activase (Werneke *et al.*, 1988a).

Identification of adenine nucleotides as a contaminant in the commercially available RuBP led to the discovery of yet another requirement for the activation of RuBPcase (Portis *et al.*, 1986). Removal of contaminating nucleotides from RuBP resulted in the loss of light-dependent stimulation of RuBPcase activity in the reconstituted system. However, activity could be restored by the addition of μM concentrations of ATP (Portis *et al.*, 1986). Machler and Nosberger (1980) observed that RuBPcase activity declines in the presence of ATP-consuming reactions suggesting that ATP is necessary for RuBPcase activation in light. Streusand and Portis (1987) demonstrated the need for ATP in the *in vitro* reconstituted system for activation of RuBPcase under physiological conditions of CO_2 and RuBP in the presence of an ATP regenerating system. In presence of ATP, the activation process was independent of light. However, ADP exhibited a strong inhibitory effect.

In vivo studies using AMP-PCP (γ, β-methylene adenosine 5′-triphosphate) and phloridzin (an energy transfer inhibitor in chloroplasts which inhibits ATP synthesis) revealed that the levels of stromal ATP were reduced to lower than that of control with a parallel decrease in the RuBPcase activity (Robinson and Portis, 1988a). Similar results were observed when isolated chloroplasts were incubated in the absence of added phosphate. A strong correlation between light activation of RuBPcase and stromal ATP level in isolated chloroplasts provides evidence for participation of RuBPcase activase in the activation of RuBPcase *in vivo* (Robinson and Portis, 1988a). Although the mechanism of action of RuBPcase activase or the function of ATP in the activation process is not clear, RuBPcase activation varies continuously with stromal ATP concentration over a physiologically meaningful range. It is also not known whether the concentration of ATP per se or the ratio of ATP to adenosine 5′-diphosphate (ADP) regulates RuBPcase activation. Changes in RuBPcase activation were much slower than changes in stromal ATP and rapid changes in ATP are not manifested by changes in RuBPcase activity (Robinson *et al.*, 1988). This indicates a relatively slow interaction of ATP with RuBPcase activase.

RuBPcase activase has now been purified (48-fold) spinach leaves by salt fractionation and ion-exchange fast protein liquid chromatography with 70 per cent recovery (Robinson *et al.*, 1988). The enzyme also exhibited ATPase activity. The purified enzyme comprised two polypeptides of 43 and 47 kD, while the holoenzyme has a molecular mass of about 200,000. The enzyme was stabilized by ATP and dithreiothreitol (DTT). The purified RuBPcase activase exhibited nucleotide specificity for ATP and was inhibited by ADP. The requirement of ATP could not be replaced by any other nucleoside triphosphate or analogs of ATP. Antibodies raised against the activase from spinach cross-reacted with two polypeptides from about 15 higher plants and with two polypeptides from extracts of *Chlamydomonas* (Salvucci *et al.*, 1987). RuBPcase activase has been obtained in partly purified form from *E. coli* (Werneke *et al.*, 1988a), which was transformed with a 16 kilobase spinach activase cDNA clone (Werneke *et al.*, 1988a). The cloned RuBPcase activase polypeptide had two consensus nucleotide binding sites (Werneke *et al.*, 1988b). The affinity for CO_2 in the activase-catalysed activation is substantially greater than the spontaneous mechanism and is well within the physiological range. The ATP-dependent activation of RuBPcase by the translation product of spinach activase gene expressed in *E. coli* confirms the presence of RuBPcase activase activity. Furthermore, RuBPcase was activated by RuBPcase activase and ATP at RuBP concentrations observed in chloroplasts of illuminated leaves. It is suggested that ATP provides energy required to either direct the carbamylation of the enzyme-RuBP complex, or bring about a conformational change in the enzyme-RuBP complex which is later stabilized by carbamylation (Werneke *et al.*, 1988a). The *in vitro* addition of RuBP to fully activated RuBPcase decreased its activity. This could, however, be restored by the addition of activase and ATP (Robinson and Portis, 1989).

RuBPcase activase has been purified and its functional requirements of ATP and RuBP well defined. However, the precise mechanism of activation process is not known. Further studies aimed at the elucidation of this mechanism will pave the way to discovering other functional requirements for the precise and optimal activation of RuBPcase in higher plants.

c) *Molecular mechanism of carboxylase and oxygenase reaction*

CO_2 and Mg^{2+} activated enzyme binds RuBP at its active site. Binding of RuBP is by its phosphate group to a lysine residue. The C_2 of RuBP has +ve charge as the bonding electrons are pulled towards the carboxy oxygen molecule. The loss of a proton from C_3 and formation of keto-enol equilibrium results in a nucleophilic enediol which allows CO_2 to react at the negatively charged C_2 of RuBP. The CO_2 molecule's being polarized and the change in the electronic structure of RuBP result in the covalent bonding of CO_2 to the C_2 of RuBP (Lawlor, 1987). The six carbon enediol intermediate is highly unstable and is hydrolysed by addition of OH^- to the C_3 of RuBP, resulting in cleavage of the intermediate into two molecules of 3-phosphoglyceric acid.

The precise mechanism of the oxygenation reaction is not yet understood. Since molecular oxygen is not charged and polarized, the mechanism of activation of O_2 is not known. Perhaps it is coordinated to the Mg^{2+} in the presence of the enediol form of RuBP leading to the formation of a hydroperoxide intermediate. This is immediately hydrolysed to form a molecule each of p-glycolate and 3-phosphoglyceric acid. The reaction mechanism is schematically represented in Fig. 2.

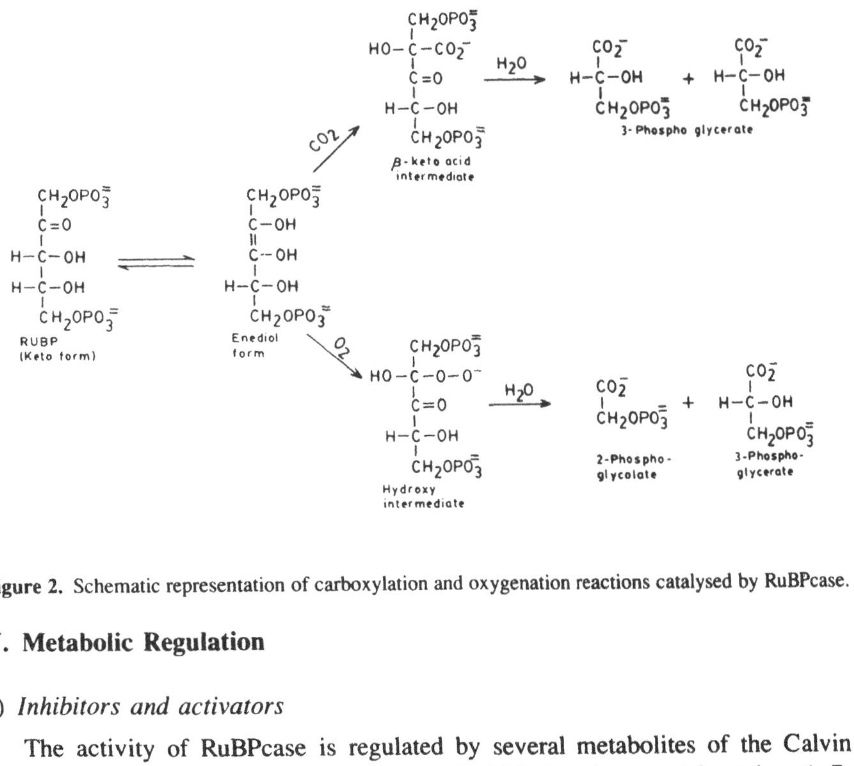

Figure 2. Schematic representation of carboxylation and oxygenation reactions catalysed by RuBPcase.

V. Metabolic Regulation

a) Inhibitors and activators

The activity of RuBPcase is regulated by several metabolites of the Calvin cycle, such as 3-phosphoglycerate, fructose-1, 6-bisphosphate, sedoheptulose-1, 7-bisphosphate, inorganic phosphate, RuBP, xylulose-1,5-bisphosphate and ribulose-5-phosphate (Badger and Lorimer, 1981; Laing and Christeller, 1976; Paulsen and Lane, 1966; Tabita and McFadden, 1972). A large number of metabolites that activate RuBPcase in vitro include ribulose-5-phosphate, xylulose-5-phosphate, erythrose-4-phosphate, dihydroxyacetone-3-phosphate, 3-phosphoglycerate, ADP, ADP-glucose, ATP and NADPH (nicotinamide adenine dinucleotide, reduced) (Buchanan and Schürman, 1973; Chu and Bassham 1975; Chollet and Anderson, 1976). It appears that phosphate esters in general are capable of affecting the activity of RuBPcase. The effect of NADPH (0.5 mM), however, seems to be specific as NAD^+ (NAD oxidized) failed to activate RuBPcase (Chu and Bassham, 1973, 1975). So far, it has not been ascertained whether these effectors have any physiological significance *in vivo* as most of the studies were carried out *in vitro* using relatively high concentrations of metabolites (e.g., 0.5 mM). The physiological levels of these metabolites are rather low in the chloroplasts. Furthermore, the concentration of RuBPcase in the stroma has been estimated to be about 4 mM, which is much higher than the concentration of the Calvin cycle intermediates (Badger and Lorimer, 1981; Miziorko and Lorimer, 1983). More recently, Pettersson and Ryde-Pettersson (1988) have studied the potential regulatory role of Calvin cycle intermediates on the activity of RuBPcase. Their findings reveal that the inhibition of RuBPcase by metabolites of Calvin cycle has no significant effect on the Calvin cycle under saturating conditions of light and CO_2. This has been ascribed to high concentrations of RuBPcase in the cell. Furthermore, 70 to 90 per cent of the intermediates are present in the enzyme-bound form and hence have little effect on the regulation of RuBPcase activity *in vivo*.

Transfer of *Chlorella* cells from light to dark resulted in a rapid decrease in the formation of 3-phosphoglyceric acid. Concomitantly, there was inhibition of phosphoribulose kinase (which converts ribulose-5-phosphate and ATP to RuBP and ADP) and RuBPcase. However, there was an increase in the levels of 6-phospho-gluconate (Pederson *et al.*, 1966). The *in vitro* addition of 6-phosphogluconate during RuBPcase assay inhibited enzyme activity (Chu and Bassham, 1972; Tabita and McFadden, 1972). Nevertheless, addition of 6-phosphogluconate (50 μM) to RuBPcase preincubated with Mg^{2+} and CO_2 brought about activation of the enzyme (Chu and Bassham, 1973). It appears that binding of RuBP is much faster than 6-phosphogluconate. Hence, binding of RuBP prior to Mg^{2+}, CO_2 and 6-phosphogluconate prevents activation of the enzyme by CO_2 and Mg^{2+}. Inhibition of the enzyme activity by 6-phosphogluconate is therefore due to competition with RuBP for its binding site. *In vivo*, in the presence of light, Mg^{2+}, and CO_2 and low concentration of 6-phosphogluconate, the enzyme can actively carboxylate RuBP. As the plant is transferred to dark, RuBP levels fall rapidly, and with the fast build-up of 6-phosphogluconate in the system, the enzyme is inhibited (Jordan *et al.*, 1983). Chu and Bassham (1973) observed that RuBP-mediated inactivation of the enzyme preincubated with HCO_3^- and Mg^{2+} persists for 20 minutes or longer and addition of 6-phosphogluconate or NADPH stimulates the enzyme activity. 6-phosphgluconate and fructose *bis*phosphate compete with RuBP when added to the enzyme along with RuBP preincubated with HCO_3^- and Mg^{2+} (K_i for 6-phosphogluconate is 20 μM; fructose diphosphate is 190 μM). NADPH or 3-phosphglyceric acid has no effect on the enzyme activity when added along with RuBP. Based on these observations, Chu and Bassham (1972) proposed that RuBPcase possesses at least four allosteric sites designated as I_1, A_1, A_2, and A_3. Binding of RuBP to site I_1 results in the inhibition of enzyme activity. In the absence of HCO_3^- and Mg^{2+}, RuBP binds to site I_1 and induces a conformational change in the enzyme that persists for about 20 minutes in the presence of bicarbonate and Mg^{2+}. The enzyme-RuBP complex has a low binding affinity for HCO_3^- and Mg^{2+}. In the absence of RuBP; HCO_3^- and

Mg^{2+} bind to site A_3 converting the enzyme into an active form. The active form of the enzyme has low affinity for RuBP at site I_1, such that, at physiological levels of RuBP, the active form is not inactivated. The active form can bind to NADPH and 6-phosphogluconate at its site A_2 during preincubation with Mg^{2+} and HCO_3^-. It has been shown that 6-phosphogluconate activates the enzyme at suboptimal concentrations of HCO_3^- and Mg^{2+} *in vitro* (Jordan *et al.*, 1983). Thus, 6-phosphogluconate could play a crucial role in the metabolic regulation of enzyme *in vivo*, since it can generate the activated from of the enzyme at physiological concentrations of CO_2 and Mg^{2+}. The precise mechanism of activation by 6-phosphogluconate is not known. It is envisaged that the eight catalytic sites on the LSU of the holoenzyme are carbamylated and bind to Mg^{2+}. Since activation can be achieved at low concentrations of 6-phosphogluconate, it is assumed that the binding of the effector (6-phosphogluconate) to one of the active sites activates the other seven catalytic sites by CO_2 and Mg^{2+} through cooperativity.

In many plant species, regulation of RuBPcase activity in light and dark is achieved through the interplay of the tight-binding inhibitor of catalysis characterized as CAlP (Seemann *et al.*, 1985; Servaites, 1985; Vu *et al.*, 1986). This inhibitor was isolated from *Solanum tuberosum* (Gutteridge *et al.*, 1986) and *Phaseolus vulgaris* (Berry *et al.*, 1987). The presence of CAlP has been observed by the loss in the CO_2-Mg^{2+} activatable RuBPcase activity from dark-treated leaves. However, this inhibitory mechanism does not operate in pea, spinach and *zea mays* (Vu *et al.*, 1984). The binding of CAlP to the catalytic site of the carbamylated form of RuBPcase inhibits the enzyme activity in dark. Structurally, CAlP is an analog of the hydrated form of 2-carboxy-3-keto-arabinitol-1, 5-*bis*phosphate which is the transition state intermediate of the carboxylation reaction (Schloss and Lorimer, 1982). Thus, CAlP is a component of a regulatory system that controls RuBPcase activity with changes in photon flux densities and associated with electron transport capacity (Seemann *et al.*, 1990).

The steady-state concentration of CAlP in *Phaseolus vulgaris* after prolonged darkness is in excess of the RuBPcase catalytic site concentration (~ 1.4 mol of CAlP/mol RuBPcase catalytic site), and regulation of RuBPcase activity is accomplished by light-dependent changes in the pool size of CAlP (Kobza and Seemann, 1988). Increased photon flux densities stimulate increased degradation of CAlP and are paralleled by the changes in RuBPcase activity (Kobza and Seemann, 1988). The degradation of CAlP has also been shown to be dependent on photosynthetic electron transport (Salvucci and Anderson, 1987; Seemann *et al.*, 1985). The bound inhibitor could be degraded by an alkaline phosphatase *in vitro*, thereby restoring the activity of RuBPcase (Seemann *et al.*, 1985). A phosphatase has been purified from tobacco chloroplast that can convert unbound CAlP to a non-inhibitory product (Salvucci *et al.*, 1988). Basically, this phosphatase converts CAlP to carboxy-arabinitol and Pi (Holbrook *et al.*, 1989). Robinson and Portis (1988b) suggested a possible role of RuBPcase activase *in vivo*, in the release of CAlP from RuBPcase although there is no concrete evidence to prove this hypothesis. A more plausible explanation is that the activase releases the inhibitor CAlP from the catalytic site and it is the phosphatase that converts it into a non-inhibitory product, 2-carboxy-D-arabinitol thereby restoring RuBPcase activity.

b) *Regulatory role of phosphorylation*

Phosphorylation of proteins has been reported in illuminated chloroplasts isolated from pea leaf tissue (Bennett, 1977). *In vivo* labelling of isolated chloroplasts with [^{32}P]-orthophosphate in light resulted in the phosphorylation of RuBPcase and light harvesting chlorophyll *a/b* binding protein. In addition, phosphorylation of thylakoid proteins is a light-dependent phenomenon. However, the mechanism by which light regulates the phosphorylation of chloroplast proteins is not understood (Bennett, 1979). Two cAMP-independent (cAMP cyclic adenosine 3', 5'-monophosphate) protein kinases have been purified from spinach chloroplasts. A low molecular weight kinase (chlorophyll protein kinase 1, 25,000) could phosphorylate the SSU of purified RuBPcase *in vitro* (Lucero Lin *et al.*, 1982). However, no mention was made of the physiological significance of phosphorylation of the SSU of RuBPcase *in vitro*.

A cAMP-independent protein kinase from the outer envelope of spinach chloroplast was purified by Soll and Buchanan (1983). This kinase was shown to phosphorylate the mature form of the SSU of RuBPcase. It was considered that phosphorylation would facilitate the transport of SSU across the chloroplast envelope. Muto and Shimogawara (1985) also reported the phosphorylation of the SSU of RuBPcase by a Ca^{2+} and phospholipid-dependent protein kinase. Phosphorylation was inhibited by EGTA, but not by antagonists of calmodulin. However, no function was assigned to the phosphorylation of SSU. Guitton and Mache (1987) isolated of Ca^{2+}-independent protein kinase associated with the stromal ribosomes of spinach chloroplasts that could phosphorylate the LSU of RuBPcase at the serine and threonine residues. No functional relevance was attributed to the phosphorylation of LSU.

To sum up, it appears that RuBPcase could be phosphorylated *in vitro* either at the SSU or at the LSU. It was, therefore, important to ascertain which of the subunits of RuBPcase is phosphorylated *in vivo*. *In vivo* labelling of the enzyme with [^{32}P]-orthophosphate has clearly demonstrated the phosphoprotein nature of RuBPcase in spinach and moss. Fractionation of the purified [^{32}P]-labelled RuBPcase on SDS-PAG revealed the selective phosphorylation of the SSU (Kaul *et al.*, 1986; Saluja *et al.*, 1986) (Fig. 3). Characterization of the acid-hydrolysed RuBPcase showed the labelling of phosphoserine residues and to a lesser extent phosphothreonine. Thus, the earlier claim that RuBPcase is phosphorylated at both the subunits does not represent the true physiological status of the enzyme. The absence of the [^{32}P] label in the LSU could not be ascribed to its dephosphorylation, as the enzyme was purified in the presence of NaF (1 mM), a potent inhibitor of phosphatases. Furthermore, immunoprecipitation of [^{32}P]-RuBPcase from crude preparations also revealed a single radioactivity band in the region of SSU on SDS-PAG (Murali, Saluja and Sachar, unpublished results).

Selective phosphorylation of SSU of RuBPcase has also been observed by us in pea leaves, *in vivo* (Murali, Saluja, and Sachar, unpublished results) (Fig. 4). However, the enzyme from pea and *Cicer* was phosphorylated at the serine and tyrosine residues (unpublished results: Murali, Saluja and Sachar; Agarwal, Saluja and Sachar).

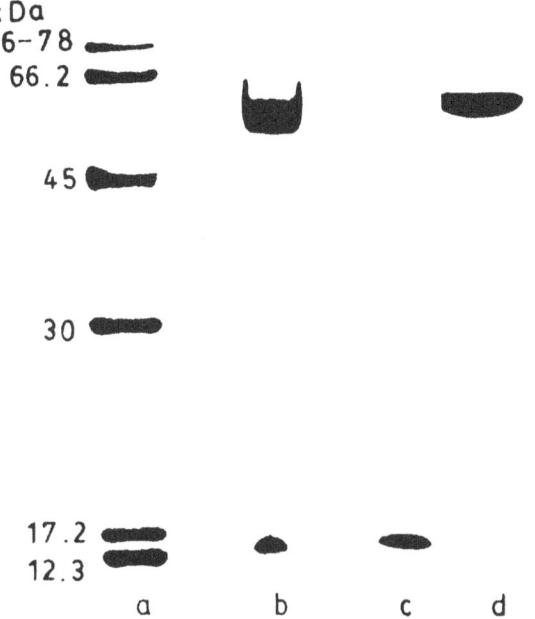

Figure 3. SDS-PAGE pattern of purified phosphorylated and dephosphorylated moss RuBPcase depicting its subunit structure and phosphorylation of the small subunit. (a) Molecular markers stained with Coomasie brilliant blue R; (b) large and small subunits of RuBPcase stained with Coomassie brilliant blue R on SDS-PAGE; (c) autoradiograph of b; dephosphorylated RuBPcase stained with Coomassie brilliant blue R (after Kaul *et al.*, 1986).

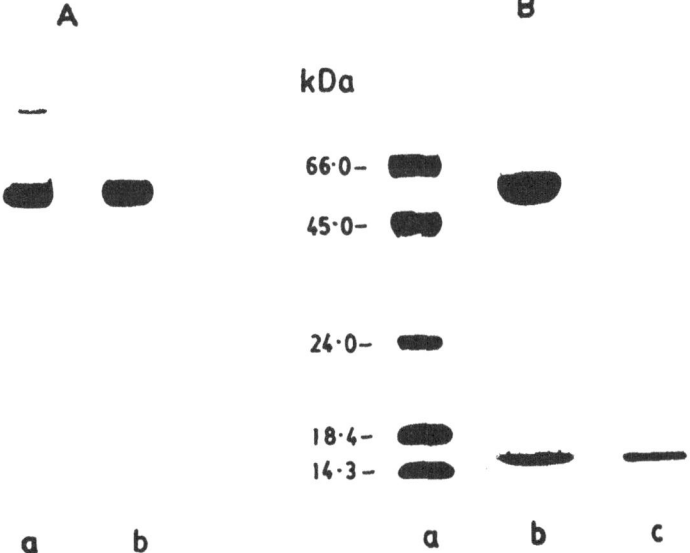

Figure 4. Electrophoretic pattern of RuBPcase purified from pea. A. Native-PAGE : lane (a) protein band of RuBPcase; lane (b) autoradiograph of (a). B. SDS-PAGE; lane (a) molecular weight markers; lane (b) protein bands of large and small subunits; lane (c) autoradiograph of (b) (unpublished data, Murali, Saluja and Sachar).

In vitro phosphorylation of RuBPcase from pea, spinach and *Cicer* was also restricted to the SSU. In contrast, our attempts to phosphorylate *Nicotiana* RuBPcase *in vivo* and *in vitro* were not successful. We therefore infer that *Nicotiana* RuBPcase is not a phosphoprotein (Agarwal, Saluja and Sachar, unpublished results).

The structure-function relationship of phosphorylation of SSU in spinach and moss was determined by dephosphorylating the purified RuBPcase with alkaline phosphatase (Kaul *et al.*, 1986; Saluja *et al.*, 1986). Dephosphorylation of moss and spinach RuBPcase resulted in dissociation of the SSUs from the catalytic LSUs. Concomitantly, there was a dramatic loss in the biological activity of the enzyme (Fig. 5). This is

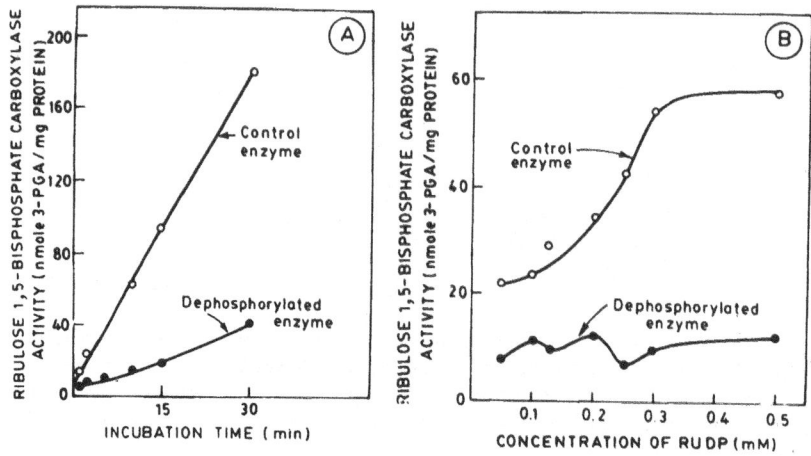

Figure 5. Differential catalytic activity of control (phosphorylated) and dephosphorylated RuBPcase in moss. Enzyme activity as a function of (A) time and (B) RuBP concentration (after Kaul *et al.*, 1986).

in agreement with the earlier observations in pea and spinach where reconstitution experiments strongly suggested that the LSU is catalytically more competent in association with SSU (Miziorko and Lorimer, 1983). Our findings clearly indicate that association of LSU with SSU in moss and spinach is inherently dependent on the phosphorylation of the SSU (Kaul *et al.*, 1986). A similar situation was observed in pea and *Cicer* where dephosphorylation of purified RuBPcase resulted in the dissociation of LSU and SSU with a concomitant loss in the enzyme activity (unpublished data: Murali, Saluja and Sachar; Agarwal, Saluja and Sachar) (Fig. 6).

It now appears that phosphorylation of RuBPcase does indeed modulate the enzyme activity in moss, pea, spinach, and *Cicer*. In all these systems, dephosphorylation of RuBPcase results in the dissociation of SSU as dimers from octameric LSU. In *Nicotiana*, RuBPcase does not appear to be a phosphoprotein and consequently treatment of purified enzyme with alkaline phosphatase does not result in the loss of the enzyme activity. It would be desirable to investigate other higher plants with a view to determine the phosphorylation status of RuBPcase. Another line of approach is to isolate and characterize protein kinases which specifically phosphorylate RuBPcase at the SSU *in vitro*. This will enable us to study the modulation of RuBPcase activity by phosphorylation *in vitro*.

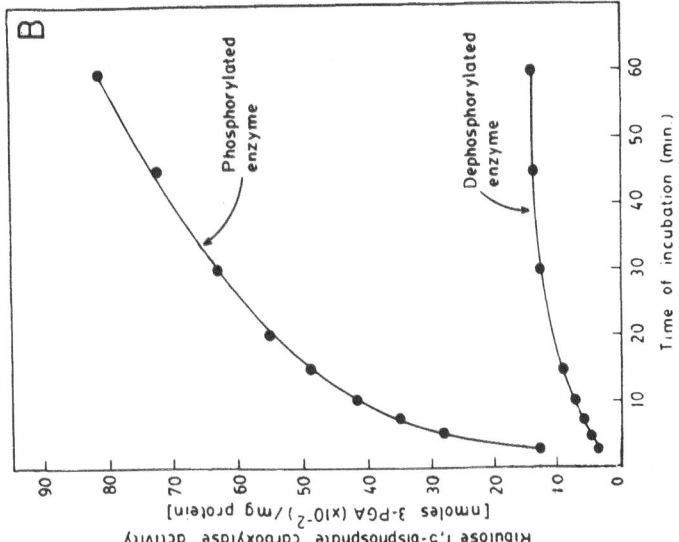

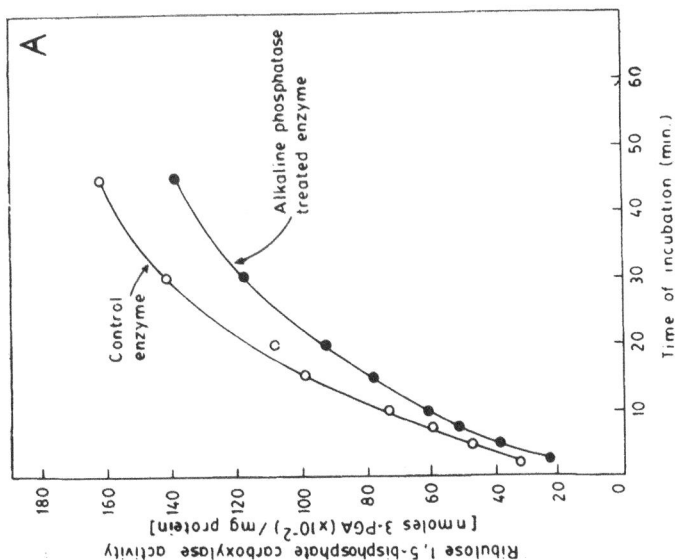

Figure 6. Differential catalytic activity of control (phosphorylated) and dephosphorylated RuBPcase from (A) *Nicotiana*, (B) *Cicer*, as a function of time (unpublished data, Agarwal, Saluja and Sachar).

306

VI. Conclusions

Many significant advances have been made in understanding the structure and function of RuBPcase. Among higher plants, this enzyme comprises eight large and eight small subunits. Catalytic activity of the enzyme resides in the LSUs. The SSU possibly has a regulatory role, but its true implication has not been ascertained. The LSU of RuBPcase is encoded in the chloroplast, while the SSU is encoded in the nucleus.

The transport mechanism of prSS from the cytosol to the chloroplast has been extensively studied. It has been convincingly shown that the transit peptide at the N′-terminus of prSS is crucial for its transport across the chloroplast membrane. Light does play a definitive role in the transport of prSS into the chloroplasts. The binding of the prSS to the envelope membrane and the translocation of the bound precursor across the membrane are light-dependent processes. Further studies have revealed that transport of prSS is an energy-dependent process and requirement for light could be sustituted by ATP. Nevertheless, the possible role of receptor(s) in the binding of prSS and the precise molecular mechanism of transport is not well understood. The assembly of LSU and SSU into holenzyme is aided by the LSU-binding protein and this process is also energy dependent. The role of the binding protein in the assembly of LSU and SSU into a holoenzyme of RuBPcase is fairly certain, but the precise molecular mechanism of the assembly process is far from clear.

There is sufficient evidence to show that the catalytic activity of RuBPcase is regulated by light. Recently, a stromal enzyme, RuBPcase activase, has been purified from spinach leaf tissue. RuBPcase activase catalyses the light-mediated activation of RuBPcase *in vivo*. It is now possible to activate RuBPcase *in vitro* with RuBPcase activase at physiological concentrations of CO_2 and Mg^{2+}. Only further studies can elucidate the precise role of activase in the metabolic regulation of RuBPcase. RuBPcase has been shown to be a phosphoprotein in some higher plants. However, some workers have claimed that phosphorylation is restricted to the SSU while others have reported phosphorylation of LSU. Labelling of RuBPcase *in vivo* with [^{32}Pi] invariably phosphorylated the small subunit in moss, pea, *Cicer* and spinach. Phosphorylation of SSU seems to play a crucial role in the regulation of RuBPcase, since dephosphorylation of the enzyme resulted in a dramatic decrease in its activity. On the contrary, RuBPcase isolated from *Nicotiana* is not a phosphoprotein. Further studies are necessary to elucidate the true significance of phosphorylation of RuBPcase.

Briefly then, a comprehensive understanding of gene expression of the two subunits of RuBPcase, transport of SSU across the chloroplast membrane and the assembly of subunits into the holoenzyme will provide strong foundations to formulate basic concepts of gene expression, transport of proteins across chloroplast membrane and assembly of multimeric proteins in plant systems.

Acknowledgements

Financial assistance from the Council of Scientific and Industrial Research to R.C. Sachar and Kanta Sachar for the scheme entitled "Metabolic regulation of ribulose-1, 5-*bis*phosphate carboxylase activity in green plants" (Sanction No: 38(704/88 EMR II) and award of SRF by Council of Scientific and Industrial Research to PM are gratefully acknowledged. We are thankful to Meenakshi Munshi and Athar for their help in preparation of this manuscript.

VII. References

Anderson LE, Price GB and Fuller RC (1968) Molecular diversity of the ribulose diphosphate carboxylase from photosynthetic microorganisms. *Science* **161**: 482–484

Anderson S and Smith SM (1986). Synthesis of the small subunit of ribulose-bisphosphate carboxylase from genes cloned into plasmids containing the SP6 promoter. *Biochem J* **240**: 709–715

Anderson T, Knight S, Schneider G, Lindquist Y, Lindquist T, Branden CI and Lorimer GH (1989) Crystal structure of the active site of ribulose bisphosphate carboxylase. *Nature* **337**: 229–234

Andrews TJ and Abel KM (1981) Kinetics and subunit interactions of ribulose bisphosphate carboxylase-oxygenase from the cyanobacterium, *Synechococcus* sp. *J Biol Chem* **256**: 8445–8451

Andrews TJ and Ballment B (1984) Active-site carbamate formation and reaction-intermediate-analog binding by ribulose bisphosphate carboxylase′ oxygenase in the absence of its small subunitss. *Proc Natl Acad Sci USA* **81**: 3660–3664

Badger MR (1980) Kinetic properties of ribulose-1, 5-bisphosphate carboxylase/oxygenase from *Anabaena variabilis*. *Arch Biochem Biophys* **201**: 247–254

Badger MR and Lorimer GH (1976) Activation of ribulose-1,5-bisphosphate oxygenase. The role of Mg^{2+} CO_2 and pH. *Arch Biochem Biophys* **175**: 723–729

Badger MR and Lorimer GH (1981) Interaction of sugar phosphates with the catalytic site of ribulose-1, 5-bisphosphate carboxylase. *Biochemistry* **20**: 2219–2225

Barcena JA and Shaw PJ (1985) Subunit arrangement of spinach ribulose-1, 5-bisphosphate carboxylase/oxygenase. *Planta* **163**: 141–144

Barraclough R and Ellis RJ (1980) Protein synthesis in chloroplasts. IX. Assembly of newly-synthesized large subunits into ribulose bisphosphate carboxylase in isolated intact pea chloroplasts. *Biochim Biophys Acta* **608**: 19–31

Bassham JA and Calvin M (1957) *The Path of Carbon in Photosynthesis*. Englewood Cliffs, Prentice Hall, New Jersey, pp 1–107

Bedbrook JR, Smith SM and Ellis RJ (1980) Molecular cloning and sequencing of cDNA encoding the precursor to the small subunit of chloroplast ribulose-1, 5-bisphosphate carboxylase. *Nature* **287**: 692–697

Bennett J (1977) Phosphorylation of chloroplast membrane polypeptides. *Nature* **269**: 344–346

Bennett J (1979) Chloroplast phosphoproteins. The protein kinase of thylakoid membrane is light-dependent. *FEBS Lett* **103**: 342–344

Berry JA, Lorimer GH, Pierce J, Seemann JR, Meeks J and Freas S (1987) Isolation, identification and synthesis of 2-carboxyarabinitol 1-phosphate, a diurnal regulator of ribulose bisphosphate carboxylase activity. *Proc Natl Acad Sci USA* **84**: 734–738

Berry JO, Nikolau BJ, Carr JP and Klessig DF (1985) Transcriptional and post-transcriptional regulation of ribulose-1, 5-bisphosphate carboxylase gene expression in light and dark-grown amaranth cotyledons. *Molec Cell Biol* **5**: 2238–2246

Berry JO, Nikolau BJ, Carr JP and Klessig DF (1986) Translational regulation of light-induced ribulose-1, 5-bisphosphate carboxylase gene expression in amaranth. *Molec Cell Biol* **6**: 2347–2353

Berry-Lowe SL, McKnight TD, Shah DM and Meagher RB (1982) The nucleotide sequence, expression and evolution of one member of a multigene family encoding the small subunit of ribulose-1, 5-bisphosphate carboxylase in soybean. *J Molec Appl Genet* **1**: 483–498

Blair GE and Ellis RJ (1973) Protein synthesis in chloroplasts. I. Light-driven synthesis of large subunit of Fraction I protein by isolated pea chloroplasts. *Biochim Biophys Acta* **319**: 223–234

Bloom MV, Milos P and Roy H (1983) Light-dependent assembly of ribulose-1, 5-bisphosphate carboxylase. *Proc Natl Acad Sci USA* **80**: 1013–1017

Bowien B, Mayer F, Spiess E, Pähler A, English U and Saenger W (1980) On the structure of crystalline ribulose bisphosphate carboxylase from *Alcaligenes eutrophus*. *Eur J Biochem* **106**: 405–410

Bradley D, van der Vies SM and Gatenby AA (1986) Expression of cyanobacterial and higher plant ribulose-1, 5-bisphosphate carboxylase genes in *Escherichia coli*. *Phil Trans R Soc Lond* **313**: 447–458

Broglie R, Coruzzi G, Lamppa G, Keith B and Chua N-H (1983) Structural analysis of nuclear genes coding for the precursor to the small subunit of wheat ribulose-1, 5-bisphosphate carboxylase. *Bio/Technology* **1**: 55–61

Brooks A and Portis AR Jr (1988) Protein bound ribulose bisphosphate correlates with deactivation of ribulose bisphosphate carboxylase in leaves. *Plant Physiol* **87**: 244–249

308

Buchanan BB and Schürmann P (1973) Ribulose 1, 5-diphosphate carboxylase: A regulatory enzyme in the photosynthetic assimilation of carbon dioxide. In: (B L Horecker and E R Standlman eds) *Current Topics in Cellular Regulation* Vol 7, Academic Press, New York, pp 1–20

Cannon S, Wang P and Roy H (1986) Inhibition of ribulose bisphosphate carboxylase assembly by antibody to a binding protein. *J Cell Biol* **103**: 1327–1335

Cashmore AR (1983) Nuclear genes encoding the small subunit of ribulose-1, 5-bisphosphate carboxylase. In: (M Kosuge and A Hollander eds) *Genetic Engineering of Plants* Plenum Press New York, pp 29–38

Cashmore AR, Szabo L, Timko M, Kausch A, Van den Broeck G, Schrejer P, Bohnert HJ, Herrera-Estrella L, Van Montagu M and Schell J (1985) Import of polypeptides into chloroplasts. *Bio/Technology* **3**: 803–808

Chan DM and Wildman SG (1972) Chloroplast DNA codes for the primary structure of the large subunit of fraction I protein. *Biochim Biophys Acta* **277**: 677–680

Charles L, Funckes-Shippy and Levine AD (1985) Cytokinin regulates the expression of nuclear genes required for photosynthesis. In: (K Steinbeck, C Arntzen, L Bogorad and S Bonitz eds) *Molecular Biology of the Photosynthetic Apparatus,* Cold Spring Harbor Laboratories Press, New York, pp 407–411

Chaudhari P and Roy H (1989) Delayed osmotic effect on *in vitro* assembly of RuBisCO. Relationship to large subunit-binding protein complex dissociation. *Plant Physiol* **89**: 1366–1371

Chollet R and Anderson LL (1976) Regulation of ribulose-1, 5-bisphosphate carboxylase-oxygenase activities by temperature pretreatment and chloroplast metabolites. *Arch Biochem Biophys* **176**: 344–351

Christeller JT and Laing WA (1978) A kinetic study of ribulose bisphosphate carboxylase from the photosynthetic bacterium *Rhodospirillum rubrum. Biochem J* **173**: 467–473

Chu DK and Bassham JA (1972) Inhibition of ribulose-1, 5-diphosphate carboxylase by 6-phosphogluconate. *Plant Physiol* **50**: 224–227

Chu DK and Bassham JA (1973) Activation and inhibition of ribulose 1, 5-diphosphate caboxylase by 6-phosphogluconate. *Plant Physiol* **52**: 373–379

Chu DK and Bassham JA (1975) Regulation of ribulose-1, 5-diphosphate carboxylase by substrates and other metabolites. Further evidence for several types of binding sites. *Plant Physiol* **55**: 720–726

Chua N-H and Schmidt GW (1978) Post-translational transport into intact chloroplast of a precursor to the small subunit of ribulose-1, 5-bisphosphate carboxylase. *Proc Natl Acad Sci USA* **75**: 6110–6114

Chua N-H and Schmidt GW (1979) Transport of proteins into mitochondria and chloroplasts. *J. Cell Biol* **81**: 461–483

Cline K, Werner-Washburne M, Andrews J and Keegstra K (1984) Thermolysin is a suitable protease for probing the surface of intact pea chloroplasts. *Plant Physiol* **75**: 675–678

Cline K, Werner-Washburne M, Lubben TH and Keegstra K (1985) Precursors to the two nuclear-encoded chloroplast proteins bind to the outer envelope membrane before being imported into the chloroplasts. *J. Biol Chem* **260**: 3691–3696

Coen DM, Bedbrook JR, Bogorad L and Rich A (1977) Maize DNA fragment encoding the large subunit of ribulose-bisphosphate carboxylase. *Proc Natl Acad Sci USA* **74**: 5487–5491

Cornwell KL and Keegstra K (1987) Evidence that a chloroplast surface protein is associated with a specific binding site for the precursor to the small subunit of ribulose 1, 5-bisphosphate carboxylase. *Plant Physiol* **85**: 780–785

Coruzzi G, Broglie R, Cashmore AR and Chua N-H (1983) Nucleotide sequences of two pea cDNA clones encoding the small subunit of ribulose-1,5-bisphosphate carboxylase and the major chlorophyll a/b-binding thylakoid polypeptide. *J Biol Chem* **258**: 1399–1402

Coruzzi G, Broglie R, Edwards C and Chua N-H (1984) Tissue specific and light-regulated expression of a pea nuclear gene encoding the small subunit of ribulose-1, 5-bisphosphate carboxylase. *EMBO J* **3**: 1671–1679

Dean C, Pichersky E and Dunsmuir P (1989) Structure, evolution and regulation of Rbc S genes in higher plants. *Annu Rev Plant Physiol Mol Biol* **40**: 415–430

Dean C, van den Elzen P Tamaki S, Dunsmuir P and Bedbrook JR (1985) Differential expression of the eight genes of the *Petunia* ribulose bisphosphate carboxylase small subunit multigene family. *EMBO J* **4**: 3055–3061

Dean C, van den Elzen P, Tamaki S, Black M, Dunsmuir P and Bedbrook JR (1987) Molecular characterization of the rbc S gene multigene family of *Petunia* (Mitchell). *Mol Gen Genet* **206**: 465–474 della-Cioppa G, Kishore GM, Beachy RN and Farley RT (1987) Protein trafficking in plant cells. *Plant Physiol* **84**: 965–968

Deng XW and Gruissem W (1987) Control of plastid gene expression during development. The limited role of transcriptional regulation. *Cell* **49**: 379–387

Donnelly MI, Hartman FC and Ramakrishnan V (1984) The shape of ribulose bisphosphate carboxylase/oxygenase in solution as inferred from small angle neutron scattering. *J Biol Chem* **259**: 406–411

Donnelly MI, Stringer CD and Hartman FC (1983) Characterization of the activator site of *Rhodospirillum rubrum* ribulose bisphosphate carboxylase/oxygenase. *Biochemistry* **22**: 4346–4352

Dron M, Rahaire M, Rochaix JD and Mets LT (1983) First DNA sequence of a chloroplast mutation. A mis-sense alteration in the ribulose bisphosphate carboxylase large subunit gene. *Plasmid* **9**: 321–324

Dunsmuir P, Smith S and Bedbrook J (1983) A number of different nuclear genes for the small subunit of RuBPcase are transcribed in *Petunia*. *Nucl Acids Res* **11**: 4117–4183

Eilers M and Schatz G (1988) Protein unfolding and the energetics of protein translocation across biological membranes. *Cell* **52**: 481–483

Eisenberg D, Baker TS, Suh SW and Smith WW (1978) Structural studies of ribulose 1, 5-bisphosphate carboxylase/oxygenase. In: (HW Siegelman and G Hind eds) *Photosynthetic Carbon Assimilation*, Vol 11, Plenum Press, New York, pp 271–281

Ellis RJ (1977) Protein synthesis by isolated chloroplasts. *Biochim Biophys Acta* **463**: 185–215

Ellis RJ (1979) The most abundant protein in the world. *TIBS* **4**: 241–244

Ellis RJ (1981) Chloroplast proteins: Synthesis, transport and assembly. *Annu Rev Plant Physiol* **32**: 111–137

Estelle M, Hanks J, McIntosh L and Somerville C (1985) Site-specific mutagenesis of ribulose-1, 5-bisphosphate carboxylase/oxygenase: Evidence that carbamate formation at Lys 191 is required for catalytic activity. *J Biol Chem* **260**: 9523–9526

Flugge UI and Hinz G (1986) Energy dependence of protein translocation into chloroplasts. *Eur J Biochem* **160**: 563–570

Fourcroy P, Klein-Ende D and Lambert C (1985) Phytochrome control of gene expression in radish seedlings. II Far-red light mediated appearance of the ribulose-1,5 bisphosphate carboxylase and the mRNA for its small subunit. *Plant Sci Lett* **37**: 235–244

Friedman AL and Keegstra K (1989) Chloroplast protein import: Quantitative analysis of receptor mediated binding. *Plant Physiol* **89**: 993–999

Gallagher TF and Ellis RJ (1982) Light stimulated transcripts of genes for two chloroplast polypeptides in isolated pea leaf nuclei. *EMBO J* **1**: 1493–1498

Gatenby AA, van der Vies SM and Bradley D (1985) Assembly in *E. coli* of a functional multisubunit ribulose bisphosphate carboxylase from a blue green alga. *Nature* **314**: 617–620

Gatenby AA, van der Vies SM and Rothstein SJ (1987) Co-expression of both the maize large and wheat small subunit genes of ribulose-bisphosphate carboxylase in *Escherichia coli*. *Eur J Biochem* **168**: 227–231

Gibson JL and Tabita FR (1977) Characterization of antiserum directed against form II ribulose-1, 5-bisphosphate carboxylase from *Rhodopseudomonas sphaeroides*. *J Bacteriol* **131**: 1020–1022

Goldschmidt-Clermont M and Rahiri M (1986) Sequence, evolution and differential expression of the two genes encoding variant small subunits of ribulose bisphosphate carboxylase/oxygenase in *Chlamydomonas reinhardtii*. *J Mol Biol* **191**: 421–432

Goloubinoff P, Gatenby AA and Lorimer GH (1989a) GroE heat-shock proteins promote assembly of foreign prokaryotic ribulose bisphosphate carboxylase oligomers in *Escherichia coli*. *Nature* **337**: 44–47

Goloubinoff P, Christeller JT, Gatenby AA and Lorimer GH (1989b). Reconstitution of active dimeric ribulose bisphosphate carboxylase from an unfolded state depends on two chaperonin proteins and Mg-ATP. *Nature* **342**: 884–889

Gordon GLR, Lawlis VB and McFadden BA (1980) 2-Carboxy-D-hexitol-1, 6-bisphosphate: An inhibitor of D-ribulose-1, 5-bisphosphate carboxylase/oxygenase. *Arch Biochem Biophys* **199**: 400–412

Grebanier AE, Champagne D and Roy H (1978) Effects of Mg^{2+} and substrates on the conformation of ribulose-1, 5-bisphosphate carboxylase. *Biochemistry* **17**: 5150–5155

Green PJ, Kay SA and Chua N-H (1987) Sequence specific interaction of nuclear factor with light-responsive elements upstream of theh rbc S-3A gene. *EMBO J* **6**: 2543–2549

Grossman A, Bartlett S and Chua N-H (1980) Energy-dependent uptake of cytoplasmically synthesized polypeptides by chloroplasts. *Nature* **285**: 625–628

Gruissem W and Zurawski G (1985a) Identification and mutational analysis of the promotor for a spinach chloroplast transfer RNA gene. *EMBO J* **4**: 1637–1644

Gruissem W and Zurawski G (1985b). Analysis of promotor regions for the spinach chloroplast rbc L, atp B and psb A genes. *EMBO J* **4**: 3375–3383

Guitton C and Mache R (1987) Phosphorylation *in vitro* of the large subunit of the ribulose-1, 5-bisphosphate carboxylase and of the glyceraldehyde-3-phosphate dehydrogenase. *Eur J Biochem* **166**: 249–254

Gutteridge S, Parry MAJ, Burton S, Keys AJ, Mudd A, Feeney J, Servaites JC and Pierce J (1986) A nocturnal inhibitor of carboxylation in leaves. *Nature* **324**: 274–276

Hall NP, Pierce J and Tolbert NE (1981) Formation of a carboxy arabinitol bisphosphate complex with ribulose bisphosphate carboxylase/oxygenase and theoretical specific activity of the enzyme. *Arch Biochem Biophys* **212**: 175–179

Hanley-Bowdoin L and Chua N-H (1987) Chloroplast promotors. *TIBS* **12**: 67–70

Hartman FC, Norton IL, Stringer CD and Schloss JV (1978) Attempts to apply affinity labeling techniques to ribulose bisphosphate carboxylase/oxygenase. In: (H W Siegelman and G Hind, eds) *Photosynthetic Carbon Assimilation*, Vol II, Plenum Press, New York, pp 245–269

Heber U, Takahawa U, Neimanis S and Shimizu-Takahawa M (1982) Transport as the basis of KOK effect. Levels of some photosynthetic intermediates and activation of light regulated enzymes during photosynthesis of chloroplasts and green leaf protoplasts. *Biochim Biophys Acta* **679**: 289–299

Hemmingsen SM and Ellis RJ (1986) Purification and properties of ribulose bisphosphate carboxylase large subunit binding protein. *Plant Physiol* **80**: 269–276

Hemmingsen SM, Woolford C, van der Vies SM, Tilly K, Dennis DT, Georgopoulos CP, Hendrix RW and Ellis RJ (1988) Homologous plant and bacterial proteins chaperone oligomeric protein assembly. *Nature* **333**: 330–334

Highfield PE and Ellis RJ (1978) Synthesis and transport of the small subunit of chloroplast ribulose bisphosphate carboxylase. *Nature* **271**: 420–424

Hinz G and Flugge GI (1988) Phosphorylation of a 51 kDa envelope polypeptide involved in protein translocation into chloroplast. *Eur J Biochem* **175**: 649–659

Holbrook GP, Bowes G and Salvucci ME (1989) Degradation of 2-carboxyarabinitol 1-phosphate by a specific chloroplast phosphatase. *Plant Physiol* **90**: 673–678

Inamine G, Nash B, Weissbach H and Brot N (1985) Light regulation of the synthesis of the large subunit of ribulose-1, 5-bisphosphate carboxylase in peas: Evidence for translational control. *Proc Natl Acad Sci USA* **82**: 5690–5694

Incharoensakdi A, Takabe T and Akazawa T (1986) Role of the small subunit of ribulose-1, 5-bisphosphate carboxylase/oxygenase in the activation process. *Arch Biochem Biophys* **248**: 62–70

Jensen RG and Bahr JT (1977) Ribulose-1, 5-bisphosphate carboxylase-oxygenase. *Annu Rev Plant Physiol* **28**: 379–400

Jordan DB and Chollet R (1983) Inhibition of ribulose bisphosphate carboxylase by substrate ribulose bisphosphate. *J Biol Chem* **258**: 13752–13758

Jordan DB, Chollet R and Ogren WL (1983) Binding of phosphorylated effectors by active and inactive forms of ribulose-1, 5-bisphosphate carboxylase. *Biochemistry* **22**: 3410–3418

Kaderbhai MA, Pickering T, Austen BM and Kaderbhai N (1988) A photoactivatable synthetic transit peptide labels 30 kDa and 52 kDa polypeptides of the chloroplast inner envelope membrane. *FEBS Lett* **232**: 313–316

Karlin-Neuman CA and Tobin EM (1986) Transit peptides of nuclear encoded chloroplast proteins share a common amino acid framework. *EMBO J* **5**: 9–13

Kaul R, Saluja D and Sachar RC (1986) Phosphorylation of small subunit plays a crucial role in the regulation of RuBPcase in moss and spinach. *FEBS Lett* **209**: 63–70

Kawashima N and Wildman SG (1970) Fraction I protein. *Annu Rev Plant Physiol* **21**: 325–358

Kawashima N and Wildman SG (1972) Studies on fraction I protein. IV. Mode of interaction of primary structure in relation to whether chloroplast or nuclear DNA contains the code for a chloroplast protein. *Biochim Biophys Acta* **262**: 42–49

Keegstra K, Olsen JO and Theg SM (1989) Chloroplastic precursors and their transport across the envelope membranes. *Annu Rev Plant Physiol Mol Biol* **40**: 491–502

Kieras FJ and Hasselkorn R (1968) Properties of ribulose-1,5-diphosphate carboxylase (carboxy-dismutase) from Chinese cabbage and photosynthetic microorganisms. *Plant Physiol* **43:** 1264–1270

Kobza I and Seemann JR (1988) Mechanisms for the light regulation of ribulose-1, 5-bisphosphate carboxylase activity and photosynthesis in intact leaves. *Proc Natl Acad Sci USA* **85:** 3815–3819

Kuhlemeier C, Green PJ and Chua N-H (1987a) Regulation of gene expression in higher plants. *Annu Rev Plant Physiol* **38:** 221–257

Kuhlemeier C, Fluhr R, Green PJ and Chua N-H (1987b) Sequences in the pea rbsc S-3A gene have homology to constitutive mammalian enhancer but function as negative regulatory elements. *Genes and Development* **1:** 247–255

Laing WA and Christeller JT (1976) A model for the kinetics of activation and catalysis of ribulose-1, 5-bisphosphate carboxylase. *Biochem J* **159:** 563–570

Langridge P (1981) Synthesis of the large subunit of spinach ribulose bisphosphate carboxylase may involve a precursor polypeptide. *FEBS Lett* **123:** 85–89

Lawlor DW (1987) *Photosynthesis: metabolism, Control and Physiology.* Longman Scientific and Technical, Harlov, England pp 1–266

Lebrun M, Washman G and Fressinet G (1987) Nucleotide sequence of a gene encoding corn ribulose-1, 5-bisphosphate carboxylase/oxygenase small subunit (rbc S). *Nucl Acids Res* **15:** 4360–4364

Lennox CP and Ellis RJ (1986) The carboxylase large subunit binding protein: photoregulation and reversible dissociation. *Biochem Soc Trans* **14:** 9–11

Lorimer GH (1979) Evidence for the existence of discrete activation and substrate sites for CO_2 on ribulose 1, 5-bisphosphate carboxylase. *J Biol Chem* **254:** 5599–5601

Lorimer GH (1981a) The carboxylation and oxgenation of ribulose-1, 5-bisphosphate: The primary events in photosynthesis and photorespiration. *Annu Rev Plant Physiol* **32:** 349–383

Lorimer GH (1981b) Ribulose bisphosphate carboxylase: Amino acid sequence of the peptide bearing the activator carbon dioxide. *Biochemistry* **20:** 1236–1240

Lorimer GH and Miziorko HM (1980) Carbamate formation on the E-amino group of a lystl residue as the basis for the activation of ribulose bisphosphate carboxylase by CO_2 and Mg^{2+}. *Biochemistry* **19:** 5321–5328

Lorimer GH, Badger MR and Andrews TJ (1976) The activation of ribulose-1, 5-bisphosphate carboxylase by carbon dioxide and magnesium ions. Equilibria kinetics, a suggested mechanism and physiological implication. *Biochemistry* **15:** 529–536

Lorimer GH, Badger MR and Heldt HW (1983) The activation of ribulose-1, 5-bisphosphate carboxylase/oxygenase. In: (H W Siegelman and G Hind, eds) *Photosynthetic Carbon Assimilation,* Plenum Press, New York, pp 283–306

Lubben TH and Keegstra K (1986) Efficient *in vitro* import of a cytosolic heat shock protein into pea chloroplasts. *Proc Natl Acad Sci USA* **83:** 5502–5506

Lubben TH, Theg SM and Keegstra K (1988) Transport of proteins into chloroplast. *Photosynth Res* **17:** 173–194

Lucero AH, Lin ZF and Racker E (1982) Protein kinases from spinach chloroplasts. II Protein substrate specificity and kinetic properties. *J Biol Chem* **257:** 12157–12160

Machler F and Nosberger J (1980) Regulation of ribulose bisphosphate carboxylase activity in intact wheat leaves by light, CO_2 and temperature. *J Exp Bot* **31:** 1485–1491

Manzara T and Gruissem W (1988) Organization and expression of the genes encoding ribulose-1, 5-bisphosphate carboxylase in higher plants. *Photosynth Res* **16:** 117–139

Mazur BJ and Chui CF (1985) Sequence of a genomic DNA clone for the small subunit of ribulose bisphosphate carboxylase-oxygenase from tobacco. *Nucl Acids Res* **13:** 2373–2386

McFadden BA (1974) The oxygenase activity of ribulose diphosphate carboxylase from *Rhodosirillum rubrum. Biochem Biophys Res Commun* **60:** 312–317

McIntosh L, Poulsen C and Bogorad L (1980) Chloroplast gene sequence for the large subunit of ribulose bisphosphate carboxylase of maize. *Nature* **288:** 556–560

Milos P and Roy H (1984) ATP-released large subunits participate in the assembly of RuBP carboxylase. *J Cell Biochem* **24:** 153–162

Mishkind ML, Wessler SR and Schmidt GW (1985) Functional determinants in transit sequences: Import and partial maturation by vascular plant chloroplast of the ribulose-1, 5-bisphosphate carboxylase small subunit of *Chlamydomonas. J Cell Biol* **100:** 226–234

Miziorko HM (1979) Ribulose-1, 5-bisphosphate carboxylase. Evidence in support of the existence of discrete CO_2 activator and CO_2-substrate sites. *J Biol Chem* **254**: 270–272

Miziorko HM and Lorimer GH (1983) Ribulose-1, 5-bisphosphate carboxylase-oxygenase. *Annu Rev Biochem* **52**: 507–535

Miziorko HM and Mildvan AS (1974) Electron paramagnetic resonance, ^1H and ^{13}C nuclear magnetic resonance studies of the interaction of manganese and bicarbonate with ribulose-1, 5-bisphosphate carboxylase. *J Biol Chem* **249**: 2743–2750

Miziorko HM, Behnke CE and Honkon EC (1982) Protein liganding to the activator cation of ribulose bisphosphate carboxylase. *Biochemistry* **21**: 6669–6674

Molt KA, Jensen RG O'Leary JW and Berry JA (1984) Photosynthesis and ribulose-1, 5-bisphosphate concentrations in intact leaves of *Xanthium strumarium* L. *Plant Physiol* **76**: 968–971

Musgrove JE, Johnson RA and Ellis RJ (1987) Dissociation of the large subunit binding protein into dissimilar subunits. *Eur J Biochem* **163**: 529–534

Muto S and Shimogawara K (1985) Calcium and phospholipid-dependent phosphorylation of ribulose-1, 5-bisphosphate carboxylase/oxygenase small subunit by a chloroplast envelope-bound protein kinase *in situ*. *FEBS Lett* **193**: 88–92

Nelson T, Harpstor MH, Mayfield SP and Taylor WC (1984) Light regulated gene expression during maize leaf development. *J Cell Biol* **98**: 558–564

Oelmuller R, Dietrich G, Link G and Mohr H (1986) Regulatory factors involved in gene-expression (subunits of ribulose-1, 5-bisphosphate carboxylase) in mustard (*Sinapis alba* L.) cotyledons. *Planta* **169**: 260–266

O'Leary MH, Jaworski RJ and Hartman FC (1979) ^{13}C nuclear magnetic resonance of the CO_2 activation of ribulose bisphosphate carboxylase from *Rhodospirillum rubrum*. *Proc Natl Acad Sci USA* **76**: 673–675

Olsen LJ, Theg SM, Selmen B and Keegstra K (1989) ATP is required for the binding of precursor proteins to chloroplast. *J Biol Chem* **264**: 6724–6729

Ostrem JA, Ramage RT and Bohnert HJ (1989) Deletion of the carboxy-terminal of the transit peptide affects processing but not import or assembly of the small subunit of ribulose-1, 5-bisphosphate carboxylase. *J Biol Chem* **264**: 3662–3666

Pain D and Blobel G (1987) Protein import into chloroplasts requires a chloroplast ATPase. *Proc Natl Acad Sci USA* **84**: 3288–3292

Pain D, Kanwar YS and Blobel G (1988) Identification of a receptor for protein import into chloroplasts and its location to envelope contact zones. *Nature* **331**: 232–237

Parry MAJ, Schmidt CNG, Cornelius MJ, Millard BN, Burton S, Gutteridge S, Dyer TA and Keys AJ (1987) Variations in properties of ribulose-1, 5-bisphosphate carboxylase from various species related to differences in amino acid sequences. *J Exp Bot* **38**: 1260–1271

Paulsen JM and Lane MD (1986) Spinach ribulose diphosphate carboxylase. I Purification and properties of the enzyme. *Biochemistry* **5**: 2350–2357

Pederson TA, Kirk M and Bassham JA (1966) Light-dark transients in levels of intermediate compounds during photosynthesis in air-adapted *Chlorella*. *Physiol Plant* **19**: 219–231

Perchorowitz JJ and Jensen RG (1983) Photosynthesis and activation of ribulose bisphosphate carboxylase in wheat seedlings. *Plant Physiol* **71**: 955–960

Pettersson G and Ryde-Pettersson R (1988) Effects of metabolite binding to ribulose bisphosphate carboxylase on the activity of the calvin photosynthesis cycle. *Eur J Biochem* **177**: 351–358

Pfisterer J, Lachmann P and Kloppstech K (1982) Transport of proteins into chloroplast. Binding of nuclear-coded chloroplast proteins to the chloroplast envelope. *Eur J Biochem* **120**: 143–148

Piechulla B, Pichersky E, Cashmore AR and Gruissem W (1986) Expression of nuclear and plastid genes for photosynthesis-specific proteins during tomato fruit development and ripening. *Plant Mol Biol* **7**: 367–376

Pierce J, Tolbert NE and Barker R (1980) Interaction of ribulose bisphosphate carboxylase/oxygenase with transition state analogs. *Biochemistry* **19**: 934–942

Polans NO, Weeden NF and Thompson WF (1985) Inheritance, organisation and mapping of rbc S and cab multigene families in pea. *Proc Natl Acad Sci USA* **82**: 5083–5087

Pon NG, Robin BR and Calvin M (1963) Mechanism of the carboxydismutase reaction. 1 The effect of the preliminary incubation of substrates, metal ions and enzyme on activity. *Biochem J* **338** 7–19

Portis AR, Salvucci ME and Ogren WL (1986) Activation of ribulose bisphoshate carboxylase/oxygenase at physiological CO_2 and ribulose bisphosphate concentrations by RuBisCO activase. *Plant Physiol* **82**: 907–971

Portis AR Jr, Chon CJ, Mosbach A and Heldt HW (1977) Fructose and sedoheptulose bisphosphatase. The sites of a possible control of CO_2 fixation by the light dependent changes of the stromal Mg^{2+} concentration. *Biochim Biophys Acta* **461**: 313–325

Poulsen C, Fluhr R, Kauffman JM, Boutry M and Chua N-H (1986) Characterization of a rbc S gene from *Nicotiana plumbaginifolia* and expression on an rbc S-CAT chimeric gene in homologous and heterologous nuclear background. *Mol Gen Genet* **205**: 193–200

Purczeld P, Chen CJ, Portis AR Jr, Heldt HW and Heber U (1978) The mechanism of the control of carbon fixation by the pH in the chloroplast stroma. Studies with nitrite-mediated protein transfer across the envelope. *Biochim Biophys Acta* **501**: 488–498

Pushkin AV, Tsurpun VL, Solovjera NA, Shubin VV, Evstigneeva ZG and Ketovich WL (1982) High molecular weight pea leaf protein similar to the groE protein of *Escherichia coli*. *Biochim Biophys Acta* **70**: 379–384

Reiss B, Wasmann CC and Bohnert HJ (1987) Regions in the transit peptide of SSU essential for transport into chloroplasts. *Mol Gen Genet* **209**: 116–121

Ritland K and Clegg MT (1987) Evolutionary analysis of plant DNA sequences. *Amer Nat* **130**: S74–S100

Robinson C and Ellis RJ (1984a) Transport of proteins into chloroplasts. Partial purification of a chloroplast protease involved in the processing of imported precursor polypeptides. *Eur J Biochem* **142**: 337–342

Robinson C and Ellis RJ (1984b) Transport of proteins into chloroplasts. The precursor of small subunit of ribulose bisphosphate carboxylase is processed to the mature size in two steps. *Eur J Biochem* **142**: 343–346

Robinson C and Ellis RJ (1985) Transport of proteins into chloroplasts. The effect of incorporation of amino acid analogs on the import and processing of chloroplast polypeptides. *Eur J Biochem* **152**: 67–73

Robinson PD, Whitman WB, Waddill F, Riggs AF and Tabita FR (1980) Isolation and sequence of the pyridoxal 5′-phosphate active site peptide from *Rhodospirillum rubrum* ribulose-1, 5-bisphosphate carboxylase/oxygenase. *Biochemistry* **19**: 4848–4853

Robinson SP and Portis AR Jr (1988a) Involvement of stromal ATP in the light activation of ribulose-1, 5-bisphosphate carboxylase/oxygenase in intact isolated chloroplasts. *Plant Physiol* **86**: 293–298

Robinson SP and Portis AR Jr (1988b) Release of the nocturnal inhibitor, carboxyarabinitol-1-phosphate from ribulose bisphosphate carboxylase/oxygenase by RuBisCo activase. *FEBS Lett* **233**: 413–416

Robinson SP and Portis AR Jr (1989) Ribulose-1, 5-bisphosphate carboxylase/oxygenase activase protein prevents the *in vitro* decline in activity of ribulose-1,5-bisphosphate carboxylase/oxygenase. *Plant Physiol* **90**: 968–971

Robinson SP, Streusand VJ, Chatfield JM and Portis AR Jr (1988) Purification and assay of RuBisCO activase from leaves. *Plant Physiol* **88**: 1008–1014

Rodermel SL and Bogorad L (1985) Maize plastid photogenes: Mapping and photoregulation of transcript levels during light-induced development. *J Cell Biol* **100**: 463–476

Roy H (1989) RuBisCO Assembly: A model system for studying the mechanism of chaperone action. *Plant Cell* **1**: 1035–1042

Roy H, Adari H and Costa KA (1979) Characterization of free subunits of ribulose-1, 5-bisphosphate carboxylase. *Plant Sci Lett* **16**: 305–318

Roy H, Costa KA and Adari H (1978) Free subunits of ribulose bisphosphate carboxylase in pea leaves. *Plant Sci Lett* **11**: 159–168

Roy H, Chaudhari P and Cannon S (1988a) Incorporation of large subunits into ribulose bisphosphate carboxylase in chloroplast extracts. Influence of added small subunits and of conditions during synthesis. *Plant Physiol* **86**: 44–49

Roy H, Hubbs A and Cannon S (1988b) Stability and dissociation of the large subunits RuBisCO binding protein complex *in vitro* and *in organello*. *Plant Physiol* **86**: 50–53

Roy H, Bloom M, Milos P and Monroe M (1982) ATP-released large subunits participate in the assembly of RuBP carboxylase. *J Cell Biol* **94**: 20–27

Ryan FJ, Jolly SO and Tolbert NE (1974) Ribulose diphosphate oxygenase. V. Presence in ribulose diphosphate carboxylase from *Rhodospirillum rubrum*. *Biochem Biophys Res Commun* **59**: 1233–1241

Saluja D, Kaul R and Sachar RC (1986) Phosphorylation plays a crucial role in the regulation of ribulose-1, 5-bisphosphate carboxylase in spinach. In: (OL Kon, MCM Chung, PLH Hway, Sai-Far Leong, KH Lok, P Thiyagarajah and PTH Wong eds) *Contemporary Themes in Biochemistry*. Proc 4th FAOB Congress, Singapore **6**: 384–385

Salvucci ME and Anderson JC (1987) Factors affecting the activation state and the level of total activity of ribulose bisphosphate carboxylase in tobacco protoplasts. *Plant Physiol* **85**: 66–71

Salvucci ME, Holbrook GP, Anderson JC and Bowes G (1988) NADPH-dependent metabolism of ribulose bisphosphate carboxylase-oxygenase inhibitor 2-carboxyarabinitol 1-phosphate by a chloroplast protein. *FEBS Lett* **231**: 197–201

Salvucci ME, Portis AR and Ogren WL (1985) A soluble chloroplast protein catalyzes ribulose bisphosphate carboxylase/oxygenase activation *in vivo*. *Photosynth Res* **7**: 193–201

Salvucci ME, Portis AR Jr and Ogren WL (1986) Light and CO_2 response of ribulose-1, 5-bisphosphate carboxylase/oxygenase activation in *Arabidopsis* leaves. *Plant Physiol* **80**: 655–659

Salvucci ME, Werneke JM, Ogren WL and Portis AR Jr (1987) Purification and species identification of RuBisCO activase. *Plant Physiol* **84**: 930–936

Sasaki Y, Nakamura Y and Matsuno R (1987) Regulation of gene expression of ribulose bisphosphate carboxylase in greening pea leaves. *Plant Mol Biol* **8**: 375–382

Sasaki Y, Tomoda Y and Kamikubo T (1984) Light regulates the gene expression of ribulose bisphosphate carboxylase at the levels of transcription and gene dosage in greening pea leaves. *FEBS Lett* **173**: 31–35

Sasaki Y, Ishiye M, Sakihama T and Kamikubo T (1981) Light induced increase in mRNA activity coding for the small subunit of ribulose-1, 5-bisphosphate carboxylase. *J Biol Chem* **256**: 2315–2320

Schloss JV and Lorimer GH (1982) The stereochemical course of ribulose bisphosphate carboxylase. Reductive trapping of the 6-carbon reaction-intermediate. *J Biol Chem* **257**:4691–4694

Schmidt GW and Mishkind ML (1986) The transport of proteins into chloroplasts. *Annu Rev Biochem* **55**: 879–912

Schnyder H, Machler F and Nosberger J (1984) Influence of temperature and O_2 concentration on photosynthesis and light activation of ribulose bisphosphate carboxylase in intact leaves of white clover (*Trifolium repens* L.). *J Exp Bot* **35**: 147–156

Seemann JR, Berry JA Freas SM and Krump MA (1985) Regulation of ribulose bisphosphate carboxylase activity *in vivo* by a light-modulated inhibitor of catalysis. *Proc Natl Acad Sci USA* **82**: 8024–8028

Seemann JR, Kobza J and Moore BD (1990) Metabolism of 2-carboxyarabinitol 1-phosphate and regulation of ribulose 1, 5-bisphosphate carboxylase activity. *Photosynth Res* **23**: 119–130

Servaites JC (1985) Binding of a phosphorylated inhibitor to ribulose bisphosphate carboxylase/oxygenase during the night. *Plant Physiol* **78**: 839–843

Sheen JY and Bogorad L (1985) Differential expression of the ribulose bisphosphate carboxylase large subunit gene in bundle sheath and mesophyll cells of developing maize leaves is influenced by light. *Plant Physiol* **79**: 1072–1076

Sheen JY and Bogorad L (1986) Expression of the ribulose-1, 5-bisphosphate carboxylase large subunit gene and three small subunit genes in two cell types of maize leaves. *EMBO J* **5**: 3417–3422

Shinozaki K and Suguira M (1982) The nucleotide sequence of the tobacco chloroplast gene for the large subunit of ribulose-1, 5-bisphosphate carboxylase/oxygenase. *Gene* **20**: 91–100

Siegel MI, Wishnick M and Lane MD (1972) Ribulose-1, 5-bisphosphate carboxylase. In: (P.D Boyer, ed) *The Enzymes*, Vol VI, Academic Press, New York, pp 169–192

Silverthorne J and Tobin EM (1984) Demonstration of transcriptional regulation of specific genes by phytochrome action. *Proc Natl Acad Sci USA* **81**: 1112–1116

Smith SM and Ellis RJ (1979) Processing of the small subunit precursor of ribulose bisphosphate carboxylase and its assembly into whole enzymes are stromal events. *Nature* **278**: 662–664

Smith SM and Ellis RJ (1981) Light-stimulated accumulation of transcripts of nuclear and chloroplast genes for ribulose bisphosphate carboxylase. *J Mol Appl Genet* **1**: 127–137

Soll J (1988) Purification and characterization of a chloroplast outer-envelope-bound, ATP-dependent protein kinase. *Plant Physiol* **87**: 898–903

Soll J and Buchanan BB (1983) Phosphorylation of chloroplast ribulose bisphosphate carboxylase/oxygenase small subunit by an envelope-bound protein kinase *in situ*. *J Biol Chem* **258**: 6686–6689

Soll J, Fischer I and Keegstra K (1988) A GTP-dependent protein kinase is located in the outer envelope membrane of pea chloroplasts. *Planta* **176**: 488–496

Somerville CR, Portis AR and Ogren WL (1982) A mutant of *Arabidopsis thaliana* which lacks activation of RuBP carboxylase *in vivo*. *Plant Physiol* **70**: 381–387

Spreitzer RJ and Mets LT (1980) Non-Mendelian mutation affecting ribulose-1, 5-bisphosphate carboxylase structure and activity. *Nature* **285**: 114–115

Spreitzer RJ, Jordan DB and Ogren WL (1982) Biochemical and genetic anaylsis on RuBP carboxylase/oxgygenase-deficient mutants and revertants of *Chlamydomonas reinhardtii*. *FEBS Lett* **148**: 117–121

Streusand VJ and Portis AR Jr (1987) RuBisCO activase mediates ATP dependent activation of ribulose bisphosphate carboxylase. *Plant Physiol* **85**: 152–154

Suguita M and Gruissem W (1987) Developmental, organ-specific and light dependent expression of the tomato ribulose-1, 5-bisphosphate carboxylase small subunit gene family. *Proc Natl Acad Sci USA* **84**: 7104–7108

Suguita W, Manzara T, Pichersky E, Cashmore AR and Gruissem W (1987) Genome organisation, sequence analysis and expression of all five genes encoding the small subunit of ribulose-1, 5-bisphosphate carboxylase/oxgygenase from tomato. *Mol Gen Genet* **209**: 247–256

Tabita FR and McFadden BA (1972) Regulation of ribulose-1, 5-diphosphate carboxylase by 6-phospho-D-gluconate. *Biochem Biophys Res Commun* **48**: 1153–1160

Tabita FR and McFadden BA (1974) One step isolation of microbial ribulose-1, 5-diphosphate carboxylase. *Arch Microbiol.* **99**: 231–240

Takabe T and Akazawa T (1973) Catalytic role of subunit A in ribulose-1, 5-diphosphate carboxylase from *Chromatium* strain D. *Arch Biochem Biophys* **157**: 303–308

Takabe T and Akazawa T (1975) Further studies on the subunit structure of *Chromatium* ribulose-1, 5-bisphosphate carboxylase. *Biochemistry* **14**: 46–50

Theg SM, Baurele C, Olsen LJ, Selmen B and Keegstra K (1989) Internal ATP is the only energy requirement for the translocation of precursor proteins across chloroplastic membranes. *J Biol Chem* **264**: 6730–6736

Thompson WF, Everett M, Polans NO, Jorgensen RA and Palmer JD (1983) Phytochrome control of RNA levels in developing pea and mung-bean leaves. *Planta* **158**: 487–500

Tobin EM (1981) Phytochrome mediated regulation of messenger RNAs for the small subunit of ribulose-1, 5-bisphosphate carboxylase and the light harvesting chlorophyll a/b protein in *Lemma gibba*. *Plant Mol Biol* **1**: 35–51

Tobin EM and Klein AO (1975) Isolation and translation of plant messenger RNA. *Plant Physiol* **56**: 88–92

Tobin EM and Silverthorne J (1985) Light regulation of gene expression in higher plants. *Annu Rev Plant Physiol* **36**: 569–593

Tomimatsu Y and Donovan JW (1981) Effect of pH, Mg^{2+}, CO_2 and mercurians on the circular dichroism, thermal stability and light sensitivity of ribulose-1, 5-bisphosphate carboxylase from alfalfa, spinach and tobacco. *Plant Physiol* **68**: 808–813

Van den Broeck G, Timko MP, Kausch AP, Cashmore AR, Van Montagu M and Herrera-Estrella L (1985) Targetting of a foreign protein to chloroplasts by fusion to the transit peptide from the small subunit of ribulose-1, 5-bisphosphate carboxylase. *Nature* **313**: 358–363

van der Vies SM, Bradley D and Gatenby AA (1986) Assembly of cyanobacterial and higher plant ribulose bisphosphate carboxylase subunits into functional homologous and heterologous enzyme molecules in *Escherichia coli*. *EMBO J* **5**: 2439–2444

Vu JVC, Allen LH and Bowes G (1984) Dark light modulation of ribulose bisphosphate carboxylase activity in plants from different photosynthetic categories. *Plant Physiol* **76**: 834–845

Vu JVC, Allen LH and Bowes G (1986) Properties of ribulose-1, 5-bisphosphate carboxylase from dark and light-exposed soybean leaves. *Plant Sci* **44**: 119–123

Wasmann CC, Reiss B, Bartlett SG and Bohnert HJ (1986) The importance of the transit peptide and the trasnported protein into chloroplast. *Mol Gen Genet* **205**: 446–453

Wasmann CC, Reiss B and Bohnert HJ (1988) Complete processing of a small subunit of ribulose-1, 5-bisphosphate carboxylase/oxygenase from pea requires the amino acid sequence Ile-Thr-Ser. *J Biol Chem* **263**: 617–619

Wasmann CC, Ramage RT, Bohnert HJ and Ostrem JA (1989) Identification of an assembly domain in the small subunit of ribulose-1, 5-bisphosphate carboxylase. *Proc Natl Acad Sci USA* **86**: 1198–1202

316

Wehmeyer B, Cashmore AR and Schafer E (1990) Photocontrol of the expression of genes encoding chlorophyll a/b binding proteins and small subunit of ribulose-1, 5-bisphosphate carboxylase in etiolated seedlings of *Lycopersicon esculentum* (L) and *Nicotiana tabacum* (L). *Plant Physiol* **93**: 990–997

Werneke JM, Chatfield JM and Ogren WL (1988a) Catalysis of ribulose bisphosphate carboxylase/oxygenase activation by the product of a RuBisCO activase cDNA clone expressed in *Escherichia coli*. *Plant Physiol* **87**: 917–920

Werneke JM, Zelinski RE and Ogren WL (1988b) Structure and expression of spinach leaf cDNA encoding RuBisCO activase. *Proc Natl Acad Sci USA* **85**: 787–791

Whitman WB, Martin MN and Tabita FR (1979) Activation and regulation of ribulose bisphosphate carboxylase/oxygenase in the absence of small subunits. *J. Biol Chem* **254**: 10184–10189

Wimpee CF, Stiekma WJ and Tobin EM (1983) Sequence heterogeneity in the RuBP carboxylase small subunit gene family of *Lemna gibba*. In: (RB Goldberg ed). *Plant Molecular Biology*. UCLA symposium on molecular and cellular biology. New Series, Vol 1–2, Alan R Liss Inc, New York, pp 391–401

Winter U and Feierabend J (1990) Multiple coordinate controls contribute to a balanced expression of ribulose-1, 5-bisphosphate carboxylase/oxygenase subunits in rye leaves. *Eur J Biochem* **187**: 445–453

Yeoh HH, Badger MR and Watson L (1980) Variation in K_m (CO_2) of ribulose-1, 5-bisphosphate carboxylase among grasses. *Plant Physiol* **66**: 1110–1112

Yeoh HH, Badger MR and Watson L (1981) Variation in kinetic properties of ribulose-1, 5-bisphosphate carboxylase among plants. *Plant Physiol* **67**: 1151–1155

Zurawski G, Perrot B, Bottomley W and Whitfeld PR (1981) The structure of the gene for the large subunit of the ribulose-1, 5-bisphosphate carboxylase from spinach chloroplast DNA. *Nucl Acids Res* **9**: 3251–3270

12

C_4 Photosynthesis and C_3-C_4 Intermediacy: Adaptive Strategies for Semiarid Tropics

A.S. Raghavendra and V.S. Rama Das

School of Life Sciences
University of Hyderabad
Hyderabad 500 134, India

CONTENTS

ABBREVIATIONS

CAM	:	Crassulacean acid metabolism
DCDP	:	3, 3-dichloro-2 (dihydroxyphosphinoyl-methyl) propenoate
DHAP	:	Dihydroxyacetone phosphate
Kcat	:	Catalytic capacity of the enzyme
Km	:	Dissociation constant of the enzyme (indicating affinity for substrate)
NAD-ME	:	NAD malic enzyme
NADP-ME	:	NADP malic enzyme
NUE	:	Nitrogen use efficiency
PEP	:	Phosphoenolpyruvate
PEP-CK	:	Phosphoenolpyruvate carboxykinase
PGA	:	Phosphoglycerate
RuBPcase	:	Ribulose bisphosphate carboxylase/oxygenase

318

ABSTRACT

The higher plants are classified into three types: C_3, C_4, and crassulacean acid metabolism (CAM). The classification is based on mechanism of photosynthetic carbon assimilation. In C_4 plants, carbon is primarily fixed into C_4 acids and subsequently metabolized through Calvin cycle. The two-step carboxylation in C_4 plants is facilitated by the intercellular compartmentation of several key enzymes involved in carbon metabolism. The enzymes necessary for formation/carboxylation of phosphoenolpyruvate (PEP) are in mesophyll while those of C_4 acid decarboxylation and CO_2 refixation are in bundle sheath. Photosynthesis in C_4 plants is optimal at high intensities of light and temperature. C_4 plants require less water or nitrogen for every unit of carbon assimilated than the C_3 species. Due to these features, C_4 plants are well adapted to grow in arid or semiarid environments and are generally distributed in tropical and subtropical regions of the world. However, the productivity of C_4 plants is quite poor in a temperate environment and may even fall below those of C_3 species. Since the discovery of C_4 photosynthesis more than 25 years ago, rapid progress has been made in our understanding of the physiology and biochemistry of C_4 plants. More research is needed to elucidate the molecular biology of gene expression and regulation in mesophyll and bundle sheath cells of C_4 plants.

There are a few plant species which tend to be neither C_3 nor C_4, but are intermediate between C_3 and C_4 types. Till now, 23 species belonging to seven genera from five families have been reported to be C_3-C_4 intermediates. These plants have intermediate values of CO_2 compensation points indicating that the process of photorespiration is much reduced. Two types of C_3-C_4 intermediates are identified. In type I, for example, *Panicum millioides* and *Moricandia arvensis*, there is no C_4 cycle but photorespiration is reduced by refixation of respired CO_2 by the concentration of organelles in the cells. At least a partial C_4 cycle operates in type II intermediates like *Flaveria* species. In both types of intermediates, the basic principle is the improvement in refixation of photorespired CO_2. The C_3-C_4 intermediates may represent an evolutionary stage in between C_3 and C_4 plants. The exact course is, however, not established. Further studies on C_3-C_4 intermediates could help in unravelling the mechanism of reduced photorespiration, operation, and evolution of C_4 photosynthesis.

I. The Discovery and After: 25 Years of C_4 Pathway

Vascular plants could be divided into three groups according to their mechanism of photosynthetic carbon fixation: C_3, C_4, and CAM. During C_4 dicarboxylic acid pathway of photosynthesis, atmospheric carbon dioxide is initially incorporated into C_4 acids (malate and aspartate) which are later decarboxylated. The released CO_2 is refixed through the conventional Calvin cycle and is assimilated into carbohydrates. Such two-step carboxylation in C_4 plants is facilitated by the existence of Kranz anatomy in the leaves, the outer mesophyll layer surrounding the inner bundle sheath cells.

The occurrence of Kranz type leaf anatomy in higher plants has long been known (Haberlandt, 1914). However, the first report of a unique C_4 type of carbon assimilation in sugar-cane leaves was made by Kortschak and his colleagues (1965). A year later, Hatch and Slack (1970) confirmed this phenomenon and presented a detailed demonstration and characterization of C_4 syndrome in several mono- and dicotyledonous plants. Rapid progress has been made in understanding and analysing the physiology and biochemistry of carbon fixation in C_4 plants. Some of the recent reviews on C_4 photosynthesis include those by Hatch (1987), Furbank and Foyer (1988), Edwards and Ku (1990), and Leegood and Osmond (1990). Readers interested in detailed literature may consult also some of the earlier reviews (Hatch and Slack, 1970; Black, 1973; Laetsch, 1974; Hatch 1977, 1978; Hatch and Osmond, 1976; Edwards and Huber, 1981) and monographs (Hatch *et al.*, 1971; Burris and Black, 1976; Edwards and Walker, 1983).

II. Taxonomic and Geographical Distribution

Out of approximately 300 families of flowering plants, the C_4 photosynthesis is known to occur in at least 18 different families of monocots and dicots (Table 1). Readers interested in compiling the check lists of C_4 plants are referred to the following literature: Downton (1975), Raghavendra and Das (1978), and Osmond *et al.* (1982).

Table 1. Distribution of C_4 plants among the families of angiosperms

Monocotyledoneae		
Cyperaceae	Liliaceae	Poaceae
Dicotyledoneae		
Acanthaceae	Aizoaceae	Amaranthaceae
Asteraceae	Boraginaceae	Capparidaceae
Caryophyllaceae	Chenopodiaceae	Euphorbiaceae
Nyctaginaceae	Polygalaceae	Portulacaceae
Scrophulariaceae	Zygophyllaceae	

The C_4 syndrome is prominent particularly in Poaceae, Cyperaceae, Chenopodiaceae, and Euphorbiaceae, which are believed to be among the most advanced families of angiosperms (Das and Raghavendra, 1980). There has been considerable interest in the distribution of C_3 and C_4 species in the families of Poaceae (Brown, 1977). Among the major subfamilies of Poaceae, all species of subfamily Eragrostoideae are C_4, while all species of subfamily Festucoideae are C_3. Both C_3 and C_4 species occur in subfamilies Panicoideae and Aristidoideae.

The taxonomic distribution of C_4 species is further associated with an interesting analogy in their geographic distribution. The C_4 species of Panicoideae and Eragrostoideae are best adapted to warm and arid environments. Because of their greater water-use efficiency and adaptation to irradiation and temperature, the C_4 species are widely distributed in tropical and subtropical regions of the world. The flora of temperate regions consist of mainly C_3 plants (Black, 1971; Osmond *et al.*, 1982).

The families containing C_4 plants (Table 1) are so far apart that it appears certain that the C_4 photosynthesis has evolved independently in several families (Smith *et al.*, 1979). The oldest known fossil C_4 plants are recorded from Miocene and Pliocene, and thus do not date back more than five to seven million years (Nambudiri *et al.*, 1978; Thomasson *et al.*, 1986). The angiosperms, on the other hand, started to appear during lower Cretaceous or upper Jurassic periods. Thus, the earliest known C_4 plants appeared approximately 130 million years after the appearance of angiosperms.

The presence of enzymes of C_4 cycle even in C_3 plants and the polyphyletic origin of C_4 pathway lead to the conclusion that only small genetic alterations involving changes in gene expression would be necessary for the transition from C_3 to C_4 photosynthesis. The formation of Kranz anatomy would be among the first of several mutation changes (Moore, 1982).

III. Physiology and Biochemistry of C_4 Photosynthesis

a. Salient features of C_4 syndrome

The operation of C_4 pathway is associated with several physiological, biochemical, anatomical, geographical, and ecological features, which are collectively called "C_4 syndrome" (Table 2). Unique physiological features include low CO_2 compensation point (0–5 ppm), low or undetectable photorespiration, no inhibition of photosynthesis by atmospheric levels of oxygen, and low discrimination of ^{13}C against ^{12}C during carbon assimilation. Furthermore, C_4 plants have a high optimal temperature of about 35°–40°C (C_3 plants have 25–30°C) and exhibit a high water-use efficiency. Photosynthesis by C_4 plants requires high light and does not get saturated even at intensities equal to full sunlight.

Table 2. Kranz syndrome: anatomical, physiological and biochemical features of leaves of C_4 plants

---

ANATOMY

Bundle sheath (Kranz) cells with several chloroplasts and mitochondria.
Mesophyll radiate, compact around bundle sheath cells.
Veins spaced very close to each other.

PHYSIOLOGY

Photosynthesis saturates at atmospheric CO_2, requires high irradiation and does not saturate at even full sunlight.
Optimal temperature of photosynthesis above 30°C.
CO_2 compensation point nears zero and unaffected by O_2.
No inhibition of photosynthesis by atmospheric levels of O_2.
No detectable photorespiration.
High quantum requirement of photosynthesis.

BIOCHEMISTRY

Operation of C_4 type of photosynthesis (initial assimilation of CO_2 into C_4 acids and their metabolism).
High levels of C_4 cycle enzymes.
Differential compartmentation of enzymes and related carbon metabolism in mesophyll and bundle sheath.
Very low turnover of photorespiratory metabolism.
Low discrimination of ^{13}C against ^{12}C as indicated by ^{13}C values.

AGRONOMY/ECOLOGY

High water-use efficiency.
High nitrogen-use efficiency.
Capable of growing in arid and semiarid climates.
Major distribution in subtropical and tropical regions.

The biochemical basis of C_4 photosynthesis (described in more detail in the next section) is a two-step carboxylation process, namely, initial fixation of CO_2 into C_4 acids by PEP carboxylase and subsequent refixation by ribulose bisphosphate carboxylase (RuBPcase) of CO_2 released from C_4 acids. Specialized Kranz anatomy of C_4 leaves allows the spatial separation of these two processes of carboxylation. β-carboxylation of PEP occurs in the outer layer of mesophyll tissue. The C_4 acids quickly move into the inner bundle sheath tissue where they are decarboxylated. The released CO_2 is fixed by RuBPcase and assimilated through Calvin cycle. The radiate mesophyll, compactly arranged around the bundle sheath, protects the RuBPcase against O_2, apart from retaining high CO_2 concentration. Further, the mesophyll tissue is able to trap and fix CO_2, if any, coming out of respiration or photorespiration.

Because of their water-use efficiency and maximal photosynthesis at high irradiation temperature, many C_4 plants grow well even in arid environments. The C_4 photosynthesis is, therefore, considered to be an appropriate ecological response to arid environments in tropical and subtropical regions. With the possible exception of maize, most of the C_4 plants are distributed in tropical and subtropical regions of the globe. Very few C_4 plants are distributed in the temperate regions.

Further detailed discussion of the physiological, biochemical, anatomical, and ecological features of C_4 plants can be found in the reviews of Black (1971, 1973), Brown (1978), Osmond et al., (1982) and Hatch (1987).

b. Carbon metabolism and subclassification of C_4 plants

In C_4 plants, the atmospheric CO_2 is initially fixed by PEP carboxylase into C_4 acids. The enzyme PEP carboxylase, which catalyses the reaction of HCO_3^- with PEP to form oxalacetate, thus forms the primary route of carbon assimilation in C_4 plants. In C_3 plants, PEP carboxylase plays only an auxiliary role. The mechanism of the C_4 pathway can be summarized as follows.

The initial fixation of CO_2 into major C_4 acids, malate and aspartate, occurs in mesophyll cells. The C_4 acids are rapidly transported into the bundle sheath and are decarboxylated to release CO_2 and pyruvate. The decarboxylation is mediated by NADP malic enzyme (NADP-ME). Aspartate is converted into oxalacetate or malate, to be decarboxylated through PEP carboxykinase (PEP-CK) or NAD malic enzyme (NAD-ME), respectively. These decarboxylations are further discussed in the following paragraph. The released CO_2 is refixed by RuBPcase and chanelled into conventional Calvin cycle. Pyruvate formed during C_4 acid decarboxylation is transported into mesophyll and converted into PEP so as to regenerate the substrate for primary carboxylation. There is a marked movement of triose phosphates between mesophyll and bundle sheath cells. A part of phosphoglycerate formed in the bundle sheath is sent into mesophyll to be reduced to dihydroxyacetone phosphate by the chloroplasts. A detailed discussion of all these reactions can be found elsewhere (Edwards and Walker, 1983; Hatch, 1987).

There are three known variations in C_4 photosynthesis depending on the key enzyme used for decarboxylation of C_4 acids: NADP-ME, NAD-ME, and PEP-CK. Accordingly, C_4 plants are subclassified into NADP-ME, NAD-ME, and PEP-CK types (Hatch, 1987). The former two classes are predominant and are known to occur in both mono- and dicotyledons, while no dicot but a few monocots are identified to be PEP-CK type (Table 3).

Table 3. Subclassification of C_4 plants into three types with suitable examples.

Group	Type		
	NADP-ME	NAD-ME	PEP-CK
Monocotyledoneae	Cyperus rotundus	Buchloe dacty-loides	Bouteloua carlipendula
	Digitaria sanguinalis	Chloris disti-chiophylla	Chloris gayana
	Pennisetum typhoides	Cynodon dactylon	Muehlenbergia sohreberi
	Setaria italica	Eleusine coracana	Panicum texanam
	Sorghum bicolor	Eragrostis ciliaris	Sporobolus poiretti
	Zea mays	Panicum miliaceum	Urochloa panicoides
Dicotyledoneae	Euphorbia maculata	Amaranthus hybridus	(None confirmed)
	Gomphrena globosa	Amaranthus retroflexus	
	Kochia childsi	Amaranthus tricolor	
	Salsola kali	Portulaca oleracea	

Unique carbon metabolism during C_4 photosynthesis is due to compartmentation of C_4 acid formation in the mesophyll and subsequent C_4 acid decarboxylation and Calvin cycle in the bundle sheath. The spatial compartmentation is achieved by the exclusive location of key enzymes. For example, PEP carboxylase and pyruvate Pi dikinase (enzyme responsible for regeneration of PEP) are located only in the mesophyll, whereas the C_4 acid decarboxylating enzymes and RuBPcase are invariably present only in the bundle sheath tissue (Furbank and Foyer, 1988).

The compartmentation of carbon metabolism extends from primary assimilation to the formation of end-products, namely, starch and sucrose. Mesophyll is the major site of sucrose formation while starch is usually produced in bundle sheath. Enzymes involved in the formation of sucrose and starch are located in the mesophyll and bundle sheath, respectively.

A marked difference in the enzyme activities is accounted for largely by differential protein composition between the mesophyll and bundle sheath cell types rather than inactive enzymes or cell-type specific inhibitors. This is demonstrated by several experiments involving either cell/protoplast separation (Aoyagi and Nakamoto, 1985; Broglie et al., 1984; Sheen and Bogorad, 1987a, b) or immunolocalization (Hattersley et al., 1977; Perrott-Rechenmann et al., 1982, 1983; Langdale et al., 1987). The reasons for differential protein composition can further be traced to the differences in mRNA accumulation and the cell-specific gene expression, as discussed here.

c. Physiological significance of C_4 cycle

C_4 cycle is essentially an additional step during normal photosynthesis by the conventional Calvin cycle. It primarily acts as a CO_2 pump to generate a several times higher concentration of CO_2 in the bundle sheath, the site of RuBPcase-mediated Calvin cycle (Hatch, 1987). In CAM plants, the CO_2 pump acts through a temporal separation of C_4 acid formation in the night followed by decarboxylation and refixation by RuBPcase in the day.

High concentration of CO_2 in the bundle sheath in turn favours the carboxylase reaction of RuBPcase, avoiding the oxygenase activity, thereby decreasing the photorespiratory outlet of carbon. In addition, the tight Kranz anatomy prevents the atmospheric oxygen from reaching the bundle sheath tissue, which is the site of photorespiratory metabolism. Thus, the spatial separation of C_4 acid formation in mesophyll and photorespiration in bundle sheath achieves two important goals: CO_2 concentration near RuBPcase and suppression of photorespiration.

Due to its sparingly soluble nature, the level of dissolved CO_2 in aquatic medium is quite low. A special mechanism, therefore, operates in aquatic plants and microorganisms (Badger, 1987). An energy-dependent pump transports CO_2 (or HCO_3^-) across the plasma membrane into the cell. A diffusion barrier minimizes the backflux of CO_2 out of the internal compartment. Finally, interconversion of CO_2 and HCO_3^- by carbonic anhydrase helps in the supply of both these forms of carbon for transport and fixation processes.

The situation in C_4 plants can be considered analogous to CO_2 concentrating mechanism in aquatic organisms. The C_4 pathway acts as the pump to concentrate CO_2 inside the bundle sheath. The exclusive localization of carbonic anhydrase in mesophyll helps in solubilizing CO_2 into HCO_3^-, the form preferred for primary carbon assimilation by PEP carboxylase. The very low permeability for CO_2 and the absence of carbonic anhydrase in bundle sheath cells make the possible leak of CO_2 minimal (Burnell and Hatch, 1988; Hatch and Burnell, 1990). Further experiments may reveal the importance of carbonic anhydrase in C_4 plants.

The operation of the C_4 pathway results also in several factors which make C_4 plants ideally suited for arid and semiarid environments. In terms of water-use at 25°C, C_4 plants are generally twice as efficient as C_3 plants. The efficiency may increase further at higher temperatures (Downes, 1969). C_4 plants are much better than C_3 species in use of nitrogen as well. The nitrogen-use efficiency (NUE) is a measure of carbon assimilation per unit of nitrogen present in either leaves or plants. On average C_4 plants are about twice as efficient as C_3 species in the use of nitrogen (Brown, 1978; Sage et al., 1987). The better NUE in C_4 plants has agronomic and evolutionary implications. It is believed to be due to the predominance of PEP carboxylase over RuBPcase and less chanelling of leaf N into the fraction I protein. Up to 50 % of soluble N in a C_3 leaf may be in RuBPcase, so that any reduction in level of N would adversely affect the levels of RuBPcase and thereby decrease photosynthetic capacity. The ability of C_4 plants to assimilate CO_2 faster than the C_3 plants is due to the combined effects of higher concentrations of CO_2 within bundle sheath cells and high specific activity of RuBPcase. There are also several reports on the higher quantum yields in leaves of C_4 plants than those in C_3 species (Edwards and Walker, 1983).

d. Regulation and development

C_4 photosynthesis is regulated by several internal as well as external factors. Among the internal factors, the levels of two key enzymes, pyruvate Pi dikinase and RuBPcase seem to be limiting the overall photosynthesis in C_4 plants. Among the ten C_4 species examined, the activities of these two enzymes were highly correlated with the maximal rates of photosynthesis (Usuda et al., 1984).

Light and temperature are among the most important environmental factors which regulate the functioning and extent of C_4 photosynthesis. Several enzymes of Calvin cycle are known to be regulated by light (Anderson, 1986). At least two key enzymes of the C_4 pathway—pyruvate Pi dikinase and NADP malate dehydrogenase—are markedly regulated by light. Recent results suggest that PEP carboxylase is also influenced by illumination (Doncaster and Leegood, 1987). The mechanism of regulation of these enzymes may, however, differ. NADP malate dehydrogenase appears to be dependent on the photosynthetic linear electron transport and the enzyme is possibly reduced on illumination (Nakamoto and Edwards, 1986). On the other hand, the activation by light of pyruvate Pi dikinase and PEP carboxylase involves phosphorylation of the enzymes (Chollet et al., 1990).

Unlike Calvin cycle enzymes, some of the key enzymes of the C_4 pathway are extremely sensitive to low temperature. A marked example is the rapid inactivation of pyruvate Pi dikinase at low temperatures (Hull et al., 1990). It has been shown that at temperatures below the optimal range, this enzyme tends to dissociate into its monomeric forms, thus becoming inactive. This inactivation appears to be one of the primary reasons for the rapid decrease in photosynthetic capacity of leaves below 10°–15°C. Sensitivity of C_4 photosynthesis to low temperature makes these plants much more photoinhibited at chilling temperatures than C_3 plants (Powles et al., 1983; Baker et al., 1990). Chilling affects primarily the mesophyll tissue in leaves of C_4 plants. The bundle sheath is less affected (Taylor and Craig, 1971).

C_4 photosynthesis develops systematically along with leaf ontogeny, particularly in monocots. Although the developmental regulation also occurs in dicots, the development is remarkably pronounced in monocots, because of polarization in cell division and expansion. Cellular differentiation in monocots proceeds in a basipetal direction. The cells at the tip of the young leaf (oldest) are photosynthetically most developed, while the cells at the base (youngest) are at a lower stage of differentiation and photosynthetic capacity (Leech, 1985).

Leaf ontogeny is a useful tool to study regulation of C_4 photosynthesis. Young leaves of maize are more C_3 in character while the older leaves are fully C_4 (Crespo et al., 1979). A similar tendency of operation of C_3 and C_4 type photosynthesis in young and old leaves, respectively, was noticed in some other C_4 plants as well (Kennedy and Laetsch 1973; Khanna and Sinha 1973; Imai and Murata 1979; Raghavendra et al., 1978; Moore et al., 1986).

A gradient of C_4 photosynthesis exists along with leaf development of monocotyledonous plants. The tip of a young maize leaf tended to be fully C_4 while the base (young in development) is more C_3 in character (Perchorowicz and Gibbs, 1980; Miranda et al., 1981a, b). Such variable C_4 function in leaves is due to the gradient of development of not only Kranz anatomy but also the levels of mRNA in the species and synthesis of related C_4 enzymes, particularly the key enzymes. Further information on the developmental regulation of photosynthesis in leaves of C_4 plants can be found in an excellent review by Nelson and Langdale (1989).

IV. C_4 Pathway in Relation to Plant Performance and Agriculture

Among the principal features of the C_4 pathway are high rates of photosynthesis, sucrose formation, assimilate export, and productivity (Edwards and Huber, 1981). The performance of the C_4 pathway is optimal at high irradiance. The C_4 cycle and Kranz anatomy bestow on these plants an improved water-use efficiency. C_4 plants are, therefore, ideally suited for arid and semiarid climates.

Because of greater photosynthetic performance, the plant communities in the tropical region, consisting of agricultural crops or fodder grasses, are known to achieve remarkable biomass yields. The productivity of C_4 crops is quite high and greater than that of C_3 species under tropical conditions (Table 4). Consequently, the economic importance of C_4 plants is considerable. They range from high-yield crops such as sugar-cane, maize, fodder and prairie grasses, *Pennisetum purpureum* (napier grass), and *Panicum virgatum* (switch grass), to the most virulent weeds, such as *Cynodon dactylon* (Bermuda grass) and *Cyperus rotundus* (nutsedge).

Table 4. Annual productivity of different C_3 and C_4 crops under tropical conditions.

Type/Crop species	Productivity (tons ha^{-1})
C_3 TYPE	
Beta vulgaris	31
Glycine max	18
Solanum tuberosum	37
Gossypium hirsutum	27
Oryza sativum	23
C_4 TYPE	
Pennisetum purpureum	85
Saccharum officinarum	67
Pennisetum typhoides	54
Zea mays	52
Sorghum vulgare, ssp. sudanensis	51

Data taken from different sources. A detailed discussion can be found in Gifford (1974), Cooper (1975), Boardman (1977) and Ludlow (1985).

Although C_4 plants are highly productive under tropical conditions, many C_4 species perform poorly in a temperate environment. There are a very few C_4 plants which can adapt to cool climates, the most notable being *Spartina* (Long, 1983). The reason is that C_4 plants are thermophilic and prefer to grow in warm areas with high irradiances (Oquist and Martin, 1986). Most of the C_4 plants are, therefore, distributed in the tropical and subtropical regions of the world. Similarly, the periods of maximum growth of C_4 plants, within a given zone, correspond to the days of high light intensity and temperature.

Another remarkable result of C_4 photosynthesis is the adaptability of plants to a variety of stress conditions. The high quantum yield of photosynthesis in C_4 plants endows a clear advantage for maintaining the growth even at limited light (Ludlow, 1985). The efficiency of C_4 plants in using water (Downes, 1969; Wong, 1979) as well as nitrogen (Brown, 1978; Sage et al., 1987) makes them capable of growing in regions deficient in water supply and poor in nitrogen.

Most of the agricultural crops are C_3 types. Only a few crops and many fodder grasses are C_4 type. A point of serious concern, however is that most of the worst weeds are C_4 species (Holm, 1969; Holm et al., 1977). In the United States, among the top 10 worst weeds affecting the field crops, the first six species are C_4 (Doresch, 1970). Accordingly, it has been logical to consider developing herbicides to inhibit C_4 cycle specifically.

Recently, Jenkins et al., (1987) reported the discovery of a specific inhibitor DCDP (3, 3-dichloro-2-(dihydroxyphosphinoyl-methyl)-propenoate) of PEP carboxylase. This new compound drastically inhibited photosynthetic carbon assimilation in C_4 plants but had only a marginal effect on CO_2 fixation in C_3 plants (Jenkins, 1989). It is too early, however, to comment on the possible application of DCDP in agriculture. Further experiments may reveal the existence of more compounds capable of affecting carboxylation in C_4 plants.

V. C_3-C_4 Intermediates

a. Discovery and occurrence

There are a few plant species which tend to be neither C_3 nor C_4 but are intermediate between C_3 and C_4 types. Among the first known C_3-C_4 intermediates are the artificial hybrids produced by crossing the C_3 and C_4 species of Atriplex. Several F_1 hybrids between Atriplex rosea (C_4-female parent) and A. triangularis (C_3-male parent) were intermediate between the parent plants in a number of anatomical, physiological, and biochemical characteristics. None of the hybrids, however, showed a complete operation of C_4 pathway. The chromosomal pairing in F_1 hybrids was highly irregular and it was difficult to reach any conclusion except that C_4 photosynthesis was not transmitted through chloroplast genome but through several nuclear genes (Osmond et al., 1980).

One of the first indications about occurrence of natural C_3-C_4 intermediates was from the extensive survey of CO_2 compensation points of several mono- and dicotyledonous species, made by Krenzer et al. (1975). They observed that Moricandia arvensis (Brassicaceae) and Panicum milioides (Poaceae) exhibited invariably CO_2 compensation points in between the typical values obtained for C_3 or C_4 species. Subsequent work established that M. arvensis and P. milioides, in fact, are C_3-C_4 intermediates. These two species exhibited intermediate status of several features connected with C_4 syndrome apart from CO_2 compensation point, such as inhibition of photosynthesis by oxygen, Kranz anatomy, or the levels of key C_4 cycle enzymes in leaves (Brown and Brown, 1975; Kanai and Kashiwagi, 1975; Apel et al., 1978).

Twenty-three plant species, including those in the genus *Alternanthera* discovered recently (Rajendrudu *et al.*, 1986), have been reported to have photosynthetic characteristics intermediate between C_3 and C_4 plants (Table 5). These species

Table 5. Naturally occurring plant species reported to be C_3-C_4 intermediates.

Monocotyledoneae
    Poaceae

Neurachne minor	*Panicum milioides*
Panicum decipiens	*Panicum schenckii*

Dicotyledoneae
    Aizoaceae

Mollugo verticillata	*Mollugo nudicaulis*

    Amaranthaceae

Alternanthera ficoides	*Alternanthera tenella*

    Asteraceae

Flaveria angustifolia	*Flaveria anomala*
Flaveria brownii	*Flaveria chloraefolia*
Flaveria floridana	*Flaveria linearis*
Flaveria oppositifolia	*Flaveria pubescens*
Flaveria ramosissima	*Flaveria sonorensis*
Flaveria vaginata	*Parthenium hysterophorus*

    Brassicaceae

Moricandia arvensis	*Moricandia spinosa*
Moricandia sinaica	

Table 6. Genera known to contain both C_3 and C_4 species, belonging to 10 families.

Monocotyledoneae

1. Cyperaceae	2. Poaceae
Cyperus	*Alloteropsis*
Scirpus	*Neurachne*
	Panicum

Dicotyledoneae

3. Aizoaceae	4. Amaranthaceae
Mollugo	*Aerva*
	Alternanthera
5. Asteraceae	6. Boraginaceae
Flaveria	*Heliotropium*
7. Chenopodiaceae	8. Euphorbiaceae
Atriplex	*Chamaesyce*
Bassia	*Euphorbia*
Kochia	
Suaeda	
9. Nyctaginaceae	10. Zygophyllaceae
Boerhaavia	*Kallstroemia*
	Zygophyllum

which include both monocots and dicots belong to several genera from five families. Most of the C_3-C_4 intermediates belong to the genera which were known for a long time to have both C_3 and C_4 species (Table 6). *Moricandia* and *Parthenium* are the only exceptions to the observation. In fact, Rajendrudu and Das (1981) observed that *P. hysterophorus* exhibited low photorespiration and subsequently the plant was confirmed to be a C_3-C_4 intermediate (Edwards and Ku, 1987). All the C_3-C_4 intermediates possess the leaf anatomy in between that of Kranz and non-Kranz, partly suppressed photorespiration compared to C_3 plants, as indicated by a reduced CO_2 compensation point, and a reduced sensitivity of net photosynthesis to oxygen. Detailed information on C_3-C_4 intermediate photosynthesis can be found in the reviews of Raghavendra (1980), Holaday and Chollet (1984), Monson *et al.* (1984), Edwards and Ku (1987), and Monson and Moore (1989).

b. Physiological characteristics

Leaves of C_3-C_4 intermediates have a basic Kranz anatomy, with bundle sheath cells containing chloroplasts. The degree of Kranz cell development, however, varies within these species. The Kranz cells may be well developed (for example, *Neurachne minor*), not as distinct as in C_4 plants (for example, *Panicum milioides*), or poorly developed (for example, *Mollugo verticillata*).

One of the most important features of all C_3-C_4 intermediates is the CO_2 compensation point, which is less than that in typical C_3 plants. The intermediate values of CO_2 compensation point (5–39 $\mu l.l^{-1}$) indicate that the level of apparent photorespiration is much reduced in the intermediate plants than in C_3 species. Considerable variation, however, exists in the CO_2 compensation points of different intermediates suggesting that the degree of reduction in photorespiration may vary. In C_4 plants, CO_2 compensation point is unaffected by ambient oxygen even up to 50 % by volume, while in C_3 species the CO_2 compensation point increases linearly with increase in CO_2 concentration. The CO_2 compensation point in the intermediate species is less sensitive to oxygen than in C_3 plants.

Most of the intermediates exhibit a biphasic response of their CO_2 compensation point to changing oxygen levels. The reasons for such biphasic response are not known. CO_2 compensation point in the intermediates is also sensitive to light intensity, increasing as the light intensity available on leaf surface is reduced from about 30 % to 5% of full sunlight. Such a response is not reported in either C_3 or C_4 plants.

Another indication of reduced photorespiration in C_3-C_4 intermediates is the extent of light-dependent release of CO_2 into CO_2-free air, after steady state photosynthesis. The extent of CO_2 evolution in light is always less in the intermediates than in C_3 plants. The absolute rates, however, vary among the different species. Similarly, the inhibition of photosynthesis by atmospheric levels of oxygen is considerably low in the intermediates when compared to C_3 species. Photosynthesis in C_3 plants is usually inhibited to about 30–35% by 21% oxygen, while there is no such inhibition in C_4 plants. The inhibition by similar oxygen levels in several intermediates ranges from 20 to 25%.

Anatomical, physiological, and biochemical features of C_3-C_4 intermediates are discussed in detail by Edwards and Ku (1987) and are summarized in Table 7.

Table 7. Major characteristics of leaves of naturally occurring C_3-C_4 intermediates

---

ANATOMY

Occurrence of Kranz cells, but the stage of development varies from poor to well-differentiated level. Mesophyll, not clearly radiate and less impact.

PHYSIOLOGY

Reduced photorespiration as indicated by:
> Intermediate levels of CO_2 compensation point (1 to 39 ppm).
> Inhibition of photosynthesis by atmospheric O_2 less than that in C_3 plants.
> Biphasic response (increase) of CO_2 compensation point to ambient oxygen.
> Increase of CO_2 compensation as the light intensity is reduced from 30 to 50% of full sunlight.
> Inhibition by oxygen of carboxylation efficiency similar to that of C_3 plants.

BIOCHEMISTRY

Dominance of C_3 type of metabolism.
Variable expression of C_4 pathway.
Incomplete compartmentation of metabolism in mesophyll and bundle sheath.
^{13}C values similar to those of C_3 plants.

ECOLOGY

Native to warm and arid climates.

Table 8. Subclassification of C_3-C_4 intermediates based on operation of C_4 pathway*

Type	Intermediate species	
TYPE I		
Without C_4 cycle	*Moricandia arvensis*	*Panicum milioides*
	Parthenium hysterophorus	*Alternanthera ficoides*
		Alternanthera tenella
TYPE II		
With partial C_4 cycle	*Flaveria anomala*	*Flaveria brownii*
	Flaveria floridana	*Flaveria linearis*
	Flaveria pubescens	*Flaveria rainocissi*
	Mollugo verticillata	*Mollugo nudicaulis*
	Neurachne minor	

* Following nine species cannot be classified in view of non-availability of evidence for operation/non-operation of C_4 cycle: *Flaveria angustifolia, F. chloraefolia, F. oppositifolia, F. sonorensis, F. vaginata, Moricandia sinaica, M. spinosa, Panicum decipiens* and *P. schenckii.*

c. Biochemistry of carbon metabolism

Operation of C_4 pathway achieves a high concentration of CO_2 in the bundle sheath and ensures a quick refixation of CO_2 released, if any. There is evidence that at least a partial C_4 cycle operates in some of the intermediates like *Flaveria brownii*. Some other intermediates like *Moricandia arvensis*, however, may not have a functional C_4 cycle. The intermediates can, therefore, be classified into two categories: type I without C_4 cycle and type II with C_4 cycle (Table 8).

In type I intermediates, photorespiration is possibly reduced by refixing photorespired CO_2 without participation of C_4 cycle. *Panicum millioides* and *Moricandia arvensis* are the best-known examples for this type. In these plants, the bundle sheath cells are enriched with not only chloroplasts but also mitochondria (Winter *et al.*, 1982; Kanai and Kashiwagi 1975; Brown *et al.*, 1983). Immunochemical studies observed that the enzyme glycine decarboxylase in *Moricandia arvensis* is located only in the bundle sheath cells (Rawsthorne *et al.*, 1988a, b). The result would be that the photorespired CO_2 released from mitochondria would be refixed by adjacent chloroplasts. As a consequence, the CO_2 compensation point is reduced.

The type I intermediates may have more PEP carboxylase in the leaves than in C_3 plants, but the activities of pyruvate Pi dikinase and C_4 acid decarboxylases are as low as in C_3 leaves. It is possible that the elevated PEP carboxylase levels can help also in the refixation of CO_2 under certain circumstances (for example, at low ambient CO_2 concentration). Such refixation by PEP carboxylase is yet to be established. Further, the intriguing points in these intermediates are the low levels of pyruvate Pi dikinase making PEP non-available for continued function of PEP carboxylase and the low activities of decarboxylating enzymes.

At least a partial C_4 pathway functions in type II intermediates, as in *Flaveria* species. The extent of operation of C_4 pathway may, however, differ in the species. For example, *Flaveria* species having the most efficient C_4 cycle first and those with the least efficient cycle last have been ranked in the following order: *F. brownii*, *F. ramosissima*, *F. floridana*, *F. pubescens*, *F. anomala*, and *F. linearis* (Monson *et al.*, 1986). All these intermediates have a clear Kranz anatomy and have substantially high activities of PEP carboxylase, pyruvate Pi dikinase, NADP malic enzyme, and NADP malate dehydrogenase. These plants can not only assimilate CO_2 into C_4 acids but also metabolize them further, to a limited extent. The major limitation on C_4 pathway in these species is the incomplete intercellular compartmentation of enzymes in mesophyll and bundle sheath. For example, PEP carboxylase is not exclusively localized in mesophyll of *F. brownii* (Cheng *et al.*, 1986). There is only a gradation of activity of PEP carboxylase, from high in the outer mesophyll layer to a low level in the inner bundle sheath.

C_4 pathway may operate to a limited extent in some of the intermediates but the experiments on quantum yield and carbon isotope discrimination indicate that the C_4 cycle does not contribute significantly to overall carbon metabolism (Edwards and Ku, 1987). Similarly, the kinetic properties of PEP carboxylase in the intermediates resemble those of the enzyme from C_3 species (Holaday *et al.*, 1981; Holbrook *et al.*, 1985).

The balance between photosynthesis and photorespiration can be modulated in C_3 plants by the kinetic properties of RuBPcase. The RuBPcase from C_4 plants is known to have a higher dissociation constant or Km (CO_2) than the enzyme from C_3 or CAM plants (Yeoh *et al.*, 1981). Further, plants capable of C_4 photosynthesis possess RuBPcase having a higher Kcat (catalytic capacity) than that found in C_3 species (Seemann *et al.*, 1984). A recent study (Wessinger *et al.*, 1989) found that the C_3-C_4 intermediate species of *Flaveria* had RuBPcase with a Km (CO_2) similar to that in C_3 species which was much less than that in C_4 species of the same genus. The mean ratio of Kcat/Km for species of each group was similar, supporting the hypothesis that changes in Km and Kcat are linked. The allocation of total soluble protein to RuBPcase was lowest in the C_4 *Flaveria* species, intermediate in the C_3-C_4 species, and highest in the C_3 species. The authors (Wessinger *et al.*, 1989) suggested that during evolution of C_4 photosynthesis, adjustments may occur in the quantity of RuBPcase prior to changes in its kinetic properties.

Photorespiratory carbon metabolism is closely related to nitrogen metabolism as well (Singh *et al.*, 1985). (See also Chapter by Kumar *et al.* this volume). A recent report indicates that a limitation on photorespiratory turnover of glycine pool may be one of the factors responsible for reduced photorespiration in *Moricandia arvensis* (Kumar and Abrol, 1990). Further experiments are required to elucidate the photorespiratory nitrogen cycle in C_3-C_4 intermediates.

d. Regulation of intermediacy

Expression and extent of C_3-C_4 intermediacy is regulated by several environmental factors but the information available is fragmentary. Growth conditions, particularly light intensity and oxygen levels, influence remarkably the photosynthetic characteristics of intermediates. When grown at low light intensities there are increases in CO_2 compensation points of *Neurachne minor* and *Panicum milioides* (Hattersley *et al.*, 1986). The influence of oxygen on CO_2 compensation point is proposed to be independent of functional C_4 cycle (Edwards and Ku, 1987). On the other hand, the model developed by Piesker and Bauwe (1984) suggests that the partial operation of a C_4 cycle can result in a biphasic response of CO_2 compensation point to ambient oxygen in C_3-C_4 intermediates. A recent report (Schuster and Monson, 1990) suggests that the C_3-C_4 intermediate photosynthesis may be an advantage over the C_3 cycle in warm environments. Further experiments are necessary to resolve the effects of oxygen on photosynthesis, photorespiration, and CO_2 compensation point in the intermediates.

Leaf age also can influence the apparent photorespiration in the intermediates. Older leaves of *Moricandia* exhibit lower CO_2 compensation point while older leaves of *Flaveria* incorporate more ^{14}C into C_4 acids than younger leaves (Apel *et al.*, 1978; Edwards and Ku, 1987).

Efforts have been made to analyse the genetics of intermediacy using hybrids between C_3 or C_4 species and the intermediates in *Panicum* (Brown *et al.*, 1985; Bouton *et al.*, 1986) and *Flaveria* (Cheng and Ku, 1985; Brown *et al.*, 1986; Cheng *et al.*, 1987). Although the mechanism of inheritance of C_3-C_4 intermediacy is not completely understood, limited information suggests that the work with *Flaveria* species could be informative since hybridization is successful and the polyploidy in this genus may not be a confusing factor.

VI. Importance of C_3-C_4 Intermediacy

a. Mechanism of reduction in photorespiration

A common feature of all the intermediates is reduced photorespiration. The most important use of the intermediates, therefore, is to understand mechanism of reduction in photorespiration. C_4 plants achieved reduction of photorespiration by limiting photorespiratory metabolism in the bundle sheath and elevating the CO_2 concentration by supplementing with C_4 acid metabolism.

As indicated in the earlier section, one group of intermediates has a partly functional C_4 cycle which helps to reduce photorespiration (for example *Flaveria brownii*). Photorespiratory CO_2 is refixed to a certain extent by the efficient PEP carboxylase during the C_4 pathway. All these plants have a Kranz anatomy but the compartmentation of photosynthetic enzymes is not as perfect as in C_4 plants.

The second group of intermediates may not have functional C_4 cycle (*Moricandia arvensis*), and seems to have evolved another strategy. Bundle sheath cells of these plants are enriched with chloroplasts as well as mitochondria. The location of key photorespiratory enzymes, like glycine decarboxylase, in bundle sheath cells results in release of CO_2 through photorespiratory metabolism mainly in the bundle sheath. The large number of chloroplasts located in close vicinity would refix the released CO_2.

In both types of intermediates, the basic principle is improvement in refixation of photorespiratory CO_2. It is also possible that the actual photorespiratory pathway itself is restricted in the intermediates but there is no clear evidence. Further experiments should be conducted to investigate if there are any alternative strategies evolved by the intermediates to reduce apparent photorespiration.

b. Evolution of C_4 plants

The intermediates evoked a lot of interest in the possible routes of evolution of C_4 plants. It is generally accepted that C_4 plants evolved later than the C_3 plants but it is not certain whether C_4 plants evolved from C_3 or represent a parallel evolution from a common ancestor. C_3-C_4 intermediates may represent an evolutionary stage in between C_3 and C_4 plants (Edwards and Ku, 1987; Monson and Moore, 1989), although the exact course is not established.

After summarizing the literature on C_3-C_4 intermediates, Edwards and Ku (1987) proposed the following hypothesis for the evolution of C_4 plants. In C_3 plants, bundle sheath, even if it exists, does not contain a significant number of organelles like chloroplasts or mitochondria. Initially, there is an enrichment of mitochondria and chloroplasts in the bundle sheath as in *Panicum laxum* and *Parthenium hysterophorus*. The level of PEP carboxylase in the leaf increases in species like *Panicum milioides* and *Moricandia arvensis*. The next stage is further development of Kranz anatomy (as in *Flaveria ramosissima*), but without a complete compartmentation of the enzymes. Finally, an improved intercellular compartmentation of the enzyme complement, associated with a fully developed Kranz anatomy, as in *Flaveria brownii*, would lead to the evolution of a typical C_4 plant. Monson and Moore (1989) also favoured the proposal of Edwards and Ku (1987).

A second alternative is evolution of Kranz anatomy without the occurrence of C_4 cycle enzymes. Elevated PEP carboxylase activity and compartmentation of C_4 acid metabolism would be the next step followed by retention of Calvin cycle and photorespiration exclusively in the bundle sheath tissue (Raghavendra, 1980).

Further experiments on C_3-C_4 intermediates, particularly *Flaveria* and *Panicum*, should be able to establish the possible course of evolution of C_4 plants from

C_3 species. Even the intermediates could have evolved through more than a single route during the evolution. There is an additional possibility that some of the intermediates evolved back from C_4 plants.

VII. Epilogue

Since the discovery of C_4 photosynthesis 25 years ago, rapid progress has been made in our understanding of the physiology and biochemistry of C_4 plants. However, detailed information on molecular aspects of differential gene expression and regulation in mesophyll and bundle sheath cells of C_4 plants is not yet available. Efforts should, therefore, be made to understand the mechanism of expression of different biochemical reactions in the leaves of C_4 plants. Another crucial area of interest is the pattern and regulation of transport of metabolites between mesophyll and bundle sheath cells. For example, attempts are being made to assess directly the diffusion of metabolites into mesophyll and bundle sheath cells, with intact plasmodesmata (Weiner et al., 1988).

C_3-C_4 intermediates offer great prospects for unravelling the mechanism of reduced photorespiration, operation, and evolution of C_4 photosynthesis. More emphasis is needed on biochemical aspects of C_3-C_4 intermediate photosynthesis. The phenomena of increased CO_2 compensation point with the decrease in light intensity and biphasic response of CO_2 compensation point to ambient oxygen are yet to be explained satisfactorily. Regulation of photosynthesis in C_3-C_4 intermediates by leaf age and environmental factors needs to be investigated in more detail. Information is lacking on the molecular biology of these intermediates. Studies on the isozyme pattern and molecular hybridization within closely related species could indicate the pattern of molecular evolution of C_4 cycle. The strategy of adaptation of C_4 plants is yet to be translated in molecular terms.

Acknowledgements

Preparation of this article was supported by grants from the Department of Science and Technology (SP/SO/A43/88) and the U.S. Dept. of Agriculture Project FG-IN-680 (IN-ARS-403).

VIII. References

Anderson LE (1986) Light/dark modulation of enzyme activity in plants. In: (JA Callow, ed) *Advances in Botanical Research*, Vol 12, Academic Press, New York, pp 1–46

Aoyagi K and Nakamoto H (1985) Pyruvate, Pi dikinase in bundle sheath strands as well as in mesophyll cells in maize. *Plant Physiol* 78: 661–664

Apel P, Ticha I and Peisker M (1978) CO_2 compensation concentrations in leaves of *Moricandia arvensis* (L.) DC at different insertion levels and O_2 concentrations. *Biochem Physiol Pflanzen* 174: 68–75

Badger MR (1987) The CO_2 concentrating mechanism in aquatic phototrophs. In: (MD Hatch and NK Boardman, eds) *The Biochemistry of Plants—A Comprehensive Treatise*, Vol 10, Academic Press, San Diego, pp 220–274

Baker NR, Nie GY, Ortiz-Lopez A, Ort DR and Long SP (1990) Analysis of chill-induced depressions of photosynthesis in maize. In: (M Baltscheffsky, ed) *Current Research in Photosynthesis*, Vol 4, Kluwer Acad Publ, Dordrecht, The Netherlands pp 565—572

Black CC (1971) Ecological implication of dividing plants into groups with distinct photosynthetic production capacity *Adv Ecol Res* 7: 87–113

334

Black CC (1973) Photosynthetic carbon fixation in relation to net CO_2 uptake. *Annu Rev Plant Physiol* **24**: 253–286

Boardman, NK (1977) The energy budget in solar energy conversion in ecological and agricultural systems. In: (R Buvet, MJ Allen and JP Massue, eds) *Living Systems as Energy Converters*. North-Holland Publishing Co., Amsterdam, pp. 307–313

Bouton JH, Brown RH, Evans PT and Jernsted JA (1986) Photosynthesis, leaf anatomy, and morphology of progeny from hybrids between C_3 and C_3/C_4 *Panicum* species. *Plant Physiol* **80**: 487–492

Broglie R, Corruzzl G, Keith B and Chua NH (1984) Molecular biology of C_4 photosynthesis in *Zea mays*: Differential localizations of proteins and mRNAs in the two leaf cell types. *Plant Mol Biol* **3**: 431–444

Brown RH (1978) A difference in N-use efficiency in C_3 and C_4 plants and its implications in adaptation and evolution. *Crop Sci* **18**: 93–98

Brown RH, Bassett CL, Cameron RG, Evans PT, Bouton JH, Black CC, Sternberg LO and DeNiro MJ (1986) Photosynthesis of F_1 hybrids between C_4 and C_3-C_4 species of *Flaveria*. *Plant Physiol* **82**: 211–217

Brown RH, Bouton JH, Rigsby LL, Rigsby M (1983) Photosynthesis of grass species differeing in carbon dioxide fixation pathways. VIII Ultrastructural characteristics of *Panicum* species in the *Laxa* group. *Plant Physiol* **71**: 425–431

Brown RH, Bouton JH, Evans PT, Malter HE and Rigsby LL (1985) Photosynthesis, morphology, leaf anatomy and cytogenetics of hybrids between C_3 and C_3/C_4 *Panicum* species. *Plant Physiol* **77**: 653–658

Brown RH and Brown WV (1975) Photosynthetic characteristics of *Panicum milioides*, 9 species with reduced photorespiration. *Crop Sci* **15**: 681–685

Brown WV (1977) The Kranz syndrome and its subtypes in grass systematics. *Mem Torrey Bot Club* **23**: 1–97

Burnell JN and Hatch MD (1988) Low bundle sheath carbonic anhydrase is apparently essential for effective C_4 pathway operation. *Plant Physiol* **86**: 1252–1256

Burris RH and Black CC (1976) eds. CO_2 *Metabolism and Plant Productivity*. University Park Press, Maryland

Cheng SH, Franseschi VR, Keefe D, Mets LJ and Ku MSB (1987) Photosynthesic characteristics of reciprocal F_1 hybrids between C_3-C_4 intermediates and C_4 *Flaveria* species. In: (J Biggins, ed) *Progress in Photosynthesis Research*, Vol 3, Martinus Nijhoff/W. Junk, The Hague, pp 637–640

Cheng SH and Ku MSB (1985) Intercellular localization of key enzymes of C_4 photosynthesis in the F_1 hybrids between C_3-C_4 intermediate and C_4 *Flaveria* species. *Plant Physiol Suppl* **77**: 90 (abstract)

Cheng SH, Moore BD and Ku MSB (1986) Unusual cellular compartmentation of photosynthetic enzymes in the C_4 species *Flaveria brownii*. *Plant Physiol Suppl* **80**: 55 (abstract)

Chollet R, Budde RHA, Jiar JA and Roeske CA (1990) Light/dark-regulation of C_4-photosynthesis enzymes by reversible phosphorylation. In: (M Baltscheffsky, ed) *Current Research in Photosynthesis*, Vol 4, Kluwer Acad Publ, Dordrecht, pp 135–142

Cooper JP (1975) Control of photosynthetic production in terrestrial systems. In: (JP Cooper, ed) *Photosynthesis and Productivity in Different Environments*, Cambridge University Press, Cambridge, pp 593–621

Crespo HM, Frean M, Cresswell CR and Tew J (1979) The occurrence of both C_3 and C_4 photosynthetic characteristics in a single *Zea mays* plant. *Planta* **147**: 257–263

Das VSR and Raghavendra AS (1980) The plurality of carbon pathway in plant photosynthesis. In: (PKK Nair, ed) *Glimpses in Plant Research*, Vol V, *Modern Methods in Plant Taxonomy*, Vikas Publishing House Pvt. Ltd., New Delhi, pp 344–351

Doncaster HD and Leegood RC (1987) Regulation of phosphoenolpyruvate carboxylase activity in maize leaves. *Plant Physiol* **84**: 82–87

Doresch R (1970) The 10 worst weeds of field crops. *Crops and Soils* **1**: 14

Downes RW (1969 Differences in transpiration rates between tropical and temperate grasses under controlled conditions. Planta **88**: 261–273

Downton WJS (1975) The occurrence of C_4 photosynthesis among plants. *Photosynthetica* **9**: 96–105

Edwards GE and Huber SC (1981) The C_4 pathway. In: (MD Hatch and NK Boardman, eds) *The Biochemistry of Plants. A Comprehensive Treatise*, Vol 8, Academic Press, New York, pp 237–281

Edwards GE and Ku MSB (1987) Biochemistry of C_3-C_4 intermediates. In: MD Hatch and NK Boardman (eds) *The Biochemistry of Plants. A Comprehensive Treatise*, Vol 10, Academic Press, New York, pp 275–325

Edwards GE and Ku MSB (1990) Regulation of the C_4 pathway of photosynthesis. In: (I Zelitch, ed) *Perspectives in Biochemistry and Genetic Regulation of Photosynthesis, Alan R. Liss Inc.*, New York, pp 175–190

Edwards GE and Walker DA (1983) *C_3, C_4: Mechanisms, and Cellular and Environmental Regulation of Photosynthesis.* Blackwell Scientific Publications, Oxford

Furbank RT and Foyer CH (1988) C_4 plants as valuable model experimental systems for the study of photosynthesis. *New Phytol* **109**: 265–277

Gifford RM (1974) A comparison of potential photosynthesis, productivity and yield of plant species with differing photosynthetic mechanism. *Aust J Plant Physiol* **1**: 107–117

Haberlandt G (1914) *Physiological Plant Anatomy.* Translated by M Drummond. Macmillan and Co, London

Hatch MD (1977) C_4 pathway photosynthesis: mechanism and physiological function. *Trends Biochem Sci* **2**: 199–202

Hatch MD (1978) Regulation of enzymes in C_4 photosynthesis. *Curr Top Cell Regul* **14**: 1–27

Hatch MD (1987) C_4 photosynthesis: A unique blend of modified biochemistry, anatomy and ultrastructure. *Biochim Biophys Acta* **895**: 81–106

Hatch MD and Burnell JN (1990) Carbonic anhydrase activity in leaves and its role in the first step of C_4 photosynthesis. *Plant Physiol* **93**: 825–828

Hatch MD and Osmond CB (1976) Compartmentation and transport in C_4 photosynthesis. In: (CR Stocking and U (Heber, eds) *Encyclopaedia of Plant Physiology*, New Series, Vol 3, Springer-Verlag, Berlin, pp 144–184

Hatch MD, Osmond CB and Slatyer RO (1971) *eds. Photosynthesis and Photorespiration.* Wiley Interscience, New York

Hatch MD and Slack CR (1970) Photosynthetic CO_2-fixation pathways. *Annu Rev Plant Physiol* **21**: 141–162

Hattersley PW, Watson L and Osmond CB (1977) *In situ* immunofluorescent labelling of ribulose-1, 5-bisphosphate carboxylase in leaves of C_3 and C_4 plants. *Aust J Plant Physiol* **4**: 523–539

Hattersley PW, Wong SC, Perry S and Roksandic Z (1986) Comparative ultrastructure and gas exchange characteristics of the C_3-C_4 intermediate, *Neurachne minor* ST Blake (Poaceae). *Plant Cell Environ* **9**: 217–233

Holaday AS and Chollet R (1984) Photosynthetic/photorespiratory characteristics of C_3-C_4 intermediate species. *Photosynth Res* **5**: 307–323

Holaday AS, Shieh YJ, Lee KW and Chollet R (1981) Anatomical, ultrastructural and enzyme studies of leaves of *Moricandia arvensis*, a C_3-C_4 intermediate species. *Biochim Biophys Acta* **637**: 334–341

Holbrook GP, Jordan DB and Chollet R (1985) Reduced apparent photorespiration by the C_3-C_4 intermediate species, *Moricandia arvensis* and *Panicum milioides*. *Plant Physiol* **77**: 578–583

Holm LG (1969) Weed problems in developing countries. *Weed Sci* **17**: 113–118

Holm LG, Plucknet DL, Pancho JV and Herberger JP (1977) *The World's Worst Weeds: Distribution and Biology.* University Press of Hawaii, Honolulu

Hull MR, Long SP and Raines CP (1990) The effects of low temperature on activities of carbon metabolism enzymes in *Zea mays* L. seedlings. In: (M Baltscheffsky, ed) *Current Research in Photosynthesis*, Vol 4, Kluwer Acad Publ, Dordrecht, pp 675–678

Imai K and Murata Y (1979) Changes in apparent photosynthesis, CO_2 compensation point and dark respiration of leaves of some Poaceae and Cyperaceae species with senescence. *Plant Cell Physiol* **20**: 1653–1658

Jenkins CLD (1989) Effects of phosphoenolpyruvate carboxylase inhibitor-3,3-dicholoro-2-(dihydroxyphosphinoylmethyl) propenoate on photosynthesis. C_4 selectivity and studies on C_4 photosynthesis. *Plant Physiol* **89**: 1231–1237

Jenkins CLD, Harris RLD and McFadden H (1987) 3, 3-dicholoro-2-dihydroxyphosphinoylmethyl-2-

propenoate, a new specific inhibitor of phosphoenolpyruvate carboxylase. *Biochem Intl* **14**:219–226

Kanai R and Kashiwagi M (1975) *Panicum milioides*, a Gramineae plant having Kranz anatomy without C_4 photosynthesis. *Plant Cell Physiol* **16**: 669–679

Kennedy RA and Laetsch WM (1973) Relationship between leaf development and primary photosynthetic products in the C_4 plant, *Portulaca oleracea* L. *Planta* **115**: 113–124

Khanna R and Sinha SK (1973) Change in the predominance from C_4 to C_3 pathway following anthesis in sorghum. *Biochem Biophys Res Commun* **52**: 121–124

Kortschak HP, Hartt CE and Burr GO (1965) Carbon dioxide fixation in sugar-cane leaves. *Plant Physiol* **40**: 209–213

Krenzer EG, Moss DN and Crookston RK (1975) Carbon dioxide compensation points of flowering plants. *Plant Physiol* **56**: 194–206

Kumar PA and Abrol YP (1990) Photorespiratory nitrogen metabolism in the C_3-C_4 intermediate species, *Moricandia arvensis* (L.) DC. *Biochem Physiol Pflanzen* **186**: 109–115

Laetsch WM (1974) The C_4 syndrome: A structural analysis. *Annu Rev Plant Physiol* **25**: 27–52

Langdale JA, Metzler MC and Nelson T (1987) The *argentia* mutation delays normal development of photosynthetic gene expression in maize. EMBO J **7**: 3643–3651

Leech RM (1985) The synthesis of cellular components in leaves. In: NR Baker, WJ Davies and CK Ong (eds) *Control of Leaf Growth*, Cambridge University Press, Cambridge, pp 93–113

Leegood RC and Osmond CB (1990) Metabolite fluxes in C_4- and CAM plants. In: Dennis and DJH Turpin, eds) *Advanced Plant Physiology and Molecular Biology*, Longman Technical Publishers, London, pp 274–298

Long SP (1983) C_4 photosynthesis at low temperatures. *Plant Cell Environ* **6**: 345–363

Ludlow MM (1985) Photosynthesis and dry matter production in C_3 and C_4 pasture plants, with special emphasis on tropical C_3 legumes and C_4 grasses. *Aust J Plant Physiol* **12**: 557–572

Miranda V, Baker NR and Long SP (1981a) Anatomical variation along the length of the *Zea mays* leaf in relation to photosynthesis. *New Phytol* **88**: 595–605

Miranda V, Baker NR and Long SP (1981b) Limitations of photosynthesis in different regions of the *Zea mays* leaf. *New Phytol* **89**: 179–190

Monson RK and Moore BD (1989) On the significance of C_3-C_4 intermediate photosynthesis to the evolution of C_4 photosynthesis. *Plant Cell Environ* **12**: 689–699

Monson RK, Edwards GE and Ku MSB (1984) C_3-C_4 intermediate photosynthesis in plants. *BioScience* **34**: 563–574

Monson RK, Moore BD, Ku MSB and Edwards GE (1986) Cofunction of C_3- and C_4- photosynthetic pathways in C_3, C_4 and C_3-C_4 intermediate, *Flaveria* species. *Planta* **163**: 493–502

Moore BD, Cheng SH and Edwards GE (1986) The influence of leaf development on the expression of C_4 metabolism in *Flaveria trinervia*, a C_4 dicot. *Plant Cell Physiol* **27**: 1159–1167

Moore PD (1982) Evolution of photosynthetic pathways in flowering plants. *Nature* **295**: 647–648

Nakamoto H and Edwards GE (1986) Light activation of pyruvate, Pi dikinase and NADP-malate dehydrogenase in mesophyll protoplasts of maize. Effect of DCMU antimycin A, CCCP and phlorizin. *Plant Physiol*. **82**: 312–315

Nambudiri EMV, Tidwell WD, Smith BN and Hebbert NP (1978) A C_4 plant from the Pliocene. *Nature* **276**: 816–817

Nelson T and Langdale JA (1989) Patterns of leaf development in C_4 plants. *Plant Cell* **1**: 3–13

Oquist G and Martin B (1986) Cold climates. In: NR Baker and SP Long (eds) *Photosynthesis in Contrasting Environments*. Elsevier Science Publishers, Amsterdam, pp 237–293

Osmond CB, Bjorkman O and Anderson CJ (1980) *Physiological Processes: Plant Ecology*. Springer-Verlag, Berlin

Osmond, CB, Winter K and Ziegler H (1982) Functional significance of different pathways of CO_2 fixation in photosynthesis. In: OL Lange, PS Nobel, CB Osmond and H Ziegler, (eds) *Encyclopaedia of Plant Physiology*, New Series, Vol 12B, *Physiological Plant Ecology II*, Springer Verlag, Berlin, pp 479–547

Peisker M and Bauwe H (1984) Modelling carbon metabolism in C_3-C_4 intermediate species. I CO_2 compensation concentration and its O_2 dependence. *Photosynthetica* **18**: 9–19

Perchorowicz JT and Gibbs M (1980) Carbon dioxide fixation and related properties in sections of developing green maize leaf. *Plant Physiol* **65**: 802–809

Perrot-Rechenmann C, Vidal J, Brulfer J, Burlet A and Gadal P (1982) A comparative immunocytochemical localization of phosphoenol pyruvate carboxylase in leaves of higher plants. *Planta* **155**: 24–30

Perrot-Rechenmann C, Jacquot JP, Gadal P, Weeden NF, Cseke C and Buchanan BB (1983) Localization of NADP-malate dehydrogenase of maize leaves by immunological methods. *Plant Sci Lett* **30**: 219–226

Powles SB, Berry JA and Bjorkman O (1983) Interaction between light and chilling temperature on the inhibition of photosynthesis in chilling sensitive plants. *Plant Cell Environ* **6**: 117–123

Raghavendra AS (1980) Characteristics of plant species intermediate between C_3 and C_4 pathways of photosynthesis: Their focus of mechanism and evolution of C_4 syndrome. *Photosynthetica* **14**: 271–283

Raghavendra AS and Das VSR (1978) The occurrence of C_4 photosynthesis: A supplementary list of C_4 plants reported during late 1974–mid-1977. *Photosynthetica* **12**: 200–208

Raghavendra AS, Rajendrudu G and Das VSR (1978) Simultaneous occurrence of C_3 and C_4 photosynthesis in relation to leaf position in *Mollugo nudicaulis*. *Nature* **273**: 143–144

Rajendrudu G and Das VSR (1981) *Parthenium hysterophorus*, L. (Asteraceae) exhibiting low photorespiration. *Curr Sci* **50**: 592–593

Rajendrudu G, Prasad JSR and Das VSR (1986) C_3-C_4 intermediate species in *Alternanthera* (Amaranthaceae). Leaf anatomy, CO_2 compensation point, net CO_2 exchange and activities of photosynthetic enzymes. *Plant Physiol* **80**: 409–414

Rawsthorne S, Hylton CM, Smith AM and Woolhouse HW (1988a) Photorespiratory metabolism and immunogold localization of photorespiratory enzymes in leaves of C_3 and C_3-C_4 intermediate species of *Moricandia*. *Planta* **173**: 283–308

Rawsthorne S, Hylton CM, Smith AM and Woolhouse HW (1988b) Distribution of photorespiratory enzymes between bundle-sheath and mesophyll cells in leaves of the C_3-C_4 intermediate species *Moricandia arvensis* (L.) DC. *Planta* **176**: 527–532

Sage RF, Pearcy RW and Seemann JR (1987) The nitrogen use efficiency of C_3 and C_4 plants. III Leaf nitrogen effects on the activity of carboxylating enzymes in *Chenopodium album* (L.) and *Amaranthus retroflexus* (L.) *Plant Physiol* **85**: 355–359

Schuster WS and Monson RK (1990) An examination of the advantages of C_3-C_4 intermediate photosynthesis in warm environments. *Plant Cell Environ* **13**: 903–912

Seemann JR, Badger MR and Berry JA (1984) Variations in specific activity of ribulose-1, 5-bisphosphate carboxylase between species utilizing differing photosynthetic pathways. *Plant Physiol* **74**: 791–794

Sheen JY and Bogorad L (1987a) Regulation or levels of nuclear transcripts for C_4 photosynthesis in bundle sheath and mesophyll cells of maize leaves. *Plant Mol Biol* **8**: 227–238

Sheen JY and Bogorad L (1987b) Differential expression of C_4 pathway genes in mesophyll and bundle sheath cells of greening maize leaves. *J Biol Chem* **262**: 11726–11730

Singh P, Kumar PA, Abrol YP and Naik MS (1985) Photorespiratory nitrogen cycle—a critical evaluation. *Physiol Plant* **66**: 169–176

Smith BN, Martin GE and Boutton TW (1979) Carbon isotopic evidence for the evolution of C_4 photosynthesis. In: (ER Klein and PD Klein, eds) *Stable Isotopes: Proceedings of Third International Conference*. Academic Press, New York, pp 231–237

Taylor AD and Craig AS (1971) Plants under climatic stress. II Low temperature, high light effects on chloroplast ultrastructure. *Plant Physiol* **47**: 719–725

Thomasson JR, Nelson ME and Zakrzewski RJ (1986) A fossil grass (Gramineae: Chloridoideae) from the Miocene with Kranz anatomy. *Science* **233**: 876–878

Usuda H, Ku MSB and Edwards GE (1984) Rate of photosynthesis relative to activity of photosynthetic enzymes, chlorophyll and soluble protein content among ten C_4 species. *Aust J Plant Physiol* **11**: 509–517

Weiner H, Burnell JN, Woodrow IE, Heldt HW and Hatch MD (1988) Metabolite diffusion into bundle sheath cells from C_4 plants. Relation to C_4 photosynthesis and plasmodesmatal function. *Plant Physiol* **88**: 815–822

Wessinger ME, Edwards GE and Ku MSB (1989) Quantity and kinetic properties of ribulose 1, 5-bisphosphate carboxylase in C_3, C_4 and C_3-C_4 intermediate species of *Flaveria* (Asteraceae). *Plant Cell Physiol* **30**: 665–671

Winter K, Usuda H, Tsuzuki M, Schmitt MR, Edwards GE, Thomas RJ and Evert RF (1982) Influence of nitrate and ammonium on photosynthetic characteristics and leaf anatomy of *Moricandia arvensis*. *Plant Physiol* **70**: 616–625

Wong SC (1979) Elevated atmospheric partial pressure of CO_2 and plant growth. I Interactions of nitrogen nutrition and photosynthetic capacity in C_3 and C_4 plants. *Oecologia* **44**: 63–74

Yeoh HH, Badger MR and Watson L (1981) Variations in kinetic properties of ribulose 1, 5-bisphosphate carboxylases among plants. *Plant Physiol* **67**: 1151–1155

13

Interaction between Carbon and Nitrogen Metabolism

P. A. Kumar, Raghuveer Polisetty and Y. P. Abrol

Division of Plant Physiology
Indian Agricultural Research Institute
New Delhi 110 012, India

CONTENTS

ABBREVIATIONS

DCMU	:	3-(3, 4-dichlorophenyl)-1, 1-dimethyl urea;
GDH	:	Glutamate dehydrogenase;
GS	:	Glutamine synthetase;
α-KG	:	α-Ketoglutaric acid/Ketoglutarate;
NADH	:	Nicotinamide adenine dinucleotide, reduced form;
NR	:	Nitrate reductase;
NUE	:	Nitrogen-use efficiency;
OAA	:	Oxalo-acetic acid;
PEP	:	Phospho-enol pyruvate;
RUBP	:	Ribulose 1, 5 bisphosphate;
RuBPcase	:	Ribulose 1, 5 bisphosphate carboxylase/oxygenase

ABSTRACT

The relationship between carbon and nitrogen metabolism in higher plants is reviewed. Photosynthetic reactions are involved in the synthesis, regulation, and maintenance of the enzymes of nitrate assimilation pathway. Chloroplast signals are needed for the synthesis of nitrate and nitrite reduction. Involvement of photorespiration in the supply of reductant is also implicated. Photorespiratory nitrogen metabolism is one of the important aspects of the interactions of carbon and nitrogen. The processes of transamination of glyoxylate, release of ammonia during glycine oxidation, and the reassimilation of photorespiratory ammonia are well understood. There is increasing evidence which indicates that the photorespiratory nitrogen cycle is not a closed one. Evidence for nitrogen import into and removal from the cycle are presented. Glutamate dehydrogenase does not appear to play a significant role in the process of reassimilation of photorespiratory ammonia. This enzyme may be involved in oxidation of glutamate. Tricarboxylic acid (TCA) cycle provides carbon cycles for amino acid biosynthesis. It operates in the light and flow of carbon from photosynthesis into TCA cycle is regulated by ammonium. Nitrate and nitrite assimilation are also implicated in the stimulation of the cycle activity. Nitrogen-use efficiency (NUE) of plants appears to be associated with the carbon assimilation pathways. Higher NUE of C_4 plants is attributed to the relatively smaller investment of nitrogen in the photosynthetic carboxylation enzymes, more efficient distribution and redistribution of nitrogen in the plant, and also perhaps the spatial separation of reactions involved in photorespiration and nitrate assimilation.

I. Introduction

Carbon and nitrogen are the two major inputs from nature into the plant system. Carbon and nitrogen metabolism in higher plants are intertwined in many ways. The important aspects of such a relationship are: (1) the role of photosynthesis and photorespiration in nitrogen metabolism; (2) the synthesis of amino acids in chloroplasts and the regulation of carbon flow from photosynthesis to amino-acid biosynthesis by ammonium; (3) the role of mitochondrial respiration in nitrate reduction; and (4) the nitrogen-use efficiency of plants. In this chapter, we dwell upon some of these aspects in detail.

II. Photosynthesis and Nitrate Assimilation

A number of investigators have implicated photosynthetic reactions in light dependent synthesis of the enzyme nitrate reductase which is the rate-limiting enzyme in the pathway of nitrate assimilation. This inference has been made on the basis of experiments with chlorophyll-deficient mutants (Sawhney et al., 1972; Warner and Kleinhoffs, 1974), etiolated seedlings treated with bleaching agents (Deane-Drummond and Johnson, 1980) or antibiotics (Sawhney and Naik, 1972). Elimination of response to light in greening or green seedlings by DCMU (3-(3, 4-dichlorophenyl)-1, 1-dimethyl urea; (Aslam et al., 1973; Sluiters-Scholten, 1975; Sihag et al., 1979) or under atmosphere free of carbon dioxide (Kannangara and Woolhouse, 1967; Sawhney and Naik, 1972) also implied dependence on photosynthetic reactions. Since the exogenously supplied sugars were not effective in replacing the photosynthetic requirement in either etiolated seedlings (Sawhney and Naik, 1972) or in illuminated DCMU-treated seedlings (Kannangara and Woolhouse, 1967), it was concluded that the enhancement of the enzyme synthesis is closely linked with some early events of photosynthesis. The adverse effects of inhibitors such as DCMU and atrazine suggested that flow of electrons via the noncyclic pathways of photosynthesis might be essential for light-dependent synthesis of nitrate reductase (Sawhney and Naik, 1972). The importance of reductant generated via photosynthetic reactions or from oxidation of carbohydrates in the dark for maintaining high nitrate reductase activity has also been deduced from studies on bean leaves (Sluiters-Scholten, 1975).

Observations against a direct involvement of photosynthetic reactions have also been documented. Synthesis of the enzyme in darkness (Travis et al., 1970; Aslam et al., 1976) or in leaves devoid of chlorophyll (Duke and Duke, 1979; Kakefuda et al., 1983) are inconsistent with the implied role of photosynthetic reactions. A number of workers believe that light per se has no direct role in the synthesis of the enzyme and provided the tissue contains adequate levels of sugar, it can be formed in the dark. The reported effect of exogenously supplied sugars in reversing the interference by DCMU (Aslam et al., 1973) or increasing the enzyme activity in the dark, which is etiolated leaves (Sihag et al., 1979) or leaves that are greened (Travis and Key, 1971), is in conformity with this view.

Recent reports indicate that a plastidic factor is involved in the synthesis of nitrate reductase and nitrite reductase (Oelmuller et al., 1988). This is similar to the proposed chloroplast signal that regulates cab gene expression (Taylor, 1989). However, Huffaker and co-workers (Ward et al., 1989) found that feeding sugars to albino iojap (maternally inherited) seedlings resulted in normal levels of nitrate reductase.

Nitrate reduction is also intimately related to carbon metabolism in terms of the supply of reducing power. The original proposal of Hageman and group (Klepper et al., 1971) argues in favour of direct supply of NADH (nicotinamide adenine nucleotide, reduced) for nitrate reductase from the oxidation of triose phosphates produced during photosynthesis. The possibility of coupling nitrate reduction with the metabolism of triose phosphates was also indicated by the work on spinach protoplasts (Rathnam, 1978). Nitrite reductase is a ferredoxin-dependent enzyme and thus is directly dependent upon photosynthetic electron transport for energy supply (Abrol et al., 1983).

We have recently proposed that glycine oxidation during photorespiration supports nitrate reduction in barley leaves (Kumar et al., 1988a) (Fig. 1). From experiments using photorespiratory metabolites and inhibitors it has been inferred that glycolate pathway is linked to nitrate reduction in C_3 plants but not in C_4 plants in terms of energy supply (Kumar and Abrol, 1989).

III. Photorespiration and Nitrogen Metabolism

Nitrogen metabolism is associated with photorespiration at several points: (1) the transamination of glyoxolate to form glycine; (2) the release of ammonia during glycine decarboxylation; and (3) the reassimilation of photorespiratory ammonia.

It is now well established that in addition to the regular flow of nitrogen through the nitrate assimilation pathway, massive recycling of nitrogen occurs during photorespiration in illuminated leaves of C_3 plants and to some extent in C_4 plant leaves (Keys et al., 1978; Kumar et al., 1984). This flow can be as much as 10-fold greater than the net assimilation of inorganic nitrogen (Wallsgrove et al., 1983).

The glycolate produced by the oxygenase activity of ribulose-1, 5 bisphosphate carboxylase/oxygenase (RuBPcase) is oxidized to glyoxylate in the peroxisomes

342

MITOCHONDRION   CYTOSOL

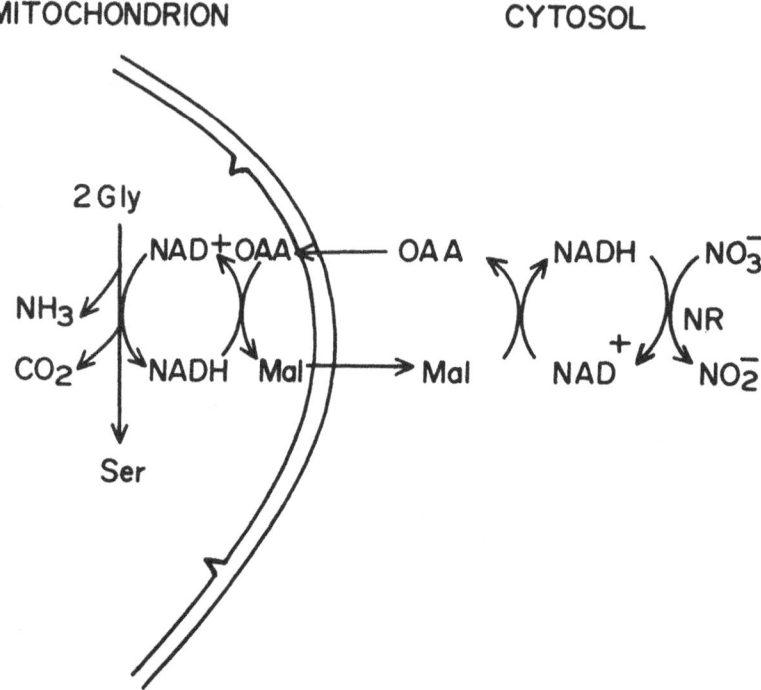

Figure 1. Interaction between glycine oxidation and nitrate reduction. (Reproduced with permission, Kumar *et al.*, 1988a).

(Husic *et al.*, 1987). Glyoxylate is then transaminated to produce glycine (Ogren, 1984). Two molecules of glycine are oxidatively decarboxylated in the mitochondria to produce serine, CO_2, NH_3, and NADH (Husic *et al.*, 1987). The reassimilation of this substantial ammonia occurs via glutamate synthase cycle. Keys *et al.* (1978) have proposed a system to explain this nitrogen recycling, termed ''photorespiratory nitrogen cycle'' (Fig. 2).

In this scheme, ammonia released in mitochondria is reassimilated via glutamine synthetase (GS) and glutamate synthase generating two molecules of glutamate, one molecule of which donates an α-amino group to glyoxylate in the peroxisomes. The second molecule is used in the synthesis of glutamine. Recent experiments suggest that photorespiratory nitrogen metabolism may not be a closed process because transamination of glyoxylate could occur using other amino donors like alanine and asparagine (Kumar and Abrol, 1990a). In labelling studies, it was found that nitrogen from ^{15}N (amino)-labelled asparagine was transferred readily into constituents of photorespiratory nitrogen cycle (glycine, serine, ammonia, and glutamine) and since this transfer decreased in low oxygen conditions and in the presence of inhibitors, it appeared to be dependent on photorespiration (Joy, 1988).

There is evidence to show that removal of amino acids (glycine, serine, and glutamate) from the photorespiratory nitrogen cycle for biosynthetic reactions occurs. Serine, glutamate, and glutamine are the major amino acids exports from the leaves to developing young tissues under normal atmospheric conditions (Madore and Grodzinski, 1984). It has been shown that increasing the concentration of oxygen

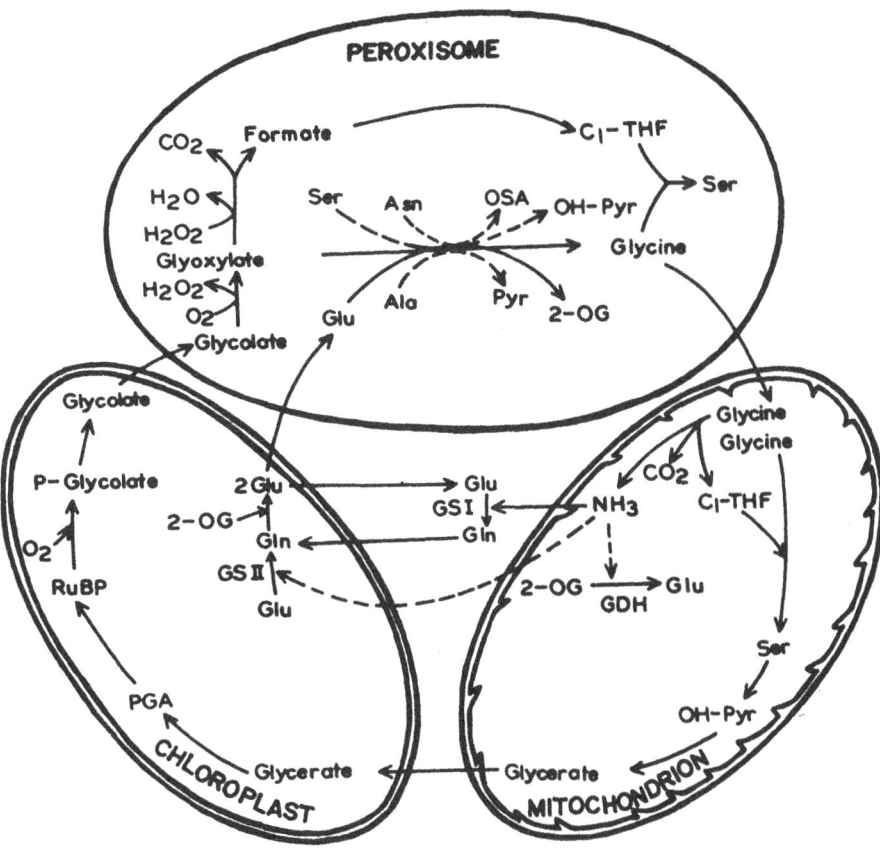

Figure 2. Photorespiratory nitrogen metabolism in higher plant cells. (Reproduced with permission, Kumar and Abrol, 1990a).

leads to a greater export of amino acids, particularly serine and glycine, from the leaves of *Salvia*. Alanine is also another important amino donor for the transamination of glyoxylate (Betsche, 1983). Experiments of Joy and his group have shown that the contribution of alanine could be higher than that from glutamate. Treatment of pea leaves with azaserine, an inhibitor of glutamate synthase, rapidly decreases the levels of alanine as well as glutamate. Stimulation of photorespiration by lowering the CO_2 concentration resulted in a rapid drop in alanine concentration while at the same time the amount of nitrogen entering alanine increased two fold (Joy, 1988). In glutamate synthase deficent barley mutants, photorespiration in air results in a considerable fall of alanine levels which decrease proportionately more rapidly than glutamate (Blackwell *et al.*, 1988). These observations show that alanine levels can change rapidly and that there is a rapid turnover related to photorespiration. The involvement of alanine in photorespiration implies that less N is recycled through glutamate than expected (Keys *et al.*, 1978).

Glutamate synthetase inhibitors like methionine sulphoximine cause rapid accumulation of ammonia, mostly photorespiratory ammonia, in C_3 plant leaves (Kumar *et al.*, 1983, 1984). This accumulation is less extensive in C_4 plants, C_3–C_4

intermediate plants, and when photorespiration is suppressed (Martin *et al.*, 1983; Kumar and Abrol, 1990b). The accumulation of ammonia is accompanied by loss by many of the soluble amino acids (Lea *et al.*, 1984; Rhodes *et al.*, 1986; Kumar *et al.*, 1990). Cullimore and Sims (1980) suggested a link between protein turnover and photorespiration in *Chlamydomonas*. Hipkin *et al.* (1982) found that ammonia generated in MSO-treated *Chlamydomonas* was related to loss of protein. Recent studies have shown that the flow of nitrogen from protein to photorespiratory ammonia occurs in *Lemna* (Rhodes *et al.*, 1986) However, it is not known how much of the amino nitrogen from protein is cycled under normal physiological conditions.

There has been some controversy whether all the glyoxylate produced from glycolate is transaminated to glycine (Singh *et al.*, 1985). Substantial CO_2 evolution by the action of H_2O_2 on glyoxylate has been reported (Grodzinski and Woodrow, 1981; Walton and Butt, 1981). The role of formate and C_1-tetrahydrofolate generated during glyoxylate decarboxylation in the synthesis of serine has been reviewed (Singh *et al.*, 1985). However, it is generally considered that under normal physiological conditions glyoxylate is transaminated to glycine, and when amino donors are not available (poor N nutrition), the glyoxylate may be decarboxylated (Husic *et al.*, 1987). This is supported by the observations of Lea and his group working with barley mutants lacking serine glyoxylate/amino-transferase (Blackwell *et al.*, 1988).

The reassimilation of photorespiratory ammonia by GS is also now well established (Kumar and Abrol, 1990a). Experiments with inhibitors, labelling studies, and use of photorespiratory mutants confirmed this view (Joy, 1988). A recent study using cDNA probe for chloroplastic GS has shown that GS_2-mRNA levels were affected by photorespiratory conditions (Edwards and Coruzzi, 1989). Plants grown under photorespiratory conditions maintained higher steady-state levels of GS_2-mRNA than plants grown in non-photorespiratory conditions.

The contention that a part of photorespiratory ammonia is reassimilated by glutamate dehydrogenase (GDH) has been put forward by Oaks and her group (Oaks and Yamaya, 1990). They found that significant levels of ^{15}N from $^{15}NH_4Cl$ or (^{15}N)-glycine were incorporated into glutamate in isolated pea and corn mitochondria (Yamaya *et al.*, 1986). However, the relative contribution of GDH to reassimilation of photorespiratory ammonia seems to be not very significant. Glutamate dehydrogenase is considered an alternative system that may operate under conditions of high ammonia concentrations usually encountered during stress or senescence (Kumar and Abrol, 1990a). Recently, ^{15}N-nuclear magnetic resonance studies on shoot-forming cultivars of white spruce buds revealed that ammonia first enters the glutamine pool and subsequently the glutamate (Thorpe *et al.*, 1989). This rules out any role of GDH in ammonia assimilation under normal conditions of growth and development. Recent studies by Robinson *et al.* (1991), with cell suspension cultures of corrot (*Daucus Carota* L. cv. Chentenay), reaffirm that primary assimilation of ammonia occurs via the glutamate synthase pathway while GDH plays a catabolic role in oxidation of glutamate and supply of carbon skeletons for effective functioning of TCA cycle.

IV. Dark Respiration and Nitrogen Metabolism

Considerable effort has been directed towards estimating dark respiration during photosynthesis because of the effect of this process on the carbon economy of the plant (Graham, 1980). Despite these efforts, the magnitude of mitochondrial dark respiration in leaves during photosynthesis is not known.

Studies of McCashin et al. (1988) in wheat leaf slices indicate that in the dark the (TCA) cycle was the major route for succinate metabolism with CO_2 as the chief end-product. In the light also, TCA cycle operates and CO_2 is released from the mitochondria. This process of CO_2 release by TCA cycle in the light is estimated at 80% of the rate occurring in the dark. It is generally agreed that TCA cycle activity is maintained in the light to provide carbon skeletons for biosynthetic reactions (Graham, 1980; Weger et al., 1988). The use of radiolabelled intermediates has provided evidence for the maintenance of TCA cycle carbon flow in photosynthesizing cells (Marsh et al., 1965). The flow of carbon through the TCA cycle increases during the assimilation of NH_4^+ and NO_3^-, to provide carbon skeletons for amino-acid biosynthesis (Kanazawa et al., 1970; Woo and Canvin, 1980; Larsen et al., 1981; and Elrifi et al., 1988). Weger et al. (1988) have shown that the assimilation of ammonium by photosynthesizing cells of Selenastrum results in increased rates of both TCA cycle CO_2 release and mitochondrial O_2 consumption. This stimulation of TCA cycle carbon flow would result in an increase in the production α-ketoglutaric acid (α-KG), needed for ammonia assimilation and net glutamate synthesis. Addition of small amounts oof ammonium (1 mM) to the medium of photosynthesizing Chlorella results in considerable stimulation of the synthesis of certain amino acids, especially glutamine, glutamate, alanine, and aspartate. This stimulation has been shown to be a consequence of an accelerated flow of carbon from photosynthesis into intermediates of TCA cycle, such as citrate (Kanazawa et al., 1970, 1972). These aspects have been discussed in detail by Bassham et al. (1981).

The assimilation of ammonium by photosynthetic organisms results in changes to both photosynthetic and respiratory carbon metabolism. An interesting observation was the suppression in the rate of photosynthesis due to inorganic nitrogen assimilation in some species of N-limited microalgae (Elrifi and Turpin, 1986). One reason could be that during ammonium assimilation Calvin cycle intermediates are exported from the chloroplast to provide carbon skeletons for N-assimilation. If the removal of triose phosphates from the Calvin cycle exceeded one-third the rate of carbon fixation, RuBp regeneration and RuBP concentrations would decrease (Elrifi et al., 1988), thus affecting the carbon fixation rates. Simultaneously, the increase in TCA cycle activity required to provide keto-acids for amino-acid synthesis coincided with an increase in NADH oxidation via the mitochondrial electron transport chain. Nitrate and nitrite assimilation stimulated TCA cycle activity to a greater degree than did the assimilation of ammonium (Weger and Turpin, 1989). A substantial proportion of the reductant generated during this process is exported to the chloroplast and used for nitrate and nitrite reduction both in the dark and during photosynthesis.

Another aspect which attracted considerable attention during the past two decades is the relationship between mitochondrial respiration and nitrate reduction. Sawhney et al. (1978) proposed that NADH generated during TCA cycle activity supports cytosolic nitrate reduction (Abrol *et al.*, 1983). This has been supported by observations that a close relationship exists between respiratory CO_2 release and nitrate reduction (Naik and Nicholas, 1981). The findings of Weger and Turpin (1989) mentioned earlier that TCA cycle reductant is exported from the mitochondrion to the chloroplast in support of transient nitrate and nitrite reduction both in the dark and during photosynthesis also supports the proposal of Sawhney *et al.* (1978).

Recently, we have proposed that NADH generated during glycine oxidation in mitochondria during photorespiration supports nitrate reduction in the cytosol (Kumar *et al.*, 1988a, b; Kumar and Abrol, 1989). Although the actual shuttle mechanism to export reducing equivalents from mitochondria is not known, it may be based upon an oxaloacetic acid malate system (Woo *et al.*, 1980; Zoglowek *et al.*, 1988). Although a significant proportion of NADH produced during photorespiratory glycine oxidation will be oxidized via the mitochondrial electron transport chain, it seems unlikely that all of the NADH from glycine could be reoxidized via the electron transport chain (Dry *et al.*, 1987). Besides, the export of some reducing equivalents from the mitochondria via substrate shuttles is probably necessary.

V. Nitrogen-use Efficiency

Plant species that fix CO_2 by C_4 cycle have higher rates of carbon dioxide uptake than species using the C_3 photosynthetic carbon reduction cycle. Greater CO_2 fixation capacity has been associated with reduced photorespiration, specialized leaf anatomy and biochemical pathways that differ in C_3 and C_4 plants. The higher photosynthetic rate of C_4 species also results in more dry matter production per unit of water transpired. Reviewing some of the earlier work published, Brown (1978) proposed a hypothesis that C_4 plants have greater NUE (biomass production per unit of N in the plant) than C_3 plants. This difference was attributed to relatively smaller investment of N in the photosynthetic carboxylation enzymes of C_4 plants than C_3 plants.

One of the characteristic differences between C_3 and C_4 species, particularly among Gramineae, is the response to increased N supply. It is postulated that C_4 species utilize N more efficiently than C_3 species. The main difference in NUE is stated to be based on partitioning of N among leaf proteins and the related carbon dioxide fixation pathways.

The greater NUE of C_4 species appears to be associated with the compartmentation of PEP carboxylase (phosphoenol pyruvate carboxylase) and RuBPcase in mesophyll and bundle sheath tissues, respectively. In C_3 species as much as 40 to 50% of the soluble leaf protein is associated with RuBPcase activity (Blenkinsop and Dale, 1974). The allotment of such a large percentage of RuBPcase appears to be inefficient use of protein.

Not only does RuBPcase account for a large percentage of leaf protein in C_3 species, but the proportion appears to rise with increased leaf protein level (Blenkinsop and Dale, 1974; Brown, 1978). The increase in percentage of RuBPcase at high leaf protein levels may indicate that its formation has a lower priority than other leaf proteins when N is limiting or that perhaps RuBPcase serves as storage protein when N is plentiful. The rapid changes in the levels of this protein under variable environmental conditions, including N nutrition, and its close relationship with photosynthesis emphasizes the importance of N availability in leaf photosynthesis (Blenkinsop and Dale, 1974).

In C_4 species, the RuBPcase is restricted to bundle sheath strands (Huber et al., 1976). Since the bundle sheath cells contain approximately 50% of the soluble leaf protein (Chen et al., 1974; Edwards and Black, 1971) and since the C_4 pathway concentrates CO_2 in these cells, a mechanism exists for a much higher NUE in CO_2 fixation. Early work of Slack and Hatch (1967) indicated that only 4–10% of soluble leaf protein of sugar-cane, corn, and sorghum was in RuBPcase compared to 50% in the C_3 species, oat, wheat, and sugar beet.

Studies of Wilson and Brown (1983) and Brown and Wilson (1983) have indicated that higher NUE may not be always associated with C_4 photosynthesis. They found *Panicum prionitis* (C_4) had NUE similar to or lower than C_3 and C_3-C_4 intermediate *Panicum* species. Studies of Brown (1985) indicate that at various low soil N levels C_4 species are likely to produce more growth and be more competitive than C_3 species. It implies C_4 species would exhibit higher NUE under low N soil conditions.

Martin et al. (1983) found that NO_3^- accumulation was significantly higher in C_3 (barley and wheat) than in C_4 (maize and sorghum) cereals. This indicated that C_3 cereals accumulated nitrate in vacuole rather than channelling it to assimilation process. Studies of Oaks et al. (1990) have indicated the apparent enhanced NUE in maize was related to faster uptake and reduction of nitrate, a more efficient distribution and redistribution of nitrogen in the plant, or the spatial separation of reactions involved in photorespiration and nitrate assimilation in the leaves of C_4 plants. These authors also postulated that spatial separation of enzymes of nitrogen assimilation seen in C_4 cereals may be contributing to the efficient functioning of C_4 cereals (Yamaya and Oaks, 1987). Maheswari et al. (1985) and Kumar (1988) found that the activity of GS is many times more in C_3 plants than in C_4 plants. It is speculated that C_4 plants invest less N in GS synthesis by virtue of their low rates of photorespiratory ammonia release.

VI. References

Abrol YP, Sawhney SK and Naik MS (1983) Reduction of nitrate and nitrite in higher plants in light and dark. *Plant Cell Environ* **6**: 595–599

Aslam M, Huffaker RC and Travis RL (1973) The interaction of respiration and photosynthesis in induction of nitrate reductase activity. *Plant Physiol* **52**: 137–141

Aslam M, Oaks A and Huffaker RC (1976) Effect of light and glucose on induction of nitrate reductase and on distribution of nitrate in etiolated barley leaves. *Plant Physiol* **58**: 588–591

Bassham JA, Larsen PO, Lawyer AL and Cornwell KL (1981) Relationships between nitrogen metabolism and photosynthesis. In: JD Bewley (ed) *Nitrogen and Carbon Metabolism*, Martinus Nijhoff, The Hague, pp 135–158

Betsche T (1983) Aminotransfer from alanine and glutamate to glycine and serine during photorespiration in oat leaves. *Plant Physiol* **71**: 961–965

Blackwell RD, Murray AJS, Lea PJ, Kendall AC, Hall NP, Turner JC and Wallsgrove RM (1988) The value of mutants unable to carry out photorespiration. *J Exp Bot* **25**: 913–926

Brown RH (1978) A difference in N-use efficiency in C_3 and C_4 plants and its implications in adaptation and evolution. *Crop Sci* **18**: 93–98

348

Brown RH (1985) Growth of C_3 and C_4 grasses under low N level. *Crop Sci* **25**: 954–957

Brown RH and Wilson JR (1983) Nitrogen response of *Panicum* species differing in CO_2 fixation pathway. II. Carbon dioxide exchange characteristics. *Crop Sci* **23**: 1154–1159

Chen TM, Dittrich P, Campbell WH and Black CC (1974) Metabolism of epidermal tissues, mesophyll cells and bundle sheath strands resolved from mature nutsedge leaves. *Arch Biochim Biophys* **163**: 246–262

Cullimore JV and Sims AP (1980) An Association between photorespiration and protein catabolism. Studies with *Chlamydomonas*. *Planta* **150**: 392–396

Deane-Drummond CE and Johnson CB (1980) Absence of nitrate reductase activity in San 9789 bleached leaves of barley seedlings. *Plant Cell Environ* **3**: 303–308

Dry IB, Bryce JH and Wiskich JT (1987) Regulation of mitochondrial respiration. In: (PK Stumpf and EE Conn, eds) *The Biochemistry of Plants*, Academic Press, and New York, pp **11**: 213–252

Duke SO and Duke SH (1979) Photosynthetic independence of light caused increase in extractable nitrate reductase from maize seedlings. *Plant Cell Physiol* **20**: 1371–1380

Edwards GE and Black CC (1971) Photosynthesis in mesophyll and bundle sheath cells isolated from *Digitaria sanguinalis* (L). scop. leaves. In: MD Hatch, CB Osmond and RC Slatyer (eds) *Photosynthesis and Photorespiration*, Wiley Interscience, New York, pp 153–168

Edwards JW and Coruzzi GM (1989) Photorespiration and light act in concert to regulate the expression of nuclear gene for chloroplast glutamine synthetase. *Plant Cell* **1**: 241–248

Elrifi IR and Turpin DH (1986) Nitrate and ammonium induced photosynthetic suppression in N-limited *Selenastrum minutum*. *Plant Physiol* **81**: 273–279

Elrifi IR, Holmes JJ, Mayo WP, Weger HG and Turpin DH (1988) RuBP limitation of photosynthetic carbon fixation: Interaction between photosynthesis, respiration and ammonium assimilation. *Plant Physiol* **87**: 395–401

Graham D (1980) Effects of light on "Dark Respiration". In: D.D. Davies (ed) *The Biochemistry of Plants*, Academic Press, New York, **2**: 525–579

Grodzinski B and Woodrow L (1981) Serine synthesis and CO_2 release from intermediates of the glycolate pathway. In: (G Akoyunoglou, ed) *Photosynthesis*, Balaban International Services, Philadelphia, **4**: 551–559

Hipkin CR, Everest SA, Rees TVA and Syrett PJ (1982) Ammonium generation by nitrogen starved cultures of *Chlamydomonas reinhardii*. *Planta* 154: 587–592

Huber SC, Hall TC and Edwards GE (1976) Differential localization of fraction 1 protein between chloroplast types. *Plant Physiol* **57**: 730–733

Husic DW, Husic HD and Tolbert NE (1987) The oxidative photosynthetic carbon cycle or C_2 cycle. *CRC Crit Rev Plant Sci* **5**: 45–400

Joy KW (1988) Ammonia, glutamine and asparagine: A carbon-nitrogen interface. *Can J Bot* **66**: 2103–2109

Kakefuda G, Duke SH and Duke SO (1983) Differential light induction of nitrate reductase in greening and photobleached soybean seedlings. *Phytochim* **19**: 2095–2097

Kanazawa T, Kirk MR and Bassham JA (1970) Regulatory effects of ammonia on carbon metabolism in synchronously growing *Chlorella pyreinoidosa*. *Biochim Biophys Acta* **205**: 401–408

Kanazawa T, Kanazawa K, Kirk MR and Bassham JA (1972) Regulatory effects of ammonia on carbon metabolism in *Chlorella pyreinoidosa* during photosynthesis and respiration. *Biochim Biophys Acta* **226**: 656–669

Kannangara CG and Woolhouse HW (1967) The role of carbon dioxide, light and nitrate in synthesis and degradation of nitrate reductase in *Perilla frutescens*. *New Phytol* **66**: 553–561

Keys AJ, Bird IF, Cornelius MJ, Lea PJ, Wallsgrove RM and Miflin BJ (1978) Photorespiratory nitrogen cycle. *Nature* (London) **275**: 741–743

Klepper LA, Flesher D and Hageman RH (1971) Generation of reduced nicotinamide dinucleotide for nitrate reduction in green leaves. *Plant Physiol* **48**: 580–590

Kumar PA (1988) *Photorespiratory ammonia assimilation in some crop species*, PhD Thesis, Post-Graduate School, Indian Agricultural Research Institute, New Delhi

Kumar PA and Abrol YP (1989) Refixation of photorespiratory CO_2 and NH_3 by the leaf slices of *Parthenium hysterophorus* L. *J Plant Physiol* **134**: 113–114

Kumar PA and Abrol YP (1990a) Ammonia assimilation in higher plants. In: YP Abrol (ed) *Nitrogen in Higher Plants*, John Wiley and Sons, New York, pp 159–179

Kumar PA and Abrol YP (1990b) Photorespiratory nitrogen metabolism in C_3-C_4 intermediate species, *Moricandia arvensis. Biochem Physiol Pflanzen* **186**: 109–115

Kumar PA, Chatterjee SR and Abrol YP (1990) Photorespiratory ammonia assimilation in the leaves of barley, sorghum and *Moricandia arvensis. Ind J Biochem Biophys* **27**: 164–166

Kumar PA, Nair TVR and Abrol YP (1983) Effect of exogenous supply of amino acids, amide, urea and ureide on free NH_4^+ level in mungbean. *Experientia* **39**: 1302–1303

Kumar PA, Nair TVR and Abrol YP (1984) Effect of photorespiratory metabolites, inhibitors and methionine sulphoximine on the accumulation of ammonia in the leaves of mungbean and *Amaranthus. Plant Sci Lett* **33**: 303–307

Kumar PA, Nair TVR and Abrol YP (1988a) Glycine supports nitrate reduction *in vivo* in barley leaves. *Plant Physiol* **88**: 1486–1488

Kumar PA, Nair TVR and Abrol YP (1988b) Nitrate reductase: Localization and source of reductant. *Proc Ind Natl Sci Acad B* **54**: 409–412

Larsen PO, Cornwell KL, Gee SL and Bassham JA (1981) Amino acid synthesis is photosynthesizing spinach cells. *Plant Physiol* **68**: 292–299

Lea PJ, Joy KW, Ramos JL and Guerrero MG (1984) Action of 2-amino-4 (methylphosphinyl) butanoic acid and its 2-oxo derivative on the metabolism of cyanobacteria. *Phytochem* **23**: 1–6

Madore M and Gordzinski B (1984) Effect of oxygen concentration on ^{14}C photoassimilate transport from leaves of *Salvia splendens* L. *Plant Physiol* **76**: 782–786

Maheswari M, Kumar PA, Nair TVR and Abrol YP (1985) Ammonia accumulation and glutamine synthetase activity in C_3 and C_4 plants. *Symposium on Nitrogen Metabolism in High Plants*, Groningen (Abstract p 23)

Marsh HV, Galmiche JM and Gibbs M (1965) Effect of light on the tricarboxylic acid cycle in *Scenedesmus. Plant Physiol* **40**: 1013–1021

Martin F, Winspear MJ, MacFarlane JD and Oaks A (1983) Effect of methionine sulphoximine on the accumulation of ammonia in C_3 and C_4 leaves. *Plant Physiol* **71**: 117–181

McCashin BG, Cossins EA and Cavin DT (1988) Dark respiration during photosynthesis in wheat leaf slices. *Plant Physiol* **87**: 155–161

Naik MS and Nicholas DJD (1981) Relation between CO_2 evolution and *in situ* reduction of nitrate in wheat leaves. *Aust J Plant Physiol* **8**: 515–524

Oaks A, HO X and Zoumadakis M (1990) Nitrogen use efficiency in C_3 and C_4 cereals. In: SK Sinha, PV Sane, SC Bharagava, PK Agarwal (eds) *Proceedings of International Congress of Plant Physiology*, New Delhi, pp 1038–1045

Oaks A and Yamaya T (1990) Nitrogen assimilation in leaves and roots—A role for glutamate dehydrogenase. In: (YP Abrol, ed) *Nitrogen in Higher Plants*, John Wiley and Sons, New York, pp 181–194

Oelmuller R, Schuster C and Mohr H (1988) Physiological characterization of a plastidic signal reguired for nitrate induced appearance of nitrate and nitrite reductase. *Planta* **174**: 75–83

Orgen WL (1984) Photorespiration: Pathways, regulation and modification. *Annu Rev Plant Physiol* **34**: 415–442

Rathnam CKM (1978) Malate and dihydroxyacetone phosphate dependent nitrate reduction in wheat (*Triticum aestivum* L.) protoplasts. *Plant Physiol* **62**: 220–223

Rhodes D, Deal L, Haworth P, Jamieson GC, Reuter CC and Erickson MC (1986) Amino acid metabolism of *Lemna minor* (L.). I Responses to methionine sulphoximine. *Plant Physiol* **82**: 1057–1062

Robinson SA, Slade AP, Fox GG, Phillips R, Ratcliffe RG and Stewart GR (1991) The role of glutamate dehydrogenase in plant nitrogen metabolism. *Plant Physiol* **95**: 509–516

Sawhney SK and Naik MS (1972) Role of light in synthesis of nitrate reductase and nitrite reductase in rice seedlings. *Plant Cell Environ* **3**: 303–308

Sawhney SK, Naik MS and Nicholas DJD (1978) Regulation of nitrate reduction by light, ATP and mitochondrial respiration in wheat leaves. *Nature* **272**: 647–648

Sawhney SK, Prakash V and Naik MS (1972) Nitrate and nitrite reductase activities in induced chlorophyll mutants of barley. *FEBS Lett* **22**: 200–202

Sihag RK, Guha-Mukherjee S and Sopory SK (1979) Effect of ammonium, sucrose and light on regulation of nitrate reductase level in *Pisum sativum. Physiol Plant* **45**: 281–287

Singh P, Kumar PA, Abrol YP and Naik MS (1985) Photorespiratory Nitrogen cycle—A critical evaluation. *Physiol Plant* **66**: 169–176

Slack CR and Hatch MD (1967) Comparative studies on the activity of carboxylase and other enzymes in relation to the new pathway of photosynthetic carbon dioxide fixation in tropical grasses. *Biochem J* **103**: 660–665

Sluiters-Scholten CM Th (1975) Photosynthesis and induction of nitrate reductase and nitrite reductase in bean leaves. *Planta* **123**: 175–184

Taylor WC (1989) Regulatory interactions between nuclear and plastid genomes. *Annu Rev Plant Physiol* **40**: 211–233

Thorpe TA, Bagh K, Cutler AJ, Dunstan DI, Mcintyre DD and Vogel HJ (1989) A ^{14}N NMR study of nitrogen metabolism in shoot-forming cultures of white spruce (*Picea glauca*) buds. *Plant Physiol* **91**: 193–198

Travis RL, Huffaker RC and Key JL (1970) Light-induced development of polyribosomes and induction of nitrate reductase in corn leaves. *Plant Physiol* **46**: 800–805

Travis RL and Key JL (1971) Correlation between polyribosomal level and ability to induce nitrate reductase in dark grown maize seedlings. *Plant Physiol* **48**: 617–620

Wallsgrove RM, Keys AJ, Lea PJ and Miflin BJ (1983) Photosynthesis, photorespiration and nitrogen metabolism. *Plant Cell Environ* **6**: 301–309

Walton NJ and Butt VS (1981) Glutamate and serine as competing donors for amination of glyoxylate in leaf peroxisomes. *Planta* **153**: 232–237

Ward MR, Grimes HD and Huffaker RC (1989) Latent nitrate reductase activity is associated with the plasmalemma of corn root. *Planta* **177**: 470–475

Warner RL and Kleinhofs A (1974) Relationships between nitrate reductase and ribulose diphosphate carboxylase activities in chlorophyll-deficient mutants of barley. *Crop Sci* **14**: 654–658

Weger HG, Birch DG, Elrifi IR and Turpin DH (1988) Ammonium assimilation requires mitochondrial respiration in the light. *Plant Physiol* **86**: 688–692

Weger HG and Turpin DH (1989) Mitochondrial respiration can support NO_3^- and NO_2^- reduction during photosynthesis. *Plant Physiol* **89**: 409–415

Wilson JR and Brown RH (1983) Nitrogen response of *Panicum* species differing in carbon dioxide fixation pathways. I Growth analysis, relative photosynthesis and carbohydrate accumulation. *Crop Sci* **23**: 1148–1153

Woo KC and Canvin DT (1980) Effect of ammonia on photosynthetic carbon fixation in isolated spinach cells. *Can J Bot* **58**: 505–510

Woo KC, Jokinen M and Canvin DT (1980) Nitrate reduction by a dicarboxylate shuttle in a reconstituted system from spinach leaves. *Aust J Plant Physiol* **7**: 123–130

Yamaya T and Oaks A (1987) Synthesis of glutamate by mitochondria: An anaplerotic function for glutamate dehydrogenase. *Physiol Plant* **70**: 749–756

Yamaya T, Oaks A, Rhodes D and Matsumoto H (1986) Synthesis of [^{15}N] glutamate from [^{15}N]H$_4^+$ and [^{15}N] glycine by mitochondria isolated from pea and corn shoots. *Plant Physiol* **81**: 754–757

Zoglowek C, Kromer S and Heldt HW (1988) Oxaloacetate and malate transport by plant mitochondria. *Plant Physiol* **87**: 109–115

14

Assimilate Partitioning within Leaves of Small Grain Cereals

Richard C. Sicher

Climate Stress Laboratory
United States Department of Agriculture
Agricultural Research Service
Beltsville Agricultural Research Center
Beltsville, MD 20705 USA

CONTENTS

ABBREVIATIONS

F2, 6BP	:	fructose 2, 6-bisphosphate
FBPase	:	fructose 1, 6-bisphosphatase
G-6-P	:	glucose 6-phosphate
PFK	:	phosphofructokinase
PPi-PFK	:	pyrophosphate-dependent, phosphofructokinase
RuBP	:	ribulose 1, 5-bisphosphate
SPS	:	sucrose phosphate synthase
triose-P	:	triose-phosphate

Names of products are included for the benefit of the reader and do not imply endoresement or preferential treatment by the US Department of Agriculture.

352

ABSTRACT

Progress occurred during the last decade with regard to understanding carbohydrate metabolism and assimilate partitioning in the *Graminaceous* plants, wheat (*Triticum aestivum* L.), barley (*Hordeum vulgare* L.) and rice (*Oryza sativa* L.). The concept that metabolic pathways are regulated at specific enzyme steps was confirmed by research performed during this period. The central importance of fructose 2,6-bisphosphate and inorganic pyrophosphate in controlling carbon fluxes in plant cells was assessed and the key role of sucrose-phosphate synthase in regulating sucrose metabolism was clearly demonstrated. It is now feasible to discuss some of the biochemical factors involved in source/sink interactions. High resolution chromatographic procedures and analytical improvements made it possible to separate and detect carbohydrates with tremendous sensitivity. Several enzymes in the leaf starch and fructan pathways have been isolated and characterized and these research areas can be expected to grow rapidly in the future. Assimilate partitioning benefited from the explosive growth in molecular biology primarily in the fields of plant development and environmental stress.

I. Introduction

The objective of this article is to review recent advances in our understanding of the distribution of assimilates in leaves of small grain cereal crops, namely wheat (*Triticum aestivum* L.), barley (*Hordeum vulgare* L.), and rice (*Oryza sativa* L.). These three species are of inestimable importance to humans and were among the earliest species subjected to cultivation (Murata and Matsushima, 1975; Briggs, 1978; Briggle, 1980). Collectively, they provide the majority of the caloric needs of the human race with the rice crop alone supporting about one-half of the world's population. Cereals are also an important source of livestock feed and they have many valuable commercial and industrial uses.

The photosynthetic properties of dicot leaves are mentioned for comparative purposes or where key information is not yet established in the small grain cereals. There are certainly many common features shared by the two leaf types and it would be difficult to discuss this subject without occasionally referring to experiments employing broad-leaf species. This is intended to be a summary article and the interested reader is encouraged to consult recent reviews and books on the subjects of carbohydrate metabolism and assimilate partitioning for a more comprehensive treatment of these subjects (Avigad, 1982; Hawker, 1985).

II. Leaves and the Strategy for Plant Survival

Wheat, rice, and barley are members of the grass family, Gramineae, and the plants possess an annual growth habit. Thus, development proceeds in stages from germination, seedling growth, tillering, stem elongation, and flowering to harvest in less than one year. Although virtually all above-ground plant parts possess some photosynthetic capacity, leaves are by far the most important autotrophic organs (Good and Bell, 1980). The cereal leaf either develops in the embryo or can be formed by the apical meristem after sprouting. The leaf initials elongate in sequence, eventually forming both a leaf blade and a leaf sheath. The leaf blade functions mainly in light interception, whereas the leaf sheath is primarily a storage and structural organ (Briggs, 1978). The younger leaves mature first and after a brief period of productivity they turn brown, wither, and die. Both minerals and nutrients are transferred from older, senescing leaves to the developing growth centers on the plant. After formation of the spike, the entire plant eventually matures and is ready for harvest.

Energy for plant growth and development is strictly derived from photosynthetic carbon assimilation. During formation of the cereal crop, reduced carbon is transported from autotrophic organs with a positive carbon balance, i.e., predominantly leaves, to various sink tissues that import carbon to support growth and respiration. During the early stages of emergence leaves are net carbon importers but before attaining one-half of full expansion, a capacity is developed to export carbohydrates and certain amino acids (Turgeon, 1989).

One intriguing problem in crop physiology has been to understand the source-path-sink relationships between various organs on a plant (Wardlaw, 1980). The capacity of a leaf to synthesize solutes from CO_2 is a major determinant of source strength and it is widely recognized that photosynthetic ability of a leaf is influenced by environmental, nutritional, and genetic parameters. Another important factor determining source strength is size of the leaf. Since leaves can be approximated as two-dimensional objects this is usually quantified as a function of leaf area. Consequently, the flag leaf of wheat is a major contributor to grain yield, whereas the smaller flag leaf on the barley plant is not (Thorne, 1965). It is generally accepted that leaves determine the timing and availability of assimilate for distribution to other organs on the plant and that sink demand determines the ultimate direction of solute movement. Unfortunately, little is known about the factors that determine sink strength or sink demand (Herold, 1980). However, there is a close correlative relationship between the proximity of an importing organ to a leaf and its ability to attract assimilates from that particular source (Cook and Evans, 1978).

There is abundant evidence to show that sinks can influence the photosynthetic capacity of a source leaf. Experiments designed to modify sink demand, such as spike removal, elevated carbohydrate levels in the leaves and stems and decreased the rate of photosynthesis (Austin and Edrich, 1975). The exact mechanism whereby a sink limitation imposes a feedback inhibition of photosynthesis is unknown, although Neales and Incoll (1968) hypothesized that increased assimilate levels in the leaf were a potential causal agent. There is no clear relationship between starch levels and the rate of photosynthesis (Peet and Kramer, 1980), but recent experiments with tobacco plants expressing a yeast invertase gene indicate that an accumulation of soluble sugars in the leaf can result in a significant inhibition of photosynthesis (Von Schaewn et al., 1990). We do not know exactly how increased levels of soluble sugars in the leaf act to reduce the rate of carbon assimilation but, as will be discussed below, some of the factors involved in this process have been identified.

III. General Characteristics of Leaf Carbohydrate Metabolism

a. Biosynthesis

The most important function of leaves with respect to whole plant partitioning is to produce photoassimilates. An outline of carbon flow during photosynthesis is diagrammed in Fig. 1. Briefly, the light reactions of photosynthesis generate ATP and NADPH, which are then consumed in the reductive pentose-phosphate pathway of the chloroplast stroma to convert atmospheric CO_2 and Pi into triose-phosphate (triose-P). Five of every six triose-P molecules must remain in the chloroplast to

354

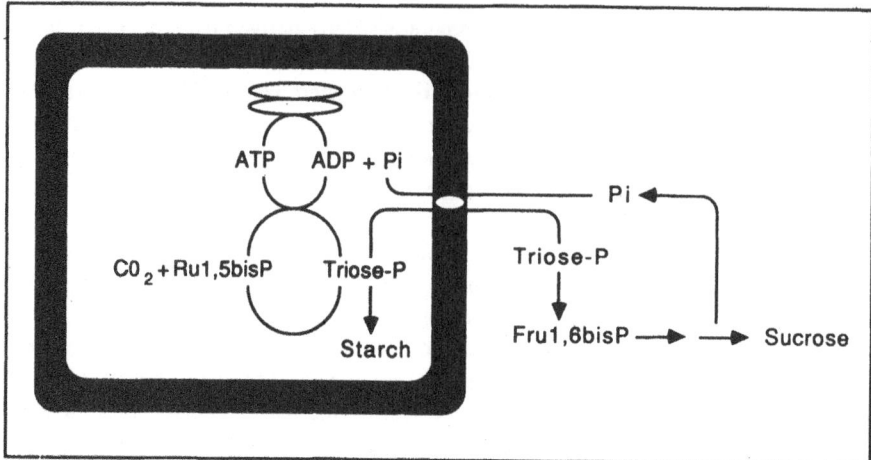

Figure 1. Diagram of carbon fluxes during photosynthesis. The compartmental separation of starch and sucrose synthesis and the recycling of Pi is shown.

remanufacture the primary CO_2 acceptor, ribulose, 1, 5-*bis*phosphate (RuBP), whereas the additional molecule is available for the biosynthesis of starch and sucrose (Sicher, 1986). These two non-structural carbohydrates serve as important temporary storage reserves in leaves. Starch is an insoluble polyglucan that is deposited in the chloroplast during the day and is degraded at night to support respiration, cellular biosynthesis, and carbon export. Sucrose is a non-reducing disaccharide that is the major transport form of carbon in most plants. The sucrose biosynthetic pathway is compartmentalized in the cytosolic fraction of mesophyll cells and it is supported by solute flux from the chloroplast. Carbon transport across the chloroplast envelope occurs on a specific membrane-bound carrier in a stoichiometric counter-exchange for Pi. The Pi liberated during sucrose synthesis can be recycled into the stroma to support photophosphorylation and to be used for CO_2 assimilation.

It is obvious that rates of sucrose biosynthesis must be controlled coordinately with rates of CO_2 assimilation in order to achieve a metabolic equilibrium (Walker and Sivak, 1986). For example, if the rate of sucrose synthesis is too low, insufficient Pi will be available for re-entry into the chloroplast and then the reductive pentose-phosphate pathway cannot be sustained. The overproduction of sucrose also cannot be maintained, since the triose-P pool will become depleted and the capacity of the chloroplast to manufacture RuBP will be impaired. Two important biochemical mechanisms for regulating the rate of sucrose synthesis have been discovered within the last decade. The first, irreversible step in the sucrose pathway is catalyzed by a cytosolic isozyme of fructose 1, 6-*bis*phosphatase (FBPase) (E.C. 3.1.3.11). This enzyme is inhibited by μM amounts of the signal metabolite fructose 2, 6-*bis*phosphate (F2, 6BP; Stitt, 1990). Experiments with differing amounts of light, CO_2, and O_2 have shown that F2, 6BP levels in the leaf vary inversely with the rate of photosynthesis. Another important observation was that F2, 6BP levels increased when export from the leaf was blocked or when soluble carbohydrate levels in the leaf were elevated (Stitt *et al.*, 1984). Clearly then, F2, 6BP is potentially an important factor in mediating the feedback inhibition of photosynthesis discussed in the preceding section. In theory, a rise in F2, 6BP should decrease the flux of carbon into the sucrose pathway and the rate of photosynthesis would then be inhibited by the resultant Pi limitation imposed on the chloroplast. This condition favors the production of starch.

The penultimate and quite probably the principal enzyme in the sucrose pathway, sucrose phosphate synthase (SPS), is also biochemically regulated (Stitt and Quick, 1989). Two levels of metabolic control have been reported and these can be identified in enzyme assays based on changes in initial rate kinetics. Widely observed light-dependent changes in SPS activity are due to an alteration in either the K_m or V_{max} properties of the enzyme (Siegl and Stitt, 1990). There is increasingly good evidence that these kinetic changes are the result of an *in vivo* enzyme phosphorylation mechanism (Walker and Huber, 1988). A second control mechanism that is based on an activation of SPS by glucose 6-phosphate (G-6-P) has been reported (Doehlert and Huber, 1984). Interaction of SPS with G-6-P probably occurs at an allosteric site and this activation is inhibited competitively by Pi. The known properties of SPS indicate that the enzyme should be most active *in vivo* when the rate of photosynthesis is high and conditions favour the production of sucrose. Inverse changes of F2, 6BP and SPS have been observed in barley primary leaves, suggesting that these two steps in sucrose biosynthesis may be coordinated (Sicher *et al.*, 1986).

In addition to starch and sucrose, wheat and barley are capable of synthesizing and storing a special class of water-soluble oligosaccharides known as "fructans" (Pollock, 1986). Fructans are composed almost exclusively of polymeric fructose residues and the cereal leaf fructans are generally low molecular weight (i.e., DP < 30). Cereal leaf fructans are predominantly of the phlein type and, like the bacterial levans, consist mostly of β-2, 6 linked fructosyl units. However, branched chain molecules containing a mixture of β-2, 1 and β-2, 6 linkages are known and small amounts of an inulin-like material with exclusively β-2, 1 fructofuranosyl linkages have been observed (Carpita *et al.*, 1989). Fructans are synthesized by successive transglycosylations that donate the fructose moiety of sucrose to either a second sucrose molecule or to an elongating oligomer. Therefore, fructan accumulation is influenced by the leaf sucrose concentration and under extreme conditions barley leaves can store fructans up to 70% of the dry weight. Large fructan concentrations also have been observed in the leaf elongation zone and in the leaf sheath (Nelson and Spollen, 1987). Both fructans and the enzymes of fructan metabolism are localized in the vacuole. The physiological function of the fructans is unknown, although it is possible that as water-soluble carbohydrates they may have a role in osmoregulation or cryoprotection. Grass species are often described as forming either starch (rice) or fructan (wheat, barley) but this generalization is probably inaccurate (Chatterton *et al.*, 1989). There is no apparent advantage to storing water-soluble carbohydrates as opposed to starch.

b. *Mobilization*

The rate at which carbon is exported from a cereal leaf can at certain times exceed the rate of sucrose synthesis. When this occurs, the various stored carbohydrate reserves are mobilized to meet demand. Two sucrose pools have been identified on the basis of radiochemical turnover patterns (Farrar and Farrar, 1986). A transport sucrose pool, which has a half-life of about 30 to 60 minutes, can be readily observed in the presence of a second storage pool with a half-life of several hours. The transport sucrose pool is located in the cytosol or in the apoplast, whereas the storage sucrose pool is thought to be present in the vacuole. About 60% of the total sucrose in barley leaves is localized in the vacuole (Wagner *et al.*, 1983).

Following the onset of darkness, the sucrose content of the leaf decreases rapidly and after a lag period of 1 to 3 hr starch mobilization is initiated. The degradation of starch in the chloroplast is dependent upon the initial attack of endohydrolases (i.e., α-amylase, E.C. 3.2.1.1) and debranching enzymes (E.C. 3.2.1.41). The soluble maltodextrins can then be hydrolyzed further by phosphorylases (E.C. 2.4.1.1). The control of starch degradation is only poorly understood because of the relative complexity of this process but it is known that high concentrations of Pi accelerate starch mobilization *in situ* (Beck and Ziegler, 1989). The carbon obtained from starch hydrolysis can be exported from the chloroplast either as triose-P units on the Pi translocator or possibly as hexose or hexose-P units on the hexose translocator. It is now relatively easy to generate mutants which are defective in starch metabolism and this may accelerate progress in this research area.

The fructan content of the cereal leaf blade is relatively low in comparison to either the stem or the culm-base. Nevertheless, diurnal fluctuations in the leaf fructan pool have been reported (Sicher and Kremer, 1986). Barley leaves contain a fructan exohydrolase that displays β-fructofuranosidase activity. This enzyme liberates terminal fructose residues from either β-2, 1 or β-2, 6 linked fructans but its activity with sucrose is negligible (Henson, 1989). The physiology of fructan hydrolysis has only been poorly characterized, although stored fructan is usually mobilized during periods of rapid plant growth. Preliminary evidence with protein synthesis inhibitors suggested that fructan exohydrolase activity increased concomitantly with fructan mobilization.

Although the majority of sucrose synthesized by the leaf is exported, it can also serve as a substrate for a number of intracellular reactions and transformations. Sucrose hydrolysis is an important starting point for glycolysis and, especially in young leaves, sucrose can be converted to starch. There are two possible pathways for sucrose to enter metabolism (Huber and Akazawa, 1986) and these are referred to as the invertase and sucrose synthase pathways, respectively (Fig. 2). The relative importance of these pathways with respect to one another is rather poorly understood and is currently receiving considerable attention. The two pathways take their names from the enzymes responsible for the initial step in sucrose cleavage and the main difference is that invertase produces free hexoses and sucrose synthase produces fructose and UDP-Glc. Subsequent use of the free hexoses requires activity of both ATP-dependent hexokinases and phosphofructokinases (PFKs), whereas the metabolism of UDP-Glc requires the involvement of UTP and pyrophosphate (PPi). The discovery of a PPi-dependent PFK, activated by F2, 6BP, provided the missing link in the sucrose synthase pathway (Stitt, 1990). The activities of sucrose synthase and PPi-dependent PFK are high in young leaves and very low in mature leaves, suggesting that this is not a major pathway in photosynthetic organs. The relative importance of PPi-dependent PFK to glycolysis is complicated by the fact that this enzyme is fully reversible. The regulatory significance of an equilibrium enzyme is understandably difficult to determine.

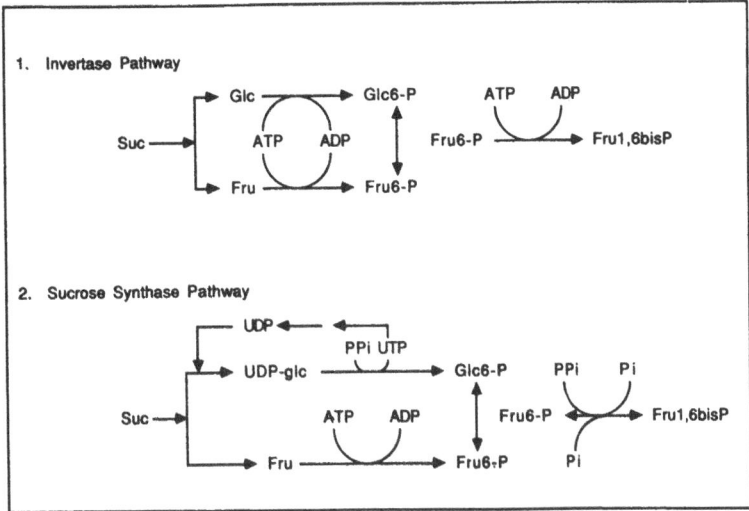

Figure 2. Invertase and sucrose synthase catalyze the principal sucrose hydrolyzing reactions of plant tissues. The utilization of fructose is similar, whereas the glucose moiety is processed by different enzymes.

IV. Environmental Effects on Assimilate Partitioning

As sessile organisms, plants are subjected to various stresses and survival is contingent upon the implementation of adaptive responses. The ability of a plant to manufacture assimilates under suboptimal conditions or to access stored carbohydrate reserves when carbon assimilation rates are below current demand clearly can have an impact upon crop performance.

a. Light

It is well recognized that light intensity, light quality, and the duration of the photoperiod can affect assimilate partitioning within cereal leaves. Barley leaves export a greater percentage of carbon fixed under high compared to low light conditions (Sicher and Kremer, 1986). Moreover, the allocation of carbon to starch and fructans is increased when light is maximal and total non-structural carbohydrate levels in the leaf are high (Pollock 1986; Chatterton *et al.*, 1989). Shortened daylengths can enhance the rate of starch accumulation, although this effect is not particularly large in species that store fructans. The red to far-red wavelength ratio can be altered when light is filtered through a canopy and this spectral change affects plant development via the phytochrome system (Hoddinott and Hall, 1982).

b. Temperature

Cereal crops maintain high rates of photosynthesis at low positive temperatures (Labate and Leegood, 1989). Translocation decreases as temperatures drop and soluble carbohydrates accumulate within the leaf. Eventually fructan synthesis is induced after sucrose levels achieve a threshold level. Elevated temperatures stimulate respiratory CO_2 losses as well as leaf expansion rates. Consequently, total non-structural carbohydrate levels in the leaf decrease as temperatures increase.

c. *Carbon Dioxide*

Insofar as CO_2 is necessary for plant growth, it is generally assumed that a doubling of atmospheric levels would enhance photosynthesis and boost plant growth. Photosynthetic productivity should increase because leaf diffusive resistances are overcome and because photorespiration is competitively inhibited. In general, experiments employing long-term exposure to enriched CO_2 support these assumptions, although enhanced dry matter production was greater in some species than others. Increased rates of photosynthesis under elevated CO_2 are often only temporary and accelerated growth rates may only occur during early growth stages. The available evidence indicates that wheat can respond positively to CO_2 enrichment (Sionit *et al.*, 1980).

d. *Nutritional Deficiencies*

Intensively cultivated crop plants are often subjected to limiting levels of mineral nutrition. Under a nutritional stress, the development and expansion of new leaves and roots is impaired and non-structural carbohydrates increase in leaves (Sicher and Kremer, 1988). Photosynthesis and protein synthesis are inhibited by N-deficiency and carbon accumulates as starch. Ostensibly this stored polyglucan can be remobilized in the event that the nutrient limitation is relieved.

e. *Anoxia*

Unlike animals, plants can survive prolonged periods of anaerobiosis. Hypoxic conditions can occur during normal agronomic practices as a result of flooding or soil compaction. The growth of barley seedlings is halted in an anoxic environment, whereas rice plants continue to grow at a reduced rate. Survival during anoxia is dependent upon the induction of fermentative metabolism leading to ethanol or lactic acid synthesis (Rivoal *et al.*, 1989). Soluble sugar levels generally decrease when O_2 is removed but little change in starch has been observed.

V. Summary

Significant progress has occurred during the last decade with regard to our understanding of plant carbohydrate metabolism. The rapid progress achieved in assessing the central importance of F2, 6BP, PPi, and PPi-dependent PFK in controlling carbon fluxes in plant cells stands out as a major advance in this field. The development of high resolution chromatographic procedures and improvements in detector sensitivity made it possible to separate and analyze carbohydrates with tremendous precision. Solid progress was achieved in the area of leaf starch research and a large number of enzymes in this pathway were isolated and characterized. The explosive growth in molecular biology has had a major impact on all fields in plant biology. The field of assimilate partitioning benefited from this technology mostly in areas of plant development and stress.

One area of future research that holds considerable promise is the production of transgenic plants. It is now possible to create genetically modified plants with increased or decreased levels of a specific enzyme and to then analyze the effects this has on carbohydrate metabolism and plant development (see Ref. Von Schwaene). It will be necessary in the future to develop better techniques for regenerating Graminaceous species from tissue culture before this approach can be applied to cereal crops.

Acknowledgements

Because of publishing constraints it was not possible to list citations for all of the research findings discussed in this article. Therefore, I would like to thank the individuals responsible for these contributions and to apologize for any inconvenience this may have caused.

VI. References

Austin RB and Edrich J (1975) Effects of ear removal on photosynthesis, carbohydrate accumulation and on the distribution of assimilated ^{14}C in wheat. *Ann Bot* **39**: 141–152

Avigad A (1982) Sucrose and other disaccharides. In: (FA Loewus and W Tanner, eds) *Encyclopedia of Plant Physiolgy: Plant Carbohydrates I, Intracellular Carbohydrates*, New Series, Vol 13A, Springer-Verlag, Berlin, pp 217–347

Beck E and Ziegler P (1989) Biosynthesis and degradation of starch in higher plants. *Annu Rev Plant Physiol Plant Mol Biol* **40**: 95–117

Briggle LW (1980) Origin and botany of wheat. In: (E Häfliger, ed) *Wheat Technical Monograph*, Documenta CIBA-GEIGY, CIBA-GEIGY Ltd., Basle, pp 6–13

Briggs DE (1978) The morphology of barley, the vegetative phase. In: (DE Briggs, ed) *Barley*, Chapman and Hall, London, pp 1–38

Carpita NC, Kanabus J and Housley TL (1989) Linkage structure of fructan oligomers from *Triticum aestivum* and *Festuca arundinacea* leaves. *J Plant Physiol* **134**: 162–168

Chatterton NJ, Harrison PA, Bennett JH and Asay KH (1989) Carbohydrate partitioning in 185 accessions of Gramineae grown under warm and cool temperatures. *J Plant Physiol* **134**: 169–179

Cook MG and Evans LT (1978) Effect of relative size and distance of competing sinks on the distribution of photosynthetic assimilates in wheat. *Aust J Plant Physiol* **5**: 495–509

Doehlert DC and Huber SC (1984) Phosphate inhibition of spinach leaf sucrose phosphate synthase as affected by glucose 6-phosphate and phosphoglucoisomerase. *Plant Physiol* **76**: 250–253

Farrar SC and Farrar JF (1986) Compartmentation and fluxes of sucrose in intact leaf blades of barley. *New Phytol* **103**: 645–657

Good NE and Bell DH (1980) Photosynthesis, plant productivity and crop yield. In: (PS Carlson, ed) *Biology of Crop Productivity*, Academic Press, New York, pp 3–51

Hawker JS (1985) Sucrose. In: (PM Dey and RA Dixon, eds) *Biochemistry of Storage Carbohydrates in Green Plants*, Academic Press, London, pp 1–51

Henson CA (1989) Purification and properties of barley stem fructan exohydrolase. *J Plant Physiol* **134**: 186–191

Herold A (1980) Regulation of photosynthesis by sink activity—The missing link. *New Phytol* **86**: 131–144

Hoddinott J and Hall LM 1982. The responses of photosynthesis and translocation rates to changes in the δ ratio of light. *Can J Bot* **60**:1285–1291

Huber SC and Akazawa T (1986) A novel sucrose pathway for sucrose degradation in cultured sycamore cells. *Plant Physiol* **81**: 1008–1013

Labate CA and Leegood RC (1989) Influence of low temperature on respiration and contents of phosphorylated intermediates in darkened barley leaves. *Plant physiol* **91**: 905–910

Murata Y and Matsushima S (1975) Rice. In: (LT Evans, ed) *Crop Physiology: Some Case Histories*. Cambridge University Press, London, pp 73–99

Neales TF and Incoll LD (1968) The control of leaf photosynthesis rate by the level of assimilate concentration in the leaf. A review of the hypothesis. *Bot Rev* **34**: 107–125

Nelson CJ and Spollen R (1987) Fructans. *Physiol Plant* **71**: 512–516

360

Peet MM and Kramer PJ (1980) Effects of decreasing source/sink ratio in soybeans on photosynthesis, photorespiration, transpiration and yield. *Plant Cell Environ* 3: 201–206

Pollock CJ (1986) Fructans and the metabolism of sucrose in vascular plants. *New Phytol* 104: 1–24

Rivoal J, Ricard B and Pradet A (1989) Glycolytic and fermentative enzyme induction during anaerobiosis in rice seedlings. *Plant Physiol Biochem* 27:43–52

Sicher RC (1986) Sucrose biosynthesis in photosynthetic tissue: Rate-controlling factors and metabolic pathway. *Physiol Plant* 67: 118–121

Sicher RC and Kremer DF (1986) Effects of temperature and irradiance on non-structural carbohydrate accumulation in barley primary leaves. *Physiol Plant* 66: 365–369

Sicher RC and DF Kremer (1988) Effects of phosphate deficiency on assimilate partitioning in barley seedlings. *Plant Sci* 57: 9–17

Sicher RC, Kremer DF and Harris WG (1986) Control of photosynthetic sucrose synthesis in barley primary leaves. Role of fructose-2, 6-bisphosphate. *Plant Physiol* 82: 15–18

Siegl G and Stitt M (1990) Partial purification of two forms of spinach leaf sucrose-phosphate synthase which differ in their kinetic properties. *Plant Sci* 66: 205–210

Sionit N, Hellmers H and Strain BR (1980) Growth and yield of wheat under CO_2 enrichment and water stress. *Crop Sci* 20: 687–690

Stitt M (1990) Fructose-2, 6-bisphosphate as a regulatory molecule in plants. *Annu Rev Plant Physiol Plant Mol Biol* 41: 153–185

Stitt M, Kürzel B and Heldt HW (1984) Control of photosynthetic sucrose synthesis by fructose 2, 6-bisphosphate. II Partitioning between sucrose and starch. *Plant Physiol* 75: 554–560

Stitt M and Quick WP (1989) Photosynthetic carbon partitioning: Its regulation and possibilities for manipulation. *Physiol Plant* 77: 633–641

Thorne GN (1965) Photosynthesis of ears and flag leaves of wheat and barley. *Ann Bot* 29: 317–329

Turgeon R (1989) The sink-source transition in leaves. *Annu Rev Plant Physiol Plant Mol Biol* 40: 119–138

Von Schaewn A, Stitt M, Schmidt R, Sonnenwald U and Willmitzer L (1990) Expression of a yeast-derived invertase in the cell wall of tobacco and *Arabidopsis* plants leads to accumulation of carbohydrate and inhibition of photosynthesis and strongly influences growth and phenotype of transgenic tobacco plants. *EMBO J* 9: 3033–3044

Wagner W, Keller F and Wiemken A (1983) Fructan metabolism in cereals: Induction in leaves and compartmentation in protoplasts and vacuoles. *Z Pflanzenphysiol*, 112: 359–372

Walker DA and Sivak MN (1986) Photosynthesis and phosphate: A cellular affair? *TIBS* 11: 176–179

Walker JL and Huber SC (1988) Regulation of sucrose phosphate synthase activity in spinach leaves by protein level and covalent modification. *Planta* 177: 116–120

Wardlaw IF (1980) Translocation and plant productivity In: (PS Carlson, ed) *Biology of Crop Productivity*, Academic Press, New York, pp 297–339

15

Source and Sink Relationship

Surma Mitra, S.N. Bhardwaj and G.C. Srivastava

Division of Plant Physiology,
Indian Agricultural Research Institute,
New Delhi 110 012, India

CONTENTS

ABBREVIATIONS

Cycocel	:	2 Chloroethyl trimethyl ammonium chloride
F6P	:	Fructose-6-phosphate;
F 1,6 BP	:	Fructose 1, 6 *bis*phosphate;
F 2,6 BP	:	Fructose 2, 6 *bis*phosphate;
F6P, 2K	:	Fructose 6 phosphate, 2 kinase;
G6P	:	Glucose 6 phosphate;
Pi	:	Inorganic phosphate;

RuBP	:	Ribulose *bis*phosphate;
RuBPcase	:	Ribulose *bis*phosphate carboxylase;
SPP	:	Sucrose phosphate phosphatase;
SPS	:	Sucrose phosphate synthase;
TP	:	Triose phosphate;
UDP	:	Uridine diphosphate
CER	:	CO_2 exchange rate

ABSTRACT

Interaction between source and sink can be envisaged as occurring due to feedback inhibition of source assimilation by sucrose in the event of its accumulation in the leaves due to low sink demand. However, conclusive evidence of sink regulation of source photosynthesis, whether by assimilates or hormones, is lacking. Sucrose loading into the phloem is driven by a pH-dependent sucrose-proton co-transport (symport) mediated by a carrier. Critical examination of the evidences in favour of and against the apoplast and symplast concepts of loading leads to the conclusion that phloem loading may occur along two pathways—apoplastic and symplastic—which are not mutually exclusive and vary with species. Unloading of assimilates at the site of utilization in the sink is the reverse of loading. The effect of environmental (light, temperature, water, and CO_2), nutritional, metabolic, and hormonal factors on the regulation of source-sink activity is discussed.

I. Introduction

Judicious balance between demand and supply of assimilates is a key to achieving higher productivity in crops where economic yield constitutes a fraction of the biomass. Opinions, however, differ on whether economic yield is enhanced through efficient partitioning and utilization of available assimilates or further improvement in net photosynthesis is required to fulfil the desired goal. It is likely that the relationship varies in different crops. Each crop might adopt its own strategy under a given environment. With this background in view, the relationship between source and sink has been examined in its various aspects, which include assimilate transport, role of vascular supply, and impact of nutritional and hormonal factors.

II. Assimilate Transport

The role of assimilates in bringing about a reduction in photosynthesis was first proposed in 1868 by Boussingault. Since then, the regulation of photosynthesis (source activity) by the level of assimilates, which, in turn, depends upon sink activity, has been evaluated from different angles (Neales and Incoll, 1968; King *et al.*, 1967; Geiger, 1976a).

Source in its broadest definition could be considered as a place from which a required assimilate is obtained, whereas *sink* constitutes the organs receiving these assimilates. In plants, supply comes from photosynthesizing organs (leaf, bract, fruit wall, etc.) designated as *source*, while demand is created in non-photosynthesizing organs (non-green plant parts like stem, root, fruit, and seeds) termed *sink* which utilize the assimilates for their growth or storage purposes. Assimilates flow within and between cells and from tissue to tissue. Vascular tissue (especially phloem) is the main channel and the flow depends on demand and supply.

Sucrose, a product of photosynthesis and the main form in which assimilates are translocated out of the leaf (Shiroya, 1977), causes feedback inhibition of photosynthesis in the event of its accumulation in the leaves due to reduction in sink demand (King *et al.*, 1967; Neales and Incoll, 1968; Geiger, 1967a; Heldt, 1976; Herold, 1980; Azc'on-Bieto, 1983). Triose phosphates (TP), the first product of photosynthesis, may move out of the chloroplast to initiate the synthesis of sucrose in the cytosol or remain in the chloroplast for starch synthesis (Fig. 1).

The rate and direction of assimilate transfer is governed by its supply from the source and the demand from developing sinks. Higher demand by sinks will, therefore,

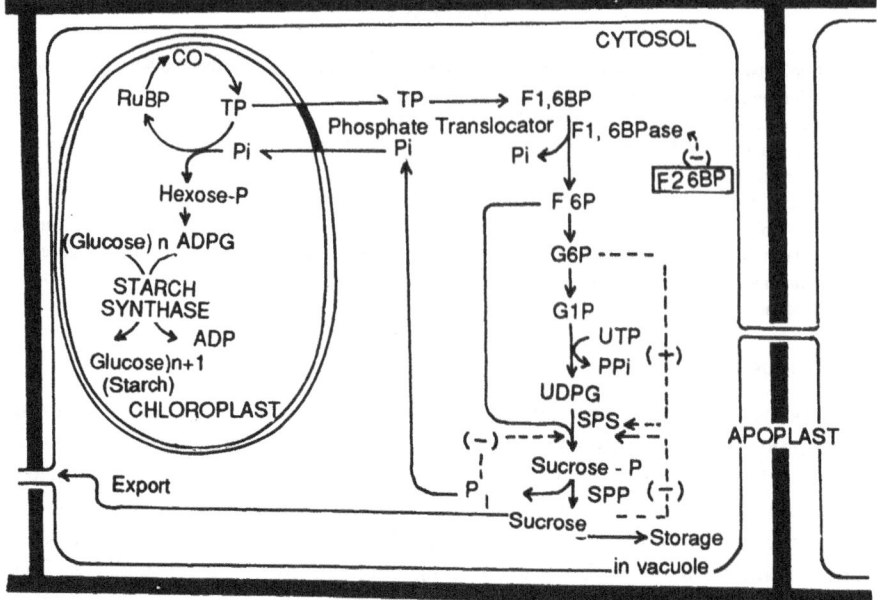

Figure 1. Diagrammatic scheme showing regulation of photosynthesis by its assimilates (Dotted line indicates suggested pathways of regulation).

increase the rate of assimilate transfer from the source and also possibly the synthesis of assimilates (photosynthesis). It is, therefore, pertinent to understand the process by which assimilates synthesized by the source are transferred towards the sink for utilization.

a) *Loading and unloading in phloem*

The first step by which assimilates are supplied to the developing sink organ is a short-distance lateral transport of assimilates from individual leaf cells to vascular bundles. This phenomenon of active and selective transport of photosynthates from the minor veins in the source region is termed *loading*. Thereafter, translocation through porous sieve plates between interconnected sieve elements of the phloem brings the assimilates to their sites of utilization (sinks) wherein they are released into the surrounding tissue from the sieve tubes by a process known as *unloading*.

b) *Apoplastic versus symplastic transport*

Phloem loading can be envisaged either as occurring directly from the apoplast through membrane transport or as occurring from the symplast of the cells surrounding the sieve element/companion cell (SE-CC) complex. Until recent years, most of the information available favoured apoplastic loading (Delrot, 1987). However, this has been challenged by recent findings which argue that apoplastic loading is not universal (Van Bel, 1987; Turgeon, 1989).

Geiger and co-workers (Geiger, 1976b) developed the symplast-apoplast-symplast concept of loading on fundamentals created by Kursanov and co-workers (Kursanov

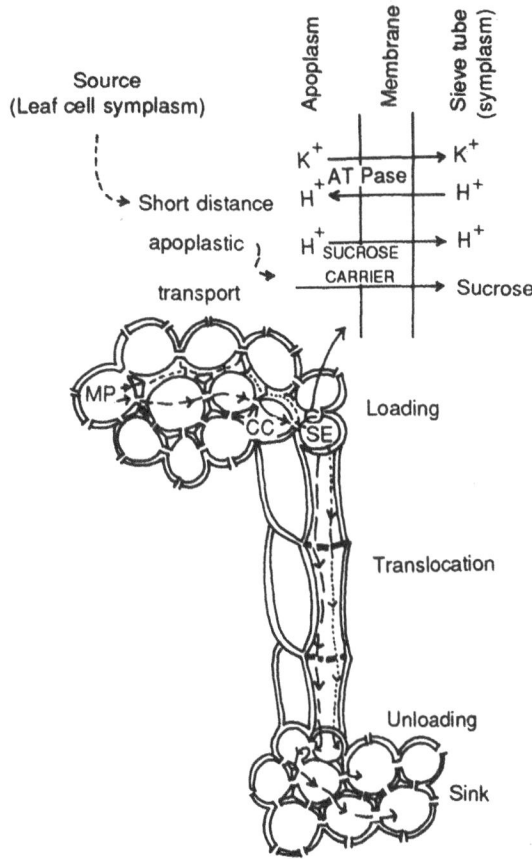

Figure 2. Diagrammatic representation of assimilate flow (based upon Giaquinta (1977) and Malik and Baker (1977).

et al., 1960) (Fig. 2). According to this concept, sucrose is released into the apoplast by leakage along the concentration gradient. In the apoplast near the minor veins, 100–200 mM sucrose has been detected. Thus, transport into the sieve tubes has to be active since it is directed against the concentration gradient. This uphill transport (phloem loading) of sucrose occurs mainly in the minor veins with transfer cells playing a major role (Pate and Günning, 1972).

Recently, there has been a spate of information on the mechanism of phloem loading, particularly of sucrose, the dominant sugar in the phloem sap of most species (Giaquinta, 1983; Delrot, 1987; Van Bel, 1987). A model was developed to explain phloem loading based on information furnished by Giaquinta (1977) and Malik and Baker (1977) (see Fig. 2). It explains that sucrose loading into the phloem is driven by a pH-dependent sucrose-proton co-transport (symport) mediated by a carrier at the plasma membrane of sieve element. The driving force for this co-transport is a separate electrogenic proton pump (H^+ efflux pump), probably a membrane-bound ATPase. The pump creates a gradient in pH and a steep electrical potential across the plasma membrane. A negative potential of −155 mV (Wright and Fisher, 1981) and a pH difference of about 2 units between the phloem sap (7.5–8.5) and the apoplasm (5.5–6.5) of leaf tissue has been observed. Evidence indicates that the movement of proton along this gradient from the apoplasm into the sieve tube cells is coupled by a carrier-mediated co-transport of sucrose. Sucrose loading is inhibited by high pH (6.0) in the apoplasm (Delrot and Bonnemain, 1981) and the transient pH increase in the apoplasm (H^+ consumption) at the onset of sucrose loading (Komor et al., 1980).

Geiger and other workers have amassed considerable evidence in favour of the apoplastic concept which may be summarized as follows:

i) Very few symplastic connections (plasmodesmata) are present between the mesophyll symplast and the SE-CC complex in plants (Fisher and Evert, 1982), thereby tilting the balance towards the apoplastic concept.

ii) Sieve tubes can absorb large quantities of photosynthate from the apoplast as evident in accumulation of ^{14}C photosynthate in minor veins in stripped leaf discs thereby indicating much higher uptake rates by sieve tubes compared to mesophyll cells (Delrot and Bonnemain, 1981).

iii) The exceedingly higher sucrose concentration in sieve tubes than in mesophyll cells makes symplastic transfer unlikely as it will have to counter a steep concentration gradient.

vi) Ample evidence is available to support the uptake of sucrose by proton-driven carriers and that sieve tubes are equipped with facilities such as a substantial transmembrane pH gradient and a high ATPase activity with a high ATP turnover. All these conditions lend excessive support to the sucrose-proton symport from the apoplast (Giaquinta, 1980 and the references cited therein).

Despite the massive support in favour of loading from the apoplast, recent reports question the universal validity of the apoplast concept (Van Bel, 1987). In fact, the concept proposed in Beta vulgaris (Giaquinta, 1976), Vicia faba (Delrot et al., 1980), and Zea mays (Heyser, 1980) was never intended to be universal and further studies are required before general conclusions can be drawn. It was further pointed out that some experiments sustaining the apoplast concept might have been interpreted erroneously. Therefore, it seems worthwhile to confront the recent findings with the arguments which have led to the apoplast concept.

Some of the recent findings which question the universal validity of apoplast transfer are mentioned. In Populus deltoides (Russin and Evert, 1985) and Cucurbita pepo (Turgeon et al., 1975), the plasmodesmatal frequency does not decline in the sieve tube area. Another anatomical feature conflicting with universal apoplast loading is the existence of two types of sieve tubes in Cucurbita pepo: abaxial sieve tubes strongly connected with the intermediary cells via the plasmodesmata and the axial ones closely linked with companion cells. Both types of sieve tubes provide an unimpeded route for symplastic transfer. A similar situation was observed in Coleus blumii leaves (Fisher, 1986). Thick-walled sieve tubes observed in several monocotyledons connected with vascular parenchyma wall by means of plasmodesmata rule out apoplastic loading (Evert et al., 1978).

Due to recent criticism of the apoplastic concept, the symplastic loading has gained in popularity and deserves serious consideration. Use of fluorescent dyes shows the existence of symplastic connections between mesophyll and vein cells (Erwee *et al.*, 1985; Madore *et al.*, 1986). Furthermore, vein cells intimately connected, via plasmodesmata, with mesophyll symplast were found to accumulate high amounts of sucrose (*see* Van Bel, 1987). The above observations and the fact that symplastic transfer has been observed to occur against the concentration gradient (Schenk, 1972) lend credibility to the symplastic concept.

In summary, phloem loading may be envisaged as occurring along two pathways—apoplastic and symplastic—which are not mutually exclusive and vary with species.

Unloading appears to be the inverse of loading in that assimilates are unloaded from the sieve tubes into the free space prior to their accumulation (Glasziov and Gayler, 1972). Unloading of sucrose into the apoplasm of the sink tissue is also facilitated by the concentration gradient created by the consumption of sucrose during growth and storage. Unloading into the apoplast appears to be an energy-dependent and possibly carrier-mediated process despite a large downhill concentration gradient from sieve tube symplast to sink apoplast (Gifford *et al.*, 1984).

III. Factors Regulating Source and Sink

a) *Environmental*

Factors affecting photosynthesis and various steps of assimilate transfer like loading, translocation, and unloading of assimilates may have a direct or indirect effect on source and sink activity. Light, temperature, water, and CO_2 are some of the primary factors which influence source and sink activity by their direct effect on photosynthesis and assimilate transfer.

i) *Light*: High light intensity stimulates photosynthesis which, in turn, yields larger quantities of assimilates for transfer. However, after prolonged and excessive illumination, assimilates might accumulate to levels sufficiently high to elicit feedback inhibition of photosynthesis. Indirectly light reduces the level of fructose 2, 6 *bis*phosphate (F 2, 6 BP), a metabolite which inhibits the enzyme *en route* sucrose biosynthesis. Another indirect regulatory effect of light manifests itself in the form of light-activation of sucrose phosphate synthase (SPS) in certain species by inducing the efflux of triose phosphates (TP) from chloroplasts into the cytosol for sucrose synthesis. Such species include maize (Sicher and Kremer, 1985; Kalt-Torres *et al.*, 1987), *Lolium temulentum* (Pollock and Housely, 1985), barley (Sicher and Kremer, 1984), and spinach (Stitt *et al.*, 1987), while no effect of light on SPS activity was observed in other species like soybean (Rufty *et al.*, 1983; Kerr *et al.*, 1985; Brown and Huber, 1987), leaves of sugar beet (Vassey, 1989), cotton (Hendrix and Huber, 1986), pea (Huber *et al.*, 1985), and sunflower (Table 1; Mitra, 1990).

Darkening the sink site enhances assimilate translocation by increased unloading (Eschrich and Eschrich, 1987).

ii) *Temperature*: Temperature directly affects several processes by virtue of its effect on the activity of several enzymes. At high temperatures, therefore, photosynthesis as well as sucrose synthesis may be affected due to disruption in enzyme activity and/or membrane structure with adverse consequences for source and sink activity.

Table 1: Effect of light and dark* treatments on SPS activity in sunflower in two separate experiments (means of three replications ± SD)

| | SPS activity (μmol sucrose/g FW/h) | |
	Light	Dark
Experiment I	71.1 ± 22.0	65.0 ± 5.2
Experiment II	77.4 ± 19.4	94.8 ± 12.5

* One-month-old plants were starved in dark for 60 hours in the dark treatment while some of these plants were exposed to light for one hour in the light treatment.
Source: Mitra, 1990.

Low temperatures, on the other hand, will result in increased storage of carbohydrates along the path of movement between the source and sink. Further lowering of temperature (possibly below 15°C for temperate species) will hamper photosynthesis and vein loading and may even cause a reduction in storage along the transport pathway (Wardlaw, 1980). He pointed out that variation in temperature can alter the proportion of different metabolites being utilized during growth. For instance, high temperatures during grain development lead to production of small grains having a high percentage of nitrogen. This may be attributed to enhanced demand for photosynthate at higher temperatures where there is a greater dependence on the lower parts of the plant for photosynthate and the relative supply of nitrogen to carbohydrates is greater.

iii) *Water*: Photosynthesis is hampered by water stress due to several reasons. The most important one is the closure of stomata at low water levels as this will directly reduce the entry of CO_2 into the leaves. Secondly, water stress can affect the activity of enzymes involved in photosynthesis. The reduction in assimilation rate with decreasing leaf water potential is often linear (Goswami and Srivastava, 1985; Khanna-Chopra, 1988). Source activity is thus regulated by water stress and this will have a bearing on translocation and sink demand. Wardlaw (1967) and Moorby *et al.* (1975) reported, on the contrary, that long-distance movement of sugar through the phloem is resistant to stress. Although this has been challenged, it does help to explain the build up of carbohydrates in the stems of plants under drought conditions as reviewed by Iljin (1957). Since translocation and grain growth are relatively insensitive to stress and photosynthesis is reduced under stress, there is a greater utilization of stored material and current photosynthate from the lower parts of the plant (Wardlaw, 1967).

iv) CO_2: CO_2 exerts its effect on source-sink activity by regulating both photosynthesis (source) and translocation. Both the processes are markedly enhanced by CO_2. As high as 50–100% increase in the photosynthetic rate was observed in several plants exposed to 1000 ppm CO_2 (Mortensen, 1984) and a four-fold increase was observed by Hesketh *et al.* (1984) in some C_3 species. In groundnut, on doubling and trebling the CO_2 concentration, an enhancement as high as 2.5-fold and 4.5-fold, respectively, was observed in the net CO_2 exchange rate (Sharma and Sengupta, 1990). A concomitant increase in translocation of ^{14}C-labelled assimilates out of the source leaves followed, as can be envisaged by decreased retention of label by 41 and 52% at doubled and trebled concentrations of CO_2, as compared to atmospheric concentrations (Sharma and Sengupta, 1990). This provided strong support to the earlier observation of Ho (1977), who observed 20 and 40% enhancement in carbon fixation and carbon transport, respectively, in CO_2-enriched plants grown at 1000 ppm CO_2. He concluded that enriched plants have a higher efficiency at elevated CO_2. (For further details see Chapter by Sicher; Sengupta and Sharma, this volume.)

b) *Nutritional*

Mineral nutrients are essential for many of the metabolic processes in plants including photosynthesis, assimilate transfer, and sink activity. As a consequence they also play a role in the regulation of source-sink activity. Some of the effects of mineral nutrients on plant growth and yield are most likely caused primarily by their influence on the phytohormones which, in turn, regulate source-sink activity.

i) *Potassium*: Several nutrients also exert their effect directly on some functions of the plant, for instance, translocation. The role of potassium ions (K^+) in enhancement of phloem loading and translocation of sucrose is well documented (Peel and Rogers, 1982) and is traced to the activation of ATPase (Giaquinta, 1980; Ho and Baker, 1982). Low K^+ concentration in the apoplasm stimulates the H^+-efflux pump, thereby facilitating sucrose loading. It could also facilitate sucrose loading indirectly by increasing the sucrose concentration in the apoplasm of the leaf tissue (Doman and Geiger, 1979). The close positive correlation between the efflux of sucrose and K^+ from wheat and tobacco mesophyll cells has led to the idea that a sucrose-H^+ co-transport operates at the plasma membrane of leaf cells (Huber and Moreland, 1981). Potassium ions when supplied to leaves caused higher SPS activity (a key enzyme in sucrose synthesis), which in turn contributed to the higher export rates in such leaves (Huber, 1984).

In addition to assimilate transport, K^+ also plays a major role in the regulation of stomatal movement. It maintains turgidity of the guard cells through stimulated uptake of water from the adjacent cells (Humble and Hsiao, 1970). Potassium ions as well affected the leaf photosynthesis through stimulation of CO_2 fixation in isolated chloroplast by maintaining favourable pH for enzyme activity and through regulation of source-sink activity. Potassium also serves as a counter ion to light-induced flux of H^+ across the thylakoid membrane resulting in the establishment of transmembrane pH gradient necessary for ATP synthesis (Lauchli and Pfluger, 1978).

ii) *Phosphorus*: It is a constituent of the energy-rich triphosphate, ATP, which is essential for photosynthesis, translocation, and many other important processes. Inorganic phosphate (Pi) is an important substrate or end product in many key enzymatic reactions including those of photosynthesis and carbohydrate metabolism. Compartmentation of Pi is, therefore, essential for the regulation of metabolic pathways in the cytoplasm and chloroplasts. Pi, at high concentration, depresses total carbon fixation (Heldt *et al.*, 1977). It also inhibits starch synthesis by inhibiting the allosteric enzyme, ADP-glucose pyrophosphorylase. Triose phosphates (TP) stimulate the enzyme. The role of Pi in transport of TP out of the chloroplast membrane is described in Section IIIc. Carbohydrate metabolism in leaves and sucrose translocation are also affected by Pi. The primary reactions in the synthesis of hexoses and sucrose require energy-rich phosphates (ATP and UTP). Sucrose-proton co-transport in phloem loading has also high requirement for Pi.

Another way in which phosphorus controls source-sink activity is by favouring flower formation in tomato (Menary and Van Staden, 1976) and wheat (Rahman and Wilson, 1977). In apple, the number of flowers per tree was found to be almost linearly related to the phosphorus content of the leaves (Bould and Parfitt, 1973). The positive correlations between the number of flowers and cytokinin level in tomato (Menary and Van Staden, 1976), on one hand, and between the phosphorus supply and the cytokinin level, on the other (Dhillon, 1978; Horgan and Wareing, 1980) provided strong evidence that cytokinin contributes to the enhancement effect of phosphorus on flower formation as concluded by Marschner (1986).

c) *Metabolic*

Several metabolic factors are known which control source-sink relationship by virtue of their effect on photosynthesis, sucrose synthesis, assimilate transfer, and utilization. Among these, factors controlling sucrose synthesis and, hence, photosynthesis are given prime importance. Before discussing these factors it is important to know how reactions occur in the chloroplast and cytosol to synthesize sucrose. Figure 1 gives a picture of these reactions. As can be observed from the TP, the first product of CO_2 fixation can either move out of the chloroplast to initiate the synthesis of sucrose in the cytosol or remain in the chloroplast for starch synthesis. The demand for sucrose by sinks controls the efflux of TP from the chloroplast via a carrier protein known as the phosphate translocator. This transport mechanism involves a strict counter exchange of one TP molecule for each Pi molecule transferred across the membrane.

The enzyme for sucrose synthesis, SPS, controls the partitioning of carbon, in the form of TP, into sucrose or starch since one forms at the expense of the other (Champigny, 1985; Foyer, 1987). Starch is, therefore, envisioned as a 'buffer' to sucrose metabolism. The first irreversible step towards sucrose synthesis is catalysed by the cytosolic enzyme, fructose *bis*phosphatase (EC 3.1.1.11), which breaks down fructose 1, 6 *bis*-phosphate (F 1, 6 BP) into fructose-6-phosphate (F6P) and Pi. The photosynthetic production of sucrose involves a two-step process (Leloir and Cardini, 1955). The first step, catalysed by SPS (EC 2.4.1.14), yields sucrose-6-phosphate and uridine disphosphate (UDP). In the next step, the free sugar is liberated by cleavage of the phosphate group by the action of sucrose phosphate phosphatase (SPP).

Pi, F 2, 6 BP, fructose *bis*phosphatase, and SPS are the main metabolites/enzymes known to regulate photosynthesis and assimilate partitioning in response to sink demand and are discussed separately.

i) *Role of Pi*

In the event of high sucrose synthesis in response to high sink demand, Pi levels in the cytosol increase and deplete the chloroplast of TP in a counter exchange. An excessive rate of carbon efflux in this manner would limit photosynthesis by draining the chloroplasts of necessary substrates for regenerating the primary CO_2 acceptor molecule, ribulose, 1,5-*bis*-phosphate (RuBP), which, in turn, would inhibit photosynthesis (Azc'on-Bieto, 1983). Conversely, if the rate of sucrose synthesis is too low, Pi levels in the cytosol fall and limit photosynthesis since Pi is needed for synthesizing ATP, which is a substrate for the reductive pentose phosphate pathway. Thus, Pi acts as the central regulatory metabolite in sucrose synthesis (Herold, 1980; Azc'on-Bieto, 1983; Sicher, 1986; Foyer, 1987).

ii) *Role of F 2, 6 BP and fructose bisphosphatase*

The main function of F 2, 6 BP, a recently discovered metabolite effector, is to mediate the flow of carbon from the chloroplast to sucrose by controlling the activity of fructose *bis*phosphatase. It is synthesized by the enzyme fructose-6-phosphate, 2 kinase (F 6P, 2K). The role of F 2, 6 BP has been well established by Sicher and Bunce (1987). The levels of F 2, 6 BP and SPS activity are interdependent in their role in the regulation of photosynthesis. As photosynthesis increases in response to increased sink demand, TP becomes available for sucrose synthesis in the cytosol and there is a decline in the F 2, 6 BP level (Stitt *et al.*, 1984). Light brings about an identical situation (Sicher and Bunce, 1987). The increased substrate and reduced inhibitor concentration activates the cytosolic fructose *bis*phosphatase. The enhanced production of hexose phosphate is, in turn, thought to stimulate SPS via a rising glucose-6-phosphate (G6P)/Pi ratio (Doehlert and Huber, 1983). On the other hand, when sucrose accumulates in the leaf it results in SPS inhibition and this leads to low concentrations of Pi and high TP concentration in the cytosol, thereby inhibiting net export of TP from the chloroplast; as a result, carbon is diverted towards starch synthesis (Herold, 1980; Huber and Israel, 1982; Champigny, 1985; Gerhardt *et al.*, 1987; Foyer, 1987; Singh, 1988).

iii) *Regulation via SPS*

Increase in sucrose above a threshold level suppresses SPS and/or fructose *bis*phosphatase activity (Plaut *et al.*, 1987). A 20–40 % increase of the overall content of hexose phosphate, UDPG, and TP in spinach leaves shows both SPS and cytosolic fructose *bis*phosphatase being inhibited *in vivo*. This caused a decline in SPS activity and two- to three-fold increase in F 2, 6 BP (Stitt *et al.*, 1984). These observations led Sicher (1986) to propose an intriguing hypothesis that a common biochemical factor, possibly cytosolic Pi level, simultaneously controls SPS and F6P, 2K activities in the leaf.

Since SPS is considered pivotal in regulating source and sink activity, its inhibitors and activators would also be expected to regulate their activity. Pi, sucrose phosphate, UDP, and F 2, 6 BP are known to inhibit SPS activity (Preiss and Greenberg, 1969; Murata, 1972; Harbron *et al.*, 1981; Amir and Preiss, 1982; Doehlert and Huber, 1983). Sucrose, however, is found to inhibit SPS only in species having low SPS activity (tobacco, pea, beans, peanut), while it failed to do so in species having high SPS (wheat, barley, and spinach) as observed by Huber (1981) (Table 2). In sunflower also, no inhibition of SPS activity by sucrose was observed (Mitra, 1990). *See* Table 3.

Table 2: Effect of sucrose on SPS activity of different species

Plant species		SPS activity extracted with Mg^{2+} + 50 mM sucrose (% of control rate)
Tobacco	(4)	45
Peanuts	(2)	0
Peas	(2)	55
Beans	(2)	70
Wheat	(4)	111
Barley	(3)	108
Spinach	(6)	102

Figures within parentheses indicate the number of separate experiments (*Source*: Huber, 1981)

Such differential behaviour of SPS in response to sucrose may be explained by the suggestion made by Salerno and Pontis (1978) that there may be isoenzymes with different sensitivities to sucrose. The possibility of such different isozymes existing in different species cannot be overruled. Mg^{2+}, citrate, and glucose-6-phosphate (G6P), on the other hand, have been observed to stimulate SPS activity (Preiss and Greenberg, 1969; Fekete, 1971; Amir and Preiss, 1982; Doehlert and Huber, 1983).

Table 3: Effect of dipping petioles in various concentrations of sucrose on SPS activity in sunflower leaves (% of control)

Sucrose (μmol/ml)	SPS activity
29.2	127
43.9	109
58.5	173

Source: Mitra, 1990.

The effect of light and sucrose on SPS activity implies that the control of carbon partitioning between sucrose and starch and SPS activity vary from species to species. Further information will be required before SPS can be considered as a regulatory enzyme controlling photosynthesis and carbon partitioning.

d) *Hormonal*

Plant growth regulators have often been implicated to exert their effects through long-distance action on source-sink activity (Wareing et al., 1968; Gieger, 1976a; Wardlaw, 1985; Warrier et al., 1987). It is suggested that sink activity may be responsible for generating an appropriate signal which is hormonal in nature and affects photosynthesis (Herold, 1980) (Fig. 3). Hypotheses implicating auxins, cytokinins, and abscisic acid have been advanced to explain the control mechanism for feedback control from the sink (Hein et al., 1979; Bangerth, 1989). Possible effects of plant growth regulators on photosynthesis, mobilization of assimilates, and source-sink interaction are discussed here.

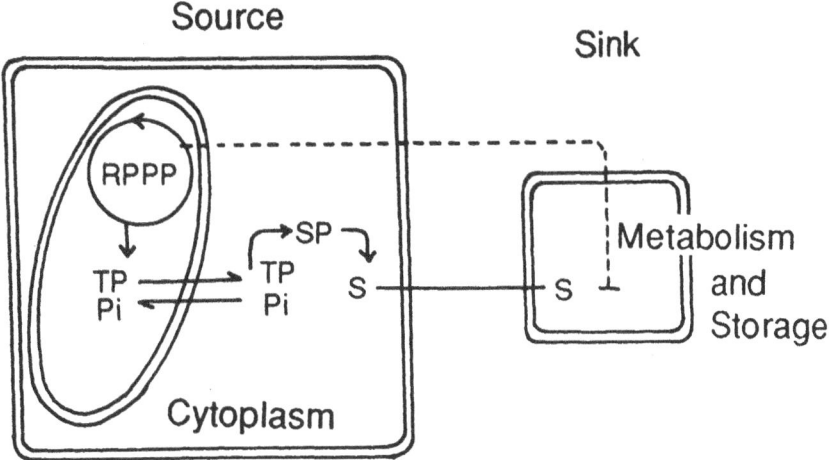

Figure 3. Hormonal signal brought about by sink activity.

i) *Photosynthesis*

Poskuta *et al.* (1975) observed increased rate of photosynthesis, photorespiration, and chlorophyll content of pea plants treated with 10 or 100 ppm GA_3. Thirty ppm GA_3 or 10 ppm NAA application increased the seed yield of peas (Doijode, 1977). Mulligan and Patrick (1979) and Turvey and Patrick (1979), however, failed to observe any increment in the rate of photosynthesis in *Phaseolus vulgaris* L. plants treated with GA_3, IAA, or kinetin. Wagner (1974) placed detached culms of barley in 5.0 ppm kinetin solution or in a mixture of kinetin, IAA, and GA each at 5.0 ppm. Rate of photosynthesis was 1.73 and 2.55 mg CO_2 per dm^2 leaf area per h in plants treated with kinetin and the mixture, respectively, in comparison to 1.38 mg CO_2 per dm^2 per h in control plants.

Effect of IAA on CO_2 fixation, electron transport, and photophosphorylation was examined by Robinson *et al.* (1978) in isolated chloroplasts of *Pisum sativum* L. and *Spinacia oleracea* L. Rate of CO_2 fixation, electron transport, and photophosphorylation did not increase significantly in response to IAA treatment. They concluded that the hormone did not affect photosynthesis in isolated chloroplasts in short-term experiments. However, increased CO_2 uptake observed in leaves following application of IAA might be due to some effect of the hormone other than a direct interaction with the chloroplasts. Foliar application of GA and IAA increased CO_2 fixation by 14 to 33% in rice plants (Anonymous, 1973, 1974). It was further shown that there was a decline in photosynthesis in GA-treated plants in the late grain filling stage while IAA-treated plants maintained higher photosynthetic rate even during this period. Similar results were also obtained in rice by Chatterjee *et al.* (1976). Starck and Stradowska (1977) observed that in radish seedlings, application of IAA (10 mg/g), GA (10 mg/g), and/or zeatin (2 mg/g) to the swollen hypocotyl stimulated $^{14}CO_2$ assimilation. In older plants, IAA + GA + zeatin did not affect $^{14}CO_2$ assimilation but treating with either of the substances separately decreased the assimilation rate. Similar results were also obtained in bean plants (Starck and Stradowska, 1978).

ii) *Mobilization of assimilates*

The concept that plant hormones may influence metabolite transport was introduced by Went (1939) in his nutrient diversion theory to explain apical dominance. Subsequently, the concept was offered as an explanation for the control of leaf senescence and the pattern of photosynthate movement in plants.

It is well established that the site of active assimilate accumulation such as grains, fruits, and shoot apices contains relatively high levels of phytohormones. The idea that these substances may act as mobilizing agents comes from the following observations:

In *intact plant observation*, relationship between endogenous level of hormones and dry matter accretion by the sink (fruit, seed, etc.) is studied.

In *application of hormone to intact plants*, endogenous plant source of hormones is removed by excising the major hormone-producing organ and assimilate movement is observed in response to the applied hormones. These studies were carried out by Bhardwaj and Verma (1985) using three wheat cultivars: HD 2009, HD 4502, and Kalyansona. Dry matter accumulation in grain tended to increase up to 21 days after anthesis. Gibberellins and auxin-like substances as well followed a similar pattern, though the cytokinin contents were the highest at seven days post-anthesis. Further evidence was deduced from exogenous hormone application, wherein it was shown that spray of 10 ppm IAA, 100 ppm GA, or 1 ppm kinetin increased ^{14}C translocation from flag leaf to the grain and consequently led to perceptible increase in grain size. Nakamura (1964) also observed that in intact *Pisum sativum* L. plant, more than 40% of the total ^{14}C accumulated in the apical region while in decapitated plants more than 70% of the total activity was distributed in basal internode. If stem stump was administered with IAA the isotope moved again to the upper internodes. Patrick (1979) reported a similar phenomenon operating in *Phaseolus vulgaris* L, observing that in a decapitated plant if auxin transport inhibitor in placed between source (leaf) and auxin applied to the decapitated apex, acropetal movement of photosynthate is checked, while in plants treated only with auxin, there is more translocation than in the control. The rate of photosynthesis of the auxin-treated plant was not different from that of control, but when partitioning of photosynthates above and below the ^{14}C-fed leaf was studied, it was found that auxin treatment favoured greater upward translocation. Auxin exerts its effect on assimilate transport by acting along the entire transport pathway. This action might be an effect on sink strength, phloem unloading, or some components of longitudinal transfer in the sieve tube throughout the length of transport pathway.

Gibberellins, another group of phytohormones, were also found to be associated with the translocation of assimilates. In developing barley grains, dry matter accumulation was parallel to the GA content in the endosperm, which was the main site of photosynthate accumulation (Mounla, 1978). Spray application of GA_3 (10 ppm) on pea led to an increase in growth by 30%. Pod number as well as average weight of pod increased (Laszlo, 1974). Banishevskaya *et al.* (1976) reported very high content of free gibberellin-like substances in the reproductive organs of pea in comparison to other vegetative parts. Sponsell and MacMillan (1978) also found (in pea) that GA_{29} reached the maximum level of about 10 µg/seed at 27 days from anthesis but declined to 1.6 µg/seed in mature seeds. Like auxins, gibberellins also stimulated the transport of photosynthate in *Phaseolus vulgaris* L. (Mulligan and Patrick, 1979; Patrick, 1979). But unlike auxins, gibberellins were reported to enhance the transport by acting locally (i.e., at the site of application). Cytokinins were also reported to be localized in their mode of action (Patrick, 1979; Turvey and Patrick, 1979).

Apart from promoters, growth inhibitors have also been known to regulate the translocation of assimilates during seed (or grain) development. Cycocel (2-chloroethyl trimethyl ammonium chloride) application during the flowering and post-flowering period considerably enhanced the translocation of carbon from the leaf to the capitulum (developing seeds) in sunflower (Pando and Srivastava, 1985). In wheat, Radley (1976) found that increased abscisic acid content in later phases of grain growth increased the cell permeability and their dehydration, which led to the cessation of photosynthate accumulation due to inactivation of starch synthesizing enzymes.

Interactions between different hormones have also been looked into. Action of cytokinins and gibberellins were found to be synergistic with auxins in stimulating metabolite transport in stem stumps of bean Mullins, 1970). This contrasted with the additive effect of these hormones in decapitated stems of *Malus sylvestris* L. (Hatch and Dowell, 1971).

IV. Source and Sink Interaction

The quantum of assimilates lost from the foliage and the rate of export from an expanded leaf are at least partly determined by the balance between leaf assimilate production and the requirements of the various sinks (Hofstra and Nelson, 1969). The onset of flowering and subsequent growth in peas leads to a rapid doubling of the photosynthesis of the whole plant (Lawrie and Wheeler, 1975). Flinn (1974) has suggested that leaflet photosynthesis in pea is regulated by the pattern of assimilate demand for the subtending pod during its development. Pods of cv. 'Onward' showed three well-defined peaks in assimilate demand. The first was due to rapid elongation of the pod; the second peak probably to inflation of the pod; and the final peak to the rapid growth of the seed. Leaflet photosynthesis fluctuates in response to demand by the developing pod but this response is least in later stage of seed growth. Flinn and Pate (1970) concluded that during this phase of growth the seed might be substantially dependent on the source of carbon from outside the node. The effect of sink removal on the source i.e. leaf photosynthesis, was studied by Pate (1975) and Lovell *et al.* (1972). Their study of pod as a provider of photosynthate showed that if seeds were removed and $^{14}CO_2$ fed to the pod. ^{14}C-labelled assimilates were exported from the pod to the other parts of the plant. Combined with this effect they found that the proportion of ^{14}C exported from a pod without seed was only one quarter of that evident when seeds were present. The result suggested that the presence of seed in a pod might have exercised a stimulatory effect on pod activity in translocation and possibly, as mentioned above, a stimulus to its photosynthetic performance.

While the importance of sink regulation of assimilation rate of source organ is now well established, there is less agreement as to how this form of control is exercised. Gastra (1963) had suggested that at saturating light intensities leaf photosynthetic rate may be the dominant factor under conditions of CO_2 limitation. Other workers have also suggested that the prime factor in sink control may be product inhibition, in which assimilate accumulating in the leaf depresses its current rate of photosynthesis (Warren-Wilson, 1966; Habeshaw, 1973). If this product inhibition hypothesis holds for pea, one might expect to find raised level of soluble hexoses in source leaflets correlated with phase of low fruit growth and depressed source photosynthesis. There is evidence from several crop species that a consuming organ, i.e. sink, can exercise a controlling influence over the production and export of assimilate by source organ. Lovell *et al.* (1972) found that rate of ^{14}C export from pea leaf is greatly increased if, 20 hours before feeding, all the other leaves are removed from the shoot. Since this export was not evident if root or shoot apices were removed at the time of defoliation, it was concluded that the demand for assimilate by these sinks sets the tempo for export.

In the sink-regulated situation described above, competition occurs. Therefore, if sink is removed, less favoured organs are expected to get more photosynthate. A similar result was obtained by Moris and Thomas (1968) where root nodules benefited due to the removal of sink.

a) *Regulation*

Crop productivity in economic terms refers to grain yield and is dependent upon photosynthesis, but evidence for a direct relationship between photosynthetic rate and economic yield is often lacking. Attempts were, therefore, made to investigate the basis for such a behaviour. Opinions differ on whether sink demand regulates photosynthetic rate. To explain the explicit relationship between assimilate transfer and sink size, experiments conducted in the authors' laboratory on wheat, pea, sunflower, and mung bean are detailed below.

i) Wheat (*Triticum aestivum*)

Studies conducted on a large number of wheat genotypes showed that higher rate of photosynthesis is not necessarily associated with higher grain yield. Since grain yield is dependent not only on the availability of assimilates but also on the ability of developing grains to accumulate and utilize them for the synthesis of the reserves, the grain yield obtained may or may not necessarily reflect the growth potential of the developing grain. Studies revealed that the diploid wheats are more efficient in photosynthetic rate than tetraploid and hexaploid wheats but their yields are rather poor. Does the poor grain growth of diploids depend on their lesser efficiency to utilize assimilates for the synthesis of reserves? The excised developing grains were cultured in ^{14}C-sucrose and their ability to synthesize reserves was determined. It appears that diploid wheat has lesser capacity to synthesize reserves than tetraploid and hexaploid wheat, thus indicating that poor grain growth of diploid wheat was due to its poor sink capacity (Ghildiyal and Sirohi, 1986). Diploid wheats possessed a relatively higher number of spikelets (per ear) and tillered profusely. Photosynthetic rate was quite high in commensuration with the demand of the sink (i.e., developing spikelets). Photosynthetic rate declined, however, with diminished sink demand during post-anthesis owing to poor grain growth and its capacity to accumulate reserve materials.

In wheat, sink (grain) size varies between cultivars and within the ear itself. Grains located in the middle spikelets and basal grains within the spikelet are generally bolder. To explain such a behaviour, translocation of externally supplied ^{14}C sucrose from the base of the ear to grains at different locations within the ear was looked into. Grain size appears to correlate well with the availability of assimilates (Fig. 4). The differences in translocation behaviour between cultivars and spikelets within an ear may be because of assimilate supply; however, under normal conditions assimilate supply may not be limiting as revealed by despikeleting and defloreting experiments (Bhardwaj and Dua, 1975; Bhardwaj and Verma, 1985). It was shown that any increase in assimilate supply through partial removal of florets or spikelets does not increase the size of the sink (grain). Differences in translocation behaviour between the cultivars are generally related with the levels of endogenous growth regulators in the developing sinks (grains) (Table 4).

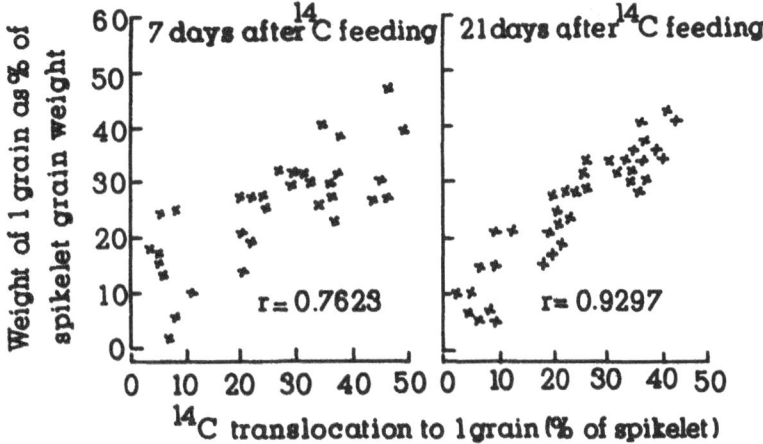

Figure. 4. Relation between ^{14}C translocation to the grain and its weight expressed as percentage of spikelet grain weight.

Table 4: Sink size and endogenous levels of growth regulators in developing wheat grains

Cultivars	1000 grain weight (g)	Auxins (ng/g fw)		Gibberellins (ng/g fw)	
		Days following post-anthesis			
		7	21	7	21
HD 2009	34.1	121	402	3.5	9.0
HD 4502	37.2	174	790	5.3	26.3
Kalyansona	26.8	116	403	3.1	21.3

Source: Bhardwaj and Dua, 1974; Bhardwaj and Verma, 1985.

ii) Pea (*Pisum sativum* var. arvensis)

Analysis of two pea cultivars for the boldness of sink (seed) size revealed that bolder seed size was found to be related to (1) the rate of flow of photosynthates to the pod rather than its capacity to fix radiocarbon, (2) the capacity of the vascular elements to translocate the assimilates, and (3) the capacity of the seed to accumulate them (Srivastava and Bhardwaj, 1987). It is, however, not clear whether the accumulation capacity of the seed or the resistance created by the vascular channel is the major determinant of seed size. Time-course analysis of movement of ^{14}C-sucrose revealed the presence of two phases of translocation. In the first phase (0–20 min), the transport was slow, while it was fast in the second phase (40–60 min). When excised seeds were allowed to accumulate ^{14}C-sucrose in sucrose media, higher accumulation of ^{14}C by seeds indicated the capacity to accumulate assimilates, whereas the amount of ^{14}C entering into insoluble fraction suggested the capacity of seeds to synthesize reserve materials like starch and lipids.

iii) Sunflower (*Helianthus annuus*)

In order to test if the photosynthetic rate is influenced by accumulation of the assimilates in the source organ (i.e., leaf), effect of shading, defoliation, and girdling on photosynthesis of sunflower leaf was investigated. Photosynthetic rate remained unaffected under the above treatments. Further, sucrose accumulation in the leaf as a result of these treatments did not cause an inhibition of SPS activity. The SPS activity was also not affected by light and does not seem to play any significant role in regulation of photosynthesis in sunflower (Mitra and Srivastava, 1992).

iv) Mung bean (*Vigna radiata*)

Studies conducted on leaf photosynthesis in deflowered plants indicate that despite starch accumulation in the source leaf, the photosynthetic rate was maintained at a higher level than the control. Exposure of dark starved plants to sunlight for different duration did not enhance the photosynthetic rate. These findings suggest the involvement of hormonal factor(s) rather than operation of feedback mechanisms (Ghildiyal and Mitra, 1988). It was further observed that defoliation or deflowering treatments enhanced the photosynthetic rate of the uppermost fully expanded leaf. It is possible that these treatments created hormonal imbalance within the plant. However, application of benzyladenine (a cytokinin) enhanced photosynthetic rate in the control plants only. It seems plausible that supply of cytokinin from the root to the leaf was augmented due to decreased competition between leaves and flowers in deflowered plants and in between leaves in defoliated plants. Increase in nodulation of deflowered plants might be based on higher availability of cytokinins (Ghildiyal and Vijaylaxmi, 1989).

b) *Manipulation*

Under saturated light intensity, photosynthetic rate of wheat leaf declined after midday. This decline appears to be independent of change in stomatal diffusive resistance of the leaf. Excision of the leaves except the uppermost fully expanded leaf sustained its photosynthetic rate at higher level during the day, while the de-eared plants (with all tillers removed) showed rapid decline in photosynthetic rate towards the evening, though it was more or less at par with the control in the morning (Fig 5). In a study conducted with three wheat cultivars (HD 2009, HD 4502, and Kalyansona), which differ in grain number per spikelet as well as per ear and single grain weight, partial despikeleting and defloreting was done. Decrease in inter-grain competition resulted in an increase in the number of grains per spikelet and single grain weight but did not compensate fully for the loss of spikelets or florets. Grain number appears to be a major yield-determining attribute. These studies indicate that (i) faster depletion of assimilates from the source leaf by a more active sink would maintain higher photosynthetic rate for a longer period during the day and (ii) sink strength determines the productivity under optimum growth conditions (Bhardwaj and Verma, 1987).

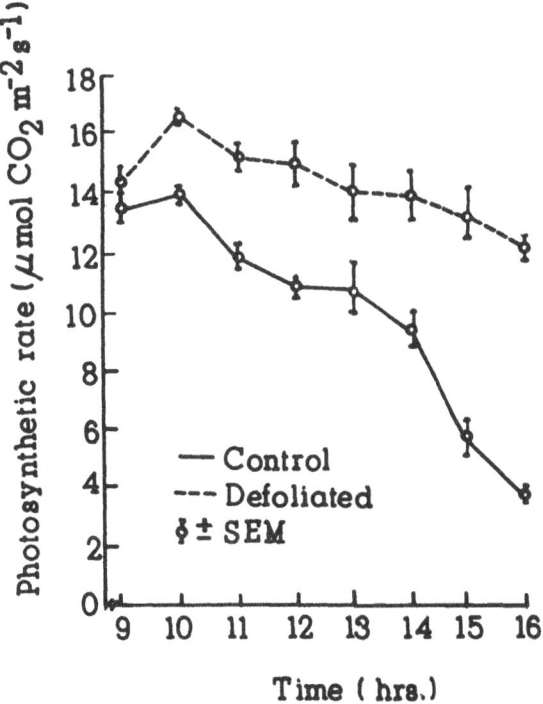

Figure 5. Photosynthetic rate in uppermost fully expanded leaf of main shoot in wheat cv. Kalyan sona three days after excision of other source leaves and in control plants.

Different photosynthesizing organs contribute differently. It has been shown in field pea (*Pisum sativum* var. arvensis) that removal of half of the leaflets had least effect on pod yield (Srivastava and Bhardwaj, 1986). The reduction in seed number was compensated for (though not fully) by increasing seed size. It seems plausible that surviving seed could get sufficient amount of assimilates even when half of the leaflets were removed. The maximum reduction in pod weight was recorded when

380

stipules and the leaflets were removed. It appears that the demand of the seed is partly compensated for by the fruit walls or by the organs located away from the fruiting node.

In sunflower, source (photosynthates) appears to be limiting in spite of the large green surface and high net assimilation rate relative to other species under full sunlight conditions. Estimations for CER indicate that leaf photosynthesis during the plant ontogeny increases from seedling to grain setting stage but declines before the seed

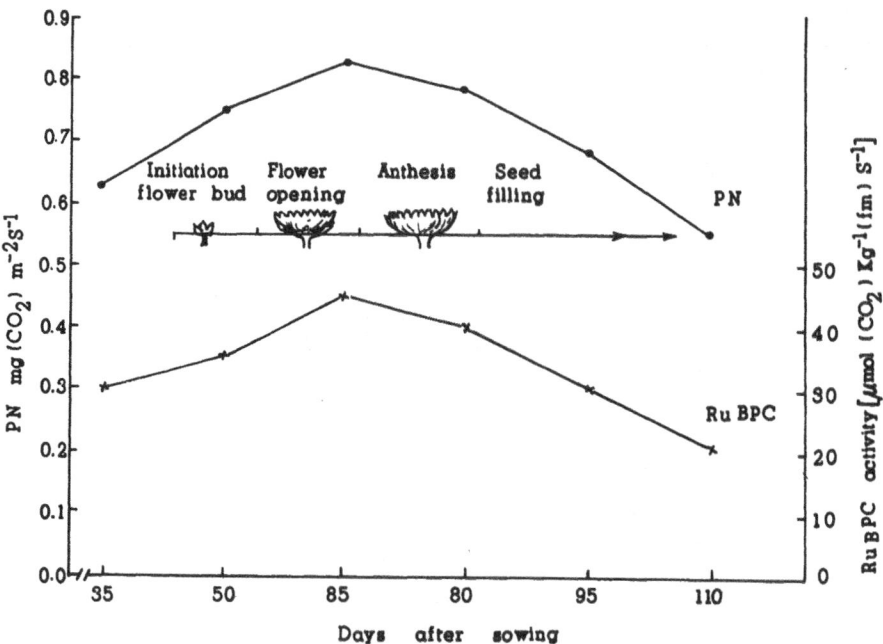

Figure 6. Photosynthetic (PN) and RuBP carboxylase activities in leaves of sunflower during plant ontogeny.

filling is complete (Fig. 6). It is plausible that most of the assimilate is utilized for building of the vegetative surface resulting in competition for the photosynthates with the developing seeds. Only the top five to three fully developed leaves and bracts in proximity to the capitulum contribute assimilates significantly towards the seed filling. The declining trend in photosynthesis during the seed filling period has been attributed to loss in ribulose *bis*phosphate carboxylase (RubPcase) enzyme activity and increase in protease activity (Table 5). Thus, it indicates that source is limiting for seed filling in sunflower.

Foliar spray of nitrogen, in the form of urea (.05%) and benzyladenine a synthetic growth regulator three times during the growing season delayed the loss in chlorophyll, RuBPcase activity and net photosynthetic carbon fixation (Goswami and Srivastava, 1988). Prolongation of active photosynthesis is likely to enhance seed yield further and the possibility of extending it through genetic manipulation needs to be explored.

Table 5: Total chlorophyll, soluble protein, RuBP carboxylase *in vivo* and protease activity in the leaves of sunflower

	Days after sowing					
Total chlorohyll	2.42	2.13	1.86	1.49	0.96	0.79
(mg/g fwt)						
Soluble protein	16.75	17.25	13.50	12.75	17.25	10.62
(mg/g fwt)	± 1.37	± 1.36	± 1.36	± 1.71	± 1.27	± 1.51
RuBP carboxylase	143.5	172.1	109.8	89.3	82.8	69.7
(μmol CO_2/h/g fwt)	± 2.9	± 3.7	± 2.2	± 2.1	± 2.1	± 2.3
Nitrate reductase	5.41	3.70	2.24	1.88	1.39	0.85
(μ mol NO_2 formed/h/g fwt)	± 0.43	± 0.37	± 0.22	± 0.21	± 0.28	± 0.09
Protease activity	11.2	14.5	16.21	16.45	17.49	18.6
(OD_{570}/g fwt/90 min)	± 1.2	± 1.8	± 1.4	± 1.8	± 1.4	± 1.2

Source: Goswami, 1985.

V. Conclusions

Crop productivity depends on photosynthesis (source) and partitioning of photosynthates to economically important parts (sink). Since these two factors may not necessarily be complementary, a question arises as to which of the two factors limits productivity. Furthermore, sink might influence the photosynthetic rate. Thus, production and utilization of dry matter within a plant would depend on each other. This relationship is influenced by numerous factors which might be environmental, nutritional, metabolic, or hormonal. It is well documented that high light intensity stimulates photosynthesis, thus providing larger amounts of assimilates for translocation, though the accumulation of assimilates may lead to feedback inhibition of photosynthesis. Change in temperature influenced both storage of assimilates and sucrose synthesis. Water stress affects stomatal regulation as well as enzyme activity, whereas increasing levels of CO_2 affect both photosynthesis and translocation of assimilates. Involvement of K^+ ions in control of stomatal movement and translocation of assimilates is a pointer in this direction. Phosphorus also plays a crucial role in regulating photosynthesis and translocation.

Most studies concerning the translocation of assimilates to different organs have been carried out using radio-assay techniques and they reveal that differences in yielding ability at cultivar level may be related to differences in the contribution(s) of photosynthesizing organs and of assimilate flow to the organs constituting yield. The concept of source response to requirement of sink envisages that the path of assimilate flow is not a limiting factor. Vascular bundles could resist the movement of assimilates but this resistance might be overcome by strong sinks. The question thus arises as to what is the basis for differences in translocation behaviour between cultivars. These differences are generally related to the levels of endogenous growth regulators in the developing sinks (grains) (Bhardwaj and Dua, 1974; Bhardwaj and Verma, 1985; Srivastava and Bhardwaj, 1987). Abscisic acid often increases membrane permeability and, therefore, probably favours unloading of sucrose in the sink (Schussler *et al.*, 1984). When the sink demand is high, the stomata are wide open to facilitate both photosynthesis and transpiration. Under such conditions, abscisic acid concentration in the source leaves is low; consequently, the stomata are fully

382

open. Support for this comes from the fact that the rates of transport of sucrose and abscisic acid from the source leaves to the sink are positively correlated (Setter *et al.*, 1980). Thus, since stomatal opening is enhanced by cytokinin and decreased by abscisic acid, changes in the abscisic acid to cytokinin ratio by different rates of import and export of these hormones seem to be an interesting hypothesis for feedback control from the sink. However, additional control mechanisms might also exist such as 'remote' hormonal signals from the developing fruits to the source leaves (Hein *et al.*, 1979). There is evidence that increased rates of auxin export in the phloem from the developing fruits to the source leaves might act as such a feedback signal (Hein *et al.*, 1984). Bangerth (1989) has implicated IAA as being this signal. His evidence is based on the observations that IAA has a direct effect on assimilate transport and is essential for vascular tissue differentiation.

VI. References

Amir J and Preiss J (1982) Kinetic characterization of spinach leaf sucrose-phosphate synthase. *Plant Physiol* **69**: 1027–1030

Anonymous (1973) Annual Report. Bose Institute, Calcutta

Anonymous (1974) Annual Report. Bose Institute, Calcutta

Azc'on-Bieto J (1983) Inhibition of photosynthesis by carbohydrates in wheat leaves. *Plant Physiol* **73**: 681–686

Bangerth F (1989) Dominance among fruits/sinks and the search for a correlative signal. *Physiol Plant* **76**: 608–614

Banishevskaya LA, Mekhtizade RM, Lustifor D Kh and Safaraliev RA (1976) Gibberellin like substances during ontogenesis of winter pea. Izvestiya Akademii Nauk Azerbaidhzhanska SSR, Biologicheskikh. Nauk. No. 1: 8–13

Bhardwaj SN and Dua IS (1974) Studies on the hormonal basis of grain yield in *aestivum wheats*. *Indian J Plant Physiol* **17**: 39–43

Bhardwaj SN and Dua IS (1975) A study of the competitive interrelationship between grain setting and growth in aestivum wheats in relation to the production of growth regulating substances. *Indian J Plant Physiol* **18**: 7–13

Bhardwaj SN and Verma V (1985) Hormonal regulation of assimilate translocation during grain growth in wheat. *Indian J Exp Biol* **23**: 719–721

Bhardwaj SN and Verma V (1987) Regulation of grain size within a developing ear of bread wheat. *Indian J Agric Sci* **57**: 710-715

Bould C and Parfitt RI (1973) Leaf analysis as a guide to the nutrition of fruit crops. X Magnesium and phosphorus sand culture experiments with apple. *J Sci Food Agric* **24**: 175–185

Boussingault JB (1968) *Agronomie, Chimie Agricole et Physiologie*, 2nd Ed: Mallet Bachelier, Paris, 1860–1874, 5 Vols, pp 236–312

Brown CS and Huber SC (1987) Photosynthesis, reserve mobilization and enzymes of sucrose metabolism in soybean (*Glycine max*) cotyledons. *Physiol Plant* **70**: 537–543

Champigny ML (1985) Regulation of photosynthetic carbon assimilation at the cellular level: A review. *Photosynthesis Res.* **6**: 273–286

Chatterjee A, Mandal RK and Sircar SM (1976) Effects of growth substances on productivity, photosynthesis and translocation of rice varieties. *Indian J Plant Physiol* **19**: 131–138

Delrot S (1987) Phloem loading: Apoplastic or symplastic? *Plant Physiol Biochem* **25**: 667–676

Delrot S and Bonnemain JL (1981) Involvement of protons as a substrate for the sucrose carrier during phloem loading in *Vicia faba* leaves. *Plant Physiol* **67**: 560–564

Dhillon SS (1978) Influence of varied phosphorus supply on growth and xylem sap cytokinin level of sycamore (*Platanus occidentalis* L.) seedlings. *Plant Physiol* **61**: 521–524

Doehlert DC and Huber SC (1983) Regulation of spinach leaf sucrose phosphate synthetase by glucose 6-phosphate, inorganic phosphate and pH. *Plant Physiol* **73**: 989–994

Doijode SD (1977) Effect of growth regulators on growth and yield of garden peas (*Pisum satium* var. hartense L.). *Mysore J Agric Sci* **11:** 114

Doman DC and Geiger DR (1979) Effect of exogenously supplied foliar potassium on phloem loading in *Beta vulgaris* L. *Plant Physiol* **64:** 528–533

Erwee MG, Goodwin PB and Van Bel AJE (1985) Cell-to-cell communication in the leaves of *Commelina cyanea* and other plants. *Plant Cell Environ* **8:** 173–178

Eschrich W and Eschrich B (1987) Control of phloem unloading by source activities and light. *Plant Physiol Biochem* **25:** 625–634

Evert RF, Eschrich W and Heyser W (1978) Leaf structure in relation to solute transport and phloem loading in *Zea mays* L. *Planta* **138:** 279–294

Fekete MAR de (1971) The regulative properties of UDP-glucose: D-fructose-6-phosphate 2-glucosyl transferase (sucrose phosphate synthetase) from *Vicia faba* cotyledons. *Eur J Biochem* **19:** 73–80

Fisher DG (1986) Ultrastructure, plasmodesmatal frequency and solute concentration in green areas of variegated *Coleus blumii* Benth leaves. *Planta* **169:**141–152

Fisher DG and Evert RF (1982) Studies on the leaf of *Amaranthus retroflexus* (Amaranthaceae): Ultrastructure, plasmodesmatal frequency, and solute concentration in relation to phloem loading. *Planta* **155:** 377–387

Flinn AM (1974) Regulation of leaflet photosynthesis by developing fruit in the pea. *Physiol Plant* **31:** 275–278

Flinn AM and Pate JS (1970) Quantitative analysis of the carbon transfer from pod and subtending leaves of ripening seeds of field peas (*P. arvensis* L.). *J Exp Bot* **21:** 71–82

Foyer CH (1987) The basis for source-sink interaction in leaves. *Plant Physiol Biochem* **25:** 649–657

Gastra P (1963) Climatic control of photosynthesis and respiration. In: (LT Evans, ed) *Environmental Control of Plant Growth*. Academic Press, New York and London, pp 113

Geiger DB (1976a) Effects of translocation and assimilate demand on photosynthesis. *Can J Bot* **54:** 2337–2345

Geiger DR (1976b) Phloem loading in source leaves. In: (IF Wardlaw and JB Passioura, eds) *Transport and Transfer Processes in Plants*. Academic Press, New York, pp 167–187

Gerhardt R, Stitt M and Heldt HW (1987) Subcellular metabolite levels in spinach leaves. Regulation of sucrose synthesis during diurnal alterations in photosynthetic partitioning. *Plant Physiol* **83:** 399–407

Ghildiyal MC (1990) Effect of urea on kinetic properties of Rubisco. *Sci Cult* **56:** 452–454

Ghildiyal MC and Mitra S (1988) Photosynthesis in relation to saccharides in mungbean leaves. *Photosynthetica* **22:** 127–128

Ghildiyal MC and Sirohi GS (1986) Photosynthesis and sink efficiency of different species of wheat. *Photosynthetica* **20:** 102–106

Ghildiyal MC and Vijaylaxmi (1989) Influence of benzyladenine on photosynthesis in mungbean under different source-sink relationships. *Photosynthetica* **23:** 667–670

Giaquinta RT (1976) Evidence for phloem loading from the apoplast. Chemical modifications of membrane sulphydryl groups. *Plant Physiol* **57:** 872–875

Giaquinta RT (1977) Phloem loading of sucrose pH dependence and selectivity. *Plant Physiol* **59:** 750–755

Giaquinta RT (1980) Mechanism and control of phloem loading of sucrose. *Ber Dtsch Bot Ges* **93:** 187–201

Giaquinta RT (1983) Phloem loading of sucrose. *Annu Rev Plant Physiol* **34:** 347–387

Gifford RM, Thorne JH, Hitz WD and Giaquinta RT (1984) Crop productivity and photoassimilate partitioning. *Science* **225:** 801–808

Glasziov KT and Gayler KR (1972) Storage of sugars in stalks of sugarcane. *Bot Rev* **38:** 471–490

Goswami BK (1985) Studies on photosynthesis and nitrogen metabolism in relation to growth and yield of sunflower (*Helianthus annuus* L.) as influenced by benzyladenine and urea. Ph.D. thesis, Indian Agricultural Research Institute, New Delhi, India p 155

Goswami BK and Srivastava GC (1985) Effect of abscisic acid and zeatin on water stress induced photosynthesis and respiration in sunflower leaves. *Indian J Plant Physiol* **28:** 350–357

Goswami BK and Srivastava GC (1988) Effect of foliar application of urea on leaf senescence and photosynthesis in sunflower (*Helianthus annuus* L.). *Photosynthetica* **22:** 99–104

Habeshaw D (1973) Translocation and the control of photosynthesis in sugarbeet. *Planta* **110:** 213–226

Harbron S, Foyer C and Walker D (1981) The purification and properties of sucrose-phosphate synthetase from spinach leaves: The involvement of this enzyme and fructose *bis*-phosphatase in the regulation of sucrose biosynthesis. *Arch Biochem* **212:** 237–246

Hatch AH and Dowell LE (1971) Hormone directed transport of ^{32}P in *Malus sylvestris* seedlings. *J Amer Soc Hort Sci* **96**: 230–234

Hein MB, Brenner ML and Brun WA (1979) Source/sink interactions in soybeans. II A possible role of IAA. *Plant Physiol* suppl. **63**: 43

Hein MB, Brenner ML and Brun WA (1984) Effects of pod removal on the transport and accumulation of abscisic acid and indole-3-acetic acid in soybean leaves. *Plant Physiol* **76**: 955–958

Heldt HW (1976) Metabolite carriers of chloroplast. In: (CR Stocking and U Heber, eds) *Encyclopedia of Plant Physiology*, Vol 3 Springer-Verlag, London, pp 137–143

Heldt HW, Chon CJ, Maronde D, Herold A, Stankovie ZS, Walker DA, Kraminer A, Kirk MR and Heber U (1977) Role of orthophosphate and other factors in the regulation of starch formation in leaves and isolated chloroplasts. *Plant Physiol* **59**: 114–1155

Hendrix DL and Huber SC (1986) Diurnal fluctuations in cotton leaf carbon export, carbohydrate content, and sucrose synthesizing enzymes. *Plant Physiol* **81**: 584–586

Herold A (1980) Regulation of photosynthesis by sink activity. The missing link. *New Phytol* **86**: 131–144

Hesketh JD, Wolley JT and Peter DB (1984) Leaf photosynthesis CO_2 exchange rates in light and CO_2 enriched environments. *Photosynthetica* **18**: 536–540

Heyser W (1980) Phloem loading in maize leaf. *Ber Dtsch Bot Ges* **93**: 221–228

Ho LC (1977) Effects of CO_2 enrichment on the rates of photosynthesis and translocation of tomato leaves. *Ann Appl Biol* **87**: 191–200

Ho LC and Baker DA (1982) Regulation of loading and unloading in long distance transport systems. *Physiol Plant* **56**: 225–230

Hofstra G and Nelson CD (1969) A comparative study of translocation of assimilated ^{14}C from leaves of different species. *Planta* **88**: 103–112

Horgan JM and Wareing PF (1980) Cytokinins and the growth response of seedlings of *Betula pendula* Roth. and *Acer pseudoplatanus* L. to nitrogen and phosphorus deficiency. *J Exp Bot* **31**: 525–532

Huber SC (1981) Interspecific variation in activity and regulation of leaf sucrose phosphate synthetase. *Z Pflanzenphysiol* **102**: 443–450

Huber SC (1984) Biochemical basis for effects of K-deficiency on assimilate export rate and accumulation of soluble sugars in soybean leaves. *Plant Physiol* **76**: 424–430

Huber SC and Israel DW (1982) Biochemical basis for partitioning of photosynthetically fixed carbon between starch and sucrose in soybean (*Glycine max* Merr.) leaves. *Plant Physiol* **69**: 691–696

Huber SC, Kerr PS and Rufty TW (1985) Diurnal changes in sucrose phosphate synthase in leaves. *Physiol Plant* **64**: 81–87

Huber SC and Moreland DE (1981) Ca-transport of potassium and sugars across the plasmalemma of mesophyll protoplasts. *Plant Physiol* **67**: 163–169

Humble GD and Hsiao TC (1970) Light-dependent influx and efflux of potassium of guard cells during stomatal opening and closing. *Plant Physiol* **46**: 483–487

Iljin WS (1957) Drought resistance in plants and physiological processes. *Annu Rev Plant Physiol* **8**: 257–274

Kalt-Torres W, Kerr PS, Usuda H and Huber SC (1987) Diurnal changes in maize leaf photosynthesis. I Carbon exchange rate, assimilate export rate, enzyme activities. *Plant Physiol* **83**: 283–288

Kerr PS, Rufty TS and Huber SC (1985) Changes in nonstructural carbohydrates in different parts of soybean (*Glycine max* L. Merr.) plants during a light/dark cycle and in extended darkness. *Plant Physiol* **78**: 576–581

Khanna-Chopra R (1988) Water stress and photosynthesis. In: (R Singh and SK Sawhney, eds) *Advances in Frontier Areas of Plant Biochemistry*. Prentice Hall of India Pvt Ltd, New Delhi, pp 32–58

King RW, Wardlaw IF and Evans LT (1967) Effect of assimilate utilization on photosynthetic rate in wheat. *Planta* **77**: 261–276

Komor E, Rotter M, Waldhauser J, Martin E and Cho BH (1980) Sucrose proton symport for phloem loading in the *Ricinus* seedling. *Ber Dtsch Bot Ges* **93**: 211–219

Kursanov AL, Brovchenko MI and Pariiskaya AN (1960) Flow of assimilates to the conducting tissue in rhubarb (*Rheum raponticum* L.) leaves. *Sov Plant Physiol* **6**: 544–551

Laszlo L (1974) Effect of GA_3 on height, number of seedy pods and average weight of pea plant (*Pisum sativum* L.) variety Petit Provencal. *Novenytermeles* **23**: 109–113

Lauchli A and Pfluger R (1978) Potassium transport through plant cell membranes and metabolic role of potassium in plants. *Proc 11th Congr Int Potash Inst Bern* pp 111–163

Lawrie C and Wheeler CT (1975) The effect of flowering and fruit formation on the supply of photosynthetic assimilates to the nodule of *Pisum sativum* L. in relation to the fixation of nitrogen. *New Phytol* **73**: 1119–1128

Leloir LF and Cardini CE (1955) The biosynthesis of sucrose phosphate. *J Biol Chem* **214**: 157–165

Lovell PH, Oo HT and Sagar CR (1972) An investigation into rate and control of assimilate movement from leaves in *Pisum sativum. J Exp Bot* **23**: 255–266

Madore MA, Dross JW and Lucas WJ (1986) Symplastic transport in *Ipomea tricolor* source leaves. Demonstration of functional symplastic connections from mesophyll to minor veins by a novel dye-tracer method. *Plant Physiol* **82**: 432–442

Malek F and Baker DA (1977) Proton contransport of sugars in phloem loading. *Planta* **135**: 297–299

Marschner H (1986) *Mineral Nutrition of Higher Plants*. Academic Press, New York, pp 674

Menary RC and Van Staden J (1976) Effect of phosphorus nutrient and cytokinins on flowering in the tomato, *Lycopersicon esculentum* Mill. *Aust J Plant Physiol* **3**: 201–205

Mitra S (1990) Studies on the regulation of photosynthesis in sunflower (*Helianthus annuus* L.). Ph.D thesis, Indian Agricultural Research Institute, New Delhi, India p 56

Mitra S and Srivastava G.C. (1992) Photosynthesis as influenced by assimilate level in sunflower (*Helianthus annuus* L.). *J. Agron Crop Sci.* in press.

Moorby J, Munns R and Walcott J (1975) Effect of water deficit on photosynthesis and tuber metabolism in potatoes. *Aust J Plant Physiol* **2**: 323–333

Moris DA and Thomas EE (1968) Distribution of ^{14}C labelled sucrose in seedling of *Pisum sativum* L. treated with indol acetic acid and kinetin. *Planta* **83**: 276–281

Mortensen LM (1984) Photosynthetic adaptation in CO_2 enriched air and the effect of intermittent CO_2 application on green-house plants. *Acta Horticulturae* **162**: 153–158

Mounla MA (1978) Gibberellin like substance in developing barley grain. *Physiol Plant* **44**: 268–272

Mulligan DR and Patrick JW (1979) Gibberellic acid-promoted transport of assimilates in stem of *Phaseolus vulgaris* L. Localized versus remote site(s) of action. *Planta* **145**: 233–238

Mullins MG (1970) Hormone directed transport of assimilates in decapitated internode of *Phaseolus vulgaris* L. *Ann Bot* **34**: 897–909

Murata T (1972) Sucrose phosphate synthetase from various plant origins. *Agric Biol Chem* **36**: 1877–1884

Nakamura E (1964) Effect of decapitation and IAA on the distribution of radio-active phosphorus in stem of *Pisum sativum*. *Plant Cell Physiol* **5**: 521–524

Mitra S and Srivastava GC (1992) Photosynthesis as influenced by assimilate level in sunflower (*Helianthus annuus* L). *J Agron Crop Sci* in press

Neales TF and Incoll LD (1968) The control of leaf photosynthesis rate by the level of assimilate concentration in the leaf: A review of the hypothesis. *Bot Rev* **34**: 107–125

Pando SB and Srivastava GC (1985) Physiological studies on seed set in sunflower III Significance of dwarfing the plant size using growth regulator. *Indian J Plant Physiol* **28**: 72–80

Pate JS (1975) Pea. In: (LT Evans, ed) *Crop Physiology*. Cambridge University Press, London, pp 191–224

Pate JS and Gunning BES (1972) Transfer cells. *Annu Rev Plant Physiol* **23**: 173–196

Patrick JW (1979) An assessment of auxin promoted transport in decapitated stems and whole shoots of *Phaseolus vulgaris* L. *Planta* **146**: 107–112

Peel AJ and Rogers S (1982) Stimulation of sugar loading into sieve elements of willow by potassium and sodium salts. *Planta* **154**: 94–96

Plaut Z, Mayoral ML and Reinhold L (1987) Effect of altered sinks: Source ratio on photosynthetic metabolism of source leaves. *Plant Physiol* **85**: 786–791

Pollock CJ and Housley TL (1985) Light-induced increase in sucrose phosphate synthetase activity in leaves of *Lolium temulentum*. *Ann Bot* **55**: 593–596

Poskuta J, Parys E and Ostrowska E (1975) Growth, CO_2 exchange rate and yield of pea (*Pisum sativum* L.) cv. "Bordi" in the field after seed pretreatment with gibberellic acid (GA_3). *Biuletyn Warzywniezy* **18**: 197–206

Preiss J and Greenberg E (1969) Allosteric regulation of uridine diphosphoglucose: D-fructose-6-phosphate-2-glucosyl transferase (EC 2.4.1.14). *Biochem Biophys Res Commun* **36**: 289–295

Radley M (1976) Development of wheat grain in relation to endogenous growth substances. *J Exp Bot* **27**: 1009–1021

Rahman MS and Wilson JH (1977) Effect of phosphorus applied as superphosphate on rate of development and spikelet number per ear in different cultivars of wheat. *Aust J Agric Res* **28**: 183–186

Robinson SP, Wiskiel JT and Paleg LG (1978) Effect of indole acetic acid on CO_2 fixation, electron transport and phosphorylation in isolated chloroplasts. *Aust J Plant Physiol* **5**: 425–432

Rufty TW, Kerr PS and Huber SC (1983) Characterization of diurnal changes in activities of enzymes involved in sucrose biosynthesis. *Plant Physiol* **73**: 428–433

Russin WA and Evert RF (1985) Studies on the leaf of *Populus deltoides* (Salicaceae): Quantitative aspects, and solute concentrations of the sieve tube members. *Am J Bot* **72**: 487–500

Salerno GL and Pontis HG Pontis (1978) Studies on sucrose phosphate synthetase. The inhibitory action of sucrose. *FEBS Lett* **86**: 263–267

Schenk E (1972) Quantitative studies on translocation in *Sagittaria graminea* Mick leaves. A new method of measurement. *Acta Bot Neerl* **21**: 231–234

Schussler JR, Brenner MI and Brun WA (1984) Abscisic acid and its relationship to seed filling in soybeans. *Plant Physiol* **76**: 301–306

Setter TL, Brun WA and Brenner ML (1980) Effect of obstructed translocation on leaf abscisic acid, and associated stomatal closure and photosynthesis decline. *Plant Physiol* **65**: 1111–1115

Sharma A and Sengupta UK (1990) Carbon dioxide enrichment effects on photosynthesis and related enzymes in *Vigna radiata* L. (Wilczek). *Indian J Plant Physiol* **33**: 340–346

Shiroya M (1977) Translocation of organic substances in sunflower 1 Downward translocation of ^{14}C-sucrose. *Plant Cell Physiol* **18**: 633–639

Sicher RC (1986) Sucrose biosynthesis in photosynthetic tissue: Rate-controlling factors and metabolic pathway. *Physiol Plant* **67**: 118–121

Sicher RC and Bunce JA (1987) Effects of light and CO_2 on fructose 2,6-bisphosphate levels in barley primary leaves. *Plant Physiol Biochem* **25**: 525–530

Sicher RC and Kremer DF (1984) Chages in sucrose-phosphate synthase activity in barley primary leaves during light/dark transitions. *Plant Physiol* **76**: 910–912

Sicher RC and Kremer DF (1985) Possible control of maize leaf sucrose-phosphate synthase activity by light modulation. *Plant Physiol* **79**: 695–698

Singh R (1988) Regulation of carbon partitioning between sucrose and starch in photosynthetic leaves of C_3 plants. In: (R Singh and SK Sawhney eds) *Advances in Frontier Areas of Plant Biochemistry*. Prentice Hall of India Pvt Ltd, New Delhi, pp 102–126

Sponsell VM and MacMillan J (1978) Metabolism of gibberellin A_{29} in seeds of *Pisum sativum* cv. Progress No. 9; Use of (^{2}H) and ^{2}H) GAs and the identification of a new GA catabolite. *Planta* **144**: 69–78

Srivastava JP and Bhardwaj SN (1986) Contribution of different photosynthesizing organs to the pod in relation to source and sink intraction in field pea (*Pisum sativum* var. *arvensis*). *Indian J Plant Physiol* **29**: 262–266

Srivastava JP and Bhardwaj SN (1987) Accumulation capacity of seed, leaf photosynthetic rate and vascular translocation in regulating the seed size in pea. *Indian J Agric Sci* **57**: 416–418

Starck Z and Stradowska M (1977) Pattern of growth and ^{14}C-assimilate distribution in relation to photosynthesis in radish plants treated with growth substances. *Acta Soc Bot Pol* **46**: 617–628

Starck Z and Stradowska M (1978) Effect of salt stress on the harmonal regulation of growth, photosynthesis and distribution of ^{14}C assimilates in bean plants. *Acta Soc Bot Pol* **47**: 245–267

Stitt M, Gerhardt R, Wilke I and Heldt HW (1987) The contribution of fructose 2,6-bisphosphate to the regulation of sucrose synthesis during photosynthesis. *Physiol Plant* **69**: 377–386

Stitt M, Herzog B and Heldt HW (1984) Control of photosynthetic sucrose synthesis by fructose 2,6-bisphosphate. I Coordination of CO_2 fixation and sucrose synthesis. *Plant Physiol* **75**: 548–553

Turgeon R (1989) The sink-source transition in leaves. *Annu Rev Plant Physiol Plant Mol Biol* **40**: 119–138

Turgeon R, Webb JA and Evert RF (1975) Ultrastructure of minor veins in *Cucurbita pepo* leaves. *Protoplasma* **83**: 217–232

Turvey PM and Patrick JW (1979) Kinetin promoted transport of asimilates in stems of *Phaseolus vulgaris* L. Localized versus remote site of action. *Planta* **147**: 151–155

Van Bel AJE (1987) The apoplast concept of phloem loading has no universal validity. *Plant Physiol Biochem* **25**: 677–686

Vassey TL (1989) Light/dark profiles of sucrose-phosphate synthase, sucrose synthase and acid invertase in leaves of sugarbeets. *Plant Physiol* **89**: 347–351

Wagner H (1974) Hormone directed transport of assimilates in barley. *Angewandte Botanik* **48**: 331–378

Wardlaw IF (1967) The effect of water stress on translocation in relation to photosynthesis and growth. 1 Effect during grain development in wheat. *Aust J Biol Sci* **20:** 25–39

Wardlaw IF (1980) Translocation and source-sink relationships. In: (PS Carlson, ed) *The Biology of Crop Productivity.* Academic Press, New York, pp 297–339

Wardlaw IF (1985) The regulation of photosynthetic rate by sink demand. In: (B Jeffcoat, AF Hawkins and AD Stead, eds) *Regulation of Sources and Sinks in Crop Plants.* Monograph 12. British Plant Growth Regulator Group, Bristol, pp 145–162

Wareing PF, Khalifa MM and Treharne CJ (1968) Rate-limiting processes in photosynthesis at saturating light intensities. *Nature* **220:** 453–457

Warren-Wilson J (1966) Effect of temperature on net assimilation rate. *Ann Bot* **30:** 755–761

Warrier A, Bhardwaj SN and Pande PC (1987) Effect of benzyladenine on grain growth in *aestivum* wheat. *Plant Cell Physiol* **28:** 735–739

Went FW (1939) Some experiments on bud growth. *Amer J Bot* **26:** 109–117

Wright JP and Fisher DB (1981) Measurement of the sieve tube membrane potential. *Plant Physiol* **67:** 845–848

16

Photosynthetic Characteristics of Fruiting Structures of Cultivated Crops

Randhir Singh

Department of Chemistry and Biochemistry
Haryana Agricultural University
Hisar 125 004, India

CONTENTS

ABBREVIATIONS

CAM	:	crassulacean acid metabolism
Chl	:	chlorophyll
OAA	:	oxaloacetate
PCR	:	photosynthetic carbon reduction
PEP	:	phosphoenol pyruvate
PGA	:	phosphoglyceric acid
RuBP	:	ribulose-1, 5-*bis*phosphate
RuBPcase	:	ribulose 1, 5-*bis*phosphate carboxylase
TCA	:	tricarboxylic acid

ABSTRACT

Yield in crop plants is built through the photosynthetic activity of leaves, stem and fruiting structures. However, the contribution of these structures varies in different crops and is influenced by factors such as moisture, nutrition, and genotype. In cereals, net photosynthesis in reproductive organs is relatively high and probably contributes as much as 50–75% of the photosynthates to developing grains. However, in legumes, the reproductive structures are generally incapable of net photosynthesis. In rapeseed, virtually 100% of the seed dry matter comes from photosynthetic CO_2 assimilation of the pod. Studies on the activities of some key enzymes of the photosynthetic carbon reduction (PCR) cycle and C_4 metabolism, rates of $^{14}CO_2$ fixation in light and dark, and initial products of photosynthetic $^{14}CO_2$ fixation conducted with fruiting structures of various crops indicated that compared to activities of ribulose-1,5-*bis*phosphate carboxylase (RuBPcase) and other photosynthetic carbon reduction (PCR) cycle enzymes, the activities of phosphoenol pyruvate (PEP) carboxylase and other enzymes of C_4 metabolism are generally much higher in fruiting structure than in the leaf. Short-term assimilation of $^{14}CO_2$ by illuminated fruiting structures produces malate as the major labelled product with less labelling in 3-phosphoglyceric acid (3-PGA), whereas the leaf shows major incorporation into 3-PGA indicating that the fruiting structures use PEP carboxylase mainly to recapture the respired or photorespired CO_2. However, the pod-wall of *Brassica* seems to be an exception, and assimilates CO_2 via the reactions of the PCR cycle. Based on the information reviewed here, a model depicting carbon assimilation in fruiting structures is proposed. Further, it is argued that fruit photosynthesis does not resemble any of the well-characterized categories of photosynthesis, namely, C_3, C_4 or crassulacean acid matabolism (CAM). It appears that fruit photosynthesis has an intermediate status between C_3, non-autotrophic tissue and C_4/CAM photosynthesis.

I. Introduction

The photosynthetic contribution of fruiting structures to their own yield or to the yield of seeds which they contain has been the subject of considerable research (Singh, 1989). Though cereals have received the greatest attention in this regard, several studies during the last few years have dealt with other crops (Singh, 1987). Estimates of photosynthetic contributions from seed heads or fruiting structures to the final harvestable yield range from negligible in green lemons and green oranges (Allison and Watson, 1966; Todd *et al.*, 1961) to values of 8–23% for rice (Enyi, 1962; Takeda and Maruta, 1956), 9–76% for barley (Buttrose and May, 1959; Frey-Wyssling and Buttrose, 1959; Thorne, 1965), 10–44% for wheat (Thorne, 1965), 13% for soybean (Thorne, 1979), 15–18% for soybean (Allison and Watson, 1966; Fischer and Wilson, 1971; Goldsworthy, 1970), 20% for chickpea (Sheoran *et al.*, 1987), 25% for field peas (Flinn and Pate, 1970), 58–63% for oats (Jennings and Shibles, 1968), and 61–100% for rapeseed (Allen *et al.*, 1971; Singh *et al.*, 1986; Sheoran *et al.*, 1991).

In crops such as barley (Thorne, 1965), wheat (Evans and Rawson, 1970), rice (Enyi, 1962, and oats (Jennings and Shibles, 1968), the net photosynthesis in reproductive organs is relatively high. As much as 50–75% of photosynthates may be translocated directly to the developing grain. However, in legumes, this does not seem to be true (Queberdeaux and Chollet, 1975), and in a number of cases the reproductive organs are incapable of CO_2 uptake (Crookston *et al.*, 1974; Khanna-Chopra and Sinha, 1974; Queberdeaux and Chollet 1975; Singh and Pandey, 1980). On the other hand, the net CO_2 fixation by pod-wall during the early stages of seed development has been reported in chickpea (Sheoran *et al.*, 1987; Singh and Pandey, 1980), pea (Atkins *et al.*, 1977), and field pea (Flinn and Pate, 1970). These studies indicate that fruiting structure possesses the photosynthetic machinery and the activity seems important in the overall process of yield buildup in these crops. However, detailed photosynthetic studies akin to those carried out in leaves have not been conducted in fruiting structures. None the less, a few reports regarding photosynthetic characteristics of fruiting structures have appeared in recent years. This information has been reviewed here, and a comprehensive picture regarding the biochemical pathway of CO_2 assimilation in these structures is presented.

II. Sites of CO_2 Exchange

Anatomical observations regarding CO_2 exchange of bean pod were made as early as 1974 by Crookston and his colleagues (Crookston *et al.*, 1974). They found the pod-wall to be about eight times as thick as the leaf blade. Its tissue appears to be composed of two major types: chlorenchyma, which makes up the outer half, and water storage parenchyma, constituting the inner half. Chlorenchyma cells near the outer surface contain very few if any chloroplasts, while those deeper within the pod tissue contain numerous plastids. Chloroplast size and frequency increase gradually from the epidermis inward. Chloroplasts occurring near the epidermis have well-developed grana, and are approximately the same size as the chloroplasts in the leaf blade, but those located near the centre of the pod-wall (vascular region) are considerably larger, mainly due to increased quantities of starch. These have very few grana, and appear more like amyloplasts than normal chloroplasts.

The occurrence of numerous amyloplast-like plastids in the cells surrounding the veins of the pod, which are located deep in the pod tissue where light penetration is minimal, suggests that the substrate for starch formation by these plastids originates externally, probably from the leaf blade, and is then translocated into the pod. The pod, therefore, appears to be more of a sink than a source of carbohydrates.

The location, frequency, and type of plastids in the pod indicate only minimal incorporation of external CO_2 by the pod. Also, stomatal and surface examinations indicate that the pod's exterior is not structured for significant atmospheric CO_2 uptake. The outer epidermis, which contains functional stomata, is underlaid by a thick sclerenchyma layer. However, the stomata occur only 25% as frequently on the pod as they do on the lower surface of the leaf blade. Collectively, the anatomy of the pod of dry bean suggests that it is not an important photosynthetic source for the developing seeds. Instead, the pod, like the seeds, is actually a storage organ for assimilates in the leaf blades.

Atkins *et al.*, (1977) used light and electron microscopy to examine the pod-wall of pea and found three distinct tissue layers, exocarp, mesocarp and endocarp. The exocarp comprises of outer epidermis of well-cutinized cells containing few chloroplasts. Stomatal density is approximately 25% of that of an adjacent stipule or leaflet. The mesocarp consists of 15 to 20 layers of parenchyma cells that are well endowed with plastids. The high degree of vacuolation suggests the bulk of the pod's solute reserve is located in these layers. The mid-region of the mesocarp is traversed by a vascular network of which the tertiary veins project downward and terminate in the lowest layers of the mesocarp. Chlorophyll density is highest in the outermost layers of the mesocarp, while the density of starch is greatest in the inner layers, especially those close to the vascular strands.

The endocarp consists of three components: an inner epidermis, a mid-region of two to three layers of thin-walled parenchyma cells, and an inner layer of sclerenchyma, two to three cells thick. Chloroplasts are abundant in the inner epidermis but do not occur in the parenchyma and fibre layers. Chloroplasts in the inner epidermis are smaller than those in the mesocarp and contain smaller starch grains. All chloroplasts of the pod possess grana. The inner epidermis in contrast with the outer epidermis has no stomata (Atkins et al., 1977; Sheoran et al., 1987). Its cells have a thin cuticle and rounded outer contours. From these structural characteristics of the pea pod, it can be deduced that the outer layer, comprising chlorenchyma of the mesocarp, captures CO_2 from the outside atmosphere whereas in the inner layer lining the pod, gas cavity is involved in assimilation of CO_2 released from respiring seeds (Singh, 1989). Besides variations in size and shape of the pods, genetic variations are also known to occur with respect to thickness of the contributing wall layers (Price and Hedley, 1980). Similarly, distribution of chloroplasts in various pod layers varies between genotypes.

In the pod of broad bean or pea, stomata in the outer epidermis or exocarp are 27% or 25% as frequent as on the abaxial surface of the respective leaflet (Atkins et al., 1977; Willmer and Johnston, 1976), i.e., 19 stomata mm^{-2} in Vicia and 40 stomata mm^{-2} in Pisum. Microscopic studies have shown 12,000–14,000 stomata per pod, with a maximum density of 30–34 stomata mm^{-2} on developing pod; after anthesis stomatal density decreases to 24–26 stomata mm^{-2} when the pod is fully expanded. In the early stages of pod development, opening and closing of these stomata are known to be as sensitive as those of the leaves (Flinn et al., 1977).

Microscopic observations made by Bain and Mercer (1964), Rhodes and Wooltorton (1967), Clijsters (1969a), de Barsy (1976), Lenz and Noga (1982), and Cordes and Blanke (1987) in apple and pear have demonstrated the presence of chloroplasts and mitochondria within these fruits. The subject has been reviewed in detail by Pantastico (1975). The parenchyma is completely free of chloroplasts, but they are present in the green tissue of the hypodermal and inner perivascular tissue. Perivascular tissue of these fruits comprises four to five layers of cells surrounding the vascular bundle which is either without chloroplasts or with a relatively small number of large chloroplasts compared with the respective leaf. Hypodermal chloroplasts present in the five to six layers below the epidermis are smaller than in the inner tissue (Phan, 1973a). They closely resemble those found in leaves and exhibit grana throughout fruit development. Such granules are generally absent in the central, bundle sheath type tissue, but are present in the hypodermal, mesophyll type tissue in the mature apple fruit (Phan, 1970, 1973a). Based on the above chloroplast dimorphism, Phan (1973a) has suggested C_3 photosynthesis for the hypodermal and C_4 for the central chloroplasts (see Chapter by Raghvendra and Das, this volume).

Cordes and Blanke (1987) have shown that inner chloroplasts of apple fruit exhibit features of both shade and sun chloroplasts, i.e., shade chloroplasts with a low chlorophyll a:b ratio of 1.0–1.1, and sun chloroplasts with a low degree of stacking, large starch granules, and few thylakoids per granum (for discussion on dynamics of chloroplast structure see Chapter by Critchley and Russell, this volume. The granal structure of the inner chloroplasts disintegrates with fruit development, light scarcity, and adaption to shade. However, PS II activity commonly associated with grana continues to be present. This was indicated by the constant chlorophyll ratio of 1.0–1.1 in the peel and core, and also by continuous starch synthesis from photosynthetic processes (Phan, 1973b). With fruit ontogeny, grana gradually disorganize as light penetration into these tissues decreases. Inner central tissues in young fruit are more active photosynthetically than in mature fruit, but less active than the hypodermis (Phan, 1973b, 1975).

In apple fruit, inner chloroplasts are larger than those of the epidermis, and are generally flattened and elongated with parallel lamellae (Phan, 1975). Starch granules disappear from inner or hypodermal chloroplasts after fruiting branches are ringed or defoliated or after fruit is stored in the dark for several days (Cordes, 1988).

In general, early developmental stages of apple chloroplasts of the central perivascular tissue resemble those found in the general PEP-carboxykinase subtype, and taken as the agranal NADP malic enzyme subtype or CAM type as grana disintegrate. Microbodies are also present adjacent to internal chloroplasts and mitochondria. The hypodermal chloroplasts are elliptical or disc-shaped and are 2–4 μm in size (Bain and Mercer, 1964), i.e., smaller than those in leaves (Rhodes and Wooltorton, 1967). Their grana comprise only four to eight thylakoids (Clijsters, 1969b) in contrast with those commonly found in leaves which possess up to 100 thylakoids per granum (Tribe and Whittaker, 1982). At full maturity of fruits, chloroplasts become disoriented while mitochondria show no signs of a breakdown.

Apple fruits contain chlorophyll a and b and the carotenoids β-carotene, lutein violaxanthin, and neoxanthin. Carotenoids may function to protect chlorophyll against photooxidation and might also help chlorophyll in maximizing light utilization in the apple peel (Downs et al., 1965; Pantastico, 1975; Phan, 1975; Gross et al., 1978; Goodheer, 1979). This pattern of pigments indicates that both the photosystems are present in apple fruit (Downs et al., 1965).

Chloroplastic pigments of the apple fruit are present primarily in the perivascular tissue of peel and core, and are less concentrated and more unevenly distributed than in the leaf (Phan, 1975). Chloroplasts retain their chlorophyll up to a very advanced stage of ripening (Clijsters, 1969 a; Pantastico, 1975; Phan, 1975; Jones, 1981). The chlorophyll $a:b$ ratio decreases with fruit development. Chlorophyllase degrades chlorophyll a more than it does chlorophyll b (Rhodes and Wooltorton, 1967).

Stomata in apple fruit are uniformly distributed over the fruit surface and function like those in the leaves. The number of stomata per fruit appears to be determined before petal fall, thereafter remaining constant throughout fruit ontogeny. However, stomatal density is maximal around petal fall, with 2–10, 10–15 (Schwerdtfeger and Buchloh, 1968) or 10–20 (Blanke and Lenz, 1985; Blanke, 1986) stomata mm^{-2}. This decreases to 6–8 stomata mm^{-2} within a week, and to less than one stomata mm^{-2} as the cuticle of the apple continues to expand (Blanke and Lenz, 1985). This suggests that an apple of 10 mm diameter would have about 600–6000 stomata per fruit, which compares more closely with CAM leaves with 11–55 stomata mm^{-2} or an average of 25 stomata mm^{-2} than with the abaxial surface of apple leaves with 320–390 stomata mm^{-2}. Hence, stomata are at least 30 times more frequent on apple leaves than on the fruit.

The glumes of wheat ears have a length of approximately 9 mm and show stomata on the abaxial side only. At the tip (first 2 mm) the chlorenchyma is arranged diffusely and the stomata are distributed irregularly (Cordes, 1988). Further down, eight green longitudinal bands, or interveinal region, of varying length and width can be observed. In the outer bands, the stomata lie in two to three rows, one beside another, and continue from the tip almost to the base. The distance between the stomata within a row decreases from the outside to the inside. In the middle part of the glume, the stomata occur in four rows and are very close to each other within a row and can be found from the tip to the base. Stomata are located on the abaxial surface of the lemma in double rows on five symmetrically distributed bands. They start, somewhat irregularly distributed, below the tip of the lemmas, and extend to 3 mm above the base on both sides and to 4.5 mm above the base in the middle row. The distance between rows of stomata is wider than on the glumes, but stomata lie very close to each other within the rows. On the adaxial side of lemmas, stomata can be seen in double rows in six bands starting 2–3 mm from the tip (inner and outer rows, respectively), and going down to 1 mm from the bottom. The distance between rows is somewhat larger than on the abaxial side. The palea has abaxial stomata in one row on the keels only. Adaxial stomata can be found in two rows at the top part of the keels, in the lower part in one row only. Stomata are distributed over almost the complete length of the palea.

Oat glumes (with a length of 20–22 mm) show 11 symmetrically distributed bands of stomata on the abaxial side with two rows of stomata each. There are more stomata in the middle than at the edges. The area at the tip (2 mm wide) contains neither chlorenchyma nor stomata. Stomata are less dense on the adaxial surface, but more evenly distributed, and tend to be smaller on the abaxial side. Adaxial stomata start 5–7 mm below the tip (occur in two rows in parallel to the middle line) with a density comparable to that of the abaxial side and end close to the base. The lemmas (with a length of 13–15 mm) have two rows of stomata in the upper third of the abaxial side in six to seven symmetrically distributed bands of high regularity and high density. Adaxially stomata occur in double rows on seven bands. In contrast to the abaxial side, stomata start 2 mm from the tip and extend to near the base. Stomata on the adaxial side are smaller and less dense than those on the abaxial side with irregular distance in between.

The palea (with a length of 13–15 mm) has stomata on the keel only. Stomata occur on two rows on the abaxial side beginning 2 mm below the tip and extending to about 5 mm above the base. Adaxially, the stomata occur in one row at the tip and in two rows further down. They extend closer to the base on the abaxial side.

Stomata of barley can mainly be found on the awns. They start near the base in single rows on two bands; further up they appear in rows of two on the two bands. The distance between the two bands is smaller than on the outer bracts of all three species. Distances between stomata within rows are comparable to those for the bracts; while the palea shows stomata in double rows on the two keels adaxially as well as abaxially no stomata can be found on the abaxial side of the lemma. Stomata on the adaxial side of the lemma are arranged irregularly at the tip. Down from there, they are distributed in four bands with rows of two stomata. Stomata in the two bands at the edge of the lemma extend to the base of the bracts, while those on the inner bands end in the middle of the bract. Stomata of the lemma seem to be smaller than those of the awns, but this may be an artifact. The frequency and distribution of stomata vary depending upon the species and the cultivar.

Knoppik *et al.* (1986) observed a less distinct opening of stomata of wheat ears at low CO_2 partial pressure than can be found in flag leaves, while stomatal conductance of leaves and ears at high CO_2 were comparable. Tschakalova and Hoffmann (1976) reported stomata opening on leaves and ears of wheat to be minimum during the course of the day with stomata of higher inserted leaves being less sensitive (because of an increasingly xeromorphic structure) than those of lower ones. Hence, ears (as the highest inserted leaves) are likely to be the most xeromorphic ones, showing the less sensitive stomata. Stomata on the awns of barley and the glumes of oat open wider than do those of other parts of the ears (Ziegler-Jöns, 1989a).

The glumes of wheat ears show bundle sheath cells surrounding the big vascular bundles. The bundles occur in the spongy parenchyma and often consist of only two tracheids and one sieve-tube element. Bundle sheath cells look less lobed than mesophyll cells, the cell walls appear to be thicker, and the chloroplasts appear less dark, with light flecks which could be interpreted as starch grains. In contrast to the Kranz anatomy in C_4 leaves, chloroplasts can be found in the bundle sheath cells of the lemmas of wheat and oat. Bundle sheath cells also are often found in the zone between spongy parenchyma and sclerenchyma. They are, however, absent in the glumes of oat and in the awns of oat and barley. In barley, the outer epidermis exhibits up to 10 stomata, whereas in wheat the number can be as high as 30 stomata per fruit (Duffus *et al.*, 1985)

In *Brassica* pod-wall, there is normal distribution of stomata in the outer epidermis (Sheoran *et al.*, 1991). They are anemocytic, but with a lower frequency compared to leaf. There is large variation in the size of stomata. The surrounding cells are parenchymatous in nature, but with outer wall comparatively thick. The inner pod-wall epidermis is a continuous layer of parenchyma cells lacking stomata. These cells are uniformly thin-walled. A transverse section of the pod-wall reveals four to six layers of parenchyma cells with intermittent small vascular bundles below the outer epidermis. Next to the inner epidermal layer there is a single layer of sclerenchyma cells which run longitudinally parallel to the axis of the fruit. The sclerenchyma cells are knitted throughout, apparently making the pod cavity an airtight compartment. At a few places, instead of sclerenchyma tissue, an extension of the thin wall can be observed. The parenchyma cells are thin-walled and contain numerous chloroplasts. The quantity of chloroplasts decreases from the outer epidermal layer (maximum) to the inner layer.

III. Light Penetration and CO_2 Diffusion

Atkins *et al.* (1977) made estimates of the irradiance flux received by endocarp and seeds of pea fruits of varying age by measuring the transmission of photosynthetically active radiation. The incident flux on the surface of the fruit was 2200 μE m^{-2} sec^{-1}. Of this, 67% was absorbed by exocarp and mesocarp and 10% by the endocarp. Illumination of the seed in the intact pod cavity was found to be 23% of incident light at 27 days after anthesis. From 27 days onward, the proportion of incident radiation reaching the endocarp and seed increased significantly, probably due to a decline in chlorophyll content of outer layers.

As already discussed, garden pea has two photosynthetically active layers in the pod-wall. The outer layer is structurally and enzymically equipped for assimilating CO_2 entering the pod from the outside atmosphere via the stomata of the outer epidermis. The high density of chloroplasts in the inner epidermis, absence of stomata, thin cuticle, and dome-shaped outer contours of the inner epidermal cells are interpreted as specializations for photoassimilation of CO_2 from the fruit cavity.

The CO_2 concentrations in the cavities of pea fruits are always many times higher than the ambient air, varying from 0.5 to 1.5% (v/v) during the life of the fruit (Flinn et al., 1977). In chickpea, this concentration is as high as 2.5% (Sheoran et al., 1987). Concentrations of CO_2 are generally lower in young than in old fruits. However, changing temperature and light conditions during the day affect the CO_2 level. Pea fruit with well-developed seeds (25–36 days after anthesis) exhibit higher levels of gas space CO_2 when the fruit is exposed to 850 $\mu E\ m^{-2}s^{-1}$ than those at 2000 $\mu E\ m^{-2}s^{-1}$. If the fruit was completely shaded during a photoperiod at 22°C, CO_2 concentration in the cavities of the fruits was always higher than that of similarly aged fruits in the night at 15°C. There is a trend during fruit development for the net daily influx of CO_2 across the inner epidermis to increase relative to that through the outer epidermis, suggesting that CO_2 is fixed within the fruit wall. This increase is most noticeable when the seeds are filling rapidly and releasing large amounts of CO_2 to the fruit gas cavity (Flinn et al., 1977).

Light, essential for synthesis of chlorophyll and starch, is diffused throughout the parenchyma of the apple fruit (Phan, 1975). It penetrates through the central cavity and illuminates the internal parts. Blanke (1986, personal communication) measured light diffusion in developing apple fruits. When young fruits were exposed to 1750 $\mu mol\ quanta\ m^{-2}s^{-1}$ above the peduncle, at core level the light diffusion was below 0.1% of the initial light intensity and decreased with fruit development from 1 to 0.1 $\mu mol\ quanta\ m^{-2}s^{-1}$; in vascular tissue, light intensities were found more intense than in parenchyma. Kliewer and Smart (1988), while conducting experiments with grape berry, suggested that the phytochrome system may be operative in fruits.

In fruits, the CO_2 molecule passes several diffusive resistances before it is fixed in the chloroplast or cytosol during photosynthesis or released to the atmosphere after mitochondrial respiration. In apple fruits, CO_2 diffuses relatively freely in the intercellular system and as the fruit grows this internal space increases up to 20–25%, v/v (Ulrich and Marcellin, 1955; Westwood, 1978). With fruit ontogeny, an internal CO_2 concentration exceeding the ambient builds up by mitochondrial respiration of predominantly imported carbon. The CO_2 gradient is, therefore, from the inner cavity to the outer epidermis, i.e., in a direction opposite to that of the leaf (Henze, 1969a, b). In young fruits, open stomata are the preferred sites of CO_2 exchange with the atmosphere (Blanke et al., 1985). However, with fruit development, the diffusive resistance to CO_2 is largely controlled by the cuticular component (Juniper and Jeffree, 1983).

Ziegler-Jöns et al. (1987) conducted detailed experiments to answer three major questions: (1) Do single parts of the ear (glumes, paleae, and lemmas) differ in structure and gas exchange characteristics? (2) Do they compete with each other for light and/or CO_2? (3) Are they adapted to their specific light conditions? To address these questions, they gave two major treatments to the ears of the spring wheat: In treatment

A, all glumes were removed, whereas in treatment B, all paleae, lemmas, and the developing grains enclosed by the glumes were removed. These treatments were given between ear emergence and anthesis or 12 weeks after anthesis. On the following day, they measured CO_2 and H_2O exchange of treated and intact ears as a response to CO_2 partial pressure (80 µbar to 2000 µbar, PPFD = 2000 µmol $m^{-2}s^{-1}$) and PPFD (0 to 2000 µmol $m^{-2}s^{-1}$; CO_2 = 330 µl L^{-1}). The CO_2 uptake for all treatments increased by the same factor (approximately three-fold) when the CO_2 partial pressure was raised from neutral to saturating level, indicating that the removal of bracts has little influence on the CO_2 environment of the remaining ones. This further indicates that stomata and mesophyll cells with chloroplasts are positioned on parts of the bracts, where no restriction of the diffusive CO_2 transport by overlapping bracts takes place. Hence, the bracts in wheat ears do not compete for CO_2. Stomata are positioned to minimize restrictions to provision with CO_2. However, shading of lemmas and paleae by glumes causes them to be under suboptimal light conditions in young wheat ears. With advancing age, as the spikelet and floret open, the provision with light is increased and the photosynthesis rate of the inner bracts is enhanced. The CO_2 uptake does not increase as much as in young ears, when the glumes are removed. Light and CO_2 response of wheat ear indicate an adaptation of the inner bracts to the lower light regime in intact ears.

IV. CO_2 Exchange

The legume fruit (pod) exhibits only small net gain of CO_2 from the atmosphere, these being restricted to 10–20 days after anthesis of its parent flower (Flinn and Pate, 1970). At other times, net daily losses of CO_2 occur although assimilation may be above the compensation point during the photoperiod. In darkness, whole fruits, pods, and seeds of field pea show substantial rates of loss of CO_2 released to the atmosphere. Harvey et al. (1976) further demonstrated that the photosynthetic and respiratory potential of the developing fruit of *Pisum sativum* depends on fruit age, light intensity, and atmospheric CO_2 concentration. Whole fruits are capable of net CO_2 uptake from the atmosphere only during the period of pod extension growth and at light intensities above 250 µE $m^{-2}s^{-1}$. Thereafter, the respiratory CO_2 due to seed development exceeds the photosynthetic capacity of the pod.

The fruit CO_2 exchange response to changes in atmospheric CO_2 concentration is complex. CO_2 concentrations above 300 ± 20 ppm markedly suppress CO_2 evolution. Conversely, CO_2 concentrations below 280 ppm result in a substantial increase in the CO_2 output of the fruit. These opposed effects are observable in either the presence or absence of light but the rate of fruit CO_2 output is always greater in darkness. This is true for almost all legumes (Atkins et al., 1977; Flinn et al., 1977). In general, pea pods in the light have net photosynthesis rates near 10% of those found in leaves or 0–0.7 mg CO_2 fruit^{-1} h^{-1} until day 29 and, thereafter, net CO_2 loss. In the dark, however, pods lose 0.2–2.6 mg CO_2 fruit^{-1}h^{-1} with maximum CO_2 loss occurring about day 21 after anthesis, a time when seeds have ceased swelling and begin to dry out.

CO_2 exchange *via* the epidermis of the developing fruit is best documented for the apple fruit, but the results vary depending upon the technique employed to determine CO_2 exchange. The CO_2 accumulation and carboxylation data imply that *in vitro* respiration of the fruit is large relative to the CO_2 uptake, i.e., routinely measured using methods such as infrared gas analysis. This is mainly because the CO_2 exchange rates are similar to *in vivo* respiration rates during the early stages of fruit development. But with fruit development the internal gas space increasingly buffers CO_2, and the permeability of epidermis to gases decreases. This leads to lowering of the CO_2 exchange relative to the *in vivo* values. To distinguish it from respiration, Clijsters (1969b) demonstrated photosynthesis of apple fruits based on: (1) a decline in respiration of fruit transferred from the dark into the light, (2) the absence of this decline after application of 3-(3, 4-dichlorophenyl)-1, 1- dimethylurea (DCMU), and (3) an increase in CO_2 photoassimilation with increasing light intensity. Unlabelled CO_2 exchange measurements of the growing apple have also been reported by a number of other workers (Kidd and West, 1925, 1947; Kurssanow, 1934; Bain and Mercer, 1964; Jones *et al.*, 1964; Hulme *et at.*, 1967; Clijsters, 1969a, b, 1975; Phan, 1970; Jones, 1981; Lenz and Noga, 1982; Noga and Lenz, 1982; Blanke, 1988a, b).

The rate of CO_2 exchange decreases by about 10-fold during fruit ontogeny, particularly in fleshy fruits such as apple or grape. After anthesis, respiration of apple fruit declines from a maximum of 4 to 0.3 in the light or from 7 to 0.7 mg CO_2 fruit^{-1} h^{-1} in the dark before ripening. This is similar to a decline from 10 to 1 mg CO_2 in darkness with grape, while the rate in light remains constant around 0.5 mg CO_2 berry^{-1} h^{-1} (Leyhe; 1987, 1988). Contrary to this, CO_2 exchange in apple increases on a per fruit basis from 0.2 to 2 mg CO_2 in light or 1 to 4 mg CO_2 fruit^{-1}h^{-1} through the 16-times smaller surface for apple. On a per fruit basis, apple fruit and grape berry both lose most of the CO_2 at maturity, irrespective of their climacteric categorization (Clijsters, 1969a; Jones, 1981; Blanke, 1988a, b).

The CO_2 exchange of developing fruit is affected by a large number of factors, such as developmental stage, temperature, light, CO_2:O_2 ratio, and carbohydrate status (Noga and Lenz, 1982). A rise in temperature from 10° to 30°C increases apple fruit respiration by four to five times (Jones, 1981), an effect that needs to be considered while calculating daily CO_2 budgets for fruits. A progressive increase in photon flux density from 0 to 1300 μmol quanta m^{-2} s^{-1}, or increase in CO_2 concentration from 0.001 to 0.09% results in a progressive decline in fruit CO_2 loss. Before the fruit matures, there is transition from respiration to net photosynthetic CO_2 uptake (Blanke and Lenz, 1989). But the respiratory CO_2 loss, although reduced, does not switch to photosynthetic CO_2 uptake on maturity (Harvey *et al.*, 1976; Jones, 1981; Noga and Lenz, 1982; Leyhe, 1987). The reductions in respiratory CO_2 loss following exposure to increased CO_2:O_2 ratio is indicative of photorespiration, although respiration of pea fruit in the dark was also suppressed following exposure to enhanced CO_2 (Harvey *et al.*, 1976). In summer, internal CO_2 concentration in fruit tissue increases from 0.03% at anthesis to several per cent. Continuous respiratory CO_2 release, accentuated by increasing temperature together with decreasing permeability of the outer epidermis, occurs during fruit ontogeny. In apple fruit, this increase is from 1% to around 8–10 % (Smith, 1947; Hall *et al.*, 1955; Burg, 1965, 1967; Marcellin, 1975; Buffer, 1976).

In wheat ears (before anthesis), the light-saturated rates of net photosynthesis of inner and outer bracts sum to be 40% higher than that of the intact ear, where the inner bracts are not light-saturated. During grain filling, rates may be only 16% higher (Ziegler-Jöns, 1989a, b). Before anthesis, inner bracts contribute nearly the same amount as glumes to the CO_2 uptake of intact ears. During grain filling, the net photosynthesis rates of glumes decrease. This is compensated, in part, by increasing CO_2 uptake of the inner bracts. Because spikelets open during grain development, paleae and lemmas receive more light, so that rates for the intact ears are partly maintained. Two weeks after anthesis, removal of the glumes gives an increase in CO_2 uptake of the inner bracts from 6.7 to 8.5 nmol s^{-1} (27%) only, compared to a 75% increase in young ears.

Rates of oat ears are considerably higher than those of wheat (Table 1), mainly because oat has twice the number of spikelets and considerably larger glumes. The contribution of glumes to the photosynthesis of the intact ear is always higher in oat than in wheat. Photosynthetic activity of intact ears of oat decreases more between the first (before anthesis) and second (two weeks after anthesis) date of measurement than of wheat ears. This is due mainly to the strong decrease in photosynthesis of glumes but also to a loss of photosynthetic activity of inner bracts, as opposed to those of wheat. In barley, awns account for 87–90 % of total CO_2 uptake of ears under atmospheric conditions. At saturating CO_2, this percentage decreases to 81–85%, which means that awns are saturated at lower intercellular partial pressures of CO_2 than are other parts of the ear.

Table 1. Net photosynthesis rate under saturating light and normal atmospheric (A^{330}) or high (A^{2000}) CO_2 concentration of ears of wheat, oat and barley

Crop	Developmental stage	Net photosynthesis (Rate (nmol s^{-1}))		A^{2000}/A^{330}
		A^{330}	A^{2000}	
Wheat	Before anthesis	12.4 ± 1.6	34.6 ± 3.6	2.8
	After anthesis	11.0 ± 0.4	31.4 ± 1.0	2.8
Oat	Before anthesis	55.4 ± 6.0	122.3 ± 9.2	2.2
	After anthesis	31.8 ± 1.9	69.6 ± 3.7	2.2
Barley	Before anthesis	27.7 ± 1.6	61.4 ± 3.7	2.2
	After anthesis	18.8 ± 2.6	41.9 ± 5.2	2.2

Most important is the high contribution of awns to the gas exchange of barley ears (87% of the total CO_2 exchange before anthesis, 90% two weeks after anthesis). This is far more than for awns of wheat (Olugbemi et al., 1976). Lemmas and paleae of barley show lower rates relative to lemmas and paleae of wheat and oat. This may result from the fact that lemmas of barley have no stomata on the abaxial side and only four bands of chlorenchyma, and as a consequence, they also have the lowest chlorophyll content of all bracts. The contribution of lemmas and paleae is high in wheat and increases after anthesis. In oat, the contribution of the inner bracts is almost constant while that of the glumes is nearly halved during the first two weeks after anthesis. Grain growth reduces uptake of CO_2 in lemmas and paleae, because of blocking of stomata facing the grains. Such stomata are unable to exchange gas with the ambient air and take up only the CO_2 respired by the grains. This effect is greater in oat, where the grains grow fast and block 70–80% of the stomata of the paleae and lemmas two weeks after anthesis. In wheat, where the longitudinal growth of the grains is slower, the corresponding values are 50–60%, respectively, However, this effect in these crops does not result in a significant reduction of photosynthesis rates of the inner bracts in intact ears, because they are partly compensated for by the better supply of light after opening of the ear structures. In barley, however, this does not happen, as there is almost no shading of lemmas and paleae by the glumes in barley ears.

In oat, photosynthesis rates of the inner parts increase 2.8-fold to 3.1-fold when CO_2 partial pressure is raised from normal atmospheric 330 µl L^{-1} to saturating 2000 µl L^{-1}. This increase is comparable to that for lemmas and paleae of barley and all parts of the wheat ears, but is significantly higher than for glumes of oat and awns of barley. The fact that the A^{2000}/A^{330} ratio is the same for glumes and inner parts of wheat ears, was interpreted by Ziegler-Jöns (1989a) as a proof of equal supply of CO_2 to all parts of wheat ears.

V. Feeding $^{14}CO_2$ to Fruiting Structures

Atkins et al. (1977) conducted feeding experiments to investigate the extent to which legume-pod layers contribute to assimilation of CO_2 absorbed either through the outer or through the inner epidermis. After 15 seconds of labelling, 42% of the ^{14}C recovered in ethanol-soluble compounds was in the endocarp, the remaining 58% in the exocarp and mesocarp. These proportions changed only slightly up to six minutes, suggesting that steady rate labelling had been achieved, and that proportion reflected the extent to which the layers participated in assimilating CO_2 from the gas cavity. In a similar study with pods conducted at 2200 µE m^{-2} s^{-1}, a value close to full sunlight, the rate of $^{14}CO_2$ fixation from the gas space was the same as at 850 µE $m^{-2}s^{-1}$, but the proportion of the ^{14}C fixed by the endocarp was increased to 66%. This suggests that under normal canopy conditions, a supply of CO_2 to the outside of fruit in a stream of air 10.035% CO_2 (v/v) resulted in the outer layers of the pod achieving the highest levels of label. By six minutes after feeding, the exocarp and mesocarp had acquired 96% of the ^{14}C recoverable in ethanol-soluble compounds; the remaining 4% was in the endocarp. About 70% of the ^{14}C in the exocarp and mesocarp was recovered from the outer 100 µm of the pod's thickness, the region where most of the chlorophyll in the outer layers was located. These results prove that in pea pods, the outer layer is responsible for the net daytime gains of CO_2 made by the fruit from the external atmosphere. The inner epidermis assimilates a major fraction of CO_2 released by seeds when a fully enlarged fruit is exposed to radiant flux equivalent to that of sunlight.

Flinn et al. (1977) indicated that carbon fixed photosynthetically in pea pod walls is translocated to the seeds. Most of this photosynthate is donated during the second half of fruit development and it relates especially to uptake of CO_2 by the pod from the fruit gas cavity. On the basis of labelling studies, photosynthate is likely to be generated in greater amount in the inner epidermis than in outer layers. The inward projecting veinlets terminating close to the endocarp might be regarded as a specialization for photosynthate retrieval from the inner epidermis, although assimilate must still traverse at least several layers of unspecialized cells, including the fibres of the endocarp in their passage to vascular elements.

Studies of the early products of $^{14}CO_2$ fixation by the inner epidermis of the pea pod have shown that a significant proportion of the ethanol soluble ^{14}C is fixed into malate and aspartate within 15 seconds after labelling (Atkins et al., 1977). Singal et al. (1986a) studied the distribution of radioactivity between malate, 3-PGA, sucrose, and glucose-6-P obtained as products of CO_2 assimilation in chickpea pods after 20, 40, 60, 120, and 300 seconds of incubation. After 20 seconds of $^{14}CO_2$ fixation in pod-wall, seed coat, and leaf, about 82, 80 and 8% was in malate, whereas 9, 10, and 77%, respectively, was in 3-PGA. The remaining label was in sucrose and glucose-6-P. The ratio of malate to 3-PGA was 9.1, 8.0, and 0.1 in pod-wall, seed coat, and leaf, respectively. Sucrose was the major labelled product in leaf after 120 seconds, but the level of labelled malate still remained higher than that of sucrose in spite of an increase in the level of labelled sucrose and a decrease in labelled malate. At 300 seconds, sucrose accounted for 70–90% of the total ^{14}C recovered.

Uptake of labelled CO_2 has been used quite extensively to assess photosynthesis in fruits (Allentoff et al., 1954; Huffaker and Wallace, 1959; Bean and Todd, 1960; Clark et al., 1961; Bean et al., 1963; Clark and Wallace, 1963; Wienicke, 1968; Young and Biale, 1968). After detached mature apples were exposed for 18 hours to 0.03, 0.05, 1, or 5% of CO_2 during darkness, 61% was found in malate, 7% in aspartate, and 2% in glutamate. Rate of CO_2 uptake were 0.05 mg CO_2 fruit^{-1} d^{-1} under ambient CO_2 whereas respiration for the same fruit was 20 mg CO_2 fruit^{-1}d^{-1}, i.e., the CO_2 uptake was 5% of the respiration. According to Wienicke (1968), the apple fruit fixes 0.01–0.05 mg CO_2 fruit^{-1} d^{-1} at earlier stages of ripening.

Willmer and Johnston (1976) conducted feeding experiments with tomato both in light and in darkness. The major compound labelled in both light and darkness was malate. In the light, the level of labelled malate decreased with time. Concomitantly, there was increase of label in sugar, mono- and diphosphates, glutamine, and alanine. After 10 second exposure to $^{14}CO_2$, about 20% of the label was found in PGA, which also tended to decrease with time. In darkness, malate, citrate, and aspartate were the only labelled compounds detectable, and the amount of label in each changed little with time. These data suggest that CO_2 is fixed by PEP carboxylase into oxaloacetate during darkness, which in turn is converted to malic acid via NADP-malate dehydrogenase.

Duffus and Watson (1988) showed immature detached cereal caryopses from barley and wheat to be capable of fixing externally supplied $^{14}CO_2$ in light and dark. They further indicated that the labelled photosynthates produced in light or darkness could be detected in the transparent layers, cross-cells, and endosperm/embryo, thus indicating that carbon can be transferred from the atmosphere to the interior of the grain. The amount of CO_2 fixed by both wheat and barley caryopses in 15 minutes is very similar. However, the distribution within the caroypses is rather different with more than 73% of the total being present in inner layers of barley but only around 40% in wheat. This may occur because the cross-cell layer in the wheat caryopses is only one cell thick but in barley it is two to three cells thick (Duffus and Cochrane, 1982). Thus, while the wheat cross-cell layer fixes as much CO_2 as barley, it may not have the same capacity for storage of assimilate, and the excess may be transferred to the transparent layer. More ^{14}C is recovered from caryopses when they are incubated in $^{14}CO_2$ without the transparent layer, thus suggesting that this layer is a barrier to the uptake of CO_2. In all cases, significant amounts of ^{14}C labelled material are found in caryopses after dark incubation with $^{14}CO_2$. Interestingly, CO_2 fixation in the chlorophyll-less mutant *Albino lemma* is significantly greater in the light than in darkness. These results indicate that intact caryopses have the ability to translocate ^{14}C–labelled assimilate derived from external CO_2 to the endosperm/embryo. Based on a number of assumptions, it is presumed that the amount of CO_2 fixed is sufficient to account for about 2% of the weight of starch found in the caryopses.

As in legumes, Singal *et al.*, (1986b) studied distribution of radioactivity between malate, 3-PGA, sucrose, and glucose-6-P obtained as products of $^{14}CO_2$ assimilation in ear parts of wheat and also of 20, 40, 60, 120 and 300 seconds of incubation in light. After 20 seconds of photosynthesis, 87, 81, 81, and 3% was in malate, whereas 8, 11.1, 11, and 55% was in 3-PGA in pericarps, glumes, awns, and flag leaves, respectively. Thus, the ratio of $C_4{:}C_3$ products was 10.9, 7.6, 7.5, and 0.05 in pericarps, glumes, awns, and flag leaves, respectively. After 120 seconds, sucrose was the major labelled product, and after 300 seconds it accounted for 89% of the total ^{14}C recovered. Recently, detailed photosynthetic experiments have been conducted with pericarp of the developing wheat grains (Caley *et al.*, 1990). Similar results as in wheat have also been obtained with rice panicles (Imaizumi *et al.*, 1990).

In *Brassica*, of the $^{14}CO_2$ fixed after 15 seconds by leaf, pod-wall, and seed, about 82, 77, and 4%, respectively, of the total radioactivity appeared in 3-PGA, whereas 7, 11, and 86%, respectively, was recovered in the C_4-acid malate (Singal *et al.*, 1987). During a chase with $^{14}CO_2$, 3-PGA in leaf and pod-wall lost radioactivity more rapidly than did malate in seeds. In all parts, however, sucrose increased with increasing chase time; the rate being relatively slow in seeds. After 120 seconds of chase, sucrose accounted for about 86, 80, and 61% of the total radioactivity in leaf, pod-wall, and seed, respectively.

VI. Enzymes of CO_2 Assimilation

The activity of RuBPcase in wheat and chickpea was higher in the leaf blade than in the fruiting structures. In *Brassica*, however, the activity was comparable in the leaf and pod-wall. The activity of $NADP^+$-glyceraldehyde-3-P-dehydrogenase and ribulose-5-P-kinase was also higher in the leaf blades than in the fruiting structures (Singal *et al.*, 1986a, b, 1987; Singh, 1987). In contrast to RuBPcase, PEP carboxylase of wheat and chickpea was more active in fruiting structures than in leaf blades. However, the activity in pod-wall and leaf of *Brassica* was more or less comparable. In fruiting structures of wheat and chickpea and in seeds of *Brassica*, the specific activity of PEP carboxylase exceeded that of RuBPcase; NADP-malate dehydrogenase and $NADP^+$-malic enzyme behaved in a manner similar to that of PEP carboxylase (Table 2).

The developing fruit of most species is largely non-autotrophic and imports most of its carbon from the adjacent leaves. Fruit fixes very little CO_2 from the ambient

Table 2. Activities of key enzymes of Calvin and C_4 metabolism in leaf and fruiting structures of wheat, chickpea and *Brassica* at 10 Days after flowering (nmol min^{-1} mg^{-1} protein)

Enzyme	Wheat				Chickpea			Brassica		
	Flag leaf	Pericarp	Awns	Glumes	Leaf	Pod-wall	Seed	Leaf	Pod-wall	Seed
RuBP carboxylase	281	8	94	45	403	73	18	1842	1149	28
NADP⁺-glyceraldephosphate dehydrogenase	1027	40	429	308	1780	257	104	1005	629	54
Ribulose-5-phosphate kinase	4149	621	2632	1765	4563	915	231	5085	3378	72
PEP carboxylase	32	76	145	55	25	74	152	28	41	138
NADP⁺-malate dehydrogenase	30	82	155	45	20	30	25	19	22	44
NADP⁺ malic enzyme	16	38	19	102	30	37	35	7	11	27

From: Singh (1987)

404

atmosphere via RuBPcase ralative to the respective leaf blade. The activity of RuBPcase in pea pod-wall is 10–100 times less than the activity of the respective leaf (Hedley *et al.*, 1975). In the green pericarp of wheat grain, aspartate, citrate, malate, and glutamate are the major labelled products of $^{14}CO_2$ feeding after one minute exposure in the light (Aoyagi and Bassham, 1984). Similar labelling pattern was obtained when fruiting structures from legume and fruit crops were fed labelled CO_2. However, the *Brassica* pod seems to be an exception to this general phenomenon (Singal *et al.*, 1987; Sheoran *et al.*, 1991).

PEP carboxylase to RuBPcase ratios on a protein basis for tomato fruit and citrus flower bud vary from 1.7 to 14 and 4 to 5, respectively (Bravdo *et al.*, 1977; Laval-Martin *et al.*, 1977; Vu *et al.*, 1985), compared with values of 0.1 for the leaf (Vu *et al.*, 1985). This shows that fruiting structures use PEP carboxylase for assimilation of CO_2 released as a result of respiration or photorespiration, and CO_2 can accumulate to levels as high as 1–8% in apple or 0.2–3.2% in legume pods. The CO_2 fixed as above could be released within the cell by the action of $NADP^+$ malic enzyme and could either be reduced via the PCR cycle or be used in the generation of carbon skeletons as shown in Fig. 1. The various enzymes from this pathway have been isolated and characterized from a number of fruit tissues (Frey-Wyssling and Buttrose, 1959; Flinn and Pate, 1970; Flinn *et al.*, 1977; Goldsworthy, 1970; Goodheer, 1979; Gross *et al.*, 1978; Dhillon *et al.*, 1985; Singal and Singh, 1986; Blanke *et al.*, 1986, 1987a, b, 1988; Gupta and Singh, 1988, 1989, 1990).

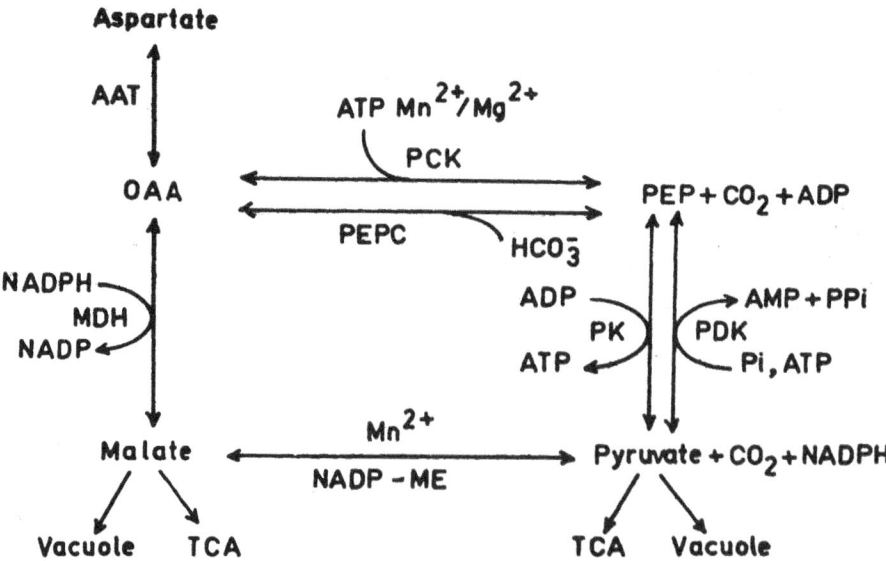

Figure 1. Malate-CO_2 shuttle and metabolic pathway of CO_2 fixation by PEP carboxylase in fruiting structures. OAA, oxalacetate; MDH, malate dehydrogenase; TCA, tricarboxylic acid cycle; PEPC, PEP carboxylase; PCK, PEP carboxykinase; PK, pyruvate kinase; PDK, pyruvate Pi dikinase; NADP-ME, NADP malic enzyme.

Table 3. Kinetics of fruit PEP carboxylase compared with that from tissues of defined photosynthesis pattern

	Photosynthesis					
	? Apple		Heterotrophic potato tuber	C_3 spinach leaf	CAM pineapple leaf	C_4 maize leaf
	Fruit	Seed				
Km (PEP)*	0.09	0.09	0.07	0.088	0.25	2.15
Km (HCO$_3$)*	0.20	0.20	0.47	0.36	0.043	0.40
Ki (HCO$_3$)*	106	115	78	81	80	86
Ki (malate)*	19	19	44	37	12	16
Vmax**	9	25	36	64	34	1470
Vmax***	19	50	18	8	95	368

* Constants expressed in mol m^{-3} substrate or inhibitor concentration, respectively
** Maximum enzyme activities expressed in nmol HCO$_3^-$ min^{-1} fresh weight^{-1} or
*** mg extracted protein
(From: Blanke *et al.* (1986); Blanke *et al.* (1987a, b); Blanke and Lenz (1989)

The enzyme PEP carboxylase has been characterized from a number of leaf and non-leaf sources (Blanke *et al.*, 1986, 1987a, b, 1988; Blanke and Notton, 1991; Singal and Singh, 1986). Apple fruit PEP carboxylase has a relatively low Km (0.20 mol m^{-3} HCO$_3^-$) and a large Ki (106–110 mol m^{-3} HCO$_3^-$) (Table 3) compared to other PEP carboxylases (Blanke *et al.*, 1987a). If the internal CO_2 is considered to be 8–10% in apple, the enzyme from this source is probably not inhibited by endogenous CO_2-bicarbonate concentrations in the fruit. PEP carboxylase from chickpea pod is activated by inorganic phosphate, and phosphate esters like glucose-6-P, α-glycerophosphate, 3-PGA, and fructose-1, 6-P$_2$, and inhibited by nucleotide triphosphates, organic acids, and the divalent cations Ca^{2+} and Mn^{+2}. Oxaloacetate and malate inhibit the enzyme non-competitively. Glucose-6-P reverses the inhibitory effects of oxaloacetate and malate. Under saturating PEP concentration, *in vitro* PEP carboxylase from apple shows a non-competitive relation between bicarbonate and PEP (Blanke *et al.*, 1987a, b).

The enzyme PEP carboxylase is present in almost all fruit tissues. However, its level and activity are not uniformly distributed. The enzyme is mainly concentrated in seeds and perivascular tissues such as the hypodermal and inner (core) perivascular tissue of apple, the exocarp and mesocarp of legume pod-wall (the mesocarp containing the vascular bundle), and the cross–cells of cereal grains which are situated between the inner pericarp, the vascular bundle, and the testa. Activities vary during the ontogeny of fruit tissues (Blanke and Lenz, 1989). Activity of PEP carboxylase increases with fruit ontogeny in apple fruit and pea cotyledons but remains constant in the grape berry. The detailed changes with ontogeny are described by Singal and his colleagues in fruit tissues of legumes (Singal *et al.*, 1986a), cereals (Singal *et al.*, 1986b), and *Brassica* (Singal *et al.*, 1987). The activity decreases if expressed on a per fruit or per g fresh weight basis as in the case of tomato and pea pod, respectively.

In tomato, PEP carboxylase activity is highest in young fruit with 0.2 μg CO_2 decreasing to 0.01 μg CO_2 fruit^{-1} h^{-1} as the fruit matures (Lavall–Martin *et al.*, 1977), whereas it remains constant around 1–1.5 μg CO_2 g^{-1} fresh weight in the grape berry (Ruffner and Kliewer, 1975). In pea, the activity increases in the cotyledons, i.e., the major part of the seeds, from 0.15 to 1.5 µg CO_2 g^{-1} fresh weight h^{-1} within 70 days from anthesis (Hedley *et al.*, 1975), whereas the activity decreases in the pod-wall from 0.5 to 0.01 µg CO_2 g^{-1} fresh weight h^{-1} (Hedley *et al.*, 1975) or 8.8–1.5 µg CO_2 g^{-1} dry matter h^{-1} (Price and Hedley, 1980). The pattern of activity in cotyledons is opposite to that found in the developing pod-wall and testa. If expressed on a pod basis, the activity is maximum (0.7–0.9 µg CO_2 h^{-1}) at 10–15 days after anthesis (Atkins *et al.*, 1977; Price and Hedley, 1980).

In cereals, the enzyme PEP carboxylase is present in all parts of the grain, i.e., glume, lemma, palea, and pericarp. In general, the activity is much higher in the pericarp (0.4–15 µg CO_2 mg^{-1} protein or chlorophyll h^{-1}) than in the glume, lemma, and palea, or in leaves with maximum activities of 4 µg CO_2 mg^{-1} protein h^{-1} (Wirth *et al.*, 1977; Duffus *et al.*, 1985).

The activity of NAD-malate dehydrogenase is much higher (10–100 times) in fruits than that of PEP carboxylase (Blanke, 1985; Laval-Martin *et al.*, 1977; Ruffner and Kliewer, 1975), varying again with developmental stage and fruit tissue or unit. In tomato, the ratio of NAD-malate hydrogenase to NADPH-malate hydrogenase is 70:1 or 50:1 in the fruit or leaf, respectively on a protein basis (Wilmer and Johnston, 1976). With NAD-malate hydrogenase activities being higher and PEP carboxylase activities lower than in leaves of C_4 plants, the malate hydrogenase to PEP carboxylase ratio is higher in fruits than in leaves of C_4 plants.

VII. Proposed Metabolic Path of CO_2 Assimilation

Based on the foregoing discussion (low RuBPcase activity and large ratio of PEP carboxylase: RuBPcase in fruiting structures compared to leaf), it could conclusively be said that these structures assimilate CO_2 by different mechanism than is known for leaves of either C_3 or C_4 plants. Accordingly, we have proposed a sequence of reactions, whereby respired or photorespired CO_2 is repeatedly fixed by PEP carboxylase into oxaloacetate which in turn is either transaminated to aspartate or converted to malate via NAD-malate hydrogenase. Malate so formed could be either stored in vacuoles or converted to pyruvate through NADP-malic enzyme. Alternatively, malate could also enter the tricarboxylic acid (TCA) cycle for the production of energy and carbon skeletons. Pyruvate could either be converted to PEP through the action of pyruvate Pi–dikinase or could give rise to acetyl CoA, which could then enter the TCA cycle. Pyruvate could also be transaminated to alanine. PEP could arise either through glycolysis or from oxaloacetate through PEP carboxykinase reaction and pyruvate through dikinase reaction. CO_2 released through the action of malic enzyme could again be refixed through PEP carboxylase reaction.

This malate-CO_2 shuttle may regulate utilization of energy, pH, aspartate synthesis and, to a limited extent, internal CO_2 concentration within the fruit. The C_3 products (PEP or pyruvate) of the decarboxylation or deacidification may also be converted to sugars via gluconeogenesis (Fig. 2). Pyruvate Pi-dikinase required to convert pyruvate to PEP is now known to occur in green grains of cereal grasses (Meyer *et al.*, 1982) and in fruits of wheat, pea, bean, plum, and castor bean (Aoyagi and Bassham, 1984). The enzyme from wheat grains resemble C_4 leaf (maize)

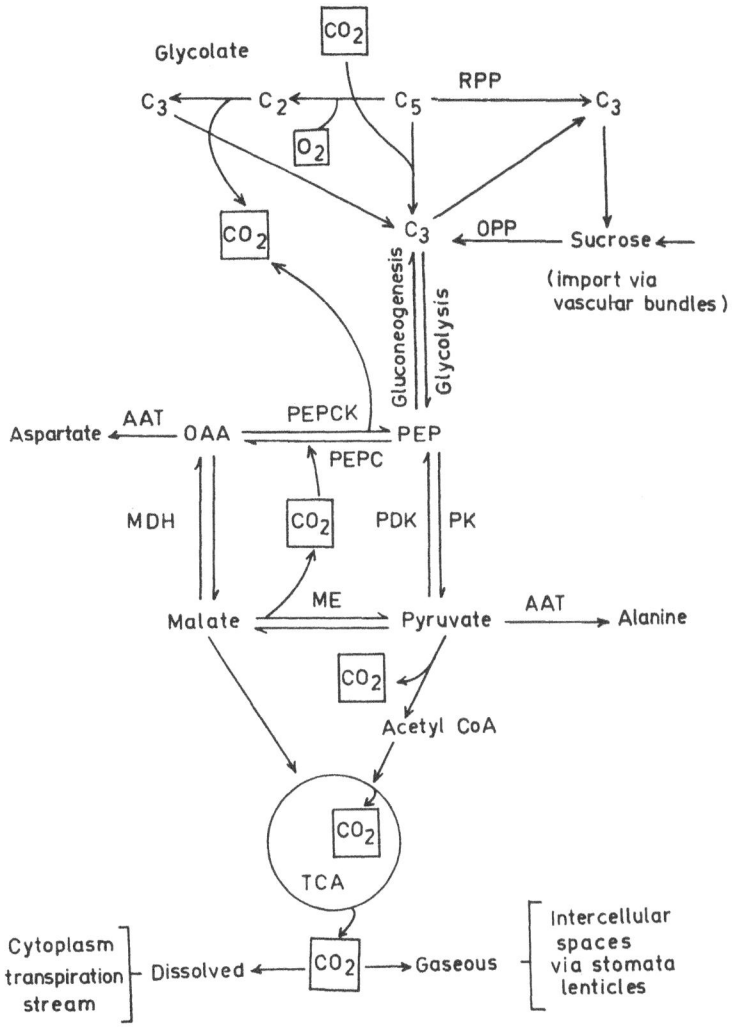

Figure 2. Proposed metabolic pathway of CO_2 assimilation in fruiting structures. AAT, aspartate aminotransferase; PEPCK, PEP carboxykinase; PEPC, PEP carboxylase; PK, pyruvate kinase; PDK, pyruvate Pi dikinase; ME, malic enzyme; AAT, alanine aminotransferase; RPP, reductive pentose phosphate pathway; OPP, oxidative pentose phosphate pathway.

enzyme with respect to pH optima and molecular weight (Aoyagi and Bassham, 1984). However, *in vitro* light activation of the fruit enzyme, as shown for leaf enzyme, is not certain. This enzyme in cereal grains has been suggested to play a role in amino-acid interconversion (Meyer *et al.*, 1982). It may also contribute to amino-acid synthesis in providing PEP for carboxylation forming oxaloacetate, which could then be aminated to form aspartate, utilizing imported alanine (Meyer *et al.*, 1982). This indicates that pyruvate Pi-dikinase in fruiting structures may not be primarily involved in photosynthesis as is the case in leaves of C_4 plants.

The reaction catalysed by NADP$^+$ malic enzyme may act as the source of NADPH in fruits (Latzko and Kelly, 1983). Pyruvate and NADH may arise through glycolysis. The activity of the two enzymes, i.e., NAD and NADP malic enzymes (components of the malate-CO_2 shuttle) in fruits increases on maturity (Hulme *et al.*, 1963, 1967), with a possible takeover from dominantly NAD malic enzyme to NADP malic enzyme. In young apple fruit, the activity of PEP carboxykinase is higher than that of PEP carboxylase (Blanke *et al.*, 1988). However, in maturing fruits the reverse is the trend. Contrary to this, grape has higher activity of PEP carboxylase.

VIII. Type of Fruit Photosynthesis

It is still not clear into which category the fruit photosynthesis falls. Factors such as the presence of β-carboxylation and a CO_2 concentrating mechanism, enhanced activity levels of the C_4 enzymes PEP carboxylase, PEP carboxykinase, malate dehydrogenase and pyruvate Pi-dikinase, affinity constants for pyruvate Pi-dikinase, and increased concentrations of C_4 organic acids such as malic acid (Blanke and Lenz, 1985) are indicative of C_4 type of photosynthesis in fruits. However, the following criteria must be fulfilled to characterize a type of photosynthesis as C_4: (1) the primary initial product of CO_2 fixation should be 4-carbon dicarboxylic acids (oxaloacetate, malate, aspartate); (2) the fixation of CO_2 as above should occur in light; (3) carbon from C_4 acids should be donated into the PCR cycle (C-4 of C_4 acids to C-1 of 3-PGA); (4) the tissue should possess the Kranz anatomy, i.e., there should be two distinct photosynthetic cell types, designated mesophyll and bundle sheath cells or Kranz cells. Kranz cells in this system are always internal to mesophyll and external to xylem and phloem.

The various properties of fruit photosynthesis are listed in Table 4. The structure of fruit chloroplasts differs from the Kranz anatomy associated with C_4 photosynthesis, but resembles those in C_3 or CAM leaves. In fruits, chloroplasts are concentrated in the perivascular tissue and partly adapt to the variation in light

Table 4. C_4, CAM, C_3, C_3/C_4-intermediate and non-autotrophic properties of fruits

C_4	CAM	C_3, C_3/C_4-intermediate or non-autotrophic
Anatomy	Chloroplast structure	
—	Low stomatal frequency	Lower chlorophyll *a:b* ratio
—	Large diffusive resistance	—
Physiology	Light—CAM	
	Internal 1–4% CO_2, exceeds ambient	High CO_2 compensation point
—	—	—
	Low transpiration:photosynthesis ratio	
Biochemistry		
—	Primary CO_2 fixation and $\delta^{13}C$	
	Large activity of C_4/CAM enzymes, β-carboxylation	Kinetics of PEP carboxylase
	Concentration of C_4 metabolites, e.g., malic acid	—
No diurnal pH fluctuation	Vacuole stores malic acid, malate is intermediate, not end-product	No diurnal pH fluctuation

penetration as the fruit develops. The stomatal frequency compares more closely with stomatal densities of CAM tissue. The chlorophyll $a:b$ ratio of around 1 in apple is lower than that in C_3 leaves with values of 2.6 (Willmer and Johnston, 1976) and C_4 leaves with 3.9 (Black, 1973). The CO_2 gradient from the inside of the fruit to the outside is in reverse order of leaves with C_3 photosynthesis where the ambient CO_2 concentration usually exceeds the internal CO_2 concentrations.

Fruit photosynthesis appears to have features resembling CAM photosynthesis as indicated by malic acid accumulation, dark CO_2-fixation, and low stomatal frequency. This should associate with diurnal pH fluctuations (Kluge and Ting, 1978), a phenomenon which does not exist in fruits (Willmer and Johnston, 1976; Blanke, 1985; Kriedemann, 1968). This indicates that in fruits cellular pH is regulated by means other than those found in CAM. Since malate rather than malic acid is the major labelled compound accounting for 61% of the total ^{14}C assimilated into apple fruit; this may be the reason for not observing any change in pH. The activities of the enzymes of malate-CO_2 shuttle, namely, PEP carboxylase, PEP carboxykinase, malic enzyme, and NAD-malate dehydrogenase, reach peak value at maturity (Blanke et al., 1982; Blanke, 1987; Hulme et al., 1967), and are comparable to the levels found in leaves of C_4 plants. However, the Km values for PEP, etc. are certainly of non-C_4 type.

From the above discussion, it is apparent that fruit photosynthesis does not resemble any of the well-characterized categories of photosynthesis, namely, C_3, C_4, or CAM. It differs from C_3 photosynthesis because of β-carboxylation and large activities of enzymes utilized therein, large internal CO_2 concentration, and large diffusive resistance of the fruit epidermis; from C_4 photosynthesis because of the absence of Kranz anatomy, reversed sequence of carboxylation, and affinity constants of fruit PEP carboxylase; and from CAM photosynthesis because of the absence of diurnal pH fluctuation and internal accumulation of CO_2 at night in the pod. From the information presented here, it appears that fruit photosynthesis has an intermediate status between C_3, non-autotrophic tissue and C_4/CAM photosynthesis.

IX. Future prospects

Stomata are scarce on the surface of fruiting structures, which imposes a question regarding their role in physiological processes. How CO_2 exchange during the light period occurs through these stomata needs to be investigated. This could be worked out by use of plant growth regulators which either close stomata in light or open them in darkness. Whether photorespiration is operative in these parts is yet to be studied. This could be done by measuring fruit respiration under variable CO_2 concentrations versus light intensities, application of photorespiratory inhibitors, or extraction of key enzymes in the glycolate pathway. Similarly, no work on light aspects of fruit photosynthesis has been carried out. This needs detailed investigations parallel to those available for leaf photosynthesis. Similarly, regulation of carbon metabolism needs to be studied so as to have a clear picture regarding the role of these plant parts in yield buildup for cultivated crops. Immunological comparisons between corresponding enzymes involved in photosynthesis of other tissues may reveal some

410

interesting differences between the structure and function of the enzymes, and help
to interpret the present data on fruit photosynthesis.

X. References

Allen EJ, Morgan DG and Ridgman WJ (1971) A physiological analysis of the growth of oilseed rape.
J Agri Sci **77**: 339–341

Allentoff N, Phillips WR and Johnston FB (1954) A ^{14}C study of carbon dioxide fixation in the apple.
J Food Sci Agric **5**: 231–238

Allison JCS and Watson DJ (1966) The production and distribution of dry matter in maize after flowering.
Ann Bot **30**: 365–381

Aoyagi K and Bassham JA (1984) Pyruvate orthophosphate dikinase of C_3 seeds and leaves as compared
to the enzyme from maize. *Plant Physiol* **75**: 387–392

Aranus JL, Alegre L, Tapia L and Calafell R (1986) Relationship between leaf structure and gas exchange
in wheat leaves at different insertion levels. *J Exp Bot* **37**: 1323–1333

Atkins CA, Kuo J, Pale JS, Flinn AM and Steele TW (1977) Photosynthetic pod-wall of pea *(Pisum sativum
L.). Plant Physiol* **60**: 779–786

Bain JM and Mercer FV (1964) Organization resistance and the respiration climacteric. *Aust J Biol Sci*
17: 78–85

Bean RC and Todd GW (1960) Photosynthesis and respiration in developing fruits. 1 $^{14}CO_2$ uptake by
young oranges in light and dark. *Plant Physiol* **35**: 425–429

Bean RC, Porter GG and Barr BK (1963) Photosynthesis and respiration in developing fruits. III Variations
in photosynthetic capacities during colour change in citrus. *Plant Physiol* **38**: 285–290

Black CC (1973) Photosynthetic carbon fixation in relation to net CO_2 uptake. *Annu Rev Plant Physiol*
24: 253–286

Blanke MM (1985) C_4/CAM photosynthese-enzyme beider CO_2 refixierung der Apfelfrucht. Dissertation,
Institut fur Obstbau, Universitat Bonn.

Blanke MM (1986) Comparative SEM study of stomata on developing quince, apple, grape and tomato
fruit. *Angewandte Botanik* **60**: 209–214

Blanke MM (1987) Comparative SEM study of stomata and surface morphology in apple. *Angewandte
Botanik* **61**: 433–438

Blanke MM (1988a) Bioenergetics of apple fruit respiration. *Plant Physiol Biochem* **26**: 245–252

Blanke MM (1988b) Wieviel CO_2 veratmet eine apfel frucht wahrend ihrer entwicklung. *Erwerbsobstbau*
30: 68–70

Blanke MM (1988c) Fruit photosynthesis In: (CJ Wright, ed) *Manipulation of Fruiting*, Proc. 47 Easter
School, University of Nottingham, Sutton Bonington, Loughborough, UK.

Blanke MM and Lenz F (1985) Spaltoffnungen, fruchtoberflache und transpiration wachsender apfelfruchte
der sorte golden delicious. *Erwerbsobstbau* **27**: 139–143

Blanke MM and Lenz F (1989) Fruit photosynthesis. *Plant Cell Environ* **12**: 31–46

Blanke MM and Notton BA (1991) Kinetics and physiological significance of photosynthetic PEP
carboxylase in avocado fruit. *J Plant Physiol* **137**: 553–558

Blanke MM, Priestley CA and Trecarne KJ (1982) Carbon conservation by apple fruits. East Malling
Research Station. Annual Report 1981, pp 146

Blanke MM, Pring RV and Goodenough PW (1985) Stomatal and cuticular respiration in fruit with C_4
photosynthesis pathways. Long Ashton Research Station. Annual Report, pp 136–137

Blanke MM, Notton BA and Hucklesby DP (1986) Physical and Kinetic properties of photosynthetic PEP
carboxylase in developing apple fruit. *Phytochemistry* **25**: 601–606

Blanke MM, Hucklesby DP, Notton BA and Lenz F (1987a) Utilization of bicarbonate by apple fruit
PEP carboxylase. *Phytochemistry* **26**: 2475–2476

Blanke MM, Hucklesby DP and Notton BA (1987b) PEP carboxylase in developing apple fruit. *J Plant
Physiol* **129**: 310–325

Blanke MM, Hucklesby DP and Notton BA (1988) Distribution and physiological significance of PEP
carboxylase in kiwi aubergine and apple fruit. *Hort Sci* **23**: 65–70

Bogin E and Wallace A (1966) CO_2 fixation in preparations of Tunisian Sweet Lemon and Eureka Lemon
fruits. *J Amer Soc Hort Sci* **88**: 298–307

Bravdo BA, Palgi A, Lurie S and Frenkel C (1977) Changing ribulose diphosphate carboxylase/oxygenase activity in ripening tomato fruit. *Plant Physiol* **60**: 309–312

Buffer G (1986) Ethylene promoted conversion of 1–amino cyclopropane–1–carboxylic acid to ethylene in peel of apple at various stages of fruit development. *Plant Physiol* **80**: 539–543

Burg SP and Burg EA (1965) Gas exchange in fruits. *Physiol Plant* **18**: 870–884

Burg SP and Burg EA (1967) Molecular requirements for the biological activity of ethylene. *Plant Physiol* **42**: 144–152

Buttrose MS and May LH (1959) Physiology of cereal grain 1 The source of carbon for the developing barley kernel. *Aust J Biol Sci* **12**: 40–52

Caley CY, Duffus CM and Jeffcoat B (1990) Photosynthesis in the pericarp of developing wheat of grains. *J Exp Bot* **41**: 303–307

Clark RB and Wallace A (1963) Dark CO_2 fixation in organic acid synthesis and accumulation in citrus fruit vescicles. *J Amer Soc Hort Sci* **83**: 322–332

Clark RB., Wallace A and Mueller RT (1961) Dark CO_2 fixation in avocado roots, leaves and fruits. *J Amer Soc Hort Sci* **78**: 161–168.

Clijsters H (1969a) On the effect of light on CO_2 exchange in developing apple fruits. *Prog Photosynthesis Res* **1**: 388–395

Clijsters H (1969b) On the photosynthetic activity of developing fruits. *Qualitas Plantarum Materiae Vegetale* **19**: 129–140

Clijsters H (1975) Relation possible entre l' activite photosynthetique de la pomme et sa maturation. *Colloques Internatiaux CNRS* **238**: 41–47

Cordes D (1988) Einfluss von Entblatterung auf Kohlen hydrat verteilung und CO_2–gasweehsel bei apfeln. *Hort Sci* **23**: 71–78

Cordes D and Blanke MM (1987) Hypodermal and core chloroplasts of developing apple fruit. Long Ashton Research Station, Annual Report, 1985

Crookston RK, Toole JO and Ojbun JL (1974) Characterization of the bean pod as a photosynthetic organ. *Crop Sci* **14**: 708–712

Das S, Sood DR, Sawhney SK and Singh R (1986) Properties of NADP$^+$–malic enzyme from pod-walls of chickpea *(Cicer arietinum)*. *Physiol Plant* **68**: 308–314

De Barsy T (1976) Le fruit de la naissance a la senescence. *La Fruit Belge* **373**: 4–10

Dhawan RS and Singh R (1983) Photosynthetic rates of leaves and pods of chickpea cultivars differing in seed weight. *Indian J Plant Physiol* **26**: 276–284

Dhillon S, Suneja SK, Sawhney SK and Singh R (1985) properties of NADP$^+$–malic enzyme from glumes of developing wheat grains. *Phytochemistry* **24**: 1657–1663

Downs RJ, Siegel- Mann HW, Butler WL and Hendricks SB (1965) Photoreceptive pigments for anthocyanin synthesis in apple skin. *Nature* **205**: 909–910

Duffus CM and Cochrane MP (1982) Carbohydrate metabolism during cereal grain development. In (AA Khan ed) *The Physiology and Biochemistry of Seed Development, Dormancy and Germination*. Elsevier Biomedical Press, Amsterdam, pp 43–66

Duffus CM, Nutbeam AR and Scragg PA (1985) Photosynthesis in the immature cereal pericarp in relation to grain growth. In: (B Jeffcoat, AF Hawkins and AD Stead, eds) *Regulation of Sources and Sinks in Crop Plants,* Monograph No. 12, British Plant Growth Regulator Group, Long Ashton, England, pp 243–256

Duffus CM and Watson PA (1988) Carbon dioxide fixation by detached cereal caryopses. *Plant Physiol* **87**: 504–509

Edwards G and Walker DA (1983) Induction. In: C_3, C_4 *Mechanisms and Cellular and Environmental Regulation of Photosynthesis*. Blackwell Scientific Publications, Oxford, pp 156–200

Enyi BAC (1962). The contribution of different organs to grain weight in upland and swamp rice. *Ann Bot* **26**: 529–531

Evans LT and Rawson HM (1970) Photosynthesis and respiration by the flag leaf and component of the ear during grain development in wheat. *Aust J Biol Sci* **23**: 245–254

Fischer KS and Wilson GL (1971) Studies of grain production in *Sorghum vulgare*. II Sites responsible for grain dry matter production during the post anthesis period. *Aust J Agric Res* **22**: 39–47

Flinn AM and Pate JS (1970) A quantitative study of carbon-transfer from pod and subtending leaf to the ripening seeds of the field pea. *J Exp Bot* **21**: 71–82

412

Flinn AM, Atkins CA and Pate JS (1977) Significance of photosynthetic and respiratory exchanges in the carbon economy of the developing pea fruit. *Plant Physiol* **60**: 412–418

Frey-Wyssling A and Buttrose MS (1959) Photosynthesis in the ear of barley. *Nature* **184**: 2031–2032

Goldsworthy PR (1970) The source of the assimilate during grain development in tall and short sorghum. *J Agric Sci* **74**: 525–531

Goodheer JC (1979) Carotenoids in the photosynthetic apparatus. *Berichte der Deutschen Botanischen Gesell Schaft 92: 427–436*

Gross J, Zachariae A, Lenz F and Eckhardt G (1978) Carotenoid changes in the peel of the golden delicious apple during ripening and storage. *Z Pflanzenphysiol* **89**: 321–332

Gupta VK and Singh R (1988) Partial purification and characterization of NADP$^+$–isocitrate dehydrogenase from immature pod-walls of chickpea. *Plant Physiol* **87**: 741–744

Gupta VK and Singh R (1989) Properties of pyruvate kinase from immature pod-wall of chickpea. *Plant Physiol Biochem* **27**: 703–711

Gupta VK and Singh R (1990) Properties of NADP$^+$–malate dehydrogenase from immature pod-wall of chickpea. *Plant Physiol Biochem.* **28**: 671–678

Hall EG, Heulin FE, Hackney FMV and Bain JM (1955) Les exchanges gazeuz dans les pommes Granny Smith. *Fruits* **10**: 149–155

Hansen P (1969) ^{14}C studies on apple trees. *Physiol Plant* **22**: 186–198

Harvey DM, Hedley CL and Kelly RJ (1976) Photosynthetic and respiratory studies during pod and seed development in *Pisum sativum* L. *Ann Bot* **40**: 993–1001

Hedley CL, Harvey DM and Kelly RJ (1975) Role of PEP carboxylase during seed development in *Pisum sativum*. *Nature* **258**: 352–354

Henze J (1969a) Muglichkeiten zur Beein flussung der Fruchtatmung und-reifung bei kernobst. *Hort Sci* **2**: 159–187

Henze J (1969b) Beziehungen Zwischen respiration und innerer Atmosphere bei CO_2 empfindliehen kernobstsorten. *Qualitas Plantarum Materiae Vegetable* **19**: 229–242

Huffaker RC and Wallace A (1959) Dark fixation of CO_2 in homogenates from citrus leaves, fruits and roots. *J Amer Soc Hort Sci* **74**: 348–357

Hulme AC Jones JD and Wooltorton LSC (1963) The respiration climacteric in apple fruits. *Proc Roy Soc Ser B* **158**: 514–535

Hulme AC, Jones JD and Wooltorton LSC (1967) The respiration climacteric in apple fruits: Some possible regulatory mechanisms. *Phytochemistry* **6**: 1343–1351

Imaizumi N, Usuda H, Nakamoto H and Ishihara K (1990) Changes in the rate of photosynthesis during grain filling and the enzymatic activities associated with the photosynthetic carbon metabolism in rice panicles. *Plant Cell Physiol* **31**: 835–842

Jennings VM and Shibles RM (1968) Genotypic differences in photosynthetic contributions of plant parts to grain yield in oats. *Crop Sci* **8**: 173–175

Jones HG (1981) CO_2 exchange of developing apple fruits. *J Exp Bot* **32**: 1203–1210

Jones HG (1983) *Plants and Microclimate*. Cambridge University Press, Cambridge.

Jones HG and Higgs KH (1982) Surface conductance and water balance of developing apple fruits. *J Exp Bot* **33**: 67–77

Jones JD, Hulme AC and Wooltorton LSC (1964) The respiration climacteric in apple fruits. *New Phytologist* **64**: 158–167

Juniper BE and Jeffree CE (1983) *Plant Surface*. Edward Arnold, London

Khanna-Chopra R and Sinha SK (1974) Photosynthesis and photosynthetic enzymes in reprodcutive organs of some crop plants. *Indian J Genet Plant Breed* **34**: 1041–1047

Kidd F and West C (1925) The course of respiratory activity throughout the life of an apple. Reports of the Food Investment Board for 1924, London, PP 27–32

Kidd F and West C (1947) A note on the assimilation of carbon dioxide by apple fruits after gathering. *New Phytologist* **46**: 274–275

Kliewer WM and Smart R (1988) Canopy manipulation for optimizing vine micro climate, crop yield and composition of grapes. In: (CJ Wright, ed) *Manipulation of Fruiting* Proc 47th Easter School, University of Nottingham Sutton Bonington, Loughborough.

Kluge M and Ting TP (1978) *Crassulacean Acid Metabolism, Springer, Berlin*

Knee M (1972) Anthocyanin, carotenoid and chlorophyll changes in the peel of Cox's orange pippin apples during ripening on and off the tree. *J Exp Bot* **23**: 84–96

Knoppik D, Selinger H, Ziegler-Jöns A (1986) Differences between the flag leaf and the ear of a spring wheat cultivar with respect to the CO_2 response of assimilation, respiration and stomatal conductance. *Physiol Plant* **68**: 451–457

Kriedemann P (1968) Observations on gas exchange in the developing Sultana berry. *Aust J Biol Sci* **21**: 907–916

Kurssanow AL (1934) Die photosynthese grunerfruchte und ihre abhangigkeit von der normalan tatigkeit der blatter. *Planta* **22**: 240–250

Latzko E and Kelly GJ (1983) The many faceted function of PEP carboxylase in C_3 plants. *Physiol Vegetale* **21**: 805–815

Laval-Martin D, Farineau J and Diamond J (1977) Light versus dark carbon metabolism in cherry tomato fruits. *Plant Physiol* **60**: 872–876

Lenz F and Blanke MM (1983) Transpiration bei apfel fruchten. Erwerbsobstbau **25**: 23–29

Lenz F and Noga G (1982) Photosynthese und atmung bei apfelfruchten. *Erwerbsobstbau* **24**: 198–200

Leyhe A (1987) Transpiration und CO_2 gaswechsel von trauben in adhangigkeit von klimafaktoren und wasserversorgung. Dissertation Institut fur Obstbau, Universitat Bonn

Leyhe A (1988) CO_2–aufnahme und abgabe sich entwickelnder trauben. *Wein Wissenschaft* **43**: 67–73

Luthra YP, Sheoran IS and Singh R (1983) Photosynthetic rates and enzyme activities of leaves, developing seeds and pod-wall of pigeonpea. *Photosynthetica* **17**: 210–215

Marcellin P (1975) Conditions physiques de la circulation des gaz respiratoires a traverse in masse des fruits et maturation. *Colloques Internationaux CNRS* **238**: 241–251

Meyer AO, Kelly GJ and Latzko E (1982) Pyruvate orthophosphate dikinase from immature grains of cereal grasses. *Plant Physiol* **69**: 7–10

Noga G and Lenz F (1982) Einfluss von verschiedenen klimajaktoren auf den CO_2–gaswechsel von apfeln wahrend der licht und dunkel periode. *Hort Sci* **47**: 193–197

Olugbemi LB, Bingham J and Austin RB (1976) Ear and flag leaf photosynthesis of awned and awnless *Triticum* species. *Ann Appl Biol* **84**: 231–240

Osmond CB, Winter K and Ziegler H (1982) Functional significance of different pathways of CO_2 fixation in photosynthesis. In: (OL Lange, PS Nobel, CB Osmond and H Ziegler eds). *Physiological Plant Ecology II Water Relations and Carbon Assimilation*, Springer, Berlin, pp 497–547

Pantastico EB (1975) *Post-harvest Physiology: Utilization of Tropical and Subtropical Fruits and Vegetables.* AVJ Westport, Conn, USA

Pearson JA and Robertson RN (1954) The physiology of growth in apple fruits. *Aust J Biol Sci* **7**: 1–17

Phan CT (1970) Photosynthetic activity of fruit tissue. *Plant Cell Physiol* **11**: 823–825

Phan CT (1973a) Chloroplasts of the peel and the internal tissue of apple fruits. *Experimentia* **29**: 1555–1557

Phan CT (1973b) A new look at the climacteric. *29th Annual Meeting of the West Canada Society of Horticulture*, pp 52–64.

Phan CT (1975) Occurrence of active chloroplasts in the internal tissues of apples. Their possible role in fruit maturation. *Colloques Internationaux CNRS* **238**: 49–55

Price DN and Hedley CL (1980) Developmental and varietal comparisons of pod carboxy levels of *Pisum sativum L.* Ann Bot **45**: 283–294

Queberdeaux B and Chollet R (1975) Growth and development of soybean pods: CO_2 exchange and enzyme studies. *Plant Physiol* **55**: 745–748

Rhodes MJC and Wooltorton LSC (1967) The respiration climacteric in apple fruits: The action of hydrolytic enzymes in peel tissue during the climacteric period in fruit detached from the tree. *Phytochemistry* **6**: 1–12

Ruffner HP and Kliewer WM (1975) PEP carboxylase activity in grape berries. *Plant Physiol* **56**: 67–71

Schwerdtfeger G and Buchloh G (1968) Die korkentwicklung in den abschlussgebeben der triebe und fruchte verschiedener apfelsorten. *Hort Sci* **33**: 77–102

Sheoran IS and Singh R (1988) Photosynthetic contribution of pod wall in seed development of chickpea. *Proc Indian Natl Sci Acad* **B53**: 531–534

Sheoran IS, Singal HR and Singh R (1987) Photosynthetic characteristics of chickpea pod wall during seed development. *Indian J Exp Biol* **25**: 843–847

414

Sheoran IS, Sawhney V, Babbar S and Singh R (1991) *In vivo* fixation of CO_2 by attached pods of *Brassica campestris* L. *Ann Bot* **67**: 425–428

Siefermann-Harms D (1981) The role of carotenoids in chloroplasts of higher plants. In: (D Mazliak, ed) *Biogenesis and Function of Plant Lipids*, pp 331–340

Singal HR, Sheoran IS and Singh R (1986a) Products of photosynthetic $^{14}CO_2$ fixation and related enzyme activities in fruiting structures of chickpea. *Physiol Plant* **66**: 457–462

Singal HR, Sheoran IS and Singh R (1986b) *in vitro* enzyme activities and products of $^{14}CO_2$ assimilation in flag leaf and ear parts of wheat. *Photosynth Res* **8**: 113–122

Singal HR, Sheoran IS and Singh R (1987) Photosynthetic carbon fixation characteristics of fruiting structures of *Brassica campestris* L. *Plant Physiol* **83**: 1043–1047

Singhal HR and Singh R (1986) Purification and properties of PEP carboxylase from immature pods of chickpea (*Cicer arietinum* L.). *Plant Physiol* **80**: 369–373

Singh R (1987) Photosynthetic carbon fixation characteristics of fruiting structures of wheat, chickpea and *Brassica* sp. *Proc Indian Natl Sci Acad* **B53**: 541–544

Singh R (1989) CO_2 fixation by PEP carboxylase in pod walls of chickpea In: (GS Singal, J Barber, RA Dilley, Govindjee, R Haselkorn and P Mohanty eds) *Photosynthesis: Molecular Biology and Energetics,* Narosa Publishing House, New Delhi, pp 315–329

Singh BK and Pandey RK (1980) Production and distribution of assimilate in chickpea. *Aust J Plant Physiol* **7**: 727–735

Singh DP, Singh P and Sharma HC (1986) Diurnal patterns of photosynthesis, evaporation and water use efficiency in mustard at different growth phases under field conditions. *Photosynthetica* **20**: 117–123

Sinha SK and Sane PV (1976) Relative photosynthetic rate in leaves and fruitwall of peas. *Indian J Exp Biol* **14**: 592–594

Smith WH (1947) A new method for the determination of the composition of the internal atmosphere of fleshy plant organs. *Ann Bot* **11**: 363–368

Takeda T and Maruta H (1956) Studies on CO_2 exchange in crop plants. IV Roles played by the various parts of the photosynthetic organs of rice plant in producing grain during ripening period. *Proc Crop Sci Soc (Japan)* **24**: 181–186

Thorne GN (1965) Photosynthesis of ears and flag leaves of wheat and barley. *Ann Bot* **29**: 317–329

Thorne JH (1979) Assimilate redistribution from soybean pod-walls during seed development. *Agron J* **71**: 812–816

Ting IP and Dugger WM (1967) CO_2 metabolism in corn roots. 1 Kinetics of carboxylation and decarboxylation. *Plant Physiol* **42**: 712–718

Todd GW, Bean RC and Propst B (1961) Photosynthesis and respiration in developing fruits. II Comparative rates at various stages of development. *Plant Physiol* **36**: 69–73

Tribe M and Whittaker P (1982) Chloroplasts and Mitochondria. *Studies in Biology* No. 31, Edward Arnold, London

Tschakalova E and Hoffmann P (1976) Strukfurelle und funktionelle grundiagen des photosynthetischen gaswechsels bei *Triticum aestivum* L. Wiss Z Humboldt Univ Berlin **25**: 723–736

Ulrich R and Marcellin P (1955) Voies et modalites des exchanges de gaz carbonique et doxygene des fruits avec 'l'atmosphere ambiante. *Recherche CNRS* **31**: 241–251

Vu JCV, Yelenowski G and Bausher MG (1985) Photosynthetic activity in the flower buds of valencia orange. *Plant Physiol* **78**: 420–423

Westwood MN (1978) *Temperate Zone Pomology*. Freeman and CO, San Francisco, USA

Wienicke J (1968) Ein bietrag zum transport ^{14}C markierter assimilate vom blatt zur Apfelfruct. *Mitteilungen Klosterneburg* **18**: 462–469

Willmer CM and Johnston WR (1976) CO_2 assimilation in some aerial plant organs and tissues. *Planta* **130**: 33–37

Wirth E, Kelly GJ, Fischbeck G and Latzko E (1977) Enzyme activities and products of CO_2 fixation in various photosynthetic organs of wheat and oat. *Z Pflanzenphysiol* **82**: 78–87

Young RE and Biale JB (1968) CO_2 effects on fruits. III Fixation of $^{14}CO_2$ in lemon in an atmosphere enriched with CO_2, *Planta* **81**: 253–263

Zelles L (1967) Untersuchungen uber den farbstoflge halt der vegetations periode und unter verschiedenen lagerbedingungen Dissertation, Institut fur dstbau, Universitat Bonn.

Ziegler-Jöns A (1989a) Gas exchange of ears of cereals in response to CO_2 and light I Relative contributions of parts of the ears of wheat, oat and barley to the gas exchange of the whole organ. *Planta* **178**: 84–91

Ziegler-Jöns A (1989b) Gas exchange of ears of cereals in response to CO_2 and light. II Occurrence of a C_3–C_4 intermediate type of photosynthesis. *Planta* **178**: 164–175

Ziegler-Jöns A, Knoppik D and Sellinger H (1987) Characteristics of the CO_2 exchange of wheat ears. In: (J Biggins, ed) *Progress in Photosynthesis Research*, Vol IV, Martinus Nijhoff, Dordrecht, The Netherlands.

STRESS; CO$_2$ ENRICHMENT

17

Effects of Water Stress on Photosynthesis of Crops and the Biochemical Mechanism

D.W. Lawlor[1] and D.C. Uprety[2]

[1]Institute of Arable Crops Research, Rothamsted Experimental Station,
Harpenden, Herts, AL5 4LA, U.K.

[2]Division of Plant Physiology, Indian Agricultural Research Institute,
New Delhi 110012, India

CONTENTS

ABBREVIATIONS

ABA	:	abscisic acid
CF_0/CF_1	:	chloroplast coupling factor
Ci	:	stomatal CO_2 concentration
g_m	:	mesophyll conductance
g_s	:	stomatal conductance
EC	:	adenylate energy charge
LAI	:	leaf area index
MPa	:	megapascal

N	:	number of moles of all solutes
NAD	:	nicotinamide adenosine dinucleotide
NADP	:	nicotinamide adenosine dinucleotide phosphate
Pg	:	gross photosynthesis
PG	:	phosphoglycolate
PGA	:	phosphoglyceric acid
Pn	:	photosynthesis per unit leaf area (net photosynthesis)
Pr	:	photorespiration
PCRC	:	photosynthetic carbon reduction cycle
R	:	gas constant
Rd	:	dark respiration
RuBPcase	:	ribulose bisphosphate carboxylase/oxygenase
ψ	:	leaf water potential
P	:	turgor potential
π	:	osmotic potential
γ	:	matric potential
ε	:	bulk modulus of elasticity
T	:	calvin temperature
V	:	volume of water

ABSTRACT

Responses of photosynthesis to water stress has been subjected to study in recent years. This review briefly describes, how stress affects photosynthesis in crops and the causes of the effects on cell metabolism. The section on 'photosynthesis under field conditions' outlines the variability of the crop species to moisture stress and refers to the differential responses of stress at different growth stages of plants. The mechanism involving adaptation of crop plants like triticale, wheat and mungbean to stress has been explained in relation to the water status, translocation and source-sink relationship in these plants. Arrangement of cellular water balance is explained on the basis of osmotic potential and cellular elasticity. Thus the adaptation to drought may be achieved by modifying metabolic processes regulating osmotic potential, cell expansion involving alterations in enzymes and changes in control system.

The causes of lowered photosynthesis under stress are analysed to identify the sites of limitations which are metabolic rather than stomatal. The major limitation to photosynthesis is considered to be impaired ATP synthesis which reduces the synthesis of RuBP and slows photosynthesis. The explanation requires critical analysis of adenylates and pyridine nucleotides in the stressed photosynthetic cell.

Two important questions concerning effects of water stress are still to be considered: (1) what are the molecular features of coupling factors which make it susceptible to increased ion concentration; (2) what stimulates the mitochondrial activity (CO_2 production, increased TCA function, carbohydrate consumption, amino acid and organic acid synthesis) as the photosynthesis process decreases with stress. There is no consensus as to the regulation of the two processes in unstressed leaves. Probably the production of ATP by unstressed chloroplast in the light suppresses mitochondrial activity. However, as the photophosphorylation decreases, the mitochondria respond to maintain the ATP concentration and energy charge of the cell. The glycolate pathway and mitochondrial electron transport possibly serve to adjust cellular energy status. The low ATP would stimulate the carbon flux from reserve carbohydrate through glycolysis to the TCA cycle and consequently increase the organic acids which along with some reductants cause accumulation of secondary metabolites. These concepts are based on experimental observations and they require formulation as specific hypotheses which can be tested, for the sequence of events in the changes of ATP, reductant, carbon-nitrogen assimilation and accumulation of products, their quantitative relations and differences between plants to understand the effects of stress on photosynthesis.

I. Introduction

Water shortage is the primary constraint to plant growth and productivity over much of the land surface (Kramer, 1983). Drought limits agricultural production, reducing the well-being of people and hindering the economic development of countries; some of the world's poorest are those from drought-prone regions. India is typical of areas where much of the land is subject to poor, erratic rainfall and to extreme evaporation. Crop failures, occurring on average once in four years in semi-arid areas, are difficult to predict (Swindale and Bidinger, 1981). Even in more humid regions, droughts often decrease yields substantially.

The two main causes of loss of production under water stress are the reduction in the crop leaf area due to fewer, smaller leaves and the lower rates of photosynthesis per unit leaf area (P_n) (Day et al., 1987; Fereres, 1985; Jordan and Ritchie, 1971; Kriedmann, 1986; Legg et al., 1979; Monteith, 1977; Vu et al., 1987; Zelitch, 1982). The response of leaves and P_n to stress differs with species and varieties of crop plants and their productivity may be affected by mechanisms which differ quantitatively. However, experiments coupled with modelling show that inhibition of growth may reduce dry matter production more than does inhibition of P_n (Legg et al., 1979; Monteith, 1977). This relative sensitivity of growth may reflect the need for plants to protect their cellular, metabolic systems (Hanson and Hitze, 1982) to sustain effective metabolism and to survive under stress. Unfortunately this strategy for survival conflicts with human demands for food. Natural vegetation of dry, hot areas with bright light is largely adapted (Kramer, 1983) but many crops are poorly adapted to such conditions in which they are often grown, although cultivated species differ substantially in growth and productivity in dry habitats (Fereres, 1985).

The responses of plants to the environment are related to the type of photosynthetic carbon metabolism (Lawlor, 1987). Plants of the C_3 group (e.g., wheat, barley, potatoes, sugar beet), which form 3-carbon compounds as the first products of photosynthesis and also photorespire (Fig. 1a), are less tolerant of hot, dry conditions than C_4 plants (e.g., sorghum, maize) (Lawlor, 1979), which form 4-carbon organic acids as the primary products and have very small rates of photorespiration (Fig. 1b). C_3 plants generally have smaller rates of light-saturated photosynthesis and larger stomatal conductances and, therefore, higher transpiration rates under given conditions than C_4 (Kramer, 1983; Lawlor, 1979). These characteristics decrease the productivity and water use efficiency of C_3 plants compared to C_4 plants in dry, hot conditions. Despite this, C_3 plants play a vital role in Indian agriculture (Swindale and Bidinger, 1981).

As water stress affects plant production, it is important to understand which parts of the photosynthetic mechanism are most sensitive in water stressed C_3 and C_4 plants (Boyer, 1983, Frederick et al., 1989; Hanson and Hitze, 1982; Kulshreshtha et al., 1987; Rawson et al., 1978; Sinha et al., 1988; Tokami et al., 1981). To decrease the sensitivity of crops to stress and to improve productivity under drought, is a long-term aim of agriculturists, physiologists and breeders. By selecting for improved characteristics (which are the result of complex biochemical and physiological processes), plants may be adapted to environments for which they were not naturally selected (Krieg, 1983). More direct intervention in the genome by genetic engineering may suppliment classical plant breeding and selection. However, current understanding of the biochemistry and physiology of plant responses to water stress is inadequate (Hanson and Hitze, 1982; Uprety and Sirohi, 1987) for intervention in the plant's genetic mechanism to achieve greater, more stable production under stress. This review briefly considers how stress affects photosynthesis in crops under field conditions and the causes of the effects on cell metabolism.

II. Photosynthesis under Field Conditions and Its Control

Crops may suffer water stress on occasions during the growing season, particularly during periods of low rainfall. The duration and severity of stress experienced depends principally on the potential of water and the hydraulic conductivity of the soil and on the evapo-transpiration from the crop and soil surface (Day et al., 1987; Hsiao and Acevedo, 1974; Jordan and Ritchie, 1971; Kramer, 1983). Plant growth and P_n differ in response to loss of cellular water; thus establish the relative susceptibility of crops to drought at different stages of growth (Boyer, 1983; Fereres, 1985; Sinha et al., 1988; Uprety and Sirohi 1985, 1987). Water stress principally affects two main processes in crops: (1) the development of leaf area, and (2) the carbon exchange rate per unit leaf area, the most important component of which is photosynthesis (Krieg, 1983; Lawlor, 1987; Zelitch, 1982). The physiology and biochemistry of the processes and the effects of drought are analysed later. Here we consider the field observations relevant to Indian conditions.

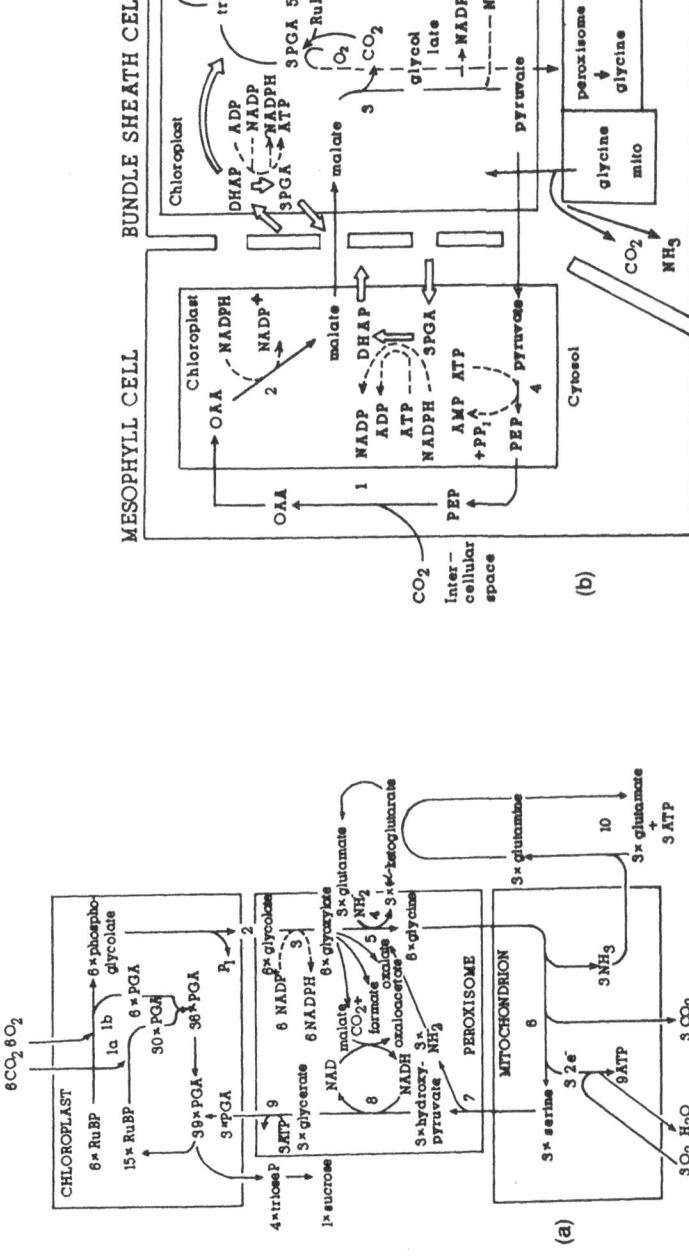

Fig. 1a. Scheme of the C_3 photosynthetic carbon reduction cycle in the chloroplast and the synthesis of glycolate, which is metabolized to glycine. Subsequent metabolism of glycine leads to the release of carbon dioxide in the light which appears as photorespiration.

Fig. 1b. In C_4 plants the carbon dioxide concentrating mechanism much decreases the production of glycolate and therefore photorespiration in comparison with C_3 plants. In both C_3 and C_4 plants, stress affects the photorespiratory processes (Lawlor, 1987).

Plants experiencing drought during vegetative growth develop smaller, often fewer leaves well before photosynthetic rates are markedly affected and the rate of leaf area growth decreases progressively with increasing drought (Aggarwal and Koundal, 1988; Day et al., 1987; Lawlor et al., 1981; Legg et al., 1979; Muchow, 1985; Sinha et al., 1988; Sung and Krieg, 1979; Zelitch, 1982). Stress effects depend strongly on the ontogenetic stage (Aggarwal and Sinha, 1984; Morgan, 1984; Tokami et al., 1981 Uprety, 1989; Uprety and Tomar, 1990; Uprety and Tomar, 1992). The number of leaves and their rate of initiation are relatively insensitive to stress in determinant (Tokami et al., 1981) compared to indeterminate plants (Frank et al., 1973), but the latter are more able to offset the effects of short-term stress. Because P_n is maintained when growth is slowed in stressed plants, inadequate assimilate supply is clearly not responsible for the short-term decrease in leaf growth with stress, and loss of turgor is the most likely cause (Kramer, 1983; Tyree and Jarvis, 1982). Recovery of turgor may lead to rapid increase in leaf growth if cells have retained the capacity to grow. The amount of assimilate required for leaf growth may be a small proportion of the total photosynthate in well-watered plants but may increase under stress (Constable and Rawson, 1980) because of the need to accumulate solutes including sugars, amino acids and organic acids (Hanson and Hitze, 1982; Kriedmann, 1986; Morgan, 1984).

Stress during the vegetative stages reduces the leaf area index (LAI) thereby decreasing light interception (Kramer, 1983; Monteith, 1977; Tokami, 1981) so that there is fall in the efficiency of production per unit of incident radiation. (Legg et al., 1979; Monteith, 1979; Zelitch, 1982). As light of higher intensity is intercepted by leaves in crops of low LAI, and P_n is less efficient in C_3 plants under such conditions, therein decline in the efficiency of dry matter production per unit of light intercepted (Fig. 2). In addition, the effect of water stress on P_n rate also reduces the photosynthetic efficiency (Frederick et al., 1989; Legg et al., 1979; Monteith, 1977; Uprety and Bhatia, 1989; Uprety and Sirohi, 1987; Zelitch, 1982). Many studies of field-grown plants have demonstrated that stress reduces the rate of P_n (Johnson et al., 1974). Generally P_n is little affected in the initial phase of water loss but falls progressively below −1 MPa in vegetative stage and −3 MPa during the reproductive stage in wheat flag leaves. However, there are marked differences in the response of varieties to stress. In Indian wheats, for example (Uprety and Sirohi, 1985, P_n decreased by 56% in cv Kalyansona but by only 10% in cv C306 at similar water potential (ψ)). The differences were related to stomatal conductance (g_s) as well as to mesophyll conductance (g_m). The wild species of wheat Triticum kotschyi (Johnson et al., 1987) is reported to have higher rates of P_n under water stress than T. aestivum partly because it maintained a larger g_m i.e., metabolism was less affected. Wild progenitors of wheats and other plants, generally have larger rates of P_n and smaller leaf area per plant than selected modern varieties; this may result in water conservation during the vegetative stage and more water during reproduction (Johnson et al., 1987) and higher water use efficiency.

In soybean, the decrease in P_n is related to smaller g_s between −1 and −1.6 MPa ψ whereas g_m fell only at −3 to −4 MPa. Cotton is similarly affected by stress (Ephrath et al., 1990; Jordan and Ritchie, 1971), g_s and P_n decrease in an exponential asymptotic relationship (Ephrath et al., 1990) between −1 and −2 MPa and g_m decreased only below −2 MPa ψ (relative water content < 75%). This study

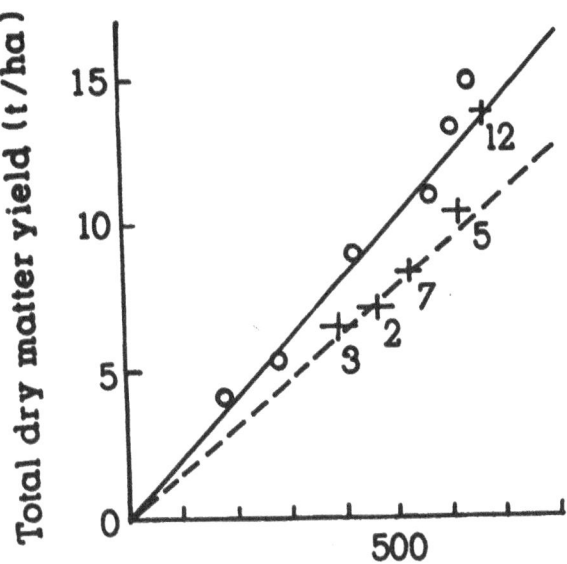

Fig. 2. Total dry-matter yield of barley crops versus the visible radiation intercepted per unit ground area by green leaves. + shows the means with standard errors from different drought treatments at Rothamsted Experimental Station, U.K. in 1976, an exceptionally dry year. 12—weekly irrigation, 2 and 3—early irrigated, late droughted, 5—late drought, 7—droughted after anthesis. Other data for experiments in the U.K. (Legg *et al.*, 1979). The solid line represents the efficiency of production without stress; the dotted line is a 24% decrease in efficiency of radiation conversion.

showed that the CO_2 concentration in the substomatal cavities (Ci) increased and that the P_n/Ci ratio decreased as g_m fell. These changes may be related to decreased RuBPcase activity (Aggarwal and Koundal, 1988); for an 0.1 MPa decrease in ψ of wheat RuBPcase activity fell by 4.4%. Photosynthesis almost stopped *i.e.*, g_m approached zero, at *ca.* −3 MPa ψ in leaves of valencia orange and the RuBPcase activity fell by 62% (Vu and Velenosry, 1988).

Generally the effects of low ψ on P_n are greater on old than young leaves; stress accelerates the natural decline in P_n with age (Morgan, 1984; Tokami *et al.*, 1981). This is accompanied by decreased g_s, decline in chlorophyll and protein (including RuBPcase content) and changed water relations in grain legumes (Uprety, 1989), pasture grasses and wheat (Aggarwal and Sinha, 1984; Ludlow and Wilson, 1971; Uprety and Sirohi, 1985). For many species in the field, mild stress seems to affect diffusion processes more than metabolism but the reverse is the case at low ψ as shown by changes in g_m and components of the photosynthetic structures.

C_4 plants frequently have larger photosynthetic rates at smaller C_i and g_s and are more productive than C_3 plants (Lawlor, 1987) under dry conditions, *e.g.*, sorghum has P_n 2.3 times greater than that of soybean and also the transpiration is less under similar, unstressed, conditions (Rawson *et al.*, 1978). Photosynthesis of soybean and sorghum ceased at very different ψ, -1.4 and -2.7 MPa, respectively (Sung and Krieg, 1979) whereas translocation of assimilates still occurred. Sorghum also has greater P_n and translocation than cotton at similar ψ. However, there are large differences between C_4 species. Maize, for example, is less drought tolerant than sorghum (Beadle et al., 1973); photosynthesis ceased at -1.2 MPa ψ but that of sorghum was still 25% of the maximum at the same ψ. Photosynthesis of maize is relatively insensitive to mild stress (0 to -0.4 MPa), decreasing in parallel with g_s as ψ decreases to -1 MPa but at lower ψ metabolic factors affect P_n (Venekamp et al., 1989).

a. Mechanisms of adaptation to drought

Maintenance of productivity in crops under drought is achieved by quantitative differences in a number of adaptive mechanisms (Swindale and Bidinger, 1981; Uprety and Bhatia, 1989; Uprety and Sirohi, 1987; Uprety and Kumar, 1990). Triticale for example, adapts by maintenance of the leaf water content through a reduction in water loss by progressive stomatal closure and through a large, efficient root system both of which keep ψ large and slow the rate of changes in water content and ψ. Rye, the male parent of triticale, exhibits similar adaptation, but with a marked ability, to recover P_n and chlorophyll compared to wheat, when irrigated after drought. Wheat, the female parent of triticale, had less marked adaptive features. The drought resistant features of triticale can be modified by the combination of the two genomes (Uprety and Kumar, 1990). Octoploid triticale generated from the hexaploid, drought resistant wheat cv C 306 was relatively less susceptible to moisture stress than hexaploid triticale, from a female tetraploid durum wheat. The octoploid maintained greater ψ but without lower g_s under moisture stress, a character which appears to be inherited from cv C 306.

Adaptation of P_n in legumes under stress (Lopez et al., 1987) is achieved by a range of mechanisms. Cowpea maintains P_n by avoidance of dehydration, with smaller g_s and paraheliotropism but without substantial osmotic adjustment. In contrast, pigeon pea adjusts osmotically at low ψ but maintains large g_s. The mung bean varieties, PS16, P105 and Pusa Baisakhi show marked variation in susceptibility of P_n and partitioning under drought, due to differences in water status (Uprety and Bhatia, 1989). The variety, P105 maintains greater ψ and RWC, which are conducive to greater P_n. The other, more susceptible varieties are less able to recover as the plants age (Uprety and Bhatia, 1989).

Variation in photosynthesis and leaf development in response to drought stress among genotypes offers opportunities for selecting plants with optimal characteristics for high P_n and productivity under drought. Such a large degree of genetic variation should be exploited. Much work remains to be done.

b. Translocation and source-sink relationships

Photosynthesis is not the only process in crop growth and productivity. Distribution of assimilates to sinks that especially yield producing organs, is also important (Patrick, 1988). Translocation of assimilates can influence the rate of P_n and is itself affected via the effects of drought on the growth of organs and also on phloem loading and unloading (Wardlaw 1967, 1969). The influence of stress on the distribution of assimilates depends on the stage of growth and the relative sensitivity of various organs. The increased root/shoot ratio in many crops under drought is attributable to greater allocation of carbon to roots (Hsiao and Acevedo, 1974). This is particularly important during vegetative growth and is possibly a consequence of the greater sensitivity of leaf expansion than photosynthesis (Wardlaw, 1969). Sink organs, such as the wheat grain, are less sensitive to stress than is photosynthesis because the spikelet adjusts osmotically more than leaves (Morgan, 1984). However, translocation of carbohydrates out of leaves was reduced during the day but increased at night in water stressed corn and soybean (Bunce, 1982). This has been ascribed to reduction in rate of transfer of sugars from mesophyll cells to the conducting tissue (Wardlaw, 1969). Velocity of transport is not affected by stress during grain filling in wheat or during vegetative growth in *Lolium*. In *Lolium* (Wardlaw, 1969) the expanding leaf was more sensitive to stress than photosynthesis of the source leaf and inhibition of leaf growth was the major factor reducing translocation. However, in wheat (Asana and Basu, 1963; Blum *et al.*, 1983), where grain growth was less sensitive to stress than P_n of the source leaf, reduced amounts of assimilates or impaired loading may have limited translocation, which can still occur (Johnson and Moss, 1976) after P_n stops; translocation of stem reserves may be important under drought, and can compensate for low P_n in early grain filling (Asana and Basu, 1963; Blum *et al.*, 1983). Drought sensitive triticale, wheat, rye (Uprety and Sirohi, 1987) and mung bean (Uprety and Bhatia, 1989) retained photosynthates in their leaves or stems and did not translocate it to sinks, whereas tolerant plants translocated assimilates under severe stress after flowering. In maize (Brevedon and Hodges, 1973), ψ inhibited translocation more than P_n, reducing vein loading. Stress between tasselling and harvest stopped P_n (McPherson and Boyer, 1977) but grain yield was decreased by ca. 40%, indicating the importance of translocation. Photosynthesis of sorghum (Sung and Krieg, 1979) was reduced by stress before translocation was affected. The velocity of translocation was unaffected upto ψ of -2.4 MPa (McPherson and Boyer, 1977)—thus the yield was limited by insufficient assimilate and reduced leaf area. However, in soybean the velocity of ^{14}C labelled photosynthate transport slowed more than P_n under stress. The reduction in P_n of plants with severe stress (-3 MPa ψ) without closure of stomates has been attributed to inhibition of transport and accumulation of assimilates due to low turgor which slows growth (Dale and Wyse, 1985); assimilates may be diverted to sinks with less negative ψ. However, such changes may lead to relatively small changes in tissue composition compared to the effects of severe stress, which are considered here later. If carbon fixation is significantly inhibited at low ψ, sink growth can be sustained by remobilization of stored assimilates from stems (Legg *et al.*, 1979; Zelitch, 1982), combined with limited growth of other non-harvested organs and increased sink activity *e.g.*, by ABA (abscisic acid) induced membrane transport of sugars (Schussler *et al.*, 1984). There is inadequate understanding of the molecular events regulating remobilization of assimilates and more research is needed to separate the direct effects of ψ on P_n from the indirect effects *via* product accumulation which is determined by translocation, sink demand and loading-unloading of assimilates.

III. Cellular Water Balance under Stress

Changes in cellular water balance are the ultimate cause of all alterations in metabolism (Kulshreshtha *et al.*, 1987; Tyree and Jarvis, 1982), growth and photosynthesis resulting from water stress (Hanson and Hitze, 1982; Kramer, 1983; Tyree and Jarvis, 1982). There is controversy over which components of the cell's water balance (Eqn. 1) are most important for specific metabolic and growth functions.

$$\psi = P + \pi + \tau \tag{1}$$

Cell water potential (ψ) is the product (eqn. 1) of turgor potential, P, osmotic potential, π, and matric potential, τ. Generally, ψ has been accepted as a measure of "stress" (Kramer 1983). It is important in determining the water potential gradient from soil to leaf and, therefore, in the water flux. Decrease in ψ correlates closely with decreased g_s, photosynthesis and organ growth (Boyer, 1983; Hanson and Hitze, 1982; Kramer, 1983). However, ψ is external to the cell membranes and, therefore to the metabolic machinery. There is little evidence that ψ *per se* affects the behaviour of enzymes, of metabolic system or of organelles within the cell (Hanson and Hitze, 1982; Kramer, 1983; Kulshreshtha *et al.*, 1987; Tyree and Jarvis, 1982). In many crops suffering from stress the range of ψ and p may be –0.5 to –2.0 MPa and 1.5 to 0 MPa, respectively. However, plants more adapted to water deficiency may have lower ψ but maintain growth through higher P by decreasing π as a result of synthesis of osmotically active solutes (Hanson and Hitze, 1982; Kramer, 1983; Lawlor and Fock, 1977a; Munns *et al.*, 1979), particularly compatible solutes such as amino acids, glycine, betaine and sugars.

The osmotic potential of cells (Eqn. 2) is generally characteristic of the species, although it may vary with nutrition and particularly with water status,

$$\pi = - \emptyset \, RTN/V \tag{2}$$

where R is the gas constant, T is the Kelvin temperature, N is the number of moles of all solutes, V is the volume of water and $\emptyset$ is an osmotic coefficient which allows for the non-ideal osmotic behaviour of the solutes; $\emptyset$ is 1 for ideal solutions but may deviate substantially from it.

As the main determinant of P, osmotic potential is clearly important for growth. Regulation of solute concentration is also important for homeostasis and efficiency of metabolic reactions. Particular concentrations of ionic species are essential to maintain the thermodynamic and structural relationships between water and macromolecules (Munns *et al.*, 1979; Stewart, 1981). With full water content and turgor, metabolic reactions (*e.g.*, enzyme processes, electron transport) are probably close to an optimum in terms of rate and types of products. Water loss increases the concentration of ions and metabolites in the cell compartments and decreases the osmotic potential (eqn. 2) (Hanson and Hitze, 1982; Kramer, 1983; Tyree and Jarvis, 1982). Alterations in concentration of substrates or regulatory components may alter the rates of metabolic processes and affect some processes more than others. This leads to inefficiency and decreased productivity of many processes including photosynthesis (Boyer, 1983; Lawlor, 1976). Thus increased concentration of ions in the cytosol may inhibit the photosynthetic enzyme RuBPcase/oxygenase (Lawlor, 1987; Lawlor and Fock, 1977a; Vu et al., 1987). The chloroplast coupling factor (CF_0-CF_1) which synthesizes ATP, requires a concentration of 1-3 mol Mg^{2+} m^{-3} in unstressed chloroplasts (Boyer, 1983; Boyer and Younis, 1983; Robinson, 1985) but in stressed tissues, with possibly double the ionic concentration, enzyme activity is inhibited. Other cellular processes have requirements and optima for particular ions, e.g., photosynthesis shows complex response to K^+ concentration (Gupta et al., 1989). Also the K^+ requirement for protein synthesis in many plants lies between 120 and 150 mol K^+ m^{-3} and changes in the concentration may affect the protein synthesizing mechanisms in the chloroplast, altering the ability to adjust to stress. For example, slower protein synthesis may impair the replacement of the rapidly turned-over D1 protein of the photosystem II reaction centre complex thus making stressed leaves more susceptible to photoinhibition (Björkman and Powles, 1984; Powles, 1984).

Accumulation of compatible solutes is common in halophytes (Hanson and Hitze, 1982); in crop plants this 'osmotic adjustment' has received much attention (Aspinall and Paleg, 1981; Hanson and Hitze, 1982; Morgan, 1984; Munns et al., 1979) as it appears to be a common adaptive mechanism to drought. It both decreases π and is a homeostatic mechanism to protect enzymes (and other macromolecules) and to maintain membrane and organelle integrity (Morgan, 1984; Munns et al., 1979; Stewart, 1981). However, the role of osmotic adjustment in maintaining the productivity of plants under water stress is more equivocal. Largely such compatible solutes accumulate at severe water deficiency, when damage is advanced (Aspinall and Paleg, 1981; Hanson and Hitze, 1982; Lawlor, 1983; Lawlor and Fock 1977, 1978; Stewart, 1981) as discussed here. The genetic differences underlying such adjustments have been examined in some studies but there is no clear picture of the way in which altering the enzyme complement of plants might improve the ability to adjust to adverse conditions, by synthesis of compatible solutes. Quantitative analysis of the fluxes of ions and metabolites in cells during the development of and adjustment to—water stress should be related to both the maintenance to water balance and to continuation of metabolic processes, which are not necessarily compatible.

Turgor (P) is the hydrostatic pressure caused by water entering cells in response to the osmotic potential of the cell contents (Kramer, 1983; Tyree and Jarvis, 1982). Turgor is the driving force for cell expansion and, therefore, for organ growth although it must be emphasised that these processes are under metabolic control and that turgor provides only the physical force for expansion. If P is small, the cell becomes flaccid and the expansion of growing cells slows, resulting in smaller organs and a reducing demand for assimilates, e.g., in cell walls and protein synthesis, a fact often overlooked. Because P changes greatly with a small decrease in cell volume (particularly within inelastic cells) and is closely coupled to growth, this mechanism provides a sensitive link between water stress in the environment and plant size (Boyer, 1983; Day et al., 1987; Fereres, 1985; Frederick et al., 1989).

The elastic modulus of cells is an important component of cellular water balance and of growth (Tyree and Jarvis, 1982). The difference in turgor potential (ΔP) is related (Eqn. 3) to the change in volume (Δv) by the bulk modulus of elasticity (ε).

$$\Delta P = \varepsilon \, \Delta \, V/V \tag{3}$$

Cells with very elastic walls and membranes change volume greatly with decrease in water content and, therefore suffer large increases in π and concentration of cellular contents and appear particularly prone to damage (Boyer, 1983; Gupta et al., 1989). Many stress resistant species have small, thick-walled, inelastic cells, so that P and ψ decrease substantially for only small changes in V and π.

Which of the components of the water balance is most important in metabolism? Particular processes may be affected by specific changes and it appears that turgor is important for cell expansion, water content for structural integrity and ionic concentration for cellular metabolism. Photosynthetic metabolism suffers irreversible damage, probably caused by the decrease in π, as ψ and π approach -1.5 to -2.0 MPa or beyond (at which level turgor is zero and cells probably plasmolysed) for plants grown with ample water and nutrients but below -3.0 MPa for plants grown under more stressed conditions. Possibly, metabolic adjustment during growth permits the development of smaller, more drought resistant cells which generally have lower π and ψ but larger P and suffer less change in volume etc (Kramer, 1983; Tyree and Jarvis, 1982). Thus adaptation to drought may be achieved modifying metabolic processes regulating π, cell expansion and these involve alterations in the amounts of enzymes, and changes in control systems; the following section considers some aspects of these processes.

IV. Photosynthetic Metabolism and Water Stress

a. Stomatal and mesophyll limitation

Decreasing ψ, π and P decrease net photosynthesis (P_n) of both C_3 (Boyer, 1983; Frederick et al., 1989; Krampitz et al., 1984; Lawlor, 1976; Lawlor and Fock, 1975) and C_4 plants (Lawlor and Fock, 1978; Pham Thi et al., 1982) as shown in Fig. 3, but the specific effects on the photosynthetic mechanism are still contested (Aspinall and Paleg, 1981; Boyer, 1983; Kramer, 1983; Lawlor, 1976; Stuhlfauth et al., 1988). There are differences in the response of species to stress depending on the conditions of growth, so it is perhaps dangerous to generalize, but there are general responses common to leaves of both C_3 and C_4 plants (Lawlor, 1979; Pham Thi et al., 1982). Despite the differences in C_3 and C_4 metabolism (Lawlor, 1987), imposition of a water deficit over periods of hours to days decreases P_n almost linearly with decreasing ψ in many plants related to increased π (Robinson, 1985) and to loss of P. Net photosynthesis decreases as P drops and almost stops at or below zero turgor (Fig. 3). The decrease in P_n is related to a decrease in stomatal conductance (g_s) and a decreased capacity of the metabolic systems to assimilate CO_2 (Boyer, 1983; Kramer, 1983; Lawlor 1976, 1983), i.e., to changed mesophyll conductance corresponding to changes in several photosynthetic processes (Krieg, 1983; Lawlor 1983, 1987).

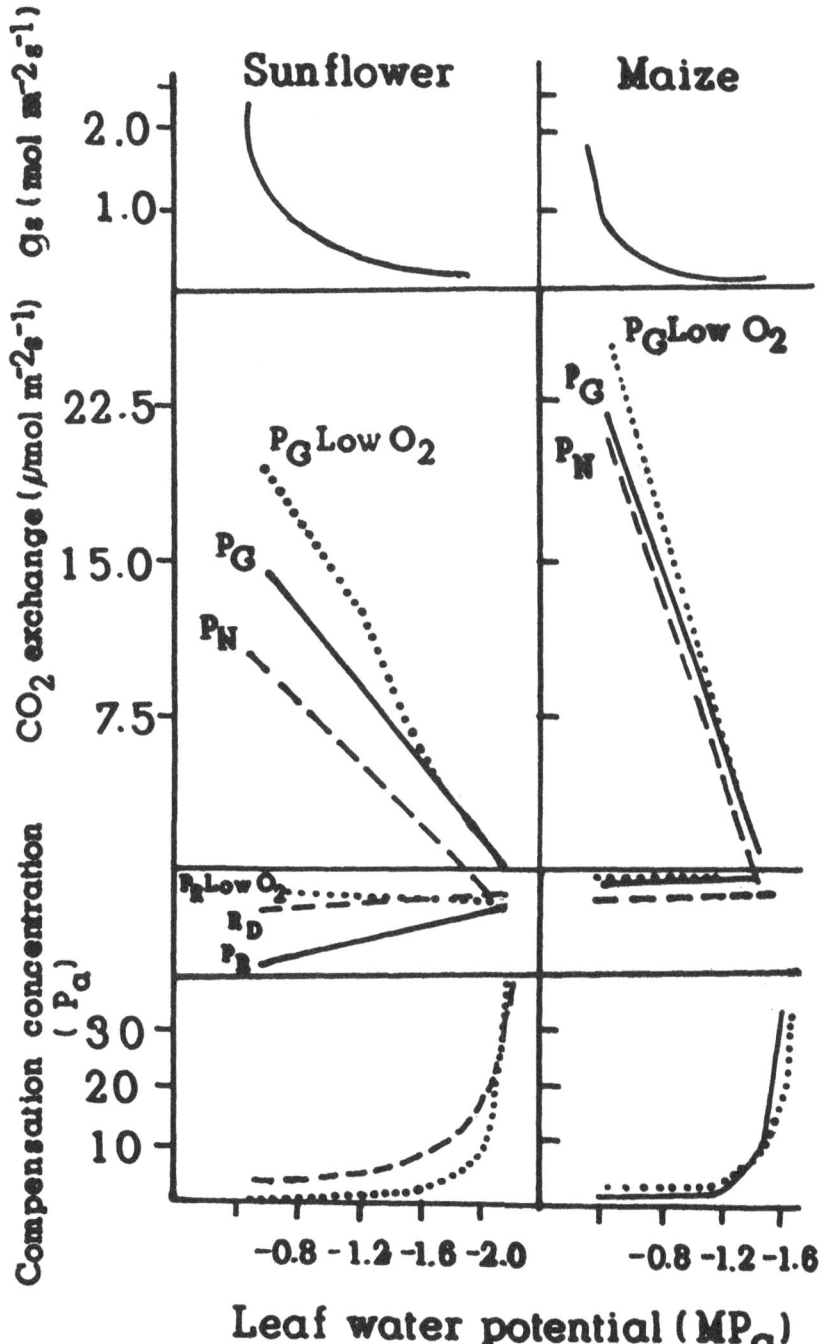

Fig. 3. Summary of the changes in stomatal conductance (g_s), gross and net photosynthesis (P_g and P_n respectively), photo- and dark-respiration (P_r and R_d) and the CO_2 compensation concentrations of C_3 (sunflower) and C_4 (maize) plants under a range of water stress (Lawlor, 1983).

Stomata close progressively with loss of turgor (Frederick *et al.*, 1989; Lawlor, 1976); this increases the resistance of the diffusion pathway, decreasing water loss and also the entry of CO_2 into leaves (Boyer, 1983; Kramer, 1983; Schulze and Hall, 1982). Depending on P_n this causes the sub-stomatal CO_2 concentration (Ci) to decrease although there is evidence that mechanisms may link assimilation rate, respiration and g_s to maintain Ci constant over a range of P_n (Farquhar and Sharkey, 1982; Krampitz *et al.*, 1984; Legg *et al.*, 1979). Stomata of plants adapted to stress may exhibit a "threshold response" (Ephrath *et al.*, 1990; Kramer, 1983), g_s remaining large as ψ decreases and then closing below a particular ψ.

The roles of g_s and mesophyll processes in determining P_n have been frequently discussed (Boyer, 1983; Farquhar and Sharkey, 1982). In earlier analyses, decreased g_s was regarded as the primary effect. However, the decreased P_n under stressed conditions is the main cause of impaired metabolism (Table 1 and Fig. 4a) (Boyer, 1983; Krieg, 1983; Lawlor, 1976; Lawlor *et al.*, 1989; Lawlor and Khanna-Chopra, 1983). Assimilation is shown as a function of Ci in Fig. 4a. The 'demand function' (Farquhar and Sharkey, 1982) is determined by the affinity of the carboxylation mechanism; the initial slope of the curve is the carboxylation efficiency and indicates the affinity of the system for CO_2 with the substrate ribulose bisphosphate (RuBP) not limiting (see Eqn. 5);. The plateau, P_{max}, (maximum assimilation) is determined by the limitation caused by inadequate supply of RuBP (Woodrow and Mott, 1989). Decreased carboxylation efficiency and P_{max} is commonly observed with decreased ψ (Fig. 4a) in leaves of both C_3 (Lawlor, 1976; Lawlor *et al.*, 1989; Lawlor and Khanna-Chopra, 1983) and C_4 plants (Kumar and Gupta, 1986; Lawlor and Fock, 1978). In contrast, the carboxylation efficiency of osmotically stressed isolated mesophyll cells (Sharkey and Badger, 1982) did not decrease although P_{max} did. The effects of stress on metabolism which are responsible for this will be considered here.

Calculaton of Ci assumes uniform g_s over the area of leaf for which P_n is determined or, if some stomata are more closed than others, that the diffusion of CO_2 in the subsequent stomatal spaces eliminates differences in Ci. If a few millimetres area have closed stomata, diffusion may not suffice. Therefore, calculation shows P_n to decrease but Ci remains relatively constant (Downton *et al.*, 1988) as observed (*e.g.*, Fig. 4a). This has been suggested to account for the decreased carboxylation efficiency with stress but $^{14}CO_2$ feeding studies suggest that assimilation rate and g_s are rather uniform in stressed leaves (Stuhlfauth *et al.*, 1988; Giminez, Young and Lawlor, unpublished).

Stomata restrict the movement of CO_2 and thereby decrease the rate of P_n of C_3 leaves in ambient CO_2 below the potential rate given by the demand function as shown in Fig. 4a where Ci depends on g_s as indicated by the arrows. The relative limitation of assimilation due to g_s is given by the ratio of P_{max} to P_n at the given CO_2. The limitation of mesophyll processes is given by $P_{max}/P_{max\ control}$. In water-stressed leaves the relative effects of g_s is generally small (Table 1) and decreases with increasing stress (Lawlor and Khanna-Chopra, 1983). Thus the decrease in P_n is largely caused by changes in the mesophyll capacity to fix CO_2 as observed in laboratory and field grown plants (Krieg, 1983). In C_4 plants, the relative stomatal control is generally smaller than in C_3 plants for the photosynthesis is saturated at

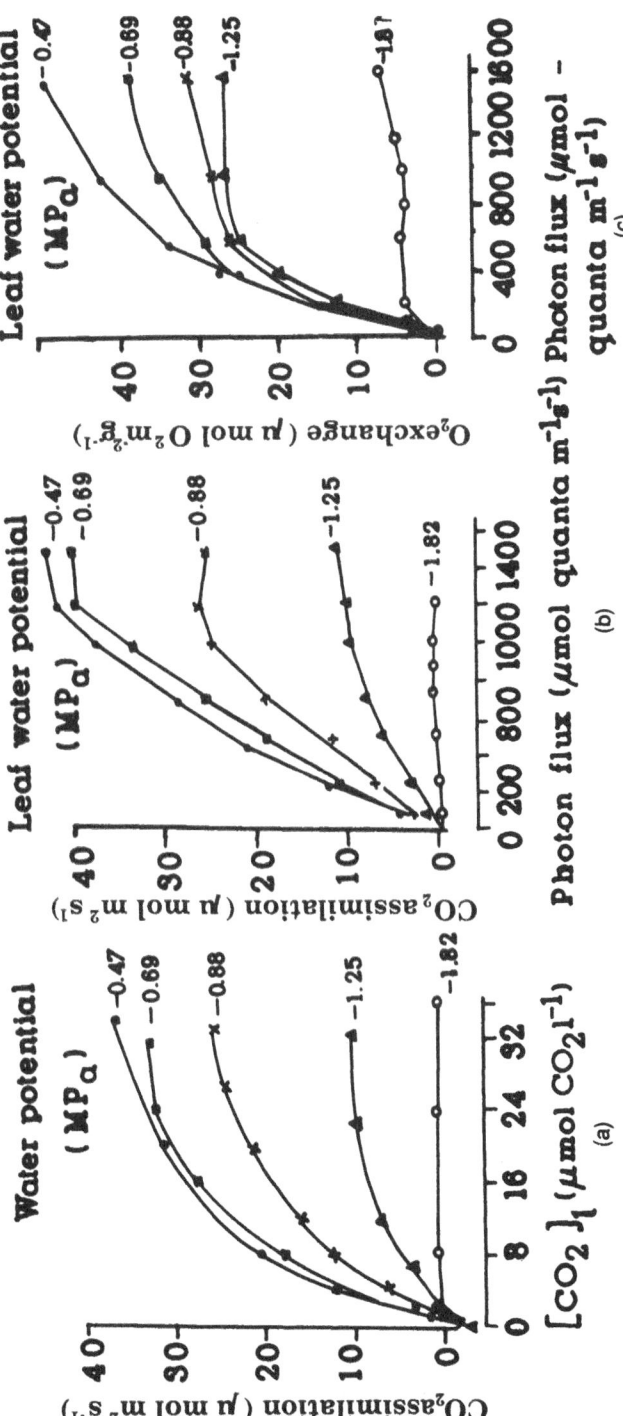

Fig. 4. The effects of stress on (a) the rate of CO_2 assimilation in relation to the calculated internal CO_2 concentration of the leaf, (b) the response of photosynthesis to the photon flux, and (c) the evolution of oxygen as a function of the photon flux in leaves of wheat (Lawlor and Khanna-Chopra, 1983).

Table 1. Effects of water stress on the stomatal and mesophyll components of net photosynthesis derived from the photosynthetic response curves of P_N to carbon dioxide in wheat shown in Fig. 4a.

Water potential (MPa)	Relative mesophyll limitation $\dfrac{P_{max}}{P_{max}\ \text{control}}$	Relative stomatal limitation $\dfrac{P_n-P_{max}}{P_{max}}$
− 0.47	1.00	0.39
− 0.69	0.88	0.34
− 0.88	0.65	0.25
− 1.25	0.25	0.09
− 1.82	0.05	0.00

ambient CO_2 (Farquhar and Sharkey, 1982; Lawlor, 1987). This analysis has important consequences for modifying photosynthesis under drought; for example, selection for smaller g_s to decrease water loss may be possible without adversely affecting assimilation in proportion. In this context assimilation of atmospheric CO_2 from the sub-stomatal spaces in leaves which have small g_s leads to enrichment of the plant dry matter in the heavy isotope of carbon, ^{13}C. The dry matter then exhibits a smaller carbon isotope discrimination in plants with large stomatal conductance. This is a characteristic difference between C_3 and C_4 plants related to their mode of CO_2 fixation and the amount of discrimination in C_3 plants depends on factors such as water stress. Because the carbon isotope discrimination develops over the life of the plant, and integrates both stomatal and mesophyll functions, it is a very useful method of selecting plants for differences in photosynthetic characteristics and indicates improvements in water use efficiency particularly under drought conditions (Farquhar and Sharkey, 1982).

b. Photosynthesis and respiration

Net photosynthesis is the difference between the gross photosynthesis (P_g) and the total respiratory loss of CO_2, i.e., photorespiration (P_r) plus dark respiration (R_d) (Lawlor, 1987).

$$P_n = P_g - (P_r + R_d) \tag{4}$$

The decreased capacity of the mesophyll may reflect inhibition of P_g or an increase of $P_r + R_d$ or both (Krampitz et al., 1984; Lawlor and Fock, 1975). It is essential to understand what controls the mesophyll's capacity to assimilate CO_2, if the mechanisms limiting photosynthesis under drought are to be modified. Clearly the effect is on P_g which decreased with stress in C_3 and C_4 plants as shown by studies using rapid measurements of $^{14}CO_2$ uptake (Lawlor, 1976; Lawlor and Fock, 1975). However, in both C_3 and C_4 plants, P_g decreased relatively less than P_n suggesting that the respiratory components increased proportionately under water stress (Hanson and Hitze, 1982; Kumar and Gupta, 1986; Lawlor and Fock, 1975).

V. Limitations to Photosynthetic Metabolism Caused by Water Stress

The limitation may occur in several parts of the mechanism (Boyer, 1983; Krieg, 1983; Lawlor, 1983): the light harvesting systems or electron transport within the chloroplast thylakoids; the energy transducing system (photophosphorylative ATP production); photosynthetic carbon reduction cycle (PCRC) in the chloroplast stroma; the utilization of assimilated carbon *e.g.*, transport out of the chloroplast. There is only a limited consensus as to the process most affected by stress and the following discussion considers the evidence.

a. Light response and electron transport

Water stress decreases P_n at all light intensities (Fig. 4b) so both apparent quantum yield (initial slope of the response curve) and P_{max} are affected. However, analysis suggests that primary capture of photon energy is insensitive to stress (Genty *et al.*, 1987). There is no information about how electron transport from the reaction centres of photosystem II to plastoquinone may be affected in the physiological range. However, the reaction centres may be affected by photoinhibition, caused by accumulation of energy which is not used by metabolism or dissipated (Björkman and Powles, 1984; Demming *et al.*, 1987; Powles, 1984). Electron transport is decreased by water stress, as shown by decreased O_2 production (Fig. 4c). However, the photosystems and electron transport are generally insensitive to changes in water balance (Keck and Boyer, 1974; Sharkey and Badger 1982; Stuhlfauth *et al.*, 1988) as expected for a process occurring in the non-aqueous environment of the lipid membranes, although the rate may depend on the conformation of the thylakoid membranes, concentration of membrane components (*e.g.*, plastoquinone) and perhaps also on ionic concentrations (Govindjee *et al.*, 1981). Water-splitting associated with PS II in the thylakoid lumen, a major process in electron transport and oxygen production, is sensitive to stress (Genty *et al.*, 1987; Govindjee *et al.*, 1981; Lawlor and Khanna-Chopra, 1983) but it is not clear if it is as a result of direct damage to the enzyme-manganese complex or to the electron transport chain.

Evidence that electron transport is maintained at a relatively large rate in stressed chloroplasts (Keck and Boyer, 1974), leading to a more reduced state of the photosynthetic system, comes from three sources: increased chlorophyll *a* fluorescence (Govindjee *et al.*, 1981; Krauser and Somersalo, 1989); maintenance of high reduction states of the pyridine nucleotides (Lawlor and Khanna-Chopra, 1983) and photoinhibition (Powles, 1984). Variable chlorophyll fluorescence increases with stress because dissipation of energy from the chlorophyll complexes is impaired. This may arise from changes in the q_Q and q_E quenching mechanisms (Krauser and Somersalo, 1989; Stuhlfauth *et al.*, 1988). q_Q and q_E are respectively the dissipation of energy as heat either directly or by transfer to other molecules and the consumption of the energy by synthesis of ATP and CO_2 assimilation. The largest increase in variable fluorescence occurs after turgor loss so that the energetics of the thylakoids are altered substantially as CO_2 assimilation is strongly decreased (Govindjee *et al.*, 1981). It also suggests that the main effect of stress is on q_E, probably because ATP synthesis is inhibited and this prevents dissipation of the proton gradient (Stuhalfauth *et al.*, 1988).

b. Photoinhibition and energy dissipation

Photoinhibition has received considerable attention (Björkman and Powles, 1984); Demmig et al., 1987) but the relationship between damage and degree of stress remains to be quantitatively assessed. One source of damage to photosynthesis involves high energy states of electron transport components. These react with, for example, molecular oxygen, producing reactive O_2 species, such as, oxygen radicals, hydroxyl ions, superoxide and peroxides, all potentially very damaging to photosynthetic components, e.g., lipids and proteins (Kacperska and Kubackaebalska, 1989; Lawlor, 1987). Lipids may be destroyed by peroxidation reactions, linked to increasing lipoxygenase activity, which is also involved in the formation of activated oxygen (Kacperska and Kubackaebalska, 1989) resulting from the over-reduced state of electron transport components. The D1 protein of the reaction centre may also be damaged. These effects are, however, secondary probably arising from damage to metabolism 'downstream' of the energy transducing process, not a direct effect of the photosystems in response to change in water balance (Hanson and Hitze, 1982; Lawlor, 1983). The importance of mechanisms dissipating high energy states and potentially damaging intermediates should be mentioned (Björkman and Powles, 1984; Demmig et al., 1987; Powles, 1984). Carotenoid quenching of high energy states of intermediates by, for example, the violaxanthin-zeaxanthin cycle, often increases in activity when tissues develop under, or are exposed to, water deficiency (Burke et al., 1985). Activity of the enzymes catalase and superoxide dismutase, which are responsible for breaking down hydrogen peroxide and superoxide respectively (Lawlor, 1987), also regulates energy metabolism, including control of the reduced pyridine nucleotides of photosynthesis. Stress may trigger the formation of the protective mechanisms (Björkman and Powles, 1984; Burke et al., 1985; Demmig et al., 1987) although if developed too rapidly it may damage the plant before protection is developed. Indeed, it is not fanciful to suggest that the genetic capacity for such protection is a key to the ability of plants to resist damage from drought.

c. ATP synthesis

Photophosphorylation results from proton flux from the thylakoid lumen to the chloroplast stroma through the ATP synthase, enzyme complex (CF_0–CF_1, or coupling factor) which spans the thylakoid membrane (Boyer and Younis, 1983; Lawlor, 1987). The driving force 'pumping' the H^+ is electron transport. Photophosphorylation is very sensitive to water stress. Boyer and co-workers (Boyer, 1983; Boyer and Younis, 1983; Keck and Boyer, 1974) suggest that increased concentration of Mg^{2+} ions, in particular, inhibits the activity of CF_0–CF_1. Experiments where Mg^{2+} increased from 3 to 9 mol Mg^{2+} m^{-3} in spinach chloroplasts decreased photophosphorylation by 50%. The ions may affect subunits of the enzyme (Younis et al.1979) by impairing ADP binding rather than ATP release from the enzyme or by affecting the flow of protons through CF_0/CF_1. This most important work should be followed up by following the energetics of thylakoids and CF_0/CF_1 activity in vivo under stress conditions.

Decreased ATP content of stressed photosynthetic cells has been observed experimentally (Hanson and Hitze, 1982; Lawlor and Khanna-Chopra, 1983)

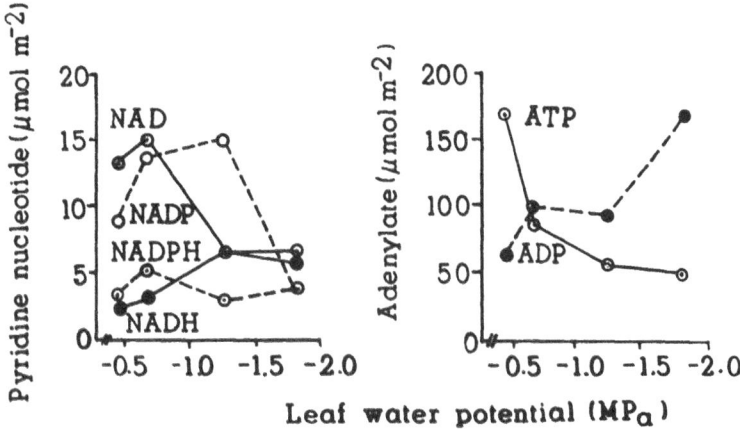

Fig. 5. The effects of short-term water stress on: (a) the pyridine nucleotides, and (b) the adenylates of illuminated wheat leaves (Lawlor and Khanna-Chopra, 1983).

(Fig. 5a), although the relation to cellular water balance is not well defined. Also the adenylate energy charge, EC, (Pradet and Raymond, 1983) decreased from 0.8 in unstressed wheat leaves to 0.57 at −1.8 MPa ψ (Lawlor et al., 1989; Legg et al., 1979; Pradet and Raymond, 1983). Such a change would greatly impair metabolism, for EC is usually controlled within narrow limits (Pradet and Raymond, 1983). Given the central role of ATP in metabolism, including regeneration of RuBP, ion transport and protein synthesis, it is not surprising that stress sufficient to decrease ATP will decrease photosynthesis.

d. RuBP production

The supply of RuBP to the carboxylation sites of RuBPcase is a most important process in photosynthesis (Eqn. 5) determining the carboxylation rate and synthesis of 3-phosphoglyceric acid (3-PGA).

$$RuBP + CO_2 = 2 \times 3\text{-PGA} \tag{5}$$

Synthesis of RuBP depends on the ATP and NADPH supply (activity of photophosphorylation and electron transport, respectively) and on the enzymes of the PCRC (Fig. 1) (Lawlor, 1987). Decreased ATP will particularly slow rates of phosphoglycerate kinase and ribulose-5-phosphate kinase reactions (Plaut and Bravdo, 1973). Enzymes of the PCRC are relatively little affected by stress (Hanson and Hitze, 1982; Plaut and Bravdo, 1973). This is based on studies without the advantage of the present knowledge of activation factors (Woodrow and Mott, 1989) and re-examination with modern methods of assay is desirable. Long-term changes in amounts of enzymes may indeed occur, and RuBPcase content was reduced in field grown plants under stress (Lawlor, unpublished). This may be a consequence of the general reduction in protein synthesis caused by impaired energy metabolism, inadequate phosphate (Lawlor et al., 1981; Legg et al., 1979) or deficient nitrogen caused by other stress-related events. Short-term effects of stress on P_n are probably not mediated through enzyme amounts. In the PCRC (Fig. 1), the control points for the operation of ions (Mg^{2+}, Pi, K^+) and products of the cycle (3-PGA, GAP) may be sites of inhibition. For example, large Pi concentration resulting from rapid loss of cell water may inhibit sedoheptulose bisphosphates and slow P_n. Conversely, 3-PGA concentration may stimulate P_n. However, understanding of such regulation is poor. One of the most studied enzymes of the PCRC is RuBPcase, which catalyses the carboxylation of RuBP (Eqn. 5) and also its oxygenation forming phosphoglycolate (PG) as well as 3PGA (Eqn. 6).

$$RuBP + O_2 = 3\text{-phosphoglyceric acid} + PG \tag{6}$$

The dual oxygenation and carboxylation function is of great importance in determining the relative rates of P_n and P_r which depends upon the CO_2/O_2 ratio at the catalytic site (Farquhar and Sharkey, 1982; Hanson and Hitze, 1982; Lawlor, 1979, 1987; Lawlor and Pearlman, 1981; Vu et al., 1987). In stressed leaves, the CO_2/O_2 ratio decreases (Lawlor and Pearlman, 1981) by only a small amount (Farquhar and Sharkey, 1982; Stuhlfauth et al., 1988) probably because of feed-back regulation by the combined effects of increased P_r and decreased P_n. Despite reduction in g_s, Ci remains relatively unaffected. This depends on the calculation of Ci; but as briefly discussed patchy stomatal closure does not generally explain the reduction of P_n in stressed leaves. A decrease in Ci with mild stress and stomatal closure (Gimenez and Lawlor, unpublished) and than to increase could reflects the balance between P_n, P_r and R_d. In modelling these processes (Lawlor and Pearlman, 1981) the value of Ci was most important for determining the ratio of CO_2/O_2 and, therefore, P_n/P_r but the model and data were indaequte to explain the carbon fluxes at severe stress. As a consequence of the characteristics of RuBPcase the decrease in P_r with decreased ψ was smaller than that in P_g. However, this is a relatively small effect and another cause of the reduced P_n is inactivation of RuBPcase in vivo, but this was not a contributing factor in sunflower (Lawlor and Gimenez, unpublished).

Stress generally decreases the carboxylation efficiency of leaves (Fig. 4a) but not of isolated cells or protoplasts. Stress may decrease the affinity of the enzyme for the limiting CO_2 or may inhibit RuBP regeneration even at very small P_n. Possibly, inhibition of CF_0-CF_1 decreases RuBP at the active sites of RuBPcase even with limiting CO_2. This suggests that ATP synthesis is coupled to the CO_2 supplies. Diffusion of CO_2 to the carboxylation site is also a potential limitation to P_n. Transport across the cell membranes and cytosol may be important if the conductivity of the pathway decreases, e.g., if carbonic anhydrase activity falls, or the solubility of CO_2 decreases. These appear to be relatively unimportant changes, although inadequately addressed.

e. Photorespiration and dark respiration

Photorespiration is CO_2 released by decarboxylation, in the mitochondria, of glycine derived from the metabolism of phosphoglycolate (PG); (Eqn. 6) by the glycolate pathway (Fig. 1a, 1b). Two molecules of glycine are decarboxylated producing one molecule of serine and releasing one molecule of ammonia and one of CO_2 (Lawlor, 1987). The P_r may be 30% of P_n of C_3 leaves in atmospheres containing 35 Pa CO_2 and 21 kPa O_2 at 25°C. Therefore, it greatly decreases the efficiency of carbon assimilation. Decreasing the O_2 pressure to 2 kPa eliminates P_r (Eqn. 6) and increases P_n by 40–50 %. Water stress decreases the absolute rate of P_r but less than it decreases P_n (Krampitz et al., 1984; Lawlor 1976, 1979; Lawlor and Fock, 1975). Consequently, the ratio of P_r/P_n increases with stress; at very small ψ, P_n may almost stop but respiration continues at a reduced rate (Fig. 3). This correleates with an increased CO_2 compensation concentration (γ) of stressed leaves (Fig. 3), suggesting that P_r may by partly responsible for the decrease in P_n. A relative increase in P_r is shown by the increased flux of C (measured by $^{14}CO_2$ techniques) into glycine and serine (Lawlor, 1976; Lawlor and Fock, 1977a) although these may be synthesised by routes other than the glycolate pathway. Amino acids form a greater proportion of assimilates in stressed leaves suggesting an increased flux of C into the glycolate pathway relative to that in photosynthesis. However, the absolute decrease in the ^{14}C in these compounds (Lawlor, 1976) and their decreased specific activity (Lawlor and Fock, 1977a) suggests synthesis by non-glycolate pathway routes.

The conclusion that P_r (i.e., CO_2 from the glycolate pathway) increases relative to P_n at low ψ is not justified by the accumulated evidence. The CO_2 release and increase in γ are insensitive to 2 kPa O_2; were the process truly photorespiratory then this should be eliminated. Therefore, a substantial part of the CO_2 released in the light from very stressed leaves cannot originate from the glycolate pathway. In addition, the much decreased P_g and P_n (Fig. 3; Fig. 4) and evidence from tracer studies (Lawlor, 1976; Lawlor and Fock, 1977a, 1977), suggest that the flux of C into the glycolate pathway is much smaller with low ψ. The sepcific activity of CO_2 released from stressed leaves using $^{14}CO_2$ exchange techniques showed that with increasing stress a larger proportion of the CO_2 originated from sources other than immediate photosynthesis (Lawlor and Fock, 1975). This may arise in two ways, either previously assimilated C is recycled (Gerland and Andre, 1987; Lawlor and Fock, 1975), and thence enters the glycolate pathway or the C released is not from that source. This respiration cannot be only P_r but include R_d, possibly from the carboxylic acid (TCA) cycle located in the mitochondria (Douce, 1985). Although forms of chloroplast respiration occur (Garab et al., 1989), utilizing the reductive pentose phosphate cycle they may be inhibited by osmotic stress and are unlikely to be the source of CO_2 (Willeford et al., 1989). Although the interaction between P_n and R_d has resisted quantitation, mitochondrial activity is probably not inhibited in the light (Douce, 1985). Rather, it may be suppressed by the large concentration of ATP or by high energy charge (Pradet and Raymond, 1983) under the normal condition in illuminated leaves. Given the decrease in ATP in stressed cells probably R_d is not inhibited as P_n decreases, rather it may increase. Consequently under mild water stress P_r may increase relative to P_n as the CO_2/O_2 ratio decreases with stomatal closure and an increased proportion of CO_2 comes from R_d and under severe stress R_d dominates (Lawlor 1983; Lawlor and Fock, 1975, 1978; Lawlor and

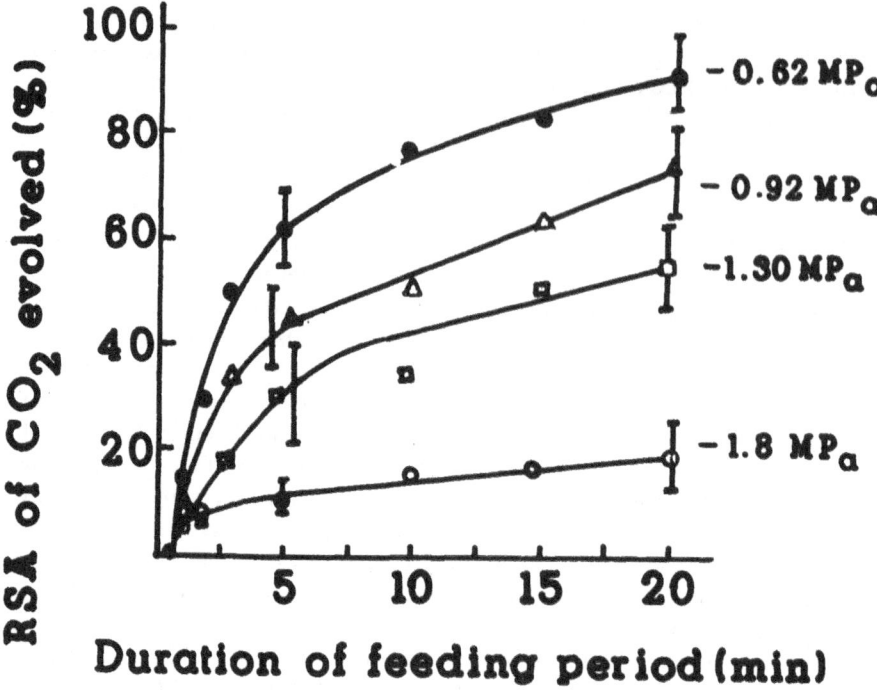

Fig. 6. The relative specific activity of the CO_2 evolved from sunflower leaves of different water potential (Lawlor and Fock, 1975).

Pearlman, 1981). This explains the smaller relative specific activity of the CO_2 released under stress (Lawlor and Fock, 1975) and the O_2 insensitivity.

C_4 plants have little photorespiration even at low ψ (Kumar and Gupta, 1986; Lawlor, 1987; Lawlor and Fock, 1978) but P_n is decreased, CO_2 is released in the light, and $(P_r + R_d)/P_n$ increases, thus increasing γ as observed in water stressed maize (Fig. 3) (Lawlor, 1976; Lawlor and Fock, 1978). This respiration is O_2 insensitive suggesting that R_d continues at low ψ. Thus in leaves with C_3 and C_4 metabolism, stress affects the mechanisms regulating photorespiratory and dark respiratory processes.

f. Summary of the causes of decreased photosynthesis in stressed leaves

Effects of stress may be explained by a mechanism based on:
— decreased stomatal conductance,
— increased ionic concentration (particularly Mg^{2+}) of the chloroplast stroma,
— progressive inhibition of ATP synthesis by the CF_0–CF_1 enzyme due to increased magnesium concentration,
— maintenance of relatively large rates of electron transport and reduction of pyridine nucleotides,
— little change in the enzyme components of photosynthesis and their characteristics.

The decrease in g_s resulting from partial loss of turgor restricts the diffusion of CO_2 and decreases P_n for leaves operating below the plateau of the P_n versus

Ci curve; there is little effect of ψ or π on P_n. Such mild stress may affect metabolism *via* the source/sink balance, *e.g.*, reduced growth combined with maintenance of P_n may increase the content of sugars and stimulate the enzymes of sucrose synthesis (Quick *et al.*, 1989). Photorespiration may decrease in absolute terms, but increase as a proportion of P_n as a result of the decrease in Ci. Further water loss may progressively inhibit photophosphorylation and the ATP content of the cells decreases. This will affect the conversion of 3-PGA to RuBP, resulting in decreased P_g; this is probably the primary stress lesion. Sucrose production will decrease (Lawlor, 1976; Lawlor and Fock, 1977a) and relatively more C will enter the glycolate pathway and other amino acid synthesizing routes.

Continuation of electron transport at relatively high rates causes over-reduction of pyridine nucleotides, leading to increased chlorophyll fluorescence, associated with a large proton gradient and reduced consumption of energy by q_e quenching in thylakoids. Consequently, the potential for photochemical damage, *i.e.*, photoinhibition, rises with decreasing P_n and decreasing cell water content. Impaired synthesis of ATP inhibits many processes essential for maintenance of cellular homeostasis, *e.g.*, transport of ions and protein synthesis. Hence, maintenance of conditions for photosynthesis will be impaired in the short- and long-term. Mechanisms for readjustment to the changed conditions clearly exist for leaves recover from often quite severe stress but are poorly understood. If the effects on metabolism are too great then the photosynthetic system breaks down irreparably.

VI. Secondary Metabolism Associated with Changes in Photosynthesis

Secondary metabolism is altered by stress (Kacperska and Kubackaebalska, 1989; Lawlor, 1983; Lawlor and Fock, 1977; Noguchi *et al.*, 1968). Yet the mechanisms by which the events, in often quite distantly related parts of metabolism, are associated with changes in the energetics and carbon assimilation of chloroplasts are unclear. Some changes associated with stress are:
— Increased accumulation of amino acids, particularly proline and glycine-betaine (Aspinall and Paleg, 1981; Hanson and Hitze, 1982; Sinha and Rajagopal, 1975; Stewart, 1981).
— Decreased carbohydrates and organic acids as primary assimilation products (Lawlor, 1979; Lawlor and Fock, 1977).
— Increased accumulation of substances with phyto-hormonal activity (abscisic acid, farnesol, polyamines and production of ethylene) (Goswami and Srivastava, 1985; Hanson and Hitze, 1982; Kacperska and Kubackaebalska, 1989; Parry and Horgan, 1990; Smith, 1985).

a. Interactions between amino acids, carbohydrates and organic acids

Accumulation of amino acids predominantly of the glutamate family, glutamate and proline (Table 2), occurs at or somewhat below zero turgor in C_3 and C_4 plants (Aspinall and Paleg, 1981; Lawlor, 1979; Lawlor and Fock 1977, 1978; Rhodes *et al.*, 1986; Stewart, 1981). Light and carbohydrates stimulate accumulation (Lawlor and Fock, 1977; Noguchi *et al.*, 1968; Stewart, 1981). The mechanism may be related to reduction of nitrate ions in photosynthetic cells, which requires reductant as

Table 2 Summary of the amounts (μmol m^{-2}) of selected metabolites in unstressed sunflower and maize leaves and the changes (%) caused by severe water stress (ca; –2.0 MPa & –1.6 MPa, respectively) (derived from data in 43, 44).

	Sunflower		Maize	
	Control	– 2.0 MPa %	Control	– 1.6 MPa %
3–PGA	37	–80	43	–90
Sucrose	1608	–100	2920	–93
Malate	896	+180	2240	–93
Glutamate	272	+175	272	+125
Glycine	15	+600	130	+530
Serine	50	+700	100	+100
Proline	40	+5400	20	+6000

ferredoxin and pyridine nucleotides. However, nitrate reductase is inhibited by stress and the activity of other enzymes of nitrogen metabolism (Sinha and Rajagopal, 1975). Therefore, it is unclear if decreased ψ affects nitrate metabolism *via* the energy and reductant supply or through enzyme activity. Possibly during the development of stress, reduction continues before inhibition of the enzyme slows the reaction. This would allow nitrate to be converted to ammonia. Even if ammonia is produced it may not be converted to glutamate by the GS/GOGAT cycle if ATP is not available (Douce, 1985; Lawlor, 1987). Ammonia may then be assimilated by glutamate dehydrogenase (GDH), located in the mitochondria (Douce, 1985; Srivastava and Singh, 1987; Yamaya and Oaks, 1987), which converts α-ketoglutarate and ammonia to glutamate (see Fig. 7). Although GDH has been regarded as of no significance for ammonia assimilation in unstressed chloroplasts, because the GS/GOGAT system functions at much lower concentrations of ammonia than GDH, lack of ATP in stressed cells may slow or inhibit the GS/GOGAT cycle (Yamaya and Oaks, 1987) and, thereore, GDH may become more important. Also GDH uses NAD(P)H, which may be available in stressed cells. Thus conditions are conducive to production of glutamate (Srivastava and Singh, 1987; Yamaya and Oaks, 1987). The stimulation of R_d, possible increase in GDH function and NADH content and organic acids of the TCA cycle point to relatively increased importance of mitochondrial metabolism (Douce, 1985) which may increase demand for carbohydrates, the content of which falls with stress (Lawlor and Fock, 1977). Organic acids are substrates for proline synthesis in bean (Venekamp *et al.*, 1989) via glutamate (Aspinall and Paleg, 1981; Douce, 1985; Noguchi *et al.*, 1968; Stewart, 1981) so the flux of carbon in stressed photosynthetic cells may be from storage carbohydrates to the TCA cycle acids, and to glutamate and proline associated with mitochondrial activity. Possibly the decrease in ATP allows glycolysis and the TCA cycle to function (as indicated in Fig. 7). However, the most important feature (and one generally ignored in attempting to link photosynthesis to so-called stress metabolites) is availability of reductant (NADPH and NADH) (Stewart, 1981). Reductant and a high reductant charge (Pradet and Raymond, 1983) are essential for synthesis of several of the stress metabolites (Fig. 7), *e.g.*, conversion of pyrroline-5-carboxylate to proline. Thus, synthesis of proline in stressed tissues may be regarded as a response to the imbalance between reductant

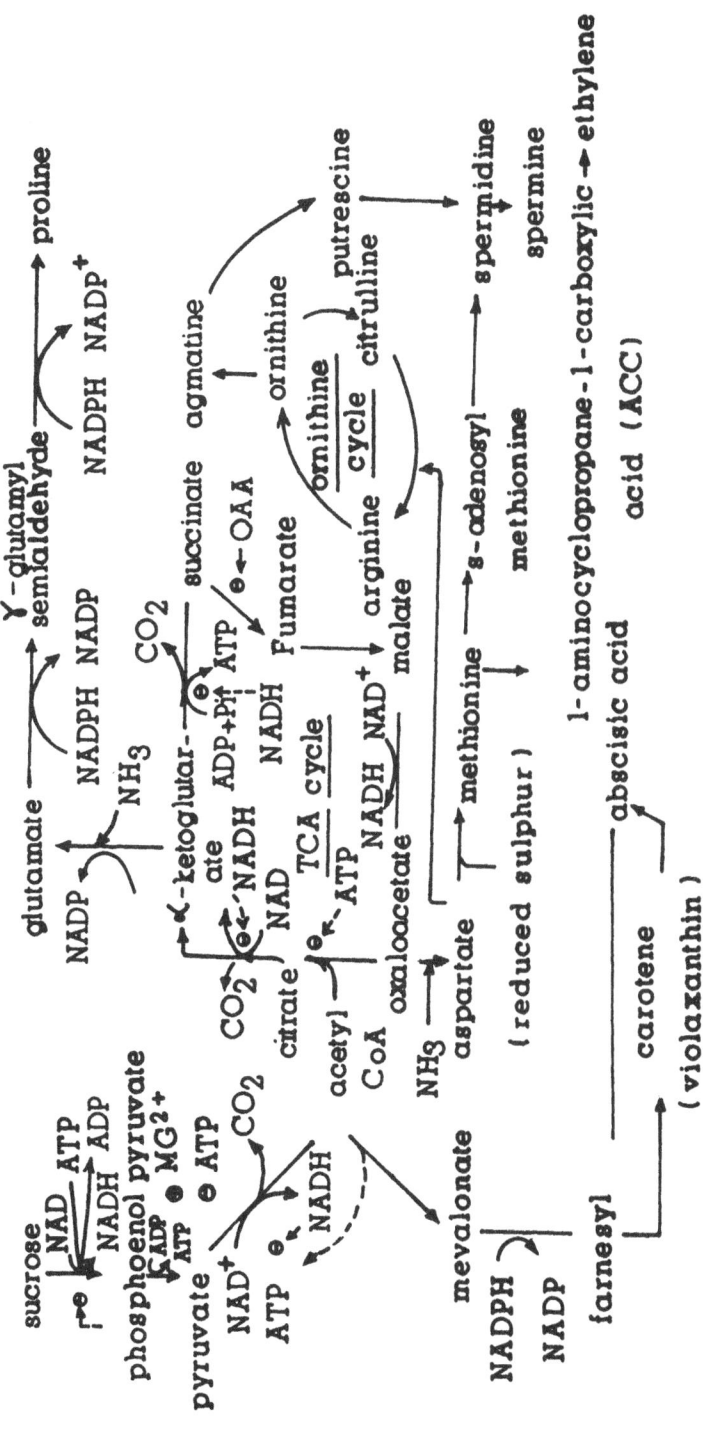

Fig. 7. A highly schematic view of the links between the metabolites which accumulate as a result of water stress, the tricarboxylic acid cycle functions and the sucrose produced in photosynthesis as the source of carbon for secondary metabolism (Douce, 1985; Hanson and Hitze, 1982; Lawlor 1979, 1983).

and ATP supply (Lawlor, 1983). Further, it is a sign of severely disturbed metabolism, associated with death of the tissue (Hanson and Hitze, 1982). Proline may play a role in protecting enzymes and membranes (Aspinall and Paleg, 1981; Hanson and Hitze, 1982) and in adjustment to stress but its value would be before the massive accumulations. Proline may also be a source of nitrogen and carbon for stressed tissue upon re-hydration (Aspinall and Paleg, 1981; Hanson and Hitze, 1982; Stewart, 1981).

Other amino acids which accumulate in stressed cells (Lawlor and Fock, 1977; Stewart, 1981) may be important substrates for synthesis of secondary products. Aspartate, formed by transamination of oxaloacetate with glutamate (Douce, 1985; Stewart, 1981), is a substrate for methionine synthesis and a source of arginine and ornithine in the ornithine cycle. From methionine and ornithine are derived the polyamines agmatine, spermine and spermidine which have growth regulatory properties (Smith, 1985), serving to reduce growth in a manner analogous to ABA. The mechanisms of these synthetic processes are poorly understood; reductant charge may be important and so also the activity of mitochondria. Although amino acid accumulation has been used as a criterion for selection of drought resistant crops, the lack of success (Hanson and Hitze, 1982) may reflect the damage caused by stress and not their role as a mechanism for minimising stress effects. The regulation of amino acid metabolism must be carefully analysed to explain the effects of stress and the links to photosynthesis (Aspinall and Paleg, 1981; Hanson and Hitze, 1982).

b. Abscisic acid, polyamines, ethylene and other hormone-like compounds

Accumulation of ABA is a significant stress response, occurring at low ψ and being partially dependent on light (Hanson and Hitze, 1982; Lawlor, 1983). A major problem is lack of clarity about the routes of ABA synthesis (Parry and Horgan, 1990). One pathway may be from pyruvate *via* acetyl CoA, with reduction by NADH to mevalonate and, in ATP demanding reactions, to farnesyl pyrophosphate and thence to ABA (Hanson and Hitze, 1982; Parry and Horgan, 1990). This is also the route for terpene synthesis and production of carotenoids such as violaxanthin and zeaxanthin. Stress generally increases the products of this pathway (Demmig *et al.*, 1987). Possibly, production of ABA is stimulated by reductant and increased acetyl CoA resulting from mitochondrial activity and reduced demand from ATP requiring processes. Abscisic acid is probably also formed by the breakdown of carotenoids (Parry and Horgan, 1990), which may be linked to the energy dissipating functions of these molecules (Björkman and Powles, 1984; Demmig *et al.*, 1987). It is tempting to speculate that a mechanism regulating growth *via* 'stress hormones' is linked to changing cell water operating through ATP synthesis and cellular energy balance, to achieve a long-term adjustment to the environment.

Ethylene is produced in stressed tissues (Kacperska and Kubackaebalska, 1989). Its synthesis is linked to the availability of methionine, formation of aminocyclopropane-3-carboxylic acid and probably to changes in the reductant state of the cell and the presence of free radicals and highly energized states of O_2 (Kacperska and Kubackaebalska, 1989). Synthesis may follow damage to membranes and the release of lipoxygenase. Thus, although not directly related to photosynthesis, production of many secondary metabolites of considerable importance in stressed cells may depend on the conditions in the chloroplast, particularly the ATP supply and the reduced redox states (Lawlor, 1983).

The energy, carbon and nitrogen fluxes have not been analysed experimentally in sufficient detail to develop a model of the interactions in stressed tissues. An experimental analysis designed to quantitate the processes and test the validity of the postulated mechanisms is essential if progress is to be achieved in selection and manipulation of plants for better productivity in dry environments.

VII. Summary

The causes of decreased P_n under stress, outlined in the first section, are analysed to identify the sites of limitation, which are metabolic rather than stomatal. The major limitation to photosynthesis is considered to be impaired ATP synthesis. This decreases synthesis of RuBP and slows P_n. From this, and the relatively high reduction state of the energy transducing systems, stems the gross changes in composition of stressed leaves. Critical analysis of the adenylates and pyridine nucleotides of stressed photosynthetic cells is required. The increase in CO_2 release in the light relative to P_n is due to progressively greater R_d rather than to P_r *sensu stricta* as ψ and P decrease and π and ionic concentration increase. Many secondary effects of stress on metabolites are related to mitochondrial activity. Two major questions concerning the effects of water stress are: what are the molecular features of coupling factor which make it susceptible to increased ion concentration? What stimulates the mitochondrial activity, increased TCA function, CO_2 production, amino acid, organic acid synthesis and carbohydrate consumption as photosynthesis decreases with stress?

Possibly, the glycolate pathway and mitochondrial electron transport serve to adjust cellular energy status and secondary metabolites may accumulate because conditions favour their synthesis rather than consumption. Low ATP may stimulate the carbon flux from carbohydrates through glycolysis to the TCA cycle. These concepts, based on experimental observations, allow development of specific hypotheses which can be tested, *e.g.*, the sequences of events in the quantitative changes of ATP, reductant, carbon and nitrogen assimilation and accumulation of products with stress. This is required in order to improve understanding of the effects of stress on photosynthesis.

VIII. References

Aggarwal PK and Sinha SK (1984) Differences in water relations and physiological characteristics in leaves of wheat associated with leaf position on plant. *Plant Physiol* 74: 1041–1045

Aggarwal PK and Koundal KR (1988) Relative sensitivity of some physiological characteristics to plant water deficits in wheat. *Plant Physiol and Biochem* 15: 161–168

Asana RD and Basu RN (1963) Studies in physiological analysis of yield. VI Analysis of the effect of water stress on grain development in wheat. *Indian J Plant Physiol* 6: 1–13

Aspinall D and Paleg LG (1981) Proline accumulation: physiological aspects. In: (LG Paleg and D Aspinall, eds) *Physiology and Biochemistry of Drought Resistance of Plant*. Academic Press, Sydney pp 203–241

Beadle CL, Stevenson RR, Neuman HH, Thurtell GW and King RM (1973) Diffusive resistance, transpiration and photosynthesis in single leaf of corn and sorghum in relation to leaf water potential. *Can J Plant Sci* 53: 537–544

446

Björkman O and Powles SB (1984) Inhibitions of photosynthetic reactions under water stress: interaction with light level. *Planta* **161**: 409–504

Blum A, Mayer J and Gozlan G (1983) Chemical desiccation of wheat plants as a stimulator of post anthesis stress: II relations to drought stress. *Field Crops Res* **6**: 147–155

Boyer JS (1983) Subcellular mechanisms of plant response to low water potential. *Agricultural Water Management* **7**: 239–248

Boyer JS and Younis HM (1983) Molecular aspects of photosynthesis at low leaf water potentials. In: (R Marcelle, H Clijsters and M van Poucke, eds) *Effects of Stress on Photosynthesis*. Martinus Nijhoff/Dr. W Junk Publishers, The Hague pp 29–33

Brevedon ER and Hodges HF (1973) Effect of moisture deficits on ^{14}C translocation in corn (*Zea mays* L.). *Plant Physiol* **52**: 436–439

Bunce JA (1982) Effect of water stress on the photosynthesis in relation to diurnal accumulation of carbohydrates in source leaves. *Can J Bot* **60**: 195–200

Burke JJ, Gamble PE, Hatfield JL and Quisenberry JE (1985) Plant morphological and biochemical responses to field water deficits I Responses of glutathione reductase activity and paraquat sensitivity. *Plant Physiol* **79**: 415–419

Constable CA and Rawson HM (1980) Carbon production and utilisation in cotton. Interference from a carbon budget. *Aust J Plant Physiol* **7**: 539–553

Dale J and Wyse RE (1985) Evidence on the mechanism of enhanced sucrose uptake at low cell turgor in leaf discs of *Phaseolus coccimius*. *Physiol Plant* **64**: 547–552

Day W, Lawlor DW and Day AT (1987) The effect of drought on barley yield and water use in two contrasting years. *Irrigation Science* **8**: 115–130

Demmig B, Winter K, Krüger A and Czygan FC (1987) Photoinhibition and zeaxanthin formation in intact leaves. *Plant Physiol* **84**: 218–224

Douce R (1985) *Mitochondria in higher plants. Structure, function and biogenesis.* Academic Press, Orlando pp 1–327

Downton WJS, Loveys BR and Grant WJR (1988) Non-uniform stomatal closure induced by water stress causes putative non-stomatal inhibition of photosynthesis. *New Phytol* **110**: 503–509

Ephrath JE, Marani A and Amibravdo B (1990) Effect of moisture stress on stomatal resistance and photosynthetic rate in cotton (*Gossypium hirsutum*) 1 Controlled levels of stress. *Field Crop Res* **23**: 117–131

Farquhar GD and Sharkey TD (1982) Stomatal conductance and photosynthesis. *Ann Rev Plant Physiol* **33**: 317–345

Fereres E (1985) Responses to water deficits in relation to breeding for drought resistance. In: (JP Srivastava, E Porceddu, E Acevedo and S. Varma, eds) *Drought tolerance in winter cereals*. John Wiley & Sons, Chicester pp. 263–274

Frank, AB, Power, JF and Willis WO (1973) Effect of temperature and plant water stress on photosynthesis, diffusion resistance and leaf water potential in spring wheat. *Agron J* **66**: 777–780

Frederick JR, Alm DM and Hesketh JD (1989) Leaf photosynthetic rates, stomatal resistances, and internal CO_2 concentrations of soybean cultivars under drought stress. *Photosynthetica* **23**: 575–584

Garab G, Lajkó F, Mustárdy L and Mórton L (1989) Respiratory control over photosynthetic electron transport in chloroplasts of higher-plant cells: evidence for chlororespiration. *Planta* **179**: 349–358

Genty B, Briantais JM and Viera Da Silva JB (1987) Effects of drought on primary photosynthetic processes of cotton leaves. *Plant Physiol* **83**: 360–364

Gerbaud A and Andre M (1987) An evaluation of the recycling in measurements of photorespiration. *Plant Physiol* **83**: 933–937

Goswami BK and Srivastava GC (1985) Effects of abscisic acid and zeatin on water stress induced photosynthesis and respiration in sunflower leaves. *Indian J Plant Physiol* **28**: 350–357

Govindjee, WJS, Downton Fork DC and Armond PA (1981) Chlorophyll *a* fluorescence transient as an indicator of water potential of leaves. *Plant Sci Lett* **20**: 191–194

Gupta AS, Berkowitz, GA and Pier PA (1989) Maintenance of photosynthesis at low leaf water potential in wheat. *Plant Physiol* **89**: 1358–1365

Hanson AD and Hitze WD (1982) Metabolic responses of mesophytes to plant water deficits. *Ann Rev Plant Physiol* **33**: 163–203

Hsiao TC and Acevedo E (1974) Plant responses to water deficits. Water use efficiency and drought resistance. *Agric Meteorol* **14**: 59–84

Johnson RR, Frey NN and Moss DN (1974) Effect of water stress on photosynthesis and transpiration of flag leaves and spikes of barley and wheat. *Crop Sci* **14**: 728–732

Johnson RC, Mornhinweg DO, Ferris DM and Heitholt JJ (1987) Leaf photosynthesis and conductance of related *Triticum* species at different water potentials. *Plant Physiol* **83**: 1014–1017

Johnson RR and Moss DN (1976) Effect of water stress on $^{14}CO_2$ fixation and translocation in wheat during grain filling. *Crop Sci* **16**: 697–701

Jordan WR and Ritche JT (1971) Influence of soil water stress on evaporation, root absorption and internal water status of cotton. *Plant Physiol* **48**: 783–788

Kacperska A and Kubackaebalska M (1989) Formation of stress ethylene depends both on ACC synthesis and on the activity of free radical-generating system. *Physiol Plant* **77**: 231–237

Keck RW and Boyer JS (1974) Chloroplast response to leaf water potentials. III Differing inhibition of electron transport and photophosphorylation. *Plant Physiol* **53**: 474–479

Kramer PJ (1983) *Water Relations of Plants*. Academic Press, New York, pp 342–389

Krampitz MJ, Klug K and Fock HP (1984) Rates of photosynthetic CO_2 evolution and dark respiration in water-stressed sunflower and bean leaves. *Photosynthetica* **18**: 322–328

Krauser GH and Somersalo S (1989) Fluorescence as a tool in photosynthesis research: application in studies of photoinhibition, cold acclimation and freezing studies. In: (DA Walker and CB Osmond, eds) *New Vistas in Measurement of Photosynthesis*. The Royal Society, London

Kriedmann PE (1986) Stomatal and photosynthetic limitations to leaf growth. *Australian J Plant Physiol* **13**: 15–31

Krieg DR (1983) Photosynthetic activity during stress. *Agricultural Water Management* **7**: 249–263

Kulshreshtha S, Mishra DP and Gupta RK (1987) Changes in contents of chlorophyll, proteins and lipids in whole chloroplasts and chloroplast membrane fractions at different leaf water potentials in drought resistant and sensitive genotypes of wheat. *Photosynthetica* **21**: 65–70

Kumar S and Gupta RK (1986) Photorespiratory CO_2 evolution and ^{14}C-glycine metabolism under different leaf water potentials in sorghum. *Photosynthetica* **20**: 401–404

Lawlor DW (1976) Water stress induced changes in photosynthesis, photorespiration, respiration and CO_2 compensation concentration in wheat. *Photosynthetica* **10**: 378–387

Lawlor DW (1976) Assimilation of carbon into photosynthetic intermediates of water-stressed wheat. *Photosynthetica* **10**: 431–439

Lawlor DW (1979) Effects of water and heat stress on carbon metabolism of plants C_3 and C_4 photosynthesis. In: (H Mussell and RC Staples, eds) *Stress Physiology in Crop Plants*, John Wiley and Sons, New York, pp. 303–306

Lawlor DW (1983) Integration of biochemical processes in the physiology of water-stressed plants. In: (R Marcele, H Clijsters and M van Pousse, eds) *The Effects of Stress on Photosynthesis*. Dr. W. Junk, Publ., The Hague, pp 35–44

Lawlor DW (1987) *Photosynthesis: metabolism, control and physiology*. Longman Scientific and Technical, Harlow, England pp 1–266

Lawlor DW, Day W, Johnston AE, Legg BJ and Parkinson KJ (1981) Growth of spring barley under drought; crop development, photosynthesis, dry matter accumulation and nutrient content. *J Agric Sci Camb* **96**: 167–186

Lawlor DW and Fock HP (1975) Photosynthesis and photorespiratory CO_2 evolution of water-stressed sunflower leaves. *Planta* **126**: 247–258

Lawlor DW and Fock HP (1977a) Photosynthetic assimilation of $^{14}CO_2$ by water-stressed sunflower leaves in two oxygen concentrations and the specific activity of products. *J Exp Bot* **28**: 320–328

Lawlor DW and Fock HP (1977) Water stress induced changes in the amounts of some photosynthetic assimilation products and respiratory metabolites of sunflower leaves. *J Exp Bot* **28**: 329–337

Lawlor DW and Fock HP (1978) Photosynthesis, respiration and carbon assimilation in water-stressed maize at two oxygen concentrations. *J Exp Bot* **29**: 579–593

Lawlor DW, Gimenez C, Ward DA and Young AT (1989) Regulation of photosynthetic carbon metabolism in water-stressed sunflower. In: (J Barber and Malkin R, eds) *Techniques and New Developments in Photosynthesis Research*. Vol *168* NATO-ASI Series A Life Science Plenum Press, New York

448

Lawlor DW and Khanna-Chopra R (1983) Regulation of photosynthesis during water stress. In: (C Sybesma, ed) *Advances in Photosynthetic Research*, Nijhoff/Junk. The Hague pp 379–382

Lawlor DW and Pearlman JG (1981) Compartmental modelling of photorespiration and carbon metabolism of water-stressed leaves. *Plant, Cell and Environ* **4**: 37–52

Legg BJ, Day W, Lawlor DW and Parkinson KJ (1979) The effects of drought on barley growth, models and measurements showing the relative importance of leaf area and photosynthetic rate. *J Agric Sci Camb* **92**: 703–716

Lopez FB, Setter TL and McDavid CR (1987) Carbon dioxide and light responses of photosynthesis in cowpea and pigeon pea during water deficit and recovery. *Plant Physiol* **85**: 990–995

Ludlow MM and Wilson GL (1971) Photosynthesis of tropical pasture plants. III Leaf age. *Aust J Biol Sci* **24**: 1077–1087

McPherson HG and Boyer JS (1977) Regulation of grain yield by photosynthesis in maize subjected to water deficiency. *Agron J* **69**: 113–117

Monteith JL (1977) Climate and the efficiency of crop production in Britain. *Phil Trans R Soc Lond B*, **281**: 277–294

Morgan JM (1984) Osmoregulation and water stress in higher plants. *Ann Rev Plant Physiol* **35**: 299–319

Muchow RC (1985) Canopy development in grain legumes grown under different water regimes in semi-arid tropical environment. *Field Crop Res* **11**: 99–109

Munns R, Brady CJ and Barlow EWR (1979) Solute accumulation and drought survival in the apex and expanding leaves of wheat during water stress. *Aust J Plant Physiol* **6**: 379–389

Noguchi M, Koiwai A, Yokoyama M and Tamaki E (1968) Studies on nitrogen metabolism in tobacco plants. IX Effect of various compounds on proline biosynthesis in the green leaves. *Plant and Cell Physiol* **9**: 35–47

Parry AD and Horgan R (1990) Carotenoids and ABA biosynthesis. *British Soc Plant Growth Regulation*, **11**: No. 2 January 1990, pp 1–9

Patrick JW (1988) Assimilation partitioning in relation to crop productivity. *Hort Sci* **23**: 33–40

Pham Thi AT, Pimentel C and Vieira da Silva J (1982) Effects of water stress on photosynthesis and photorespiration of *Atriplex nummularia*, a C_4 plant. *Photosynthetica* **16**: 334–342

Plaut Z and Bravdo B (1973) Response of carbon dioxide fixation to water stress. Parallel measurements on isolated chloroplasts and intact spinach leaves. *Plant Physiol* **52**: 28–32

Powles SB (1984) Photoinhibition of photosynthesis induced by visible light. *Ann Rev Plant Physiol* **35**: 15–44

Pradet A and Raymond P (1983) Adenine nucleotide ratios and adenylate energy charge in energy metabolism. *Ann Rev Plant Physiol* **34**: 1–19

Quick P, Siegl G, Neuhaus E, Feil R and Stitt M (1989) Short-term water stress leads to a stimulation of sucrose synthesis by activating sucrose phosphate synthase. *Planta* **177**: 535–546

Rawson HM, Turner NC and Begg JE (1978) Agronomic and physiological responses of soybean and sorghum crops to water deficits. 4 Photosynthesis, transpiration and water use efficiency. *Aust J Plant Physiol* **5**: 195–209

Rhodes D, Hander S and Bressan RA (1986) Metabolic changes associated with adaptation of plant cells to water stress. *Plant Physiol* **82**: 890–903

Robinson SP (1985) Osmotic adjustment by intact isolated chloroplasts in response to osmotic stress and its effect on photosynthesis and chloroplast volume. *Plant Physiol* **79**: 996–1002

Schulze ED and Hall AE (1982) Stomatal responses, water loss and CO_2 assimilation rates of plants in contrasting environments. Physiological plant Ecology II Water relations and carbon assimilation. *Encyclo Plant Physiol New Series* **12(B)**: 181–230

Schussler JR, Brenner ML and Brun WA (1984) Abscisic acid and its relationship to seed filling in soybean. *Plant Physiol* **76**: 301–306

Sharkey RD and Badger MR (1982) Effects of water stress on photosynthetic electron transport, photophosphorylation and metabolite levels of *Xanthium stumarium* mesophyll cells. *Planta* **156**: 193–203

Sihna SK, Khanna-Chopra R, Aggarwal PK, Chaturvedi GS and Koundal KR (1988) Effect of drought on shoot growth: significance of metabolism to growth and yield. In: *Principles and methods of crop improvement for drought resistance on rice. Special International Symp* IRRI los Banos, Laguna, Philippines. pp 153–170

Sinha SK and Rajagopal V (1975) Effect of moisture stress on nitrate reductase activity and accumulation of proline in sorghum In: *Crop Response to Environmental Stresses*. Vivekananda Laboratory for Hill Agriculture, Almora, U.P.

Smith TA (1985) Polyamines. *Ann Rev Plant Physiol* **36**: 117–143

Srivastava HS and Singh RP (1987) Role and regulation of glutamate dehydrogenase activity in higher plants. *Phytochemistry* **26**: 597–610

Stewart, CR (1981) Proline accumulation: biochemical aspects. In: (LG Paleg and D Aspinall, eds) *Physiology and Biochemistry of Drought Resistance in Plants*, Academic Press, Sydney pp 243–259

Stuhlfauth T, Sültemeyer DF, Weinz S and Fock HP (1988) Fluorescence quenching and gas exchange in a water stressed C_3 plant, *Digitails lanata*. *Plant Physiol* **86**: 246–250

Sung FJM and Krieg DR (1979) Relative sensitivity of photosynthetic assimilation and translocation of ^{14}carbon to water stress. *Plant Physiol* **64**: 852–856

Swindale LD and Bidinger FR (1981) Introduction: the human consequences of drought and crop research priorities for their alleviation. In: (LG Paleg and D Aspinall, eds) *Physiology and Biochemistry of Drought Resistance in Plants*, Academic Press Sydney

Tokami S, Turner NC and Rawson HM (1981) Leaf expansion of four sunflower (*Helianthus annuus* L.) cultivars in relation to water deficits. I Pattern during plant development. *Plant Cell Environ* **4**: 399–407

Tyree MT and Jarvis PG (1982) Water in tissues and cells. In: (OL Lange, PS Nobel, CB Osmond and H Ziegler, eds) *Encyclopaedia of plant physiology*, New Series, Vol 12B, Physiological Plant Ecology II, Water relations and carbon assimilation, Springer-Verlag, Berlin, pp 35–37

Uprety DC (1989) Photosynthetic capacity of green gram (*Phaseolus radiatus*) leaves under influence of water stress. *Indian J Agric Sci* **59**: 404–407

Uprety DC and Bhatia A (1989) Effect of water stress on photosynthesis, productivity and water status of mungbean (*Vigna radiata* L.) Wilczek. *J Agron and Crop Sci* (Germany), **163**: 115–123

Uprety DC and Kumar R (1990) Influence of parental genotypes on drought resistance in triticales. *Science and Culture* **56**: 458–460

Uprety DC and Sirohi GS (1985) Effect of water stress on photosynthesis and water relations of wheat varieties. *Indian J Plant Physiol* **28**: 107–114

Uprety DC and Sirohi GS (1987) Comparative study on the effect of water stress on photosynthesis and water relations of triticale, rye and wheat. *J Agron and Crop Sci (Germany)* **159**: 349–355

Uprety DC and Tomar VK (1990) Comparative study on the moisture stress susceptibility of uniculm and three culmed (oligoculm) wheats for their photosynthesis, productivity and water status. *Proc. Indian National Sci. Academy* (**B**)**56**: 469–475

Uprety DC and Tomar VK (1992) Effect of cytoplasm on the drought resistance and photosynthesis of *Brassica carinata*. *Photosynthetica* **26**: (in press).

Venekamp JH, Lampe JEM and Koot JTM (1989) Organic acids as sources for drought-induced proline synthesis in field bean plants, *Vicia faba* L. *J Plant Physiol* **133**: 654–659

Vu JC, Allen LH and Bowes G (1987) Drought stress and elevated CO_2 effects on soybean ribulose bisphosphate carboxylase activity and canopy photosynthetic rates. *Plant Physiol* **83**: 573–578

Vu CVJ and Velenosry G (1988) Water deficit and associated changes in some photosynthetic parameters in leaves of Valencia orange (*Citrus sinensis* L. Osback). *Plant Physiol* **88**: 375–378

Wardlaw IF (1967) The effect of water stress on translocation in relation to photosynthesis and growth. I Effect during grain development in wheat. *Aust J Biol Sci* **20**: 25–39

Wardlaw IF (1969) The effect of water stress on translocation in relation to photosynthesis and growth II Effect during leaf development in *Lolium temulentum*. *Aust J Biol Sci* **22**: 1–16

Willeford KO, Ahluwalia, KJK and Gibbs M (1989) Inhibition of chloroplastic respiration by osmotic dehydration. *Plant Physiol* **89**: 1158–1160

Woodrow IE and Mott KA (1989) Rate limitation of non-steady state photosynthesis by ribulose-1, 5-bisphosphate carboxylase in spinach. *Aust J Plant Physiol* **26**: 487–500

Yamaya T and Oaks A (1987) Synthesis of glutamate by mitochondria—an anaplerotic function for glutamate dehydrogenase. *Physiol Plant* **70** 749–756

Younis HM, Boyer JS and Govindjee (1979) Conformation and activity of chloroplast coupling factor exposed to low chemical potential of water in cells. *Biochim Biophys Acta* **548**: 328–340

Zelitch I (1982) The close relationship between net photosynthesis and crop yield. *Bioscience* **32**: 796–802

18

Effect of Heavy Metals on Photosynthesis in Higher Plants

I.S. Sheoran* and Randhir Singh

Department of Chemistry and Biochemistry
Haryana Agricultural University
Hisar 125 004 India

CONTENTS

ABBREVIATIONS

ABA	:	Abscisic acid
ADPG	:	Adenosine diphosphate glucose
ALA	:	5-Amino levulinic acid
Chla	:	Chlorophyll *a*
DHAP	:	Dihydroxyacetone phosphate
DGDG	:	Digalactosyl diacylglycerol
DPC	:	Diphenyl carbazide
ET	:	Electron transport
FFA	:	Free fatty acid
F-6-P	:	Fructose-6-phosphate
F-1, 6-P2	:	Fructose-1, 6-*bis*phosphate
G-6-P	:	Glucose-6-phosphate
LHC	:	Light harvesting complex
MGDG	:	Monogalactosyl diacylglycerol
MV	:	Methyl viologen
PA	:	Phosphatidic acid
PCR cycle	:	Photosynthetic carbon reduction cycle

* Present Address: Seed Technology Centre, Haryana Agricultural University, Hisar, India

PC	:	phosphatidyl choline
PG	:	phosphatidyl glycerol
3-PGA	:	3-Phospho-glyceric acid
PEP	:	Phosphoenol pyruvate
PSI	:	Photosystem I
PS II	:	Photosystem II
RuBP	:	Ribulose-1, 5-*bis*phosphate
RuBPCase	:	Ribulose-1, 5-*bis*phosphate carboxylase/oxygenase
SC	:	Semicarbazide
SL	:	Sulfoquinovosyl diacylglycerol

ABSTRACT

Heavy metals have adverse effects on plant growth and metabolism. Photosynthesis is quite sensitive to heavy metal toxicity and both *in vivo* and *in vitro* photosynthetic CO_2 fixation are affected by heavy metals. Most of the elements studied are inhibitory to photosystem (PS) II, PS I being less sensitive in isolated chloroplasts. A common site of action seems to be at the oxidizing side of PS II. Photophosphorylation and/or enzyme activity are also possible sites of inhibition. Inhibition *in vivo* seems to be due to multiple effects of these metals. The immediate effect is on stomatal closure followed by chloroplastic changes. Long-term exposure results in reduced leaf growth, decreased photosynthetic pigments, changed chloroplast structure, and decreased enzyme activities for CO_2 assimilation.

I Introduction

Increasing development of fossil fuel resources, application of sludges to agricultural lands, and continuing release of industrial wastes is undoubtedly resulting in redistribution of trace metals in the environment throughout the world (Davis, 1984). Increased concentrations of these trace elements, particularly the heavy metals such as Pb^{2+}, Hg^{2+}, Cd^{2+}, Ni^{2+}, Cr^{2+} and Zn^{2+}, in the soil result in concomitant increase in their concentrations in plants and possibly in the food chain. Many of these heavy metals are found in food products and are toxic to both animals and human beings.

Plants for their normal growth and development require a number of inorganic ions, known as essential elements. These ions regulate several vital physiological processes of plants. Any change in the level of these essential elements in the soil or in the environment results in manifestation of either deficiency symptoms or inhibitory effects, which finally lead to reduced plant growth and development (Epstein, 1972; Mohanty and Mohanty, 1988). On the other hand, the presence of heavy metals such as Pb^{2+}, Cd^{2+}, Ni^{2+}, Hg^{2+} and Cr^{2+} in the soil always results in reduced plant growth (Kleinen-Hammans *et al.*, 1976; Agarwala *et al.*, 1977; Allinson and Dzialo, 1981; De Filippis *et al.*, 1981; Woolhouse, 1983). These heavy metals when present at an elevated level in the environment are taken up by the root systems of the plant and are accumulated in different plant parts, thereby leading to reduced plant growth and impaired metabolism (Mishra and Kar, 1974; Bingham *et al.*, 1975; Allinson and Dzialo, 1981; Lepp, 1981; Bradshaw, 1984; Chugh *et al.*, 1992).

Laboratory experiments have indicated that photosynthesis, the vital process by which green plants prepare their food, is highly sensitive to heavy metals, more so than any other metabolic process. Photosynthetic CO_2 fixation in higher plants is inhibited by Hg^{2+}, Pb^{2+}, Cd^{2+}, Ni^{2+}, Zn^{2+}, Co^{2+}, Cr^{2+} and Cu^{2+} when supplied to the nutrient medium in a number of plant species (Bazzaz and Govindjee, 1974; Bazzaz *et al.*, 1974a, b; Carlson *et al.*, 1975; Austenfeld, 1979; Baszynski *et al.*, 1980; Van Assche *et al.*, 1980; Van Assche and Clijsters, 1983; Clijsters and Van Assche, 1985; Weigel, 1985; Stilborova *et al.*, 1986; Krupa *et al.*, 1987; Jana *et al.*, 1987; Mohanty and Mohanty, 1988; Malik, 1989; Bhardwaj and Mascarenhas, 1989; Sheoran *et al.*, 1990 a, b). In most of these studies, the effect of these metals has been assessed in *in vitro* system, such as intact chloroplasts or chloroplast fragments, and epidermal strips. Few reports are available regarding their effects in *in vivo* system. The most widely studied heavy metal is cadmium. The exact mechanism of action of these heavy metals on photosynthesis is, however, still not clearly understood. We present here the current status of our understanding of the problem with major emphasis on Cd^{2+} and Ni^{2+}.

II. General Effects of Heavy Metals

a. Growth

A big group of ions such as Hg^{2+}, Pb^{2+}, Cd^{2+}, Co^{2+}, Ni^{2+}, Cr^{2+}, Cu^{2+}, Zn^{2+} and Al^{3+} are toxic to plants when present at an elevated level in the soil. Cu^{2+} and Zn^{2+} are essential elements for plant growth but excess amounts inhibit growth (Epstein, 1972; Reilly and Reilly, 1973; Agarwala *et al.*, 1977; Taylor and Foy, 1985; Stroinski and Szczotka, 1989; Sheoran *et al.*, 1990b). Ni^{2+} has also been shown to be an essential element (Brown *et al.*, 1987), but a high concentration inhibits growth (Mishra and Kar, 1974; Sheoran *et al.*, 1990b). Hg^{2+}, Pb^{2+}, Cd^{2+} and Ni^{2+} are phytotoxic even at a very low level (Bazzaz *et al.*, 1974a; Foy *et al.*, 1978; Weigel and Jager, 1980; Woolhouse, 1983; Mohanty and Mohanty, 1988). Toxicity of these ions causes reduction in germination (Aggarwal *et al.*, 1991; Dua and Sawhney, 1991; Mittal and Sawhney, 1990), reduction in plant growth, yellowing of leaves (chlorosis), fewer roots, and lesser dry matter and grain yield (Lepp, 1981; Woolhouse, 1983). The major effect of these metal ions is on the root growth followed by the leaf growth.

A 50% growth reduction in corn and barley was observed with 1.2 and 5.6 ppm cadmium solution, respectively (Page *et al.*, 1972). However, the effect of heavy metals on growth and yield depends on the type of metal ion, type of growth media (soil), and species and growth stages of plants. In wheat, growth and yield are affected almost to the same extent by Cd^{2+} and Ni^{2+} but in pigeon pea, Cd^{2+} is more depressive (Aggarwal *et al.*, 1991). Also, the adverse effect of Cd^{2+} decreases with the type of rooting medium; it is maximum with river sand followed by dune sand and sandy soil (farm soil). The effect decreases with the increase in clay content of the soil (Narwal *et al.*, 1990).

b. Metabolism

Heavy metals affect plant metabolism in many ways. Cadmium inhibits nitrogen fixation and nitrogen metabolism in legumes (Porter and Sheridan, 1981; Brockup and Capone, 1985; Sawhney *et al.*, 1990; Chugh *et al.*, 1992; Mittal and Sawhney, 1990). Nitrate reduction is also affected adversely in crop plants (Muthuchelian *et al.*, 1988; Singh *et al.*, 1988). Translocation of sugars and their metabolism are impaired (Rauser, 1977; Malik, 1989). Pb^{2+}-induced growth reduction is mainly due to decreased concentration of auxin (Lane *et al.*, 1978). Mitochondrial electron transport (Bittel *et al.*, 1974), enzymes of pentose phosphate pathway (Huang *et al.*, 1974), and respiratory activities are inhibited by heavy metal application in various plant species (Lee *et al.*, 1976; Lamoreaux and Chaney, 1978; De Filippis *et al.*, 1981). These metals also decrease the protein content by affecting protein synthesis as well as affecting hydrolytic enzymes (Lepp, 1981; Malik, 1989; Sheoran *et al.*, 1990a; Dua and Sawhney, 1991). Assimilatory sulphate reduction increases under Cd^{2+} treatment in *Zea mays* (Nussbaum *et al.*, 1988). Activity of SH-containing enzymes decreases by various heavy metals because of their strong SH-antagonistic activity (Vallee and Ulmer, 1972).

c. Metal tolerance

A great deal of work has been done on metal tolerance by plants (Bradshaw, 1984; Baker, 1987). Plants are able to live and grow on soils with varying heavy metal contents, indicating that they have the capacity to develop a specific detoxification system to cope to some degree with poisonous heavy metals. In mammals and some fungi, this is accomplished by metallothioneins (Kagi and Vallee, 1961). There have been a number of reports in recent years that both differentiated plants and plant cells grown in culture produce heavy metal binding complexes when exposed to these metal ions (Gruenhagel et al., 1985; Steffens et al., 1986; Fujita and Kawanishi, 1987; Grill et al., 1987; Reese and Wagner, 1987; Scheller et al., 1987). In general, these complexes have been poorly characterized and their structures unknown. However, it has been shown recently that the most abundant heavy metal binding complex in a number of higher plants comprise a family of sulphur-rich peptides that are structurally related to glutathione (Steffens et al., 1986). These peptides, termed "phytochelatin", have the structures (r-Glu-Cys)$_n$-Gly (n = 2-10) and are analogous to metallothioneins in that they are induced by and bind heavy metals (Grill et al., 1987. The mechanism of action of these two groups of compounds appears to be similar in their use of cysteine-SH groups to bind heavy metals. This subject has recently been reviewed by Steffens (1990).

The structure of phytochelatins indicates that these peptides are synthesized enzymically and are not primary translation products of mRNA (Grill et al., 1987). However, enhanced gene expression has been shown in tobacco culture under Cd^{2+} (Hirt et al., 1989). The structural similarity between phytochelatins and glutathione and inhibitor studies suggest that glutathione is involved in the synthesis of these heavy metal binding peptides (Scheller et al., 1987). Whether such peptides and/or Cd^{2+}-binding proteins are synthesized in the chloroplast is not yet known.

III. Effect on Photosynthesis

Photosynthesis has been shown to be the most sensitive process affected by heavy metals. Heavy metals are known to interfere with a number of photosynthetic functions. Major work on this aspect has been done on isolated systems. Less emphasis has been given to studies on intact plants.

a. Effect on isolated chloroplasts

Metal ions are involved in both the light and the dark reactions of CO_2 fixation in chloroplasts. Ionic analysis shows that K^+ is the major intrachloroplastic cation followed by Mg^{2+} (Nakatani et al., 1979). Metal ions such as Mg^{2+}, Mn^{2+}, Cu^{2+}, Fe^{2+} and Ca^{2+} form an integral part of several photosynthetic components. Besides their participation as integral constituents, these ions play structural as well as functional roles in regulating the primary photosynthetic processes. (See Govindjee and Coleman, Chapter 3)

Because of environmental concern, the effect of heavy metal ions such as Pb^{2+}, Hg^{2+}, Cd^{2+}, Zn^{2+}, Cu^{2+}, Ni^{2+} and Co^{2+} on chloroplast activity has been studied by several workers (Hampp et al., 1976; Tripathy and Mohanty, 1980; Krupa et al., 1987; Mohanty and Mohanty, 1988; Mohanty et al., 1989; Murthy et al., 1989). Many studies

456

carried out with isolated chloroplasts indicate that heavy metals affect the light reactions much more than the dark enzymic reactions of CO_2 fixation. Photosynthetic electron transport chain has been found to be the most sensitive to these metals. Many studies with isolated chloroplasts showed that PS II is very sensitive to these metals (Bazzaz and Govindjee, 1974; Li and Miles, 1975; Tripathy and Mohanty, 1980; Tripathy et al., 1981, 1983; Mohanty and Mohanty, 1988; Mohanty et al., 1989), particularly the oxidizing side of PS II. The exact mechanism is not known. Various studies have shown that heavy metals such as Pb^{2+}, Cd^{2+}, Zn^{2+} and Ni^{2+} inhibit electron transport by blocking the flow of electrons to PS II at the water splitting side. Based on various studies a number of sites of action are proposed (Fig. 1).

Recovery of PS II activity by semicarbazide or diphenylcarbazide suggests that the major site of action of heavy metals is at the oxidizing side (site I). However, various other sites of inhibition (sites II, III, and IV) have also been proposed. Li and Miles (1975) indicated that the inhibition of chloroplast electron transport by cadmium is complex. When short treatment times or low concentrations of Cd^{2+} are used, the inhibition can be reversed to a limited extent by PS II electron donors. If amounts of restoration are considered significant, it would indicate the inhibitory site to be at the point of O_2 evolution (site I). However, when slightly higher concentrations of Cd^{2+} or somewhat longer incubation time is used, the restoration of electron transport by these donors is reduced to zero. In the same chloroplasts. PS I is fully functional when a PS I electron donor is provided. These data can be interpreted to mean that the ultimate site of Cd^{2+} inhibition rests with the primary electron donor or the reaction centre of PS II (Possibly site II in Fig. 1). The loss in chlorophyll, particularly long wave absorbing form of Chl a, suggests that Cd^{2+} alters the reaction centre chlorophyll of PS II. Hg^{2+} interferes at the reducing side of PS II; it possibly substitutes for Cu^{2+} in plastocyanin, disturbing electron transport between two photosystems (site III in Fig. 1) (Clijsters and Van Assche, 1985; Murthy et al., 1989). A light-dependent interaction of Zn^{2+} between PS II and PS I (possibly at plastoquinone level) is also observed (Van Assche and Clijsters, 1983). Co^{2+}, Ni^{2+} and Zn^{2+} have been shown to impair PS II activity at the level of secondary quinone electron acceptor of PS II (Mohanty et al., 1989).

Addition of Zn^{2+} (Tripathy and Mohanty, 1980, 1981), Cd^{2+} (Van-Duijvendi et al., 1975), Ni^{2+} (Tripathy et al., 1981) or Co^{2+} (Tripathy et al., 1983) to isolated chloroplasts inhibited electron transport at the same site in vitro and application of the same electron donors could overcome this effect in a similar way. Moreover,

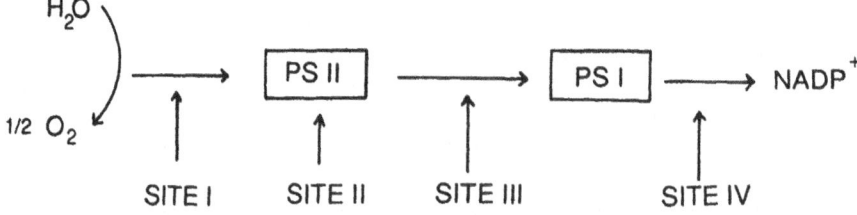

Figure 1. **Possible sites of inhibition of electron transport by various heavy metals.** Possible sites are indicated by arrow. Thick arrow means most probable site of action whereas dotted (thin) arrow means least probable site of action.

PS II inhibition by Zn^{2+} and Cd^{2+} at the water splitting side was also found in *Euglena gracilis* (De Filippis *et al.*, 1981) in a quasi *in vivo* approach where artificial electron transport dyes were taken up by the living alga. Hill activity is reduced by heavy metals in *Anacystis nidulans* (Singh and Singh, 1987). The *in vitro* inhibitory action of Pb^{2+} is Mn^{2+}-independent and directly related to PS II reaction centre (Miles *et al.*, 1972). Failure of restoration of activity by addition of Mn^{2+} in Zn^{2+} treated chloroplast indicates that in barley, inhibition of chloroplast function is not due to loss of Mn^{2+} and inactivation or loss of Mn^{2+} containing H_2O splitting enzyme (Tripathy and Mohanty, 1980, 1981). Washing of chloroplasts results in restoration of O_2 evolving activity by 76%. However, Ni^{2+} inhibition is due either to inactivation of PS II reaction centre or alteration of membrane structure of the photosynthetic apparatus as the activity could not be restored by exogenous electron donors such as $MnCl_2$, benzidine, or NH_2OH. The inhibition could not be recovered by washing the chloroplasts (Tripathy *et al.*, 1981, 1983). Copper-mediated lipid peroxidation, causing loss of chloroplast membrane function, is another mechanism of inhibition proposed for this element (Sandmann and Böger, 1980).

Interaction of heavy metals with functional SH-group is generally proposed as the probable mechanism of inhibition for several of the reactions affected. This applies to the electron transport effects which are not restorable by electron donors. At the reducing side of PS I, Cu^{2+}, Zn^{2+}, Cd^{2+} and Hg^{2+} affect the NADP-oxidoreductase, an enzyme with functional SH-group (site IV in Fig. 1) (Lucero *et al.*, 1976; De Filippis *et al.*, 1981; Clijsters and Van Assche, 1985).

PS I activity with methyl viologen as an electron acceptor is generally more resistant to heavy metal toxicity. The major inhibitory effects are reported with Zn^{2+}, Cd^{2+} and Hg^{2+} in *Euglena* (De Filippis *et al.*, 1981), Cd^{2+} in lettuce (Weigel, 1985), Co^{2+}, Ni^{2+} and Zn^{2+} in barley (Tripathy *et al.*, 1981, 1983; Tripathy and Mohanty, 1980) and Pb^{2+} (Wong and Govindjee, 1976). Inhibition of PS I activity is observed only at higher concentration of heavy metals. In many cases an increase in PS I activity has been observed under heavy metal treatments (Bhardwaj and Mascarenhas, 1989; P. Mohanty, personal communication).

Non-cyclic photophosphorylation is very sensitive to heavy metals *in vitro*. Cu^{2+} and Cd^{2+} inhibit both cyclic and non-cyclic photophosphorylation at low concentrations (Clijsters and Van Assche, 1985; Lucero *et al.*, 1976). These reports clearly indicate that in an isolated system, heavy metal ions reduce the production of photosynthetic adenosine triphosphate (ATP) and $NADPH_2$. Shortage of these reducing equivalents (ATP and $NADPH_2$) could affect the photosynthetic CO_2 reduction *in vivo* but significant inhibition of the specific activity of several Calvin cycle enzymes (Ernst, 1980) is also observed in leaf homogenates from plants treated with Zn^{2+} (Van Assche and Clijsters, 1983), Cd^{2+} and Ni^{2+} (Sheoran *et al.*, 1990a), and Cd^{2+} (Malik, 1989).

Weigel (1985) used the mesophyll protoplasts from lettuce, a situation close to *in vivo*, for understanding the mechanism of inhibition of CO_2 fixation by Cd^{2+}. He observed 40–60% inhibition of photosynthesis by pre-treatment for 10 minutes with 2 mM Cd^{2+}. This inhibition is independent of the light intensity to which protoplasts are exposed. Chlorophyll fluorescence studies revealed that primary photochemical reactions associated with thylakoid membranes are not affected by the metal ions.

Light activation of ribulose bisphosphate carboxylase (RuBPcase) is also not inhibited by Cd^{2+} under rate limiting CO_2 concentrations; inhibition of CO_2 fixation is smaller than V_{max} of CO_2 reduction indicating that carboxylation reaction of the Calvin cycle is not susceptible to Cd^{2+} (for detailed discussion on RuBPcase, see Chapter 11 this volume). Measurement of relative concentrations of (^{14}C) labelled Calvin cycle intermediates showed that Cd^{2+} causes a decrease in 3-PGA (phosphoglyceric acid) to triose phosphate ratio and an increase in the triose phosphate/ribulose-1, 5 *bis*phosphate ratio. Weigel concluded that in protoplast, Cd^{2+} affects photosynthesis mainly at the level of dark reactions and that the site of inhibition may be localized in the regenerative phase of Calvin cycle.

b. Effect on whole plant photosynthesis

Conclusions drawn from *in vitro* studies have to be considered with care for *in vivo* effects at least for the following reasons. First, although we know that Cd^{2+}, Pb^{2+} and Zn^{2+} tend to accumulate in the chloroplast, actual concentration in this organelle *in vivo* is very scarce. Secondly, the resulting effect on the ions balance in the chloroplast can have important consequences on its *in vivo* activity. Studies carried out by Weigel (1985) raised a question as to whether the effect of heavy metals will be at the same sites as observed with isolated chloroplasts. At the whole plant level, heavy metals can affect photosynthesis at stomata, mesophyll, pigment synthesis, light reactions, and dark reactions and, indirectly, by affecting various other metabolic pathways.

c. Effect on stomatal conductance

Heavy metals may impair leaf transpiration and CO_2 fixation by decreasing leaf conductance to CO_2 diffusion as a result of stomatal closure. A direct effect on *in vivo* stomatal regulation has been postulated for Cd^{2+}, Ni^{2+} and Pb^{2+} (Carlson *et al.*, 1975) or for Al^{3+} in experiments with isolated epidermal strips, floating on heavy metal solutions and showing stomatal closure under these conditions (Bazzaz *et al.*, 1974a; Carlson *et al.*, 1975). Cd^{2+} inhibited net photosynthesis and transpiration and decreased stomatal and internal conductance to CO_2 diffusion in excised silver maple leaves (Lamoreaux and Chaney, 1978). The linear relationship between net photosynthesis and inhibition of transpiration suggests that the response of Cd^{2+} might be due to stomatal closure (Bazzaz *et al.*, 1974b; Huang *et al.*, 1974; Carlson *et al.*, 1975; Lamoreaux and Chaney, 1978; Malik, 1989; Sheoran *et al.*, 1990a, b). Transpiration rates are also affected by these metals but to a different extent (Table 1). The decrease in stomatal conductance and transpiration are of lower magnitude compared to inhibition of photosynthesis (Table 1), which indicates that stomatal conductance cannot be the sole reason for decreased photosynthesis (Sheoran *et al.*, 1990a, b). In wheat, photosynthesis is reduced more by Ni^{2+} than by Cd^{2+} but stomata are closed to the same extent (Table 2). Inhibition of photosynthesis in wheat leaves by Cd^{2+} and Ni^{2+} depends upon the leaf position as well (Table 2).

The mechanism of stomatal closure is considered to be related to abscisic acid accumulation (ABA). Application of Cd^{2+} increases ABA content and affects water relations of expanding bean leaves, which in turn might result in stomatal closure

Table 1. Effect of Cd^{2+} and Ni^{2+} on photosynthesis (μmol m^{-2} s^{-1}), transpiration (μmol m^{-2} s^{-1}) and stomatal conductance (μmol m^{-2} s^{-1}) in wheat and pigeon pea. Observations recorded 10 days after treatment.

Treatments (mM)	Wheat			Pigeon pea		
	Photosynthesis	Transpiration	Stomatal conductance	Photosynthesis	Transpiration	Stomatal conductance
Control	14.54 ± 1.1	3.14 ± 0.2	270 ± 18	15.72 ± 1.1	8.91 ± 0.8	345 ± 28
Cd 5	12.13 ± 1.1	3.00 ± 0.3	249 ± 21	7.92 ± 0.6	7.51 ± 0.7	213 ± 14
10	10.28 ± 0.9	3.06 ± 0.2	214 ± 18	3.57 ± 0.2	4.49 ± 0.5	100 ± 11
15	8.31 ± 0.6	2.76 ± 0.3	168 ± 14	d	d	d
20	8.53 ± 0.8	2.63 ± 0.1	167 ± 14	d	d	d
Ni 5	10.11 ± 1.3	3.36 ± 0.2	234 ± 21	7.92 ± 0.4	6.73 ± 0.5	361 ± 19
10	8.65 ± 0.4	3.05 ± 0.1	191 ± 13	5.11 ± 0.4	5.52 ± 0.2	145 ± 14
15	5.66 ± 0.5	3.11 ± 0.1	202 ± 18	5.21 ± 0.4	5.60 ± 0.5	150 ± 10
20	4.52 ± 0.4	3.14 ± 0.2	163 ± 12	4.40 ± 0.2	4.58 ± 0.3	110 ± 10

* In wheat and pigeon pea, treatments were given 60 and 45 days after sowing, respectively, through the rooting medium in sand-grown plants. In wheat, observations were recorded on the 2nd leaf from top and in pigeon pea on the 4th leaf.

d = dried.

Table 2. Effect of Cd^{2+} and Ni^{2+} on photosynthesis (μmol m^{-2} s^{-1}) and stomatal conductance (μmol m^{-2} s^{-1}) in wheat leaves on day 10 after treatment. Treatment was given 60 days after sowing. Observations are the means of four replicates ± SE

Treatments (mM)	Leaf position from top					
	1		2		3	
	Photosynthesis	Stomatal conductance	Photosynthesis	Stomatal conductance	Photosynthesis	Stomatal conductance
Control	17.0 ± 1.1	297 ± 14	15.0 ± 1.6	301 ± 21	12.61 ± 0.8	213 ± 11
Cd 5	15.64 ± 1.2	267 ± 19	10.68 ± 0.9	267 ± 10	9.54 ± 0.9	200 ± 9
10	12.64 ± 1.2	240 ± 19	10.95 ± 1.0	234 ± 14	7.22 ± 0.9	168 ± 8
15	10.34 ± 1.0	151 ± 12	11.0 ± 1.2	214 ± 19	3.58 ± 0.4	100 ± 12
20	9.63 ± 0.8	180 ± 13	11.18 ± 1.4	203 ± 19	4.86 ± 0.8	122 ± 11
Ni 5	8.85 ± 1.0	188 ± 14	11.97 ± 1.0	315 ± 11	10.53 ± 1.0	210 ± 18
10	6.06 ± 0.7	179 ± 14	11.0 ± 1.0	222 ± 18	9.0 ± 0.8	172 ± 14
15	1.53 ± 0.2	164 ±12	9.27 ± 0.8	265 ± 18	6.13 ± 0.5	179 ± 14
20	1.08 ± 0.1	80 ± 10	8.69 ± 1.0	218 ± 5	4.66 ± 0.2	192 ± 11

under heavy metal treatments (Poschensieder *et al.*, 1989). Excess of Ni^{2+}, Co^{2+} and Zn^{2+} increase the ABA content without any adverse effect on plant water relations and hence stomatal closure (Rauser and Dumbroff, 1981). Our results with wheat and pigeon pea indicate that Cd^{2+} induces water deficiency in leaves, whereas Ni^{2+} has no effect. Stomatal closure is, however, observed under both these treatments (Table 1). It is difficult, therefore, to explain the mechanism of stomatal closure under heavy metal treatments. Zn^{2+} toxicity affects the premature loss of stomatal regulation (Van Assche *et al.*, 1980), possibly reflecting early onset of senescence as observed for Cd^{2+} (McLilley and Walker, 1974; Baker, 1987; Malik, 1989; Sheoran *et al.*, 1990b). It has yet to be proved that heavy metals directly interfere with K^+ fluxes in the guard cells.

In general, heavy metals decrease leaf expansion, resulting in more compact leaf structure and increased stomatal resistance (Clijsters and Van Assche, 1985; Barcelo *et al.*, 1988; Aggarwal *et al.*, 1991). Based on these observations, it can be said that changes in leaf architecture and stomata are side effects of heavy metals, the primary site of action being at some other level.

d. Effect on photosynthetic pigments

Heavy metals application causes yellowing of leaves which points towards a reduction in pigments status and consequently reduction in photosynthesis in higher plants. The chlorophyll content of leaf tissue has been used as a criterion for metal toxicity (Burton *et al.*, 1986). Reduction in chlorophyll content under heavy metal treatment has been reported by various workers in different crops (Baszynski *et al.*, 1980; Lepp, 1981; Woolhouse, 1983; Burton *et al.*, 1986; Bhardwaj and Mascarenhas, 1989; Malik, 1989; Sheoran *et al.*, 1990a, b). In tomato plants treated with Cd^{2+}, CO_2 fixation rate per unit chlorophyll rises, but this activity decreases when expressed on a leaf area basis. The authors concluded that Cd^{2+} primarily affected the photosynthetic pigment, before photosynthetic CO_2 fixation (Baszynski *et al.*, 1980). Reduction in chlorophyll content has been observed to be due to inhibition of chlorophyll biosynthesis in barley seedlings under Cd^{2+} (Stobart *et al.*, 1985). Cd^{2+} inhibits the synthesis of 5 amino levulinic acid (ALA) and the protochlorophyllide reductase ternary complex with its substrates. It has no effect on the constituent enzymes that catalyse the synthesis of free protochlorophyllide from ALA. These results clearly indicate that Cd^{2+} binding with thiol groups of both protochlorophyllide reductase protein and enzyme(s) involved in the light-dependent synthesis of ALA are responsible for reduced chlorophyll biosynthesis. Cu^{2+} and Cd^{2+} also affected the biosynthesis of chlorophyll in *Phaseolus vulgaris* L. and *Vigna sinensis* (Rani *et al.*, 1987; Muthuchelian *et al.*, 1988).

Time course studies with wheat and pigeon pea indicated that reduction in chlorophyll starts only after the inhibition of photosynthesis (Sheoran, unpublished data). Similarly, in pigeon pea and wheat, reduction in photosynthesis could not be explained on the basis of reduced chlorophyll content under Cd^{2+} and Ni^{2+} (Malik, 1989; Sheoran *et al.*, 1990a, b). In general, carotenoids are less affected by heavy metals compared to chlorophyll in higher plants (Baszynski *et al.*, 1980; Clijsters and Van Assche, 1985). Therefore, reduction in photosynthesis pigment concentration is not responsible for the short-term effects of heavy metals on photosynthesis. Under long-term conditions, however, chlorophyll content decreases as a consequence of the general senescing effect of heavy metals.

e. Effect on photosynthetic apparatus

Long-term Cd^{2+} application causes disorganization of chloroplast fine structure similar to that of senescence response (Baszynski et al., 1980). The principal symptoms of Cd^{2+} effects are the occurrence of large plastoglobules and a disorganization of lamellar structure, mainly grana stacks (for review on leaf senescence see Grover and Mohanty, Chapter 9 of this book). The effects of Cd^{2+} can be reversed to some extent by application of Mn^{2+} to the growth medium (Baszynski et al., 1980). Cadmium-treated bush bean plants show disorganized thylakoids and disrupted chloroplast envelopes in some cells (Barcelo et al., 1988). The etioplasts of dark-grown soybean plants fail to develop a prolamellar body under 100 µM cadmium treatment. Severe disruption of grana and irregularly spaced thylakoids are observed in soybean but not in corn plants treated with cadmium and exposed to light for 48 hours (Ghoshroy and Nadakavukaren, 1990).

Application of Cd^{2+} to wheat seedlings has no qualitative effect on thylakoid membrane protein composition but lipid composition is affected to a large extent (Malik et al., 1992). Thylakoid total lipids, total glycolipids, total phospholipids and total neutral lipids decrease by 22, 23, 12 and 25%, respectively. Glycolipids of thylakoid membranes show three major constituents: MGDG, DGDG, and SL. The level of MGDG and DGDG decreased by 32 and 27%, respectively, under Cd^{2+} treatment. PG and phosphatidyl choline contents decreased by 57 and 31%, respectively, but PA and FFA content increased (Malik et al., 1992). These results clearly indicate that Cd^{2+} treatment reduces the photosynthesis in vivo by affecting the architecture of thylakoid membranes which in turn affect the light reactions of photosynthesis adversely, as has been reported in isolated chloroplasts. However, little information is available regarding the in vivo effects of heavy metals on light reactions. It has been reported in pigeon pea, wheat, and tomato that application of heavy metals to whole plants affects the PS II activity of isolated chloroplasts adversely without any effect on PS I activity (Baszynski et al., 1980; Bhardwaj and Mascarenhas, 1989; Malik, 1989). Increase in PS I activity possibly due to uncoupling of light harvesting complex from PS II complex resulting in increase in energy transfer from light harvesting complex to PS I and concomitant increase in PS I activity in wheat grown under Cd^{2+} has been reported (Bhardwaj and Mascarenhas, 1989).

f. Effect on carbon metabolism

Reduced photosynthesis (in vivo) under heavy metal treatment could also be due to their effects on various enzymes and metabolites of carbon reduction cycle. Heavy metal toxicity in plants has been attributed to the binding of these metal ions to enzymes resulting in the alteration of their catalytic functions (Eicchan et al., 1969; Ernst, 1980; Clijsters and Van Assche, 1985). Decreased activity of key enzymes of CO_2 fixation (RuBPcase, PEP-carboxylase) has been observed in various crops under heavy metal treatments (Clijsters and Van Assche, 1985; Stilborova and Leblora, 1985; Malik, 1989; Sheoran et al., 1990 a). Decreased RuBPcase/oxygenase activity ratio has been observed under Zn^{2+} treatments (Van Assche and Clijsters, 1983). Metal substitution might be the reason for this effect. Uptake of heavy metals induces metal substitution in the metallo-proteins, impairing enzyme activity. Substitution of Zn^{2+} and Mn^{2+} by Cd^{2+} was proved to be the inhibitory mechanism of Cd^{2+} toxicity for carbonic anhydrase (Ernst, 1980) and for the PS II water splitting enzyme (Baszynski et al., 1980), respectively. In vitro substitution of Mg^{2+} by Mn^{2+} or Co^{2+} and Ni^{2+} (Wildner and Henkel, 1979) decreased the RuBPcase/oxygenase activity ratio. The changed ratio is reflected in increased CO_2 compensation and accumulation of glycolate. This aspect needs detailed investigation. Zn^{2+}, Cd^{2+}, Pb^{2+} and Hg^{2+}, powerful SH-antagonist, affect enzyme activity (Ernst, 1980; De Filippis et al., 1981). Due to competition for uptake by the roots, heavy metal toxicity can induce deficiency of other essential elements. This deficiency could cause inhibition of metal-dependent enzymatic reactions.

A detailed study in pigeon pea has shown that the activity of various enzymes of PCR-cycle namely RuBPcase, 3-PGA kinase, NAD- and NADP-glyceraldehyde phosphate dehydrogenases, aldolase and fructose 1, 6-bisphosphatase gets reduced only by 10–40% when photosynthesis is reduced to 80% under 10 mM Cd^{2+} application. Ni^{2+} affects these activities and photosynthesis to a lesser extent. The effect of these metal ions depends on the plant growth stage, ion concentration, and duration of exposure to these ions (Sheoran et al., 1990a). A similar response is recorded in wheat, when exposed to different Cd^{2+} concentrations and at three growth stages (Malik, 1989). Cd^{2+} drastically affects the enzyme functioning related to carbon metabolism. The activity of ADPG-pyrophosphorylase was most sensitive to this treatment. Levels of various Calvin cycle intermediates, namely, 3-PGA, DHAP, F-1, 6-P_2, F-6-P, ribulose-1, 5-P_2 and G-6-P, were reduced to different extents by Cd^{2+} application (Table 3). Surprisingly, reduction in the activity of various enzymes and metabolites could not explain the drastic reduction in in vivo CO_2 fixation by these plants. This indicates that reduction in photosynthesis at whole plant level is regulated by the various mechanisms and not by a single factor.

Photosynthate partitioning has been shown to be affected by heavy metals. Labelled studies (using $^{14}CO_2$) indicate that in pigeon pea, a lower amount of carbon is transported towards roots under cadmium treatments (Jitender, personal communication). However, assimilate transport from the fully developed leaf is more under low concentration (3 mM) but less under high concentration (6 mM) of cadmium. The level of total soluble carbohydrates and starch is not affected by these concentrations but sucrose content as well as the activity of sucrose phosphate synthase decreases (Sheoran et al., 1991). These results indicate that carbon partitioning between starch and sucrose is not affected by cadmium but sucrose biosynthesis is inhibited.

It is clear that in vitro and in vivo photosynthetic CO_2 fixation is affected by several heavy metals. All elements studied are inhibitory to PS II, PS I is less sensitive in isolated chloroplasts. A common site of action at the oxidizing side of PS II is proposed. Photophosphorylation or enzyme activity are also possible sites of heavy metal inhibition. In vivo inhibition seems to be a multisite effect of heavy metals. The immediate effect is on stomatal closure followed by light and dark reactions.

Table 3. Effect of Cd^{2+} on the levels of various metabolites (nmol mg chl^{-1}) at different days after treatment at vegetative stage of wheat*

Metabolites	Treatment	Days after treatment		
		1	7	10
3-Phosphoglyceric acid	C	153	140	141
	T	141 (94)	67 (48)	43 (31)
Dihydroxyacetone-phosphate	C	45.6	51.5	51.5
	T	39.9 (80)	33.1 (66)	25.2 (60)
Fructose 1, 6-*bis*phosphate	C	36.8	39.9	39.9
	T	36.6 (99)	19.2 (48)	16.8 (41)
Fructose-6-phosphate	C	32.5	29.9	29.9
	T	29.8 (92)	18.1 (61)	15.7 (53)
Ribulose-1, 5-*bis*phosphate	C	136	143	143
	T	132 (97)	102 (72)	91 (64)
Glucose-6-phosphate	C	56	53	54
	T	49 (88)	22 (42)	17 (32)
Fructose-2, 6-*bis*phosphate	C	0.36	0.39	0.39
	T	0.35 (97)	0.33 (84)	0.31 (80)
Inorganic pyrophosphate	C	0.72	0.73	0.73
	T	0.69 (96)	0.52 (72)	0.51 (70)
Inorganic phosphate	C	436	500	500
	T	430 (99)	243 (49)	221 (44)
Photosynthesis (mg CO_2 dm^{-2} h^{-1})	C	21.2	21.5	21.5
	T	15.3 (72)	12.6 (59)	11.6 (54)

* Plants were grown in dune sand and treatment was given 50 days after sowing.
C = control; T = treatment (20 mM Cd^{2+})
Values in parentheses are per cent of control.

Long-term exposure results in reduced leaf growth, decreased photosynthetic pigments, changed chloroplast structure, and decreased activity of enzymes of CO_2 fixation.

Acknowledgements

Financed in part by a USDA grant under Cooperative Agriculture Research Grant Program (Grant No. 687).

IV. References

Agarwala SC, Bisht SS and Sharma C (1977) Relative effectiveness of certain heavy metals in producing toxicity and symptoms of iron deficiency in barley. *Can J Bot* **55**: 1299–1307

Aggarwal N, Singh J and Sheoran IS (1991) Effect of cadmium and nickel on germination, early seedling growth and photosynthesis of wheat and pigeon pea. *Int J Trop Res*: in press

Allinson DW and Dzialo C (1981) The influence of Pb^{2+}, Cd^{2+} and Ni^{2+} on the growth of rye grass and oats. *Plant Soil* **62**: 81–89

Austenfeld GA (1979) Effect of Ni^{2+}, Co^{2+} and Cr^{2+} on net photosynthesis of primary and secondary leaves of *Phaseolus vulgaris* L. *Photosynthetica* **13**: 434–438

Baker AJM (1987) Metal tolerance. *New Phytol* **106**: 93–111

Barcelo J, Vazquez, MD and Poschensieder C (1988) Structural and ultrastructural disorders in Cd-treated bush bean plants (*Phaseolus-vulgaris* L.) *New Phytol* **108**: 37–49

Baszynski T, Wadja L, Krol M, Wolinska D, Kurpa Z and Tukendorf A (1980) Photosynthetic activities of Cd-treated tomato plants. *Physiol Plant* **48**: 365–370

Bazzaz FA, Carlson RW and Rolfe GL (1974a) The effect of heavy metals on plants. 1 inhibition of gas exchange in sunflower by Pb^{2+}, Cd^{2+}, Ni^{2+} and Ti^{2+}. *Environ Pollut* **7**: 241–246

Bassaz FA, Rolfe GL and Carlson RW (1974b) Effect of Cd^{2+} on photosynthesis and transpiration of excised leaves of corn and sunflower. *Physiol Plant* **32**: 372–376

Bazzaz MB and Govindjee (1974) Effect of cadmium nitrate on spectral characteristic and light reactions of chloroplasts. *Environ Lett* **6**: 1–12

Bhardwaj R and Mascarenhas C (1989) Cadmium induced inhibition of photosynthesis *in vivo* during development of chloroplast in *Triticum aestivum* L. *Plant Physiol Biochem* **16**: 40–48

Bingham FT, Page AL, Mahler RJ and Ganje T (1975) Yield and cadmium accumulation of plants grown on soil treated with cadmium enriched sewage. *J Environ Qual* **4**: 207–211

Bittel JE, Kolppe DE and Miller JR (1974) Sorption of heavy metal cations by corn mitochondria and the effects on electron and energy transfer reaction. *Physiol Plant* **30**: 226–230

Bradshaw AD (1984) In: (D Evered and GM Collins, eds) *Origin and Development of Adaptation*, Pitman, London, pp 4–19

Brockup I and Capone DG (1985) The effect of several metals and organic pollutants on nitrogen fixation by roots and *Rhizobium* of *Zostera marina* L. *Environ Expt Bot* **25**: 145–151

Brown PH, Welch RM and Cary EE (1987) Nickel—A micronutrient essential for higher plants. *Plant Physiol* **85**: 801–804

Burton KW, King JB and Morgan E (1986) Chlorophyll as an indicator of the upper critical tissue concentration of Cd^{2+} in plants. *Water Air Soil Pollut* **27**: 147–154

Carlson RW, Bazzaz FA and Rolfe GL (1975) The effect of heavy metals on plants parts II Net photosynthesis and transpiration of whole corn and sunflower plants treated with Pb^{2+}, Cd^{2+}, Ni^{2+} and Ti^{2+}. *Environ Res* **10**: 113–120

Chugh LK, Gupta VK and Sawhney SK (1992) Effect of cadmium on enzymes of nitrogen metabolism in pea seedlings. *Phytochemistry* **31**: 395–400

Clijsters H and Van Assche F (1985) Inhibition of photosynthesis by heavy metals. *Photosynth Res* **7**: 31–40

Davis RD (1984) Cadmium—A complex environmental problem II Cadmium in sludges used as fertilizer. *Experientia* **40**: 117–126

Dua A and Sawhney SK (1991) Effect of chromium on activities of hydrolytic enzymes in germinating pea seeds. *Environ Expt Bot.* **31**: 133–139

De Filippis LE, Hampp R and Zeigler H (1981) The effect of sub-lethal concentrations of zinc, cadmium and mercury on *Euglena* II respiration, photosynthesis and photochemical activities. *Arch Microbiol* **128**: 407–411

Eicchan GL, Clark P and Tarien E (1969) The interaction of metal ion with polynucleotides and related compounds. *J Biol Chem* **244**: 937–942

Epstein E (1972) *Mineral Nutrition of Plants: Principles and Perspectives*. John Wiley & Sons, New York

Ernst WHO (1980) Biochemical aspects of cadmium in plants. In: (JO Nriagu, ed) *Cadmium in the Environment Part I*, John Wiley & Sons, New York, pp 639–653

Foy CD, Chaney RL and White MC (1978) The physiology of metal toxicity in plants. *Annu Rev Plant Physiol* **29**: 511–566

Fujita M and Kawanishi T (1987) Cd^{2+}-binding complexes from the root tissues of various higher plants cultivated in Cd^{2+}-containing medium. *Plant Cell Physiol* **28**: 379–382

Ghoshroy S and Nadakavukaren MJ (1990) Influence of cadmium on the ultrastructure of developing chloroplasts in soybean and corn. *Environ Expt Bot* **30**: 187–192

Grill E, Winnacker EL and Zenk MN (1987) Phytochelatins—A class of heavy metal binding peptides from plants are functionally analogous to metallothioneins. *Proc Natl Acad Sci USA* **84**: 439–444

Gruenhagel L, Weigel HJ, Ilge D and Jaeger JH (1985) Isolation and partial characterization of cadmium binding protein from *Pisum sativum. J Plant Physiol* 125: 275–283

Hampp R, Beulich K and Ziegler H (1976) Effect of zinc and cadmium on photosynthetic CO_2 fixation and Hill activity of isolated chloroplasts. *Z Pfln Physiol* 77: 336–344

Hirt H, Casari C and Barta A (1989) Cd-enhanced gene expression in suspension culture cells of tobacco. *Planta* 179: 414–420

Huang C, Bazzaz FA and Vanderhoef LN (1974) The inhibition of soybean metabolism by cadmium and lead. *Plant Physiol* 54: 122–124

Jana S, Dalal T and Barua B (1987) Effect and relative toxicity of heavy metals on *Cuscuta reflexa. Water Air Soil Pollut.* 33: 23–27

Kagi J and Vallee B (1961) Metallothionein: A cadmium and zinc containing protein from equine renal cortex. *J Biol Chem* 236: 2435–2442

Kleinen-Hammans JW, Rabon LPLM and Pietersen HQ (1976) Participation of pigment complexes in uptake and incorporation of mercury ions by *Anacystis. Photosynthetica* 10: 440–446

Krupa, Z, Skorzynska E, Maksymic W and Baszynski T (1987) Effect of Cd^{2+} treatment on the photosynthetic apparatus and its photochemical activities in greening radish seedlings. *Photosynthetica* 21: 156–164

Lamoreaux RJ and Chaney WR (1978) The effect of cadmium on net photosynthesis, transpiration and dark respiration of excised Silver Maple leaves. *Physiol Plant* 43: 231–236

Lane SD, Martin ES and Garrod JF (1978) Lead toxicity effects on indole-3-acetic acid induced cell elongation. *Planta* 144: 79–84

Lee KC, Cunningham BA, Paulsen GM, Liang GH and Moore RB (1976) Effects of cadmium on respiration rate and activities of several enzymes in soybean seedlings. *Physiol Plant* 36: 4–6

Lepp NW (1981) Effect of heavy metal pollution on plants Vol. 1 and II, Applied Science Publisher, London

Li EH and Miles CD (1975) Effects of cadmium on photoreaction II of chloroplasts. *Plant Sci Lett* 5: 33–40

Lucero H, Andreo C and Vallyos RH (1976) Sulphydryl groups in photosynthetic energy conservation III. Inhibition of photophosphorylation in spinach chloroplasts by $CdCl_2$. *Plant Sci Lett* 6: 309–313

Malik D (1989) Effect of cadmium on photosynthetic efficiency of wheat (*Triticum aestivum* L.) Ph. D. Thesis, Haryana Agricultural University, Hisar

Malik D, Sheoran IS and Singh R (1992) Effect of cadmium on lipid composition of wheat (*Triticum-aestivum* L.) thylakoid membranes. *Indian J. Biochem Biophys*: in press

McLilley R and Walker DA (1974) An improved spectrophotometric assay for ribulose bisphosphate carboxylase. *Biochim Biophys Acta* 358: 226–229

Miles CD, Brandle JR, Daniel DJ, Chu-Der O, Schnare PD and Uhlik DJ (1972) Inhibition of PS II in isolated chloroplasts by lead. *Plant Physiol* 49: 820–825

Mishra D, and Kar M (1974) Nickel in plant growth and metabolism. *Bot Rev* 40: 395–452

Mittal S and Sawhney SK (1990) Influence of lead on enzymes of nitrogen metabolism in germinating pea seeds. *Plant Physiol and Biochem* 17: 75–81

Mohanty N and Mohanty P (1988) Cation effects on primary processes of photosynthesis. In: R Singh and SK Sawhney eds) *Advances in Frontier Areas of Plant Biochemistry.* Prentice-Hall, New Delhi, pp 1–18

Mohanty N, Vass I and Demeter (1989) Impairment of PS II activity at the level of secondary quinone electron acceptor in chloroplasts treated with Co^{2+}, Zn^{2+}, Ni^{2+} ions. *Physiol Plant* 76: 386–390

Murthy SDS, Sabat SC and Monanty P (1989) Mercury induced inhibition of PS II activity and changes in emission of fluorescence from phycobilisomes in intact cells of the cyanobacterium, *Spirulina plantensis. Plant Cell Physiol* 30: 1153–1157

Muthuchelian K, Rani SMV and Paliwal K (1988) Differential action of Cu^{2+} and Cd^{2+} on chlorophyll biosynthesis and NRA in *Vigna sinensis. Indian J Plant Physiol* 31: 169–173

Nakatani HY, Barber J and Minski MJ (1979) The influence of the thylakoid membrane surface properties on the distribution of ions in chloroplasts. *Biochim Biophys Acta* 358: 226–229

Narwal RP, Singh M and Dahiya DJ (1990) Effect of Cd^{2+} on plant growth and heavy metal content of corn (*Zea mays* L.). *Crop Res* 3: 13–20

Nussbaum S, Schmutz D and Brunoi C (1988) Regulation of assimilatory sulfate reduction by cadmium in *Zea mays* L. *Plant Physiol* 88: 1407–1410

Page AL, Bingham FT and Nelson C (1972) Cadmium absorption and growth of various plant species as influenced by solution cadmium concentrations. *J Environ Qual* **1**: 288–291

Porter JR and Sheridan A (1981) Inhibition of nitrogen fixation in alfalfa by arsenate, heavy metals, fluoride, and simulated acid rains. *Plant Physiol* **68**: 143–148

Poschen-Srieder CH, Gunse B and Barcelo J (1989) Influence of Cd^{2+} on water relations, stomatal resistance and ABA content in expanding bean leaves. *Plant Physiol* **90**: 1365–1369

Rani SMV, Mothuchelian K and Paliwal K (1987) Differential toxicity of Cu^{2+} and Cd^{2+} on chlorophyll biosynthesis and NRA in *Phaseolus vulgaris* L. *Ann Plant Physiol* **1**: 126–135

Rauser WE (1977) Early effects of phytotoxic burdens of cadmium, cobalt, nickel and zinc in white beans. *Can J Bot* **56**: 1744–1749

Rauser WE and Dumbroff EB (1981) Effect of excess cobalt, nickel and zinc on the water relations of *Phaseolus vulgaris* L. *Environ Exp Bot* **21**: 249–255

Reese RN and Wagner GJ (1987) Properties of tobacco (*Nicotiana tobaccum*) cadmium binding peptide (s). Unique non-metallothionein cadmium ligenda. *Biochem J* **241**: 641–647

Reilly A and Reilly C (1973) Zinc, lead and copper tolerance in the grass, *Stereochlaena cameronii* (Stapf) Clayton. *New Phytol* **72**: 1041–1046

Sandmann G and Boger P (1980) Copper-mediated lipid peroxidation processes in photosynthetic membranes. *Plant Physiol* **66**: 797–800

Sawhney V, Sheoran IS and Singh R (1990) Nitrogen fixation, photosynthesis and enzymes of ammonia assimilation and ureide biogenesis in nodules of mung bean (*Vigna radiata*) grown in presence of cadmium. *Indian J. Exp. Biol.* **28**: 883–886

Scheller HV, Huang B, Hatch E and Goldserough PB (1987) Phytochelatin synthesis and glutathione levels in response to heavy metals in tomato cells. *Plant Physiol* **85**: 1031–1035

Sheoran IS, Singal HR and Singh R (1990a) Effect of cadmium and nickel on photosynthesis and the enzymes of photosynthetic carbon reduction cycle in pigeon pea. *Photosynth Res* **23**: 345–351

Sheoran IS, Aggarwal N and Singh R (1990b) Effect of cadmium and nickel on *in vivo* carbon dioxide exchange rate of pigeon pea (*Cajanus cajan* L.) *Plant Soil* **129**: 243–249

Sheoran IS, Gupta VK, Laura JS and Singh R (1991) Photosynthetic carbon fixation, translocation and metabolite levels in pigeon pea (*Cajanus cajan* L.) leaves exposed to excess cadmium. *Indian J. Exp Biol* **29**: 857–861

Singh DP and Singh SP (1987) Action of heavy metals on Hill activity and O_2 evolution in *Anacystis nidulans*. *Plant Physiol* **83**: 12–14

Singh DN, Srivastava HS and Singh RP (1988) Nitrate assimilation in pea leaves in the presence of Cd^{2+}. *Water Air Soil Pollut* **42**: 1–5

Steffens JC (1990) The heavy metal-binding peptides of plants. *Annu Rev Plant Physiol Plant Mol Biol* **41**: 553–575

Steffens JC, Hunt DF and Williams BG (1986) Accumulation of non-protein metal binding polypeptides (r-glutamyl-cysteinyl) glycine in selected cadmium resistant tomato cells. *J Biol Chem* **261**: 13879–13882

Stilborova M, Doudravova, M, Breezinova A, and Friedsich A (1986) Effect of heavy metal ions on growth and biochemical characteristics of photosynthesis of barley. *Photosynthetica* **20**: 418–425

Stilborova M and Leblora S (1985) Heavy metal inactivation of maize PEP-carboxylase isoenzyme. *Photosynthetica* **19**: 500–503a

Stobart AK, Griffiths WT, Ameen-Sukhari I and Sherwood RP (1985) The effect of Cd^{2+} on the biosynthesis of chlorophyll in leaves of barley. *Physiol Plant* **63**: 293–298

Stroinski A and Szczotka Z (1989) Effect of Cd^{2+} and *Phytophthora infestans* on polyamine levels in potato leaves. *Physiol plant* **77**: 244–246

Taylor GJ and Foy CD (1985) Differential uptake and toxicity of ionic and chelated copper in *Triticum aestivum*. *Can J Bot* **63**: 1271–1275

Thurman DA (1981) Mechanisms of metal tolerance in higher plants. In: (NW Lepp, ed) *Effect of Heavy Metal Pollution on Plants*, Vol II, Applied Science Publishers, London, pp. 239–251

Tirpathy BC, Bhatia B and Mohanty P (1981) Inactivation of chloroplast photosynthetic electron transport activity by Ni^{2+}. *Biochim Biophys Acta* **638**: 217–224

Tripathy BC, Bhatia B and Mohanty P (1983) Cobalt ions inhibit electron transport activity of photosystem II without affecting photosystem I. *Biochim Biophys Acta* **722**: 88–93

468

Tripathy BC and Mohanty P (1980) Zinc inhibited electron tranport of photosynthesis in isolated barley chloroplast. *Plant Physiol* **66:** 1174–1178

Tripathy BC and Mohanty P, (1981) Stabilization by glutaraldehyde fixation of chloroplast structure and function against heavy metal ion induced damage. *Plant Sci Letts,* **22:** 253-261

Vallee BL and Ulmer DD (1972) Biochemical effects of mercury, cadmium and lead. *Annu Rev Biochem* **41:** 91–128

Van-Assche F, Ceulemans R and Clijsters H (1980) Net photosynthesis in *Phaseolus vulgaris* L. Photosynth Res **1:** 171–180

Van-Assche F and Clijsters H (1983) Multiple effects of heavy metal toxicity on photosynthesis. In: (R Marcelle, H Clijsters and Van Pouckem, eds) *Effects of Stress on Photosynthesis,* Martinus-Nijhoff, The Hague, pp 371–382

Van-Duijvendi JK, Matteoli MA and Desmet GM (1975) On the inhibitory action of cadmium on the donor side of PS II in isolated chloroplasts. *Biochim Biophys Acta* **408:** 164–169

Weigel HJ (1985) The effect of Cd^{2+} on photosynthetic reaction of mesophyll protoplasts. *Physiol Plant* **63:** 192–200

Weigel HJ (1985) Inhibition of photosynthetic reactions of isolated chloroplasts by Cd^{2+}. *J Plant Physiol* **119:** 179–189

Weigel HJ and Jager HJ (1980) Subcellular distribution and chemical form of cadmium in bean plants. *Plant Physiol* **65:** 480–482

Wildner GF and Henkel J (1979) The effect of divalent metal ions on the activity of Mg^{2+} depleted ribulose-1, 5-diphosphate oxygenase. *Planta* **46:** 223–228

Wong D and Govindjee (1976) Effects of lead ions on PS I in isolated chloroplasts. Studies on the reaction center P 700. *Photosynthetica* **10:** 241–254

Woolhouse HW (1983) Toxicity and tolerance in the responses of plants to metals. In: (A Parson and MH Zimmermann eds) *Encyclopedia of Plant Physiology,* New Series 12C, Springer-Verlag, Berlin, pp 245–300

19

The Role of Carotenoids in Protection against Photoinhibition

Prabhat K. Sharma* and David O. Hall

Division of Biosphere Sciences
King's College London
Campden Hill Road
London W8 7AH, U.K.

CONTENTS

ABBREVIATIONS

DTT	:	Dithiothreitol
^{1}H NMR	:	Proton nuclear magnetic resonance
LHC	:	Light harvesting complex
PS I	:	Photo System I
PS II	:	Photo System II
PPFD	:	Photosynthesis Photo Flux density

* Present Address: Goa University, Taleigaun Plateau, P.O. Bambolin, Goa 403 202.

470

ABSTRACT

Recent achievements in the investigation of the properties of the carotenoids in relation to photoprotection are discussed. Focus is on the presence of carotenoids in the reaction centre and light harvesting complex, and their functions in photoprotection. Emphasis is given to the xanthophyll cycle and β-carotene, which play a significant role in photoprotection by dissipating the excessive excitation energy.

I. Introduction

Prolonged exposure of leaves to light levels where the excitation energy exceeds the capacity for orderly dissipation by the photosynthetic system can lead to photoinhibition (Bjorkman, 1987a; Krause et al., 1990; Oquist et al., 1987). Photoinhibition is characterized by a sustained decrease in the efficiency of photon utilization by photosystem II (PS II) photochemistry (Bjorkman and Demmig, 1987) and is observed when (i) shade-adapted leaves or plants grown at low light are exposed to high light (Bjorkman, 1987a; Sharma and Hall 1990, 1991) and (ii) sun leaves acclimated to natural sunlight are exposed to additional environmental stresses such as unfavourably low and high temperature (Richards and Hall, 1987; Masojidek et al., 1991; Mishra et al., 1991), water deficit (Genty et al., 1987), and salt stresses (Sharma and Hall, 1991). Under these conditions a given light level which was previously not excessive becomes inhibitory because the utilization of energy through photosynthesis is decreased by the additional stress conditions (Bjorkman and Powles, 1984; Ludlow and Powles, 1988; Genty et al., 1987; Sharma and Hall, 1990).

Under normal conditions the light energy absorbed by the pigments organized in chlorophyll-protein complexes of PS II and photo system I (PS I) is utilized in a controlled manner leading to the generation of NADPH and ATP. However, under photoinhibitory conditions this process is not able to dissipate all the available energy resulting from the decreased efficiency of the photosynthetic system in utilizing the energy. There are other built-in mechanisms of energy dissipation (carotenoids, photorespiration, Mehler reaction, reducing substances, superoxide dismutase, etc.) which play a role in providing some protection to the PS II reaction centres against photoinhibition. In this review, we discuss the role of carotenoids (xanthophyll cycle) in this process. Readers interested in the broader field of carotenoids in photosynthesis may refer to reviews by Cogdell and Frank (1987), Frank et al. (1991), Krinsky (1979), Mathis and Schenck (1982), Mimuro and Katoh (1991), Siefermann-Harms (1987), Truscott (1990).

Carotenoids function as accessory light harvesting pigments (Davidson and Cogdell, 1981) in addition to the important protection role in the chloroplasts (Demmig et al., 1987b; Koyama, 1991; Krinsky, 1979; Sharma and Hall, 1990). The group of primary plant carotenoids can be divided into the O_2-free carotenoids and the xanthophylls which contain O_2. The carotenoids of chloroplasts include β-carotene (25–40%), lutein (40–57%), violaxanthin (9–12%) and neoxanthin (5–13%) as major and regular components of thylakoids of higher plants; their amount varies within the range depending upon the growth conditions and stress factors (Britton, 1990).

Photoprotective Functions: There are two proven photoprotective reactions of carotenoids: one is to dissipate the excessive energy by transferring energy from the triplet to ground state and the other is to scavenge the O_2 radicals generated under photoinhibitory conditions.

$$Chlorophyll \longrightarrow {}^3Chl* \tag{1}$$

$${}^3Chl* + {}^1Car \longrightarrow {}^1Chl + {}^3Car* \tag{2}$$

$${}^3Car* \longrightarrow {}^1Car + Heat \tag{3}$$

The quenching of ${}^3Chl*$ by carotenoids to prevent the generation of singlet O_2 through triplet sensitization is a protective mechanism. Equations (1), (2) and (3) indicate that the relative energies of the components which make the above reactions possible.

$$Carotenoid \xrightarrow[dark]{hv, O_2} Epoxy\text{-}carotenoids \tag{4}$$

Equation (4) shows how carotenoids remove singlet O_2 generated during photoinhibition by a mechanism of epoxidation. This process is carried out by the xanthophylls, which are O_2-containing carotenoids. During photoinhibitory treatment violaxanthin is de-epoxidized to zeaxanthin via antheraxanthin, which is reversed in the dark (Fig. 1a). There are reports (Demmig et al., 1988) which suggest that

(A)   (B)

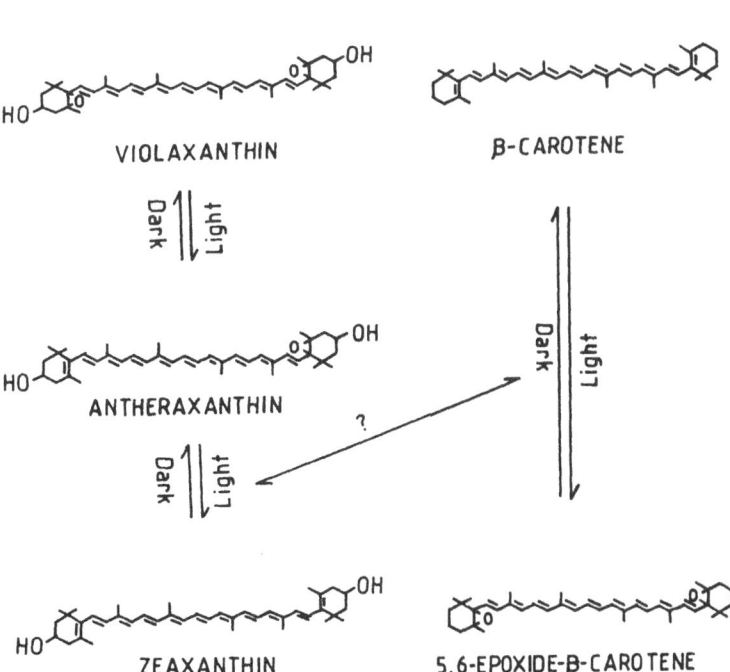

Figure 1. The xanthophyll cycle (a) and β-carotene cycle (b).

β-carotene can also be converted to zeaxanthin by a process shown in Fig. 1b. Whether or not antheraxanthin (an intermediate product of xanthophyll cycle) has a function similar to that of zeaxanthin cannot clearly be determined from the work published so far (Demmig-Adams, 1990). However, a contribution of antheraxanthin to energy dissipation cannot be ruled out.

Light harvesting function: It has been established that carotenoids function as supplementary light harvesters (Davidson and Cogdell, 1988). The light harvesting reactions of carotenoids include the absorption of light energy by ^1Car to generate an excited singlet state ($^1Car*$) and its energy transfer to ground state chlorophyll (^1Chl).

$$^1Car + h\nu \longrightarrow {}^1Car* \tag{5}$$

$$^1Car + {}^1Chl \longrightarrow {}^1Car + {}^1Chl* \tag{6}$$

The mechanism of energy transition either by an electron exchange mechanism (Cogdell, 1985) or by a dipole mechanism still needs to be determined.

II. Structure of Carotenoids

Proton nuclear magnetic resonance (^1H NMR) spectroscopy and Raman spectroscopy of the carotenoids revealed that carotenoids present in the reaction centre have an all-*cis* configuration while carotenoids present in the light harvesting complex (LHC) have an all-*trans* configuration (Koyama *et al.*, 1988; Koyama *et al.*, 1990).

a. Reaction centre

Carotenoids in the reaction centre are present in both bound and free form. ^1H NMR spectroscopy of carotenoids extracted from the reaction centres from *Rhodobacter sphaeroides* has provided conclusive evidence for a configuration assignment of 15-*cis* (Koyama *et al.*, 1988; Koyama, 1990; Koyama *et al.*, 1990; Lutz *et al.*, 1987). Raman spectroscopy has demonstrated that the free and bound carotenoids in the reaction centres have essentially the same 15-*cis* configuration (Lutz *et al.*, 1987; Koyama *et al.*, 1988), thus facilitating energy dissipation in the reaction centre. X-ray crystallography was used to determine the location of carotenoid molecules in the reaction centre of *Rhodobacter sphaeroides* (Yeats *et al.*, 1988). A central mono-*cis* configuration was observed. However, detailed conformation still remains to be determined.

b. Light harvesting complex

Comparison of the Raman spectra of the LHC bound and free carotenoids indicates that carotenoids in the LHCs of *Rhodobacter sphaeroides*, *Rhodospirillum rubrum*, and *Rhodopseudomonas viridis* have an all-*trans* configuration (Lutz *et al.*, 1978). Raman spectrum of the LHC bound carotenoids detects two different types of conformation, i.e., twisted and planar forms (Iwata *et al.*, 1985). The twisting is not ascribed to the type of carotenoid, but to the specific interaction between the carotenoid and apoprotein (Hayashi *et al.*, 1989), and it is also correlated with efficiency of energy transfer (Noguchi *et al.*, 1990).

III. Photoprotective Function of Carotenoids

Are the carotenoids present in the reaction centre and/or in the LHC involved in energy dissipation? Under reducing conditions (i.e., in a plant grown at excessive light) a charge recombination takes place at the special pair chlorophyll in the PS II reaction centre and $^3Chl^*$ is generated. The main function of reaction centre bound carotenoid, as described in equations (1), (2), and (3), is to accept the triplet energy and then to dissipate it to the surroundings as heat.

It has been proposed that the dissipation process may occur at several sites within or around the PS II and PS I reaction centres. Dissipation of excitation energy has been suggested to occur via a back reaction within the reaction centre of PS II thereby leading to a recombination after the initial charge separation (Weis and Berry, 1987; Weis *et al.*, 1987). However, a diversion of significant amounts of excitation energy away from the photochemical reaction centres has been suggested to occur within the LHC associated with PS II and PS I (Bjorkman, 1987b; Demmig and Bjorkman, 1987; Horton, 1990; Rees and Horton, 1990). The dissipation process within the pigment complex is thought to involve the carotenoid zeaxanthin (Demmig *et al.*, 1987a; Demmig-Adams *et al.*, 1989). Here we present evidence suggesting that the zeaxanthin is involved in the dissipation of excessive light energy.

High energy state quenching or pH-dependent quenching (decreasing both flourescence parameters F_m and F_o) reflects a dissipation process which can apparently persist for very different lengths of time. For this type of fluorescence quenching (*i.e.*, non-photochemical quenching) a correlation was found between the calculated activity of a radiationless energy dissipation process and the content of the carotenoid zeaxanthin in the leaves of *Nerium oleander* (Demmig *et al.*, 1987a). Such a correlation was found to exist between non-photochemical quenching and zeaxanthin formation. The observation of such a correlation led to the proposal that zeaxanthin may be involved in this dissipation process in the chlorophyll pigment-protein complex.

Zeaxanthin acts as a competitor for the excitation energy under excessive light, since an increase in radiationless energy dissipation, which Demmig-Adams *et al.* (1990) suggested to be associated with zeaxanthin, results in a decrease in photochemical efficiency (Bjorkman, 1987a, 1987b). The capacity for a rapid removal of this competitor (zeaxanthin) for the excitation energy facilitates a rapid increase in the efficiency of photochemical energy conversion upon return to conditions under which light is no longer excessive. Evidence for the photoprotective function of zeaxanthin is based on experiments carried out using either an inhibitor of zeaxanthin formation or an organism which lacks the xanthophyll cycle.

Bilger *et al.* (1989) reported that dithiothreitol (DTT), an inhibitor of violaxanthin de-epoxidation, could be administered through the cut petiole of leaves. In such DTT-treated leaves zeaxanthin formation was completely inhibited without affecting photosynthetic O_2 evolution. It was observed that DTT-treated (zeaxanthin-free) leaves had an approximately 30% greater reduction state of PS II compared to that of control leaves containing zeaxanthin. This indicates that an increased amount of energy reaches the PS II reaction centre in the DTT-treated (zeaxanthin-free) leaves because excessive energy could not be dissipated by the mechanism depicted in Fig. 1.

Certain cyanobacteria naturally lack the xanthophyll cycle, for example, those found in symbiotic association with lichens. Demmig-Adams *et al.* (1990) compared the response of such cyanobacteria with those of green algae which possess the xanthophyll cycle. A comparative study of chlorophyll fluorescence characteristics as well as the tolerance of photosynthesis to high irradiance was carried out in zeaxanthin-free cyanobacteria and zeaxanthin-containing green algae symbiotically associated with lichens. Upon exposure to excess light, the green algae associated with lichens formed zeaxanthin rapidly from violaxanthin and also exhibited the type of rapid fluorescence quenching associated with a quenching of F_o. The reduction state (over-energization) of PS II centres was maintained at a low level when the degree of non-photochemical fluorescence quenching was high resulting in lesser degree of damage to photosynthesis. In contrast, the cyanobacteria associated with lichens which lack a xanthophyll cycle did not form any zeaxanthin and also did not exhibit rapidly developing and relaxing non-photochemical quenching, indicative of radiationless energy dissipation. There was also no quenching of F_o and the reduction state (excitation state) of PS II centres was high under excess irradiance. The tolerance to longer exposure at high irradiance was considerably greater in zeaxanthin-containing green algae than in the zeaxanthin-free cyanobacteria. These results resembled that of DTT-treated leaves (Demmig-Adams, 1990).

Sharma (1990) observed that there may be a certain limitation to the role of the xanthophyll cycle as a protection mechanism against photoinhibition. This was concluded because the zeaxanthin content after a photoinhibition treatment at 1600 μmol m^{-2} s^{-1} at 5°C was not further increased when plants were photoinhibited at a higher PPFD (2500 μmol m^{-2} s^{-1}), although photoinhibition itself increased (Table 1). In addition, greater increase in zeaxanthin and decrease in violaxanthin content were observed at 20°C where the photoinhibition observed was not as great as at 5°C. This suggested that: (1) Changes in the xanthophyll cycle are temperature dependent. Under chilling conditions the dissipation of excessive radiation energy by carotenoids is less than at 20°C resulting in the greater damage to photosynthesis at 5°C; (2) With higher irradiances (greater than 1600 μmol m^{-2} s^{-1}) no further

Table 1. The effect of photoinhibition on carotenoid (xanthophyll cycle) composition in sorghum leaves

Carotenoids	Control	Photoinhibition at 5°C and 1600 PAR	Photoinhibition at 5°C and 2500 PAR	Photoinhibition at 20°C and 1600 PAR	Photoinhibition at 20°C and 2500 PAR
Violaxanthin	26.4	7.1	8.7	5.7	6.4
Antheraxanthin	1.5	2.6	1.7	4.1	2.6
Zeaxanthin	0.0	14.6	13.7	19.7	19.8
V + A + Z	27.9	24.3	24.1	29.1	28.8

Data shown are percentages of total carotenoids. PAR: photosynthetic active radiation, $\times$ μmol m^{-2} s^{-1}.
V + A + Z: violaxanthin + antheraxanthin + zeaxanthin.

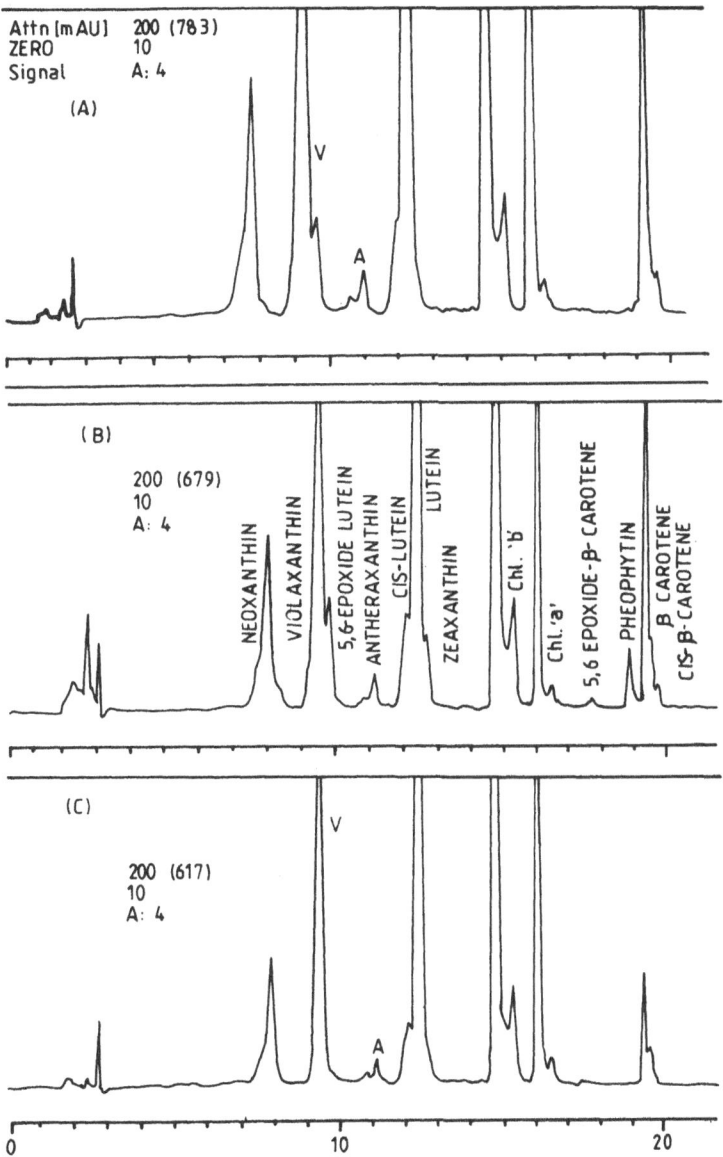

Figure 2. HPLC analysis of 13-day-old sorghum leaves: (A) control (no photoinhibition); (B) photoinhibition at 5°C and irradiance of 2500 μmol m^{-2} s^{-1} for four hours; (C) 48 hours recovery from photoinhibitory treatment. A—antheraxanthin, V—violaxanthin; Z—zeaxanthin.

increase in the zeaxanthin content suggests a limitation in the dissipation of excessive energy by the xanthophylls (Fig. 2).

The detailed mechanism of the triplet energy transfer and dissipation remains to be determined. Experimental data now available indicate a route of triplet energy transfer and a possible mechanism of energy dissipation in which isomerization of triplet carotenoid plays an essential role (Koyama, 1991). The fact that zeaxanthin is the carotenoid which is formed exclusively under an excess of light and removed under limiting light or dark (Demmig et al., 1988; Sharma and Hall, 1990) is also consistent with the assumption that the singlet excited state of chlorophyll, which is normally used for photochemistry, may interact with zeaxanthin under an excess of light in a process leading to de-excitation.

IV. Conclusions

Much information has been accumulated in recent years on the location and structure of carotenoids in the pigment-protein complexes (reaction centres and light harvesting complexes of photosynthetic membranes). However, there is still uncertainty as to the detailed mechanism of photoprotection. The main lines of evidence for a relationship between zeaxanthin and radiationless energy dissipation are:
(1) The inhibition of the dissipation process in the chlorophyll pigment complex in organisms treated with an inhibitor of the formation of zeaxanthin-free violaxanthin which does not affect the rate of photosynthesis or other dissipation processes.
(2) The lack of radiationless energy dissipation in the chlorophyll pigment-protein complexes in organisms which lack the xanthophyll cycle.
(3) The zeaxanthin-associated energy dissipation process in the chlorophyll-protein complexes can occur alone or in combination with other dissipation mechanisms within or around the PS II centres.

A photoprotective function for zeaxanthin and for the radiationless energy dissipation process in the chlorophyll-protein complex is supported by the finding that a variety of zeaxanthin-containing organisms: (1) exhibit the ability to prevent overexcitation of the PS II centres more efficiently than the corresponding zeaxanthin-free organisms, and (2) possess the capacity for a rapid and complete recovery from high light stress, whereas the corresponding zeaxanthin-free organisms exhibit sustained depression in the efficiency of photochemical and photosynthetic energy conversion.

V. References

Bilger W, Bjorkman O and Thayer SS (1989) Light-induced spectral absorbance changes in relation to photosynthesis and the epoxidation state of xanthophyll cycle components in cotton leaves. Plant Physiol 91: 542–551

Bjorkman O (1987a) High irradiance stress in higher plants and interaction with other stress factors. In: (J Biggins, ed) Progress in Photosynthesis Research, Vol 4. Martinus Nijhoff, Dordrecht, The Netherlands, pp 11–18

Bjorkman O (1987b) Chlorophyll fluorescence in leaves and its relation to photon yield of photosynthesis. In: (J Kyle, CB Osmond and CJ Arntzen, eds) Photoinhibition. Elsevier, Amsterdam, pp 123–144

Bjorkman O and Demmig B (1987) Photon yield of O₂ evolution and chlorophyll fluorescence (77K) and photon yield of O₂ evolution in leaves of higher plants. Planta 171: 489–504

Bjorkman O and Powles SB (1984) Inhibition of photosynthetic reaction under water stress: Interaction with light level. Planta 161: 490–504

Britton G (1990) Biosynthesis of chloroplast carotenoids. In: (M. Baltscheffsky, ed) Current Research in Photosynthesis, Vol 4. Kluwer Acad Publ, Dordrecht, The Netherlands, pp 827–834

Cogdell RJ (1985) Carotenoids in photosynthesis. Pure Appl Chem 57: 723–728

Cogdell RJ and Frank HA (1987) How carotenoids function in photosynthetic bacteria. Biochim Biophys Acta 895: 63–79

Davidson E and Cogdell RJ (1981) Reconstitution of carotenoids into light harvesting pigment-protein complexes from the carotenoid less mutant of *Rhodopseudomonas spheroides* R26. *Biochim Biophys Acta* **635**: 295–303

Demmig B and Bjorkman O (1987) Comparison of effect of excessive light on chlorophyll fluorescence (77K) and photon yield of O_2 evolution in the leaves of higher plants. *Planta* **171**: 171–184

Demmig B, Winter K, Kruger A and Czygen FC (1987a). Photoinhibition and zeaxanthin formation in intact leaves. A possible role of the xanthophyll cycle in the dissipation of excess light. *Plant Physiol* **84**: 218–224

Demmig B, Winter K, Kruger A and Czygen FC (1987b) Zeaxanthin and heat dissipation of excess light energy in *Nerium oleander* exposed to a combination of high light and water stress. *Plant Physiol* **87**: 17–24

Demmig-Adams B, Winter K, Kruger A, Czygen FC (1987) Zeaxanthin and the heat dissipation of excess light energy in *Nerium oleander* exposed to a combination of high light and water stress. *Plant Physiol* **87**: 17–24

Demmig-Adams B, Adams III WW, Winter K, Meyer A, Schereiber U, Pereira JS, Kruger A, Czygen FC, Lange OL (1989) Photochemical efficiency of PS II photon yield of O_2 evolution, photosynthetic capacity and carotenoid composition during the midway depression of net CO_2 uptake in *Arbutus unedo* plants growing in Portugal. *Planta* **177**: 377–387

Demmig-Adams B (1990) Carotenoids and photoprotection in plants: A role for the xanthophyll, zeaxanthin. *Biochim Biophys Acta* **1020**: 1–24

Demmig-Adams B, Adams WW III, Czygen FC, Schreiber U and Lange OL (1990) Difference in the capacity for radiationless energy dissipation in the photochemical apparatus of green and blue green algal lichens associated with differences in carotenoid composition. *Planta* **180**: 582–589

Frank HA, Violette CA, Trautman JK, Shreve A, Owens TG and Albrecht AC (1991) Carotenoids in photosynthesis: Structure and photochemistry. *Pure Appl Chem* **63**: 109–114

Genty B, Briantais JM and Da Silva JBV (1987) Effects of drought on primary photosynthetic processes of cotton leaves. *Plant Physiol* **83**: 360–364

Hayashi H, Noguchi T and Tasumi M (1989) Studies on the interrelationship among the intensity of a Raman marker band of carotenoids, polyene chain structure and efficiency of the energy transfer from carotenoids to bacteriochlorophyll in photosynthetic bacteria. *Photochem Photobiol* **49**: 337–343

Horton P (1990) Regulation of light harvesting by metabolic events. In: (M Baltscheffsky, ed) *Current Research in Photosynthesis*, Vol 4. Kluwer Acad Publ Dordrecht, The Netherlands, pp 111–118

Iwata K, Hayashi H and Tasumi M (1985) Resonance Raman studies on the conformation of all-trans carotenoids in LHC of photosynthetic bacteria. *Biochim Biophys Acta* **810**: 269–273

Koyama Y, Kanaji M and Shimamura T (1988) Configuration of neurosporene isomers isolated from the reaction centre and the LHC of *Rhodobacter sphaeroides* Glc. A resonance Raman electronic absorption and ^{1}H NMR study. *Photochem Photobiol* **48**: 107–114

Koyama Y (1990) Natural selection of carotenoid configuration by the reaction centre and the LHC of photosynthetic bacteria. In: (NI Krinsky, MM Mathews-Roth and RF Tayler, eds) *Carotenoids: Chemistry and Biology*. Plenum, New York, pp 207–222

Koyama Y, Takatsuka I, Kanaji M, Tomomoto K, Kito M, Shimamura T, Yamashita J, Saiki K and Tsukida K (1990) Configuration of carotenoids in the reaction centre and LHC of *Rhodospirillum rubrum*. Natural selection of carotenoid configuration by pigment-protein complexes. *Photchem Photobiol* **51**: 119–128

Koyama Y (1991) Structure and functions of carotenoids in photosynthetic systems. *J Photochem Photobiol B: Biol* **9**: 265–280

Krause GH, Somersalo S, Zumbusch E, Weyers B and Laasch H (1990) On the mechanism of photoinhibition in chloroplasts relationship between changes in fluorescence and activity of PS II. *J Plant Physiol* **136**: 472–479

Krinsky NI (1979) Carotenoid protection against oxidation. *Pure Appl Chem* **51**: 649–660

Ludlow MM and Powles SB (1988) Effect of photoinhibition induced by water stress on growth and yield of grain of sorghum. *Aust J Plant Physiol* **15**: 179–194

Lutz M, Agalidis I, Hervo G, Cogdell RJ and Reiss-Husson F (1978) On the state of carotenoids bound to reaction centres of photosynthetic bacteria, a resonance Raman study. *Biochim Biophys Acta* **503**: 287–303

478

Lutz M, Szoonarski W, Berger G, Robert B and Neumann J-M (1987) The stereoisomerism of bacterial reaction centre-bound carotenoids revisited: An electronic absorption, resonance Raman and ^{1}H NMR study. *Biochim Biophys Acta* **894**: 423–433

Masojidek J, Trivedi S, Halshaw L, Alexiou A and Hall DO (1991) The synergistic effect of drought and light stresses in sorghum and pearl millet. *Plant Physiol* **96**: In press

Mathis P and Schenck CC (1982) The function of carotenoids in photosynthesis. In: (G. Britton and TW Goodwin, eds) *Carotenoids: Chemistry and Biochemistry*. Pergamon, Oxford, pp 339–347

Mimuro M and Katoh T (1991) Carotenoids in photosynthesis absorption, transfer and dissipation of light energy. *Pure Appl Chem* **63**: 123–130

Mishra S, Subrahmanyam D and Singhal GS (1991) Interrelationship between salt and light stress on primary processes of photosynthesis. *J Plant Physiol* **138**: 92–96

Noguchi T, Hayashi H and Tasumi M (1990) Factors controlling the efficiency of energy transfer from carotenoids to bacterio-chlorophyll in purple photosynthetic bacteria. *Biochim Biophys Acta* **1017**: 280–290

Oquist G, Greer DH and Ogren E (1987) Light stress and low temperature. In: (DJ Kyle, CB Osmond and CJ Arntzen, eds) *Photoinhibition*. Elsevier, The Netherlands, pp 67–88

Rees D and Horton P (1990) The mechanisms of changes in PS II efficiency in spinach thylakoids. *Biochim Biophys Acta* **1016**: 219–227

Richards GE and Hall DO (1987) Photoinhibition at chilling in intact leaves and isolated chloroplasts. In: (J Biggins, ed) *Progress in Photosynthesis Research*, Vol 4. Martinus Nijhoff, Dordrecht, The Netherlands, pp 38–42

Sharma PK (1990) Effects of environmental stresses (light, temperature and salt) on photosynthesis in sorghum. Ph.D. Thesis, King's College, University of London, London pp 225–229

Sharma PK and Hall DO (1990) The effect of photoinhibition, salt stress and their interaction on photosynthesis in sorghum. In: (M Baltscheffsky, ed) *Current Research in Photosynthesis*, Vol 2. Kluwer Acad Publ, Dordrecht, The Netherlands, pp 487–490

Sharma PK and Hall DO (1991) Interaction of salt stress and photoinhibition on photosynthesis in barley and sorghum. *J Plant Physiol* **138**: 614–619

Siefermann-Harms D (1987) The light-harvesting and protective functions of carotenoids in photosynthetic membranes. *Physiol Plant* **69**: 561–568

Truscott TG (1990) The photophysics and photochemistry of carotenoids. *J Photochem Photobiol B: Biol* **6**: 359–371

Weis E and Berry JA (1987) Quantum efficiency of PS II in relation to energy dependent quenching of chlorophyll fluorescence. *Biochim Biophys Acta* **894**: 198–208

Weis E, Ball JT and Berry JA (1987) Photosynthetic control of electron transport in leaves of *Phaseolus vulgaris*—Evidence for regulation of PS II by proton gradient. In: (J Biggins, ed) *Progress in Photosynthesis Research*, Vol 2. Martinus Nijhoff, Dordrecht, The Netherlands, pp 553–556

Yeats TO, Koyama H, Chirino A, Rees DC, Allen JP and Feber G (1988) Structure of reaction centre from *Rhodobacter sphaeroides* R-26 and 2.4.1: Protein-cofactor (Bacteriochlorophyll, bacteriopheophytin and carotenoid) interactions. *Proc Natl Acad Sci USA* **85**: 7993–7997

20

Carbon Dioxide Enrichment Effects on Photosynthesis and Plant Growth

U.K. Sengupta and Aruna Sharma

Division of Plant Physiology,
Indian Agricultural Research Institute,
New Delhi 110012, India

CONTENTS

ABBREVIATIONS

ABA	:	abscisic acid
Ca	:	ambient CO$_2$ concentration
CAM	:	crassulacean acid metabolism
CER	:	carbon dioxide exchange rate
Ci	:	internal CO$_2$ concentration
LAI	:	leaf area index
NAR	:	net assimilation rate
PEP	:	phosphoenol-pyruvate
RLGR	:	relative leaf growth rate
RuBPcase	:	ribulose-bisphosphate carboxylase/oxygenase

480

ABSTRACT

Increasing atmospheric carbon dioxide directly stimulates photosynthesis and plant growth. However, other environmental variables affect the relation between CO_2 concentration and photosynthesis and thus the links between carbon assimilation, growth and yield. Limitations to photosynthesis under short-term enrichment conditions are quite unlike those found in plants grown in high CO_2 for a longer time. The effects of high CO_2 concentrations on the rate of photosynthesis under both short-term and long-term enrichment conditions are discussed. Acclimation of photosynthesis, i.e., an increase or decrease in the rate under long-term CO_2 enrichment is examined with emphasis on factors regulating photosynthesis at the metabolic level. Apart from stimulatory effect on photosynthesis, high CO_2 concentrations affect plant growth and biomass partitioning. The role of elevated CO_2 concentrations in enhancement of growth and yield are discussed. Finally, we discuss the interaction of high CO_2 with environmental variables like temperature, water and nutrient in context of photosynthesis and growth in order to assess the present state of knowledge and to identify areas for future research.

I. Introduction

The record of carbon dioxide measurements in the atmosphere which began in Antarctica in 1957 and Mauno Loa in Hawai in 1958 indicate clearly that the concentration of CO_2 in the atmosphere is increasing rapidly. There is a consensus that prior to industrial revolution, about 130 years ago, the global CO_2 concentration was between 270 and 280 ppm. Present day concentration is about 353 ppm and is increasing at the rate of 1.2 ppm (Houghton *et al.*, 1990). Increase in CO_2 concentration is expected to cause global warming due to the fact that CO_2, though reasonably transparent to short wave radiation, absorbs longer wavelength thermal radiation from the Earth's surface. The increased CO_2 concentration will thus increase the surface air temperature and affect the associated climatic factors such as cloudiness and precipitation. This is referred to as 'climatic effect' of CO_2 enrichment.

CO_2 also has a 'biological effect' primarily because it serves as a substrate for photosynthetic carbon assimilation. Concomitantly, increase in CO_2 results in decline in photorespiratory activity and alters stomatal aperture as well. These changes are expected to enhance plant growth and yield, and affect the global eco-systems (Bazzaz, 1990). Environmental variables such as light, water, nutrients and temperature interact strongly with CO_2 and thus modify the response of plants to increased CO_2 concentration.

In this chapter, the present state of knowledge regarding the physiological effects of CO_2 enrichment on photosynthesis, growth and biomass partitioning are reviewed and future areas of research are identified.

II. CO_2 Enrichment and Rate of Photosynthesis

The reductive photosynthetic carbon cycle or C_3 pathway is the primary pathway for net CO_2 fixation and formation of sugar phosphates in all photosynthetic organisms. Increased atmospheric CO_2 will, of course, not change the principle photosynthetic pathway. However, the rate at which CO_2 is assimilated and the levels of endogenous end products may be altered by high CO_2 level. This is because increase in ambient CO_2 concentration (C_a) increases internal CO_2 concentration (C_i), resulting in more available CO_2 at the carboxylating site (Tolbert and Zelitch, 1983). Many data are available on the effects of CO_2 concentration on photosynthesis. In some studies carbon dioxide exchange rate (CER) has been measured in leaves of plants grown at ambient or control level of CO_2 and measured at elevated CO_2 concentrations after enriching for a few hours, while in others, measurements were made on leaves grown at elevated CO_2 concentrations for at least a week. Under short-term enrichment, limitations to CER are unlike those found in plants grown for long in high CO_2 concentrations. When we consider short-term experiments, a large increase in the rate of photosynthesis occurs with increasing CO_2 concentrations, at least at high irradiance. As much as four times increase in photosynthetic rate has been reported (Table 1).

Table 1. Effect of short-term CO_2 enrichment on net carbon exchange rate (CER): Response of plant species

Crop Type	CO_2 concentration	Enrichment period	Increase in CER	References
C₃ species				
Tomato	0.1%	–	2.4 times	Bishop and Whittingham, 1968
Tomato	750 and 1000 ppm	6 hr	70%	Ho, 1977
Lolium	720 ppm	6 hr	Doubled	Wardlaw, 1982
Soybean	650 ppm	–	150%	Huber *et al.*, 1984
Soybean	1670 ppm	few minutes	4 times	Brun and Cooper, 1967
Strawberry	600 ppm	6-7 hr	45%	Campbell and Young, 1986
Mungbean	600 and 900 ppm	6 hr	1.8 and 2.4 times	Sharma and Sengupta, 1990
C₄ species				
Sorghum	720 ppm	6 hr	25%	Wardlaw, 1982
Conifer				
Ponderosa pine	500 ppm	2 days	84%	Green and Wright, 1977

Under conditions of moderate to high light intensities and temperatures, the rate of CO_2 fixation in C_3 plants roughly doubles with doubling of present atmospheric CO_2 concentration (Fig. 1). The C_3 pathway, found in most plants, is generally strongly limited by the low CO_2 concentration in the present-day atmosphere. There is a further limitation imposed by the high atmospheric concentration of O_2 which causes photorespiration. Plants possessing the C_4 dicarboxylic pathway for CO_2 fixation (C_4 plants) differ from C_3 plants in their response to increased CO_2 since C_4 pathway is an adaptation for concentrating the internal CO_2 concentration (for details see Chapter by Raghavendra and Das, this volume). In addition, photosynthesis in C_4 plants is not limited by O_2. The results obtained so far for C_4 plants are contradictory. Some authors report that C_4 photosynthesis is not affected by CO_2 enrichment (Wong, 1979; Morison and Gifford, 1983; Rogers *et al.*, 1983), while others have shown a positive correlation between CO_2 concentration and net photosynthesis (Akita and Moss, 1973; Van Bavel, 1974; Wardlaw, 1982; Potvin and Strain, 1985). Most of the C_4 plants studied show an increase in photosynthetic rate in the range of 20 to 30%.

The influence of high CO_2 concentrations on photosynthesis in crassulacean acid metabolism (CAM) plants is not well known. The dark CO_2 fixation characteristic

482

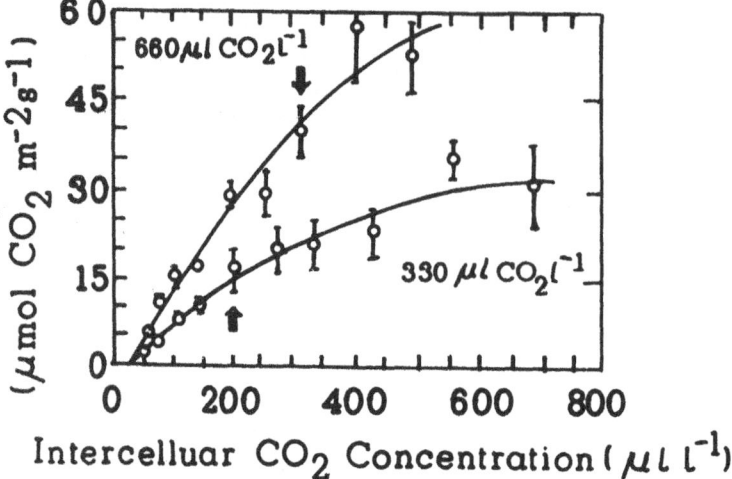

Figure 1. Leaflet photosynthetic rates versus intercellular CO_2 concentration for soybean plants grown at 330 and 660 ppm CO_2. The SD of the data is shown as vertical bars through each symbol. Arrows indicate the rates measured at the respective growth CO_2 concentration. Measurements were made 56 to 60 days after planting (from Campbell et al., 1988).

of this pathway involves phosphoenol pyruvate (PEP) carboxylase and fixation of CO_2 into C_4 acid. Osmond and Bjorkman (1975) studied the response of *Kalanchoe daigremontiana*, a species capable of both CAM and C_3 photosynthesis. The study revealed a dark CO_2 uptake response with a zero CO_2 compensation point, little increase in CO_2 uptake rates above 200 ppm intercellular CO_2 concentration and no O_2 effect on CO_2 uptake, a response similar to C_4 assimilation. Doubling of CO_2 concentration enhanced CO_2 uptake rates in dark only by 10%, but by 50% in the light. Minor physiological response to elevated CO_2 has been reported in the CAM plant, *Agave vilmoriana* (Szarek et al., 1987). Negligible differences in stomatal conductance and CO_2 assimilation rates were observed over an entire 24 hr period between plants grown in elevated (750 pm) CO_2 and normal CO_2 indicating saturated or near-saturated PEP carboxylase activity at ambient level of CO_2 concentration.

Measurements made over short periods of time do not necessarily provide reliable information concerning what occurs when plants are grown at high CO_2 concentrations for a longer period. Analysis of response of eight C_3 and two C_4 species showed that in the overall crop response to doubled CO_2 concentration the weighted average short-term CER response was 52%, whereas acclimated CER response was of the order of only 29% (Cure and Acock, 1986). Soybean and cotton are the two species studied extensively under both short- and long-term exposure to different CO_2 concentrations (Mauney et al., 1978; Radin et al., 1987; Cure et al., 1989). In most other crops, short-term CER data and acclimated CER data were collected by different investigators under different conditions and, therefore, comparisons cannot be made.

Data have accumulated on the effects on CER of growing plants in different concentrations of CO_2 (Table 2); there is generally an increase in CER in plants

Table 2. Response of plant species to long-term CO_2 enrichment in terms of CER and plant growth

Crop	CO_2 concentration	Period of enrichment	CER	Growth	Reference
C₃ species					
Cucumber	300-5400 ppm	From seeding to 3 weeks	Initial high rate but below control by 15 days	–	Aoki and Yabuki, 1977
Tobacco	1000 ppm	35 days after transplanting	70-80% of the control	Increase in leaf area	Raper and Peedin, 1978
Cotton	640 ppm	From emergence, rate measured at 40 days	1.5 times increase	Two fold increase in short dry weight, 60% increase in leaf area	Wong, 1979
Soybean	1000 ppm	Pod set to maturity	Almost double but less in plants grown at 1000 ppm and measured at 350 ppm	Increased dry weight of vegetative and reproductive tissues	Clough *et al.*, 1981
Soybean	650 ppm	From emergence	25% increase	60-70% increase in leaf dry weight	Heber *et al.*, 1984
Soybean	660 ppm (outdoor chamber)	Grown from seed	150% greater at seed filling stage	–	Campbell *et al.*, 1984
C₄ species					
Maize	640 ppm	From emergence, rate measured	Negligible increase	No increase in shoot dry weight, less than 10% increase in leaf area	Woug, 1979

grown in high CO_2 concentrations. Though the initial rate was large and later decreased, it was always greater than the CER measured in plants grown in ambient CO_2 concentration (Wong, 1979; Clough et al., 1981; Campbell et al., 1988). There are some reports where CER measured at high CO_2 concentration was similar to or less than that of plants grown at 'ambient' CO_2 (Havelka et al., 1984a; Peet et al., 1986). Non-acclimation of photosynthetic efficiency to increased CO_2 concentration has been reported in case of many C_3 plants (Sionit et al., 1984; Huber et al., 1984 Jones et al., 1985; Baysdorter and Bassham, 1985; Campbell et al., 1988). Mauney et al., (1978) grew four species of plants for 12 weeks in air conditioned glass houses with 350 and 660 ppm compared to 330 ppm of CO_2 (Table 3). The rate of photosynthesis per unit of leaf area, measured at frequent intervals on single leaves of plants grown at 660 ppm compared to 330 ppm of CO_2 showed increases of 41% for soybean and 15% for cotton, but only 7% for sunflower and 2% for sorghum. This clearly shows that species do differ in their response to CO_2 enrichment. Moreover, the net assimilation rate (NAR) of young plants was increased by high CO_2 but not that of mature plants. Radin et al., (1987) found a very large response to CO_2 enrichment in cotton with assimilation rate not saturated at all internal CO_2 concentrations, no decrease in photosynthetic capacity for most of the season, and maintenance of very high internal CO_2 concentration as the basis.

Several workers have observed that plants grown in elevated CO_2 show a decline in photosynthetic rate with time. Aoki and Yabuki (1977) grew cucumber plants for up to three weeks in chambers exposed to the sun and subjected to CO_2 concentration ranging form 300 to 5400 ppm. Although initial rates of photosynthesis in 5400 and 2400 ppm were twice those at 300 ppm, the rates fell below that of control by the fifth day. In tobacco, grown at 1000 ppm concentration (Raper and Peedin, 1978), photosynthetic rates were between 70 to 80% of the rate of plants kept at 400 ppm; although the leaf area was greater in high CO_2 than in low CO_2, photosynthetic rate per unit leaf area declined. Lower photosynthetic rate has been reported for *Pharbitis nil* (Hicklenton and Jolliffe, 1980), soybean (Clough and Peet, 1981), *Desmodium paniculatum* (Wulff and Strain 1982), cotton (DeLucia et al., 1985), and tomato and carnation (Bruggnick, 1984).

Table 3. Effects of 660 ppm of CO_2 on rate of photosynthesis, leaf area and dry weight accumulation after 12 weeks

Species	CER	Leaf area	Dry weight
Cotton	15	91	109
Soybean	41	180	382
Sunflower	7	142	60
Sorghum	1.5	15	18

Values are percentage increase above values for plants grown in 330 ppm (adapted from Mauney et al., 1978).

Variation in photosynthetic response to high CO_2 levels exists not only between species but between experimental treatments with single species. Sionit et al. (1984) found that container-grown soybean showed greater response to high CO_2 than field-grown plants. In addition, it is difficult to draw conclusions about response to CO_2 due to great variability in experimental conditions. Data obtained from greenhouse studies are very different from those obtained from the field even for the same crop. Cotton acclimates to increased CO_2 by decreasing photosynthetic capacity correlated with increased starch content in controlled environmental conditions (Mauney et al., 1978). In the field, in open top chambers, there was no decrease in photosynthetic efficiency in spite of a great increase in starch. It has been suggested that high sink demand due to early maturity of fruits in the field was responsible for non-acclimation of CER under field conditions (Radin et al., 1987). Another factor leading to variability in the data is that CER response categories include leaf as well as canopy measurement and this could be misleading. Gifford (1977) showed, in a growth chamber study, that photosynthesis of wheat plants grown at 490 ppm CO_2 were 56 % for the flag leaf, 40 % for the plant canopy, and 238 % on a unit ground area basis compared to those grown at 290 ppm CO_2.

Several reasons have been attributed to the decline in the photosynthetic rate, although the underlying mechanisms are not well understood. These are: (1) decrease in carboxylating efficiency due to decrease in amount and activity of ribulose bisphosphate carboxylase (RuBPcase) (Sage and Pearcy, 1987; Fetcher et al., 1988; Sage et al., 1989); (2) suppression of sucrose synthesis due to inhibition of the triose-P-carrier and reduction in the activity of sucrose phosphate synthase, resulting in accumulation of starch (Huber et al., 1984) and feedback inhibition of photosynthesis (Mauney et al., 1978; Azcon-Bieto, 1983) and/or physical damage at chloroplast level (Madsen, 1975; Sasek et al., 1985; Wulff and Strain 1982); (3) reduction in RuBPcase activity, regeneration of RuBP and rate of photosynthetic electron transport (von Cammerer and Farquhar, 1984); (4) reduction in inorganic phosphate concentration in chloroplast at high internal CO_2 concentration (Herold, 1980; Sharkey and Badger, 1984); and (5) increase in carbonic anhydrase activity, which causes direct inhibition of photophosphorylation via increased bicarbonate concentration (Chang, 1975).

Some conclusions on the CER that can be drawn from CO_2 enrichment studies are: (1) plants with the C_3 pathway usually show a greater increase in rate than plants with C_4 pathway; (2) elevated CO_2 concentrations reduce or completely eliminate photorespiration; (3) response to CO_2 is more pronounced under high level of other resources, especially water, nutrient, and light; (4) there are important differences between species in response to enhanced CO_2; (5) with long-term exposure to high concentrations of CO_2, many species show only an initially temporary increase in photosynthetic rate, which later falls below those of plants kept at ambient CO_2 concentration; and (6) adjustment of photosynthesis during growth under high CO_2 depends on the species as well as the resource availability, and, therefore, species even of the same community may differ in their response to CO_2.

III. Regulation of Photosynthesis under CO_2 Enrichment

The response of leaf photosynthesis to elevated CO_2 concentrations depends on various external and internal limits to CO_2 uptake, their dependence on CO_2 concentration, and their interactions with environmental factors. CO_2 must be physically transported from the atmosphere to the carboxylating site in the chloroplast, before carboxylation takes place. Raschke (1979) coined two separate terms for these two determinants. The transport has been termed the "supply function" and the carboxylation, the "demand function". The components of either process can limit the rate of CO_2 uptake to varying extent depending on environmental factors and leaf characteristics. The path of CO_2 into the leaf can be considered as a series of conductances associated with the leaf boundary layer, the stomata, the intercellular air space, the cell wall, and intercellular liquid phase. The stomata are responsible for the largest drop in CO_2 concentration along the diffusion path. Though the CO_2 concentration drop across the stomata is different in C_3 and C_4 plants C_i, in both, remains remarkably constant over a wide range of stomatal conductances. Many factors like leaf water status, leaf age, nitrogen nutrition, and light regime during growth affect the stomatal conductance (Wong, 1979). It is found that stomata frequently respond to changes in C_a so that the C_a/C_i ratio is maintained at 0.6 to 0.8 in C_3 species and 0.3 to 0.5 in C_4 species. This suggests a close coordination of photosynthetic activity (i.e., in the mesophyll) and the stomatal conductance as C_a increases.

a. Stomatal response and water-use efficiency

Increase in net photosynthesis due to increase in CO_2 concentration can be attributed to the steeper CO_2 gradient between ambient atmosphere and carboxylating site in the chloroplast and availability of CO_2 for carboxylation. Therefore, the ability of the additional carbon dioxide to enter the leaf is a critical factor that determines the response of the plant to increased CO_2 levels.

Response of stomata to CO_2 concentration is well documented. Morison (1985a) recorded stomatal response in over 50 species, including gymnosperms, with C_3, C_4 and CAM photosynthesis. The literature reveals great variation in the response from no change (Berhuizen and Slayter, 1964) to a tenfold change in stomatal conductance with increasing CO_2 from 300 to 500 ppm (Waritt et al., 1980). In general, partial stomatal closure occurs at higher than ambient CO_2 concentration and opening at lower than ambient concentration. It was widely thought that stomata of C_3 species are less sensitive to CO_2 than those of C_4 species. The evidence for this is rather slender and is based on the results of early workers where direct response of stomata to humidity was not accounted for (Pallas 1965; Akita and Moss, 1973). The data are insufficient to conclude unequivocally that stomata of C_3 and C_4 plants differ (Morison, 1985b). Morison and Gifford (1983) studied two C_3 and two C_4 grass species grown and examined under identical conditions. They found that the sensitivity of stomatal conductance to either C_a or C_i was similar. The calculated sensitivity to CO_2 concentration was linearly related to the size of the conductance in the species examined. There appears to be no significant difference between the CO_2 sensitivity of stomata of C_3 species and that of C_4 species.

It has been suggested that C_i remains fairly constant when other environmental factors are changed. Constancy of C_i is indicative of a linear relationship between photosynthetic rate and stomatal conductance. Diurnal studies of leaf assimilation suggest that C_i does vary within limits with variation in temperature, light, and humidity. In many laboratory experiments, when C_a was increased, C_i changed and only in very few cases was the CO_2 sensitivity of stomata enough to maintain constant C_i. In all other cases, it was C_a/C_i which was constant and not C_i itself, although there is yet no convincing evidence that stomata are able to sense the ratio of C_a/C_i. This clearly indicates that stomatal sensitivity to CO_2 is not a limitation to the transport of CO_2 to the carboxylating centre.

It appears that the importance of the CO_2 response in normal stomatal behaviour is overemphasized. Stomatal conductance decreases with increase in the ambient CO_2 concentration, but for CO_2 diffusion to be a limiting factor, stomatal closure must be complete as CO_2 diffusion remains high and nearly uniform until stomata are competely closed (Ting and Loomis, 1963; El-Sharkawy and Hesketh, 1964) (Fig. 2). The fact that photosynthetic rate and stomatal conductance are linearly related and that C_i in leaves is regulated at near constant level has led to the concept that C_i some metabolite other than CO_2 regulates stomatal behaviour and stomatal mesophyll coordination. Cowan *et al.* (1982) suggested abscisic acid (ABA) as that metabolite. Radin *et al.* (1988) found that in cotton plants grown at 650 ppm CO_2, photosynthetic capacity was strongly correlated with stomatal conductance even though the procedure adopted discounted the effect of variable C_i (Table 4). They suggested that this "residual" coordination implied the existence of a messenger in CO_2-enriched plants. Enriched CO_2 greatly increased stomatal response to ABA and it appears that in CO_2-enriched atmospheres, ABA may function as such a messenger. Enhancement of stomatal closure in response to ABA by CO_2 enrichment has been reported by Raschke (1975). Synergistic action of CO_2 and

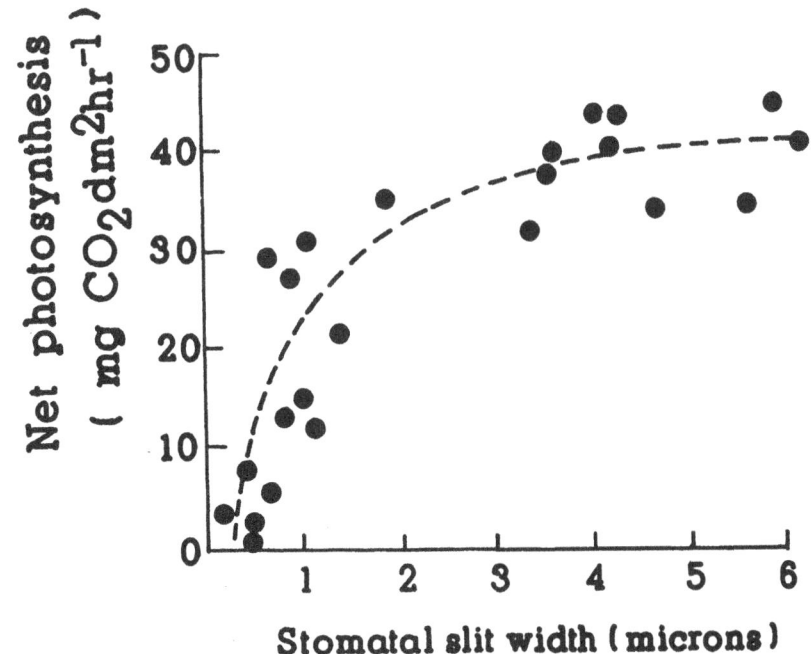

Figure 2. The effect of stomatal opening on the rate of photosynthesis of sunflower leaves (from El-Sharkawy and Hesketh, 1964).

Table 4. Variability and correlations among gas-exchange properties of cotton leaves

Treatment	Range of leaf properties			A versus g
	Assimilation rate (A)	Stomatal conductance (g)	C_i	
	$\mu mol \ m^{-2}s^{-1}$	$mol \ m^{-2}s^{-1}$	$\mu l/l$	
Ambient CO_2 (350 ppm)				
Well-watered	13.7–29.7	0.20–0.75	184–305	0.41**
Water-stressed	10.3–30.0	0.17–0.77	164–304	0.45**
Enriched CO_2 (650 ppm)				
Well-watered	16.9–44.9	0.13–0.58	337–552	0.49**
Water-stressed	16.4–61.8	0.15–0.56	377–552	0.46**

** highly significant

Source : Radin et al., 1988

ABA has been reported in cotton by Radin and Ackerson (1981). Research done so far indicates that though there is a definite effect of CO_2 on stomata, whole leaf gas exchange studies are not indicative of its importance in stomatal regulation. More work in this area is necessary to elucidate the role of CO_2 response in photosynthesis.

Though there is still ambiguity about the role of changes in C_a in stomatal behaviour, high C_a does reduce stomatal conductance in C_3 and C_4 species and also increases assimilation rate at least in C_3 plants. It is, therefore, expected that increased CO_2 concentration will increase water-use efficiency. Increases in instantaneous water-use efficiency with high CO_2 are well documented (Morison, 1985a). Plants grown with long-term enrichment of CO_2 show responses of assimilation rate and stomatal conductance which are subject to secondary responses and feedback controls when moving from the level of photosynthetic metabolism to crop yields (Warwick and Gifford, 1985). Both C_3 and C_4 species experience increase in water-use efficiency via reduced transpiration and hence increased growth under conditions of limited water (Morison, 1985b). In C_3 species, water-use efficiency doubled due to both increased assimilation rate and decreased stomatal conductance, while in C_4 species a greater decline in stomatal conductance was responsible for increased water-use efficiency rather than the small increase in assimilation rate.

Morison (1985b) in their review have quoted increases of about 60 to 160% in transpiration efficiency in both C_4 and C_3 plants. However, when plant growth is considered, increases in water-use effficiency in terms of dry matter produced per unit of water lost are not as large as those of transpiration efficiency. More experimental work at larger scale is required to predict the impact of increasing CO_2 concentrations.

b. *Carboxylation reaction*

The supply function is primarily determined by the stomata, while the demand function is determined by carboxylation reaction and RuBP supply. The enzyme responsible for this reaction, RuBP carboxylase/oxygenase, is the most abundant protein in the biosphere and ubiquitous in photosynthetic tissues. Though much studied (extensively and intensively), its regulation, activation, and reaction mechanism in the leaf are still poorly understood (see chapter by Sachar, Saluja and Murali, this volume). Since this enzyme exhibits two functions, competition between the two gaseous substrates, CO_2 and O_2, for the RuBP at the same enzyme site determines the flow of carbon between the reductive and the oxidative modes of the photosynthetic carbon cycle (Ogren, 1977; Lorimer, 1981).

CO_2 is not only a substrate for RuBPcase but is also required for the activation of this enzyme. The kinetics of this enzyme involves activation by slow binding with an activated CO_2 followed by rapid binding of Mg^{++} to form the activated enzyme complex (Lorimer, 1981). Upon binding of the substrate RuBP, the complex then reacts either with a molecule of the substrate, CO_2, or O_2. In addition, the CO_2 level in the chloroplast is very low and is estimated to be only 5 µM. Although the true K_m values of this enzyme are about 15 µM CO_2 and 450 µM O_2, the apparent K_m (CO_2) in air level of O_2 is about 26 µM. Therefore, if otherwise active, this enzyme may only be functioning at less than 0.1 V_{max} as carboxylase, while the rest of its potential activity either is unused or acts in part as oxygenase. Thus, excess carboxylase is available in the plant and with more CO_2, carboxylation reaction can occur faster per unit time per unit leaf area.

CO_2 uptake rate at low CO_2 concentration in the chloroplast is linearly related to the CO_2 concentration, the slope being determined by the carboxylase activity in the leaf (Fig. 3). At these levels, RuBP concentrations are considered to be saturating

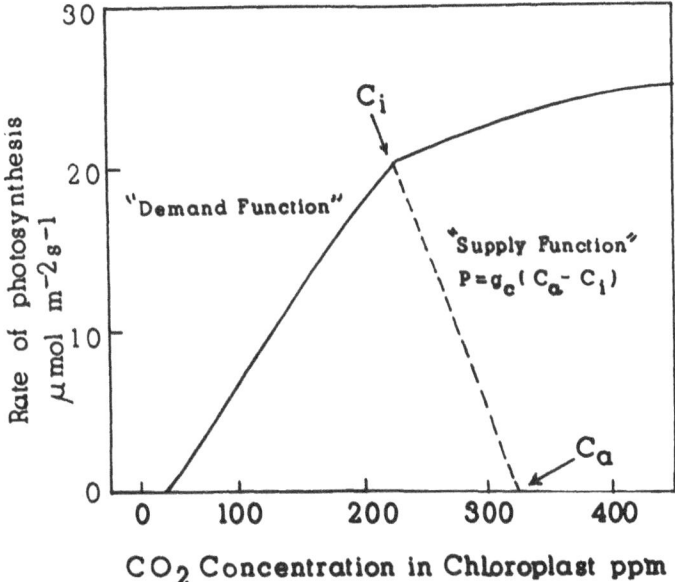

Figure 3. General relationship between the rate of photosynthesis and CO_2 concentration in the chloroplast for a C_3 plant. The supply function is represented by the broker line, while the demand function is given by the solid line (from Pearcy and Bjorkman, 1983; redrawan from Farquhar and Sharkey, 1982).

for RuBPcase. Farquhar and Sharkey (1982) have developed a model for CO_2 fixation by air-grown plants (Fig. 3). They suggested that at moderate temperatures and light levels, the transition from a RuBP-saturated to a RuBP-limited state occurs at about 220–240 ppm C_i (found in C_3 plant at ambient CO_2 concentration of 340 ppm). After the sharp break in the rate, further increase at a lower rate with further rise in CO_2 concentration suggests that other factors like electron transport capacity and RuBP supply become rate limiting. During short-term exposure to elevated CO_2 concentrations the rate of photosynthesis in light-saturated conditions is primarily limited by the capacity to regenerate P_i from phosphorylated photosysnthetic intermediates (Sage and Sharkey, 1987; Labate and Leegood, 1988), while at sub-saturating light intensity, the capacity of the light harvesting and electron transport system to regenerate RuBP limits photosynthesis at high CO_2 (von Cammerer and Farquhar, 1984). Short-term CO_2 enhancement disrupts the balance between RuBPcase and RuBP regeneration and possibly P_i regeneration. In response, the capacity of processes like RuBP carboxylation is regulated to balance the limiting processes like P_i or RuBP regeneration (Sage et al., 1988; Sharkey et al., 1988). Under long-term enrichment, however, mechanisms by which photosynthesis is limited are different. Many species show acclimation in photosynthesis under growth CO_2 concentrations which is not observed under short-term CO_2 enrichment.

The lack of response of RuBPcase activation state in the light to short-term changes in CO_2 concentration has been observed by several workers. No response of activation state has been reported in *Arabidopsis* by Salvucci et al. (1986), in *Triticum aestivum* seedlings by Perchorowicz and Jensen (1983). Campbell et al. (1988) studied the effect of CO_2 concentration on RuBPcase activity and amount in relation to photosynthesis in soybean leaves. Short-term enrichment of about one hour to a range of CO_2 concentrations(110–880 ppm) did not have any effect on activaton state of RuBPcase *in vivo* (initial activity) in eight-week-old plants, though a marked increase in the photosynthetic rate was observed. Sharma and Sengupta (1990) also reported no response of initial RuBPcase activity to CO_2 concentrations in mungbean. A single leaf was exposed to 600 and 900 ppm CO_2 concentrations for five hours. Although there was tremendous increase in photosynthetic rate (2.5 times in 900 ppm CO_2), there was no change in RuBPcase activity. Campbell et al. (1988) have also studied the activation and amount of RuBPcase in eight-week old plants grown in three CO_2 concentrations from 160 to 900 ppm CO_2. At high irradiance levels, no difference in initial activity of RuBPcase on a leaflet area or soluble protein basis was observed among plants grown at these CO_2 concentrations. At saturating irradiance, and varying CO_2 concentrations, leaflets grown at elevated CO_2 always had greater photosynthetic rates.

The observation that at high irradiance, changes in CO_2 did not cause a response in RuBPcase activation is consistent with low K(act), k (CO_2) of RuBPcase *in vivo* as reported by Portis et al. (1986). In fact, soybean RuBPcase activation *in vivo* is effectively CO_2-saturated at C_a less than 110 ppm (Campbell et al., 1988) while 50% RuBPcase activation has been reported in *Raphanus sativus* at as low as 10 ppm by von Cammerer and Edmondson (1986). Campbell et al. (1988) also measured leaflet soluble protein and the percentage C_a of soluble protein comprising

RuBPcase and found that there remained unchanged in plants in CO_2 concentrations either below or above 350 ppm CO_2. The apparent K_m, however, decreased slightly as a result of CO_2 enrichment. Increased photosynthesis rate was attributed to availability of substrate CO_2 and reduction of photorespiration but not to activation of RuBPcase enzyme (Table 5). Yelle *et al.* (1989a) on the other hand found that in case of tomato, 900 ppm growth CO_2 concentration increased initial RuBPcase activity for the first five weeks of treatment, but the difference did not persist during the last weeks. Though the activation of RuBPcase was always higher under elevated CO_2, there was a sharp decline of photosynthetic rate indicating that long-term decline of photosynthesis could not be attributed to the deficiency in the activation of RuBPcase.

Table 5. Soluble protein, RuBP carboxylase protein, apparent K_m (CO_2) and V_{max} of RuBP carboxylase from crude extracts of soybean leaflets grown at six CO_2 concentrations

Growth CO_2 concentration	Leaf soluble protein	RuBP carboxylase protein	Apparent K_m (CO_2)	V_{max}
ppm	gm^{-2}	%	µM	µmol CO_2 min^{-1} mg^{-1} protein
160	2.53 ± 0.01	56	9.7 ± 0.3	1.38 ± 0.01
220	3.23 ± 0.02	54		
280	2.58 ± 0.02		9.7 ± 0.7	1.60 ± 0.03
330	2.31 ± 0.01	57	9.4 ± 0.7	1.51 ± 0.03
660	2.28 ± 0.04	54	8.8 ± 0.7	1.29 ± 0.03
990	2.29 ± 0.01	55	8.4 ± 0.5	1.54 ± 0.02

Source: Campbell *et al.*, 1988.

RuBPcase activity has been reported to be low in CO_2-enriched plants which showed acclimation in photosynthesis rate in bean in studies by Porter and Grodzinski (1984), in cotton and maize by Wong (1979), in soybean, by Vu *et al.* (1983) and in cucumber by Peet *et al.* (1986). Slightly lower RuBPcase activity with similar photosynthetic rate has been reported for wheat (Havelka *et al.*, 1984a), while higher RuBPcase activity and photosynthesis has been reported for CO_2-enriched tomato by Hicklenton and Jolliffe (1978, 1980). Yelle *et al.* (1989a) found higher initial activities of RuBPcase in tomato plants grown in elevated concentrations of CO_2. The activation of RuBPcase under high CO_2 levels was high and the decline in photosynthetic rate was attributed to the decline of activated RuBPcase in CO_2-enriched plants (Fig. 4). They suggest that the lowering of activated RuBPcase can be due to decrease of RuBPcase protein or the presence of specific inhibitors that bind to the enzyme under high CO_2 causing incomplete activation of the enzyme. The enzyme content was found to be reduced by almost 50%. Decreased synthesis of RuBPcase protein could be related to high carbohydrate status of the enriched leaves. Reduction in the amount of RuBPcase protein has been reported in *Nerium oleander* grown under high CO_2 levels by Dowton *et al.* (1980). Campbell *et al.* (1988) have reported no change in RuBPcase protein in soybean leaves grown in elevated CO_2 concentrations. Reduction in the amount of RuBPcase protein may

492

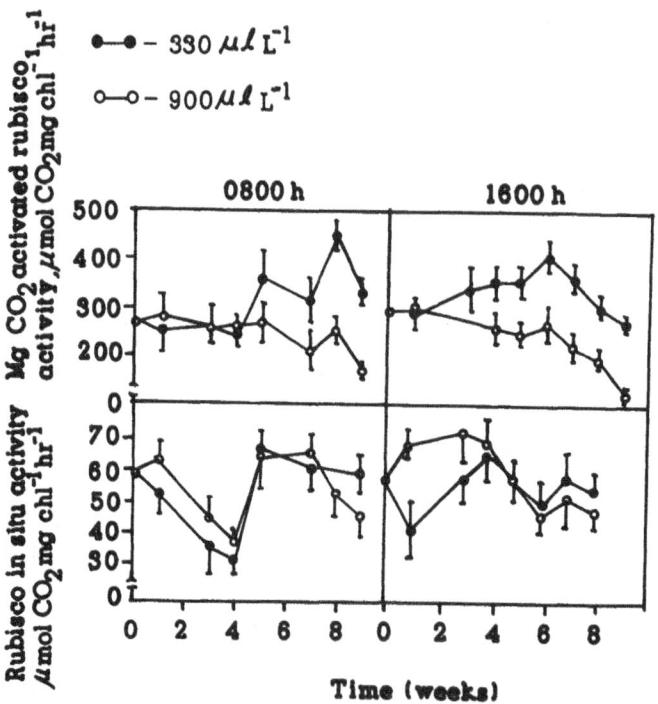

Figure 4. Initial and Mg^{2+} CO_2-activated RuBPcase activity for the fifth (top to bottom) leaf of tomato grown for a 10-week period. Each point represents the mean of four values ± SE (from Figs. 1 and 2 Yelle *et al.*, 1989).

be a response to balance all components of photosynthetic apparatus. Bloom *et al.* (1985) and Field and Mooney (1986) have suggested that plants maximize resource-use efficiency by allocating resources, principally N, to maintain a balance between photosynthetic and non-photosynthetic process. Reduced carbon exchange rate may reflect reallocation of N away from RuBPcase and into light harvesting, electron transport, and P_i regeneration processes. Since RuBPcase constitutes the single largest sink for N in the photosynthetic apparatus (Sage *et al.*, 1987), changes in the content will have greatest effect on N partitioning within leaves.

Much work has been done on the effect of CO_2 concentration on photosynthesis and RuBPcase enzyme. However, there are some points which require consideration. Short-term increases in carbon dioxide concentration not only show differences in the magnitude of enhancement of carbon dioxide assimilation rate, but also disrupt the balance between RuBPcase activity, RuBP regeneration, P_i regeneration, and electron transport in a way which is not found in plants grown under CO_2 enrichment. In such studies, it has been noted that many species show an acclimation of photosynthesis to increased concentrations of CO_2. As discussed earlier, several causes have been put forward for this. There is no conclusive evidence that reduction in RuBPcase activity, amount, or activation is responsible for photosynthetic acclimation. The work of Sage *et al.* (1989) elucidates this point. They studied five C_3 species which responded differently to increased CO_2 concentration in acclimation of photosynthesis. In two species, initial slope was unaffected but the photosynthetic rate at high CO_2 increased. In one species, initial slope decreased but CO_2-saturated rate was little affected, while two other species showed decrease in both initial slope and CO_2-saturated rate. However, all five species showed reduced activation state, and only two species showed reduction in RuBPcase amount. Since RuBPcase capacity and RuBP regeneration both limit photosynthesis, their balance may not persist under high CO_2. A better understanding of how these two capacities change with increase in CO_2 is needed along with difference in response among species and cultivars.

c. *Chemical partitioning of leaf carbon*

CO_2 concentrations above ambient not only increase the rate of CO_2 uptake, but also alter the partitioning of the newly fixed carbon into end products. The reductive C_3 photosynthetic carbon cycle is responsible for synthesis of triose phosphate from CO_2 and regeneration of RuBP. In C_4 pathway, CO_2 is trapped as malate or asparate, which are decarboxylated in bundle sheath cells, resulting in increased CO_2 concentration at the carboxylating site. Triose phosphates are transported out of the chloroplast via the triose-P-shuttle against P_i along with reducing equivalents. Sucrose, the major product to be translocated out of the leaf, is synthesized in the cytosol, whereas starch synthesis takes place in the chloroplast. For details see chapters by Sicher; Mitra *et al.*, this volume.

Under short-term enrichment, partitioning of carbon has been studied in several plants. There are conflicting reports as to the extent to which the extra carbon fixed is used for export (sucrose synthesis) compared to storage (starch synthesis) in leaves (Huber *et al.*, 1984). Ho (1977) has reported increased rate of carbon transport per unit fresh weight of leaf (40–45%) in tomato plants grown in 1000 ppm CO_2 even though there was increased carbon accumulation. It has also been reported that pool size of amino acids, particularly glycine and serine, were lower under CO_2 enrichment (Madore and Grodzinski, 1985; Sengupta, 1988). There was more export of currently labelled photosynthates and less labelling of starch.

Partitioning of extra photosynthates into storage rather than to translocation has been reported in *Lolium* by Wardlaw (1982). In soybean and wheat, additional photosynthates accumulated as starch (Lee and Whittingham, 1974; Sengupta, 1988). Finn and Brun (1982) observed a 30% increase in starch content within 36 hours of transferring soybean plants from ambient to 1000 ppm. Huber *et al.* (1984) reported that under both short-term and long-term CO_2 enrichment, there was increase in photosynthetic rates and rate of starch accumulation. Leaf sucrose concentration increased slightly and there was no increase in the activity of sucrose phosphate synthase (Table 6).

Since instantaneous photosynthetic rate depends on several factors, CO_2 enrichment may and does affect partitioning of newly fixed carbon between export and storage. Long-term CO_2 enrichment causes a definite increase in starch content.

Table 6. Long-term CO_2 enrichment effects on rates of carbon fixation, export and sucrose phosphate synthase activity in soybean leaves

CO_2	Rate (mg CH_2O $dm^{-2}h^{-1}$)				Sucrose concentration (mg dm^{-2})		SPS activity (μmol dm^{-2} h^{-1})
	CER	Dry weight accmulation	Starch accumulation	Export	9.00 hr	12.00 hr	
Ambient	30.3 ± 0.8	13.0 ± 1.1	11.2 ± 0.4	17.3 ± 1.8	6.5 ± 0.3	10.5 ± 0.9	57 ± 5
300 ppm above ambient	37.1 ± 1.2	21.7 ± 1.6	19.3 ± 2.5	15.4 ± 2.8	12.0 ± 0.7	21.0 ± 1.2	57 ± 6

All values are means ± SE on 12 determinations (from Huber et al., 1984).

As early as 1902, Brown and Escombe found that starch content of leaves from species such as *Fuschia, Cucurbita,*and *Impatiens* was higher in plants grown in enhanced CO_2 concentrations. Higher starch content has been reported in several plant species (Madsen, 1975; Thomas et al., 1975; Mauney et al., 1979; Sasek et al., 1985; Wong, 1990). A strong negative correlation between leaf starch concentration and photosynthesis has been observed for cotton (Mauney et al., 1978), wheat (Azcon-Bieto, 1983), tobacco (Thomas et al., 1975), and tomato (Madsen, 1975). However, long-term decline of photosynthetic efficiency could not be attributed to starch or sugar accumulation in tomato species grown at 900 ppm CO_2 (Yelle et al., 1989b). Delucia et al. (1985) noted an increase in starch in leaves on a diurnal basis, leading to feedback inhibition of photosynthesis in cotton grown in 675 ppm and 1000 ppm CO_2.

Excessive starch accumultion has been known to disrupt chloroplast structure. Leaves of plants grown at 1000 ppm CO_2 accumulated up to 94% more starch than those grown at 350 ppm. The total chlorophyll content as well as chlorophyll *a/b* ratio was also lower (Cave et al., 1981). Deformed chloroplasts leading to diminished photosynthesis has been reported by several workers (Madsen, 1968; 1975; Wulff and Strain, 1982; Tolbert and Zelitch, 1983; Sasek et al., 1985). Starch-induced physical damage at subcellular level needs to be further investigated. Electron micrographs of chloroplasts from *Trifolium* leaves grown at 1000 ppm CO_2 showed the development of very large aberrant starch granules (Cave et al., 1981). Large starch granules and reduced grana formation were also observed in *Desmodium* (Wulff and Strain, 1982). Reduction not only in chlorophyll content but also in chlorophyll *a/b* ratio (Cave et al., 1981; Wulff and Strain, 1982; Delucia et al., 1985) supports the view that damage can be at grana or photosynthetic unit level.

Vu et al. (1989) could not show any visual chlorotic symptoms during different stages of development in soybean plants grown at high CO_2. In fact, chlorophyll content at 39 days was somewhat higher in the leaves of CO_2-enriched plants. Starch accumulation did not disrupt chloroplast structure despite a threefold increase in starch content. An increase in the number of palisade cells resulting in an increase in number of chloroplasts per leaf was observed. This resulted in more storage sites for starch, avoided chloroplast disruption, and was responsible for non-acclimation in photosynthetic rate in soybean. Though continuous and excessive starch accumulation has been reported under CO_2 enrichment, the extent of starch accumulation varies with developmental stages of the plant. Where sink demand is high, there is less starch accumulation (Pharr et al., 1985; Sharma, 1986). Conversely reduction in sink strength increases starch accumulation (Clough, et al., 1981). Accumulated starch may not allow the light to reach chloroplast (Warren, 1966) or increase the diffusion path length of carbon dioxide during internal CO_2 transport (Nafziger and Koller, 1976).

Accumulation of starch as a result of CO_2 enrichment may or may not cause decreases in photosynthetic rates, often observed under long-term CO_2-enrichment conditions. Starch accumulation and its dark mobilization are not only linked to carbon exchange rate but are also responsible for eventual gain of dry matter in various organs and therefore for increased growth and biomass.

IV. Growth, Biomass Partitioning and Economic Yield under CO_2 Enrichment

Exposure of plants to high CO_2 concentrations produces several growth effects in addition to those caused by increase in CER. Increases in leaf size as a result of CO_2 enrichment have been reported (Ford and Thorne, 1967; Hardy and Havelka, 1977; Ho, 1977; Riechers and Strain, 1988). It is accompanied by increase in leaf area index (LAI) and leaf weight (Paez et al., 1983; Sasek and Strain, 1989). Newton (1965) has reported that in seedlings of cucumber exposed to high CO_2 concentrations, expansion rate and leaf size increased in the first few leaves only, and subsequent increase in growth rate was due to production of new primordia. Cure et al. (1989) also found that in soybean, at high CO_2, emergence and expansion rates of leaves on main stem increased but the areas of individual leaves was affected very little. Leaf development on later branches accounted for more than 40% of the total increase in leaf area. Increase the leaf expansion rate can be due to increased availability of photosynthate as well as to improved turgor due to changes in water status caused by increased stomatal conductance (Pearcy and Bjorkman, 1983; Sasek and Strain, 1989).

High CO_2 concentrations increased branching in both woody and herbaceous plants and decreased shoot/root ratio. In *Pinus* and *Picea* plants, shoot/root ratio declined approximately 16% after 12 months' growth at 1200 compared to 325 ppm CO_2 (Tinus, 1972). Exposure of roots of potato plants to high CO_2 greatly increased tuber formation (Arteca et al.,1979). These changes in the pattern of growth indicate a significant modification in the partitioning of photosynthates to organs.

The time of flowering is affected differently by high CO_2 in various kinds of plants. Hesketh and Hellmers (1973) found great delay in floral initiation in sorghum under 1000 ppm CO_2 but only a slight delay in corn, sunflower, and cotton. Hastening of flower and bud initiation and flowering has been reported in Alaska pea (Paez et al., 1980). According to Hicklenton and Jolliffe (1980), high concentrations of CO_2 caused *Pharbitis* to flower under non-inductive, long day conditions, but caused delayed flowering under short day conditions. Under non-limiting water and nutrients, increase in CO_2 concentrations usually is accompanied by at least a temporary increase in growth. Allen (1979) has summarized considerable data on the effects of enhanced CO_2 concentration on plant growth: in most experiments growth increased, but in some it decreased. Tinus (1972) reported that exposure of pine seedlings to 1200 ppm of CO_2 for 12 months resulted in nearly a 50% increase in dry weight above controls grown at 325 ppm. Substantial increase in dry matter production by herbaceous plants grown for several weeks to several months in high CO_2 concentrations have been reported by Cooper and Brun (1967) in two varieties of soybean, grown in 1350 ppm CO_2. Mauney et al. (1978) found that, in soybean, cotton and sunflower plants grown to maturity in 660 ppm CO_2 showed an increase in dry weight of 382%, 110% and 60%, respectively, while sorghum showed no statistically significant increase compared to plant grown in 330 ppm CO_2. Idso et al. (1987) found that biomass and seed yield increased by 82% and 87%, respectively, in cotton when CO_2 was doubled in open-top chambers. Dry matter accumulation and seed yield increased by 22% and 7%, respectively, in soybean (Rogers et al., 1986). Havelka et al. (1984a) observed an increase of 20% in dry matter and 17% in grain yield of wheat with enrichment to 1200 ppm CO_2. Maize, a C_4 plant, showed an increase of 49% and 55% in total dry matter and yield due to CO_2 doubling (Rogers et al., 1983). Many of the yield and dry matter increase values differ in various studies from the mean responses of dry matter and yield to CO_2 doubling as given by Cure and Acock (1986). This is because of the great variability of conditions in the field. It has been suggested that determinate plants like sorghum and sunflower do not show larger increases in dry weight than indeterminate types like soybean and cotton (Kramer, 1981). C_3 plants respond more than C_4 plants in terms of dry matter accumulation. Cure and Acock (1986) concluded from a survey of data on CO_2 on responses to elevated CO_2 for 10 leading crop species that the effect of CO_2 doubling on biomass accumulation among C_3 grasses appears to be about 28% but data on C_3 broad-leaf species is sparse. The effect is, however, more than in C_3 grasses. Biomass response to CO_2 doubling was low in C_4 species: 9% and 4% in corn and sorghum, respectively.

Kimball (1983) analysed 430 observations and concluded that the mean yield response is 33% more with a doubling of atmospheric CO_2 concentration. It has been noted that the beneficial effect of CO_2 enrichment on yield is due more to increase in the number of structures than to increase in their size. In soybean, the seed number increased greatly resulting in increased yield (Havelka et al., 1984b; Rogers et al., 1983). The increase in seed number was mainly due to increase in the number of pods and not in seeds per pod (Ackerson et al., 1984). Cure and Acock (1986) in their analysis of crop response to CO_2 doubling, found decrease in harvest index in case of soybean only. This suggests that extra growth induced by increasing CO_2 is not always evenly distributed between organs. For bean, elevated CO_2 increased the number of pods, but at the same time decreased seeds per pod without affecting mean seed weight (Gustafson, 1984). Increase in yield, in case of wheat has been attributed to an increase in number of grains, rather than grain weight (Havelka et al., 1984a; Fischer and Aguilar, 1976). Cure and Acock (1986) found an increase in yield as well as in harvest index in an analysis of crop resposnse to CO_2 doubling in crops like wheat, barley, rice, corn, sorghum, and potato. Since photosynthesis of C_4 crops such as corn responds relatively less to an increase in CO_2, yields of C_4 plants are expected to increase less than C_3 crops. Kimball (1986) showed an average yield increase of about 14% with a doubling of CO_2 for C_4 species.

Biomass and yield increases are highly responsive to CO_2 enrichment during exponential seedling growth (Thomas *et al.*, 1975; Patterson and Flint, 1980; Havelka *et al.*, 1984a) but diminish with growth and development to such an extent that plant age and CO_2 can show large positive interactions (Kriedemann and Wong, 1984). Neales and Nicholls (1978) have cited instances where wheat grown beyond 21 days of age with 600 ppm CO_2 actually grew more slowly than non-enriched controls. Relative growth rate was depressed by 44%. Kriedemann and Wong (1984) found that in some C_3 species, the increased final biomass in plants enriched with 1350 ppm CO_2, although partly due to high NAR, was more a result of faster leaf expansion during early exponential growth. This response diminished with age and relative leaf growth rate (RLGR) reached lower values than control (Table 7). CO_2

Table 7. Relative leaf growth rate ($cm^2\ cm^{-2}\ day^{-1}$), net assimilation rate ($mg\ dm^{-2}\ wk^{-1}$), and relative growth rate ($mg\ mg^{-2}\ day^{-1}$) for CO_2-enriched *Cucumber* and *Brassica*

Age	Interval (days)	Ambient (330 ppm)			CO_2-Enriched (1350 ppm)		
		RLGR	NAR	RGR	RLGR	NAR	RGR
A	0–21	.164	485	.107	.215	928	.173
	22–40	.093	608	.138	.122	902	.147
	40–52	.061	229	.036	.035	406	.051
B	0–18	.23	490	.195	.960	2110	.258
	18–24	.297	684	.307	.222	829	.291
	24–40	.130	465	.155	.120	651	.147
	40–52	.061	567	.114	.020	466	.066

Adapted from Kriedemann and Wong, 1984

concentration of 1000 ppm did not increase total fruit weight, leaf area, or relative growth rates beyond the first 16 days after seeding. Fruit number increased more than fruit weight (Peet, 1986). The effect of CO_2 enrichment on yield is dependent on the stage of growth at which it is applied. In wheat, maximum yield response was observed when plants were exposed to 1200 ppm CO_2 during the period from jointing to anthesis (Havelka *et al.*, 1984a). Ackerson *et al.* (1984) found that in soybean, enrichment from early pod development until maturity and from emergence to maturity could only increase yield by 27% and 36%. Exposure to high CO_2 from emergence to anthesis or from anthesis to early pod development had no effect on the yield. In spring wheat, CO_2 enrichment increased yield most during tillering (Fischer and Aguilar, 1976). These findings suggest that additional carbohydrates are needed during early flower and seed development (Ackerson *et al.*, 1984) and are not used during grain filling (Havelka *et al.*, 1984a).

Rates of photosynthesis are not necessarily closely linked with growth and increase in biomass. There are numerous intermediate steps between the carbohydrate production in the leaf and growth and dry matter accumulation in the roots, stem, flower, fruit, and seed. Growth and yield depend on many other factors in addition to rate of photosynthesis, such as dark respiration, partitioning of photosynthates, and nitrogen metabolism as well as environmental factors such as water, mineral nutrition, temperature and light. Mauney *et al.* (1978) have shown that for soybean, 12 weeks of CO_2 enrichment caused 41% increase in photosynthetic rate but leaf area and dry weight increased by 180% and 382%. Considerable variation among species in the relationship between rate of photosynthesis, leaf area, and dry weight accumulation has been reported by Patterson and Flint (1980) (Table 8). Although

Table 8. Effect of enhanced CO_2 on net assimilation rate, leaf area duration, and dry weight accumulation for the period 24 to 45 days after germination.

Species	Net assimilation rate		Leaf area duration		Dry weight accumulation	
	600 ppm	1000 ppm	600 ppm	1000 ppm	600 ppm	1000 ppm
Corn	4	−11	3	−1	−1	−13
Itchgrass	−3	−8	23	7	18	−18
Soybean	17	34	10	28	22	23
Velvet Leaf	24	35	7	13	32	32

Data are for values at 600 and 1000 ppm expressed as percentage increases above values for 350 ppm (adapted from Patterson and Flint, 1980).

with long-term exposure to increased CO_2 many plant species are unable to maintain high photosynthetic rates, in most cases the plant growth increased significantly; whether this increase in biomass is due to higher photosynthesis per unit leaf area or to increased plant canopy there by leading to increased total photosynthesis is not very clear. However, it appears that higher CO_2 concentrations are beneficial to most plants in terms of the effect on growth and yield.

V. Interaction with Other Environmental Factors

The interaction of carbon dioxide with other environmental factors such as light, temperature, soil moisture, and nutrient availability is widely demonstrated. Several workers have shown that under limited nitrogen and phosphorus supply, CO_2 enhancement fades away (Wong, 1979; Goudriaan and de Ruiters, 1983; Brown and Higginbotham, 1986). Inadequate nitrogen limits the formation of new tissue and, therefore, the demand for photosynthate. Hocking and Meyer (1985) observed a consistent increase in growth (18%) of cocklebur to high CO_2 over a range of nitrate concentrations with little change in nitrate uptake relative to controls. Also in plants exposed to elevated CO_2, a large portion of the plant N content was located in the roots. On the other hand, increased nitrogen uptake by soybean plant in high CO_2 has been reported by Cure *et al.* (1988). They found that the increased nitrogen uptake was associated with larger root system. Increase in whole plant growth and seed yield due to high nitrate levels was substantially greater at high CO_2. The stimulation of growth by CO_2 enrichment at different NO_3^- treatments was consistently associated with increased whole plant photosynthetic capacity, as NAR and leaf area increased (Fig. 5) with increased N utilization efficiency in the production of dry matter.

Goudriaan and de Ruiters (1983) studying the effect of CO_2 doubling under nitrogen and phosphorus shortage in several plant species found that except for faba

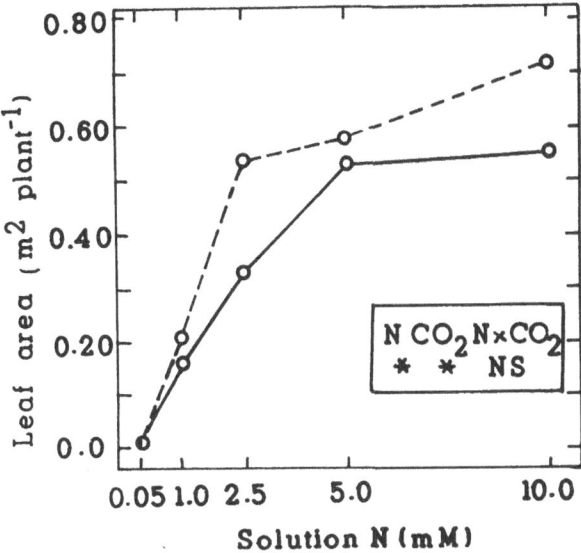

Figure 5. Total canopy leaf area for soybean plants grown in 350 (closed symbols) and 700 (open symbols) ppm CO_2 and five NO_3^- and CO_2 effects from the regression analysis is included (from Cure *et al.*, 1988).

bean, CO_2 did not increase growth under phosphorus shortage. The interaction of CO_2 with shortage of phosphorous appears to be fully governed by the law of limiting factors and the biochemical role of phosphorus could not be overcome by increased CO_2. Lack of response of photosynthesis to an increase in CO_2 concentration under sulphur deficiency has been reported in sugarbeet by Thomas and Hill (1949).

It is well documented that plant growth is decreased as the water supply is reduced (Boyer, 1971; Sionit *et al.*, 1980), as is the increase in growth due to CO_2 enrichment (Sionit *et al.*, 1981a). Increased CO_2 is thought to ameliorate the detrimental effects of drought stress on plant growth (Sionit *et al.*, 1981a; Paez *et al.*, 1984). Drought stress may be avoided or lessened by elevated CO_2 in two ways. With decreased stomatal conductance less water is used as water potential remains high and the onset of stress is delayed. This allows continued growth during drought, and increasing overall biomass production. Also, under stress high CO_2 maintains greater photosynthetic rates than ambient CO_2.

Paez *et al.* (1983) have reported decrease in total leaf potential under water stress in alaska pea, with a slower decrease under high CO_2 regime, partly due to reduced stomatal conductance. High CO_2 counteracted the reduction in height and leaf area that was observed in low CO_2 unwatered plants throughout the drought period. Maintenance of higher photosynthetic rates up to the end in non-watered aster plants at 650 ppm CO_2 has been reported by Wray and Strain (1986). The per cent reduction in stomatal conductance was, however, the same at 650 ppm and 380 ppm CO_2 (Fig. 6).

500

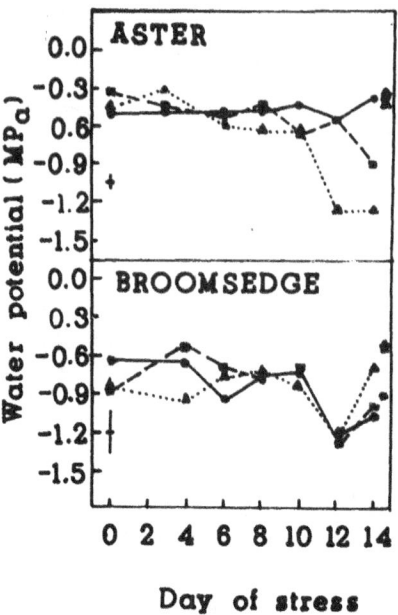

Figure 6. Time of course of leaf water potentials of aster and broomsedge during a period without watering when grown at 380 ppm CO_2 (....), 500 ppm CO_2 (----), or 650 ppm CO_2 (—). Symbols at far right are well-watered controls at the end of the drought. Error bars indicate average ± SE of the means. N = 5. (From Wray and Strain, 1986).

Maintenance of higher photosynthetic rates at enriched levels of CO_2 by several C_4 grass species has been reported by Sionit and Patterson (1985). Wray and Strain (1986) found a differential response to drought stress in case of a C_3 (aster) and a C_4 (broomsedge) plant under CO_2 enrichment. Unwatered, CO_2-enriched aster plants had greater leaf potential, greater photosynthetic rates, and more total dry weight than non-watered plants at 380 ppm CO_2. The response of broomsedge to drought was similar in enriched and unenriched plants and there was no significant interaction of CO_2 enrichment and drought.

Water stress decreases the expansion of organs more than it slows photosynthesis, and, therefore, under high CO_2, carbohydrate accumulation may increase although quality in terms of nitrogen content may suffer. Although stimulation of photosynthesis by CO_2 become smaller with increase in the degree of stress, the combined effect of improved water balance and increased substrate will probably increase the productivity of C_3 plants under water stress by maintaining source and sink development and function and thus partially offsetting the damage (Lawlor, 1991). Gifford (1979) showed yield enhancement of two cultivars of wheat when water-stressed plants were grown with increased CO_2.

The interaction of CO_2 with temperature is critical to the response of plants to climate change. Pearcy and Bjorkman (1983) have reviewed the response of a C_3 and C_4 plant to leaf temperature at 330 and 1000 ppm CO_2 and found that for C_3 plants, enhancement is least at low temperature. Probably due to decline of the affinity of RuBPcase for CO_2 and increased affinity for O_2 with increased temperature, the temperature optimum for photosynthesis is generally higher at high CO_2 than at low CO_2 concentrations. Temperature responses of C_4 shrub were not altered by high CO_2. This can have important implications in warm environments. Radin et al. (1987) found substantial differences in photosynthetic response to C_i between mid-summer and late summer in cotton exposed to elevated CO_2 levels in open-top chambers in the field. They found that CO_2-enriched plants had lower mesophyll conductance to CO_2 in September, which had lower temperature than July, compared to those in ambient CO_2. Decreasing temperatures may increase the solubility of CO_2 relative to O_2 (Ku and Edwards, 1977) and increase the affinity of RuBPcase for CO_2 (Laing et al., 1974). An increase of 2.5 times in CER and of temperature optima from 25° to 37°C has been reported with increase in CO_2 concentration form 320 to 1900 ppm (Jurik et al., 1984).

The interaction of CO_2 with temperature manifests itself in the growth responses also. Cure (1985) found that at high temperature a doubling of CO_2 generally caused greater increase in growth. Kimball (1985) has noted that relative growth stimulation is rather constant with temperature over the temperature range at which plants normally grow. There are some experiments showing limited response to CO_2 concentration at low temperatures. Sionit et al. (1981b) observed that in okra, CO_2 could greatly stimulate growth at lower than normal temperatures. Below 26/20°C day/night temperature the plants did not survive at 350 ppm but could thrive even at 20/14°C at CO_2 concentrations of 450, 675 and 1000 ppm. Contrary to this, Idso et al. (1987) found that elevated CO_2 stimulated growth only over the range of 19° to 34°C for five C_3 plant species. Below 19°C, elevated CO_2 concentrations had a negative effect on growth. Cool temperature (20°C) has been reported to promote leaf injury to CO_2-enriched plants of Phaseolus vulgaris (Ehret and Jolliffe, 1985). Perhaps many plants may not be able to thrive in higher CO_2 concentrations if temperature increases.

VI. Conclusions

Studies conducted on the direct effects of high CO_2 on plants clearly indicate that the present level of CO_2 is limiting for photosynthesis. Some general patterns of the physiological responses of plants are now apparent. High CO_2 enhances photosynthesis more in C_3 plants than C_4 plants and within those groups, there are important differences among species. Plants grown continuously in high CO_2 show an acclimation of photosynthesis which has been attributed to such factors as reduced RuBP activation, amount, or activity, limitation of P_i regeneration, and feedback inhibition due to starch accumulation. No clear picture regarding the causes and mechanisms leading to acclimation has, however, emerged. Clearly, high CO_2 reduces stomatal conductance. However, stomatal sensitivity is not a limitation to the transport of CO_2. Reduction in stomatal conductance coupled with enhanced photosynthetic rate results in the increase in water-use efficiency in both C_3 and C_4 plants. In addition to photosynthesis, high CO_2 has a stimulatory effect on growth and number of structures, resulting in increased dry matter and yield. An average of 33% increase in yield has been observed by CO_2 doubling.

Experimental data for CO_2 enrichment on various crop species are poorly represented in the literature with the exception of soybean and cotton; information on other key economic crops such as wheat, rice, legumes, and oilseeds is totally inadequate. Since CO_2 is increasing gradually, studies of the effects of small increments of CO_2 concentrations are needed. Emphasis is required on exposure to high CO_2 at different growth stages, as plant response changes with growth due to shift in priority of sinks.

CO_2 interacts strongly with environmental variables like light, temperature, water, and nutrient. Due to enhanced water-use efficiency, CO_2 enrichment ameliorates the harmful effects of water stress but the effects are diminished under nitrogen and phosphorus limitation. Overall plant responses are strongly stimulated under higher level of these resources. Most of the data on CO_2 enrichment are from studies conducted in greenhouses. The predicted doubling of CO_2 by the middle of the next century is expected to be accompanied by changes in climatic factors like temperature and precipitation. Field studies, in combination with environmental variables, and close association with predicted models need to be done for proper optimization of crops to future climate.

VII. References

Ackerson RC, Havelka UD and Boyle MG (1984) CO_2 enrichment effects on soybean physiology. II Effects of stage specific CO_2 exposure. *Crop Sci* **24**: 1150–1154

Akita S and Moss DN (1973) Differential stomatal response between C_3 and C_4 species to atmospheric CO_2 concentration and light. *Crop Sci* **13**: 234–237

Allen LH Jr (1979) Potential for carbon dioxide enrichment. In: (BJ Barfield and JF Gerber, eds) *Modification of the Aerial Environment of Plants*. American Society of Agricultural Engineers Monograph, St. Joseph, Michigan pp 500–519

Aoki M and Yabuki K (1977) Studies on the carbon dioxide enrichment for plant growth. VII Changes in dry matter production and photosynthetic rate of cucumber during carbon dioxide enrichment. *Agric Meteorol* **18**: 475–485

Arteca RN, Pooviah BW and Smith OE (1979) Changes in carbon fixation, tuberization and growth induced by CO_2 application to the root zone of potato plants. *Science* **21**: 1279–1280

Azcon-Bieto J (1983) Inhibition of photosynthesis by carbohydrates in wheat leaves. *Plant Physiol* **73**: 681–686

Baysdorter C and Bassham JA (1985) Photosynthate supply and utilization of alfalfa. A developmental shift from a source to a sink limitation. *Plant Physiol* **77**: 313–317

Bazzaz FA (1990) The response of natural ecosystems to the rising global CO_2 levels. *Annu Rev Ecol Syst* **21**: 167–196

Berhuizen JF and Slatyer RO (1964) Photosynthesis of cotton leaves under a range of environmental conditions in relation to internal and external diffusive resistances. *Aust J Biol Sci* **17**: 348–359

Bishop PM and Whittingham CP (1968) The photosynthesis of tomato plants in carbon dioxide enriched atmosphere. *Photosynthetica* **2**: 31–38

Bloom AJ, Chapin FS III and Mooney HA (1985) Resource limitation in plants—An economic analogy. *Annu Rev Ecol Syst* **16**: 363–392

Boyer JS (1971) Recovery of photosynthesis in sunflower after a period of low leaf water potential. *Plant Physiol* **47**: 816–820

Brown HT and Escombe F (1902) Influence of varying amount of carbon dioxide in the air on the photosynthetic process of leaves and on the mode of growth of plants. *Proc Roy Soc Lond* **70**: 397–413

Brown K and Higginbotham KO (1986) Effects of carbon dioxide enrichment and nitrogen supply on growth of boreal tree seedlings. *Tree Physiol* **2**: 223–232

Bruggnick GT (1984) Effect of CO_2 concentration on growth and photosynthesis of young tomato and carnation plants. *Acta Horticulture* **162**: 279

Brun WA and Cooper RL (1967) Effects of light intensity and carbon dioxide concentration on photosynthetic rate of soybean. *Crop Sci* **7**: 451–454

Campbell WJ, Allen LH Jr and Bowes G (1988) Effects of CO_2 concentration on Rubisco activity, amount and photosynthesis in soybean leaves. *Plant Physiol* **88**: 1310–1316

Cambell DE and Young R (1986) Short term CO_2 exchange response to temperature, irradiance and CO_2 concentration in strawberry. *Photosynth Res* **8**: 31–40

Cave G, Tolley LC and Strain BR (1981) Effect of carbon dioxide enrichment on chlorophyll content, starch content and starch grain structure in *Trifolium subterraneum* leaves. *Physiol Plant* **51**: 171–174

Chang CW (1975) Carbon dioxide and senescence in cotton plants. *Plant Physiol* **55**: 515–519

Clough JM and Peet MM (1981) Effects of intermittent exposure to high atmospheric CO_2 on vegetative growth in soybean. *Physiol Plant* **53**: 565–569

Clough JM, Peet MM and Kramer PJ (1981) Effects of high atmospheric CO_2 and sink size on rates of photosynthesis of a soybean cultivar. *Plant Physiol* **67**: 1007–1010

Cooper RL and Brun WA (1967) Response of soybeans to a carbon dioxide enriched atmosphere. *Crop Sci* **7**: 455–457

Cowan IR, Raven JA, Hartung W and Farquhar GD (1982) A possible role of abscisic acid in coupling stomatal conductance and photosynthetic carbon metabolism in leaves. *Aust J Plant Physiol* **9**: 489–498

Cure JD and Acock B (1986) Crop responses to carbon dioxide doubling: A literature survey. *Agric For Meteorol* **38**: 127–145

Cure JD (1985) CO_2 doubling response. A crop survey. In: *Direct Effects of Increasing CO_2 on Vegetation*. DOE/ER-0238. US Dept of Energy, Carbon Dioxide Res Div, Washington DC, pp 99–116

Cure JD, Israel DW and Rufty TW Jr (1988) Nitrogen stress effects on growth and seed yield of non-nodulated soybean exposed to elevated carbon dioxide. *Crop Sci* **28**: 671–677

Cure JD, Rufty TW Jr and Israel DW (1989) Alterations in soybean leaf development and photosynthesis in a CO_2-enriched atmosphere. *Bot Gaz* **150**: 337–345

Delucia EH, Sasek TW and Strain BR (1985) Photosynthetic inhibition after long-term exposure to elevated levels of atmospheric carbon dioxide. *Photosynth Res* **7**: 175–184

Dowton WJS, Bjorkman O and Pike CS (1980) Consequences of increased atmospheric concentration of carbon dioxide for growth and photosynthesis of higher plants. In: (GI Pearman, ed) *Carbon Dioxide and Climate*. Australian Research, Australian Academy of Science, Canberra, pp 143–151

Ehret DL and Jolliffe PA (1985) Leaf injury to bean plants grown in carbon dioxide enriched atmosphere. *Can J Bot* **63**: 2015–2020

El-Sharkawy MA and Hesketh JD (1964) Effects of temperature and water deficit on leaf photosynthetic rates of different species. *Crop Sci* **9**: 514–517

Farquhar GD and Sharkey TD (1982) Stomatal conductance and photosynthesis. *Annu Rev Plant Physiol* **33**: 317–345

Fetcher N, Jaeger CH, Strain BR and Sionit N (1988) Long term elevation of atmospheric CO_2 concentration and the carbon exchange rates of saplings of *Pinus taeda* L. and *Liquidambar styraciflua* L. *Tree Physiol* **4**: 255–262

Field C and Mooney HA (1986) The photosynthesis nitrogen relationship in wild plants. In: (TA Givinish, ed) *On the Economy of Plant Form and Function*. Cambridge University Press, London, pp 25–55

Finn GA and Brun WA (1982) Effect of atmospheric CO_2 enrichment on growth, non-structural carbohydrate content and root nodule activity in soybean. *Plant Physiol* **69**: 327–331

Fischer RA and Aguilar M (1976) Yield potential in a dwarf spring wheat and the effect of carbon dioxide fertilization. *Agron J* **68**: 749–752

Ford MA and Thorne GN (1967) Response of CO_2 concentration on growth of sugarbeet, barley, kale and maize. *Ann Bot* **31**: 629–644

Gifford RM (1977) Growth pattern, carbon dioxide exchange and dry weight distribution in wheat growing under differing photosynthetic environments. *Aust J Plant Physiol* **4**: 99–110

Gifford RM (1979) Growth and yield of CO_2 enriched wheat under water limited conditions. *Aust J Plant Physiol* **6**: 367–378

504

Goudriaan J and de Ruiters HE (1983) Plant growth in response to CO_2 enrichment, at two levels of nitrogen and phosphorus supply. 1. Dry matter, leaf area and development. *Neth J Agric Sci* **31**: 157–169

Green K and Wright R (1977) Field response of photosynthesis to CO_2 enhancement in ponderosa pine. *Ecology* **58**: 687–692

Gustafson SW (1984) Effects of CO_2 enrichment during flowering and podfill on net photosynthesis, dry matter accumulation and yield of beans *Phaseolus vulgaris* L. *Diss Abs Int* **44(10)**: 2954

Hardy R and Havelka UD (1977) Possible routes to increase the conversion of solar energy to food and feed by grain legumes and cereal grains (crop production) In: (MA Miyachi, S San Pietro Pieto and A Tamura, eds) *Biological Solar Energy Conversion*. Academic Press, London, New York, pp 299–322

Havelka UD, Wittenbach VA and Boyle MG (1984a) CO_2 enrichment effects on wheat yield and physiology. *Crop Sci* **24**: 1163–1168

Havelka UD, Ackerson RC, Boyle MG and Wittenbach VA (1984b) CO_2 enrichment effects on soybean physiology. I. Effects of long-term CO_2 exposure. *Crop Sci* **24**: 1146–1150

Herold A (1980) Regulation of photosynthesis by sink activity—The missing link. *New Phytol* **86**: 131–144

Hesketh JD and Hellmers H (1973) Floral initiation in your plant species growing in CO_2 enriched air. *Environ Control Biol* **111**: 51–53

Hicklenton PR and Jolliffe PA (1978) Effects of greenhouse CO_2 enrichment on the yield and photosynthetic physiology of tomato plants. *Can J Plant Sci* **58**: 801–817

Hicklenton PR and Jolliffe PA (1980) Carbon dioxide and flowering in *Pharbits nil* Choisy. *Plant Physiol* **66**: 13–17

Ho LC (1977) Effects of CO_2 enrichment on the rates of photosynthesis and translocation of tomato leaves. *Ann Appl Biol* **87**: 191–200

Hocking PJ and Meyer CP (1985) Responses of Noogoora Burr (*Xanthium occidentale* Bertol) to nitrogen supply and carbon dioxide enrichment. *Ann Biol* (London) **55**: 835–844

Houghton JT, Jenkins GJ and Ephramms JJ (1990) *Editors* Climate change: The IPCC Scientific Assessment. Cambridge University Press, Cambridge pp 26

Huber SC, Rogers H and Israel DW (1984) Effects of CO_2 enrichment on photosynthesis and photosynthate partitioning is soybean (*Glycine max*) leaves. *Physiol Plant* **62**: 95–101

Idso SB, Kimball BA and Mauney JR (1987) Atmospheric CO_2 enrichment effects on cotton midday foliage temperature: Implications for plant water use and crop yield. *Agron J* **79**: 667–672

Jones P, Allen LH Jr, Jones JW and Valle R (1985) Photosynthesis and transpiration responses of soybean canopies to short and long term CO_2 treatments. *Agron J* **77**: 119–126

Jurik TW, Weber JA and Gates DM (1984) Short term effects of CO_2 on gas exchange of leaves of bigtooth aspen (*Populus grandidentata*) in the field. *Plant Physiol* **75**: 1022–1026

Kimball BA (1983) Carbon dioxide and agricultural yield: An assemblage and analysis of 430 prior observations. *Agron J* **75**: 779–788

Kimball BA (1985) Adaptation of vegetation and management practices to a higher carbon dioxide world. In: (BR Strain and JD Cure, eds) *Effects of Increasing Carbon Dioxide on Vegetation*, US Department of Energy, Washington DC, pp 185–204

Kimball BA (1986) Influence of elevated CO_2 on crop yield. In: (HZ Enoch and BA Kimball, eds) *Carbon Dioxide Enrichment of Green House Crops*, Vol 2, *Physiology Yield and Economics*, CRC Press, Boca Raton, Florida, pp 105–115

Kramer PJ (1981) Carbon dioxide concentration, photosynthesis and dry matter production. *Bioscience* **31**: 29–33

Kriedemann PE and Wong SC (1984) Growth response and photosynthetic adaptation to carbon dioxide: Comparative behaviour in some C_3 species. In: (C sybesma, ed) *Advances in Photosynthetic Research*, Vol IV, Martinus Nijhoff/Dr W Junk Publishers, The Hague, Boston, Lancaster, pp 209–212

Ku SB and Edwards GE (1977) Oxygen inhibition of photosynthesis. I. Temperature dependence and relation to O_2/CO_2 solubility ratio. *Plant Physiol* **59**: 986–990

Labate CA and Leegood RC (1988) Limitation of photosynthesis by changes in temperature. Factors affecting the response of carbon dioxide assimilation to temperature in barley leaves. *Planta* **168**: 84–93

Laing WA, Ogren WL and Hageman RH (1974) Regulation of soybean net photosynthetic CO_2 fixation by the interaction of CO_2, O_2 and ribulose 1, 5-diphosphate carboxylase. *Plant Physiol* **54**: 678–685

Lawlor DW (1991) Response of plants to elevated carbon dioxide: The role of photosynthesis, sink demand ad environmental stress. In: (YP Abrol, PN Wattal, A Gnanam, Govindjee, DR Ort and AH Teramura, eds) *Impact of Global Climatic Changes on Photosynthesis and Plant Productivity*. Proceedings: Indo US Workshop, New Delhi. Oxford & IBH Publishing CO Pvt Ltd, New Delhi, pp 431–446

Lee RB and Whittingham CP (1974) The influence of partial pressures in tomato leaf. *J Exp Bot* 25: 277–287

Lorimer GH (1981) The carboxylation and oxygenation of ribulose-1, 5-bisphosphate: The primary event in photosynthesis and photorespiration. *Annu Rev Plant Physiol* 32: 379–384

Madore M and Grodzinski B (1985) Effects of CO_2 enrichment on growth and photoassimilate on transport in dwarf cucumber (*Cucumus sativus* I). *J Plant Physiol* 121: 51–71

Madsen E (1968) Effects of CO_2 concentration on the accumulation of starch and sugar in tomato leaves. *Physiol Plant* 21: 168–175

Madsen E (1975) Effects of CO_2 enrichment on growth, development, fruit production and fruit quality in tomato from a physiological point of view. In: (N de Bilderling and P Chouard, eds) *Phytotrons in Horticultural Research*, Goulhier-Villars, Paris, pp 318–330

Mauney JR, Fry KE and Guinn G (1978) Relationship of photosynthetic rate to growth and fruiting of cotton, soybean, sorghum and sunflower. *Crop Sci* 18: 259–263

Mauney JR, Guinn G, Fry KE and Hesketh JD (1979) Correlation of photosynthetic carbon dioxide uptake and carbohydrate accumulation in cotton, soybean, sorghum and sunflower. *Photosynthetica* 13: 260–266

Morison JIL (1985a) Intercellular CO_2 concentration and stomatal response to CO_2. In: (E Zeiger, I Cowan and GD Farquhar, eds). *Stomatal Function*, Stanford University Press, Stanford, pp 229–252

Morison JIL (1985b) Sensitivity of stomata and water-use efficiency to high CO_2. *Plant Cell and Environ* 8: 467–474

Morison JIL and Gifford RM (1983) Stomatal sensitivity to carbon dioxide and humidity—A comparison of two C_3 and two C_4 grass species. *Plant Physiol* 71: 789–796

Nafziger ED and Koller HR (1976) Influence of leaf starch concentration on CO_2 assimilation in soybean. *Plant Physiol* 57: 560–563

Neales TF and Nicholls AO (1978) Growth responses of young wheat plants to a range of ambient CO_2 levels. *Aust J Plant Physiol* 5: 45–59

Newton P (1965) Growth of *Cucumis sativus* variety Butcher's Disease Resister with two concentrations of carbon dioxide. *Ann Appl Biol* 56: 55–64

Ogren WL (1977) Increasing carbon fixation by crop plants. In: (G Akoyunoglou, ed) *Fourth International Congress on Photosynthesis*, Balaban Int Sci services, Philadelphia, pp 721–733

Osmond CB and Bjorkman O (1973) Pathways of CO_2 fixation in the CAM plant *Kalanchoe diagremontiana* II. Effects of O_2 and CO_2 concentration on light and dark fixation. *Aust J Plant Physiol* 2: 155–162

Paez A, Hellmers H and Strain BR (1980) CO_2 effects on apical dominance in *Pisum sativum*. *Physiol Plant* 50: 43–46

Paez A, Hellmers H and Strain BR (1983) CO_2 enrichment, drought stress and growth of alaska pea plant (*Pisum sativum*). *Physiol Plant* 58: 161–165

Paez A, Hellmers H and Strain BR (1984) Carbon dioxide enrichment and water stress interaction on growth of two tomato cultivars. *J Agric Sci Cam* 102: 687–693

Pallas JE Jr (1965) Transpiration and stomatal opening with changes in carbon dioxide content of air. *Science* 147: 171–173

Patterson DT and Flint EP (1980) Potential effects of global atmospheric CO_2 enrichment on the growth and competitiveness of C_3 and C_4 weed and crop plants. *Weed Sci* 28: 71–75

Pearcy RW and Bjorkman O (1983) Physiological effects. In: (ER Lemon, ed). *CO_2 and Plant: The response of plant to Rising Level of Atmospheric Caron Dioxide*. Westview Press, Boulder, Colorado, pp 65–105

Peet MM (1986) Acclimation to high CO_2 in monoecious cucumbers. I Vegetative and reproduction growth. *Plant Physiol* 80: 59–62

Peet MM, Huber SC and Patterson DT (1986) Acclimation to high CO_2 in monoecious cucumbers. II Carbon exchange rates, enzyme activities and starch and nutrient concentrations. *Plant Physiol* 80: 63–67

506

Perchorowicz JT and Jensen RG (1983) Photosynthesis and activation of ribulose bisphosphate carboxylase in wheat seedlings. Regulation by CO_2 and O_2. *Plant Physiol* **71**: 955–960

Pharr DM, Huber SC and Sox HN (1985) Leaf carbohydrate status and enzymes of translocate synthesis in fruiting and vegetative plants of *Cucumis sativus* L. *Plant Physiol* **77**: 104–108

Porter MA and Grodzinski B (1984) Acclimation to high CO_2 in bean. Carbonic anhydrase and ribulose biophosphate carboxylase. *Plant Physiol* **74**: 413–416

Portis AR Jr, Salvucci ME and Ogren WL (1986) Activation of ribulose bisphosphate concentrations by rubisco activase. *Plant Physiol* **82**: 967–971

Potvin C and Strain BR (1985) Effects of CO_2 enrichment and temperature on growth in two C_4 weeds, *Echinochloa crusgalli* and *Eleusine indica*. *Can Bot* **63**: 1495–1497

Radin JW and Ackerson RC (1981) Water relations of cotton plants under nitrogen deficiency. III Stomatal conductance, photosynthesis, and abscisic acid accumulation during drought. *Plant Physiol* **67**: 115–119

Radin JW, Kimball BA, Hendrix DL and Mauney JR (1987) Photosynthesis of cotton plants exposed to elevated levels of carbon dioxide in the field. *Photosynthesis Research* **12**: 191–203

Radin JW, Hartung W, Kimball BA and Mauney JR (1988) Correlation of stomatal conductance with photosynthetic capacity of cotton only in a CO_2-enriched atmosphere: Mediation by abscisic acid? *Plant Physiol* **88**: 1058–1062

Raper CD Jr and Peedin GF (1978) Photosynthetic rate during steady-state growth as influenced by carbon dioxide concentration. *Bot Gaz* **139**: 147–149

Raschke (1975) Simultaneous requirement of carbon dioxide and abscisic acid for stomatal closing in *Xanthium strumarium* L. *Planta* **125**: 243–259

Raschke (1979) Movement of stomata. In: (W Haupt and ME Feinlab, eds) *Encyclopedia Plant Physiology* New Series 7, pp 383–441

Riechers GD and Strain BR (1988) Growth of blue grama (*Boutelous gracilis*) in response to atmospheric carbon dioxide enrichment. *Can Bot* **66**: 1570–1573

Rogers HH, Bingham GE, Cure JD, Smith JM and Surano KA (1983) Response of selected plant species to elevated carbon dioxide in the field. *Environ Qual* **12**: 569–574

Rogers HH, Cure JD and Smith JM (1986) Soybean growth and yield response to elevated carbon dioxide. *Agric Ecosyst Environ* **16**: 113–128

Sage RF and Pearcy RW (1987) The nitrogen use efficiency of C_3 and C_4 plants. I Leaf nitrogen, growth and biomass partitioning in *Chenopoduim album*(L) and *Amaranthus retraflexus* (L). *Plant Physiol* **84**: 954–958

Sage RF and Sharkey TD (1987) The effect of temperature on the occurrence of O_2 and CO_2 insensitive photosynthesis in field grown plants. *Plant Physiol* **84**: 658–664

Sage RF, Pearcy RW and Seemann JR (1987) The nitrogen use efficiency of C_3 and C_4 plants. III Leaf nitrogen effects on the activity of carboxylating enzymes in *Chenopodium\album* L and *Amaranthus retroflexus* L. *Plant Physiol* **85**: 355–359

Sage RF, Sharkey TD and Seemann JR (1988) The *in vivo* response of the ribulose-1, 5 bisphosphate carboxylase activation state and the pool sizes of photosynthetic metabolites to elevated CO_2 in *Phaseolus vulgaris* L. *Planta* **174**: 407–416

Sage RF, Sharkey TD and Seemann JR (1989) Acclimation of photosynthesis to elevated CO_2 of five C_3 species. *Plant Physiol* **89**: 590–596

Salvucci ME, Portis AR Jr and Ogren WL (1986) Light and CO_2 response of ribulose-1, 5 bisphosphate carboxylase oxygenase activation in *Arabidopsis* leaves. *Plant Physiol* **80**: 655–659

Sasek TW and Strain BR (1989) Effects of carbon dioxide on the expansion and size of Kurzu (*Pueraria lobata*) leaves. *Weed Sci* **76**: 23–28

Sasek TW, Delucia EH and Strain BR (1985) Reversibility of photosynthetic inhibition in cotton after long-term exposure to elevated CO_2 concentrations. *Plant Physiol* **78**: 619–622

Sengupta UK (1988) Effect of increasing CO_2 concentration on photosynthesis and photorespiration in wheat leaf. *Curr Sci* **57**: 145–146

Sharkey TD and Badger MR (1984) Factors limiting photosynthesis as determined from gas exchange characteristics and metabolite pool size. In: (C Sybesma, ed) *Advances in Photosynthesis Research*, Vol 7, Martinus Nijhoff Dr W Junk Publishers, The Hague, pp 325–328

Sharkey TD, JA Berry and Sage RF (1988) Regulation of photosynthetic electron transport in *Phaseolus vulgaris* L. as determined by room temperature chlorophylla *a* fluorescence. *Planta* **176**: 415–424

Sharma A (1986) Studies on relationship of carbon dioxide assimilation and translocation of assimilates in mungbean. PhD Thesis, PG School, Indian Agricultural Research Institute, New Delhi

Sharma A and Sengupta UK (1990) Carbon dioxide enrichment effects on photosynthesis and related enzymes in *Vigna radiata* L (Wilczek). *Indian J Plant Physiol* 33: 340–346

Sionit N and Patterson DT (1985) Response of C_4 grasses to atmospheric CO_2 enrichment. II Effects of water stress. *Crop Sci* 25: 533–537

Sionit N, Hellmers H and Strain BR (1980) Growth and yield of wheat under CO_2 enrichment and water stress. *Crop Sci* 20: 687–690

Sionit N, Strain BR, Hellmers H and Kramer PJ (1981a) Effects of atmospheric CO_2 concentrations and water stress on water relations of wheat. *Bot Gaz* 142: 191–196

Sionit N, Strain BR and Beckford RA (1981b) Environmental control on the growth and yield of ckra. I Effect of temperature and CO_2 enrichment at cool temperature. *Crop Sci* 21: 885–888

Sionit N, Rogers HH, Bingham GE and Strain BR (1984) Photosynthesis and stomatal conductance with CO_2 enrichment of container and field grown soybeans. *Agron* 76: 447–451

Szarek SR, Holthe PA and Ting IP (1987) Minor physiological response to elevated CO_2 by the CAM plant *Agava vilmoriniana*. *Plant Physiol* 83: 938–940

Thomas JD and Hill GR (1949) Photosynthesis under field conditions. In: (J Franck and WE Loomis, eds). *Photosynthesis in Plants*, Iowa State College Press, Ames, pp 19–52

Thomas JF, Raper CD Jr, Anderson CE and Dowas RJ (1975) Growth of young tobacco plants as affected by carbon dioxide and nutrient variables. *Agron* 67: 685–689

Ting I and Loomis WE (1963) Diffusion through stomates *Amer J Bot* 30:866–872

Tinus RW (1972) CO_2 enriched atmosphere speeds growth of ponderosa pine and blue spruce seedlings. *Tree Plant Notes* 23: 12–15

Tolbert NE and Zelitch I (1983) Carbon metabolism. In: (ER Lemon, ed). *CO₂ and Plants: The Response of Plants to Rising Levels of Atmospheric Carbon Dioxide*, Westview Press, Boulder, Colorado, pp 21–64

Van Bavel MCH (1974) Antitransparent action of carbon dioxide on intact sorghum plants. *Crop Sci* 14: 208–212

von Cammerer S and Edmondson DL (1986) Relationship between steady state gas exchange, *in vivo* ribulose bisphosphate caroxylase activity and some carbon reduction cycle intermediates in *Raphanus sativus*. *Aust J Plant Physiol* 13: 669–688

von Cammerer S and Farquhar GD (1984) Effects of partial defoliation, changes of irradiance during growth, short term water stress and growth at enhanced P (CO_2) on the photosynthetic capacity of leaves of *Phaseolus vulgaris* L. *Planta* 160: 320–329

Vu JCV, Allen LH and Bowes G (1983) Effects of light and elevated atmospheric CO_2 on the ribulose bisphosphate carboxylase activity and ribulose bisphosphate level of soybean leaves. *Plant Physiol* 73: 729–734

Vu JCV, Allen LH and Bowes G (1989) Leaf ultrastructure, carbohydrates and protein of soybeans grown under CO_2 enrichment. *Environ Exp Bot* 29: 141–147

Wardlaw IF (1982) Assimilate movement in *Lolium* and *Sorghum* leaves, III Carbon dioxide concentration effects on the metabolism and translocation of photosynthates. *Aust Plant Physiol* 9: 705–715

Waritt B, Landsberg JJ and Thorpe MR (1980) Responses of apple leaf stomata to environmental factors. *Plant Cell Environ* 3: 13–22

Warren WJ (1966) An analysis of plant growth and its control in arctic environments. *Ann Bot* (London) 30: 383–402

Warwick RA and Gifford RM (1985) Climatic change and agriculture: Assessing the reponse of global agriculture to increasing CO_2 In: UNEP/WMO/ICSU-SCOPE. *International Assessment of the Impact of an Increased Anthropogenic Input of Carbon Dioxide on the Environment*, Chap 8, pp 1–24

Wong SC (1979) Elevated atmospheric partial pressure of CO_2 and plant growth. I Interactions of nitrogen nutrition and photosynthetic capacity in C_2 and C_3 plants. *Oecologia* 44: 68–74

Wong SC (1990) Elevated atmospheric partial pressures of CO_2 and plant growth. II Non structural carbohydrate content in cotton plants and its effects on growth parameters. *Photosynth Res* 23: 171–180

Wray SM and Strain BR (1986) Response of two old field perennials to interactions of CO_2 enrichment and drought stress. *Ame J Bot* 73: 1486–1491

508

Wulff R and Strain BR (1982) Effects of caron dioxide enrichment of growth and photosynthesis in *Desmodium paniculatum. Can Bot* **73**: 1486–1491

Yelle S, Benson RC, Trudel MC Jr and Gosselin A (1989a) Acclimation of two tomato species to high atmospheric CO_2 II. Ribulose-1, 5-bisphosphate carboxylase oxygenase and phosphoenolpyruvate carboxylase. *Plant Physiol* **90**: 1473–1477

Yelle S, Beeson RC Jr, Trudel MJ and Gosselin A (1989b) Acclimation of two tomato species to high atmospheric CO_2. I Sugar and starch concentration. *Plant Physiol* **90**: 1465–1472

21

The Influence of Atmospheric CO₂ Enrichment on Allocation Patterns of Carbon and Nitrogen in Plants from Natural Vegetations

I. Stulen, J. den Hertog and Katrien Jansen

Department of Plant Biology
University of Groningen
PO Box 14, 9750 AA Haren (Gn)
The Netherlands

CONTENTS

ABBREVIATIONS

LAR	:	leaf area ratio
LWR	:	leaf weight ratio
NAR	:	net assimilation rate
PS	:	photosynthesis
RGR	:	relative growth rate
RR	:	root respiration
RuBPcase	:	Ribulose *bis*phosphate carboxylase/oxygenase
SLA	:	specific leaf area
SR	:	Shoot respiration
S/R	:	shoot/root ratio
SWR	:	shoot weight ratio

510

ABSTRACT

The physiological response of plant species to elevated atmospheric CO_2 concentrations varies greatly. This paper examines the extent to which experimental conditions influence the growth response of plants from natural vegetations to elevated CO_2 concentrations. It is concluded that nitrogen availability is an important factor in explaining the variable shoot/root responses as regards dry matter partitioning. At optimum nitrogen availability, shoot-root ratio is not changed due to high CO_2. At limiting nitrogen availability, relatively more dry matter is invested in the root in some cases. However, changes in light interception of the shoot can greatly influence the results.

Stimulation of growth and photosynthesis by an elevated CO_2 level is usually temporary. To what extent the transient nature of the stimulation can be explained by changes in concentration and partitioning of non-structural carbohydrates in the leaf is discussed. From the data presented in this paper, it is concluded that changes in shoot morphology rather than changes in carbohydrate partitioning are important in explaining the transient nature of stimulation of growth and photosynthesis.

I. Introduction

Over the past few decades investigations on the effects of elevated atmospheric CO_2 concentrations on plant growth and productivity have been numerous (For details see chapter by Sengupta and Sharma). Many of these studies used crop species showing enhanced dry matter production upon doubling of the current atmospheric CO_2 concentration (Wittwer, 1983; Kimball, 1983, 1986; Acock and Pasternak, 1986; Cure and Acock, 1986; Poorter, 1992). Comparatively little research has been carried out with species from natural vegetations. Knowledge of the physiological response of these species is important in predicting ecosystem responses to elevated CO_2 concentrations (Bazzaz 1990; Mooney et al., 1991).

The physiological response of plant species to elevated CO_2 concentrations varies greatly and depends on experimental conditions. Nitrogen and water availability may restrict the stimulation of dry matter production by an elevated CO_2 concentration (Gifford, 1979; Wong, 1979; Patterson and Flint, 1982; Goudriaan and de Ruiter, 1983). Whether an elevated CO_2 concentration changes the partitioning of dry matter between shoot and root largely depends on the availability of nutrients as well as water (Stulen and den Hertog, 1992).

The aim of this paper is to assess the influence of elevated atmospheric CO_2 concentrations on the allocation pattern in plants from natural vegetations. Attention will be paid to allocation of carbon and nitrogen between and within shoot and roots in relation to photosynthetic capacity and acquisition of nitrogen. The consequences of changes in the pattern of allocation for the functioning of the plant in its natural habitat will be discussed.

II. Partitioning of Dry Matter and Carbon

Information on dry matter partitioning between shoot and root is needed to estimate the partitioning of resources between photosynthetic tissue and tissue for acquisition of water and nutrients. An optimal allocation pattern is important to balance resource acquisition (Schulze et al., 1983; Bloom et al., 1985; Pearcy et al., 1987).

a. *Partitioning of dry matter between shoot and root*

Literature data on the influence of an elevated CO_2 concentration on dry matter partitioning between various plant parts, measured as shoot/root ratio (S/R) or leaf weight ratio (LWR), revealed a variable response: an increase, a decrease, or no effect at all. The assumption that an increase in atmospheric CO_2 concentration generally leads to a relatively greater allocation of dry matter to the root (Bazzaz, 1990; Enoch, 1990) needs critical reconsideration. By analysing the data in relation to experimental conditions as regards nutrient and water availability, developmental stage of the plant, phenology, and sink strength, some general trends were found (Stulen and den Hertog, 1992). Under well-fertilized conditions the increase in dry matter by an elevated CO_2 concentration was partitioned in proportion over the vegetative parts of the plants for part of the experimental period. In some cases an increase in S/R during prolonged exposure to an elevated CO_2 concentration was found, which was ascribed to an increase in self-shading, resulting in a relatively higher investment of dry matter in photosynthetic tissue (Poorter et al., 1988). Analysis of experiments, set up to investigate the influence of a limited nutrient supply in combination with an elevated CO_2 concentration, showed no change in dry matter partitioning, or a relatively higher investment in root dry matter, resulting in a decrease in S/R. Oberbauer and co-workers (1986), in experiments with tundra species in which the smallest S/R values were consistently found in the high CO_2, low nutrient treatment, concluded that the enhancement of carbon allocation to roots at an elevated CO_2 level may be important, especially for species that naturally occur in nutrient-limited ecosystems.

Analysis of the available data on the influence of an elevated CO_2 concentration on S/R further revealed that the developmental stage of the plant during the experimental period could offer further explanation for the variation in response to an elevated CO_2 concentration between species. It was concluded that changes in dry matter partitioning over the plant components as a result of an elevated CO_2 concentration might depend on changes in relative sink strength of the various plant organs. A direct or indirect effect of an elevated CO_2 concentration on morphological characteristics might play a role in assimilate partitioning as well (Stulen and den Hertog, 1992). The effect of an elevated CO_2 concentration on the partitioning of dry matter between above-ground and below-ground plant organs is influenced by processes such as earlier cessation of stem growth, reduction in leaf number (Mousseau and Enoch, 1989), and distinct genetic growth patterns affecting the chronological events in root and shoot growth (Kaushal et al., 1989).

Changes in accumulation of non-structural carbohydrates by an elevated CO_2 concentration can interfere in the interpretation of changes in S/R based on total dry weight. Wong (1990) demonstrated that differences in S/R were much smaller when expressed on the basis of structural dry weight rather than on total dry weight. Therefore, it is useful to compare growth parameters on the basis of structural dry weight with those on a total dry weight basis.

Stulen and den Hertog (1992) concluded that the response of S/R to an elevated CO_2 concentration is not in accordance with the functional equilibrium theory (Brouwer, 1962; Thornley, 1972; Wilson, 1988). The effect of an elevated CO_2 concentration on carbon partitioning might be better described in more physiological terms by measuring the ratio between non-structural carbohydrates and free amino acids (Campagna and Margolis, 1989). Campagna and Margolis (1989) concluded that this ratio correlated well with the partitioning of carbon between shoots and roots in *Picea mariana* Mill. grown at different levels of CO_2. They suggested that measurement of this ratio in the shoot might be useful in predicting the dry matter allocation between above-ground and below-ground organs in field studies.

b. *Partitioning of dry matter within the leaf*

Growth analysis of a number of species from natural vegetations showed that the initially observed stimulation of relative growth rate (RGR) by an elevated CO_2 level did not last the whole experimental period (Poorter *et al.*, 1988; den Hertog and Stulen, 1990). A transient stimulation of the RGR has been reported for many other species (Bazzaz, 1990). The length of the period of growth stimulation appears to vary between experiments and species. In seedlings of *Plantago major* ssp. *pleiosperma* and *Urtica dioica*, grown at a quantum flux density of 550 μmol m^{-2} s^{-1}, and a non-limiting supply of nutrients, the stimulation of RGR due to an increase in CO_2 concentration from 350 μl l^{-1} to 700 μl l^{-1} only lasted the first 10 days of the treatment. In *P. major* ssp. *major*, grown under similar conditions except for the quantum flux density (250 μmol m^{-2} s^{-1}), RGR was stimulated for four weeks (Poorter *et al.*, 1988). A comparative study with two ecotypes of *P. lanceolata*, one from a hayfield with erect rosettes with few, long and slender leaves, and another from a pasture with small, flat rosettes with many small leaves, showed that the ecotypes responded differently. Relative growth rate of the pasture plants was stimulated during the first four weeks only, while RGR of the hayfield plants was stimulated for more than six weeks (Table 1).

Table 1: Relative growth rate (mg g^{-1} day^{-1}) in ecotypes of *Plantago lanceolata* (see text), grown at 350 (low) or 700 (high) μl l^{-1} CO_2, a quantum flux density of 260 μmol m^{-2} s^{-1}, and a non-limiting nitrate supply. The CO_2 treatment was started at day 0, and RGR ($\pm$ SD) was calculated by linear regression from ln transformed dry weight data.

P. lanceolata type	Period (days)	RGR (mg g^{-1} day $^{-1}$)	
		low	high
exp 1			
hayfield	14–47	110 ± 10	140 ± 10*
pasture	12–45	90 ± 10	90 ± 10 ns
exp 2			
pasture	6–28	100 ± 10	140 ± 10*

ns: not significant
*significant at the 0.05 level.

The transient nature of the stimulation of RGR by an elevated CO_2 concentration might be caused by changes in the pattern of biomass allocation within the leaf. A detailed growth analysis showed that in *P. major* ssp. *pleiosperma* and *U. dioica* the effect on RGR was due to an increase in net assimilation rate (NAR) right from the beginning of the exposure to an elevated CO_2 concentration; after 10 days both RGR and NAR returned to the control level. In these experiments the positive effect of NAR on RGR was partly counteracted by a 10 per cent decrease in specific leaf area (SLA) (den Hertog and Stulen 1990; Table 2). The effect on SLA could be partly

Table 2: Changes in the leaves of *P. major* ssp. *pleiosperma* grown at 350 (low) or 700 (high) $\mu l\ l^{-1}$ CO_2 ($\pm$ SE), a quantum flux density of 550 $\mu mol\ m^{-2}\ s^{-1}$, and a non-limiting nitrate supply. The CO_2 treatment was started at day 0. Pairwise comparisons has been made using Student t tests. For further experimental conditions see den Hertog *et al.* (1992).

Day	Dry wt (%)		Carbon (mgC gDw⁻¹)		RGR # (mg g⁻¹ day⁻¹)		SLA (m² kg⁻¹)		Photosynthesis (μmol m⁻² s⁻¹)	
	low	high	low	high	low	high	low	high	low	high
0	12.2	11.8 ns	394	398**	299	329	40.0	40.8*	9.7	
6			393	399**	261	313				22.4
7	11.0	11.4 ns			260	303	33.1	30.3**		
10	11.9	12.9 ns			259	271	30.5	25.0**		
11					260	262			11.9	32.1
12	12.1	12.7 ns	405	417**	261	254	26.0	22.9**		
14					262	247			12.3	14.4
15			417	430**	263	248				
17	12.9	13.1**			261	262	24.3	20.5**		

\# RGR was calculated from fitted curves (den Hertog *et al.*, 1991).

ns: not significant.

* significant at the 0.05 level.

**significant at the 0.01 level.

explained by the higher dry matter concentration in the leaves (Table 2). In these experiments there was no effect on dry matter allocation between shoot and root, so that the leaf area ratio (LAR), the product of SLA and LWR, was affected to the same degree as SLA (den Hertog et al., 1992). A reduction in SLA was also found for *P. major* ssp. *major*. At the end of the experiment, however, when RGR was no longer stimulated, the reduction in SLA had also disappeared (Poorter et al., 1988). The pasture ecotype of *P. lanceolata* showed a different response. The SLA was not affected during the period of growth stimulation, while a decrease was found when growth stimulation had disappeared (Table 3).

Table 3: Photosynthesis and SLA in *P. lanceolata* pasture ecotype (see text), grown at 350 (low) or 700 (high) $\mu l \; l^{-1} \; CO_2$, a quantum flux density of 260 $\mu mol \; m^{-2} \; s^{-1}$, and a non-limiting nitrate supply. The CO_2 treatment was started at day 0. For details of photosynthesis measurement see den Hertog et al. (1992). Pairwise comparison of the data was made using a Student t test.

Day	Photosynthesis $(\mu mol \; CO_2 \; m^{-2} \; s^{-1})$		SLA $(m^2 \; kg \; Dw^{-1})$	
	low	high	low	high
14	8.3	12.7*	21.2	21.2 ns
21	10.8	13.1*	19.3	19.3 ns
28	7.2	12.4**	18.6	19.9 ns
35	8.1	10.6 ns	19.7	16.7*

ns: not significant
*significant at the 0.05 level.
**significant at the 0.01 level.

From these experiments with related *Plantago* species we conclude that in general SLA is reduced as a result of an elevated CO_2 concentration, but that the beginning and duration of the response differ. A reduction in SLA by an elevated CO_2 concentration was found for other species as well (Acock and Pasternak, 1986). The SLA appears to be the only morphological characteristic that responds consistently to an elevated CO_2 concentration. It is, therefore, concluded that plants can change the investment of dry matter in relation to their leaf area, resulting in a smaller leaf area per leaf weight at elevated CO_2 concentrations. As a consequence of the reduction in SLA, RGR does not respond as much as would be expected from the increase in photosynthesis by high CO_2.

c. *Partitioning of carbon compounds within the leaf in relation to photosynthesis*

The stimulation of RGR by an elevated CO_2 concentration is caused by an increased photosynthesis rate (Wong, 1979; Wulff and Strain, 1982; Poorter et al., 1988; Bazzaz, 1990; den Hertog et al., 1991). The time course of the stimulation, however, is not always known. When investigated, it appears to vary with experiment and species (Bazzaz, 1990). In *P. major* ssp. *major* seedlings, grown at a quantum flux density of 250 $\mu mol \; m^{-2} \; s^{-1}$, photosynthesis was stimulated for at least six weeks (Poorter et al., 1988). The stimulation of photosynthesis in *P. major* ssp. *pleiosperma*, grown at 550 $\mu mol \; m^{-2} \; s^{-1}$, was much shorter. In these seedlings photosynthesis decreased and nearly reached the control level after two weeks' exposure to high CO_2. In the pasture ecotype of of *P. lanceolata*, grown at 250 μmol m^{-2} s^{-1}, photosynthesis was stimulated for four weeks before the control level was reached (Table 3).

The transient nature of the stimulation of photosynthesis could have several reasons. A decline in the activity of the carboxylating enzyme ribulose-1,5-bisphosphate carboxylase/oxygenase, (RuBPcase), due to changes in the amount or activation state of the enzyme (Sage and Pearcy, 1987; Sage *et al.*, 1989), might play a role. This aspect will be dealt with in relation to changes in nitrogen partitioning later in this paper. Feedback inhibition or even physical damage by starch accumulation in the chloroplast might be another important factor in explaining the transient nature of the stimulation of photosynthesis by high CO_2 (Wulff and Strain, 1982; DeLucia *et al.*, 1985) together with changes in sink strength (Clough *et al.*, 1981; Koch *et al.*, 1986).

Wulff and Strain (1982) found for *Desmodium paniculatum* that the decrease in photosynthesis on a leaf area basis in plants exposed to 1000 μl l^{-1} CO_2 and a quantum flux density of 650 μmol m^{-2} s^{-1} was correlated with an increase in starch accumulation and reduced grana formation in chloroplasts. Poorter and co-workers (1988) measured changes in sugar and starch concentration in leaves of *P. major* ssp. *major*, grown at a quantum flux density of 250 μmol m^{-2} s^{-1} and exposed to 350 and 700 μl l^{-1}. The difference in starch accumulation of the leaves at the end of their experiments was less than in the beginning. A similar trend was also observed in our experiments with *U. dioica* and *P. lanceolata* ecotypes, grown at a quantum flux density of 550 and 250 μmol $m^{-2}s^{-1}$, respectively, and the same CO_2 concentrations as used by Poorter and co-workers (1988) (Table 4). Starch concentration in plants grown at an elevated CO_2 concentration tended to decrease during the experimental period. In *U. dioica*, starch concentration in the high CO_2 plants was higher than in the control plants throughout the experiment, while the concentration of soluble sugars was higher at the last harvest only. The *P. lanceolata* ecotypes showed a different response. In the hayfield type, starch concentration was not increased due to an elevated CO_2 concentration, while the concentration of soluble sugars increased during the whole experimental period. In the pasture type, soluble sugar concentration increased as well, while starch concentration increased at the end of the experiment only (Table 4). These experiments show that changes in soluble carbohydrate concentration as a result of an elevated CO_2 concentration can vary to a great extent. In the *P. lanceolata* ecotypes, the stimulation of both photosynthesis and RGR correlates with a relative accumulation of soluble sugars and starch at that time. The absolute starch concentration at that time, however, is less than in the beginning when both RGR and photosynthesis are stimulated. From the data obtained with the various *Plantago* species grown at different quantum flux densities, we conclude that it is unlikely that the disappearance of the stimulation of photosynthesis under these conditions is associated with the accumulation of starch.

d. *Shoot architecture in relation to photosynthesis*

An explanation for the transient nature of the stimulation of RGR and photo-

516

Table 4: Carbohydrate concentration and stimulation of RGR in *U. dioica* and *P. lanceolata* ecotypes. Carbohydrates were measured after Fales (1951). The plants were grown at 350 (low) or 700 (high) μl l^{-1} CO_2, a quantum flux density of 550 and 250 μmol m^{-2} s^{-1}, respectively, and a non-limiting nitrate supply. The CO_2 treatment was started at day 0. Pairwise comparisons have been made using Student t tests. For further experimental conditions see den Hertog and Stulen (1990).

Day	Sugar (mg gDw^{-1})		Starch (mg gDw^{-1})		RGR stimulation by high CO_2 (%)
	low	high	low	high	
U. dioica					
8	70.54	72.96 ns	84.82	124.56*	20
17	65.60	83.68*	67.57	86.08*	0
P. lanceolata hayfield					
15	41.1	45.1*	80.3	96.1 ns	present
40	46.6	56.0*	77.7	80.2 ns	present
pasture					
15	40.4	45.9*	83.7	91.8 ns	present
40	36.6	54.8**	67.4	79.6*	absent

ns: not significant
*significant at the 0.05 level.
**significant at the 0.01 level.

synthesis might be found in changes in the shoot architecture of the plants. In the case of *P. major* ssp. *major*, a rosette plant, the decrease in photosynthesis was a result of an increase in self-shading during the development of the plant (Poorter *et al.*, 1988). Enlargement of the projected leaf area was more important in enhancing photosynthesis than an increase in total leaf area. To what extent this explanation holds for more species is still an open question. Our preliminary experiments with the pasture ecotype of *P. lanceolata* showed no changes in the ratio of real to projected leaf area after 35 days in elevated CO_2, so that differences in shading in this case can be excluded. Reekie and Bazzaz (1989) investigated the response of seedlings of five tropical trees, grown both individually and in competition with each other, at ambient (350) and two levels of elevated CO_2 (525 and 700 μl l^{-1}). CO_2 influenced canopy structure of the individual plants in a different way and affected the competitive outcome through its effect on canopy architecture. They concluded that changes in light interception as a result of changes in shoot architecture might play an important role under competitive conditions in the field.

e. Carbon budget of the whole plant

Den Hertog and co-workers (1992) investigated to what extent an elevated CO_2 concentration influences the carbon budget of *P. major* ssp. *pleiosperma*. The influence of an elevated CO_2 concentration (700 μl l^{-1}) on the carbon budget was calculated from measurements on carbon fixation by photosynthesis, carbon incorporation into different plant parts, and carbon losses in shoot and root respiration (Table 1, Fig. 1). As stated in the previous section, the initial stimulation of photosynthesis

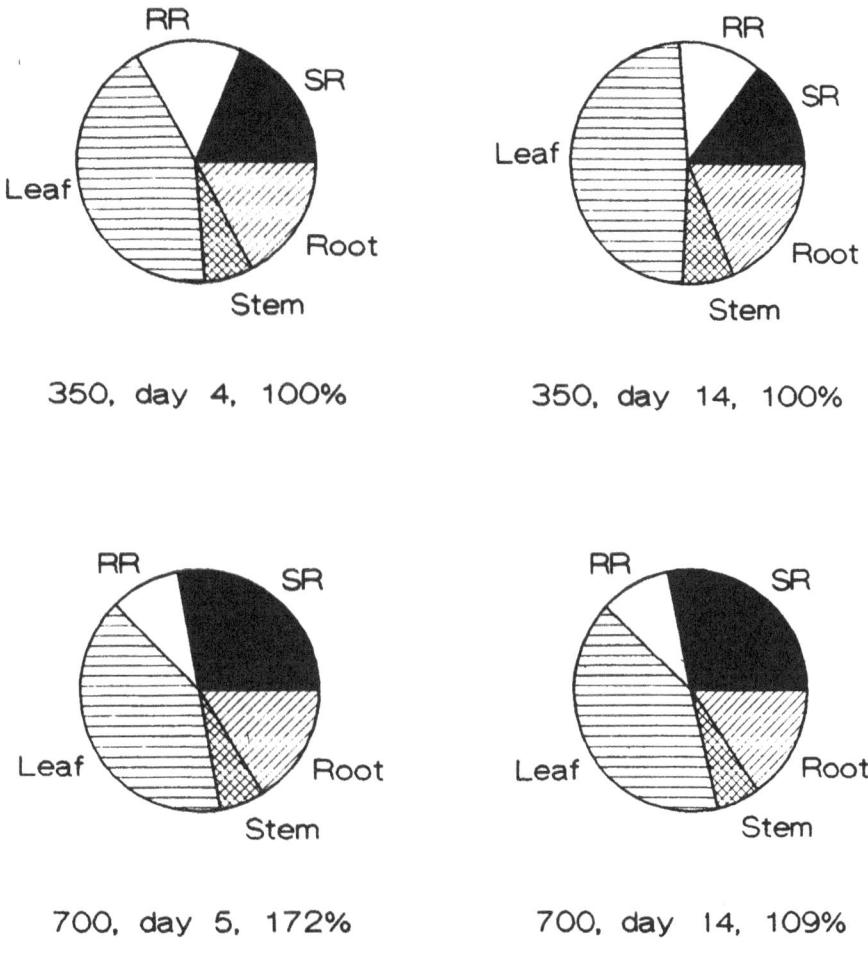

Figure. 1. Carbon budget of *Plantago major* ssp. *pleiosperma* grown at 350 and 700 µl l⁻¹ CO_2 on day 4 and 5 and day 14. The numbers under the circles indicate the photosynthesis rates on that day as a percentage of that of control plants. The segments indicate the proportion of C used in shoot respiration (SR), root respiration (RR), and in leaf, stem, and root growth. For calculations and experimental data see den Hertog *et al.* (1992).

by high CO_2 gradually decreased. At the beginning of the experiment the carbon concentration was the same in roots and shoot. Carbon concentration in the roots remained at the same level for both treatments. During the experimental period carbon concentration in the shoot increased for both treatments, but relatively more at the high CO_2 treatment. The rate of root respiration was not affected by high CO_2, in contrast to that of the shoot, which increased at high CO_2 (den Hertog *et al.*, 1992). From these data the carbon budget for both treatments was calculated at two points: during the early stage when RGR is stimulated by high CO_2 and at the end of the experiment when RGR is the same for both treatments (Fig. 1). Shifts in the total carbon budget as a result of high CO_2 are mainly due to the effects on shoot respiration. The relative contribution of shoot respiration to the carbon budget was larger at high CO_2. This effect lasted the whole experimental period. As a result, a lower percentage of the incoming carbon was invested in dry matter in the high CO_2 treatment.

The stimulation of shoot respiration by high CO_2 in this experiment (den Hertog et al., 1992) might be associated with a higher carbohydrate concentration at the elevated CO_2 concentration (Table 4). Such a positive correlation between respiration rate and sugar concentrations is often found (Lambers, 1985). Poorter and co-workers (1988) found that shoot respiration was stimulated by high CO_2 during the first part of the experiment only. In their experiments, the increase in starch accumulation in the leaves as a result of an elevated CO_2 level was also highest at the beginning of the experiment.

In the experiments of den Hertog and co-workers (1992) with *P. major* ssp. *pleiosperma* the rate of root respiration was not influenced by the CO_2 treatment. Poorter and co-workers (1988), however, found an initial increase in root respiration at high CO_2, followed by a return to control level. Literature data obtained with crop species also show a variable response. In *Triticum aestivum*, high CO_2 caused a decrease in root respiration, but in *Vigna radiata* and *Helianthus annuus* (Gifford et al., 1985) root respiration was not affected. Increased carbohydrate availability as a consequence of an elevated CO_2 concentration might increase respiration through the alternative pathway. In *P. major* ssp. *pleiosperma*, however, no shift in the contribution of cytochrome and alternative pathway to total respiration was observed (den Hertog, unpublished results). In contrast, in *T. aestivum* the engagement of the alternative pathway decreased (Gifford et al., 1985)

The response of both shoot and root respiration to an elevated CO_2 concentration clearly varies between experiments and species. Differences in experimental conditions as quantum flux density and nutrient availability, resulting in differences in RGR and soluble carbohydrate concentration, might play an important role in explaining the varibale responses (Amthor, 1991). A more detailed study of the influence of an elevated CO_2 concentration on the components of root respiration, i.e., for growth, maintenance, and ion uptake (van der Werf et al., 1988), might help in understanding the differences in response of root respiration. A high CO_2 concentration may influence uptake and allocation of nutrients as nitrogen, as discussed later in this paper. Since the costs of ion uptake can amount to as much as 40 per cent of respiratory ATP (van der Werf et al., 1988; Poorter et al., 1992), a shift in the balance between root respiration for growth and for ion uptake, leading to changes in total respiration, can be expected, if nutrient uptake is changed as a result of an elevated CO_2 concentration.

III. Partitioning of Nitrogen

Nitrogen availability is one of the most important factors in determining growth rate and productivity of plants (Bradshaw et al., 1964). It is well documented that internal partitioning of carbon can be altered by nitrogen stress (Rufty et al., 1984).

a. Between shoot and root

The effect of an elevated CO_2 concentration on nitrogen concentration in the plant was studied at various levels of nitrogen nutrition. In a comparative study with *P. major* ssp. *pleiosperma* and *U. dioica* grown at a non-limiting nitrate supply, shoot nitrogen concentration of *P. major* was reduced soon after the plants were transferred to high CO_2, whereas that of *U. dioica* was unchanged. Root nitrogen concentration was unchanged by the CO_2 treatment. Since in this experiment shoot weight ratio (SWR) was not changed by an elevated CO_2 concentration, relatively less nitrogen was allocated to the shoot of *P. major* (Table 5).

Table 5. N concentration in shoot and root and SWR on a dry weight basis in *P. major* ssp. *major* and *U. dioica* grown at low (350 $\mu l\ l^{-1}$) or high (700 $\mu l\ l^{-1}$) CO_2. The CO_2 treatment was started at day 0. Pairwise comparisons have been made using Student t tests. For further experimental conditions see den Hertog and Stulen (1990).

Day	Shoot N ($mgN\ gDw^{-1}$)		Root N ($mgN\ gDw^{-1}$)		SWR	
	low	high	low	high	low	high
P. major						
0	51.4		47.1		0.741	0.738 ns
7	54.5	46.8**	46.0	46.1 ns	0.735	0.754*
14	53.4	41.5**	47.6	42.4**	0.727	0.718 ns
17	50.7	39.5**	44.7	45.4 ns	0.726	0.735 ns
U. dioica						
0	45.3	43.7 ns	48.8		0.878	0.879 ns
7	48.4	46.9 ns	60.0	57.5 ns	0.788	0.772 ns
14	50.5	50.3 ns	54.2	54.5 ns	0.764	0.759 ns
22	52.1	47.3*	53.6	55.4 ns	0.770	0.783 ns

ns: not significant.
*significant at the 0.05 level.
**significant at the 0.01 level.

Norby and co-workes (1986) investigated the influence of an elevated CO_2 concentration on the uptake and partitioning of nitrogen in *Quercus alba* seedlings grown in nutrient-poor soil. The high CO_2 treatment resulted in a greater root size. Total uptake of nitrogen was not affected but the limited pool of nitrogen was allocated more to metabolically active tissue as fine roots and leaves. The authors suggested that the fact that relatively more nitrogen was invested in fine roots could explain the increased uptake of other nutrients, as P and K was increased at high CO_2 in this experiment.

b. *Within the leaf in relation to photosynthesis*

Many studies have shown that at an elevated CO_2 concentration total nitrogen concentration on a dry weight basis in the above-ground parts is reduced, leading to an increase in C/N ratio (Hocking and Meyer, 1985; Overdieck and Reining 1986; Larigauderie *et al.*, 1988; Overdieck *et al.*, 1988; Wong, 1979; 1990). A lowered C/N ratio of the leaf may decrease the nutritive value to herbivores, and the decay of the litter (Overdieck *et al.*, 1988; Bazzaz, 1990; Lambers, 1992).

There is evidence that nitrate-N concentration in the leaf is decreased by elevated CO_2 levels (Hucklesby and Blanke, 1990; Yelle *et al.*, 1987). The fact that nitrogen concentration in the leaf in *P. major* ssp. *pleiosperma* decreased, but was constant in *U. dioica* (Table 5) can possibly be explained by changes in the nitrate fraction.If a smaller fraction of total leaf nitrogen accumulates as nitrate, the decreased shoot nitrogen content can be more effectively used for growth (Freijsen and Veen, 1989).

The partitioning of nitrogen between leaf proteins is an important factor in determining the N use efficiency, i.e., the amount of biomass that can be produced per unit of N (Brown, 1978). RuBPcase accounts for a large percentage of soluble leaf protein in C_3 species (50 per cent), but less in C_4 species (see chapter by Kumar *et al.*, this volume). Changes in RuBPcase level, therefore, will have the greatest effect on N partitioning within the leaf (Sage *et al.*, 1989). For a C_3 species, a decrease in RuBPcase level at an elevated CO_2 level would lead to a more efficient use of the nitrogen invested in the leaf; under nutrient limitation less nitrogen would be needed per unit dry matter increment. Hocking and Meyer (1985) concluded from their experiments with *Xanthium* grown under nitrogen limitation that the increased growth of the nitrogen-stressed plants at an elevated CO_2 level was not caused by an increase in nitrogen uptake by the root, but resulted from a decrease in the RuBPcase content in the leaf, which enabled the plant to use the available nitrogen for other growth processes. The activity of RuBPcase decreased in a number of species due to an elevated CO_2 concentration (Hinkelton and Jolliffe, 1980; Peet *et al.*, 1985). Sage and co-workers (1989) investigated the photosynthetic response to an elevated CO_2 concentration of five species. The RuBPcase content was lower in two of the five species only, but in all species the RuBPcase activation state was reduced. It is interesting to note that the weedy species *Chenopodium album* showed the most economical response to an elevated CO_2 concentration: its RuBPcase level decreased while at the same time the rate of photosynthesis increased. Koch and co-workers (1986) found different results with two citrus species. In *Carrizo citrange*, no clear difference in RuBPcase level was found between the control and the high CO_2 treatment while in citrumelo RuBPcase level remained higher at the high CO_2. According to the authors these contrasting results might be explained by species differences, the time of sampling, or a leaf-age effect.

c. *In relation to growth rate*

In the experiments of Overdieck and co-workers (1988), dry matter production was still stimulated by an enhanced CO_2 level when N concentration in the leaf declined to about 2 per cent of dry matter. At a very low nitrogen availability, however, growth was no longer stimulated by CO_2 (Goudriaan and de Ruiter, 1983). The data of Larigauderie and co-workers (1988) demonstrated that under severe nitrogen stress photosynthesis can be reduced in plants exposed to an elevated CO_2 concentration and that this reduction can be attributed to the low nitrogen concentration found in these plants.

Lambers and co-workers (1989) concluded that information on nitrogen productivity or NAR alone is not sufficient to understand the physiological basis of variation in RGR. Knowledge of the rates of photosynthesis and of the partitioning of nitrogen and biomass over the various plant organs is also required to explain variation in growth rate either between species or depending on environmental conditions. Clearly more experiments on the effect of an elevated CO_2 concentration on changes in partitioning of nitrogen in combination with a detailed growth analysis are needed to fully understand the impact of an elevated CO_2 level on RGR.

IV. Concluding Remarks

Nitrogen availability is an important factor determining the growth response to an elevated atmospheric CO_2 level. The partitioning of dry matter between shoot and root does not seem to be influenced when nitrogen is not limiting, while root growth is relatively favoured when nitrogen is limiting. The latter could be of importance for situations in the field where plants often have to cope with nutrient limitation.

The growth stimulation by high CO_2 disappears during the experimental period. To what extent changes in soluble carbohydrate partitioning in the leaf are responsible for the transient nature of growth stimulation and photosynthesis cannot be clarified with the current knowledge. From our experiments, however, it seems unlikely that starch accumulation is involved. Changes in shoot architecture, resulting in increased self-shading, might play an important role in explaining the changes in growth stimulation and photosynthesis.

Specific leaf area was reduced right from the start of the CO_2 treatment in a number of cases, thus partly counteracting the increased photosynthesis. It is important to include morphological parameters in models predicting the response of plants to an elevated atmospheric CO_2 concentration. Knowledge on the influence of an elevated CO_2 concentration on carbon losses by the plant in root and shoot respiration should be expanded, in order to be able to draw conclusions on the carbon budget of the whole plant.

Nitrogen partitioning between shoot and root may be changed as a result of an elevated CO_2 concentration, resulting in a lower leaf nitrogen concentration. Due to more efficient use of the nitrogen invested in RuBPcase, overall growth and photosynthesis are enhanced. Only at very low nitrogen concentrations an elevated CO_2 concentration does not stimulate growth and photosynthesis.

Acknowledgements

We are grateful to Sander Pot for carrying out part of the experiments. Thanks are due to P.J.C. Kuiper and L.J. de Kok for their comments on the manuscript and Marion Cambridge for her linguistic corrections.

V. References

Acock B and Pasternak D (1986) Effects of CO_2 concentration on composition, anatomy and morphology of plants. In: (HZ Enoch and BA Kimball, eds) *Carbon Dioxide Enrichment of Greenhouse Crops.* II *Physiology, Yield and Economics* CRC Press, Boca Raton, pp 41–53

Amthor JS (1991) Respiration in a future, higher-CO_2 world. *Plant Cell* and *Environ* **14**: 13–20

Bazzaz FA (1990) The response of natural ecosystems to the rising global levels. *Annu Rev Ecol Syst* **21**: 167–196

Bloom AJ Chapin FS III and Mooney HA (1985) Resource limitation in plants—An economic analogy. *Annu Rev Ecol Syst* **16**: 363–392

Bradshaw AD, Chadwick MJ, Jowett D and Snaydon RW (1964) Experimental investigations into the mineral nutrition of several grass species. IV Nitrogen level. *J Ecol* **52**: 665–677

Brouwer R (1962) Functional equilibrium: Sense or nonsense? *Neth J Agric Sci* **31**: 335–348

Brown RH (1978) A difference in N use efficiency in C_3 and C_4 plants and its implications in adaptation and evolution. *Crop Sci* **18**: 93–98

Campagna MA and Margolis HA (1989) Influence of short-term atmospheric CO_2 enrichment on growth, allocation patterns, and biochemistry of black spruce seedlings different stages of development. *Can J For Res* **19**: 773–782

Clough JM, Peet MM and Kramer PJ (1981) Effects of high atmospheric CO_2 and sink size on rates of photosynthesis of a soybean cultivar. *Plant Physiol* **67**: 1007–1010

Cure JD and Acock B (1986) Crop responses to carbon dioxide doubling: A literature survey. *Agric For Meteorol* **38**: 127–145

den Hertog J and Stulen I (1990) The effects of an elevated atmospheric CO_2 concentration on dry matter and nitrogen allocation In: (J Goudriaan, H van Keulen and HH van Laar, eds) *The Greenhouse Effect and Primary Productivity in European Agro-ecosystems*, Pudoc, Wageningen, pp 27–30

den Hertog J, Stulen I and Lambers H (1992) Assimilation and allocation of carbon in *Plantago major* as affected by atmospheric CO_2 levels: A case study. *Vegetatio*: in press.

DeLucia EH, Sasek TW and strain BR (1985) Photosynthetic inhibition after long-term exposure to elevated levels of atmospheric carbon dioxide. *Photosynth Res* **7**: 175–184

Enoch HZ (1990) Crop responses to aerial carbon dioxide. *Acta Horticulturae* **268**: 17–33

Fales FW (1985) The assimilation and degradation of carbohydrates by yeast cells. *J Biol Chem* **193**: 113–124

Freijsen AHJ and Veen BW (1989) Phenotypic variation in growth as affected by N-supply: Nitrogen productivity. In: (H Lambers, ML Cambridge, H Konings and TL Pons, eds) *Cases and Consequences of Variation in Growth Rate and Productivity of Higher Plants*, SPB Academic Publishing, The Hague, pp 1–19

Gifford RM (1979) Growth and yield of CO_2-enriched wheat under water-limited conditions. *Aust J Plant Physiol* **6**: 367–378

Gifford RM, Lambers H and Morison JIL (1985) Respiration of crop species under CO_2 enrichment. *Physiol Plant* **63**: 351–356

Goudriaan J and de Ruiter HE (1983) Plant growth in response to CO_2 enrichment, at two levels of nitrogen and phosphorus supply. I Dry matter, leaf area and development. *Neth J Agric Sci* **31**: 157–169

Hinkelton PR and Jolliffe PA (1980) Alterations in the physiology of CO_2 exchange in tomato plants grown in CO_2 atmospheres. *Can J Bot* **58**: 2181–2189

Hocking PJ and Meyer CP (1985) Responses of Noogoora Burr (*Xanthium occidentale* Bertol.) to nitrogen supply and carbon dioxide enrichment. *Annu Bot* **55**: 835–844

Hucklesby DP and Blanke MM (1990) Limitation of nitrogen assimilation in plants. II Effect of CO_2, GA_3 and photoperiod on nitrate accumulation in spinach. *Gartenbauwissenschaft* **55**: 159–162

Kaushal P, Guehl JM and Aussenac G (1989) Differential growth response to atmospheric carbon dioxide enrichment in seedlings of *Cedrus atlantica* and *Pinus nigra* ssp. *laricio* var. *corsicana. Can J For Res* **19**: 1351–1358

Kimball BA (1983) Carbon dioxide and agricultural yield: An assemblage and analysis of 430 prior observations. *Agron J* **75**: 779–789

Kimball BA (1986) CO_2 stimulation of growth and yield under environmental restraints. In: (HZ Enoch and BA Kimball, eds) *Carbon dioxide enrichment of greenhouse crops. II Physiology, yield and economics*. CRC Press, Boca Raton, pp 53–67

Koch KE, Jones PH, Avigne WT and Allen LH (1986) Growth, dry matter partitioning, and diurnal activities of RuBP carboxylase in citrus seedlings maintained at two levels of CO_2. *Physiol Plant* **67**: 477–484

Lambers H (1985) Respiration in intact plants and tissues: Its regulation and dependence on environmental factors, metabolism and invaded organisms. In: (R Douce and DA Day, eds) *Encyclopedia of Plant Physiology*, NS Vol 18, Springer-Verlag, Berlin, pp 418–473

Lambers H (1992) Rising CO$_2$, secondary plant metabolism, plant herbivore interactions and litter decomposition. Theoretical considerations. *Vegetatio*: in press.

Lambers H, Freijsen N, Poorter H, Hirose T and Van der Werff A (1989) In: (H Lambers, ML Cambridge, H Konings and TL Pons, eds) *Causes and Consequences of Variation in Growth Rate and Productivity of Higher Plants*, SPB Academic Publishing, The Hague, pp 1–17

Larigauderie A, Hilbert DW and Oechel WC (1988) Effects of CO$_2$ enrichment and nitrogen availability on resource acquisition and resource allocation in a grass, *Bromus mollis*. *Oecologia* **77**: 544–549

Mousseau M and Enoch HZ (1989) Carbon dioxide enrichment reduces shoot growth in sweet chestnut seedlings (*Castanea sativa* Mill.). *Plant Cell Environ* **12**: 927–934

Mooney HA, Drake BG, Luxmoore RJ, Oechel WC and Pitelka LF (1991) Predicting ecosystem responses to elevated CO$_2$ concentrations *Bio Science* **41**: 96–105

Norby RJ, O'Neill EG and Luxmoore RJ (1986) Effects of atmospheric CO$_2$ enrichment on the growth and mineral nutrition of *Quercus alba* seedlings in nutrient-poor soil. *Plant Physiol* **82**: 83–89

Oberbauer SF, Sionit N, Hasting SJ and Oechel WC (1986) Effects of CO$_2$ enrichment on growth, photosynthesis, and nutrient concentration of Alaskan tundra species. *Can J Bot* **64**: 2993–2998

Overdieck D and Reining E (1986) Effect of atmospheric CO$_2$ enrichment on perennial ryegrass (*Lolium perenne* L.) and white clover (*Trifolium repens* L.) competing in managed model ecosystems. *Acta Oecologia* **7**: 367–378

Overdieck D, Reid CH and Strain (1988) The effects of preindustrial and future CO$_2$ concentrations on growth, dry matter production and the C/N relationship in plants at low nutrient supply: *Vigna unguiculata* (cowpea), *Abelmoschus esculentus* (okra) and *Raphanus sativus* (radish). *Angew Botanik* **62**: 119–134

Patterson DT and Flint EP (1982) Interacting effects of CO$_2$ and nutrient concentration. *Weed Sci* **30**: 389–394

Pearcy RW, Bjorkman O, Caldwell MM, Keeley, JJ, Monsin AK and Strain BB (1987) Carbon gain by plants in natural environments. *BioScience* **37**: 21–29

Peet MM, Huber SC and Patterson DT (1985) Acclimation to high CO$_2$ concentrations in two monoecious cucumbers. II Alterations in gas exchange rates, enzyme activities, and starch and nutrient concentrations. *Plant Physiol* **80**: 63–67

Poorter H (1992) Effect of elevated atmospheric CO$_2$ on growth, photosynthesis and respiration. *Vegetatio*: in press

Poorter H, Pot CS and Lambers H (1988) The effect of an elevated atmospheric CO$_2$ concentration on growth, photosynthesis and respiration in *Plantago major*. *Physiol Plant* **73**: 553–559

Poorter H, van der Werf A, Atkin OK and Lambers H (1992) Respiratory energy requirements of roots vary with the potential growth of a plant species. *Physiol Plant*: in press

Reekie EG and Bazzaz FA (1989) Competition and patterns of resource use among seedlings of five tropical trees grown at ambient and elevated CO$_2$. *Oecologia* **79**: 212–222

Rufty TW, Jr Raper CD and Huber SC (1984) Alterations in internal partitioning of carbon in soybean plants in response to nitrogen stress. *Can J Bot* **62**: 501–508

Sage RF and Pearcy RW (1987) The nitrogen use efficiency of C$_3$ and C$_4$ plants. I Leaf nitrogen, growth and biomass partitioning in *Chenopodium album* (L.) and *Amaranthus retroflexus* (L.). *Plant Physiol* **84**: 954–958

Sage RF, Sharkey TD and Seeman JR (1989) The acclimation of photosynthesis to elevated CO$_2$ in five C$_3$ species. *Plant Physiol* **89**: 590–596

Schulze E-D Schilling K and Nagarajah S (1983) Carbohydrate partitioning in relation to whole plant production and water use of *Vigna unguiculata* (L.) Walp. *Oecologia* **58**: 169–177

Stulen I and den Hertog J (1992) Root development and functioning under atmospheric CO$_2$ enrichment. *Vegetatio*: in press

Thornley JHM (1972) A balanced quantitative model for root:shoot ratios in vegetative plants. *Ann Bot* **36**: 431–441

Van der Werf A, Kooijman A, Welschen R and Lambers H (1988) Respiratory energy costs for the maintenance of biomass, for growth and for ion uptake in roots of *Carex diandra* and *Carex acutiformis*. *Physiol Plant* **56**: 33–37

Wittwer SH (1983) Rising atmospheric CO$_2$ and crop productivity. *Hortscience* **18**: 667–673

Wong SC (1979) Elevated atmospheric partial pressure of CO_2 and plant growth. I Interactions of nitrogen nutrition and photosynthetic capacity in C_3 and C_4 plants. *Oecologia* **44**: 68–74

Wong SC (1990) Elevated atmospheric partial pressure of CO_2 and plant growth. II Non-structural carbohydrate content in cotton plants and its effect on growth. *Photosynth Res* **23**: 171–180

Wulff RD and Strain BR (1982) Effects of CO_2 enrichment on growth and photosynthesis in *Desmodium paniculatum*. *Can J Bot* **60**: 1084–1091

Yelle S, Gosselin A, Trudel MJ (1987) Effect of atmospheric CO_2 concentration and root-zone temperature on growth, mineral nutrition, and nitrate reductase activity of greenhouse tomato. *J Amer Soc Hort Sci* **112**: 1036–1040

GENETIC VARIATION; PRODUCTIVITY

22

Genetic Variation in Photosynthetic Characteristics in Wheat: Causes and Consequences

K.C. Bansal[1], D.C. Uprety[2] and Y.P. Abrol[2]

[1]Department of Cellular & Developmental Biology
Harvard University, Cambridge, MA 02138 USA
[2]Division of Plant Physiology
Indian Agricultural Research Institute, New Delhi 110012, India

CONTENTS

ABBREVIATIONS

ATP	:	Adenosine triphosphate
CER	:	Carbon exchange rate
FeCN	:	Ferricyanide
NADP	:	Nicotinamide adenine dinucleotide phosphate
PS I	:	Photosystem I
PS II	:	Photosystem II
rbc S	:	Ribulose *bis*phosphate carboxylase small subunit
RuBPcase	:	Ribulose *bis*phosphate carboxylase

ABSTRACT

Genetic variation in photosynthetic characteristics in *Triticum aestivum* and related species, which include leaf photosynthetic rate at tillering and anthesis, electron transport and ribulose bisphosphate carboxylase activities are described. Causes of this variation in these characteristics are discussed at anatomical, physiological and biochemical level. It is suggested that this genetic variability, particularly high rates of leaf photosynthesis in wild diploid and tetraploid wheat species, can be exploited for increasing biomass production of modern wheat cultivars through interspecific hybridization. Selection for higher leaf photosynthate rate throughout crop growth duration could be used as a potential physiological tool in selecting genotypes producing higher biomass and grain yield.

I. Introduction

"Wheat is physiologically and genetically capable of much higher productivity and photosynthetic efficiency than has been recorded in a field environment" (Bugbee and Salisbury, 1988). The efficiency of photosynthesis has, however, remained stable with the rise in grain yield potential (Carver *et al.*, 1989). The physiological basis of improved grain yield of present-day cultivars of wheat (*Triticum aestivum*) has been the greater dry-matter partitioning to grains or, in other words, the increased harvest index (Austin *et al.*, 1980; Evans, 1981, 1984; Gifford *et al.*, 1984; Deckard *et al.*, 1985). On the contrary, Waddington *et al.* (1986) showed a close association of grain yield of Mexican wheat genotypes with biomass yield rather than with higher harvest index. On the basis of these observations it is plausible to speculate that an improvement in above-ground biomass can bring about increased grain yield of wheat. In general, the biomass production and grain yield of a particular cereal crop are largely determined by an array of complex physiological and biochemical processes including photosynthesis. Thus, in principle, the genetic improvement in photosynthetic characteristics of individual leaves or canopy can be the physiological basis of enhanced crop productivity and grain yield of wheat. To meet the above objective, genetic diversity in photosynthetic rates of flag leaves and early-formed leaves could be used. A substantial variation has been reported in photosynthetic rates of flag leaves at anthesis (Evans and Dunstone 1970; Austin *et al.*, 1982) and early-formed leaves at tillering (Johnson *et al.*, 1987b). Certain wild species possess photosynthetic rates considerably higher than the present-day hexaploid wheat, *T. aestivum*. The use of alien species as sources of physiological traits (e.g., photosynthesis) for wheat improvement has been suggested by Fedak (1985).

Recently, Carver and Nevo (1990) conducted experiments to identify potential sources of high photosynthetic efficiency for future wheat improvement. They were successful in identifying some accessions of *T. dicoccoides* having high photosynthetic capacity with no significant reduction in leaf size. Earlier studies had shown a strong association of higher photosynthetic rate with reduction in leaf size (Evans and Dunstone, 1970). Since both leaf size and photosynthetic capacity play a pivotal role in determining productivity, the benefits of such an accession possessing higher photosynthetic rate without significant reduction in leaf size in a wheat improvement-breeding programme would be enormous. However, many of the studies aimed at identifying genetic variation in photosynthetic rates in different crops have not been directly concerned with its consequences in relation to crop productivity. This could be due to the lack of correlation between photosynthetic rate and grain yield (Rawson *et al.*, 1983). Dornhoff and Shibles (1970) observed a relationship between leaf net CO_2 exchange and seed yield of soybean. This was later confirmed by Harrison *et al.* (1981), who found that the seed yield of soybean and canopy photosynthesis were significantly related. However, Curtis *et al.* (1969) reported no correlation between photosynthesis and yield of 36 cultivars of soybean. Babu *et al.* (1985) found a positive association of net CO_2 exchange after anthesis with total dry matter and pod yield of black gram. More recently, Kelly (1988) found a relationship between increased canopy photosynthesis and an increase in grain yield of maize. These studies, therefore, present a correlative evidence for an improved biomass and harvest index with an increase in the rate of leaf photosynthesis. On the other hand, in an excellent review, Gifford (1987) presented a comprehensive account of the constraints on selecting for higher carbon exchange rates while aiming at increasing the crop productiivity. However, the fact remains that crop biomass production is by and large determined by the photosynthetic capacity of leaves. Why thus, should an increased leaf photosynthetic rate not be used as a potential genetic tool to improve crop yields if the appropriate partitioning of increased dry matter to grains is maintained (Wardlaw, 1990)?

Genetic variation in photosynthetic characteristics of leaves and its causes are discussed in this review. Emphasis is given on exploring the possibility of using this variation as a potential genetic source for wheat improvement.

II. Genetic Variation in Photosynthetic Characteristics

Genetic variation in the photosynthetic capacity of leaves, particularly flag leaves at anthesis, is well documented in wheat (Evans and Dunstone, 1970; Khan and Tsunoda, 1970; Austin *et al.*, 1982, 1984; Bansal and Abrol, 1990). Originally, the variation in leaf photosynthesis of wheat became known when Evans and Dunstone (1970) and Khan and Tsunoda (1970), respectively, studied the physiological aspects of evolution and evolutionary trends in leaf photosynthesis of wheat species and its wild relatives. Both groups used the wild progenitors available and the present-day cultivated forms of wheat. The various wheat genotypes, wild as well as cultivated, belonged to three levels of ploidy: wild diploids (2n = 14), wild and cultivated tetraploids (2n = 28), and cultivated hexaploids (2n = 42). With the shift from wild to cultivated forms and with the increase in ploidy, the grain and leaf size increased, whereas the photosynthetic rate per unit leaf area declined significantly. Enlargement of leaf area was found associated with a decline in leaf nitrogen content. The number of stomata per unit leaf surface area and the assimilation rate per unit chlorophyll *a* decreased while the chlorophyll *a* content in leaves increased in parallel with increase in ploidy. Finally, the capacity to import by ears and developing grains improved with increase in ploidy and has been associated with enhanced remobilization of dry matter to grains. However, total dry matter (above-ground biomass) did not increase with increase in ploidy (Austin *et al.*, 1982, 1986; Uprety *et al.*, 1987).

The origin of different genomes has been investigated in the present-day hexaploid *T. aestivum* (ABD), an allopolyploid species. The details of their origin and the evolutionary pattern are presented diagrammatically (Figure 1). While *T. boeoticum*, a diploid wild species, is the source of A genome, a species closely related to *T. speltoides* (Riley *et al.*, 1958; Dvorak and Zhang, 1990) or *Aegilops longissima*

530

(Ogihara and Tsunewaki, 1982) is considered to be the donor of B genome. Through hybridization of the two species, a tetraploid *T. dicoccoides* with genome AB came into existence. *Aegilops squarrosa* provided the D genome, which hybridized with *T. dicoccoides* (AB) to give rise to *T. aestivum* (ABD). The addition of the D genome on *T. aestivum* conferred on this species baking characteristics as well as wide climatic adaptation which allowed its cultivation from the sub-humid to the semiarid areas (Zohary et al., 1969). *Triticum aestivum* is thus the only form of wheat cultivated in most locations in the world, although at some locations tetraploid species *T. durum* and *T. dicoccum* are preferred.

Group	Species	Genome
	Triticum boeoticum	A
Diploid	*Aegilops speltoides*	B
(2n = 14)	*Aegilops squarrosa*	D
	Triticum monococcum	A
	Triticum dicoccoides	AB
Tetraploid	*Triticum dicoccum*	AB
(2n = 28)	*Triticum durum*	AB
Hexaploid	*Triticum spelta*	ABD
(2n = 42)	*Triticum aestivum*	ABD

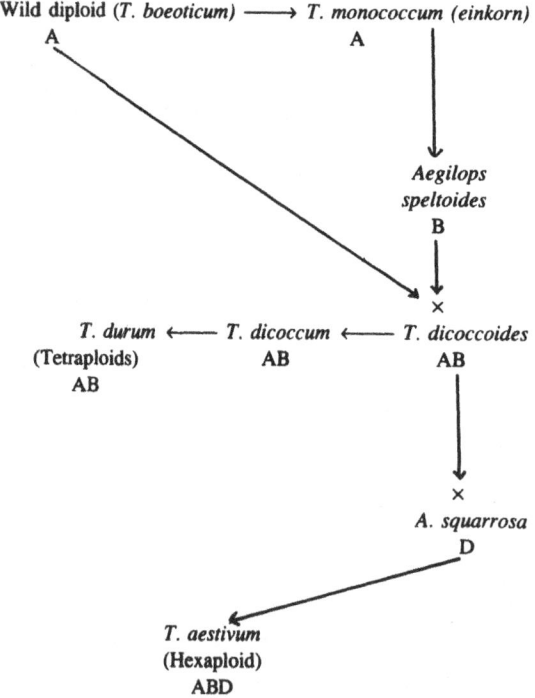

Figure 1. Diagrammatic presentation of evolution in wheat and genomic configuration

a. Leaf photosynthetic rates at anthesis

The differences in leaf photosynthetic rates at saturating light intensities, earlier observed by Evans and Dunstone (1970) and Khan and Tsunoda (1970), were later confirmed by Austin *et al.* (1982) and Uprety *et al.* (1987). Austin *et al.* (1982) measured the rates of net photosynthesis of flag leaves of 15 genotypes of *T. aestivum* and related wild diploid and tetraploid species. Intact leaves on field-grown plants were used and the photosynthetic rates were measured throughout their life. While the mean rates of photosynthesis per unit leaf area increased to a maximum at about two to three weeks after flag leaf emergence in diploids and hexaploids, the rates when expressed on per unit chlorophyll basis were constant over time, particularly in diploids. In both diploids and hexaploids, the rates of photosynthesis declined after attaining a maximum. However, the decline was more sharp in diploids than in hexaploids. The tetraploids were intermediate in these respects. Similar results were reported by Ghildiyal and Sirohi (1986). On an average of each ploidy group, the maximum net photosynthetic rates were 38, 32, and 28 mg CO_2 dm^{-2} h^{-1}, respectively, in diploids, tetraploids, and hexaploids. These rates were negatively correlated with flag leaf area as has been observed earlier by Evans and Dunstone (1970). However, Austin *et al.* (1982) found that *T. urartu* and *T. thaoudar* had higher photosynthetic rates than would be expected on the basis of their flag leaf area. This could possibly be related to their more efficient dark reactions involved in the fixation and assimilation of CO_2 than light reactions involved in generating ATP and NADPH. We have observed that the mean thylakoid Hill activity in seedling leaves of diploids gets saturated at relatively low light intensity as compared to the activity of either tetraploids or hexaploids (Bansal *et al.*, 1991b). In case of flag leaves, Dunstone *et al.* (1973) have earlier shown that net photosynthetic rates in diploids increased with increase in light intensity, whereas the rates in tetraploids and hexaploids reached their maxima at intermediate light intensities. These comparisons highlight the differential response of isolated thylakoids and intact leaves to changing light intensities.

On studying a large number of species or genotypes in each ploidy group, the differences in photosynthetic rates within a given ploidy level became evident (Austin *et al.*, 1984). The diploids maintained the highest rates of photosynthesis. The rates for individual diploids varied from 7.9 to 9.5, for tetraploids from 6.0 to 9.4, and for hexaploids from 6.1 to 8.3 CO_2 mg^{-1} chl h^{-1}. *T. urartu, T. boeoticum, T. thaoudar*, and *T. aegilopoides* showed consistently the highest rates of photosynthesis. In yet another experiment, Austin *et al.* (1986) compared the photosynthetic rates among species and found that *T. urartu* had a significantly higher rate than other diploids. Furthermore, photosynthetic rates of organs other than leaves were also greater for wild diploids than hexaploids. The flag leaves of wild diploid species exhibited higher rates of photosynthesis than those of tetraploids and hexaploids when the various species were grown at different temperature and photoperiod (Bansal and Abrol, 1990). While comparing the photosynthetic efficiency of cultivable and wild forms of diploid, tetraploid, and hexaploid wheats, Uprety *et al.* (1987) observed that the wild genotypes within ploidy level had higher rate of photosynthesis than cultivable ones. On the other hand, the leaf area and harvest index were higher in

Table 1. Variability for photosynthesis and yield characters in cultivated and wild forms of *Triticum* species within ploidy level.

	Photosynthesis (μmol CO_2dm^{-2} hr^{-1})	Stomatal resistance (s/cm)	Shoots/ plant	Dry matter g/8 plants	Harvest index (%)	Grain yield (g/8 plants)
Triticum species						
Tetraploid						
(2n = 42)						
T. durum (cultivated)	355	2.20	11.0	255.00	25.40	64.86
T. dicoccum (wild)	397	2.35	15.3	267.00	19.83	52.76
T. dicoccoides (wild)	422	1.34	18.0	236.00	10.76	25.47
Hexaploid						
(2n = 42)						
T. aestivum (cultivated)	287	3.23	11.0	168.50	34.86	58.57
T. compactum (wild)	330	2.02	16.0	234.50	19.83	44.17
CD at 5% P	32.20	0.50	3.75	27.30	3.63	10.71

Source: Uprety *et al.* (1987).

cultivable forms. The possible explanation suggested by them for the fall in photosynthesis rate of cultivated wheat was that the number of chloroplasts did not increase in proportion to leaf and cell size. An increase in leaf area was observed in domesticated form within ploidy level. The wild form of tetraploid and hexaploid species had profuse tillering compared to domesticated forms. This study indicated that there was enough sink requirement at this stage to be commensurate with high rate of photosynthesis of these wild genotypes. The increase in yield in domesticated forms would have been derived either from reduction in wastage of whole plant assimilates or increased translocation to grains. Uprety et al. (1987) demonstrated that the reduction in assimilate wastage occurred in terms of significant decrease in tillers per plant in domesticated tetraploids and hexaploids compared to their wild forms. The lesser number of tillers in the domesticated forms resulted in increased biomass of individual tillers. This may possibly provide increased potential in the individual tillers to fulfill the demands of increasing sink size of the cultivable tetraploids and hexaploids (Table 1). Similar findings have been reported by other workers (Planchon and Fesquet, 1982; Haour-Lurton and Planchon, 1985; Ghildiyal and Sirohi, 1986).

b. Leaf photosynthetic rates at tillering

In contrast to the findings of Dunstone et al. (1973), who reported that differences in photosynthetic rates of Triticum species were absent in early-formed leaves, Johnson et al. (1987) noticed variation in leaf photosynthesis at early vegetative growth stages. Though the importance of variation in flag leaf photosynthesis is beyond dispute, the differences in early-formed leaves are of practical significance. As pointed out by Johnson et al. (1987), differences in photosynthesis detected during early vegetative growth stages could hasten the process of progeny identification with high photosynthesis in a segregating population.

Studies concerning the variation in photosynthesis at early growth stages are relatively few (Johnson et al., 1987b, 1988; Carver et al., 1989; Carver and Nevo, 1990). On six-week-old plants of four wild diploids, six wild tetraploids, and two cultivated hexaploids grown in a growth chamber, Johnson et al. (1987) observed that net assimilation rates for diploids and tetraploids were comparable, i.e., 27.2 and 27.0 μmol CO_2 m^{-2} s^{-1}, respectively, whereas the hexaploids exhibited lower assimilation rate, 22.6 μmol CO_2 m^{-2} s^{-1}. Individually, the wild diploid T. speltoides showed the highest leaf photosynthesis (31.4 μmol CO_2 m^{-2} s^{-1}) and one of the accessions of T. dicoccoides had the lowest value of 20.7, close to those of two cultivars of T. aestivum (cv. Study, 20.9 and cv.TAM W-101, 24.3 μmol CO_2 m^{-2} s^{-1}). Johnson et al. (1987) related higher assimilation rate to a higher capacity for mesophyll photosynthesis rather than stomatal conductance per se. In addition, they found that T. dicoccoides accession PI 428109 had significantly higher assimilation rate than PL 428042, in both the early-formed leaves at tillering and flag leaves at anthesis. However, the two accessions did not differ in leaf area, internal leaf CO_2 concentration, or water-use efficiency. On the basis of genetic analysis, Johnson et al. (1988) suggested that genes governing high assimilation were not cytoplasmic in origin and that the inheritance of A was through additive gene action. Earlier,

T. urartu was characterized as the only wild diploid species having higher leaf photosynthesis rate at both seedling and anthesis stage. The above reports are in contrast to the results of Rawson *et al.* (1983), who found no correlation between the assimilation rates of seedlings and flag leaves.

c. Photosynthetic electron transport activity

Besides the leaf photosynthetic rate per se, a possible manipulable component of photosynthesis is the light harvesting system associated with photosynthetic electron transport. Genotypic differences in Hill activity and photophosphorylation were reported by Kleese (1966) and Miflin and Hageman (1966). This variation in photochemical activity in wheat has been suggested as a potential source for improving photosynthetic efficiency by Zelenski *et al.* (1978), Miginiac-Maslow *et al.* (1979), Hieke (1984), Morgan and Austin (1986), Watanabe *et al.* (1988), Bansal and Abrol (1990) and Bansal *et al.* (1991b).

Zelenski *et al.* (1978) reported that flag leaves of diploid species had 42 % greater uncoupled Hill reaction rates than those of hexaploids. We measured the photosystem II activity ($H_2O \rightarrow$ FeCN) in isolated chloroplasts from flag leaves of different wheat species and noticed about 48% greater activity in diploids as compared to hexaploids (Bansal and Abrol, 1990). Austin *et al.* (1982), however, failed to detect any difference between ploidy levels in the rates of oxygen evolution associated with carbon fixation from isolated protoplasts or chloroplasts. According to Miginiac-Maslow *et al.* (1979), the protoplasts from seedling leaves of diploid species, had 40 % higher Hill reaction rate than the hexaploids. On the contrary, Hieke (1984) reported a tendency of the primary leaves of diploid species to have lower photochemical activity than that of tetra-and hexaploid species. This was the case when electron transport activity was calculated on total chlorophyll basis; the activity calculated on the basis of dry matter reached high values in tetraploid species but values remained low in *T. aestivum*. Watanabe *et al.* (1988) measured photosynthetic O_2 evolution in mesophyll protoplasts isolated from seedling leaves of wheat species. They found that O_2 evolution per unit number of protoplasts increased with increase in ploidy but was not proportional to ploidy. In a relatively large number of species, eight diploid, five tetraploid, and four hexaploid, the uncoupled ferricyanide-dependent Hill activity of isolated thylakoids was not absolutely related to ploidy (Bansal *et al.*, 1991b). An increase in the activity was noticed with a transition from diploid to tetraploid but the latter and hexaploids were comparable. In addition, we studied the Hill activity in response to changing light intensity. *T. boeoticum*, a diploid wild species, became saturated at relatively low intensity compared to *T. dicoccoides* and *T. aestivum*. The latter two species did not show saturation with increasing intensity. The results suggested that pool size of different electron transport components in these species may be quite different. According to Morgan and Austin (1986) higher light-saturated photosynthetic rates were related to higher density of photosynthetic reaction centres. However, a detailed analysis of the species is required to establish the differences in various subcomponents of photosynthetic electron-transport.

d. RuBPcase and photosynthetic carbon metabolism

Sirohi and Ghildiyal (1975) reported a correlation between leaf carbon exchange rate and carboxylase activity while Murthy and Singh (1979) found a correlation between flag leaf RuBPcase per unit area and grain yield for a range of wheat cultivars. Although the causes and the physiological significance of such relationships have not been addressed adequately, the importance of increased RuBPcase activity in relation to an improvement in leaf photosynthesis cannot be disregarded. For example, Randall et al. (1977) found that a decaploid tall fescue with higher RuBPcase activity had almost twice the carbon exchange rate, estimated per unit leaf area or per unit leaf dry weight. Their study suggests that despite the unavoidable oxygenase activity associated with RuBPcase, there is a possibility of improving leaf carbon exchange rate through a higher RuBPcase activity. The genetic variation available in wheat RuBPcase specific activity (Evans and Austin, 1986) or content (Bansal et al., 1991a) may be used to breed cultivars with higher enzyme content or activity and subsequently higher leaf CER (carbon exchange rate). A significant variation in RuBPcase (carboxylase as well as oxygenase) activity has been reported in the flag leaves of T. monococcum, T. urartu, T. dicoccum, T. dicoccoides, and T. aestivum (Holbrook et al., 1984). However, the ratio of carboxylase to oxygenase activity did not differ significantly among the species.

Evans and Seemann (1984) found that the in vitro RuBPcase activity for hexaploid wheat T. aestivum was 30% higher than for the diploid T. monococcum, although the two species did not differ in their respective RuBPcase content for a given leaf nitrogen content. The higher RuBPcase activity in T. aestivum was attributed to the large subunit coded by B genome in the chloroplast.

Ghildiyal and Sirohi (1986) also noticed higher RuBPcase activity per unit leaf area in flag leaves of hexaploid and tetraploid species than in flag leaves of diploid T. monococcum. Although in vitro enzyme activity could not be correlated with leaf photosynthetic rate in this study, there are evidences that light-saturated rates of photosynthesis are commensurate with RuBPcase content per unit leaf area (Bjorkman, 1968; Van Caemmerer and Farquhar, 1981). The concentration of RuBPcase in the chloroplasts has been reported to be less in diploid species than in hexaploids (Dean and Leech, 1982). We detected a significant genetic variation in RuBPcase content per unit leaf area as well as per unit leaf weight (Bansal et al., 1991a). Diploid species exhibited, on an average, the lowest enzyme amount; however, T. urartu had values similar to those of tetraploid and hexaploid species. Total soluble leaf protein on fresh weight basis did not differ among three ploidy levels but on leaf area basis it was the highest in tetraploids. Interestingly, the proportion of total soluble protein-in RuBPcase increased with increase in ploidy in contrast to the observations of Evans and Austin (1986). This suggested that further increase in the proportion of total soluble protein in RuBPcase would be achieved only through better N-fertilization that might increase the existing level of total soluble protein (Lawlor et al., 1989). Increasing the amount of RuBPcase at the expense of other soluble proteins has not been associated with an increase in CO_2 assimilation in wheat (Dean and Leech, 1982), though such an association was observed in tall fescue (Joseph et al., 1981). Increase in total soluble protein as a result of better N-fertilization, therefore, appears to be more applicable in wheat with regard to an increase in RuBPcase level.

Attempts have also been made to compare wheat species at different ploidy levels in their rate of photorespiration and photosynthetic carbon metabolism (Dunstone et al., 1973; Champigny and Moyse, 1979; Holbrook et al., 1984). The role of photorespiration in carbon recovery during carbon metabolism has always been questioned in different crops. However, on the basis of a positive correlation between leaf CER and rate of photorespiration in wheat (Dunstone et al., 1973), it seems more likely that photorespiration is a useful process. Champigny and Moyse (1979) indeed observed that glycolate pathway provided 40 to 75% of the sucrose-^{14}C, when ^{14}C was supplied either through glycolate pathway or by the Calvin cycle intermediates. They found that in A. squarrosa and T. aestivum, the greater capability to synthesize sucrose was associated with the greater participation of photorespiration and the glycolate pathway. The wild diploid, T. aegilopoides had better participation of Calvin cycle intermediates in the synthesis of sucrose.

III. Causes of Variation in Photosynthetic Characteristics

As noted in the preceding section, the hexaploid wheat species show lower leaf photosynthetic rate, particularly in flag leaves, than the allied diploid and tetraploid species. However, the physiological or biochemical basis of the differences in photosynthesis among species has not been fully resolved. The addition of D genome to hexaploid wheat (ABD) has been considered to be one of the possible causes for low rate of photosynthesis, since the expression of D genome has an inhibiting effect on gene expression of A and B genomes (Champigny and Moyse, 1979). According to Haour-Lurton and Planchon (1985), chromosome 7D in a hexaploid species appeared responsible for the decrease in net photosynthesis of flag leaves developed under high irradiances. Earlier work had reported a deleterious effect of 7D chromosome on RuBPcase activity. On the contrary, Watanabe et al. (1988) failed to assign any major role to D genome in the photosynthetic oxygen evolution of protoplasts. The absence of photosynthesis-enhancing genes from the wild A genome could also be partly responsible for the low rate of photosynthesis in hexaploid species (Maan and Lucken, 1970). Apart from this genetic basis, the anatomical, physiological, and biochemical bases of differences in photosynthetic characteristics among species have been investigated and are discussed below.

a. Anatomical

The rate of photosynthesis of a crop species is determined by various anatomical features of the leaf. There are many ways in which the anatomical characteristics may contribute to the higher or lower rate of photosynthesis. In diploid species, as reported by Khan and Tsunoda (1971), the mesophyll cells are arranged more compactly than in hexaploids, and are arranged radially around the closely spaced veins. This "panicoid"-like leaf structure was found related to higher photosynthetic rate in diploid species. Parker and Ford (1982) found that in hexaploids, the veins were widely spaced, giving a larger pathway for the movement of photosynthates and water between veins and chloroplasts. Thus, widely spaced veins in hexaploids might result in the lower rate of photosynthesis. The evidence in favour of this came from Austin et al. (1982), who showed a positive correlation between vein density and maximum photosynthetic rates in 15 genotypes of wheat. The other important anatomical feature affecting photosynthesis is the degree of exposure of mesophyll cell surface to the air-filled space inside the leaf. The greater the cell surface exposed to the air-filled space per unit leaf area, the higher the rate of photosynthesis, due to more diffusion of carbon dioxide into the mesophyll cells (Dunstone and Evans, 1974). The wild diploid *T. urartu* has this anatomical feature (Parker and Ford, 1982) and has been reported to have the highest rate of photosynthesis among 15 wheat genotypes studied by Austin *et al.* (1982). Austin *et al.* (1982) studied the relationship of flag leaf photosynthetic rate with several anatomical features. Since the various anatomical features were interrelated, the relative importance of each could not be ascertained in explaining the differences in photosynthesis among species. However, differences in their internal anatomical features appeared to be the main cause of differences in photosynthesis. Jellings and Leech (1984) failed to detect a relationship between the number of chloroplasts per unit leaf area and photosynthetic capacity of first leaves of nine *Triticum* genotypes. They concluded that cell size with respect to the size of the nuclear genome was of overriding importance in determining the rate of photosynthesis on unit leaf area basis and that the influence of cell size was due to intracellular rather than extracellular characteristics. The variation in photosynthesis per unit chlorophyll could be accounted for up to 42% by mesophyll cell plan area alone (Austin *et al.*, 1984). The remaining variation could possibly be explained by other anatomical characteristics including the ratio of mesophyll cell surface area to mesophyll cell volume. The higher ratio will be expected to result in a reduced diffusion resistance in the pathway of carbon dioxide uptake to the carboxylation site inside the chloroplasts. However, the correlation of photosynthetic capacity per unit chlorophyll with the above ratio was low (r = + 0.501) in 34 genotypes of field-grown wheat.

There exists a negative correlation betwen the number of mesophyll cells per leaf area and the ploidy level (Dean and Leech, 1982; Lieckfeldt, 1989). There are approximately 394,000 and 124,000 mesophyll cells cm^{-2} in fully expanded first leaves of *T. monococcum* and *T. aestivum*, respectively. Accordingly, hexaploids have larger cells than diploids. Despite the bigger cell size and more abundant nuclear DNA, the hexaploids show a lower rate of photosynthesis than the diploids. It thus appears that in diploids, the smaller cell size is related to their greater photosynthetic capacity. The mechanism, however, of such a relationship remains largely unknown. Chloroplast number per cell increases proportionately with increase in ploidy, whereas the ratio of total chloroplast plan area per cell to mesophyll cell plan area decreases. The decrease in the latter ratio might explain to some extent the lowered rate of photosynthesis in hexaploids. Jellings and Leech (1984) found a relationship between the chloroplast cover per cell and the photosynthetic capacity per unit leaf area. It was, however not significant. The differences among species in the number of chloroplasts per unit leaf area also could not account for the differences in the photosynthetic capacity. It could be argued on the basis of the above discussion that factors other than anatomical features might as well play a significant role in deciding the rate of photosynthesis of different species.

b. Physiological

In order to determine the physiological basis of variation in photosynthesis among species, many physiological characters including stomatal conductance, stomatal diffusive resistance, mesophyll photosynthesis, chlorophyll level, and photochemical activity of isolated chloroplasts have been studied. Net leaf photosynthetic rate depends on (1) the availability of CO_2 that has diffused through stomata to the mesophyll cells and (2) the mesophyll capacity for photosynthesis. If higher stomatal conductance and higher capacity for mesophyll photosynthesis are associated with each other (see Johnson et al., 1987a), a positive correlation between leaf photosynthetic rate and stomatal conductance will be expected (K.C. Bansal, unpublished). In a field study with a large number of genotypes, Austin et al. (1982) showed that higher flag leaf photosynthetic rates of diploids were not absolutely associated with differences in stomatal conductance. In contrast, Johnson et al. (1987a), while studying photosynthetic differences among species in early-formed leaves at tillering, found a positive correlation between stomatal conductance and photosynthesis. However, they did not completely relate the higher photosynthetic rates of diploids to stomatal conductance. The high rate was coupled with higher mesophyll capacity. The importance of the latter in maintaining high leaf photosynthetic rates became evident when Johnson et al. (1987b) found that higher leaf photosynthesis in T. kotschyi, a tetraploid desert annual, was related to the higher capacity for mesophyll photosynthesis under water-deficit conditions. However, under well-watered conditions, stomatal conductance alone could account for higher leaf photosynthesis of T. kotschyi compared to T. aestivum. According to Ghildiyal and Sirohi (1986), stomatal diffusion resistance accounted for the differences in photosynthetic rates of flag leaves rather than the RuBPcase activity.

In general, no correlation between leaf photosynthesis and amount of chlorophyll has been observed in crop species. This is also true for both modern wheat (Murthy and Singh, 1979) and wild species of different ploidy levels. However, when such a relationship was worked out during late grain filling period, Ellison et al. (1983) found a positive correlation between flag leaf photosynthetic rates and chlorophyll levels. This suggests that chlorophyll content might limit the full expression of photosynthesis, particularly during grain filling. Austin et al. (1982) did not find such a relationship in various species of different ploidy levels. On the other hand, Heike (1984) reported that diploid species had higher chlorophyll content than tetra- and hexaploids. She, however, did not measure the rate of photosynthesis. If a relationship were sought between the chlorophyll content and already reported rates of leaf photosynthesis in different species, a positive correlation would be evident. Thus, the differences in photosynthesis rates could be explained on the basis of chlorophyll content per fresh matter as against per unit area (Austin et al., 1982).

Malkin et al. (1980) found that the density of photosystem II (PS II) reaction centres was well correlated with light-saturated rates of photosynthesis. Austin et al. (1984) concluded that differences in the density of PS II reaction centres may account partly for the differences in leaf photosynthesis rates among species. For example, T. urartu, a diploid species with high rate of photosynthesis, had a greater density of PS II reaction centres than T. aestivum, a hexaploid species with comparatively low rate of photosynthesis. Furthermore, the diploid species had significantly higher rates of CO_2-dependent O_2 evolution from isolated chloroplasts and protoplasts (per unit chlorophyll) than hexaploid wheat. This further reflects the importance of chlorophyll molecules associated with PS II in explaining the observed differences in leaf photosynthesis among species. In a preliminary study, Heike (1984) found a relatively greater area density of PS I reaction centres in *T. durum*, a cultivated tetraploid species, than in *T. monococcum*. But the latter had more chlorophyll content per fresh matter than the former. This indicated a larger photosynthetic unit size in *T. monococcum* than in *T. durum*. It may be concluded that size of photosynthetic unit is important rather than density of PS I reaction center with regard to the possible fractional cause of variation in photosynthetic rates among different species. A more detailed investigation is required to support the above view.

Austin *et al.* (1982) failed to detect a relationship between the rates of O_2 evolution (per unit chlorophyll) associated with carbon fixation from isolated protoplasts or chloroplasts and flag leaf photosynthetic rates of species exhibiting variability in photosynthesis rates. Similarly, in rye grass genotypes, Treharne (1972) observed no relationship between *in vitro* chloroplast Hill activity expressed either per unit leaf area or per unit chlorophyll and leaf carbon exchange rate.

We found that the enhanced basal electron transport rate of uncoupler-treated thylakoids was minimum in diploids whereas the rate of photosynthesis in seedling leaves was maximum in species of this ploidy group (Bansal *et al.*, 1991a, b). This suggests the efficient utilization of ATP and NADPH by diploid species during subsequent fixation of CO_2. The genotypic differences observed in leaf photosynthesis could also be explained on the basis of differences in ATP synthesizing capacity of chloroplasts of different species (Sinha and Khanna, 1972).

c. Biochemical

Holbrook *et al.* (1984) conducted experiments to test whether the high rates of photosynthesis in diploid species could be explained by the presence of C_4 pathway or lack of photorespiration. However, their results suggested rapid photorespiration and an absence of C_4 pathway for CO_2 fixation in all the species. Furthermore, the various species did not differ significantly in their mean ratio of carboxylase to oxygenase activities of RuBPcase. The *in vitro* carboxylase activity of RuBPcase also failed to explain the faster rate of leaf photosynthesis in diploids when Evans and Seemann (1984) reported 30% less activity for the diploid, *T. monococcum* than for *T. aestivum*. The faster rate of photosynthesis in diploids despite much less carboxylase activity than in hexaploids might reflect the differences among species in the level of activated RuBPcase. Very recently, Quick *et al.* (1991) investigated the impact of decreased RuBPcase on photosynthesis in transgenic tobacco transformed with "antisense" rbcS. They found that decreased level of RuBPcase was associated with a compensatory increase in RuBPcase activation, since the photosynthesis got inhibited by only 6% in transformed tobacco plants when RuBPcase was reduced to about 60% of the amount in the wild type. On the basis of their results it could be hypothesized that the level of activated RuBPcase must be significantly higher in diploid species than in hexaploids. Evans and Austin (1986) reported significant variation among species in *in vitro* specific activity. However, the difference in *in vitro* specific activity was not associated with *in vivo* leaf gas exchange (Evans, 1986). The species that showed high specific activity of RuBPcase had either B-type or S-type of cytoplasmic genome encoding the large subunit of RuBPcase. The relationship between the rate of photosynthesis and RuBPcase content has been reported. Lieckfeldt (1989) reported higher rates of photosynthesis in diploids and tetraploids than in hexaploids, and higher amounts of RuBPcase in diploids than in hexaploids, on a leaf area basis. Furthermore, when the calculations were made per leaf, the higher rates of photosynthesis in *T. dicoccoides* and *T. dicoccum* were evidently related to the higher RuBPcase content. In barley also, a strong genotypic correlation between gross photosynthesis rate and RuBPcase content has been reported recently (Ecochard *et al.*, 1991). However, we failed to detect such a relationship in wheat (Bansal *et al.*, 1991b). In our experiment conducted with seven diploid, four tetraploid and three hexaploid species, the rate of photosynthesis (area basis) decreased, whereas RuBPcase content (per leaf weight or area) increased with increase in ploidy, although the members within a ploidy group exhibited variation. In light of the recent investigation of Quick *et al.* (1991), estimation of the activation state of RuBPcase might explain the known differences in the rates of photosynthesis of different species.

IV. Use of High Photosynthesis Rate of Wild Wheat in Improving the Grain Yield Potential of Domestic Wheat

While the basic understanding of factors controlling plant productivity is being attempted (Ogren, 1991; Zelitch, 1991), the exploitation of genetic variation in leaf photosynthesis rates in breeding improved wheat varieties is of immediate concern (Fedak, 1985). Plant productivity is determined by the amount of light intercepted and by the conversion-efficiency of intercepted light to photosynthetic products. Thus, the physiological basis of improved plant productivity is the improved rate of leaf photosynthesis. The genetic improvement in grain yield of present-day wheat cultivars has been the result of improved dry-matter partitioning to grains (Gifford *et al.*, 1984) whereas the dry-matter production per se has remained more or less constant (Austin *et al.*, 1982). Further improvements in grain yields as a result of increase in dry-matter partitioning to grains seems impractical due to theoretical limits of harvest-index. This raises the strong possibility of enhancing dry-matter production per se through improved rate of leaf photosynthesis which in turn might result in higher grain yields. For obvious reasons, dry-matter production before and after anthesis determines the grain yield potential of wheat. However, recently Slafer *et al.* (1990) reported that the genetic improvement of wheat has not been associated with biomass produced during pre-anthesis, rather, it has been related to the improved partitioning of biomass to spikes. Increased partitioning of photosynthates to spikes during pre-anthesis stage might well increase the number of "sinks" in terms of number of grains. Grain number per m^2 has been shown to be most closely related with grain yield (Fischer, 1985; Bansal and Sinha, 1991). Improvement in the biomass production and partitioning to spikes during pre-anthesis period might be expected to occur in wheat genotypes bred using the species, possessing high rate of photosynthesis in early-formed leaves (Johnson *et al.*, 1987). *T. dicoccoides* (PI 428109), a wild tetraploid species, has been shown to exhibit higher rate of photosynthesis in early-formed leaves as well as in flag leaves. Johnson *et al.* (1988) attempted the cross between PI 428109 and hexaploid wheat TAM W-101. The reciprocal F_1 progenies resulting from the cross showed similar leaf photosynthesis rates suggesting that nuclear genome, rather than chloroplast genome, regulated the higher rate of photosynthesis in PI 428109. They did not, however, measure the biomass produced at anthesis or at maturity of the resulting F_1 progenies. Nevertheless, the possibility of transferring nuclear genes from PI 428109 to hexaploid wheat exists via recombination and selection. The introgression of genes from PI 428109 can be another attractive approach for raising the leaf photosynthetic capacity of hexaploid wheat (Austin *et al.*, 1984; Carver *et al.*, 1989). Recently, Carver and Nevo (1990) demonstrated a wide range of genetic diversity among native populations of *T. dicoccoides*. Some accessions were identified with exceptionally high rate of photosynthesis combined with larger leaf size. Apparently, the relationship between leaf area and rate of photosynthesis is not always negative and thus points strongly to the exploitation of *T. dicoccoides* accessions identified by Carver and Nevo (1990) in improving the grain yield potential of domestic wheat. Besides *T. dicoccoides*, *T. urartu*, a diploid wild species, is another potential source for the genes of high rates of photosynthesis. This diploid species has been shown to possess relatively high rates of photosynthesis in early-formed as well as flag leaves (Austin *et al.*, 1984). Austin *et al.* (1985) determined photosynthesis in lines derived from the cross *T. aestivum* × *T. urartu* with *T. urartu* as pollen parent. They found that some lines possessed the high photosynthetic genes of *T. urartu* and exhibited higher rate of leaf photosynthesis. On the basis of a cross between *T. durum* and *T. urartu*, it was suggested that lower rates of photosynthesis in hexaploids were possibly due to the presence of D genome and that by replacing D genome of hexaploids by A genome, an increased rate of photosynthesis could be observed despite the increase in ploidy (Austin *et al.*, 1984).

From the foregoing, it is evident that the traits that confer not only specific disease or pest resistance, but also high rate of leaf photosynthesis can be transferred from alien wheat species to domesticated hexaploid wheats. That the alien species are crossable with hexaploid wheat has been earlier demonstrated by numerous investigators. Recurrent backcrossing, screening, and cytogenetic analyses will form the basis of evolving the improved wheat cultivars possessing higher biomass and grain yield. The manipulation of photosynthetic genes at the molecular level to modify photosynthesis in transgenic plants appears to be another attractive alternative. The knowledge gathered by Chao *et al.* (1989) on the chromosomal location and copy number of genes coding for photosynthetic enzymes in wheat and its closely related species will prove useful in this regard.

V. Conclusions

Photosynthetic efficiency is the key factor determining plant productivity of different crops including wheat. However, there have been instances where photosynthetic efficiency could not be related to productivity, in particular to grain yield of wheat. The improvement in grain yield of wheat has been basically associated with dry-matter partitioning to the grains. Further improvement is likely to be achieved by increased photosynthetic efficiency combined with the maintenance of larger leaves, and increased number of "sinks" with the capacity to utilize the additional photosynthates. Significant genotypic variation exists among different wheat species with regard to photosynthetic efficiency, although higher rates of photosynthesis have been associated with smaller leaf size in wild diploid species. Nevertheless, species have been identified (e.g., *T. dicoccoides,* accession PI 428109) that possess high photosynthetic efficiency as well as larger leaves that are morphologically similar to *T. aestivum.* Furthermore, there are species available (e.g., *T. urartu, T. dicoccoides*) that exhibit higher photosynthetic efficiency throughout the plant growth duration. The inhibitory effect of D genome and the positive effect of A genome on photosynthesis have also been demonstrated at hexaploid level. It is thus suggested that interspecific hybridization using the wild species mentioned above as the potential genetic source for the improvement of photosynthetic efficiency can bring about further improvements in the existing level of wheat grain yields. Manipulation of photosynthetic genes at molecular level to modify the photosynthetic efficiency and enhance production in transgenic plants is also gaining importance.

Acknowledgements

The authors thank Dr. Ellen Kearns for going through the manuscript and Julie O'Neil, Ellie Valminuto and Joan White for meticulous typing.

VI. References

Austin RB, Bingham J, Blackwell RD, Evans LT, Ford MR, Morgan CL and Taylor M (1980) Genetic improvements in winter wheat yields since 1900 and associated physiological changes. *J Agric Sci Camb* **94:** 675–689

Austin RB, Ford MR, Morgan CL, Kamenski A and Miller TE (1984) Genetic constraints on photosynthesis and yield in wheat. In: (Sybesma, C. ed) *Advances in Photosynthesis Research*, Vol 4., Ed. Martinus Nijhoff/Dr W Junk Publishers The Hague pp 103–110

Austin RB, Morgan CL and Ford MA (1986) Dry matter yields and photosynthetic rates of diploid and hexaploid *Triticum* species. *Ann Bot* **57:** 847–858

Austin RB, Morgan CL, Ford MA and Bhagwat SG (1982) Flag leaf photosynthesis of *Triticum aestivum* and related diploid and tetraploid species. *Ann Bot* **49:** 177–189

Babu RC, Srinivasan PS, Natraja Ratnam N and Rangasamy SRS (1985) Relationship between leaf photosynthetic rate and yield in Blackgram (*Vigna mungo* (L.) Hepper) genotypes. *Photosynthetica* **19:** 159–163

Bansal KC and Abrol YP (1990) Photochemical activity of isolated chloroplast in relation to flag leaf photosynthesis in *Triticum aestivum* L. In: (M. Baltscheffsky, ed) *Current Research in Photosynthesis,* Kluwer Acad Publ, Dordrecht, The Netherlands, pp 941–943

Bansal KC and Sinha SK (1991) Assesment of drought resistance in *Triticum aestivum* L. and related species. 1 Total dry matter and grain stability. *Euphytica* **56:** 7–14

Bansal KC, Abdin MZ, Sivasanker A and Abrol YP (1991a) Genotypic variation in RuBPcase and its relationship with the leaf photosynthesis in *Triticum* and *Aegilops* species. *Photosynthetica* **25:** 303–306

Bansal KC, Sabat SC and Abrol YP (1991b) Thylakoid Hill activity in relation to ploidy in wheat. *Photosynthetica* **25:** 307–311

Bjorkman O (1968) Carboxydismutase activity in shade adapted and sun-adapted species of higher plants. *Physiol Plant* **21:** 1–10

Bugbee BG and Salisbury FB (1988) Exploring the limits of crop productivity. 1 Photosynthetic efficiency of wheat in high irradiance environments. *Plant Physiol* **88:** 869–878

Carver BF, Johnson RC and Rayburn AL (1989) Genetic analysis of photosynthetic variation in hexaploid and tetraploid wheat and their interspecific hybrids. *Photosynth Res* **20:** 105–118

Carver BF and Nevo E (1990) Genetic diversity of photosynthetic characters in native populations of *Triticum dicoccoides. Photosynth Res* **25:** 119–128

Champigny ML and Moyse A (1979) Photosynthetic carbon metabolism in wild primitive and cultivated forms of wheat at 3 levels of ploidy: Role of the glycolate pathway. *Plant Cell Physiol* **20**: 1167–1178

Chao S, Raines CA, Longstaff M, Sharp PJ, Gale MD and Dyer TA (1989) Chromosomal location and copy number in wheat and some of its close relatives of genes for enzymes involved in photosynthesis. *Mol Gen Genet* **218**: 423–430

Curtis PE, Ogren WL and Hageman RH (1969) Varietal effects in soybean photosynthesis and photorespiration. *Crop Sci* **9**: 323–327

Dean C and Leech RM (1982) Genome expression during manual leaf development 2 Direct correlation between Ribulose bisphosphate carboxylase content *Plant Physiol* **70**: 1605–1608

Deckard EL, Busch RH and Kofoid KD (1985) Physiological aspects of spring wheat improvement. In: (J Hasper, Schreder and R Howell, eds) *Exploitation of Physiological and Genetic Variability to Enhance Crop Productivity*. ASPP, Rockland MD pp 45–54

Dornhoff GM and Shibles RM (1970) Varietal differences in net photosynthesis of soybean leaves. Crop Sci. **10**: 42–45

Dunstone RL, Gifford RM and Evans LT (1973) Photosynthetic characteristics of modern and primitive wheat species in relation to ontogeny and adaptation to light. *Aust J Biol Sci* **26**: 295–307

Dunstone RL and Evans LT (1974) Role of changes in cell size in the evolution of wheat. *Aust J Plant Physiol* **1**: 157–165

Dvorak J and Zhang HB (1990) Variation in repeated nucleotide sequences sheds light on the phylogeny of the wheat B and C genome. *Proc Natl Acad Sci, USA* **87**: 9640–9644

Ecochard R, Cavalie G, Nicco C, Piquemal M and Sarrafi A (1991) Rubisco content and specific activity in barley (*Hordeum vulgare* L.). I Genetic variability. *J Exp Bot* **42**: 39–43

Ellison F, Derera NF and Pederson DG (1983) Inheritance of physiological characters associated with yield variation in bread wheat. *Euphytica* **32**: 241–255

Evans JR (1986) The relationship between carbon dioxide limited photosynthetic rate and RuBPcase content in two nuclear cytoplasm substitution lines of wheat and the coordination of RuBP carboxylation and electron transport capacities. *Planta* **167**: 344–350

Evans JR and Seemann JR (1984) Differences between wheat genotypes in specific activity of RuBP case and the relationship to photosynthesis. *Plant Physiol* **74**: 759–765

Evans LT (1981) Yield improvement in wheat: Empirical or analytical? In: (LT Evans and WJ Peacock, eds) *Wheat Science-Today and Tomorrow'* Cambridge University Press, Cambridge. pp 203–222

Evans LT (1984) Physiological aspects of varietal improvement. In: (J. Gustafson, ed) *Gene Manipulation in Plant Improvement*. University of Missouri, Columbia. pp 121–146

Evans LT and Dunstone RL (1970) Some physiological aspects of evolution in wheat. *Aust J Biol Sci* **34**: 673–680

Fischer RA (1985) Number of kernels in wheat crops and the influence of solar radiation and temperature. *J. Agric Sci Camb* **105**: 447–461

Ghildiyal MC and Sirohi GS (1986) Photosynthesis and sink efficiency of different species of wheat. *Photosynthetica* **20**: 102–106

Gifford RM (1987) Barriers to increasing crop productivity by genetic improvement in photosynthesis. In: (J, Biggens, ed) *Progress in Photosynthesis Research*, Vol IV Martinus Nijhoff Publishers, Dordrecht pp 377–385

Gifford RM, Thorne JH. Hitz WD and Giaquinta RT (1984) Crop productivity and photoassimilate partitioning. *Science* **225**: 801–808

Haour-Lurton B and Planchon C (1985) Role of D-genome chromosomes in photosynthesis expression in wheats. *Theor Appl Genet* **69**: 443–446

Harrison SA, Boerma HR and Ashley DA (1981) Heritability of canopy—apparent photosynthesis and its relationship to seed yield in soybeans. *Crop Sci* **21**: 222–226

Hieke B (1984) Photosynthetic electron transport of isolated chloroplasts and its relation to shoot biomass production in seedlings of selected evolutionary forms of wheat. *Photosynthetica* **17**: 578–589

Holbrook GP, Keys AJ and Leech RM (1984) Biochemistry of photosynthesis in species of *Triticum* of differing ploidy. *Plant Physiol* **74**: 12–15

Jellings AJ and Leech RM (1984) Anatomical variation in 1st leaves of 9 *Triticum* genotypes and its relationship to photosynthetic capacity. New Phytol **96**: 371–382

Johnson RC, Kebede H, Mornhinweg DW, Carver BF, Rayburn AL and Nguyen HT (1987) Photosynthetic differences among *Triticum* accessions at tillering. *Crop Sci* **27**: 1046–1050

Johnson RC, Carver BF, Mornhinweg DW, Kebede H, Ferris DM and Rayburn A (1988) Photosynthetic variation in *Triticum dicoccoides* accessions: physiology and genetics. *Plant Physiol Biochem* **26**: 439–444

Joseph MC, Randall DD and Nelson CJ (1981) Photosynthesis in polyploid tall Fescue. II Photosynthesis and ribulose-1, 5-bisphosphate carboxylase of polyploid tall fescue. *Plant Physiol* **68**: 894–898

Kelly H (1988) Corn bred successfully for higher photosynthetic rate. *Agric Res* **36**: 13–14

Khan MA and Tsunoda S (1970) Evolutionary trends in leaf photosynthesis and related leaf characters among cultivated wheat species and its wild relatives. *Japan J Breed* **20**: 133–140

Khan MA and Tsunoda S (1971) Comparative leaf anatomy of cultivated wheats and wild relatives with reference to their leaf photosynthetic rates. *Japan J Breed* **21**: 143–150

Kleese RA (1966) Photophosphorylation in barley. *Crop Sci* **6**: 524–527

Lawlor DW, Konturii M and Young AT (1989) Photosynthesis by flag leaves of wheat in relation to protein, RuBPcase activity and nitrogen supply. *J Exp Bot* **40**: 43–52

Lieckfeldt E (1989) Importance of leaf anatomy for characterization of primary leaf photosynthetic efficiency in different genotypes of wheat triticum. *Photosynthetica* **23**: 63–70

Maan SS and Lucken KA (1970) Interaction of *Triticum boeoticum* cytoplasm and genomes of *T. aestivum* and *T. durum*, restoration of male fertility and plant vigour. Euphytica **19**: 498–508

Malkin S, Armond PA, Mooney HA and Fork DC (1980) Photosystem II photosynthetic unit sizes from fluorescence induction in leaves: Correlation to photosynthetic capacity. *Plant Physiol* **67**: 570–579

Miflin BJ and Hageman RH (1966) Activity of chloroplasts isolated from maize inbred and their F_1 hybrids. *Crop Sci* **6**: 185–187

Miginiac-Maslow M, Hoarua A and Moyse A (1979) Hill reaction studies with protoplasts from cultivated wheats and their wild relatives. *Z Pflanzenphysiol* **95**: 95–104

Morgan CL and Austin RB (1986) Analysis of fluorescence transients of DCMU-treated leaves of *Triticum* species to provide estimates of the densities of photosystem II reaction centres. *Photosynthesis Res* **7**: 203–219

Murthy KK and Singh M (1979) Photosynthesis, chlorophyll content and ribulose diphosphate carboxylase activity in relation to yield in wheat genotypes. *A. agric Sci* **93**: 7–11

Ogihara Y and Tsunewaki K (1982) Molecular basis of the genetic diversity of the cytoplasm in *Triticum* and *Aegilops*. I. Diversity of the chloroplast genome and its lineage revealed by the restriction pattern of t. DNAS. *Japan J Genet* **57**: 371–396

Ogren WL (1991) Control of plant productivity by photosynthetic carboxylation. *Int. Conf. on the Optimization of Plant Productivity*, March 7–9, 1991, Beach, Florida (Meeting Abstr.) p 2

Parker ML and Ford MA (1982) The structure of the mesophyll of flag leaves in 3 triticum species. *Ann Bot* **49**: 165–176

Planchon C and Fesquet J (1982) Effect of the D genome and of selection on photosynthesis in wheat. *Theor Appl Genet* **61**: 359–365

Quick WP, Schurr U, Scheibe R Schulze ED, Rodermel SR, Bogorad L and Stitt M (1991) Decreased ribulose-1, 5-bisphosphate carboxylase-oxygenase in transgenic tobacco transformed with "antisense" abcs. I Impact on photosynthesis in ambient growth conditions. *Planta* **183**: 542–554

Randall DD Nelson CJ and Assay KH (1977) Ribulose bisphosphate carboxylase: Altered genetic expression in tall fescue. *Plant Physiol* **59**: 38–41

Rawson HM, Hindmarsh JH, Fischer RA and Stockman YM (1983) Changes in leaf photosynthesis with plant anatomy and relationships with yield per ear in wheat cultivars and 120 progeny. *Aust J Plant Physiol* **10**: 503–514

Riley R, Unrau J and Chapman V (1958) Evidence on the origin of the B genome of wheat. *J Heredity* **49**: 91–98

Sinha SK and Khanna R (1972) Developmental changes of photophosphorylation rate in wild and cultivated wheats. *Photosynthetica* **6**: 195–196

Sirohi GS and Ghildiyal MC (1975) Varietal differences in photosynthetic carboxylases and chlorophylls in wheat varieties. *Indian J. Exp Biol* **13**: 42–44

Slafer GA, Andrade FH and Satorre EH (1990) Genetic improvement effects on pre-anthesis physiological attributes related to wheat grain yield. *Field Crop Res* **23**: 255–264

Treharne KJ (1972) In: (AR Rees, ed) *Crop Processes in Controlled Environments*. Academic Press, New York pp 285–303

Uprety DC, Ghildiyal MC and Sirohi GS (1987) Photosynthesis among cultivated and wild genotypes within ploidy levels in wheat. *Indian J Agric Sci* **37**: 601–604

Von Caemmerer S and Farquhar GD (1981) Some relationships between the biochemistry of photosynthesis and the gas exchange of leaves. *Planta* **153**: 376–387

Waddington SR, Ransom JK, Osmanzai M and Saunders DA (1986) Improvement in the yield potential of bread wheat adapted to north west Mexico. *Crop Sci* **26**: 698–703

Wardlaw IF (1990) The control of carbon partitioning in plants. *New Phytol* **116**: 341–381

Watanabe N, Kawajiri T and Nishikawa K (1988) Contribution of D genome to the photosynthetic oxygen evolution in mesophyll protoplasts isolated from leaves of wheat seedlings. *Plant Physiol Biochem* **26**: 421–426

Zelenski MI, Mogileva G, Shitova I and Fattakhova R (1978) Hill reaction of chloroplasts from some species, varieties and cultivars of wheat. *Photosynthetica* **12**: 428–435

Zelitch I (1991) Control by regulation of photorespiration. *Int. Conf. on the Optimization of plant productivity*, March 7–9, 1991, Cocoa Beach, Florida (Meeting Abstr.). p 3

Zohary D, Harlan JR and Vardi A (1969) The wild diploid progenitors of wheat and their breeding value. *Euphytica* **18**: 58–65

23

The Significance of Light-limiting Photosynthesis to Crop Canopy Carbon Gain and Productivity—A Theoretical Analysis

S.P. Long

Department of Biology
University of Essex
Colchester, CO4 3SQ UK

CONTENTS

See Appendix to this chapter for explanation of abbreviations.

ABSTRACT

Analyses of the photosynthetic rates of CO_2 uptake (A) by crop leaves have centred on light-saturated rates (A_{sat}). Little attention has been given to the apparent quantum yield (ø), which determines the initial slope of the response of A to incident photon flux (Q), or to the convexity coefficient (Θ), which determines the duration of the transition from light-limited to light-saturated photosynthesis as light is increased. To assess the quantitative significance of these parameters of the leaf photosynthetic response to photosynthesis at the crop level, a mechanistic mathematical model was constructed which relates the individual leaf light response to the daily crop CO_2 uptake. Computer simulations of the model were conducted for canopy sizes, architectures, and light levels typical for some major tropical cereals. The simulations suggest ø rather than A_{sat} to be the major determinant of crop CO_2 uptake, under a majority of conditions. Sensitivity analysis shows that crop CO_2 uptake is only more sensitive to A_{sat} than ø under the highest light levels, and then only when leaf area index is relatively low. The simulations also suggest that if ø and Θ are decreased in parallel, as has been suggested for photoinhibition, then canopy CO_2 uptake is strongly decreased for all combinations of leaf area index and photon flux.

I. Introduction

Much research of the leaf photosynthetic rates of crops has concerned maximum rates achieved under light-saturated conditions. Although this has proved a useful criterion for identifying more productive genotypes in a few cases, in many situations maximum leaf photosynthetic rates have proved to lack any correlation with crop dry matter production (Beadle et al., 1985; Long, 1985; Long and Hällgren, 1985).

From a theoretical viewpoint this poor correlation is not surprising. In dense crop canopies, a large proportion of the leaves will remain shaded for all or most of the day. The upper exposed leaves will only receive saturating sunlight for part of the day—during the evening, in the early morning, and on cloudy days, light levels are unlikely to be sufficient to saturate photosynthesis. Even exposed leaves at the top of the canopy can receive insufficient light for saturation if they are oriented at right angles to the sun or their inclination is such that the sun's rays strike them at a shallow angle. Figure 1 illustrates the proportion of a canopy of leaf area index (L) 5 which would be sunlit on a day with a clear sky at the equator in June, assuming that the leaves within that canopy are inclined and orientated at random. Even at midday, less than 40 % of the canopy is sunlit, while less than one-quarter of the canopy is sunlit for the first and last three hours of daylight. Such considerations suggest that canopy photosynthesis and hence crop carbon gain will not only be determined by A_{sat}, but also by capacity for CO_2 uptake under light-limiting conditions.

The response of photosynthetic CO_2 uptake (A) at the leaf level to incident photon flux (Q) is commonly described by a non-rectangular hyperbola (Thornley, 1976; Baker et al., 1988; Leverenz et al., 1990). The form of the response of A to Q is determined by three parameters (Fig. 2a): A_{sat}, which determines the plateau of the hyperbola (0–50 µmol m^{-2} s^{-1}); ø, the initial and maximum slope of the response of A to Q, i.e., the apparent quantum yield (dimensionless, range 0–0.125); and Θ, the convexity coefficient for the transition from the light-limited to light-saturated photosynthesis (dimensionless, range 0–1). Photosynthetic rate under strictly light-limited conditions will depend on ø, but the extent to which ø will influence

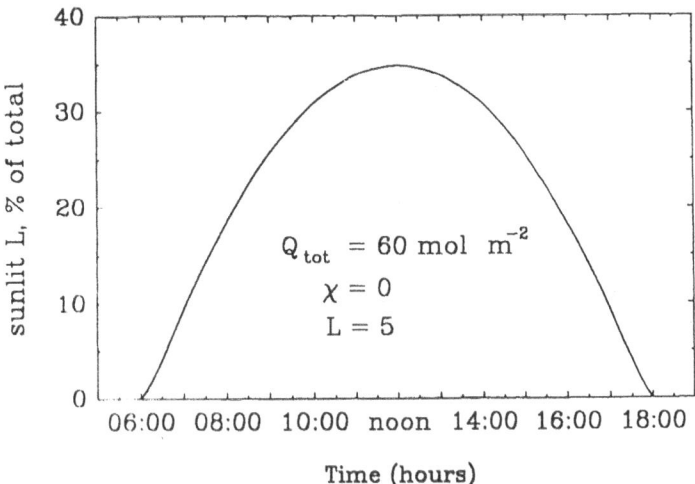

Figure 1. The percentage of leaf area within a canopy which would receive direct sunlight at different times of the day, calculated from the equations of Forseth and Norman (1992). Calculations illustrated are for a canopy of leaf area index (L) 5, leaf inclination index (χ) 0, and a clear sky day at 0° latitude with a total light receipt of 60 mol m^{-2} over the day.

A at higher photon fluxes will also depend on Θ. If $\Theta \rightarrow 1$ the curve consists of two linear phases, an initial slope $\emptyset$ followed by a sharp transition to a plateau A_{sat}. As Θ decreases, the range of photon fluxes in which there is a transition between domination of the curve by $\emptyset$ and by A_{sat} is extended.

Figure 2b illustrates the effects of variation in these parameters on the shape of the response of A to Q for an individual leaf. When A_{sat} is reduced by 50% from 30 to 15 μmol m^{-1} s^{-1}, it can be seen that A at low photon fluxes (< 500 μmol m^{-2} s^{-1}) is little affected. However, above 500 μmol m^{-2} s^{-1}, the effect of a decrease in A_{sat} becomes increasingly pronounced. By contrast, when $\emptyset$ is decreased by 50% from .06 to .03, then A is halved at low photon fluxes, but at photon fluxes close to full sunlight, A is only slightly reduced. However, if Θ, in addition to $\emptyset$, is decreased by 50% from 0.9 to 0.45, then the effect of the decrease in maximum quantum yield is evident at far higher photon fluxes.

A_{sat} is well known to vary markedly with environmental conditions and genotype (Beadle *et al.*, 1985; Long, 1985). Its magnitude is commonly co-limited by the amount of active carboxylase, the diffusion of CO_2 to the site of carboxylation, and the rate of regeneration of acceptor for carboxylation, which in turn is commonly limited by the maximum rate of electron transport *in vivo* (Long, 1985). Apparent quantum yield ($\emptyset$) is determined by the pathway of photosynthetic CO_2 assimilation (and ratio of photorespiration to photosynthesis), the efficiency of transduction of the light energy intercepted by the leaf, and the absorptance of the leaf. At normal atmospheric CO_2 concentrations the apparent quantum yield can also be reduced by closure of stomata

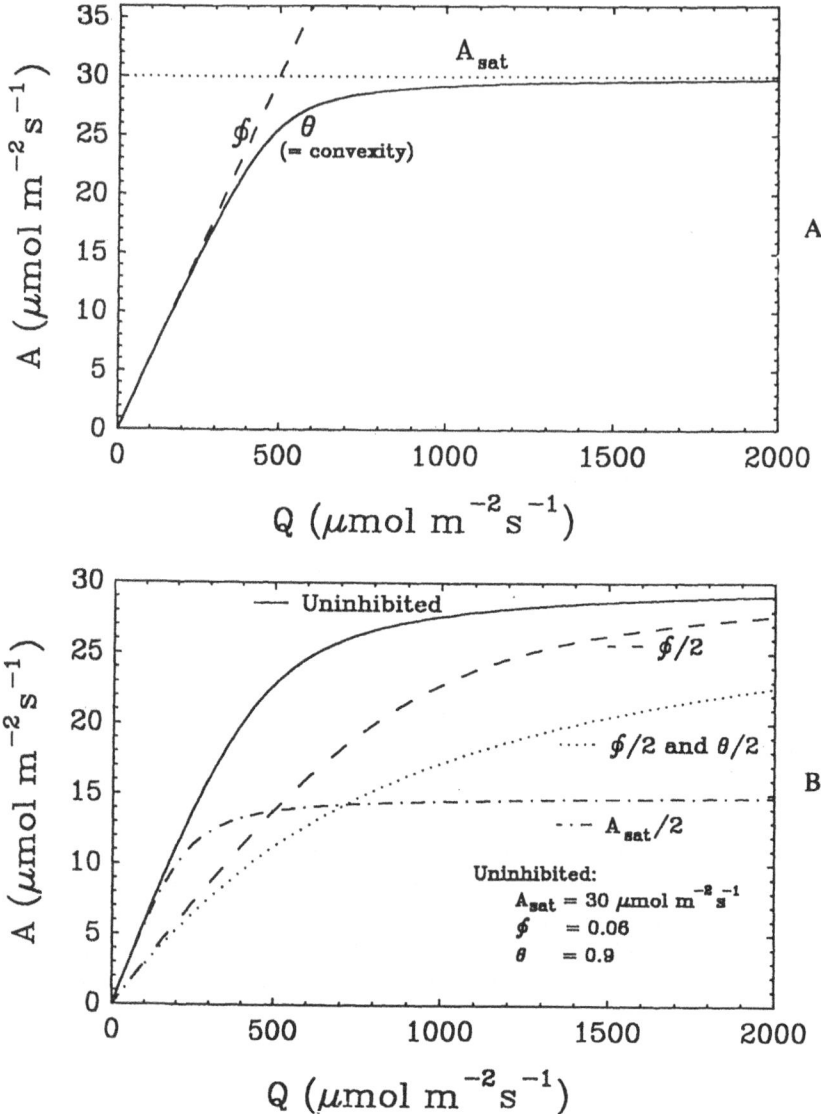

Figure 2A. An illustration of the non-rectangular form of the response of CO_2 uptake by an individual leaf (A) to incident photon flux (Q) as predicted by equation (B), indicating the three parameters determining the response: A_{sat}, ø and Θ.

2B. Upper curve, labelled uninhibited, the light response obtained from equation 2 when A_{sat} = 30 μmol $m^{-2} s^{-1}$, Θ = 0.9 and ø = 0.06. The other curves illustrate the effect of a 50% decrease in each of the parameters on the shape of the response curve.

in patches (Terashima *et al.*, 1988). Although many potential mechanisms that could alter the efficiency of energy transduction have been noted (Baker *et al.*, 1988), ø is suggested to be remarkably constant between healthy plants of different species (Ehleringer and Björkman, 1977; Pearcy and Ehleringer, 1984; Björkman and Demmig,

1987). However, photoinhibition is one mechanism which can produce marked decreases in ø (Baker *et al.*, 1988), while smaller decreases are caused by variation in absorptivity (Björkam and Demmig, 1987). Factors determining the convexity coefficient (Θ) are less well understood. The anatomical arrangement of the photosynthetic tissues will influence Θ, but photoinhibition has also been shown to decrease Θ (Leverenz *et al.*, 1990; Ögren & Sjöström, 1990).

The significance of these parameters, determining the response A to Q, to the daily integral of photosynthetic carbon assimilation by a crop ($A_{c,tot}$), will depend on the size and organization of the crop canopy and the quantity and angular distribution light received.

The objective of this analysis is to quantitatively assess the relative important of variation in A_{sat}, Θ, and ø at the individual leaf level to crop canopy photosynthesis and net carbon gain in the field, with the particular aim of assessing the importance of light-limited photosynthesis to crop carbon gain.

II. Theoretical Considerations

The relationship of A to Q at the leaf level is described by a non-rectangular hyperbola. Thornley (1976) developed a quadratic equation for this relationship incorporating A_{sat}, ø, and Θ. This equation is solved below for its positive root:

$$A = \frac{\phi \cdot Q + A_{sat} - \sqrt{[(\phi \cdot Q + A_{sat})^2 + 4\Theta \cdot A_{sat} \cdot Q]}}{2\Theta} \tag{1}$$

To assess the significance of the three parameters of this relationship to CO_2 uptake at the crop level, a computer model for the calculation of canopy rates of CO_2 gain was developed. Solar elevation was computed for 15 min intervals for June 21 at latitude 0° using the equations of Ross (1975). The incident radiation and ratios of diffuse to direct radiation were determined from solar elevation, the solar constant, and atmospheric transmissivities, the latter being varied to produce photon fluxes ranging from those expected on a day with a clear sky to that expected on a day with sufficient cloud to decrease the total incident light (Q_{tot}) by 80%. The proportion of radiation in the photosynthetically active waveband was calculated from the tables of Ross (1975) and the mean energy of photons in this waveband for daylight assumed to be 2500 kJ mol^{-1} (Kubin, 1971). The canopy was assumed to have a foliar inclination index (X) of 0, i.e., a random distribution of leaves with respect to inclination giving a mean inclination angle of 45°. Foliar inclination index (X) follows the definition of Ross (1975). Leaf area indices (L) were varied in the simulations from 0.5 to 10. Given Q_{tot}, solar angle, ratio of diffuse to direct light, L and X, the proportions of the canopy which were sunlit and shaded could be computed. The L_{sun} and L_{shade} and the mean photon fluxes of these two categories (Q_{sun} and Q_{shade}) of leaf within the canopy were calculated for 15 min intervals through the day by the procedures of Norman (1980) and Forseth and Norman (1991). From the estimated Q_{sun} and Q_{shade} and A for both categories of leaf was estimated from equation (1). The daily total of gross photosynthetic CO_2 fixation ($P_{c,tot}$) was then obtained by integrating over the day:

$$P_{c,tot} = \int_{0}^{t=24h} fQ_{sun} \cdot L_{sun} + fQ_{shade} \cdot (L - L_{sun}) \qquad (2)$$

where f indicates A as a function of Q calculated from equation (1).

Integration was conducted numerically, using the Euler method (Jeffers, 1978), with 15 min interval steps.

Respiratory losses (R) over the same 24 hr period were computed following the rationale of McCree and von Bavel (1975). Here respiration is assumed to be a function of gross carbon assimilation and mass of the crop. The former constitutes "growth respiration", the latter "maintenance respiration"; only the latter shows a temperature dependence (Charles-Edwards, 1982). In these calculations R was determined by:

$$R = a \cdot P_{c,tot} + b \cdot L \cdot F^{-1} \qquad (3)$$

Values of a and b at 25°C were those derived by McCree and von Bavel (1975) for a *Sorghum* crop and F was assumed to be 0.01 m^2 g^{-1}. To correct daily canopy carbon gain, net of respiratory losses:

$$A_{c,tot} = P_{c,tot} - R \qquad (4)$$

By combining equations (2), (3) and (4), $A_{c,tot}$ was determined directly:

$$A_{c,tot} = -b \cdot L \cdot F^{-1} + (1-a) \int_{0}^{t=24h} fQ_{sun} \cdot L_{sun} + fQ_{shade} \cdot (L-L_{sun}) \qquad (5)$$

To examine the relative significance of changes in ø, Θ, and A_{sat}, sensitivity analyses were conducted. Values of the three parameters were decreased in sequence and in combination by 50% to numerically assess their significance to $A_{c,tot}$ for different canopy sizes (L) and different total photon fluxes (Q_{tot}). Realistic values for healthy and uninhibited leaves were assumed to be ø = 0.06 Θ = 0.9, and A_{sat} = 30 µmol m^{-2} s^{-1} (Pearcy and Eheringer, 1984; Long, 1985; Leverenz et al., 1990; Long et al., 1990 Long and Drake, 1991). The simulation assumed that all leaf and canopy parameters remained constant through the day and that temperature within the canopy was constant. Senstivity(s) of total canopy carbon gain ($A_{c,tot}$) to change in ø, Θ, and A_{sat} was assessed:

$$s = \frac{A'_{c,tot} - A_{c,tot}}{A_{c,tot}} \qquad (6)$$

The effect of variation in L on $A_{c,tot}$ was assessed for a clear sky day with a Q_{tot} of 60 µmol m^{-2} s^{-1} and a dull day with 20 µmol m^{-2} s^{-1}, these values being at the extremes of photon flux receipts for the equator at this time of year. The effect of variation in Q_{tot} was assessed by assuming an L of 5, which would be typical of a well-developed crop under conditions of good nutrition and water supply (Beadle *et al.*, 1985). Values of L and X selected are within the range that could be expected of tropical crops such as maize, sorghum, and rice in the tropics (Ross, 1975; Beadle *et al.*, 1985).

All simulations of the model were conducted via a program written in QuickBASIC v.4 using double precision real numbers and executed on 80386 based microcomputer (DeskPro 386/20, Compaq).

III. Specific Case Studies

The effect of indiviual decreases in ø, Θ and A_{sat} on the instantaneous rate of canopy photosynthesis over the course of a day are illustrated (Fig. 3). This simulation

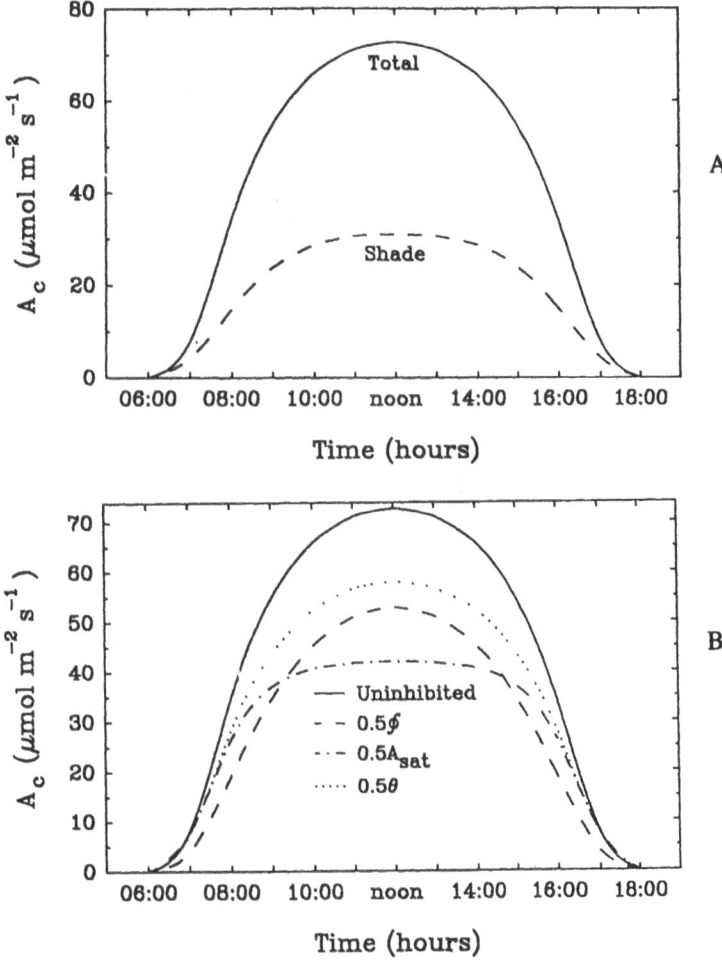

Figure 3A. Simulated diurnal course of CO_2 uptake rate by a crop canopy on a unit ground area basis (A_c); L = 3. The simulation is for day with a total photon flux of 40 mol m^{-2} at the equator.
3B. Indicates the contribution of the shaded leaves within the canopy and the upper curve the total for shaded and unshaded leaves combined. But indicates the effect that a 50% decrease in A_{sat}, Θ, or ø would have on canopy photosynthesis.

is for a leaf area index (L) of 3 and a day with a photon flux of 40 mol m^{-2} (around 20 MJ m^{-2}), which might be typical of this time of year at the equator (e.g., Piedade *et al.*, 1991). Figure 3a shows that even under these sunny conditions, shaded leaves contribute 45% of total canopy photosynthesis over the course of the whole day, and considerably more during the early and later parts of the day. A 50% decrease in A_{sat} has little effect on the instantaneous rate of canopy photosynthesis (A_c) in the early and late part of the day. However, it does result in a marked depression of A_c at midday (Fig. 3b). By contrast a 50% decrease in ø produces the larger decrease in A_c early and late in the day, but because a significant proportion of the canopy is shaded throughout the day, it also produces a substantial depression of A_c in the middle of the day (Fig. 3b). The importance of the convexity of the transition between light-limited and light-saturated photosynthesis is illustrated by the effect of a 50% decrease in the convexity coefficient (Θ), without any change in either ø or A_{sat}. This decreases A_c throughout the day, and the loss integrated over the course of the day is 19% relative to that which would be assimilated by a canopy in which Θ is unchanged. A 50% decrease in A_{sat} produces a 34% decrease in assimilation over the day compared to a 32% decrease for a 50% decrease in ø. Thus, under these conditions, canopy photosynthesis appears similarly sensitive to A_{sat} and ø. Clearly, with increasing L and decreasing Q, ø will assume greater importance—this is illustrated in Figs. 4 and 5.

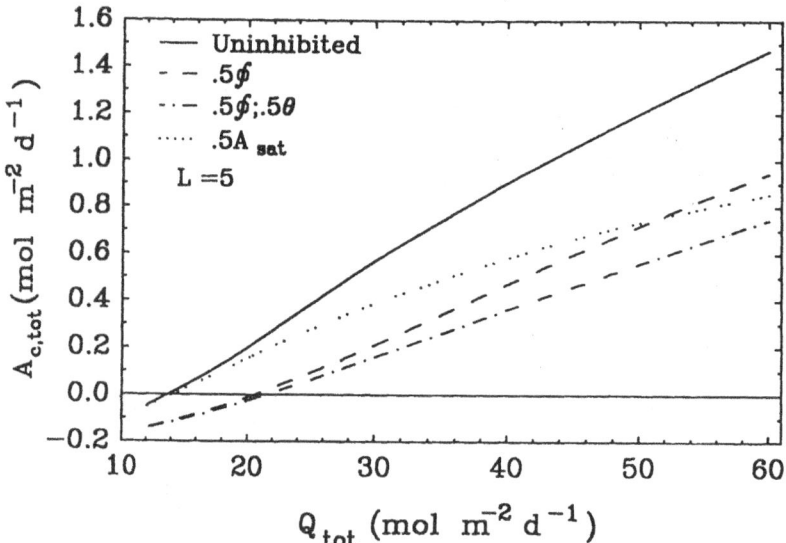

Figure 4. The response of net crop carbon exchange per unit ground area ($A_{c,tot}$) to the daily total of photon flux (Q_{tot}). The leaf area index is 5. As in previous simulations, "uninhibited" assumes that for all leaves $A_{sat} = 30$ µmol m^{-2} s^{-1}, $\Theta = 0.9$, and ø $= 0.06$. The effect of a 50% decrease in each of these parameters on the relationship of $A_{c,tot}$ with Q_{tot} is illustrated.

For L = 3 and the highest light levels likely at 0° of latitude in June, sensitivity analysis of $A_{c,tot}$ shows variation in A_{sat} to be the more important determinant.

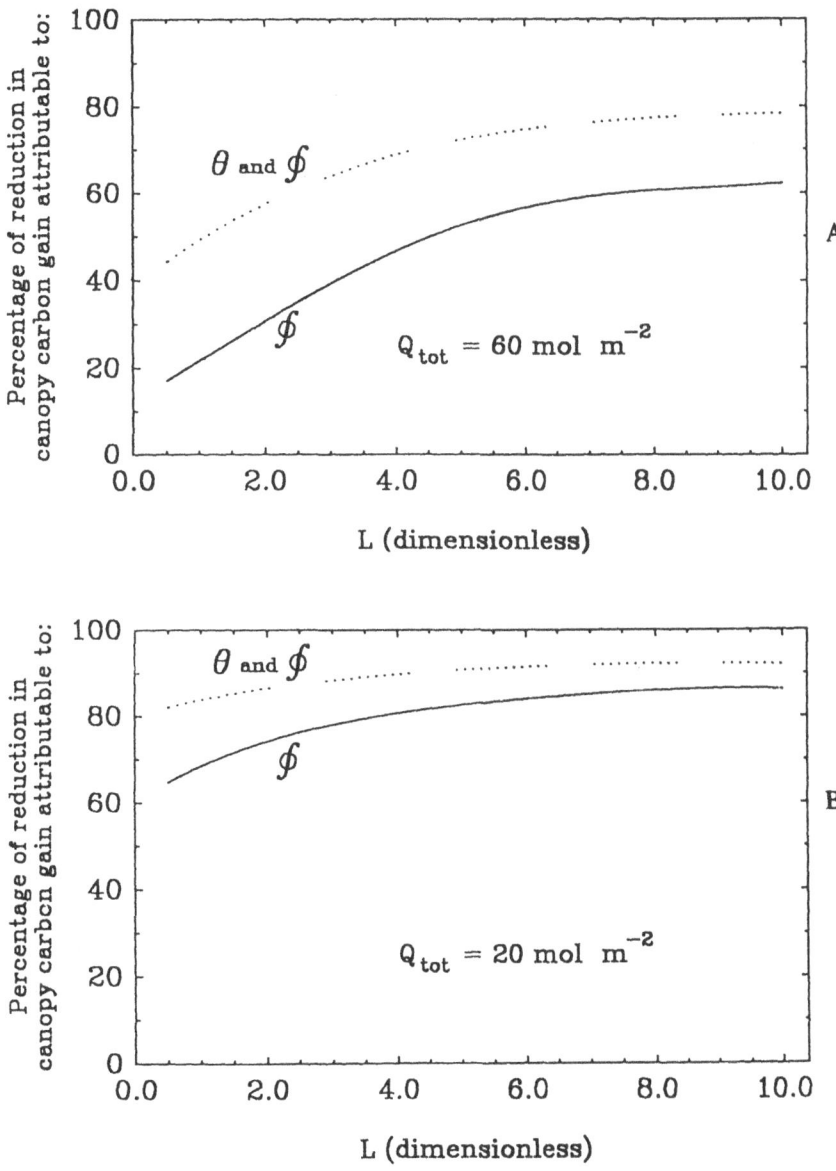

Figure 5A. The proportion of the decrease in $A_{c,tot}$ produced by a 50% decrease in A_{sat}, Θ and ϕ that can be attributed to a decrease in ϕ alone or to combined decrease in the ϕ and Θ in absence of change in A_{sat} as a function of leaf area index (L). Q_{tot} for this simulation was 60 mol m^{-2}.
5B. as for a., but Q_{tot} = 20 mol^{-2}.

However, even under these high light conditions sensitivity to A_{sat} is only around 50% greater than to ϕ (Table 1). Under light levels equivalent to dull days at the equator, $A_{c,tot}$ is affected 2.5 times more by variation in ϕ than by variation in A_{sat} (Table 1).

Table 1. A sensitivity analysis of canopy carbon gain $A_{c,tot}$ to variation in the three parameters of the leaf light response curve.

	ø	Θ	A_{sat}
Clear sky	0.31	0.19	0.45
Cloud cover	0.71	0.23	0.27

Figure 4 illustrates the effect of decreases in ø, Θ, and A_{sat} at the leaf level on $A_{c,tot}$ over the range of likely light levels for a canopy of L = 5. At the lowest daily photon fluxes (10 mol m^{-2} d^{-1}), a 50% decrease in A_{sat} has virtually no effect on $A_{c,tot}$, and only at the highest light levels is a 50% decrease in A_{sat} more significant to $A_{c,tot}$ than a 50% decrease in ø. If a decrease in ø is combined with a decrease in Θ, as may occur in photoinhibition *in vivo* (Leverenz *et al.*, 1990; Ögren & Sjöström, 1990), then the decrease in $A_{c,tot}$ is more marked, and is greater than the effect of a 50% decrease in A_{sat} at all photon fluxes. This greater importance of a depression in ø and Θ is also apparent if L rather than Q_{tot} is varied (Fig. 5).

If ø, Θ, and A_{sat} are all decreased by 50% how much of the decrease is attributable to light-limited photosynthesis, i.e., ø? This may be assessed by:

$$p = \frac{A_{c,tot} - A'_{c,tot}}{A_{c,tot} - A''_{c,tot}} \tag{7}$$

where p is the proportion of the decrease that would occur with a decrease in ø alone; $A_{c,tot}$ is the net CO_2 gain with unchanged values of ø, Θ, and A_{sat}; $A''_{c,tot}$ is the net CO_2 gain with a 50% decrease in ø, Θ and A_{sat} and $A'_{c,tot}$ is the net CO_2 gain with a 50% decrease in ø

Since a decrease in Θ will increase the influence of variation of ø at higher photon fluxes, the proportion of the total decrease that could be attributed to a combined decrease in ø and Θ was also assessed. Figure 5a shows that under high light conditions, ø does not account for more than 50% of the total decrease in $A_{c,tot}$ until L → 5, but when a combined decrease in ø and Θ is considered then a 50% decrease in $A_{c,tot}$ is effected at L < 1, even though A_{sat} has not been decreased. At L = 8 combined decrease of ø and Θ accounts for 80% of the decrease in $A_{c,tot}$, even under these high light conditions. At a low photon flux, decrease in ø accounts for over 60% of the decrease in $A_{c,tot}$ even in an open canopy of L = 0.5, while at L ≥ 5 it accounts for over 80% of the decrease in $A_{c,tot}$. When decrease in ø and Θ are considered in combination they account for more than 80% of the decrease in $A_{c,tot}$ at all values of L, indicating that variation in A_{sat} under these conditions is of little significance (Fig. 5b).

These simulations suggest that capacity for photosynthesis under light-limiting conditions is a major factor and often the dominant factor determining whole crop photosynthetic carbon gain. Simulations were conducted for photon fluxes typical within the tropics; at lower photon fluxes, the influence of light-limited photosynthesis could be expected to be even greater. Although June was chosen for these simulations, the range of Q_{tot} values used vary little from those obtained in practice at the equator at other times of year (e.g., Piedade *et al.*, 1991). The simulation did not take account of the diurnal course of temperature. However, in an equatorial climate this would be small, and given a mean value of 25°C, variation around this would be expected to have only a small effect on A_{sat} and ∅ (Long, 1985). The foliar inclination index of 0 is similar to the value found for all but the most erect-leaved cultivars of the major cereals. Crops such as sunflower, which show lower mean inclination angles, could be expected to show an even greater dependence on ∅ at similar values of L, since here light will be more strongly absorbed by the upper layers of the canopy.

As noted in the inroduction, $∅_{abs}$ is remarkably constant between species. However, variability in absorptivity of leaves increases the variability of ∅ (Björkman and Demmig, 1987). Lack of variability in ∅ and the critical role of ∅ in determining $A_{c,tot}$ could be linked. Figure 4 shows that $A_{c,tot}$ is critically affected by decrease in ∅ at low photon fluxes. A canopy of leaves in which A_{sat} is decreased by 50% shows a light compensation point for $A_{c,tot}$, i.e., a positive carbon gain over 24 hr, at $Q_{tot} = 13.5$ mol m^{-2} d^{-1}, and only 0.2 mol m^{-2} d^{-1} greater than the compensation point for a canopy in which the parameters have not been reduced. However, if ∅ is decreased by 50% then the compensation point rises to 21.6 mol m^{-2} d^{-1}. It is likely that there are strong selective pressures against any increase in light compensation point since this will determine the point below which a plant is no longer able to maintain a net positive carbon balance and therefore survive. Thus, while a decrease in A_{sat} may decrease productivity, a decrease in ∅ will not only decrease productivity, but directly influence an individual's survival. This may provide one evolutionary explanation of the wide variation in A_{sat} between species compared to the narrow variation in ∅.

While ∅ varies little among plants grown under optimal conditions, variation in ∅ is induced by suboptimal growth conditions, in particular those producing photoinhibition (Baker *et al.*, 1988). This situation provides empirical evidence of the significance of ∅ to crop carbon gain. Photoinhibition is apparently common where low temperatures and relatively high light levels coincide. In crops of *Zea mays* and *Brassica napus* grown in southern England, decrease in ∅ under low temperature conditions has been shown to be closely correlated ($r^2 > 0.88$) with decrease in the efficiency with which the crop utilizes intercepted radiation in the production of new dry matter (Long *et al.*, 1990). These responses, however, were obtained in crops of low L growing during periods of relatively low light receipt. In denser canopies photoinhibition will not be evenly distributed through the canopy, affecting the upper leaves to the greatest extent. Decrease in ∅ in the upper leaves will be of less significance to canopy photosynthesis than to the lower leaves. However, if the convexity of the light response curve of photosynthesis is simultaneously decreased by photoinhibition (e.g., Ögren and Sjöström, 1990), then the effect of a decreased ∅ would extend to higher photon fluxes. Relatively little is known of variability in Θ and the exact factors determining its magnitude. The results suggest that improved understanding of the factors maintaining a high ∅ and Θ, and hence capacity for CO_2 uptake under light-limiting conditions, will be critical to understanding the potential for engineering crops with improved potential for canopy`carbon gain.

558

IV. Appendix—Abbreviations

a   : growth respriation constant (dimensionless)

A   : rate of CO_2 uptake per unit projected area of leaf (μmol m^{-2} s^{-1})

A_{sat} : A at light-saturation (μmol m^{-2} s^{-1})

A_c : instantaneous rate of canopy photosynthetic carbon uptake per unit of ground area (μmol m^{-2} s^{-1})

$A_{c,tot}$: daily integral of canopy photosynthetic CO_2 uptake net of respiratory losses (mol m^{-2} d^{-1})

$A'_{c,tot}$: $A_{c,tot}$ computed with a 50% decrease in either A_{sat}, ø, or Θ

$A''_{c,tot}$: $A_{c,tot}$ computed with a 50% decrease in A_{sat}, ø, and Θ

b   : maintenance respiration constant (d^{-1})

F   : leaf area ratio, i.e., the ratio of leaf area to dry weight of a crop (g m^{-2})

L   : leaf area index, i.e., projected leaf area per unit ground area (dimensionless)

L_{shade} : the L shaded from direct sunlight at any point in time

L_{sun} : the L illuminated by direct sunlight at any point in time

$P_{c,tot}$: daily integral of canopy photosynthetic CO_2 uptake uncorrected for respiratory losses (mol m^{-2} d^{-1})

Q   : photosynthetically active photon flux (μmol m^{-2} s^{-1})

Q_{sun} : mean Q over the sunlit leaves of a canopy

Q_{shade} : mean Q over the shaded leaves of a canopy

Q_{tot} : the photon flux accumulated over one day (mol m^{-2} d^{-1})

R   : dark respiration rate per unit ground area of crop (μmol m^{-2} s^{-1})

s   : proportionate change in $A_{c,tot}$ given a 50% decrease in either A_{sat}, ø, or Θ

X   : foliar inclination index (dimensionless)

ø   : the maximum apparent quantum yield, i.e., ratio of CO_2 molecules absorbed per photon incident on the leaf under light-limiting conditions (dimensionless)

$ø_{abs}$: ø calculated on an absorbed light basis

Θ : the convexity coeffecient of the response of A to Q (dimensionless)

V. References

Baker NR, Long SP and Ort DR (1988) The effects of temperature on photosynthesis. In: (SP Long and FI Woodward, eds.) *Plants and Temperature*, Cambridge University Press, Cambridge, pp 347–375

Beadle CL, Long SP, Imbamba SK, Hall DO and Olembo RJ (1985) *Photosynthesis in Relation to Bioproductivity*, UNEP/Tycooly International, Oxford. pp 291

Björkman O and Demmig B (1987) Photon yield of O_2 evolution and chlorophyll fluorescence characteristics at 77K among vascular plants of diverse origins. *Planta* **170:** 489–504

Charles-Edwards DA (1982) *Physiological Determinants of Crop Growth*, Academic Press, Sydney

Ehleringer JR and Björkman O (1977) Quantum yields for CO_2 uptake in C_3 and C_4 plants. *Plant Physiol* **59:** 86–90

Forseth I and Norman JM (1992) Light and plant canopies. In: (DO Hall, JMO Scurlock, HR Bolhar-Nordenkampf, RC Leegood and SP Long, eds) *Techniques in Bioproductivity and Photosynthesis*, Chapman and Hall, London: in press

Jeffers JNR (1978) *An Introduction to Systems Analysis: With Ecological Application*, Edward Arnold, London

Kubin S (1971) Measurement of radiant energy. In: (Z Sestak, J Catsky, PG Jarvis, eds) *Photosynthetic Production: Manual of Methods*, Dr. W. Junk, The Hague, pp 702–763

Leverenz JW, Falk S, Pilström C-M and Samuelsson G (1990) The effects of photoinhibition on the photosynthetic light-response curve of green plant cells (*Chlamydomonas reinhardtii*). *Planta* **182:** 345–357

Long SP (1985) Leaf gas exchange. In: (J Barber and NR Baker, eds) *Photosynthetic Mechanisms and the Environment*, Elsevier, Amsterdam, pp 453–499

Long SP and Drake BG (1991) The effect of the long-term elevation of CO_2 concentration in the field on the quantum yield of photosynthesis of the C_3 sedge: *Scirpus olneyii*. *Plant Physiol* **96:** 221–226.

Long SP, Farage PK, Groom Q, Macharia JMN and Baker NR (1990) Damage to photosynthesis during chilling and freezing, and its significance to the photosynthetic productivity of field crops. *Curr Res Photosynth* **4:** 853–842

Long, SP and Hällgren J-E (1985) Measurement of CO_2 assimilation by plants in the field and the laboratory. In: (J Coombs, DO Hall, SP Long, JMO Scurlock, eds) *Techniques in Bioproductivity and Photosynthesis*, ED 2, Pergamon, Oxford, pp 62–94

McCree KJ and van Bavel, CHM (1977) Respiration and crop production: A case study with two crops under water stress. In: (JJ Landsberg, CV Cutting, eds) *Evironmental Effects on Crop Physiology*, Academic Press, London, pp 199–216

Norman, JM (1980) Interfacing leaf and canopy light interception models. In: (JD Hesketh and JW Jones, eds) *Predicting Photosynthesis for Ecosystem Models*, Vol 2, CRC Press, Boca Raton, pp 49–67

Ögren E and Sjöström M (1990) Estimation of the effect of photoinhibition on carbon gain in leaves of a willow canopy. *Planta* **181**: 560–567

Pearcy RW and Ehleringer J (1984) Comparative ecophysiology of C_3 and C_4 plants. *Plant Cell Environ* **7**: 1–13

Piedade MTF, Junk WJ and Long SP (1991) The productivity of the C_4 grass. *Echinochloa polystachya* on the Amazon floodpain. *Ecology* **72**: 1456–1463

Ross J. (1975) Radiative transfer in plant communities. In: (JL Monteith, ed) *Vegetation and the Atmosphere* Vol I, Academic Press, London, pp 13–56

Terashima I, Wong S-C Osmond CB and Farquhar GD (1988) Characterisation of non-uniform photosynthesis induced by abscisic acid leaves having different mesophyll anatomies. *Plant Cell Physiol;* **29**: 385–394

Thornley JHM (1976) *Mathematical Models in Plant Physiology*, Academic Press, London, p. 315

24

Leaf Photosynthesis in Rice in Relation to Grain Yield

Ryuichi Ishii

Laboratory of Crop Science
Faculty of Agriculture
The University of Tokyo, Yayoi
Bunkyo-ku, Tokyo 113, Japan

CONTENTS

ABBREVIATIONS

CPS	:	Canopy photosynthesis
LAI	:	Leaf area index
LPS	:	Leaf photosynthesis, or apparent rate of photosynthetic CO_2 uptake per unit leaf area
SLW	:	Specific leaf weight
RuBP	:	Ribulose-1, 5 bisphosphate
RuBPcase	:	Ribulose-1, 5 bisphosphate carboxylase
RGR	:	Relative growth rate
Rs	:	Stomatal resistance
Rm	:	Mesophyll resistance
Rc	:	Carboxylation resistance
Rt	:	Transportation resistance
SLA	:	Specific leaf area

562

ABSTRACT

Differences in leaf photosynthesis and their relationship to grain yield in *Oryza* species are discussed. During the evolutionary process, improvement of biochemical reactions of photosynthesis took place toward the direction of increase in CO_2 fixing reaction catalyzed by RuBPcase. Most advanced *Oryza* species acquired the characteristics of high leaf photosynthesis for longer duration and efficient partitioning of photosynthetic products to the grains. Amongst the different groups of varieties, higher yield was associated with high respsonse of photosynthesis to nitrogen and maintenance of high photosynthetic activity of the flag leaf during the grain filling period. Studies on F_1 heterosis in photosynthesis suggest that high yields could be associated with higher integrated photosynthesis of the flag leaf during the ripening period.

I. Introduction

It can be assumed that crop growth rate in the field is in parallel with the net CO_2 gain or CO_2 balance of the crop population. The net CO_2 gain is defined as the difference between the apparent photosynthetic CO_2 uptake (LPS) in the day and the respiratory CO_2 release in the night by the canopy. The apparent canopy photosynthesis (CPS) is the important process which determines crop growth rates and eventually the crop yield. The CPS depends on the following three factors:

(1) Leaf area index (LAI).
(2) Canopy architecture influencing light penetration.
(3) Apparent rate of photosynthetic CO_2 uptake per unit leaf area.

The maintenance of high LAI has been realized in the modern high-yielding varieties tolerant to lodging under adequate fertility and high plant density. Canopy structure was also improved by development of varieties with erect leaves so that incident solar radiation can penetrate into the deep level of canopy.

It is not yet concluded, however, whether LPS has been improved through the breeding process of rice. Many crop scientists in the world, of course, have been interested in the improvement of LPS, expecting that it would contribute to the increase of crop growth or crop yield. Evans and Dunstone (1970) reported that in the domestication process of wheat (*Triticum aestivum*), LPS showed a decreasing trend, although LAI showed an increasing one in the flag leaves. However, Yoshida (1973) reported in rice plants that if LPS, particularly in the grain filling stage, was activated by the CO_2 enrichment in the air, the grain yield increased through the increased filled grain percentage. This suggests that there is a close relationship between LPS and yield in rice, or that LPS can be a breeding target for the high yielding varieties.

II. Differences in Leaf Photosynthesis and Yield in *Oryza* Species

a. *Leaf Photosynthesis*

Cook and Evans (1983) studied the photosynthetic characteristics of wild and cultivated species of genus *Oryza* and found that, in general, the LPS at the seedling stage was high in the Asian cultivated species, *O. sativa* as compared to the Asian and African wild species, or to the African cultivated species, *O. glaberrima*. Between the subspecies in *O. sativa*, upland indica rice and javanica rice showed a relatively low LPS.

Table 1. Photosynthetic rates and specific leaf weight in *Oryza* species (Cook and Evans, 1983)

Category	Photosynthesis mg dm^{-2} $^{-1}$	SLW (mg cm^{-2})
O. rufipogon	35.2 ± 1.5	3.17 ± 0.11
O. nivara	33.4 ± 0.6	2.66 ± 0.06
O. spontanea	37.8 ± 2.9	3.46 ± 0.44
O. sativa		
indica		
primitive	37.6 ± 1.3	3.16 ± 0.08
upland	34.2 ± 0.5	2.76 ± 0.05
intermediate	41.0 ± 1.5	3.25 ± 0.09
"Taiwan"	39.8 ± 1.5	2.94 ± 0.08
advanced	39.2 ± 1.2	2.92 ± 0.07
indica × japonica	43.7 ± 1.2	3.39 ± 0.06
japonica		
old	41.1 ± 0.6	2.98 ± 0.09
advanced	40.5 ± 2.0	3.02 ± 0.07
javanica	34.9 ± 0.7	3.00 ± 0.03
O. barthii	34.8 ± 0.5	2.65 ± 1.69
O. stapfii	30.2 ± 2.3	2.51 ± 0.17
O. glaberrima	30.8 ± 0.8	2.63 ± 0.03

They further suggested that the interspecific difference of LPS in *Oryza* species could be attributed to the difference in nitrogen content per unit leaf area, and to the specific leaf weight (SLW), an index of leaf thickness. Apparently, therefore, it is possible to achieve in thicker leaves which contain more nitrogen per unit leaf area. Saka (1985) found that ribulose 1,5-bisphosphate carboxylase (RuBPcase) activity was high in the cultivated and its closely related species, compared to other wild *Oryza* species. He also found that there was practically no difference in the ratio of RuBP carboxylase to oxygenase between the *Oryza* species, suggesting that the relative rate of photorespiration to photosynthesis did not change through the domestication process. Makino *et al.* (1987) compared the concentration and kinetic characteristics of RuBPcase in *Oryza* species. As shown in Table 2, the ratio of RuBPcase against total soluble protein was 52 % in *O. sativa* and its ancestor, *O. perennis*, while the ratio was as low as 42–49 % in *O. glaberrima*, *O. breviligulata* and all other wild species. They also found that k_m (CO_2) was low in *O. glaberrima*, compared to other *Oryza* species.

These observations imply that in the evolutionary process of *Oryza*, the improvement of biochemical reaction of photosynthesis took place toward the direction of the increase in CO_2 fixing reaction catalysed by RuBPcase.

b. *Relation with Grain Yield*

Cook and Evans (1983) examined the relationship between LPS and growth rate in *Oryza* species, and found that during vegetative stage the relative growth rate (RGR) showed a negative correlation with LPS, while it showed a positive correlation

564

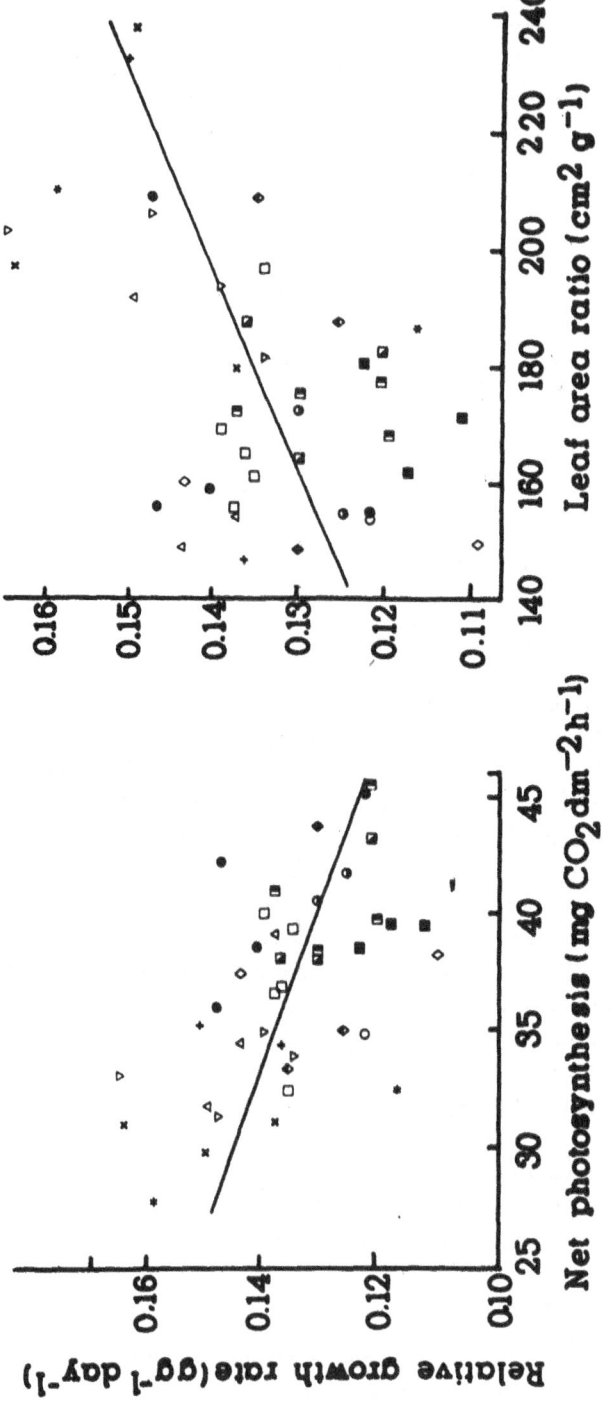

Figure 1. Correlations between relative growth rate and leaf photosynthesis (left), and leaf area ratio (right) in *Oryza* species (Cook and Evans, 1983).

Symbols: cultivars examined: *O. rufipogon* (Δ), *O. nivara* (▽), *O. spontanea* (◐), primitive (θ), upland (◊), intermediate (▣), "Taiwan" (o) and advanced (o) indicas, old (o) and advanced ( ) japonicas, *indica* × *japonica* (◆), *javanica* (o), *O. barthii* (+), *O. stapfii* (*) and *O. glaberrima* (×).

Table 2. Variations in the kinetic constants of RuBPcase (25°C, pH 8.0), and the ratio of enzyme protein to soluble protein among *Oryza* species (Makino *et al., 1987*)

Species	Genome	$K_m(CO_2)$ (µM)	V_{max} (units mg^{-1} of carboxylase)	Enzyme per sol protein (%)	Number of varieties
O. sativa L.	AA	10.2 ± 1.0	1.72 ± 0.13	52 ± 2	25
O. perennis Moench	AA	10.6 ± 1.0	1.84 ± 0.11	52 ± 1	2
O. glaberrima Steud.	A^sA^s	7.8 ± 0.3	1.66 ± 0.06	49 ± 0	3
O. breviligulata A. Chev. et Roehr.	A^sA^s	8.3 ± 1.2	1.67 ± 0.08	47 ± 2	2
O. punctata Korschy et Steud.	BB	12.0 ± 1.2	1.68 ± 0.07	44 ± 4	2
O. minuta J.S. Presl. ex Chandl.	BBCC	9.6	1.98	49	1
O. eichingeri A. Peter	CC	10.2 ± 0.7	1.95 ± 0.04	42 ± 0	2
O. officinalis Wall. wx Watt.	CC	11.6 ± 0.7	2.23 ± 0.03	45 ± 3	2
O. latifolia Desv.	CCDD	9.7 ± 0.1	1.84 ± 0.11	45 ± 3	2
O. grandiglumis (Doell) Prod.	CCDD	10.0	1.79	47	1
O. australiensis Domin	EE	11.3 ± 0.4	1.79 ± 0.24	46 ± 3	2
O. brachyantha A. Chiev. et Roehr	FF	11.7 ± 0.8	1.77 ± 0.13	47 ± 0	2

The kinetic constants were determined according to Wilkinson (1961).
±SE of the difference in the same species.

with leaf area ratio (Fig. 1). This would mean that the species maintain high RGR in the vegetative stage by developing the large leaf area, rather than by exploiting high LPS. On the other hand a positive correlation between LPS at flag leaf after heading and the grain yield (Fig. 2) indicates the importance of the current photosynthetic activity in determining the genotypic difference in grain yield. This contention is supported by the observation on *O. sativa* which maintains high LPS for longer duration after heading than other *Oryza* species. This finding showed that *O. sativa* also had high ripening percentage and high harvest index.

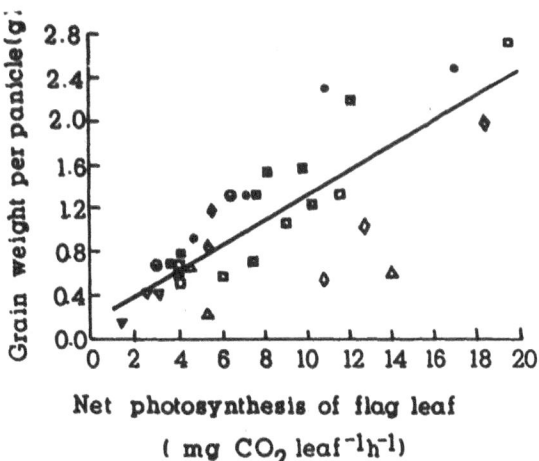

Figure 2. Relationship between grain yield per panicle and flag leaf photosynthesis one to two weeks after anthesis in *Oryza* species (Cook and Evans, 1983). Symbols are the same as in Fig. 1.

It could be concluded that *O. sativa,* the most advanced species in *Oryza,* acquired the characteristics of high LPS for longer duration and efficient partitioning of the photosynthetic products to the grain, through the domestication process.

III. Varietal Difference of Leaf Photosynthesis and Its Mechanism in *Oryza sativa*

a. *Leaf Photosynthesis*

Many rice researchers in Japan have been interested in the varietal differences in LPS, expecting that this can be a breeding criterion for achieving high yields. Osada and Murata (1962a, b, 1965a, b) compared LPS between the nitrogen tolerant and nitorgen intolerant varieties and found that the heavy application of nitrogen fertilizer increased LPS more in the nitrogen-tolerant than in the nitrogen-intolerant varieties without affecting respiration, and resulting in increased ratio of photosynthesis to respiration in such varieties. The japonica-indica hybrid varieties recently developed in Korea, which were known for their high response to nitrogen, also showed a large response of LPS to nitrogen (Cho and Murata, 1980; Cho *et al.*, 1980). These data suggest that the breeding for high-yielding varieties by endowing them with high responsiveness to nitrogen fertilizer was accompanied by improvement in LPS.

Sasaki *et al.* (1986) examined differences in LPS in 32 varieties which were bred in the 19th century in Japan, and found that there was no significant difference in LPS of flag leaf between old, middle, and new variety groups at the heading time when the flag leaf showed the maximum potential LPS (Table 3). However, during the grain filling period, LPS of the flag leaf showed large difference between the groups, with the highest value in the new variety group and the lowest in the old group. It was also interesting to observe that nitrogen top-dressing prevented decline in LPS due to delay in leaf senescence during the grain filling period in the middle and new varieties, although in the old varieties such an effect of nitrogen top-dressing was not observed. Further experiments (Sasaki and Ishii, 1992) showed that difference in during grain filling time were not maintained between these groups in different years. However, LPS at the heading time was comparatively stable.

Table 3. Leaf photosynthetic rate of the flag leaves in old, middle, and new variety groups of rice bred in Japan.

Variety group	LPS ($mgCO_2/dm^2/hr$)				
	Heading	Grain filling		Harvesting	
		$-N$	$+N$	$-N$	$+N$
Old	38.3	27.4	27.5	20.9	19.6
Middle	38.5	28.9	33.1	19.1	21.3
New	39.2	31.3	36.3	19.7	20.6

+N, top-dressing of N at the heading time; –N, no topdressing.
Variety groups old, middle, and new consist of 7 varieties bred between 1882 and 1913, 10 varieties bred between 1921 and 1940, and 15 varieties bred between 1949 and 1976, respectively.
The values of LPS are expressed as the mean of the varieties of each group.

b. *Leaf Morphological Characteristics in relation to photosynthesis*

Tsunoda (1959) examined, in detail, the morphological characteristics of leaf and found that the specific leaf area (SLA) was smaller and hence leaves were thicker in the nitrogen-tolerant varieties than in the intolerant ones. The thick leaves naturally lead to narrow and erect leaves which have been found useful for breeding high-yielding varieties in Japan. This trend is also observed in indica varieties (Murty *et al.*, 1973; Ohno, 1976).

Many studies show a positive correlation between LPS and leaf nitrogen content *vis-a-vis* the soluble protein content which usually corresponds to enzymic protein. Since about 50% of the soluble protein in the leaf is RuBPcase (Makino *et al.*, 1984), varietal differences in LPS could be attributed to its content. Thick leaves can contain relatively more chlorophyll and photosynthetic enzymes per unit leaf area (Tsunoda, 1960).

Leaf photosynthesis is said to be limited by two resistances, the stomatal (Rs) and mesophyll (Rm) resistance. The latter resistance involves CO_2 diffusion process from the stomatal cavity to the chloroplast and the CO_2 fixation process at the chloroplasts. Sasaki and Ishii (1990) attempted to separate the Rm into two components, CO_2 diffusion or CO_2 transportation resistance (Rr), and CO_2 fixation, or carboxylation resistance (Rc). They estimated the relative magnitude of each resistance to the total one (Rt) in 31 Japanese rice varieties and found that Rs, Rr and Rc occupied 20–37%, 4–17%, and 55–67% against Rt, respectively. This suggests that the most important determinant in the varietal difference of LPS is carboxylation resistance, which is considered to be regulated by RuBPcase activity. The high correlation observed between LPS and RuBP carboxylase in rice varieties supports this suggestion (Saka, 1985).

c. *The Grain Yield and Leaf Photosynthesis*

Ohno (1976) reported that varietal difference in LPS is very small and the contribution of LPS accounts the 30% varietal difference in grain yield, and attributed the remaining 70% to difference in leaf area. The duration of the leaf area shows a positive correlation with grain yield, but LPS in the senescing period shows a comparatively good correlation. Thus, maintaining longer duration of leaf area after heading would increase the grain yield potential.

IV. Genetics of Leaf Photosynthesis

Hayashi *et al.* (1977) reported that LPS is controlled by the single major gene, and low LPS might be dominant (Fig. 3). In soybean, similar results were obtained by Ojima *et al.* (1969). These data are preliminary and need further elucidation.

Recently, several researchers found F_1 heterosis in photosynthesis of rice. Murayama *et al.* (1987) reported, in the crossing trial of many rice varieties, such an evident 57% heterosis in rice photosynthesis compared to the midparent, and 51% against the higher parent. On the other hand, Xuan *et al.* (1989) failed to find the heterosis in potential photosynthesis in the flag leaves of F_1 hybrid, in the experiment using male sterile, maintenance, and restoration lines of Chinese rice.

568

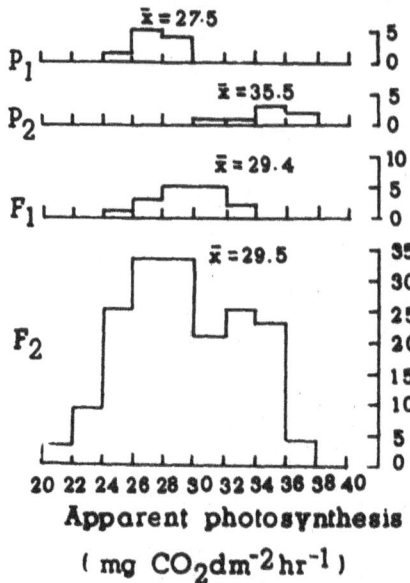

Figure 3. Frequency distribution of leaf photosynthesis in the parents (P1 and P2), and their F1 and F2 progenies of rice (Hayashi *et al.*, 1977). P1, Nakate-shinsenbon; P2, CP-SLO.

They found, however an evident heterosis in LPS in the senescing period of the flag leaves, and consequently the integrated LPS of the flag leaves through the senescing period was larger in F_1 hybrids than in the parents. From this, they inferred that the high yield of F_1 hybrids could be attributed to large integrated LPS of the flag leaves during the ripening period.

V. Conclusions

It is still debatable whether LPS can serve as criterion for breeding high-yielding varieties. On the basis of our present findings one can conclude that LPS in the flag leaf during grain filling period can be used as one of the criteria. Studies need to be conducted on the mechanism of progress of leaf senescence. It is reported that enzymatic protein degradation is regulated by some plant hormones like cytokinins, which are formed in the roots. We need to examine the possibility of maintaining high root activity during grain filling period.

VI. References

Cho DS and Murata Y (1980) Studies on the photosynthesis and dry matter production of rice plants. I Varietal differences in photosynthetic activity induced by nitrogen top-dressing. *Jpn J Crop Sci* **49**: 88–94 (in Japanese with English summary)

Cho DS, Yokoi S and Murata Y (1980) Studies on photosynthesis and dry matter production of rice plants. II. Varietal differences in the relationship between the content of nitrogenous constituents in the leaf and its photosynthetic activity induced by nitrogen top dressing. *Jpn J Crop Sci* **49**: 608–614 (in Japanese with English summary)

Cook MG and Evans LT (1983) Some physiological aspects of the domestication and improvement of rice (*Oryza* spp.). *Field Crops Res* **6**: 219–238

Evans LT and Dunstone RL (1970) Some physiological aspects of evolution in wheat. *Aust J Biol Sci* **23**: 725–741

Hayashi K, Yamamoto T and Nakagahra M (1977) Genetic control for leaf photosynthesis in rice. *Oryza sativa* L. *Japan J Breed* **27**: 49–56

Makino A, Mae T and Ohira K (1984) Relation between nitrogen and ribulose-1, 5-bisphosphate carboxylase in rice leaves from emergence through senescence. *Plant Cell Physiol* **25**: 429–437

Makino A, Mae T and Ohira K (1987) Variation in the contents and kinetic properties of ribulose-1, 5-bisphosphate carboxylases among rice species. *Plant Cell Physiol* **28**: 799–804

Murayama S, Miyazato K and Nose A (1987) Studies on dry matter production of F1 hybrid in rice. I Heterosis in the single leaf photosyntheetic rate. *Jpn J Crop Sci* **56**: 198–203

Murty KS, Nayak SK and Sahu G (1973) Photosynthetic efficiency in rice varieties. *ISNA Newsletter* **2**: 5–6

Ohno Y (1976) Varietal differences of photosynthetic efficiency and dry matter production in indica rice. *Tech Bull TARC* **9**: 1–72

Ojima M, Kawashima R and Mikoshiba K. (1969) Studies on the seed production of soybean. VII The ability of photosynthesis in F1 and F2 generations. *Proc Crop Sci Soc Jpn* **38**: 693–699 (in Japanese with English summary)

Osada A and Murata Y (1962a) Studies on the relationship between photosynthesis and varietal adaptability for heavy manuring in rice plant. I The relationship in the case of medium-maturing varieties. *Proc Crop Sci Soc Jpn* **30**: 220–223 (in Japanese with English summary)

Osada A and Murata Y (1962b) Studies on the relationship between photosynthesis and varietal adaptability for heavy manuring in rice plant. II The relationship in the case of early maturing varieties. *Proc Crop Sci Soc Jpn* **30**: 224–227 (in Japanese with English summary)

Osada A and Murata, Y (1965a) Studies on the relationship between photosynthesis and varietal adaptability for heavy manuring in rice plant. III Effects of photosynthetic characteristics on dry matter production and ripening of rice varieties. *Proc Crop Sci Soc Jpn* **33**: 460–466 (in Japanese with English summary)

Osada A and Murata Y (1965b) Varietal difference in the rate of photosynthesis of rice plant and its relation to dry matter production. *Proc Crop Sci Soc Jpn* **33**: 454–459 (in Japanese with English summary)

Saka H (1985) Variation in the activities of several photosynthetic enzymes during the growth stages in several genotypes and species of genus *Oryza*. *Bull Natl Inst Agric Sci* **D36**: 247–282 (in Japanese with English summary)

Sasaki H and Ishii R (1990) $\delta^{13}C$ analysis to approach the mechanism of varietal difference of photosynthetic rate in rice plants. In: (M Baltscheffsky, ed) *Current Research in Photosynthesis*, Vol IV, Kluwer Acad Publ, Dordrecht. The Netherlands, pp 895–898

Sasaki H and Ishii R (1992) Cultivar differences in leaf photosynthesis of rice (*Oryza sativa* L.) bred in Japan. *Photosynthesis Res:* in press.

Sasaki H, Ishii R and Kumura A (1986) Studies on varietal difference of leaf photosynthesis in rice. I The leaf photosynthesis in different growth stages. Presentation in 182nd Ann Meet Crop Sci Soc Jpn

Tsunoda S (1959) A developmental analysis of yielding ability in varieties of field crops. III The depth of green colour and the nitrogen content of leaves. *Jpn J Breed* **10**: 39–42

Xuan SN, Kanda A, Yamagishi T and Ishii R (1989) Studies on grain yield, dry matter production and photosynthesis in F1 hybrid rice. III Photosynthetic characteristics of F1 hybrid rice. Presentation in 188th Ann Meet Crop Sci Soc Jpn

Yoshida S (1973) Effects of CO_2 enrichment at different stages of panicle development on yield components and yield of rice (*Oryza sativa* L.). *Soil Sci Plant Nutr* **19**: 311–316

25

Photosynthesis Improvement as a Way to Increase Crop Yield

H. Medrano and J. Vadell

Lab de Fisiología Vegetal
Dept. de Biología Ambiental
Institut d'Estudis Avançats.- Universitat de les Illes Balears
07071 Palma de Mallorca
Spain

CONTENTS

ABBREVIATIONS

LAI	:	leaf area index
P_{max}	:	photosynthesis at saturated light intensities
Γ	:	CO_2 compensation point

572

ABSTRACT

Attempts to breed crop varieties with higher rates of photosynthesis have met with no success although considerable genetic variation in photosynthesis rates exist in several crop species and a positive correlation between leaf photosynthesis and productivity is reported in a number of experiments. An efficient partitioning of assimilated carbon seems to be more critical in determining plant productivity. Selection for low dark respiration or enhanced light interception have successfully increased net carbon gain by the plant. Selection by survival under low CO_2 atmosphere in tobacco haploids and on *Lolium multiflorum* L. Italian ryegrass cultivar RvP population showed a significant increase in plant productivity even when the leaf photosynthesis rate was not high. For Italian ryegrass RvP population, the surviving plants under low CO_2 had had significant increase in initial dry weight but the difference disappeared in the second regrowth. No single character of *ryegrass* or tabacco genotypes could account for their survival under low CO_2. Even though photosynthesis and plant production are closely related, large environmental and ontogenic-induced variations in leaf photosynthesis rate make it difficult to achieve a good estimate of its contribution to the entire plant carbon economy.

I. Introduction

It has been argued by several research that the productivity of some crops has attained a plateau (Fig. 1) and that further improvement may be difficult to achieve by the presently available technologies and breeding methods. The phenomenal improvement in crop productivity over the last few decades has been attributed primarily to an efficient partitioning of resources to the economic parts. This is evident also from the fact that there is virtually no improvement in the biomass of the improved varieties compared to their wild progenitors. Recently, efforts have been made to understand the constraints, if any, in improving the total biomass productivity.

We discuss several issues concerned with the problem of improving biomass productivity and the methods developed to screen genotypes with high potential

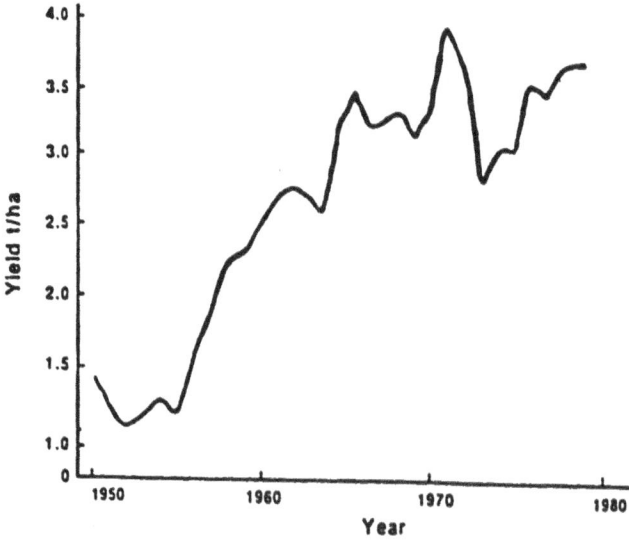

Figure 1. Sorghum grain yields in the USA from 1950 to 1980. (Fehr, 1984).

biomass production capabilities. First, while we accept that the net photosynthetic rate constitutes the primary component in biomass production, we show that use of this parameter may often be incorrect as it is associated with several confounding variables. We briefly review earlier work in this area and evaluate the success and attempts to delineate the confounding variables. Secondly, we consider a few alternatives such as selection for enhanced light interception and low dark respiration as possible approaches for attaining higher net productivity. Finally, we discuss a more recent approach, that of selection at low CO_2 for identifying genotypes of plants that essentially have a low CO_2 compensation point. Preliminary results of work in this direction conducted in our laboratory are presented.

II. Selection for Photosynthetic Efficiency—Pitfalls and Prospects

Selection for photosynthetic efficiency to improve biomass productivity was chiefly inspired by two facts: (1) considerable genetic variation for photosynthetic efficiency was reported both within and between species; and (2) the photosynthetic rates differ markedly between the C_3 and C_4 plant species.

Incoporation of the C_4 syndrome was suggested as a possible method to improve the photosynthetic yield of C_3 plant species. Björkman and co-workers (1971) attempted to produce C_3-C_4 hybrids of *Atriplex* to see whether these exhibit anatomical, physiological and biochemical C_4 characterstics. None of these hybrids showed complete C_4 syndrome. However, a few of the hybrids did show certain anatomical and biochemical features akin to the C_4 plants. In contrast, results of hybridization of the C_3 and C_4 types in *Flaveria* have indicated that there is a possibility of improving crop biomass through the improvement of photosynthetic syndrome (Huber *et al.*, 1989).

Though considerable variation in photosynthetic rate has been demonstrated in several crop species, only occasionaly has such variation been reflected in the variations in biomass productivity. For example, in *Lolium perenne* (Wilson, 1975, 1981) and in *Festuca arundinacea* (Assay *et al.*, 1974), considerable intraspecific variation in photosynthetic rate has been demonstrated at saturated light intensities (P_{max}). However, selection of genotypes with higher photosynthetic rate did not lead to higher productivity (Nelson *et al.*, 1975a). In *Triticum*, Evans and Dunstone (1970), showed that the highest rates of P_{max} corresponded to the diploid with poor crop productivity compared to the hexaploid (For details see chapter by Bansal, Uprety and Abrol, this volume.). In view of these observations it is doubtful if selection for high P_{max} will lead to higher biomass.

Gifford and Evans (1981) and Elmore (1981) argued that biomass productivity might depend more on the efficiency of storage and translocation rate than on CO_2 assimilation rate curtiva per se. Horst *et al.*, (1978) also emphasized that carbon partitioning efficiency seems to be more critical in determining productivity than P_{max}. From a study on tall fescue grass, Wilhelm and Nelson (1978) suggested that genetic selection for both source and sink activity is needed to improve the proportion of photosynthates going to the derived sink. They suggested it may open the way to an economic use of the additional photosynthates obtained by the increase in photosynthetic rate (cf. Nelson, 1988). By crossing a high P_{max} decaploid with a normal hexaploid, they obtained an octoploid progeny intermediate in photosynthetic rate with higher yield than their parents. The better investment in leaf growth of the hybrid seemed to be the basis of this yield improvement.

Reporting a number of experiments demonstrating a positive correlation between P_{max} and biomass productivity, Zelitch (1982) argued that lack of correlation observed by earlier workers may be because of several methodological problems and confounding variables. Such problems are compounded by the narrow differences among the genotypes and the large intra-plant variations. It is conceded that measurement of photosynthetic rate on single leaf basis does not reflect the efficiency of entire plant performance. This is because the whole plant productivity depends upon the effective light interception by the leaves mediated through leaf morphology, size inclination, and total plant leaf area.

Nevertheless the genetic variability in photosynthetic rate has been related to variations in leaf thickness, its size, cell size and ploidy status (Nelson, 1988). The reported variability in photosynthetic rate could be due to these characters which indirectly modulate the rates of photosynthesis and not due to genetic variations in the photosynthetic rate itself. Leaf age and the ontogentic developmental phases of the plant are also known to influence significantly the photosynthetic rate. We have also shown wide variation in photosynthetic rate during leaf development along plant profile and through the day in tobacco, ryegrass, and subterranean clover (Fig. 2). Apart from these intrinsic features, estimates of photosynthetic rate are affected by extraneous factors such as temperature, day length, light intensity, air and soil humidity, and seasonal and diurnal fluctuations.

Finally, respiration also contributes to confounding the relationship between the P_{max} and biomass productivity. Thus, the lack of correlation appears to be more a problem due to inadequate estimation coupled with the confounding variables and probably not an indication of lack of relationship. However, in any case, as Gifford (1987) points out, selection for leaf photosynthetic rate may not be an effective way to obtain new cultivars of higher productivity. No cultivars with improved yield have yet been generated by selecting for higher photosynthetic rate.

III. Alternate Selection Methods for Improving Biomass Productivity

Several traits have been identified that may serve as selection criteria for improved biomass production, such as higher light interception by the canopy and an efficient carbohydrate utilization.

Selection of *Lolium perenne* genotypes for leaf and tiller insertion angle and more rigid leaves showed 30 per cent increase in dry matter production with respect to the initial population (Rhodes, 1973). All the characteristics involved in this selection process were related with light interception. Crop with more erect and rigid leaves can develop higher leaf cell index which facilitates light penetration in the canopy thus allowing higher crop growth rates. Gutschik (1988) proposed that opimization of specific leaf weight improves crop growth rate and water use efficency. While Le Cain and co-workers (1989) reported positive correlation between specific leaf weight and P_{max}, Nelson (1988) reported the contrary.

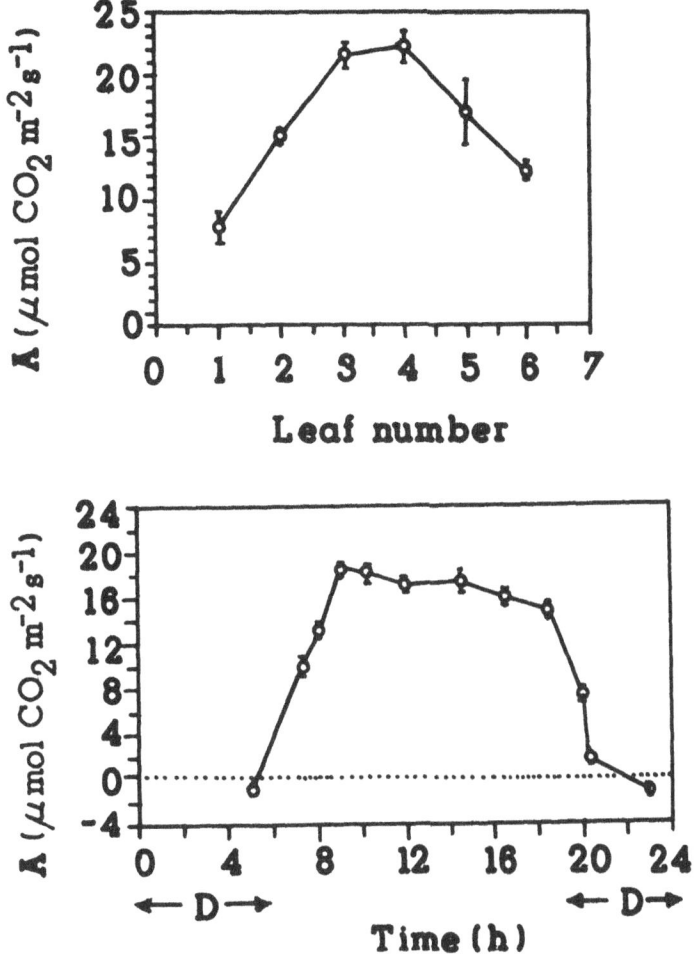

Figure 2. CO₂ Assimilation rate (A) Subterranean clover leaves: Variations in respect to leaf age (number) and day time, D = dark period.

There are also some reports on the effect of reduction in respiratory losses on plant production. Wilson (1975) identified genetic variation in respiration rates of young, fully expanded leaves of S23 perennial ryegrass and demonstrated negative correlation between dry matter production and respiration rate in simulated swards under growth room and glasshouse conditions. After a selection programme for low respiration rate, the selected genotypes showed substantial reduction in respiratory losses that was reflected in a higher dry matter production (13 to 20 per cent) under field conditions (Wilson, 1982; Wilson and Jones, 1982).

Since photorespiration represents a waste of assimilated carbon, the reduction of photorespiratory losses has also been shown as a means to improve photosynthesis balance and dry matter production. Zelitch and Day (1973) found tobacco genotypes with low photorespiration rate, which have also higher levels of CO₂ assimilation.

The progeny obtained by self-pollination of such genotypes, however, did not maintain such higher assimilation levels.

Considering the need to evolve suitable methodology that takes into account both the assimilatory and the respiratory rates, selection of plants for their survival under low CO_2 levels has been proposed (Nasyrov, 1978).

IV. Survival under Low CO_2—A New Selection Criterion for Improving Crop Biomass Production

Leaf photosynthetic rate represents the balance between assimilatory processes and respiration which take place simultaneously in the leaf. Differences in photorespiration between C_3 and C_4 plants are clearly reflected in the differences for the CO_2 compensation point (Γ) which is close to zero for C_4 and varies from 30 to 80 ppm CO_2 for C_3 leaves, at high light intensities, optimal temperature and 21 per cent O_2 in the atmosphere. At the CO_2 compensation point, leaf CO_2 assimilation exactly balances respiratory plus photorespiratory losses, so that low Γ values represent better balance between gains and losses of carbon, and are commonly associated also with a higher photosynthetic rate. In fact, a clear negative correlation has been reported between Γ values and maximal photosynthesis rate in different species (Downton and Tregunna, 1968; Krenzer et al., 1975). This parameter has been considered suitable to discriminate genotypes with higher photosynthetic rate and/or reduced photorespiratory losses (Smith et al., 1976; Somerville and Somerville, 1986; Nasyrov, 1978). It may, however, be mentioned that CO_2 compensation point value as a selection criterion is difficult to use because its determination for each genotype needs a lot of time and effort as well as expensive equipment.

At CO_2 compensation point there would be no net assimilation of carbon, a plant maintained under this CO_2 concentration value will die in a few days. Menz and co-workers (1969) and Cannel and co-workers (1969) placed large populations of soybean plantlets of representative genotypes in closed illuminated chambers accompanied by some C_4 plants. With time, C_4 plants decreased the CO_2 concentration level below the compensation point of the C_3 ones, resulting in mortality of C_3 plants. However, from 2458 different soybean genotypes tested for survival in these chambers, about 5 per cent of the genotypes which died early and about 5% which died late did not show statistically significant difference between their CO_2 compensation point values. Nelson and co-workers (1975b) subjected tall fescue seedlings to a sub-Γ atmosphere in a chamber under continuous illumination and concluded that the method was not effective to select genotypes of higher leaf photosynthetic rate or reduced photorespiration. However, the genotypes which survived for a longer period showed higher dark respiration rates, a factor that might be associated with their increased vigour.

Although these reports do not encourage the use of this method for screening germplasm for high photosynthetic rate, the simplicity of the method and the possibility of a large population screening are clear advantages that overcome the main technical limitations of direct measurement of leaf photosynthesis. On these grounds, the method was considered suitable for screening for photorespiratory characteristics by Nasyrov (1978). We report below two specific situations where selection under low CO_2 was carried out.

a. Tobacco haploids

We have developed a screening chamber in which haploid tobacco plants obtained from *in vitro* anther culture were first acclimatised to the growth conditions (27°C, 600 μE $m^{-2}sec^{-1}$, 12 hrs light/12 hrs dark cycles in a 350 ppm CO_2 atmosphere, and Hoagland's solution). After acclimation, the circulating atmosphere was modified to maintain 60 ppm CO_2 in the chamber. Populations of haploid small plants were subjected to this low CO_2 atmosphere for 45 days. When survival was about 10 per cent, the surviving plants were allowed to grow in a greenhouse. Some of the plants were diploidized by colchicine treatment to obtain seeds (Figure 3). The selected genotypes exhibited clear advantage over the rest of the haploids as well as over the genotype Wisconsin 38 from which anthers were obtained (Medrano and Primo-Millo, 1985) with regard to early growth, plant production, and photosynthesis rate. Successive field assays showed a consistent advantage of selected genotypes with respect to the source cultivar in plant production (Medrano *et al.*, 1989). Field evaluation also showed an increase in leaf area per plant as one of the features of selected genotypes that might explain to some extent the significant increase observed in dry matter production (Table 1). Nevertheless, measurements of leaf photosynthesis rate showed no significant differences among selected genotypes and the control (Delgado and Medrano, 1990). However, dark respiration rate of mature leaf on dry matter basis was consistently lower in these selected genotypes (Delgado *et al.*, 1990).

In tobacco, leaf being the economically important component, genotypes with larger leaves were favoured and selected by conventional breeding programmes. Screening tobacco plants under low CO_2 atmosphere enabled us to select genotypes with higher productivity, but did not show otherwise any clear advantage in either instantaneous leaf photosynthetic rates or CO_2 compensation points measured in different environments.

b. Italian ryegrass

Italian ryegrass is a forage crop very common in the Mediterranean area and, as in tobacco, the leaves are the economically important part. We have followed the same procedure to screen Italian ryegrass genotypes for survival under low CO_2 atmosphere. Since it was not possible to develop haploids, the source population for screening was cultivar RvP, commonly used as base population in breeding programmes because of its high plant to plant variability. The CO_2 concentration in the circulating air during the selection procedure was adjusted to maintain slightly higher than the CO_2 compensation point for this species. The duration of the treatment period was altered to achieve around 10 per cent survival (Medrano and Pol, 1986). The surviving genotypes were planted in pots after the selection procedure was accomplished. Preliminary data on growth and dry matter in pot experiments

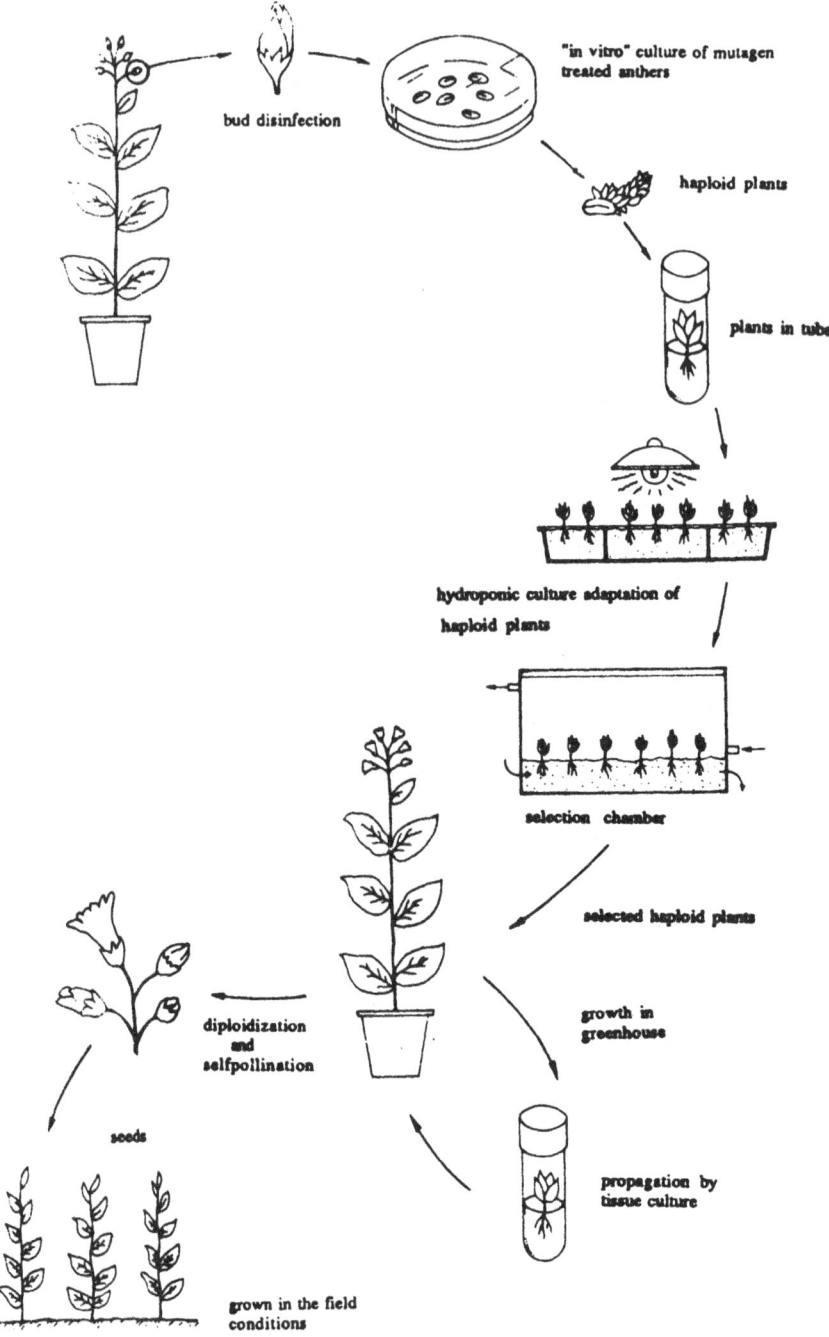

Figure 3. Diagram of experimental procedure to select haploid *Nicotiana tabacum* plants by survival in a low CO_2 atmosphere chamber.

showed a clear significant advantage for the selected ones in comparison to the RvP population. Nevertheless, this difference decreased in the second regrowth and almost

Table 1. Leaf area and plant production in field-grown plants of tobacco genotypes selected by low CO_2 survival

Genotype Experiment	Leaf area per plant m^2	Number of leaves per plant	Leaf dry weight (g plant^{-1})	Total dry weight (g plant^{-1})
CONTROL				
1	—	22.7 ± 1.2	72.4 ± 5.2	135.8
2	1.25 ± 0.48	26.1 ± 0.9	96.6 ± 6.7	219.0 ± 15.9
3	2.59 ± 0.22	28.6 ± 2.3	117.0 ± 16.9	235.3 ± 31.7
4	1.00 ± 0.46	29.0 ± 1.1	88.8 ± 4.8	142.0 ± 6.0
SP355				
1	—	24.9 ± 1.0	123.0 ± 4.8	264.6
2	1.60 ± 0.48	25.4 ± 1.0	123.5 ± 5.8	272.5 ± 14.0
3	—			
4	1.09 ± 0.46	27.2 ± 0.7	97.6 ± 6.0	162.6 ± 10.9
SP422				
1	—	20.5 ± 0.9	68.3 ± 5.8	150.8
2	1.43 ± 0.69	26.5 ± 1.8	124.2 ± 10.1	296.2 ± 21.5
3	2.88 ± 0.20	32.0 ± 1.7	126.6 ± 16.8	264.6 ± 16.8
4	1.12 ± 0.60	28.0 ± 2.0	105.1 ± 5.0	185.0 ± 12.5
SP432				
1	—	25.4 ± 0.5	124.7 ± 6.2	267.7 ±
2	1.40 ± 0.61	23.9 ± 0.3	111.3 ± 5.8	239.4 ± 11.4
3	—			
4	1.00 ± 0.50	24.9 ± 0.9	84.1 ± 5.0	142.0 ± 10.6
SP435				
1	—	25.2 ± 0.9	125.4 ± 9.2	289.1
2	1.43 ± 1.27	28.3 ± 1.9	139.6 ± 13.0	319.0 ± 29.8
3	—			
4	1.03 ± 0.60	26.4 ± 0.8	93.4 ± 6.1	150.4 ± 10.3
SP451				
1	—	23.9 ± 0.6	127.6 ± 7.8	267.6
2	1.50 ± 0.60	26.6 ± 1.3	131.2 ± 12.5	280.8 ± 25.2
3	3.24 ± 1.77	29.0 ± 3.4	160.7 ± 10.9	321.7 ± 17.3
4	1.15 ± 0.70	28.0 ± 0.9	110.1 ± 6.3	179.0 ± 10.9

1. Pol, 1984.   2. Medrano et al., 1989.   3. Delgado and Medrano, 1990.   4. Delgado, 1990

disappeared in the third regrowth of the plants (Fig. 4). Leaf photosynthesis rates, measured on attached last fully expanded leaves with a portable equipment, showed no differences among average values of surviving genotypes and those of RvP original population (Pol, unpublished data). The lack of correspondence between these results and the ones reported from selected tobacco haploids could be explained in terms of a different adaptive pattern of each species to low CO_2 partial pressures. Furthermore, the extent of genetic diversity of initial screening material of haploids could also influence the response to low CO_2 levels. Tobacco haploids from mutagen-treated anthers show a lot of variability in morphological characters (Medrano et al., 1986), which facilitates greater genetic variability for selection of lines of greater productivity.

In the foregoing analysis involving two species when the entire plant is exposed to low CO_2 concentration or little higher than CO_2 compensation point, the survival

580

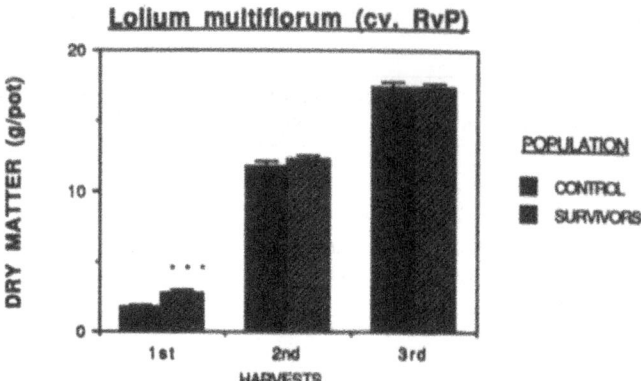

Figure 4. Plant production in unselected and low CO_2 survival genotypes of *Lolium multiflorum*.

of the plants depends on the improved carbon economy of the entire plant rather than on the low CO_2 compensation point. The sum of small contributions from several morphological and physiological parameters in different parts of plants could result in improvement of net carbon gain without any substantial increase in instantaneous leaf photosynthetic rate or leaf CO_2 compensation point.

V. Summary

The information discussed here indicates that the results did not agree with the expectations. In those cases where leaf photosynthesis rate was improved, there was no parallel response in plant growth. On the other hand, when selection was made for related characters of photosynthesis, the plant production was increased without any increase in the instantaneous leaf photosynthetic rates. Nevertheless, the efforts devoted to improve photosynthesis has been compensated with a significant progress in the knowledge of relationship between leaf photosynthesis and plant growth. Photosynthesis and plant production are obviously closely related, but it is difficult to establish the relationship because of the large variability in photosynthetic rate induced by leaf age, plant development, source sink ratio, and environmental factors. Because of these variations, importance of maximum photosynthesis rate appears to be much lower than expected. The variations also make it difficult to achieve a good estimate of accurate photosynthetic rate and its actual contribution to the entire plant carbon economy. More data on photosynthetic performance under field conditions and a better knowledge of leaf age, plant development, and environmental factors on leaf characters is necessary to establish the relationship between leaf photosynthesis and productivity. Financial support by the CICYT-Progr. Nac. de Inv. Agraria-Project no. AGR 89 0729, in gratefully acknowledged.

VI. References

Assay KH, Nelson CJ and Horts GL (1974) Genetic variability for net photosynthesis in tall fescue. *Crop Sci* **14**: 571–574

Björkman O, Nobs M, Pearcy R, Boynton J and Berry J (1971) Characteristics of hybrids between C$_3$ and C$_4$ *Atriplex*. In: (MD Hatch, CB Osmond and RO Slayter eds) *Photosynthesis and Photorespiration*, Wiley-Interscience, Sydney, pp 105–119

Cannell RQ, Bruns WA and Moss DN (1969) A search for high net photosynthetic rate among soybean genotypes. *Crop Sci* **9**: 840–841

Delgado E, Azcón Bieto J and Medrano H (1989) Photosynthesis, dark respiration and plant production in *Nicotiana tabacum* genotypes derived from haploids selected by low CO_2 survival. In: (M Baltscheffsky ed) Current Research in Photosynthesis, Vol IV, Kluwer Academic Publishers, Dordrecht, The Netherlands, pp 39–42

Delgado E (1990) Caracterization fotosintética de líneas de *Nicotiana tabacum* L seleccionadas en cámara de bajo contenido en CO_2. Doctoral Thesis, Universitat de les Illes Balears, Palma de Mallorca, pp. 308.

Delgado E and Medrano H (1991) Field performance and leaf characteristics of *Nicotiana tabacum* genotypes selected by low CO_2 survival. *Photosynthetica* **25**: in press

Delgado E, Azcon-Bieto J, Aranda X, Palazón J and Medrano H (1992) Leaf photosynthesis and respiration of high CO_2 grown tobacco plants selected for survival under CO_2 compensation point conditions. *Plant Physiol*: in press

Downton WJS and Tregunna ED (1986) Carbon dioxide compensation. Its relation to photosynthetic carboxylation reactions, systematics of the Gramineae and leaf anatomy. *Can J Bot* **46**: 207–215

Dunstone RL and Evans LT (1974) Role of changes in cell size in the evolution of wheat. *Aust J Plant Physiol* **1**: 157–165

Elmore CD (1981) The paradox of no correlation between leaf photosynthesis rates and crop yields. In: (JD Hesketh and JW Jones, eds) *Predicting Photosynthesis for Ecosystems*, CRC Press, Boca Raton, pp 155–167

Evans LT and Dunstone RL (1970) Some physiological aspects of evolution in wheat. *Aust J Biol Sci* **23**: 725–741

Fehr WR (1984) Genetic contributions to yield gains of five major crop plants. CSSA Special Pub No p. 101

Gifford RM (1987) Barriers to increasing crop productivity by genetic improvement in photosynthesis. (In: J Biggens, ed) *Progress in Photosynthesis Research*, Vol 4, M Nijhoff Publishers, Dordrecht, The Netherlands p 377–384

Gifford, RM and Evans LT (1981) Photosynthesis, carbon partitioning, and yield. *Annu Rev Plant Physiol* **32**: 485–509

Gutschik VP (1988) Optimization of specific leaf mass, internal CO_2 concentration, and chlorophyll content in crop canopies. *Plant Physiol Biochem* **26**: 525–537

Horst GL, Nelson CJ and Assay KH (1978) Relationship of leaf elongation to forage yield of tall fescue genotypes. *Crop Sci* **18**: 1715–719

Huber WE, Brown RH, Bouton JH and Sternberg LO'R (1989) CO_2 exchange, cytogenetics, and leaf anatomy of hybrids between photosynthetically distinct *Flaveria* species. *Plant Physiol* **89**: 839–844

Krenzer EG, Moss DN, Crookston RK (1975) Carbon dioxide compensation points of flowering plants. *Plant Physiol* **56**: 194–206

Le Cain DR, Morgan JA and Zerbi G (1989) Leaf anatomy and gas exchange in nearly isogenic semidwarf and tall winter wheat. *Crop Sci* **29**: 1264–1251

Medrano H and Pol A (1986) Selección de gramíneas forrajeras por eficacia fotosintética. *Rev Pastos* **XVI**: 191–204

Medrano H, Pol A and Delgado E (1989) Plant production, photosynthesis rate and related characters in doubled-haploid lines of *Nicotiana tabacum* selected by photosynthetic efficiency. In: (J Barber and R Malkin eds) *Techniques and New Developments in Photosynthesis Research*, Plenum Press London, pp 481–484

Medrano H and Primo-Millo E (1985) Selection of *Nicotiana tabacum* haploids of high photosynthetic efficiency. *Plant Physiol* **79**: 505–508

Medrano H, Primo-Millo E and Guerri J (1986) Ethylmethane-sulphonate effects on anther cultures of *Nicotiana tabacum. Euphytica* **35**: 161–168

Menz KM, Moss DN, Cannell RQ and Brun WA (1969) Screening for photosynthetic efficiency. *Crop Sci* **9**: 692–694

Nasyrov YS (1978) Genetic control of photosynthesis and improving of crop productivity. *Annu Rev Plant Physiol* **29**: 215–237

Nelson CJ (1988) Genetic associations between photosynthetic characteristics and yield: Review of the evidence. *Plant Physiol Biochem* **26**: 543–554

582

Nelson GJ, Asay KH and Horst GL (1975a) Relationship of leaf photosynthesis to forage yield of tall fescue. *Crop Sci* **15**: 476–478

Nelson CJ, Asay KH and Patton LD (1975b) Photosynthesis responses of tall fescue to selection for longevity below the CO_2 compensation point. *Crop Sci* **15**: 629–633

Pearcy RW (1990) Sunflecks and photosynthesis in plant canopies. *Annu Rev Plant Physiol Plant Mol Biol* **41**: 421–453

Pol A. (1984) Caracterització i hibridació de línies diplohaploides de *Nicotiana tabacum*, L seleccionades per eficacia fotosintética. Tesis Licenciatura. Univesitat de les Illes Balears. Palma de Mallorca, pp. 139

Rhodes I (1973) The relationship between productivity and some components of canopy structure in ryegrass (*Lolium* spp.). III Spaced plant characters, their heritabilities and relationship to sward yield. *J Agric Sci* **80**: 171–176

Smith EW, Tolbert NE and Ku HS (1976) Variables affecting the CO_2 compensation point. *Plant Physiol* **58**: 143

Somerville CR and Somerville SC (1986) Regulation of photorespiration. In: (CA Neyra, ed) *Biochemical Basis of Plant Breeding* Vol I CRC Press Boca Raton, pp 89–131

Wilhelm WW and Nelson CJ (1978) Irradiance response of tall fescue genotypes with contrasting levels of photosynthesis and yield. *Crop Sci* **18**: 405–408

Wilson D (1975) Variation in leaf respiration in relation to growth and photosynthesis of *Lolium*. *Ann Appl Biol* **80**: 323–328

Wilson D (1981) Breeding for morphological and physiological traits. In: (JF Kenneth, ed) *Plant Breeding* Iowa State University Press, Iowa pp 325–346

Wilson D (1982) Response to selection for dark respiration rate of mature leaves in *Lolium perenne* and its effects on growth of young plant and simulated swards. *Ann Bot* **49**: 313–312

Wilson D and Jones JG (1982) Effect of selection for dark respiration rate of mature leaves on crop yields of *Lolium perenne* cv S 23. *Ann Bot* **49**: 313–320

Zelitch I (1982) The close relationship between net photosythesis and crop yield. *BioScience* **32**: 796–802

Zelitch I and Day PR (1973) The effect on net photosynthesis of pedigree selection for low and high rates of photorespiration in tobacco. *Plant Physiol* **52**: 33–37

AUTHOR INDEX

584

586

SUBJECT INDEX

The manufacturer's authorised representative in the EU is Springer
Nature Customer Service Centre GmbH, Europaplatz 3, 69115 Heidelberg,
Germany. If you have any concerns regarding our products, please
contact ProductSafety@springernature.com

Printed and bound by CPI Group (UK) Ltd, Croydon, CR0 4YY

23/04/2026

02095593-0014